Just the Essentials of Elementary Statistics

11E

Johnson | Kuby

CENGAGE
Learning

Australia • Brazil • Japan • Korea • Mexico • Singapore • Spain • United Kingdom • United States

CENGAGE
Learning™

**Just the Essentials of Elementary Statistics,
11e**

Executive Editors:
Maureen Staudt
Michael Stranz

Senior Project Development Manager:
Linda deStefano

Marketing Specialist:
Courtney Sheldon

Senior Production/Manufacturing Manager:
Donna M. Brown

PreMedia Manager:
Joel Brennecke

Sr. Rights Acquisition Account Manager:
Todd Osborne

Cover Image:
Getty Images*

*Unless otherwise noted, all cover images used by Custom
Solutions, a part of Cengage Learning, have been supplied
courtesy of Getty Images with the exception of the Earthview
cover image, which has been supplied by the National
Aeronautics and Space Administration (NASA).

Elementary Statistics, 11th Edition
Johnson | Kuby

For product information and technology assistance, contact us at
Cengage Learning Customer & Sales Support, 1-800-354-9706

For permission to use material from this text or product,
submit all requests online at **cengage.com/permissions**
Further permissions questions can be emailed to
permissionrequest@cengage.com

This book contains select works from existing Cengage Learning resources and
was produced by Cengage Learning Custom Solutions for collegiate use. As such,
those adopting and/or contributing to this work are responsible for editorial
content accuracy, continuity and completeness.

Compilation © 2011 Cengage Learning

ISBN-13: 978-1-133-27014-0

ISBN-10: 1-133-27014-X

Cengage Learning
5191 Natorp Boulevard
Mason, Ohio 45040
USA
Cengage Learning is a leading provider of customized learning solutions with
office locations around the globe, including Singapore, the United Kingdom,
Australia, Mexico, Brazil, and Japan. Locate your local office at:
international.cengage.com/region.

Cengage Learning products are represented in Canada by Nelson Education, Ltd.
For your lifelong learning solutions, visit **www.cengage.com/custom.**
Visit our corporate website at **www.cengage.com.**

Printed in the United States of America

Custom Contents

Elementary Statistics

Preface

Over the years since it was first published, *Elementary Statistics* has developed into an exceptionally readable and trusted introductory textbook that promotes learning, understanding, and motivation by presenting statistics in a real-world context without sacrificing mathematical rigor. Along the way, discipline after discipline has evolved to recognize that statistics is a highly valuable tool for them and that statistics reaches into multiple areas of daily life, resulting in at least one statistics course being recommended for students at most schools. As they have from the start but now more than ever to support current curricula, the applications, examples, and exercises in this text appropriately contain data from a wide variety of areas of interest, including the physical and social sciences, public opinion and political science, business, economics, and medicine. In *Elementary Statistics*, Eleventh Edition, we continue to strive for ever greater readability and a common sense tone that appeals to students who are increasingly more interested in application than theory.

Overview of What's New in and to This Edition

Instructors familiar with the text will notice the following changes in this edition.

New Chapter-Opening Vignettes

More than 50% of the book's chapter-opening vignettes, each of which focuses on an everyday aspect of life, are new. Illustrated with statistical information, each chapter opener provides a relevant, familiar context for students' initial step into the concepts covered in the chapter.

New Applied Examples

Nearly 20% of the text's applied examples are new or updated to help engage student interest. Key statistical concepts are presented with enhanced step-by-step solutions.

Over 20% New and Updated Exercises

Many of the exercises are new or updated to reflect current events and other timely topics. The text's more than 1700 exercises provide a wealth of practice problems, with each exercise set including a range of exercise types that progress from basic recall to multi-step to items requiring critical thinking. As always, most exercises can be calculated either by hand or by using technology.

Completely Rewritten Normal Probability Distribution Coverage

Chapter 6, "Normal Probability Distributions," has been completely rewritten to present the Standard Normal Distribution utilizing the cumulative approach, incorporating a more intuitive idea with respect to the total area under a curve and following more closely to the format used with graphing calculators and statistical software. To support this change, a new, corresponding two-page Table 3, "Cumulative Areas of the Standard Normal Distribution," is included among the tables at the back of the text.

New Visuals Throughout

In addition to new chapter-opening photos and graphics, new and redesigned art appears throughout Examples, Applied Examples, and Exercises.

Dynamic New Online Teaching and Learning Resources

See pages xvii–xviii for details about the Eleventh Edition's instructor and student supplements.

Tour of the Eleventh Edition

Continuing, new, and updated features include the following:

Emphasis on Interpretation of Statistical Information and Real Applications

Immediately in Chapter 1, when students learn foundational key terms and procedures; in Chapter 4, "Probability," where analysis rather than formula is highlighted; and continuing throughout the text, the authors emphasize the role of interpretation in statistical analysis. Examples and exercises feature real applications of statistics, and chapter-opening vignettes enhance the relevance of the material for students. Critical thinking exercises throughout the chapters further support the book's practical, proven approach.

NEW AND UPDATED Chapter Openers

Chapter outlines with a brief description of what's covered in each major section now appear on the first page of every chapter to help orient students and better prepare them for the instruction to come. Extended, engaging examples again open each chapter to illustrate a familiar situation using statistics in a relevant, approachable manner for the student. New chapter openers focus on students' average weekday time expenditure (Chapter 2), number of automobiles per household in USA (Chapter 5), and the relationship between the

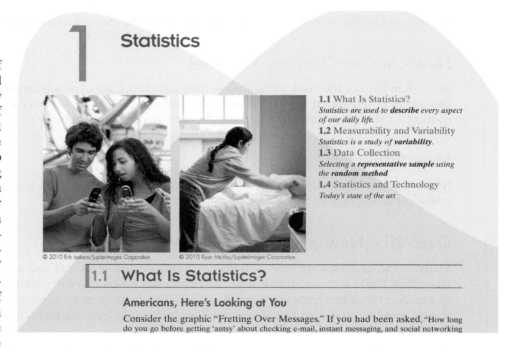

1 Statistics

1.1 What Is Statistics?
*Statistics are used to **describe** every aspect of our daily life.*
1.2 Measurability and Variability
*Statistics is a study of **variability**.*
1.3 Data Collection
*Selecting a **representative sample** using the **random method***
1.4 Statistics and Technology
Today's state of the art

© 2010 Erik Isakson/Jupiterimages Corporation   © 2010 Ryan McVay/Jupiterimages Corporation

1.1   What Is Statistics?

Americans, Here's Looking at You
Consider the graphic "Fretting Over Messages." If you had been asked, "How long do you go before getting 'antsy' about checking e-mail, instant messaging, and social networking

length and weight of a fish (Chapter 3). Chapter 9's "Floor to Door," Chapter 10's "Battle of the Sexes—Commute Time," and Chapter 12's "The Morning Rush" are also among those with updated openers.

NEW AND UPDATED Examples

Throughout the text, examples that present the step-by-step solution process for key statistical concepts and methods have updated or replaced to ensure currency and relevance. New and updated examples focus on statistical factors pertaining to topics such as finding the area under the normal curve using the cumulative normal distribution (Chapter 6) and applying that technique in Chapters 7 and 8. A basic probability experiment involving golf balls is presented in Chapter 4.

NEW AND UPDATED Applied Examples

Relevant Applied Examples incorporate statistical concepts to demonstrate how statistics work in the real world. New and updated data relevant to areas such as sports (Chapter 1), growth charts (Chapter 2), SUVs (Chapter 3), ceramic floor tiles (Chapter 9), and microchips (Chapter 10) will capture students' attention.

APPLIED EXAMPLE 1.6

THE BIG PAYCHECK

World's Highest Paid Athletes

Forbes' list of the highest-paid athletes looks at earnings derived from salaries, bonuses, prize money, endorsements, and licensing income between June 2008 and June 2009 and does not deduct for taxes or agents' fees.
Here are the Top 5:

Rank	Athlete	Sport	Earnings
1	Tiger Woods	Golf	$110 million
2	Kobe Bryant	Basketball	$45 million
2	Michael Jordan	Basketball	$45 million
2	Kimi Raikkonen	Auto Racing	$45 million
5	David Beckham	Soccer	$42 million

Source: http://www.getlisty.com/preview/highest-paid-athletes/

Left: Image copyright cloki, 2012. Used under license from Shutterstock.com;
Left center: Image copyright Charlene Bayerle, 2012. Used under license from Shutterstock.com;
Right center: Image copyright Rafa Irusta, 2012. Used under license from Shutterstock.com;
Right: Image copyright hanzl, 2012. Used under license from Shutterstock.com

Let's face it; most of us dream about making this much in our whole lifetime. If one is gainfully employed every year from 21 years old to 62 and makes $1 million each year, that would be $42 million for one's whole life. Most of us cannot even wrap our heads around that concept. You probably are thinking, "Sign me up to be a superstar athlete!"

Let's see how we can apply our new terminology to the "Big Paycheck." First, the general population of interest would be professional athletes. Furthermore, the information in the above table demonstrates the various types of variables. The athlete's name is generally not considered to be a variable; it is for the purpose of identification only. The other three kinds of information are variables:

1. Rank is qualitative and an ordinal variable since it incorporates the concept of ordered position.

DID YOU KNOW?

Yellowstone contains approximately one-half of the world's hydrothermal features. There are over 10,000 hydrothermal features, including over 300 geysers, in the park.

UPDATED Did You Know? and FYI Sidebars

The strategically placed Did You Know? sidebars present brief histories and fun facts to provide an informative and entertaining look at related concepts or methods being presented in the corresponding section of a given chapter. Similarly, the FYI segments provide helpful tips and insights at key points in each chapter.

FYI An ogive of these exam grades would graphically determine these same percentiles, without the use of formulas.

3.84 As summer heats up, Americans turn to ice cream as a way to cool off. One of the questions asked as part of a July 2009 Harris Poll was, "What is your favorite way to eat ice cream?" The study included 2177 U.S. adults.

FAVORITE WAY TO EAT ICE CREAM

Base: All adults who eat ice cream

Favorite Way	Male, %	Female, %
Cup	50	41
Cone	24	34
Sundae	17	18
Sandwich	2	2
Other	8	5
Total	**101**	**100**

Image copyright © M. Unal Ozmen. Used under license from Shutterstock.com.

NEW AND UPDATED Exercises

With nearly 25% new and updated exercises, the Eleventh Edition of *Elementary Statistics* provides instructors with up-to-date, relevant homework sets geared toward students' interests. Additionally, more than 300 classic exercises, as well as the solutions to odd-numbered exercises, are available online. With more than 2000 exercises total, instructors have great choice when creating assignments, and students have ample opportunity for practice.

The accompanying "Favorite Way to Eat Ice Cream" graphic lists in percentages the distributions of the ways both genders prefer to eat their ice cream.

a. Identify the population, the variables, and the type of variables.

b. Construct a bar graph showing the two distributions side by side.

c. Do the distributions seem to be different for the genders? Explain.

NEW Visuals Throughout

Done in an updated style, new and redesigned art appears throughout Examples, Applied Examples, and Exercises. The two samples that follow show the art's new style.

2.164 This statistical offering is a rather clever graphic using artistic license along with bills as the bars of a bar graph. An "A for effort," as you have heard before, but the scale aspects of the graph have been compromised.

What do you Plan to Spend your Tax Refund on?

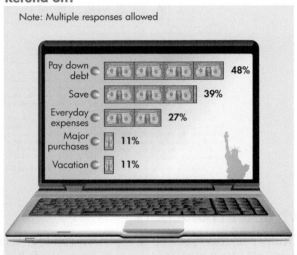

Source: National Retail Federation 2009 Tax Returns Consumer Intentions and Actions survey of 8,426 consumers. Margin of error ±1 percentage point.

Numerous Skillbuilder Applet Exercises

Found in Section and Chapter Exercises, Skillbuilder Applet Exercises help students "see" statistical concepts and allow hands-on exploration of statistical concepts and calculations. Skillbuilder Applet Exercises are easy to spot in the book and direct students to access the applets online.

3.5 "The perfect age" graphic shows the results from a 9×2 contingency table for one qualitative and one quantitative variable.

"The Perfect Age"

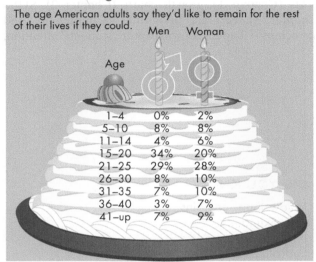

Source: Data from Cindy Hall and Genevieve Lynn, *USA TODAY*; IRC Research for Walt Disney. © 1998 *USA TODAY*, reprinted by permission.

a. Identify the population and name the qualitative and quantitative variables. adults; gender; age would like to remain rest of life

b. Construct a bar graph showing the two distributions side by side.

c. Does there seem to be a big difference between the genders on this subject? no

4.30 Skillbuilder Applet Exercise demonstrates the law of large numbers and also allows you to see if you have psychic powers. Repeat the simulations at least 50 times, guessing between picking either a red card or a black card from a deck of cards.

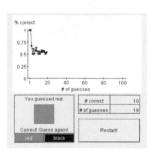

a. What proportion of the time did you guess correctly?

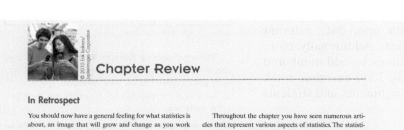

Chapter Review

In Retrospect

You should now have a general feeling for what statistics is about, an image that will grow and change as you work your way through this book. You know what a sample and a population are and the distinction between qualitative (attribute) and quantitative (numerical) variables. You should also have an appreciation for and a partial understanding of how important random samples are in statistics.

Throughout the chapter you have seen numerous articles that represent various aspects of statistics. The statistical graphics display a variety of information about ourselves as we describe ourselves and other aspects of the world around us. Statistics can even be entertaining. The examples are endless. Look around and find some examples of statistics in your daily life (see Exercises 1.77 and 1.78, p. 30).

Streamlined Chapter Reviews

The Chapter Review for each chapter includes the following pedagogically sound elements:

- **In Retrospect**, a summary of concepts covered that points out the relationships between each.

- **Vocabulary and Key Concept List**, which shows students at a glance what was covered and provides a page reference so that they can check their understanding.

- **Learning Outcomes**, intended to complement the Vocabulary and Key Concept lists, these summary listings outline the key concepts presented in the chapter and provide references to relevant pages and corresponding review exercises to help ensure that students understand chapter material.

Vocabulary and Key Concepts

attribute variable (p. 6)	infinite population (p. 5)	sample (p. 5)
biased sampling method (p. 16)	judgment sample (p. 19)	sample design (p. 19)
categorical variable (p. 6)	multistage random sampling (p. 22)	sampling frame (p. 18)
census (p. 18)	nominal variable (p. 7)	sampling method (p. 16)
cluster sample (p. 22)	numerical data (p. 7)	simple random sample (p. 20)
continuous variable (p. 8)	numerical variable (p. 6)	single-stage sampling, (p. 19)
convenience sample (p. 17)	observational study (p. 18)	statistic (p. 6)
data (p. 5)	ordinal variable (p. 7)	statistics (p. 1)
data collection (p. 16)	parameter (p. 5)	strata (p. 22)
data value (p. 5)	population (p. 4)	stratified random sample (p. 22)
descriptive statistics (p. 1)	probability sample (p. 19)	survey (p. 18)
discrete variable (p. 8)	proportional stratified sample (p. 22)	systematic sample (p. 21)
experiment (p. 5)	qualitative variable (p. 6)	unbiased sampling method (p. 16)
finite population (p. 5)	quantitative variable (p. 6)	variability (p. 15)
haphazard (p. 20)	representative (p. 19)	variable (p. 5)
inferential statistics (p. 1)	response variable (p. 5)	volunteer sample (p. 17)

Learning Outcomes

• Understand and be able to describe the difference between descriptive and inferential statistics.	p. 1, Ex. 1.7, 1.8, 1.67
• Understand and be able to identify and interpret the relationships between sample and population, and statistic and parameter.	pp. 4–6, EXP 1.5
• Know and be able to identify and describe the different types of variables.	pp. 6–8, Ex. 1.33, 1.34
• Understand how convenience and volunteer samples result in biased samples.	pp. 16–17, Ex. 1.45
• Understand the differences among and be able to identify experiments, observational studies, and judgment samples.	pp. 18–19
• Understand and be able to describe the single-stage sampling methods of "simple random sample" and "systematic sampling."	pp. 19–21
• Understand and be able to describe the multistage sampling methods of "stratified sampling" and "cluster sampling."	pp. 22–23
• Understand that variability is inherent in everything and in the sampling process.	p. 14–15, Ex. 1.38

Chapter Exercises

1.65 We want to describe the so-called typical student at your college. Describe a variable that measures some characteristic of a student and results in

a. Attribute data

b. Numerical data

1.66 A candidate for a political office claims that he will win the election. A poll is conducted and 35 of 150 voters indicate that they will vote for the candidate, 100 voters indicate that they will vote for his opponent, and 15 voters are undecided.

a. What is the population parameter of interest?

b. What is the value of the sample statistic that might be used to estimate the population parameter?

c. Would you tend to believe the candidate based on the results of the poll?

1.67 A researcher studying consumer buying habits asks every 20th person entering Publix Supermarket how many times per week he or she goes grocery shopping. She then records the answer as *T*.

a. Is *T* = 3 an example of (1) a sample, (2) a variable, (3) a statistic, (4) a parameter, or (5) a data value?

Suppose the researcher questions 427 shoppers during the survey.

(continue on page 28)

- **Chapter Exercises** offer practice on all of the concepts found in the chapter while also tying in comprehensive material learned from earlier chapters. Answers to selected exercises are provided at the back of the book.

- **Chapter Practice Test**, which provides a formal self-evaluation of the student's mastery of the material before being tested in class. Answers to test questions are provided at the back of the book.

Chapter Practice Test

PART I: Knowing the Definitions

Answer "True" if the statement is always true. If the statement is not always true, replace the words printed in bold with words that make the statement always true.

1.1 **Inferential** statistics is the study and description of data that result from an experiment.

1.2 **Descriptive statistics** is the study of a sample that enables us to make projections or estimates about the population from which the sample is drawn.

1.3 A **population** is typically a very large collection of individuals or objects about which we desire information.

1.4 A statistic is the calculated measure of some characteristic of a **population**.

1.5 A parameter is the measure of some characteristic of a **sample**.

1.6 As a result of surveying 50 freshmen, it was found that 16 had participated in interscholastic sports, 23 had served as officers of classes and clubs, and 18 had been in school plays during their high school years. This is an example of **numerical data**.

1.7 The "number of rotten apples per shipping crate" is an example of a **qualitative** variable.

1.8 The "thickness of a sheet of sheet metal" used in a manufacturing process is an example of a **quantitative** variable.

1.9 A **representative** sample is a sample obtained in such a way that all individuals had an equal chance of being selected.

1.10 The basic objectives of **statistics** are obtaining a sample, inspecting this sample, and then making inferences about the unknown characteristics of the population from which the sample was drawn.

PART II: Applying the Concepts

The owners of Corner Convenience Store are concerned about the quality of service their customers receive. In order to study the service, they collected samples for each of several variables.

1.11 Classify each of the following variables as nominal, ordinal, discrete, or continuous:

a. Method of payment for purchases (cash, credit card, check)

Updated Technology Instructions for MINITAB, Excel, and TI-83/84 appear throughout each chapter and are now color coded for easy reference. Provided alongside corresponding material, these instructions allow instructors to easily choose which statistical technology, if any, they want to incorporate into their courses.

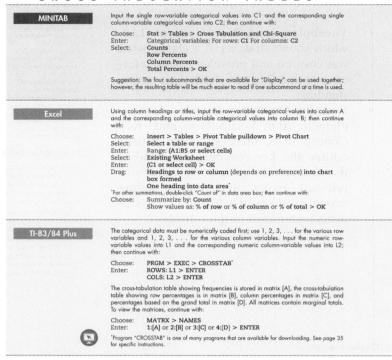

TECHNOLOGY INSTRUCTIONS:
CROSS-TABULATION TABLES

MINITAB

Input the single row-variable categorical values into C1 and the corresponding single column-variable categorical values into C2; then continue with:

Choose: **Stat > Tables > Cross Tabulation and Chi-Square**
Enter: Categorical variables: For rows: **C1** For columns: **C2**
Select: **Counts**
         Row Percents
         Column Percents
         Total Percents > OK

Suggestion: The four subcommands that are available for "Display" can be used together; however, the resulting table will be much easier to read if one subcommand at a time is used.

Excel

Using column headings or titles, input the row-variable categorical values into column A and the corresponding column-variable categorical values into column B; then continue with:

Choose: **Insert > Tables > Pivot Table pulldown > Pivot Chart**
Select: **Select a table or range**
Enter: Range: **(A1:B5 or select cells)**
Select: **Existing Worksheet**
Enter: **(C1 or select cell) > OK**
Drag: **Headings to row or column** (depends on preference) into chart box formed
         One heading into data area°
°For other summations, double-click "Count of" in data area box; then continue with:
Choose: Summarize by: **Count**
         Show values as: **% of row** or **% of column** or **% of total > OK**

TI-83/84 Plus

The categorical data must be numerically coded first; use 1, 2, 3, . . . for the various row variables and 1, 2, 3, . . . for the various column variables. Input the numeric row-variable values into L1 and the corresponding numeric column-variable values into L2; then continue with:

Choose: **PRGM > EXEC > CROSSTAB**°
Enter: **ROWS: L1 > ENTER**
         COLS: L2 > ENTER

The cross-tabulation table showing frequencies is stored in matrix [A], the cross-tabulation table showing row percentages is in matrix [B], column percentages in matrix [C], and percentages based on the grand total in matrix [D]. All matrices contain marginal totals. To view the matrices, continue with:

Choose: **MATRX > NAMES**
Enter: **1:[A] or 2:[B] or 3:[C] or 4:[D] > ENTER**

°Program "CROSSTAB" is one of many programs that are available for downloading. See page 35 for specific instructions.

NEW AND UPDATED Data sets, totaling more than 400 and ranging from small to large, give students ample opportunity to practice using their statistical calculator or computer.

Technology Manuals offer additional instruction in these various statistical technologies. The following manuals are available online:

- MINITAB Manual by Diane L. Benner and Linda M. Myers, Harrisburg Area Community College
- Excel Manual by Diane L. Benner and Linda M. Myers, Harrisburg Area Community College
- TI-83/84 Manual by Kevin Fox, Shasta College

Note : These manuals are available in print as well as online. Instructors, contact your Cengage Learning sales representative or our customer service group to learn about how these manuals can be customized for your course.

This edition's additional valuable assets, changes, and improvements include

- Enhanced ogive coverage and discussion to improve overall usefulness and student comprehension (Chapter 2)
- Early introduction and coverage of bivariate data to ensure a logical progression of topics (Chapter 3)
- Increased focus on analysis and comprehension as opposed to a formula-driven approach to probability (Chapter 4)
- Timely association between the U.S. 2010 Census and sampling distributions (Chapter 7)

Quantitative Data

One major reason for constructing a graph of **quantitative data** is to display its *distribution*.

> **Distribution** The pattern of variability displayed by the data of a variable. The distribution displays the frequency of each value of the variable.

One of the simplest graphs used to display a distribution is the *dotplot*.

- Pedagogical flexibility with side-by-side *p*-value and classical approaches to hypothesis testing (Chapters 9–14)
- Reorganization of selected in-chapter sections for increased clarity with respect to the connection of topics (Chapter 5)
- Relevant forms of statistical terms/formulas added to supplement various chapters
- In-chapter running definitions now even easier to spot

Related Teaching and Learning Resources

UPDATED **Student Solution Manual.** Written by Patricia Kuby, this resource contains fully worked out solutions for all odd-numbered exercises, giving students a way to check their answers and ensure that they took the correct steps to arrive at an answer. It also provides hints, tips, and additional interpretation for specific exercises.

Annotated Instructor's Edition, *Elementary Statistics*, 11th Edition

This instructor version of the text features annotated exercise answers on pages with exercise sets, including exercises for which solutions are not provided in the answer key in the back of the book.

PowerLecture™ for Elementary Statistics, 11th Edition

This disc provides the instructor with dynamic media tools for teaching, including Microsoft® PowerPoint® lecture slides and figures from the book. Create, deliver, and customize tests (both print and online) in minutes with ExamView® Computerized Testing, featuring algorithmic equations. You'll also find a link to the Solution Builder online solutions manual, allowing you to easily build solution sets for homework or exams.

ExamView Assessment Suite for Elementary Statistics, 11th Edition

Available on the PowerLecture™ disc and featuring automatic grading, ExamView® testing software allows instructors to quickly create, deliver, and customize tests for class in print and online formats. The program includes a test bank with hundreds of questions customized directly to the text, with all questions also provided in PDF and Microsoft® Word formats for instructors who opt not to use the software component.

NEW Solution Builder for Elementary Statistics, 11th Edition

This online instructor database offers complete worked solutions to all exercises in the text, allowing you to create customized, secure solutions printouts (in PDF format) matched exactly to the problems you assign in class. www.cengage.com/solutionbuilder

NEW Statistics CourseMate

The **Statistics CourseMate** site for Johnson/Kuby's *Elementary Statistics* 11th Edition brings chapter topics to life with interactive learning, study, and exam preparation tools, including quizzes and flashcards for the Vocabulary and Key Concepts listed at the end of each chapter. The site also provides an **eBook** version of the text with highlighting and note taking capabilities. Throughout the chapters, the CourseMate icon 🖥 flags concepts and examples that have corresponding interactive resources such as **video** and **animated tutorials** that demonstrate, step by step, how to solve problems; **datasets** for exercises and examples; **Skillbuilder Applets** to help students better understand concepts; **technology manuals**; and software to download including **Data Analysis Plus** and **TI-83/84 Plus** programs. For instructors, this text's CourseMate also includes Engagement Tracker, a first-of-its-kind tool that monitors student engagement in the course. Go to login.cengage.com to access these resources.

UPDATED AND IMPROVED Book Companion Website

This resource offers book- and course-specific resources such as data sets for exercises and self quizzes. Students access this site's resources through cengage-brain.com. Instructors access password-protected resources by signing into their accounts through login.cengage.com.

NEW Aplia™

Aplia™ for Elementary Statistics, Eleventh Edition, is an online interactive learning solution that improves comprehension and outcomes by increasing student effort and engagement. Founded by a professor to enhance his own courses, Aplia provides automatically graded assignments with detailed, immediate explanations on every question, and innovative teaching materials. Our easy to use system has been used by more than 1,000,000 students at over 1800 institutions.

NEW Enhanced WebAssign

Exclusively from Cengage Learning, Enhanced WebAssign® offers an extensive online program for Statistics to encourage the practice that's so critical for concept mastery. The meticulously crafted pedagogy and exercises in this proven text become even more effective in Enhanced WebAssign, supplemented by multimedia support and immediate feedback as students complete their assignments. You can assign as many as 1,000 homework problems that match this text's section exercises. Students benefit from a interactive eBook with search and highlighting features, a "practice another version" feature on many problems (activated at your discretion), and links to solution videos, interactive tutorials, and even live online help. An Enhanced WebAssign® Start Smart Guide is available for students (see below).

NEW Enhanced WebAssign® Start Smart Guide for Students

The Enhanced WebAssign® Student Start Smart Guide helps students get up and running quickly with Enhanced WebAssign so they can study smarter and improve their performance in class.

Note: Students access all complimentary and premium online study and homework resources through cengagebrain.com. Numerous package options are available. Please see the additional information provided on the inside front cover or contact your sales representative or Cengage Learning customer service group.

Acknowledgments

It is a pleasure to acknowledge the aid and encouragement we have received form students and colleagues at Monroe Community College throughout the development of this text. Likewise, we are grateful to the reviewers and survey respondents who offered invaluable guidance as we planned this new edition:

Roger Abernathy, Orange Coast College
Lisa W. Kay, Eastern Kentucky University
Francis Nacozy, Miracosta Community College and Palomar Community College
Phillip Nissen, Georgia State University

Maureen Petkewich, University of South Carolina
Mehdi Safaee, Southwestern College and Grossmont College
Heyday Zahedani, California State University at San Marcos

Finally, we would again like to thank the reviewers who read and offered suggestions for previous editions:

Nancy Adcox, *Mt. San Antonio College*
Paul Alper, *College of St. Thomas*
William D. Bandes, *San Diego Mesa College*
Matrese Benkofske, *Missouri Western State College*
Tim Biehler, *Fingerlakes Community College*
Barbara Jean Blass, *Oakland Community College*
Austin Bonis, *Rochester Institute of Technology*
Nancy C. Bowers, *Pennsylvania College of Technology*
Shane Brewer, *College of Eastern Utah, San Juan Campus*
Robert Buck, *Slippery Rock University*
Louis F. Bush, *San Diego City College*
Ronnie Catipon, *Franklin University*
Rodney E. Chase, *Oakland Community College*
Pinyuen Chen, *Syracuse University*
Wayne Clark, *Parkland College*
David M. Crystal, *Rochester Institute of Technology*
Joyce Curry and Frank C. Denny, *Chabot College*
Larry Dorn, *Fresno Community College*
Shirley Dowdy, *West Virginia University*
Thomas English, *Pennsylvania State University, Erie*
Kenneth Fairbanks, *Murray State University*
Dr. William P. Fox, *Francis Marion University*
Joan Garfield, *University of Minnesota General College*
Monica Geist, *Front Range Community College*
David Gurney, *Southeastern Louisiana University*
Edwin Hackleman
Carol Hall, *New Mexico State University*
Silas Halperin, *Syracuse University*
Noal Harbertson, *California State University, Fresno*
Hank Harmeling, *North Shore Community College*
Bryan A. Haworth, *California State College at Bakersfield*
Harold Hayford, *Pennsylvania State University, Altoona*
Jim Helms, *Waycross College*
Marty Hodges, *Colorado Technical University*
John C. Holahan, *Xerox Corporation*
James E. Holstein, *University of Missouri*
Soon B. Hong, *Grand Valley State University*
Robert Hoyt, *Southwestern Montana University*
Peter Intarapanach, *Southern Connecticut State University*
T. Henry Jablonski, Jr., *East Tennessee State University*
Brian Jean, *Bakersfield University*
Jann-Huei Jinn, *Grand Valley State University*
Sherry Johnson
Meyer M. Kaplan, *The William Patterson College of New Jersey*
Michael Karelius, *American River College*
Anand S. Katiyar, *McNeese State University*
Jane Keller, *Metropolitan Community College*
Gayle S. Kent, *Florida Southern College*
Andrew Kim, *Westfield State College*
Amy Kimchuk, *University of the Sciences in Philadelphia*
Raymond Knodel, *Bemidji State University*
Larry Lesser, *University of Northern Colorado*

Natalie Lochner, *Rollins College*
Robert O. Maier, *El Camino College*
Linda McCarley, *Bevill State Community College*
Mark Anthony McComb, *Mississippi College*
Carolyn Meitler, *Concordia University Wisconsin*
John Meyer, *Muhlenberg College*
Jeffrey Mock, *Diablo Valley College*
David Naccarato, *University of New Haven*
Harold Nemer, *Riverside Community College*
John Noonan, *Mount Vernon Nazarene University*
Dennis O'Brien, *University of Wisconsin, LaCrosse*
Chandler Pike, *University of Georgia*
Daniel Powers, *University of Texas, Austin*
Janet M. Rich, *Miami-Dade Junior College*
Larry J. Ringer, *Texas A & M University*
John T. Ritschdorff, *Marist College*
John Rogers, *California Polytechnic Institute at San Luis Obispo*
Neil Rogness, *Grand Valley State University*
Thomas Rotolo, *University of Arizona*
Barbara F. Ryan and Thomas A. Ryan, *Pennsylvania State University*
Robert J. Salhany, *Rhode Island College*
Melody Smith, *Dyersburg State Community College*
Dr. Sherman Sowby, *California State University, Fresno*
Roger Spalding, *Monroe County Community College*
Timothy Stebbins, *Kalamazoo Valley Community College*
Howard Stratton, *State University of New York at Albany*
Larry Stephens, *University of Nebraska-Omaha*
Paul Stephenson, *Grand Valley State University*
Richard Stockbridge, *University of Wisconsin, Milwaukee*
Thomas Sturm, *College of St. Thomas*
Edward A. Sylvestre, *Eastman Kodak Co.*
Gwen Terwilliger
William K. Tomhave, *Concordia College, Moorhead, MN*
Bruce Trumbo, *California State University, Hayward*
Richard Uschold, *Canisius College*
John C. Van Druff, *Fort Steilacoom Community College*
Philip A. Van Veidhuizen, *University of Alaska*
John Vincenzi, *Saddleback College*
Kenneth D. Wantling, *Montgomery College*
Joan Weiss, *Fairfield University*
Mary Wheeler, *Monroe Community College*
Barbara Whitney, *Big Bend Community College*
Sharon Whitton, *Hofstra University*
Don Williams, *Austin College*
Rebecca Wong, *West Valley College*
Pablo Zafra, *Kean University*
Yvonne Zubovic, *Indiana University Purdue University, Fort Wayne*

Robert Johnson
Patricia Kuby

1 Statistics

© 2010 Erik Isakson/Jupiterimages Corporation

© 2010 Ryan McVay/Jupiterimages Corporation

1.1 What Is Statistics?
*Statistics are used to **describe** every aspect of our daily life.*

1.2 Measurability and Variability
*Statistics is a study of **variability**.*

1.3 Data Collection
*Selecting a **representative sample** using the **random method***

1.4 Statistics and Technology
Today's state of the art

1.1 What Is Statistics?

Americans, Here's Looking at You

Consider the graphic "Fretting Over Messages." If you had been asked, "How long do you go before getting 'antsy' about checking e-mail, instant messaging, and social networking sites?" how would you have answered? Do you believe that the diagram accurately depicts your answer? Now, don't you wonder how and where this information was obtained?

How often is your bed made? Find your answer on the graphic "Making the Bed." Does the frequency of bed making shown on the graph seem to suggest what you believe actually happens with all beds?

These "statistical" tidbits come from a variety of sources and present only a tiny sampling of what can be learned about Americans.

Fretting Over Messages

Are you fretting about messages?
How Wi-Fi users responded when asked how long they go before they get "antsy" about checking e-mail, instant messaging and social networking sites:

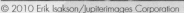

One hour or less **47%**

One day **46%**

One week **7%**

Source: Impulse Research for Qwest Communications online survey of 1,063 adult Wi-Fi users in April 2009.

Making the Bed

How often is your bed made?
Four percent of women say never, and 2% say only when company comes. Other responses:

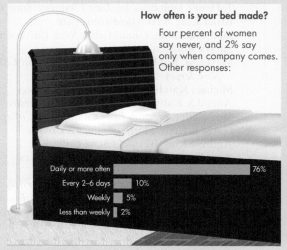

Daily or more often **76%**
Every 2–6 days **10%**
Weekly **5%**
Less than weekly **2%**

Source: Consumer Reports National Research Center survey of 1,008 women. Margin of error ±3.2 percentage points.

FYI A great source of information on Americans is the *Statistical Abstract of the United States* published annually by the U.S. Census Bureau (http://www.census.gov/). From the 1000+ page book or the website, you can find a statistical insight into many of the most obscure and unusual facets of our lives. Consider: "How many hours do we work and play?" "How much do we spend on snack food?" or "What source is one of the largest consumers of renewable energy?" The questions, data, and statistics stretch far and wide!

As you study statistics, you will learn how to read and analyze many types of statistical measures so that you can then arrive at appropriate conclusions.

So as we embark on our journey into the study of statistics, we must begin with the definition of *statistics* and expand on the details involved.

Statistics has become the universal language of the sciences. As potential users of statistics, we need to master both the "science" and the "art" of using statistical methodology correctly. Careful use of statistical methods will enable us to obtain accurate information from data. These methods include (1) carefully defining the situation, (2) gathering data, (3) accurately summarizing the data, and (4) deriving and communicating meaningful conclusions.

Statistics involves information, numbers, and visual graphics to summarize this information, and its interpretation. The word *statistics* has different meanings to people of varied backgrounds and interests. To some people it is a field of "hocus-pocus" in which a person attempts to overwhelm others with incorrect information and conclusions. To others it is a way of collecting and displaying information. And to still another group it is a way of "making decisions in the face of uncertainty." In the proper perspective, each of these points of view is correct.

The field of statistics can be roughly subdivided into two areas: descriptive statistics and inferential statistics. **Descriptive statistics** is what most people think of when they hear the word *statistics*. It includes the collection, presentation, and description of sample data. The term **inferential statistics** refers to the technique of interpreting the values resulting from the descriptive techniques and making decisions and drawing conclusions about the population.

Statistics is more than just numbers; it is data, what is done to data, what is learned from the data, and the resulting conclusions. Let's use the following definition:

Statistics The science of collecting, describing, and interpreting data.

Before we begin our detailed study, let's look at a few illustrations of how and when statistics can be applied.

APPLIED EXAMPLE 1.1

FISH'S AGE

Jose Luis Pelaez/Photographer's Choice/Getty Images

HOW OLD IS MY FISH
Average age by length of largemouth bass in New York State.

Length, in.	8	9	10	11	12	13	14
Age, yrs.	2	3	3	4	4	5	5

Source: NYS DEC Freshwater Fishing Guide

Forget about my dad's and my grandpa's age, I just want to know, "How old is my fish?" How does that work? **Statistics!** You will learn about "averages" in Chapter 2. This information also seems to imply that if the fish's length is measured, the fish's age is then known. Additional statistical techniques can be used to describe the relationship between the fish's age based on the fish's length, and as a result age can be estimated. You will learn about the statistical method for data like this in Chapter 3.

APPLIED EXAMPLE 1.2

OH, THE CONVENIENCE OF TECHNOLOGY

Do you have a cell phone? Are you talking or texting at times you shouldn't be? Consider the teens and young drivers who were surveyed below. Are they focused appropriately while in class or while driving? Do you see yourself in either of these situations?

Many Teens Use Phones in Class

84% teens have cell phones

16% teens do not

An average of 440 text messages sent per week –110 of them during class. Works out to be 3 text messages per class period.

Source:
Common Sense Media survey of 1013 teens May/June 2009

Busy Behind the Wheel

Most drivers ages 16–20 admit to risky driving habits.

16–20 year olds who say they have driven and done this:

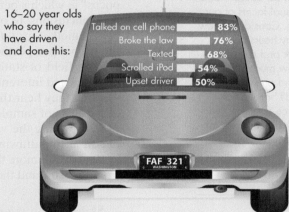

Talked on cell phone	83%
Broke the law	76%
Texted	68%
Scrolled iPod	54%
Upset driver	50%

FAF 321
WASHINGTON

Source: National Organization for Youth Safety, Allstate Foundation online survey of 605 drivers ages 16–20. (6/16/09)

Much information is given in these graphics about teens and young drivers. A vast majority of teens have cell phones and are using them all the time, including when they should not be, in the classroom and on the road. Consider what information was collected to formulate these graphics—first and foremost, cell phone status; number of text messages per week; number of text messages during class per week; and types of activities while driving. How would the organizations responsible for the surveys use that collected information to come up with the 84% and 83% shown in the above graphics?

Always take note of the source for published statistics (and any other provided detail); it will tell you much about the information being presented. In these cases, both are national organizations. Allstate is a funding partner for the National Youth Health and Safety Coalition, and Common Sense Media is a respected leader on children and media issues. These "source" details can give you a clue as to the quality of the information. Also note the type of survey used, if given, as it can offer additional insight about the quality. What is an online survey? How do they work? Are the results reliable?

The print media publish graphs and charts telling us how various organizations or people think as a whole. Do you ever wonder: How much of what you think is directly influenced by the information you read in these articles? Do you ever ponder the concern, "Is this information biased?"

APPLIED EXAMPLE 1.3

Employers Look for Positive Attitude

What do employers look for in a seasonal employee?

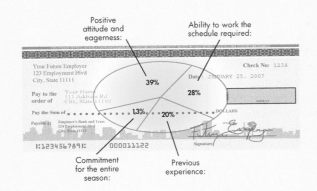

Positive attitude and eagerness:

Ability to work the schedule required:

39%

28%

13%   20%

Commitment for the entire season:

Previous experience:

Source: SnagAjob.com survey of 1,043 hiring managers. Margin of error: ±3 percentage points

WHAT EMPLOYERS ARE LOOKING FOR

This graphic on the left reports that 39% of employers consider a positive attitude and eagerness as the top qualities for a seasonal employee. Where did this information come from? Is it true? Note the source, SnagAJob.com. Notice that the organization conducted a survey of 1,043 hiring managers. How was this information collected? How was the collected information converted into the information reported? A margin of error is also reported as ± 3 percentage points. Based on this additional detail, the 39% on the graph becomes "between 36% and 42% of employers look for a positive attitude and eagerness in their seasonal employees."

You will learn about margin of error in Chapter 8.

APPLIED EXAMPLE 1.4

SHARK ATTACK

Consider the **International Shark Attack File** (ISAF), which is a compilation of all known shark attacks that is administered by the American Elasmobranch Society and the Florida Museum of Natural History and is shown in the graph and chart below.

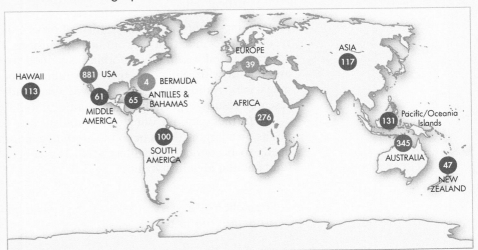

Source: http://www.flmnh.ufl.edu/fish/sharks/statistics/GAttack/World.htm

Territory	Total Attacks	Fatal Attacks	Last Fatality
USA (w/out Hawaii)	881	38	2005
Australia	345	135	2006
Africa	276	70	2004
Asia	117	55	2000
Pacific/Oceania Islands (w/out Hawaii)	131	50	2007
Hawaii	113	15	2004
South America	100	23	2006

Territory	Total Attacks	Fatal Attacks	Last Fatality
Antilles & Bahamas	65	19	1972
Middle America	61	31	1997
New Zealand	47	9	1968
Europe	39	19	1984
Bermuda	4	0	
Unspecified	20	6	1965
WORLD	2,199	470	2007

Common sense? Using common sense while reviewing the graph, one would certainly stay away from the United States if he or she enjoys the ocean. The United States has two-fifths of the world's shark attacks! U.S. waters must be full of sharks and the sharks must be mad!

Common sense—remember? What is going on with this graph? Is it a bit misleading? What else could be influencing the statistics shown here? First, one must take into consideration how much coastline of a country or continent comes in contact with an ocean.

Secondly, who is keeping track of these attacks? Notice the source of the map and chart, the Florida Museum of Natural History, a museum in the United States. Apparently, the United States is trying to keep track of unprovoked shark attacks. What else is different about the United States as compared to the other areas? Is the ocean a recreational area in the other places? What is the economy of these other areas and/or who is keeping track of their shark attacks?

FYI Statistics Is a Tricky Business
"One ounce of statistics technique requires one pound of common sense for proper application."

Remember to consider the source when reading a statistical report. Be sure you are looking at the complete picture.

The uses of statistics are unlimited. It is much harder to name a field in which statistics is not used than it is to name one in which statistics plays an integral part. The following are a few examples of how and where statistics are used:

- In education, descriptive statistics are frequently used to describe test results.
- In science, the data resulting from experiments must be collected and analyzed.
- In government, many kinds of statistical data are collected all the time. In fact, the U.S. government is probably the world's greatest collector of statistical data.

A very important part of the statistical process is that of studying the statistical results and formulating appropriate conclusions. These conclusions must then be communicated accurately—nothing is gained from research unless the findings are shared with others. Statistics are being reported everywhere: newspapers, magazines, radio, and television. We read and hear about all kinds of new research results, especially in the health-related fields.

To further continue our study of statistics, we need to "talk the talk." Statistics has its own jargon, terms beyond *descriptive statistics* and *inferential statistics*, that needs to be defined and illustrated. The concept of a population is the most fundamental idea in statistics.

Population A collection, or set, of individuals, objects, or events whose properties are to be analyzed.

The population is the complete collection of individuals or objects that are of interest to the sample collector. The population of concern must be carefully defined and is considered fully defined only when its membership list of elements is specified. The set of "all students who have ever attended a U.S. college" is an example of a well-defined population.

Typically, we think of a population as a collection of people. However, in statistics the population could be a collection of animals, manufactured objects, whatever. For example, the set of all redwood trees in California could be a population.

There are two kinds of populations: finite and infinite. When the membership of a population can be (or could be) physically listed, the population is said to be **finite**. When the membership is unlimited, the population is **infinite**. The books in your college library form a finite population; the OPAC (Online Public Access Catalog, the computerized card catalog) lists the exact membership. All the registered voters in the United States form a very large finite population; if necessary, a composite of all voter lists from all voting precincts across the United States could be compiled. On the other hand, the population of all people who might use aspirin and the population of all 40-watt light bulbs to be produced by General Electric are infinite. Large populations are difficult to study; therefore, it is customary to select a *sample* and study the data in that sample.

> **Sample** A subset of a population.

A sample consists of the individuals, objects, or measurements selected from the population by the sample collector.

> **Variable (or response variable)** A characteristic of interest about each individual element of a population or sample.

A student's age at entrance into college, the color of the student's hair, the student's height, and the student's weight are four variables.

> **Data value** The value of the variable associated with one element of a population or sample. This value may be a number, a word, or a symbol.

For example, Bill Jones entered college at age "23," his hair is "brown," he is "71 inches" tall, and he weighs "183 pounds." These four data values are the values for the four variables as applied to Bill Jones.

> **Data** The set of values collected from the variable from each of the elements that belong to the sample. Once all the data are collected, it is common practice to refer to the set of data as the sample.

The set of 25 heights collected from 25 students is an example of a set of data.

> **Experiment** A planned activity whose results yield a set of data.

An experiment includes the activities for both selecting the elements and obtaining the data values.

> **Parameter** A numerical value summarizing all the data of an entire population.

The "average" age at time of admission for all students who have ever attended our college and the "proportion" of students who were older than 21 years of age when they entered college are examples of two population parameters. A parameter is a value that describes the entire population. Often a Greek letter is used to symbolize the name of a parameter. These symbols will be assigned as we study specific parameters.

FYI Parameters describe the population; notice that both words begin with the letter **p**. A statistic describes the sample; notice that both words begin with the letter **s**.

For every parameter there is a *corresponding sample statistic.* The statistic describes the sample the same way the parameter describes the population.

> **Statistic** A numerical value summarizing the sample data.

The "average" height, found by using the set of 25 heights, is an example of a sample statistic. A statistic is a value that describes a sample. Most sample statistics are found with the aid of formulas and are typically assigned symbolic names that are letters of the English alphabet (for example, $\bar{x}$, s, and r).

E X A M P L E   1 . 5

 APPLYING THE BASIC TERMS

A statistics student is interested in finding out something about the average dollar value of cars owned by the faculty members of our college. Each of the eight terms just described can be identified in this situation.

1. The *population* is the collection of all cars owned by all faculty members at our college.
2. A *sample* is any subset of that population. For example, the cars owned by members of the mathematics department is a sample.
3. The *variable* is the "dollar value" of each individual car.
4. One *data value* is the dollar value of a particular car. Mr. Jones's car, for example, is valued at $9400.
5. The *data* are the set of values that correspond to the sample obtained (9400; 8700; 15,950; . . .).
6. The *experiment* consists of the methods used to select the cars that form the sample and to determine the value of each car in the sample. It could be carried out by questioning each member of the mathematics department, or in other ways.
7. The *parameter* about which we are seeking information is the "average" value of all cars in the population.
8. The *statistic* that will be found is the "average" value of the cars in the sample.

FYI Parameters are fixed in value, whereas statistics vary in value.

Note: If a second sample were to be taken, it would result in a different set of people being selected—say, the English department—and therefore a different value would be anticipated for the statistic "average value." The average value for "all faculty-owned cars" would not change, however.

There are basically two kinds of variables: (1) variables that result in *qualitative* information and (2) variables that result in *quantitative* information.

> **Qualitative, or attribute, or categorical, variable** A variable that describes or categorizes an element of a population.
>
> **Quantitative, or numerical, variable** A variable that quantifies an element of a population.

Video tutorial available—logon and learn more at cengagebrain.com

A sample of four hair-salon customers was surveyed for their "hair color," "hometown," and "level of satisfaction" with the results of their salon treatment. All three variables are examples of qualitative (attribute) variables because they describe some characteristic of the person, and all people with the same attribute belong to the same category. The data collected were {blonde, brown, black, brown}, {Brighton, Columbus, Albany, Jacksonville}, and {very satisfied, satisfied, somewhat satisfied, not satisfied}.

The "total cost" of textbooks purchased by each student for this semester's classes is an example of a quantitative (numerical) variable. A sample resulted in the following data: $238.87, $94.57, $139.24. [To find the "average cost," simply add the three numbers and divide by 3: $(238.87 + 94.57 + 139.24)/3 = 157.56.]

Note: Arithmetic operations, such as addition and averaging, are meaningful for data that result from a quantitative variable.

Each of these types of variables (qualitative and quantitative) can be further subdivided as illustrated in the following diagram.

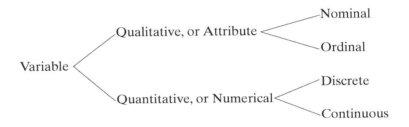

Qualitative variables may be characterized as nominal or ordinal.

> **Nominal variable** A qualitative variable that characterizes (or describes, or names) an element of a population. Not only are arithmetic operations not meaningful for data that result from a nominal variable, but an order cannot be assigned to the categories.

In the survey of four hair-salon customers, two of the variables, "hair color" and "hometown," are examples of nominal variables because both name some characteristic of the person and it would be meaningless to find the sample average by adding and dividing by 4. For example, (blonde + brown + black + brown)/4 is undefined. Furthermore, color of hair and hometown do not have an order to their categories.

> **Ordinal variable** A qualitative variable that incorporates an ordered position, or ranking.

In the survey of four hair-salon customers, the variable "level of satisfaction" is an example of an ordinal variable because it does incorporate an ordered ranking: "Very satisfied" ranks ahead of "satisfied," which ranks ahead of "somewhat satisfied." Another illustration of an ordinal variable is the ranking of five landscape pictures according to someone's preference: first choice, second choice, and so on.

Quantitative or numerical variables can also be subdivided into two classifications: *discrete* variables and *continuous* variables.

> **Discrete variable** A quantitative variable that can assume a countable number of values. Intuitively, the discrete variable can assume any values corresponding to isolated points along a line interval. That is, there is a gap between any two values.
>
> **Continuous variable** A quantitative variable that can assume an uncountable number of values. Intuitively, the continuous variable can assume any value along a line interval, including every possible value between any two values.

In many cases, the two types of variables can be distinguished by deciding whether the variables are related to a count or a measurement. The variable "number of courses for which you are currently registered" is an example of a discrete variable; the values of the variable may be found by counting the courses. (When we count, fractional values cannot occur; thus, gaps can occur between the values.) The variable "weight of books and supplies you are carrying as you attend class today" is an example of a continuous random variable; the values of the variable may be found by measuring the weight. (When we measure, any fractional value can occur; thus, every value along the number line is possible.)

When trying to determine whether a variable is discrete or continuous, remember to look at the variable and think about the values that might occur. Do not look at only data values that have been recorded; they can be very misleading.

Consider the variable "judge's score" at a figure-skating competition. If we look at some scores that have previously occurred, 9.9, 9.5, 8.8, 10.0, and we see the presence of decimals, we might think that all fractions are possible and conclude that the variable is continuous. This is not true, however. A score of 9.134 is impossible; thus, there are gaps between the possible values and the variable is discrete.

Note: Don't let the appearance of the data fool you in regard to their type. Qualitative variables are not always easy to recognize; sometimes they appear as numbers. The sample of hair colors could be coded: 1 = black, 2 = blonde, 3 = brown. The sample data would then appear as {2, 3, 1, 3}, but they are still nominal data. Calculating the "average hair color" [(2 + 3 + 1 + 3)/4 = 9/4 = 2.25] is still meaningless. The hometowns could be identified using ZIP codes. The average of the ZIP codes doesn't make sense either; therefore, ZIP code numbers are nominal too.

Let's look at another example. Suppose that after surveying a parking lot, I summarized the sample data by reporting 5 red, 8 blue, 6 green, and 2 yellow cars. You must look at each individual source to determine the kind of information being collected. One specific car was red; "red" is the data value from that one car, and red is an attribute. Thus, this collection (5 red, 8 blue, and so on) is a summary of nominal data.

Another example of information that is deceiving is an identification number. Flight #249 and Room #168 both appear to be numerical data. However, the numeral 249 does not describe any property of the flight—late or on time, quality of snack served, number of passengers, or anything else about the flight. The flight number only identifies a specific flight. Driver's license numbers, Social Security numbers, and bank account numbers are all identification numbers used in the nominal sense, not in the quantitative sense.

Remember to inspect the individual variable and one individual data value, and you should have little trouble distinguishing among the various types of variables.

APPLIED EXAMPLE 1.6

THE BIG PAYCHECK

World's Highest Paid Athletes

Forbes' list of the highest-paid athletes looks at earnings derived from salaries, bonuses, prize money, endorsements, and licensing income between June 2008 and June 2009 and does not deduct for taxes or agents' fees.

Here are the Top 5:

Rank	Athlete	Sport	Earnings
1	Tiger Woods	Golf	$110 million
2	Kobe Bryant	Basketball	$45 million
2	Michael Jordan	Basketball	$45 million
2	Kimi Raikkonen	Auto Racing	$45 million
5	David Beckham	Soccer	$42 million

Source: http://www.getlisty.com/preview/highest-paid-athletes/

Let's face it; most of us dream about making this much in our whole lifetime. If one is gainfully employed every year from 21 years old to 62 and makes $1 million each year, that would be $42 million for one's whole life. Most of us cannot even wrap our heads around that concept. You probably are thinking, "Sign me up to be a superstar athlete!"

Let's see how we can apply our new terminology to the "Big Paycheck." First, the general population of interest would be professional athletes. Furthermore, the information in the above table demonstrates the various types of variables. The athlete's name is generally not considered to be a variable; it is for the purpose of identification only. The other three kinds of information are variables:

1. Rank is qualitative and an ordinal variable since it incorporates the concept of ordered position.

2. Sport is qualitative and a nominal variable since it describes the athlete's sport.

3. Earnings is quantitative and a continuous variable as it is a measure of the athlete's income. Amounts of money are generally considered continuous since fractional parts of dollars are possible, even though the amount is generally rounded to the nearest dollar or penny.

SECTION 1.1 EXERCISES

1.1 Postyour.info is a worldwide service where Internet users from arround the world can take part in questionnaires. [http://postyour.info/] Below is a graph depicting the combined summary of how users answered one of the posted questions. Results given in Percent (count).

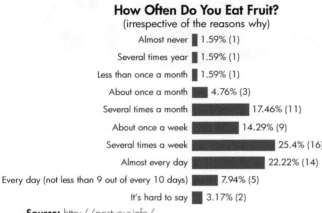

How Often Do You Eat Fruit?
(irrespective of the reasons why)

Almost never	1.59% (1)
Several times year	1.59% (1)
Less than once a month	1.59% (1)
About once a month	4.76% (3)
Several times a month	17.46% (11)
About once a week	14.29% (9)
Several times a week	25.4% (16)
Almost every day	22.22% (14)
Every day (not less than 9 out of every 10 days)	7.94% (5)
It's hard to say	3.17% (2)

Source: http://postyour.info/

a. What question was asked and answered in order to gather the information presented on this graph?

b. Who was asked the question?

c. How many people answered the question?

d. Verify the percentages 1.59%, 17.46%, and 25.4%.

e. Are the percentages reported in this graph likely to be representative of all people? Explain why or why not.

1.2 Do you work hard for your money? Java professionals think they do, reporting long working hours at their jobs. Java developers from around the world were surveyed about the number of hours they work weekly. Listed here are the average number of hours worked weekly in various regions of the United States and the world.

Region	Hours Worked	Region	Hours Worked
U.S.	48	California	50
Northeast	47	Pacific NW	47
Mid-Atlantic	49	Canada	43
South	47	Europe	48
Midwest	47	Asia	47
Central Mt	51	South America and Africa	49

Source: Jupitermedia Corporation

a. How many hours do you work per week (or anticipate working after you graduate)?

b. What happened to the 40-hour workweek? Does it appear to exist for the Java professional?

c. Does the information in this chart make a career as a Java professional seem attractive?

1.3 a. Each of the statistical graphics presented on the first page of this chapter seems to suggest that information is about what population? Is that the case? Justify your answer.

b. Describe the information that was collected and used to determine the statistics reported in "Are you fretting about messages?"

c. "47%—One hour or less" was one specific statistic reported in "Are you fretting about messages?" Describe what that statistic tells you.

1.4 a. Consider the graphic "How often is your bed made?" If you had been asked, how would you have responded? What does the percentage associated with your answer mean? Explain.

b. How do you interpret the "5%—Weekly" reported in "How often is your bed made?"

1.5 a. Write a 50-word paragraph describing what the word "statistics" means to you right now.

b. Write a 50-word paragraph describing what the word "random" means to you right now.

c. Write a 50-word paragraph describing what the word "sample" means to you right now.

1.6 *Statistics* is defined on page 1 as "the science of collecting, describing, and interpreting data." Using your own words, write a sentence describing each of the three statistical activities. Retain your work for Exercise 1.79.

1.7 Determine which of the following statements is descriptive in nature and which is inferential. Refer to "How old is my fish?" in Applied Example 1.1 (p. 1).

a. All 9-inch largemouth bass in New York State are an average of three years old.

b. Of the largemouth bass used in the sample to make up the *NYS DEC Freshwater Fishing Guide,* the average age of 9-inch largemouth bass was three years.

1.8 Determine which of the following statements is descriptive in nature and which is inferential. Refer to "Many teens use phones in class" in Applied Example 1.2 (p. 2).

a. 84% of the surveyed teens in May and June of 2009 had cell phones.

b. In May/June 2009, 16% of all teens did not have a cell phone.

1.9 Refer to the graphic "Setting a Date for Date Night."

Setting a Date for Date Night

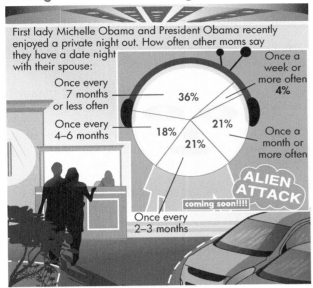

Source: Frigidaire Motherload Index survey of 1,170 married women, ages 25–50 who have two or more children.

a.   What group of people were polled?

b.   How many people were polled?

c.   What information was obtained from each person?

d.   Explain the meaning of "18% say once every 4–6 months."

e.   How many people answered "Once every 4–6 months"?

1.10 International Communications Research (ICR) conducted the 2008 Spring Cleaning Survey for the Soap and Detergent Association. ICR questioned 777 American adults who spring-clean about which spring-cleaning chore they would like to hire someone to do for them. The "chore" results were: 47% washing windows, 23% cleaning the bathroom, 12% cleaning the kitchen, 8% dusting, 7% mopping, 3% other. The survey has a margin of error of plus or minus 3.52%.

a.   What is the population?

b.   How many people were polled?

c.   What information was obtained from each person?

d.   Using the information given, estimate the number of surveyed adults who would gladly hire someone to wash the windows, if they could.

e.   What do you think the "margin of error of plus or minus 3.52%" means?

f.   How would you use the "margin of error" in estimating the percentage of all adults who would like to hire out the spring-cleaning chore of "Cleaning the kitchen"?

1.11 Opinion Research Corporation conducted the 2009 Lemelson-MIT Invention Index survey of 501 teens, ages 12–17. The teens were asked what everyday invention they thought would be obsolete in five years. Refer to the graphic "In One Day, Out the Next."

In One Day, Out the Next

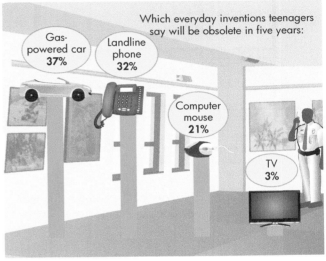

Source: 2009 Lemelson-MIT Invention Index survey of 501 teens, ages 12–17, by Opinion Research Corp. Margin of error ± 4.3 percentage points.

a.   What is the population?

b.   How many people were polled?

c.   What information was obtained from each person?

d.   Estimate the number of surveyed teens who thought the computer mouse will be obsolete in five years.

e.   What do you think the "margin of error" of ± 4.3 percentage points means?

f.   How would you use the "margin of error" in estimating the percentage of all teens who thought the computer mouse will be obsolete in five years?

1.12 Refer to the graphic on next page "What do you Plan to Spend your Tax Refund on?" (page 12).

a.   Describe the population of interest.

b.   Describe the sample most likely used for this report.

c.   Identify the variables used to collect this information.

(continue on page 12)

What do you Plan to Spend your Tax Refund on?

Note: Multiple responses allowed

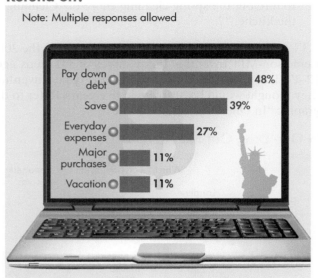

Pay down debt 48%
Save 39%
Everyday expenses 27%
Major purchases 11%
Vacation 11%

Source: National Retail Federation 2009 Tax Returns Consumer Intentions and Actions survey of 8,426 consumers. Margin of error ±1 percentage point.

d. What were the majority of people going to do with their tax refund? How does this majority show on the graph?

1.13 During a radio broadcast a few years ago, David Essel reported the following three statistics: (1) the U.S. divorce rate is 55%, and when married adults were asked if they would remarry their spouse, (2) 75% of the women said yes, and (3) 65% of the men said yes.

a. What is the "stay married" rate?

b. There seems to be a contradiction in this information. How is it possible for all three of these statements to be correct? Explain.

1.14 A working knowledge of statistics is very helpful when you want to understand the statistics reported in the news. The news media and our government often make a statement such as, "Crime rate jumped 50% in your city."

a. Does an increase in rate from 4 to 6 represent an increase of 50%? Explain.

b. Why would anybody report an increase from 4 to 6 as a "50% rate jump"?

1.15 Of the adult U.S. population, 36% has an allergy. A sample of 1200 randomly selected adults resulted in 33.2% having an allergy.

a. Describe the population.

b. What is the sample?

c. Describe the variable.

d. Identify the statistic and give its value.

e. Identify the parameter and give its value.

1.16 Find a recent newspaper article that illustrates an "apples are bad" type of report.

1.17 In your own words, explain why the parameter is fixed and the statistic varies.

1.18 Is a football jersey number a quantitative or a categorical variable? Support your answer with a detailed explanation.

1.19 a. Name two customer-related attribute variables that a newly opened department store might find informative to study.

b. Name two customer-related numerical variables that a newly opened department store might find informative to study.

1.20 a. Name two customer-related nominal variables that a newly opened department store might find informative to study.

b. Name two customer-related ordinal variables that a newly opened department store might find informative to study.

1.21 Skillbuilder Applet Exercise simulates taking a sample of size 10 from a population of 100 college students. Take one sample and note the outcome.

	64 inches
	73 inches
	64 inches
	70 inches
	70 inches
	69 inches
	68 inches
	64 inches
	63 inches
	68 inches

POPULATION: Mean = 66.9 inches   SAMPLE: Mean = 67.3 inches
Percent Female = 64.0%   Percent Female = 80.0%

a. Name the attribute variable involved in this experiment. Is it nominal or ordinal?

b. Name the numerical variable involved in this experiment. Is it discrete or continuous?

1.22 a. Explain why the variable "score" for the home team at a basketball game is discrete.

b. Explain why the variable "number of minutes to commute to work" is continuous.

1.23 The severity of side effects experienced by patients who are being treated with a particular medicine is under study. The severity is measured on a scale of none, mild, moderate, severe, very severe.

a. Name the variable of interest.

b. Identify the type of variable.

1.24 A nationwide poll of adults on cell phone use and driving was conducted during May 2009 by the Harris Poll.

Their responses to "How dangerous is it for a driver to use a cell phone while driving?" were categorized as "very dangerous," "dangerous," "somewhat dangerous," "slightly dangerous," or "not dangerous at all."

a.   Name the variable of interest.

b.   Identify the type of variable.

1.25 Students are being surveyed about the weight of books and supplies they are carrying as they attend class.

a.   Identify the variable of interest.

b.   Identify the type of variable.

c.   List a few values that might occur in a sample.

1.26 A drug manufacturer is interested in the proportion of persons with hypertension (elevated blood pressure) whose condition can be controlled by a new drug the company has developed. A study involving 5000 individuals with hypertension is conducted, and it is found that 80% of the individuals are able to control their hypertension with the drug. Assuming that the 5000 individuals are representative of the group that has hypertension, answer the following questions:

a.   What is the population?

b.   What is the sample?

c.   Identify the parameter of interest.

d.   Identify the statistic and give its value.

e.   Do we know the value of the parameter?

1.27 The admissions office wants to estimate the cost of textbooks for students at our college. Let the variable x be the total cost of all textbooks purchased by a student this semester. The plan is to randomly identify 100 students and obtain their total textbook costs. The average cost for the 100 students will be used to estimate the average cost for all students.

a.   Describe the parameter the admissions office wishes to estimate.

b.   Describe the population.

c.   Describe the variable involved.

d.   Describe the sample.

e.   Describe the statistic and how you would use the data collected to calculate the statistic.

1.28 A quality-control technician selects assembled parts from an assembly line and records the following information concerning each part:

X: defective or nondefective

Y: the employee number of the individual who assembled the part

Z: the weight of the part

a.   What is the population?

b.   Is the population finite or infinite?

c.   What is the sample?

d.   Classify the three variables as either attribute or numerical.

1.29 Select 10 students currently enrolled at your college and collect data for these three variables:

X: number of courses enrolled in
Y: total cost of textbooks and supplies for courses
Z: method of payment used for textbooks and supplies

a.   What is the population?

b.   Is the population finite or infinite?

c.   What is the sample?

d.   Classify the three variables as nominal, ordinal, discrete, or continuous.

1.30 A study was conducted by Aventis Pharmaceuticals Inc. to measure the adverse side effects of Allegra™, a drug used for the treatment of seasonal allergies. A sample of 679 allergy sufferers in the United States was given 60 mg amounts of the drug twice a day. The patients were to report whether or not they experienced relief from their allergies as well as any adverse side effects (viral infection, nausea, drowsiness, etc.).

Source: *Good Housekeeping*, February 2005

a.   What was the population being studied?

b.   What was the sample?

c.   What were the characteristics of interest about each element in the population?

d.   Were the data being collected qualitative or quantitative?

1.31 On the next page is a small sample of the 165 2009 pickup trucks listed on MPGoMatic.com and available to the buying public. Refer to the table for this exercise (1.31) on page 14.

a.   What was the population from which this sample was taken?

b.   How many individuals were in the population? In the sample?

c.   How many variables?

(continue on page 14)

Table for Exercise 1.31

Manufacturer	Model	Drive	Engine Size (no. cylinders)	Engine Size, Displacement (liters)	Transmission	City MPG	Hwy MPG
CHEVROLET	COLORADO	2WD	4	2.9	Manual	18	24
GMC	CANYON	2WD	5	3.7	Auto	17	23
HUMMER	H3T	4WD	8	5.3	Auto	13	16
MITSUBISHI	RAIDER	4WD	8	4.7	Auto	9	12
SUZUKI	EQUATOR	2WD	4	2.5	Auto	17	22
TOYOTA	TACOMA	4WD	6	4.0	Manual	14	19

Source: http://www.mpgomatic.com/2009/

d.   Name the qualitative/categorical variables.

e.   Which of the qualitative variables are nominal? Ordinal?

f.   Name the quantitative variables.

g.   Which of the quantitative variables are discrete? Continuous?

1.32 Skillbuilder Applet Exercise simulates taking a sample of size 10 from a population of 100 college students. Take a sample of size 10.

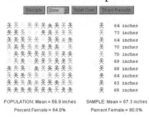

a.   What is the population?

b.   Is the population finite or infinite?

c.   Name two parameters and give their values.

d.   What is the sample?

e.   Name the two corresponding statistics and give their values.

f.   Take another sample of size 10. Which of the items above remain fixed and which changed?

1.33 Identify each of the following as examples of (1) attribute (qualitative) or (2) numerical (quantitative) variables:

a.   The breaking strength of a given type of string

b.   The hair color of children auditioning for the musical *Annie*

c.   The number of stop signs in towns of fewer than 500 people

d.   Whether or not a faucet is defective

e.   The number of questions answered correctly on a standardized test

f.   The length of time required to answer a telephone call at a certain real estate office

1.34 Identify each of the following as examples of (1) nominal, (2) ordinal, (3) discrete, or (4) continuous variables:

a.   A poll of registered voters as to which candidate they support

b.   The length of time required for a wound to heal when a new medicine is being used

c.   The number of televisions within a household

d.   The distance first-year college women can kick a football

e.   The number of pages per job coming from a computer printer

f.   The kind of tree used as a Christmas tree

1.35 Suppose a 12-year-old asked you to explain the difference between a sample and a population.

a.   What information should your answer include?

b.   What reasons would you give him or her for why one would take a sample instead of surveying every member of the population?

1.36 Suppose a 12-year-old asked you to explain the difference between a statistic and a parameter.

a.   What information should your answer include?

b.   What reasons would you give him or her for why one would report the value of a statistic instead of the value of a parameter?

1.2 Measurability and Variability

Within a set of measurement data, we always expect variation. If little or no variation is found, we would guess that the measuring device is not calibrated with a small enough unit. For example, we take a carton of a favorite candy bar and weigh each bar individually. We observe that each of the 24 candy bars weighs 7/8 ounce, to the nearest 1/8 ounce. Does this mean that the bars are all identical in weight? Not really! Suppose we were to weigh them on an analytical balance that weighs to the nearest ten-thousandth of an ounce. Now the 24 weights will most likely show **variability**.

It does not matter what the response variable is; there will most likely be variability in the data if the tool of measurement is precise enough. One of the primary objectives of statistical analysis is measuring variability. For example, in the study of quality control, measuring variability is absolutely essential. Controlling (or reducing) the variability in a manufacturing process is a field all its own—namely, statistical process control.

S E C T I O N   1 . 2   E X E R C I S E S

1.37 Suppose we measure the weights (in pounds) of the individuals in each of the following groups:

>   Group 1: cheerleaders for National Football League teams
>   Group 2: players for National Football League teams

For which group would you expect the data to have more variability? Explain why.

1.38 Suppose you are trying to decide which of two machines to purchase. Furthermore, suppose the length to which the machines cut a particular product part is important. If both machines produce parts that have the same length on the average, what other consideration regarding the lengths would be important? Why?

1.39 Consumer activist groups for years have encouraged retailers to use unit pricing of products. They argue that food prices, for example, should always be labeled in $/ounce, $/pound, $/gram, $/liter, etc., in addition to $/package, $/can, $/box, $/bottle, etc. Explain why.

1.40 A coin-operated coffee vending machine dispenses, on the average, 6 oz of coffee per cup. Can this statement be true of a vending machine that occasion-

ally dispenses only enough to barely fill the cup half full (say, 4 oz)? Explain.

1.41 Teachers use examinations to measure students' knowledge about a subject. Explain how "a lack of variability in the students' scores might indicate that the exam was not a very effective measuring device."

1.42 Skillbuilder Applet Exercise simulates sampling from a population of college students.

a.   Take 10 samples of size 4, keeping track of the sample averages of hours per week that students study. Find the range of these averages by subtracting the lowest average from the highest average.

b.   Take 10 samples of size 10, keeping track of the sample averages of hours per week that students study. Find the range of these averages by subtracting the lowest average from the highest average.

c.   Which sample size demonstrates more variability?

d.   If the population average is about 15 hours per week, which sample size demonstrates this most accurately? Why?

1.3 Data Collection

Because it is generally impossible to study an entire population (every individual in a country, all college students, every medical patient, etc.), researchers typically rely on *sampling* to acquire the information, or *data,* needed. It is important to obtain "good data" because the inferences ultimately made will be based on the statistics obtained from these data. These inferences are only as good as the data.

Although it is relatively easy to define "good data" as data that accurately represent the population from which they were taken, it is not easy to guarantee that a particular sampling method will produce "good data." We need to use sampling (**data collection**) methods that will produce data that are representative of the population and not *biased*.

> **Sampling method** The process of selecting items or events that will become the sample.
>
> **Biased sampling method** A sampling method that produces data that systematically differ from the sampled population. Repeated sampling will not correct the bias.
>
> **Unbiased sampling method** A sampling method that is not biased and produces data that are representative of the sampled population.

APPLIED EXAMPLE 1.7

A MODERN HIGH-TECH VOLUNTEER SAMPLE

Public Survey—Let's surprise NBC!
In December 2008, NBC posted the question below on their website to survey the public.

Live Vote   March 16, 2009, with 12,810,699 responses tallied

IN GOD WE TRUST
Should the motto "In God We Trust" be removed from U.S. currency?

Yes. It's a violation of the principle of separation of church and state
14%

No. The motto has historical and patriotic significance and does nothing to establish a state religion.
86%

At the same time, the e-mail below was being circulated to help "get out the vote."

> Here's your chance to let the media know where the people stand on our faith in God, as a nation. NBC is taking a poll on "In God We Trust" to stay on our American currency.
>
> Please send this to every Christian you know so they can vote on this important subject. Please do it right away, before NBC takes this off the web page.
>
> This is not sent for discussion; if you agree forward it, if you don't, delete it. By me forwarding it, you know how I feel. I'll bet this was a surprise to NBC.

No meaningful statistical conclusions can be drawn from this survey. The sampling process is severely flawed, and the results were very likely to be strongly biased and not representative of the American population. Can you give at least two reasons why the results of this survey do not represent good statistical practices? See Exercise 1.46.

Two commonly used sampling methods that often result in biased samples are the *convenience* and *volunteer samples.*

A **convenience sample,** sometimes called a *grab* sample, occurs when items are chosen arbitrarily and in an unstructured manner from a population, whereas a **volunteer sample** consists of results collected from those elements of the population that chose to contribute the needed information on their own initiative.

Did you ever buy a basket of fruit at the market based on the "good appearance" of the fruit on top, only to later discover that the rest of the fruit was not as fresh? It was too inconvenient to inspect the bottom fruit, so you trusted a convenience sample. Has your teacher used your class as a sample from which to gather data? As a group, the class is quite convenient, but is it truly representative of the school's population? (Consider the differences among day, evening, and/or weekend students; type of course; etc.)

Have you ever mailed back your responses to a magazine survey? Under what conditions did (would) you take the time to complete such a questionnaire? Most people's immediate attitude is to ignore the survey. Those with strong feelings will make the effort to respond; therefore, representative samples should not be expected when volunteer samples are collected.

The Data-Collection Process

The collection of data for statistical analysis is an involved process and includes the following steps:

1. Define the objectives of the survey or study.
   Examples: compare the effectiveness of a new drug to the effectiveness of the standard drug; estimate the average household income in the United States.
2. Define the variable and the population of interest.
   Examples: length of recovery time for patients suffering from a particular disease; total income for households in the United States.
3. Define the data collection and data-measuring schemes.
   This includes sampling frame, sampling procedures, sample size, and the data-measuring device (questionnaire, telephone, and so on).
4. Collect your sample. Select the subjects to be sampled and collect the data.
5. Review of the sampling process upon completion of collection.

Often an analyst is stuck with data already collected, possibly even data collected for other purposes, which makes it impossible to determine whether the data are "good." Using approved techniques to collect your own data is much preferred. Although this text is concerned chiefly with various data analysis techniques, you should be aware of the concerns of data collection.

The following example describes the population and the variable of interest for a specific investigation:

APPLIED EXAMPLE 1.8

POPULATION AND VARIABLE OF INTEREST

The admissions dean at our college wishes to estimate the current "average" cost of textbooks per semester, per student. The population of interest is the "currently enrolled student body," and the variable is the "total amount spent for textbooks" by each student this semester.

Two methods commonly used to collect data are *experiments* and *observational studies.* In an **experiment**, the investigator controls or modifies the environment and observes the effect on the variable under study. We often read about laboratory results obtained by using white rats to test different doses of a new medication and its effect on blood pressure. The experimental treatments were designed specifically to obtain the data needed to study the effect on the variable. In an **observational study**, the investigator does not modify the environment and does not control the process being observed. The data are obtained by sampling some of the population of interest. **Surveys** are observational studies of people.

APPLIED EXAMPLE 1.9

EXPERIMENT OR OBSERVATIONAL STUDY?

SURGICAL INFECTION IS A MATTER OF TIME

Many surgical patients fail to get timely doses of the right medications, raising the risk of infection, researchers write in the Archives of Surgery. Of 30 million operations performed each year in the USA, about 2% are complicated by an on-site infection, the report says. The study of 34,000 surgical patients at nearly 3,000 hospitals in 2001 found that only 56% got prophylactic medications within an hour of surgery, when they can be effective.

Source: *USA Today,* February 22, 2006

This study is an example of an observational study. The researchers did not modify or try to control the environment. They observed what was happening and wrote up their findings.

If every element in the population can be listed, or enumerated, and observed, then a **census** is compiled. However, censuses are seldom used because they are often difficult and time-consuming to compile, and therefore very expensive. Imagine the task of compiling a census of every person who is a potential client at a brokerage firm. In situations similar to this, a *sample survey* is usually conducted.

When selecting a sample for a survey, it is necessary to construct a *sampling frame.*

Sampling frame A list, or set, of the elements belonging to the population from which the sample will be drawn.

Ideally, the sampling frame should be identical to the population, with every element of the population included once and only once. In this case, a census would become the sampling frame. In other situations, a census may not be so easy to obtain because a complete list is not available. Lists of registered voters or the telephone directory are sometimes used as sampling frames of the general public. Depending on the nature of the information being sought, the list of registered voters or the telephone directory may or may not serve as an unbiased sampling frame. Because only the elements in the frame have a chance to be selected as part of the sample, it is important that the sampling frame be **representative** of the population.

Once a representative sampling frame has been established, we proceed with selecting the sample elements from the sampling frame. This selection process is called the **sample design**. There are many different types of sample designs; however, they all fit into two categories: *judgment samples* and *probability samples.*

> **Judgment samples** Samples that are selected on the basis of being judged "typical."

When a judgment sample is collected, the person selecting the sample chooses items that he or she thinks are representative of the population. The validity of the results from a judgment sample reflects the soundness of the collector's judgment. This is not an acceptable statistical procedure.

> **Probability samples** Samples in which the elements to be selected are drawn on the basis of probability. Each element in a population has a certain probability of being selected as part of the sample.

The inferences that will be studied later in this textbook are based on the assumption that our sample data are obtained using a probability sample. There are many ways to design probability samples. We will look at two of them, single-stage methods and multistage methods, and learn about a few of the many specific designs that are possible.

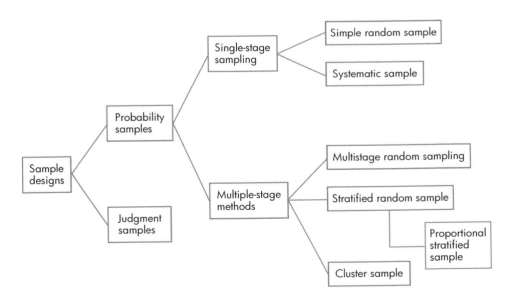

Single-Stage Methods

> **Single-stage sampling** A sample design in which the elements of the sampling frame are treated equally and there is no subdividing or partitioning of the frame.

One of the most common single-stage probability sampling methods used to collect data is the *simple random sample.*

> **Simple random sample** A sample selected in such a way that every element in the population or sampling frame has an equal probability of being chosen. Equivalently, all samples of size *n* have an equal chance of being selected.

Note: Random samples are obtained either by sampling with replacement from a finite population or by sampling without replacement from an infinite population.

Inherent in the concept of randomness is the idea that the next result (or occurrence) is not predictable. When a random sample is drawn, every effort must be made to ensure that each element has an equal probability of being selected and that the next result does not become predictable. The proper procedure for selecting a simple random sample requires the use of random numbers. Mistakes are commonly made because the term *random* (equal chance) is confused with **haphazard** (without pattern).

To select a simple random sample, first assign an identifying number to each element in the sampling frame. This is usually done sequentially using the same number of digits for each element. Then using random numbers with the same number of digits, select as many numbers as are needed for the sample size desired. Each numbered element in the sampling frame that corresponds to a selected random number is chosen for the sample.

EXAMPLE 1.10

USING RANDOM NUMBERS

The admissions office at our college wishes to estimate the current "average" cost of textbooks per semester, per student. The population of interest is the "currently enrolled student body," and the variable is the "total amount spent for textbooks" by each student this semester. Because a random sample is desired, Mr. Clar, who works in the admissions office, has obtained a computer list of this semester's full-time enrollment. There were 4265 student names on the list. He numbered the students 0001, 0002, 0003, and so on, up to 4265; then, using four-digit random numbers, he identified a sample: 1288, 2177, 1952, 2463, 1644, 1004, and so on, were selected. (See the *Student Solutions Manual* for a discussion of the use of random numbers.)

A simple random sample is our first step toward an unbiased sample. Random samples are required for most of the statistical procedures presented in this book. Without a random design, the conclusions we draw from the statistical procedures may not be reliable.

EXAMPLE 1.11

PROCESS FOR COLLECTING DATA

Consider the graphic "Employers look for positive attitude" on page 3 and the five steps of the data-collection process.

1. *Define the objectives of the survey or experiment.* Determine the opinion of employers regarding what qualities they look for when hiring seasonal employees.
2. *Define the variable and the population of interest.* The variable is the opinion or response to a question concerning qualities or characteristics. The population of interest is all U.S. hiring managers.
3. *Define the data-collection and data-measuring schemes.* Based on the graphic itself, it can be seen that the source for the percentages presented was SnagAJob.com. Upon further investigation, IPSOS Public Affairs, a third-party research firm, conducted the survey on behalf of the "hourly job website" SnagAJob.com between February 20 and 25, 2009. It was an online survey of 1043 hiring managers with responsibility for hiring summer and seasonal employees by the hour.
4. *Collect the sample.* The information collected from each hiring manager was his or her single "most" essential quality/characteristic that a seasonal employee should possess.
5. *Review the sampling process upon completion of collection.* Since the sampling process was an online survey, were only hiring managers that conduct their business online aware of this survey? Were various areas of the country and types of businesses represented? Perhaps you can think of additional concerns.

In concept, the simple random sample is the simplest of the probability sampling techniques, but it is seldom used in practice because it often is an inefficient technique. One of the easiest-to-use methods for approximating a simple random sample is the *systematic sampling method.*

> **Systematic sample** A sample in which every kth item of the sampling frame is selected, starting from a first element, which is randomly selected from the first k elements.

To select an x percent (%) systematic sample, we will need to randomly select 1 element from every $\frac{100}{x}$ elements. After the first element is randomly located within the first $\frac{100}{x}$ elements, we proceed to select every $\frac{100}{x}$th item thereafter until we have the desired number of data values for our sample.

For example, if we desire a 3% systematic sample, we would locate the first item by randomly selecting an integer between 1 and 33 ($\frac{100}{x} = \frac{100}{3} = 33.33$, which when rounded becomes 33). Suppose 23 was randomly selected. This means that our first data value is obtained from the subject in the 23rd position in the sampling frame. The second data value will come from the subject in the 56th (23 + 33 = 56) position; the third, from the 89th (56 + 33); and so on, until our sample is complete.

The systematic technique is easy to describe and execute; however, it has some inherent dangers when the sampling frame is repetitive or cyclical in nature. For example, a systematic sample of every kth house along a long street might result in a sample disproportional with regard to houses on corner lots. The resulting information would likely be biased if the purpose for sampling is to learn about support for a proposed sidewalk tax. In these situations the results may not approximate a simple random sample.

Multistage Methods

When sampling very large populations, sometimes it is necessary to use a *multistage sampling* design to approximate random sampling.

> **Multistage random sampling** A sample design in which the elements of the sampling frame are subdivided and the sample is chosen in more than one stage.

Multistage sampling designs often start by dividing a very large population into subpopulations on the basis of some characteristic. These subpopulations are called **strata**. These smaller, easier-to-work-with strata can then be sampled separately. One such sample design is the *stratified random sampling method.*

> **Stratified random sample** A sample obtained by stratifying the population, or sampling frame, and then selecting a number of items from each of the strata by means of a simple random sampling technique.

A stratified random sample results when the population, or sampling frame, is subdivided into various strata, usually some already occurring natural subdivisions, and then a subsample is drawn from each of these strata. These subsamples may be drawn from the various strata by using random or systematic methods. The subsamples are summarized separately first and then combined to draw conclusions about the entire population.

When a population with several strata is sampled, we often require that the number of items collected from each Stratum be proportional to the size of the strata; this method is called a *proportional stratified sampling.*

> **Proportional stratified sample** A sample obtained by stratifying the population, or sampling frame, and then selecting a number of items in proportion to the size of the strata from each strata by means of a simple random sampling technique.

A convenient way to express the idea of proportional sampling is to establish a quota. For example, the quota "1 for every 150" directs you to select 1 data value for each 150 elements in each strata. That way, the size of the strata determines the size of the subsample from that strata. The subsamples are summarized separately and then combined to draw conclusions about the entire population.

Another sampling method that starts by stratifying the population, or sampling frame, is a *cluster sample.*

> **Cluster sample** A sample obtained by stratifying the population, or sampling frame, and then selecting some or all of the items from some, but not all, of the strata.

The cluster sample is a multistage design. It uses either random or systematic methods to select the strata (clusters) to be sampled (first stage) and then uses either random or systematic methods to select elements from each identified cluster (second stage). The cluster sampling method also allows the possibility of selecting all of the elements from each identified cluster. Either way, the subsamples are summarized separately and the information then combined.

To illustrate a possible multistage random sampling process, consider that a sample is needed from a large country. In the first stage, the country is divided into smaller regions, such as states, and a random sample of these states is selected. In the second stage, a random sample of smaller areas within the selected states (counties) is then chosen. In the third stage, a random sample of even smaller areas (townships) is taken within each county. Finally in the fourth stage, if these townships are sufficiently small for the purposes of the study, the researcher might continue by collecting simple random samples from each of the identified townships. This would mean that the entire sample was made up of several "local" subsamples identified as a result of the several stages.

Sample design is not a simple matter; many colleges and universities offer separate courses in sample surveying and experimental design. The topic of survey sampling is a complete textbook in itself. It is thus intended for the preceding information to provide you with an overview of sampling and put its role in perspective.

S E C T I O N   1 . 3   E X E R C I S E S

1.43 *USA Today* regularly asks readers, "Have a complaint about airline baggage, refunds, advertising, customer service? Write: . . ."

a. What kind of sampling method is this?

b. Are the results likely to be biased? Explain.

1.44 *USA Today* conducted a survey asking readers, "What is the most hilarious thing that has ever happened to you en route or during a business trip?"

a. What kind of sampling method is this?

b. Are the results likely to be biased? Explain.

1.45 In a survey about families, Ann Landers—a well-known advice columnist—asked parents if they would have kids again; 70% responded "No." An independent random survey asking the same question yielded a 90% "Yes" response. Give at least one explanation why the resulting percent from the Landers survey is so much different from the random sample's percent.

1.46 Describe two reasons why the results of the "In God We Trust" survey in Applied Example 1.7 on page 16 should not be expected to be representative of the population.

1.47 We all know that exercise is good for us. But can exercise prevent or delay the symptoms of Parkinson's disease? A recent study by the Harvard School of Public Health studied 48,000 men and 77,000 women who were relatively healthy and middle-aged or over. During the course of the study, 387 people developed the disease. The study found that men who had participated in some vigorous activity at least twice a week in high school, college, and up to age 40 had a 60% reduced risk of getting Parkinson's. The study found no such reduction for women. What type of sampling does this represent?

Source: "Exercise may prevent Parkinson's," *USA Today*, February 22, 2005

1.48 A wholesale food distributor in a large metropolitan area would like to test the demand for a new food product. He distributes food through five large supermarket chains. The food distributor selects a sample of stores located in areas where he believes the shoppers are receptive to trying new products. What type of sampling does this represent?

1.49 Consider a simple population consisting of only the numbers 1, 2, and 3 (an unlimited number of each). There are nine different samples of size 2 that could be drawn from this population: (1, 1), (1, 2), (1, 3), (2, 1), (2, 2), (2, 3), (3, 1), (3, 2), (3, 3).

a. If the population consists of the numbers 1, 2, 3, and 4, list all the samples of size 2 that could possibly be selected.

b. If the population consists of the numbers 1, 2, and 3, list all the samples of size 3 that could possibly be selected.

1.50 a. What is a sampling frame?

b. What did Mr. Clar use for a sampling frame in Example 1.10, page 20?

c. Where did the number 1288 come from, and how was it used?

1.51 An article titled "Surface Sampling in Gravel Streams" (*Journal of Hydraulic Engineering,* April 1993) discusses grid sampling and arial sampling. Grid sampling involves the removal by hand of stones found at specific points. These points are established on the gravel surface through the use of a wire mesh or by using predetermined distances on a survey tape. The material collected by grid sampling is usually analyzed as a frequency distribution. An arial sample is collected by removing all the particles found in a predetermined area of a channel bed.

(continue on page 24)

The material recovered is most often analyzed as a frequency distribution by weight. Would you categorize these sample designs as judgment samples or probability samples?

1.52 A random sample could be very difficult to obtain. Why?

1.53 Why is the random sample so important in statistics?

1.54 Sheila Jones works for an established marketing research company in Cincinnati, Ohio. Her supervisor just handed her a list of 500 four-digit random numbers extracted from a statistical table of random digits. He told Sheila to conduct a survey by calling 500 Cincinnati residents on the telephone, provided the last four digits of the phone number matched one of the numbers on the list. If Sheila follows her supervisor's instructions, is he assured of obtaining a random sample of respondents? Explain.

1.55 Describe in detail how you would select a 4% systematic sample of the adults in a nearby large city in order to complete a survey about a political issue.

1.56 a. What body of the federal government illustrates a stratified sampling of the people? (A random selection process is not used.)

  b. What body of the federal government illustrates a proportional sampling of the people? (A random selection process is not used.)

1.57 Suppose that you have been hired by a group of all-sports radio stations to determine the age distribution of their listeners. Describe in detail how you would select a random sample of 2500 from the 35 listening areas involved.

1.58 Explain why the polls that are so frequently quoted during early returns on Election Day TV coverage are an example of cluster sampling.

1.59 The telephone book might not be a representative sampling frame. Explain why.

1.60 The election board's voter registration list is not a census of the adult population. Explain why.

1.61 Replace incandescent light bulbs with compact fluorescent bulbs that use up to 75% less energy and last up to 10 times as long. From "Simple Ways to Save Energy," *NYSEG Energy Lines*, February 2009.

Image copyright Ossile, 2012. Used under license from Shutterstock.com

a. What two claims are made in the above statement by the New York State Electric and Gas Company? State them in terms of a statistical parameter.

b. Do you feel the two statements by NYSEG are reasonable and likely to be true? Explain.

c. If you feel a claim is reasonable and likely true, would you be driven to find evidence to verify its truth? Explain.

d. If you feel a claim is not reasonable and likely not true, would you be driven to find evidence to verify it to be incorrect? Explain.

e. Would you be more likely to research the situation described in part c or d? Explain.

f. How would you proceed to attempt to verify "up to 75% less energy"?

g. How would you proceed to attempt to verify "last up to 10 times as long"?

1.4 Statistics and Technology

In recent years, electronic technology has had a tremendous impact on almost every aspect of life. The field of statistics is no exception. As you will see, the field of statistics uses many techniques that are repetitive in nature: calculations of numerical statistics, procedures for constructing graphic displays of data, and procedures that are followed to formulate statistical inferences. Computers and calculators are very good at performing these sometimes long and tedious operations. If your computer has one of the standard statistical packages or if you have a statistical calculator, then it will make the analysis easy to perform.

Throughout this textbook, as statistical procedures are studied, you will find the information you need to have a computer complete the same procedures using MINITAB and Excel software. Calculator procedures will also be demonstrated for the TI-83/84 Plus calculator.

An explanation of the most common typographical conventions that will be used in this textbook follows. As additional explanations or selections are needed, they will be given.

TECHNOLOGY INSTRUCTIONS: BASIC CONVENTIONS

MINITAB

FYI For information about obtaining MINITAB, check the Internet at http://www .minitab.com.

Choose: tells you to make a menu selection by a mouse "point and click" entry.
For example: **Choose: Stat > Quality Tools > Pareto Chart** instructs you to, in sequence, "point and click on" **Stat** on the menu bar, "followed by" **Quality Tools** on the pull-down, and then "followed by" **Pareto Chart** on the second pull-down.

Select: indicates that you should click on the small box or circle to the left of a specified item.

Enter: instructs you to type or select information needed for a specific item.

Excel

FYI Excel is part of Microsoft Office and can be found on many personal computers.

Choose: tells you to make a menu or tab selection by a mouse "point and click" entry.
For example: **Choose: Insert > Scatter > 1st graph picture** instructs you to, in sequence, "point and click on" the **Insert** tab, followed by **Scatter** under "Charts" section, followed by **1st graph picture** on the Chart subtype.

Select: indicates that you should click on the small box or circle to the left of a specified item. It is often followed by a "point and click on" **Next, Close,** or **Finish** on the dialog window.

Enter: instructs you to type or select information needed for a specific item.

TI-83/84 Plus

FYI For information about obtaining TI-83/84 Plus, check the Internet at http://www.ti.com/calc.

Choose: tells you which keys to press or menu selections to make.
For example: **Choose: Zoom > 9:ZoomStat > Trace > > >** instructs you to press the **Zoom** key, followed by selecting **9:ZoomStat** from the menu, followed by pressing the **Trace key; > > >** indicates that you should press arrow keys repeatedly to move along a graph to obtain important points.

Enter: instructs you to type or select information needed for a specific item.

Screen Capture: gives pictures of what your calculator screen should look like with chosen specifications highlighted.

Additional details about the use of MINITAB and Excel are available by using the Help system in the MINITAB and Excel software. Additional details for the TI-83/84 are contained in its corresponding *TI-83/84 Plus Graphing Calculator Guidebook.* Specific details on the use of computers and calculators available to you need to be obtained from your instructor or from your local computer lab person.

Your local computer center can provide you with a list of what is available to you. Some of the more readily available packaged programs are MINITAB, JMP-IN, and SPSS (Statistical Package for the Social Sciences).

Note: *There is a great temptation to use the computer or calculator to analyze any and all sets of data and then treat the results as though the statistics are correct. Remember the adage: "Garbage-in, garbage-out!" Responsible use of statistical methodology is very important. The burden is on the user to ensure that the appropriate methods are correctly applied and that accurate conclusions are drawn and communicated to others.*

S E C T I O N   1 . 4   E X E R C I S E S

1.62 How have computers increased the usefulness of statistics for professionals such as researchers, government workers who analyze data, statistical consultants, and others?

1.63 How might computers help you in statistics?

1.64 a. Have you ever heard someone say, "It must be right, that's what my calculator told me"? Explain why the calculator may or may not have given the correct answer.

b. What is meant by the phrase "Garbage-in, garbage-out!" and how have computers increased the probability that studies might be victimized by the adage?

Chapter Review

In Retrospect

You should now have a general feeling for what statistics is about, an image that will grow and change as you work your way through this book. You know what a sample and a population are and the distinction between qualitative (attribute) and quantitative (numerical) variables. You should also have an appreciation for and a partial understanding of how important random samples are in statistics.

Throughout the chapter you have seen numerous articles that represent various aspects of statistics. The statistical graphics display a variety of information about ourselves as we describe ourselves and other aspects of the world around us. Statistics can even be entertaining. The examples are endless. Look around and find some examples of statistics in your daily life (see Exercises 1.77 and 1.78, p. 30).

CourseMate The **Statistics CourseMate** site for this text brings chapter topics to life with interactive learning, study, and exam preparation tools, including quizzes and flashcards for the Vocabulary and Key Concepts that follow. The site also provides an **eBook** version of the text with highlighting and note taking capabilities. Throughout chapters, the CourseMate icon 🖥 flags concepts and examples that have corresponding interactive resources such as **video** and **animated tutorials** that demonstrate, step by step, how to solve problems; **datasets** for exercises and examples; **Skillbuilder Applets** to help you better understand concepts; **technology manuals**; and software to download including **Data Analysis Plus** (a suite of statistical macros for Excel) and **TI-83/84 Plus** programs—logon at **www.cengagebrain.com**.

Vocabulary and Key Concepts

attribute variable (p. 6)
biased sampling method (p. 16)
categorical variable (p. 6)
census (p. 18)
cluster sample (p. 22)
continuous variable (p. 8)
convenience sample (p. 17)
data (p. 5)
data collection (p. 16)
data value (p. 5)
descriptive statistics (p. 1)
discrete variable (p. 8)
experiment (p. 5)
finite population (p. 5)
haphazard (p. 20)
inferential statistics (p. 1)

infinite population (p. 5)
judgment sample (p. 19)
multistage random sampling (p. 22)
nominal variable (p. 7)
numerical data (p. 7)
numerical variable (p. 6)
observational study (p. 18)
ordinal variable (p. 7)
parameter (p. 5)
population (p. 4)
probability sample (p. 19)
proportional stratified sample (p. 22)
qualitative variable (p. 6)
quantitative variable (p. 6)
representative (p. 19)
response variable (p. 5)

sample (p. 5)
sample design (p. 19)
sampling frame (p. 18)
sampling method (p. 16)
simple random sample (p. 20)
single-stage sampling, (p. 19)
statistic (p. 6)
statistics (p. 1)
strata (p. 22)
stratified random sample (p. 22)
survey (p. 18)
systematic sample (p. 21)
unbiased sampling method (p. 16)
variability (p. 15)
variable (p. 5)
volunteer sample (p. 17)

Learning Outcomes

- Understand and be able to describe the difference between descriptive and inferential statistics.   p. 1, Ex. 1.7, 1.8, 1.67

- Understand and be able to identify and interpret the relationships between sample and population, and statistic and parameter.   pp. 4–6, EXP 1.5

- Know and be able to identify and describe the different types of variables.   pp. 6–8, Ex. 1.33, 1.34

- Understand how convenience and volunteer samples result in biased samples.   pp. 16–17, Ex. 1.45

- Understand the differences among and be able to identify experiments, observational studies, and judgment samples.   pp. 18–19

- Understand and be able to describe the single-stage sampling methods of "simple random sample" and "systematic sampling."   pp. 19–21

- Understand and be able to describe the multistage sampling methods of "stratified sampling" and "cluster sampling."   pp. 22–23

- Understand that variability is inherent in everything and in the sampling process.   pp. 14–15, Ex. 1.38

Chapter Exercises

1.65 We want to describe the so-called typical student at your college. Describe a variable that measures some characteristic of a student and results in

a. Attribute data

b. Numerical data

1.66 A candidate for a political office claims that he will win the election. A poll is conducted and 35 of 150 voters indicate that they will vote for the candidate, 100 voters indicate that they will vote for his opponent, and 15 voters are undecided.

a. What is the population parameter of interest?

b. What is the value of the sample statistic that might be used to estimate the population parameter?

c. Would you tend to believe the candidate based on the results of the poll?

1.67 A researcher studying consumer buying habits asks every 20th person entering Publix Supermarket how many times per week he or she goes grocery shopping. She then records the answer as T.

a. Is $T = 3$ an example of (1) a sample, (2) a variable, (3) a statistic, (4) a parameter, or (5) a data value?

Suppose the researcher questions 427 shoppers during the survey.

(continue on page 28)

b. Give an example of a question that can be answered using the tools of descriptive statistics.

c. Give an example of a question that can be answered using the tools of inferential statistics.

1.68 A researcher studying the attitudes of parents of preschool children interviews a random sample of 50 mothers, each having one preschool child. He asks each mother, "How many times did you compliment your child yesterday?" He records the answer as *C*.

a. Is *C* = 4 an example of (1) a data value, (2) a statistic, (3) a parameter, (4) a variable, or (5) a sample?

b. Give an example of a question that can be answered using the tools of descriptive statistics.

c. Give an example of a question that can be answered using the tools of inferential statistics.

1.69 Consider the June 8, 2009, *USA Today* article titled "Credit card delinquencies rise."

Credit card delinquencies rise

The delinquency rate for bank-issued credit cards rose 11% in the first three months of the year, according to credit-reporting agency TransUnion. The delinquency rate jumped to 1.32% this year, from 1.19% in the first three months of 2008, TransUnion said. The statistic measures the percentage of card holders who are three months or more past due on their payments for Master-Card, Visa, American Express and Discover cards. The average total debt on bank cards also rose, jumping to $5,776 from $5,548 last year. Balances typically rise in the first quarter, as holiday spending comes due, said Ezra Becker, director of consulting and strategy in TransUnion's financial services group. But retail sales results showed holiday spending took a steep drop. That likely means higher balances reflect consumers using credit cards to pay for necessities, he said.

Source: Credit Card Delinquencies Rise," *USA Today*, June 8, 2009. Copyright © 2009, *USA Today*. Reprinted with permission.

a. What is the population?

b. Name at least 3 variables that must have been used.

c. Classify all the variables of the study as either attribute or numerical.

1.70 Harris Interactive conducted an online poll of U.S. adults during August 2008 in anticipation of September being Library Sign-up Month.

Libary Card Use

These are some of the results of a Harris Interactive poll of 2,710 U.S. adults conducted online between August 11 and 17, 2008.

68% of Americans currently own a library card.
Certain groups are more likely to have a library card than others—Echo Boomers (those 18–31) are more likely to have one over other age categories (70% versus 68–65%); women over men (73% versus 62%); Hispanics over African Americans and Whites (72% versus 67% and 66%); Midwesterners (72%) over Easterners (65%) and Southerners (63%).
Politically there is also a difference as Democrats are more likely to have a library card over Republicans and Independents (71% versus 67% and 61%).
Over one-third (35%) of people with a library card have used the library 1 to 5 times in the past year and 15% have used it more than 25 times in the past year.

Source: http://www.harrisinteractive.com/harris_poll/

a. What is the population?

b. Name at least 3 variables that must have been used.

c. Classify all the variables of the study as either attribute or numerical.

1.71

Do you Support Using Cameras to Identify Red-light Runners?

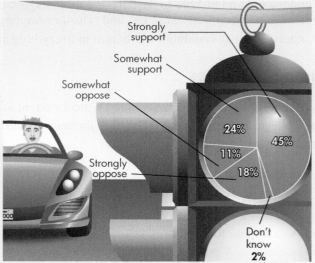

Source: Public Opinion Strategies survey of 800 likely voters, April 2009

The graphic above shows how 800 likely voters in April 2009 felt about using cameras to identify red-light runners. Would you classify the data collected and used to determine these percentages as qualitative (nominal or ordinal) or quantitative (discrete or continuous)? Why?

1.72 Consider the May 12, 2009, *USA Today* article titled "Simulated acupuncture eases pain."

Simulated acupuncture eases pain

Acupuncture brought more relief to people with back pain than standard treatments, whether it was done with a toothpick or a real needle, a study finds, but how acupuncture works remains unclear. In the study, 638 adults with chronic low back pain were divided into four groups and received standardized acupuncture treatment; individually prescribed acupuncture treatment; simulated acupuncture treatment using a toothpick in a needle guide tube that did not pierce the skin as regular acupuncture does but was targeted at correct acupuncture points; or standard medical treatment (medication and physical therapy). After eight weeks, 60% of those who got any type of acupuncture reported significant improvement compared with those who got standard care alone, says the study in this week's *Archives of Internal Medicine*.

Source: Nanci Hellmich, "Simulated Acupuncture Eases Pain," *USA Today*, May 12, 1999. Copyright © 1999, *USA Today*. Reprinted with permission.

a. What is the population?

b. What is the sample?

c. Is this a judgment sample or a probability sample?

d. If this study is a probability sample, what type of sampling method do you think was used?

1.73 "Pull Down the Shades," an article in the July 2009 *Good Housekeeping* magazine, presented the results of a study of 5000 people in Hawaii done by the University of Hawaii in Manoa. Data collected at beaches, parks, and pools in sunny Honolulu revealed that only 4 in 10 adults wear sunglasses to protect their eyes.

a. Was this study an experiment or an observational study?

b. Identify the parameter of interest.

c. Identify the statistic and give its value.

1.74 The 700 Club: Barry Bonds played for the San Francisco Giants and near the end of his career was on pace to become baseball's home run king. He joined Hank Aaron and Babe Ruth as the only Major League players to have hit more than 700 home runs in their careers. The graphic below is a look at how they amassed their totals.

a. Describe and compare the overall appearance of the three graphs. Include thoughts about such things as length of career, when the most home runs per year were hit and their relationship to the aging process, and any others you think of.

b. Does it appear that one of them was more consistent with annual home run production?

c. From the evidence presented here, who do you think should be called the "Home Run King"?

(continue on page 30)

Graphic and data for Exercise 1.74

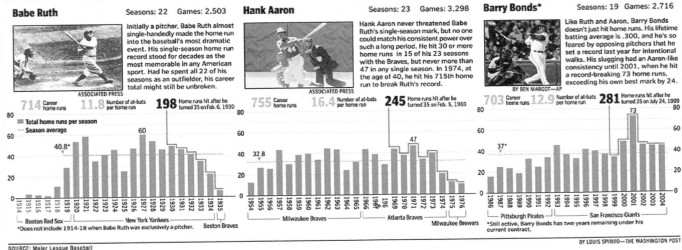

Source: The *Washington Post*

d. Was Barry Bonds' 73 home runs in one season a fluke?

e. If you were a team owner and interested in home run production over the next several years, you would like very much to sign a player for your team that would duplicate which one of these players? Let's say you were signing him at age 21. At age 35.

1.75 Describe in your own words and give an example of each of the following terms. Your examples should not be ones given in class or in the textbook.

a. variable   b. data   c. sample

d. population   e. statistic f. parameter

1.76 Describe in your own words and give an example of the following terms. Your examples should not be ones given in class or in the textbook.

a. random sample   b. probability sample

c. judgment sample

1.77 Find an article or an advertisement in a newspaper or magazine that exemplifies the use of statistics.

a. Identify and describe one statistic reported in the article.

Chapter Practice Test

PART I: Knowing the Definitions

Answer "True" if the statement is always true. If the statement is not always true, replace the words printed in bold with words that make the statement always true.

1.1 **Inferential** statistics is the study and description of data that result from an experiment.

1.2 **Descriptive statistics** is the study of a sample that enables us to make projections or estimates about the population from which the sample is drawn.

1.3 A **population** is typically a very large collection of individuals or objects about which we desire information.

1.4 A statistic is the calculated measure of some characteristic of a **population**.

1.5 A parameter is the measure of some characteristic of a **sample**.

1.6 As a result of surveying 50 freshmen, it was found that 16 had participated in interscholastic sports, 23 had served as officers of classes and clubs, and 18 had been in school plays during their high school years. This is an example of **numerical data**.

b. Identify and describe the variable related to the statistic in part a.

c. Identify and describe the sample related to the statistic in part a.

d. Identify and describe the population from which the sample in part c was taken.

1.78 a. Find an article in a newspaper or magazine that exemplifies the use of statistics in a way that might be considered "entertainment" or "recreational." Describe why you think this article fits one of these categories.

b. Find an article in a newspaper or magazine that exemplifies the use of statistics and presents an unusual finding as the result of a study. Describe why these results are (or are not) "newsworthy."

1.79 In Exercise 1.6, you were asked to write a sentence for each of the three statistical activities given in the definition of *statistics*. Now that you have completed the chapter, review your work. Again, using your own words, change and/or enhance your work to complete a paragraph on the definition of *statistics*.

1.7 The "number of rotten apples per shipping crate" is an example of a **qualitative** variable.

1.8 The "thickness of a sheet of sheet metal" used in a manufacturing process is an example of a **quantitative** variable.

1.9 A **representative** sample is a sample obtained in such a way that all individuals had an equal chance of being selected.

1.10 The basic objectives of **statistics** are obtaining a sample, inspecting this sample, and then making inferences about the unknown characteristics of the population from which the sample was drawn.

PART II: Applying the Concepts

The owners of Corner Convenience Store are concerned about the quality of service their customers receive. In order to study the service, they collected samples for each of several variables.

1.11 Classify each of the following variables as nominal, ordinal, discrete, or continuous:

a. Method of payment for purchases (cash, credit card, check)

b. Customer satisfaction (very satisfied, satisfied, not satisfied)

c. Amount of sales tax on purchase

d. Number of items purchased

e. Customer's driver's license number

1.12 The mean checkout time for all customers at Corner Convenience Store is to be estimated using the mean checkout time for 75 randomly selected customers. Match the items in column 2 with the statistical terms in column 1.

1	2
_____ data value	(a) the 75 customers
_____ data	(b) the mean time for all customers
_____ experiment	(c) 2 minutes, one customer's checkout time
_____ parameter	(d) the mean time for the 75 customers
_____ population	(e) all customers at Corner Convenience Store
_____ sample	(f) the checkout time for one customer
_____ statistic	(g) the 75 times
_____ variable	(h) the process used to select 75 customers and measure their times

PART III: Understanding the Concepts

Write a brief paragraph in response to each question.

1.13 The population and the sample are both sets of objects. Describe the relationship between them and give an example.

1.14 The variable and the data for a specific situation are closely related. Explain this relationship and give an example.

1.15 The data, the statistic, and the parameter are all values used to describe a statistical situation. How does one distinguish among these three terms? Give an example.

1.16 What conditions are required for a sample to be a random sample? Explain and include an example of a sample that is random and one that is not random.

2 Descriptive Analysis and Presentation of Single-Variable Data

© 2010 Alys Tomlinson/Jupiterimages

© 2010 Chris Whitehead/Jupiterimages

GRAPHIC PRESENTATION OF DATA

2.1 Graphs, Pareto Diagrams, and Stem-and-Leaf Displays
*A **picture** is worth a thousand words.*

2.2 Frequency Distributions and Histograms
***Graphical methods** for larger sets of data*

NUMERICAL DESCRIPTIVE STATISTICS

2.3 Measures of Central Tendency
*The mean, median, mode, and midrange are **average** values.*

2.4 Measures of Dispersion
*Measuring the **amount of spread** in a set of data*

2.5 Measures of Position
***Comparing** one data value to the set of data*

2.6 Interpreting and Understanding Standard Deviation
*The length of a **standardized yardstick***

2.7 The Art of Statistical Deception
*"**Tricky**" graphs and **insufficient information** mislead.*

2.1 Graphs, Pareto Diagrams, and Stem-and-Leaf Displays

Students, Here's Looking at You

Consider all the information in the graph, specifically called a pie chart or circle graph. Does your day break up into the categories shown on the next page? Or do you have an extra category or two? Maybe even fewer categories? Now consider the time given for each activity. On the

Time Use on an Average Weekday for Full-time University and College Students

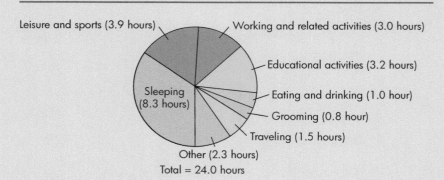

Leisure and sports (3.9 hours)
Working and related activities (3.0 hours)
Educational activities (3.2 hours)
Eating and drinking (1.0 hour)
Grooming (0.8 hour)
Traveling (1.5 hours)
Sleeping (8.3 hours)
Other (2.3 hours)
Total = 24.0 hours

NOTE: Data include individuals, ages 15 to 49, who were enrolled full time at a university or college. Data include non-holiday weekends and are averages for 2003–07.
Source: Bureau of Labor Statistics

FYI ATUS is a federally administered, continuous survey on time use in the United States sponsored by the Bureau of Labor Statistics and conducted by the U.S. Census Bureau.

average, how does the amount of time you spend compare? Perhaps you have entirely different categories. Are you wishing you had the 8.3 hours of sleep, on the average? We do!

Can you imagine all this information written out in sentences? Graphical displays can truly be worth a thousand words. This one pie chart summarizes "Time Use" information from a 2003–2007 American Time Use Survey (ATUS) of over 50,000 Americans. Since it is a cross-sectional survey, realize that this graph included only the full-time college students that participated.

Now that you know the source and see the overall sample size, you may feel that these data portray a relatively accurate picture of a college student's day. Perhaps you may be looking more closely at one of the categories. Do you have questions about the average 0.8 hour per day on grooming? Did you think there may be a gender difference? It gets you thinking, doesn't it?

FYI There is no single correct answer when constructing a graphic display. The analyst's judgment and circumstances surrounding the problem play major roles in the development of the graphic.

As demonstrated with the graph on page 32, one of the most useful ways to become acquainted with data is to use an initial exploratory data analysis technique that will result in a pictorial representation of the data. The display will visually reveal patterns of behavior of the variable being studied. There are several graphic (pictorial) ways to describe data. The type of data and the idea to be presented determine which method is used.

Qualitative Data

Pie charts (circle graphs) and bar graphs Graphs that are used to summarize **qualitative,** or attribute, or categorical **data.** Pie charts (circle graphs) show the amount of data that belong to each category as a proportional part of a circle. Bar graphs show the amount of data that belong to each category as a proportionally sized rectangular area.

E X A M P L E   2 . 1

GRAPHING QUALITATIVE DATA

Table 2.1 lists the number of cases of each type of operation performed at General Hospital last year.

TABLE 2.1 Operations Performed at General Hospital Last Year [TA02-01]

Type of Operation	Number of Cases
Thoracic	20
Bones and joints	45
Eye, ear, nose, and throat	58
General	98
Abdominal	115
Urologic	74
Proctologic	65
Neurosurgery	23
Total	498

The data in Table 2.1 are displayed on a pie chart in Figure 2.1, with each type of operation represented by a relative proportion of a circle, found by dividing the number of cases by the total sample size, namely, 498. The proportions are then reported as percentages (for example, 25% is 1/4 of the circle). Figure 2.2 displays the same "type of operation" data but in the form of a bar graph. Bar graphs of attribute data should be drawn with a space between bars of equal width.

FIGURE 2.1
Pie Chart

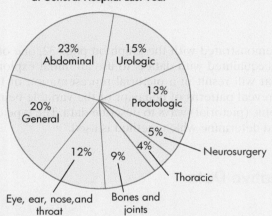

Operations Performed at General Hospital Last Year

FIGURE 2.2
Bar Graph

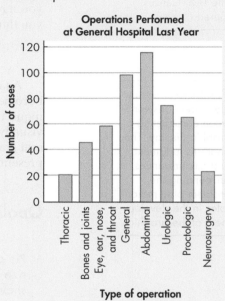

Operations Performed at General Hospital Last Year

FYI All graphic representations need to be completely self-explanatory. That includes a descriptive, meaningful title and proper identification of the quantities and variables involved.

TECHNOLOGY INSTRUCTIONS: PIE CHART

Input the categories into C1 and the corresponding frequencies into C2; then continue with:

Choose: **Graph > Pie Chart . . .**
Select: **Chart values from a table**
Enter: **Categorical variable: C1 Summary variables: C2**
Select: **Labels > Title/Footnotes** Enter: Title: **your title**
Select: **Slice Labels** > Select desired labels > **OK > OK**

Input the categories into column A and the corresponding frequencies into column B; activate both columns of data by highlighting and selecting the column names and data cells, then continue with:

Choose: **Insert > Pie > 1st picture** (usually)
Choose: Chart Layouts—**Layout 1**
Enter: Chart title: **Your title**

To edit the pie chart:

Click On: Anywhere clear on the chart—use handles to size
              Any cell in the category or frequency column and type in
              different name or amount > ENTER

TI-83/84 Plus

Input the frequencies for the various categories into L1; then continue with:

Choose: PRGM > EXEC > CIRCLE*
Enter: LIST: L1 > ENTER
           DATA DISPLAYED?: 1:PERCENTAGES
                    OR
                        2:DATA

*The TI-83/84 Plus program "CIRCLE" and others are available for downloading through cengagebrain.com. The TI-83/84 Plus programs and data files may be in zipped, or compressed, format. If so, save the files and uncompress them using a zip utility. Download the programs to your calculator using TI-Graph Link software.

When the bar graph is presented in the form of a *Pareto diagram*, it presents additional and very helpful information.

> **Pareto diagram** A bar graph with the bars arranged from the most numerous category to the least numerous category. It includes a line graph displaying the cumulative percentages and counts for the bars.

The Pareto diagram is popular in quality-control applications. A Pareto diagram of types of defects will show the ones that have the greatest effect on the defective rate in order of effect. It is then easy to see which defects should be targeted to most effectively lower the defective rate.

E X A M P L E   2 . 2

PARETO DIAGRAM OF HATE CRIMES

The FBI reported the number of hate crimes by category for 2003 (http://www.fbi.gov/). The Pareto diagram in Figure 2.3 shows the 8715 categorized hate crimes, their percentages, and cumulative percentages.

FIGURE 2.3
Pareto Diagram

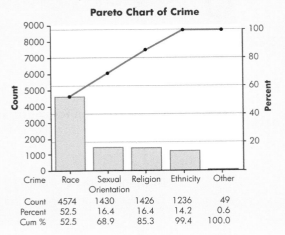

Pareto Chart of Crime

Crime	Race	Sexual Orientation	Religion	Ethnicity	Other
Count	4574	1430	1426	1236	49
Percent	52.5	16.4	16.4	14.2	0.6
Cum %	52.5	68.9	85.3	99.4	100.0

Video tutorial available—logon and learn more at cengagebrain.com

T E C H N O L O G Y   I N S T R U C T I O N S : P A R E T O   D I A G R A M

MINITAB

Input the categories into C1 and the corresponding frequencies into C2; then continue with:

Choose: **Stat Chart > Quality Tools > Pareto**
Select: **Chart defects table**
Enter: Defects or attribute data in: **C1**
            Frequencies in: **C2**
Select: **Options**
Enter: Title: **your title > OK > OK**

Excel

Input the categories into column A and the corresponding frequencies into column B (column headings are optional); then continue with:
    First, sorting the table:

Activate both columns of the distribution
Choose: **Data > AZ / ZA Sort**
Select: Sort by: **frequency column**
            Order: **Largest to Smallest > OK**
Choose: **Insert > Column > 1st picture** (usually)
Choose: Chart Layouts—Layout 9
Enter: Chart title: **your title**
            Category (x) axis title: **title for x-axis**
            Value (y) axis title: **title for y-axis**

To edit the Pareto diagram:

Click on: Anywhere clear on the chart
            –use handles to size
            Any title name to change
            Any cell in the category column and type in a name > **Enter**

Excel does not include the line graph.

TI-83/84 Plus

Input the numbered categories into L1 and the corresponding frequencies into L2; then continue with:

Choose: **PRGM > EXEC > PARETO***
Enter: **LIST: L2 > ENTER**
Ymax: **at least the sum of the frequencies > ENTER**
Yscl: **increment for y-axis > ENTER**

*Program "PARETO" is one of many programs that are available for downloading. See page 35 for specific instructions.

Quantitative Data

One major reason for constructing a graph of **quantitative data** is to display its *distribution*.

> **Distribution** The pattern of variability displayed by the data of a variable. The distribution displays the frequency of each value of the variable.

One of the simplest graphs used to display a distribution is the *dotplot*.

> **Dotplot display** Displays the data of a sample by representing each data value with a dot positioned along a scale. This scale can be either horizontal or vertical. The frequency of the values is represented along the other scale.

E X A M P L E   2 . 3

DOTPLOT OF EXAM GRADES

Table 2.2 provides a sample of 19 exam grades randomly selected from a large class.

TABLE 2.2
Sample of 19 Exam Grades [TA02-02]

76	74	82	96	66	76	78	72	52	68
86	84	62	76	78	92	82	74	88	

FIGURE 2.4 is a dotplot of the 19 exam scores.

FIGURE 2.4
Dotplot

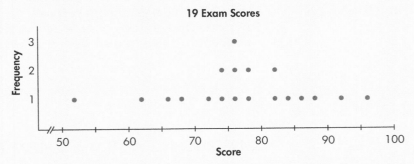

Notice how the data in Figure 2.4 are "bunched" near the center and more "spread out" near the extremes.

The dotplot display is a convenient technique to use as you first begin to analyze the data. It results in a picture of the data that sorts the data into numerical order. (To *sort* data is to list the data in rank order according to numerical value.)

T E C H N O L O G Y   I N S T R U C T I O N S :   D O T P L O T

MINITAB

Input the data into C1; then continue with:

Choose: **Graph > Dotplot . . . > One Y, Simple > OK**
Enter: **Graph Variables: C1 > OK**

Excel

The dotplot display is not available, but the initial step of ranking the data can be done. Input the data into column A and activate the column of data; then continue with:

Choose: **Data > AZ ↓ (Sort)**
Use the sorted data to finish constructing the dotplot display.

TI-83/84 Plus	Input the data into L1; then continue with:

Choose: **PRGM > EXEC > DOTPLOT***
Enter: **LIST: L1 > ENTER**
          Xmin: at most the lowest x value
          Xmax: at least the highest x value
          Xscl: 0 or increment
          Ymax: at least the highest frequency

*Program "DOTPLOT" is one of many programs that are available for downloading. See page 35 for specific instructions.

In recent years a technique known as the *stem-and-leaf display* has become very popular for summarizing numerical data. It is a combination of a graphic technique and a sorting technique. These displays are simple to create and use, and they are well suited to computer applications.

> **Stem-and-leaf display** Displays the data of a sample using the actual digits that make up the data values. Each numerical value is divided into two parts: The leading digit(s) becomes the stem, and the trailing digit(s) becomes the leaf. The stems are located along the main axis, and a leaf for each data value is located so as to display the distribution of the data.

E X A M P L E   2 . 4

CONSTRUCTING A STEM-AND-LEAF DISPLAY

FIGURE 2.5A
Unfinished Stem-and-Leaf Display

19 Exam Scores

5	2
6	6 8 2
7	6 4 6 8 2 6 8 4
8	2 6 4 2 8
9	6 2

FIGURE 2.5B
Final Stem-and-Leaf Display

19 Exam Scores

5	2
6	2 6 8
7	2 4 4 6 6 6 8 8
8	2 2 4 6 8
9	2 6

Let's construct a stem-and-leaf display for the 19 exam scores given in Table 2.2 on page 37.

At a quick glance we see that there are scores in the 50s, 60s, 70s, 80s, and 90s. Let's use the first digit of each score as the stem and the second digit as the leaf. Typically, the display is constructed vertically. We draw a vertical line and place the stems, in order, to the left of the line.

```
5 |
6 |
7 |
8 |
9 |
```

Next we place each leaf on its stem. This is done by placing the trailing digit on the right side of the vertical line opposite its corresponding leading digit. Our first data value is 76; 7 is the stem and 6 is the leaf. Thus, we place a 6 opposite the stem 7:

$$7 \mid 6$$

The next data value is 74, so a leaf of 4 is placed on the 7 stem next to the 6.

$$7 \mid 6 \ 4$$

FIGURE 2.6

Stem-and-Leaf Display

19 Exam Scores

(50–54) 5	2
(55–59) 5	
(60–64) 6	2
(65–69) 6	6 8
(70–74) 7	2 4 4
(75–79) 7	6 6 6 8 8
(80–84) 8	2 2 4
(85–89) 8	6 8
(90–94) 9	2
(95–99) 9	6

The next data value is 82, so a leaf of 2 is placed on the 8 stem.

$$
\begin{array}{c|c}
7 & 6\ 4 \\
8 & 2
\end{array}
$$

We continue until each of the other 16 leaves is placed on the display. Figure 2.5A shows the resulting stem-and-leaf display; Figure 2.5B shows the completed stem-and-leaf display after the leaves have been ordered.

From Figure 2.5B, we see that the grades are centered around the 70s. In this case, all scores with the same tens digit were placed on the same branch, but this may not always be desired. Suppose we reconstruct the display; this time instead of grouping 10 possible values on each stem, let's group the values so that only 5 possible values could fall on each stem, as shown in Figure 2.6. Do you notice a difference in the appearance of Figure 2.6? The general shape is approximately symmetrical about the high 70s. Our information is a little more refined, but basically we see the same distribution.

TECHNOLOGY INSTRUCTIONS: STEM-AND-LEAF DISPLAY

MINITAB

Input the data into C1; then continue with:
Choose: **Graph > Stem-and-Leaf . . .**
Enter: **Graph variables: C1**
             Increment: stem width (optional) **> OK**

Excel

Input the data into column A; then continue with:
Choose: **Add-Ins > Data Analysis Plus* > Stem and Leaf Display > OK**
Enter: Input Range: **(A2:A6 or select cells)**
             Increment: **Stem Increment**

*Data Analysis Plus is a collection of statistical macros for Excel and one of many programs available for downloading through cengagebrain.com.

TI-83/84 Plus

Input the data into L1; then continue with:
Choose: **STAT > EDIT > 2:SortA(**
Enter: **L1**

Use sorted data to finish constructing the stem-and-leaf diagram by hand.

It is fairly typical of many variables to display a distribution that is concentrated (mounded) about a central value and then in some manner dispersed in one or both directions. Often a graphic display reveals something that the analyst may or may not have anticipated. Example 2.5 demonstrates what generally occurs when two populations are sampled together.

E X A M P L E   2 . 5

OVERLAPPING DISTRIBUTIONS

A random sample of 50 college students was selected. Their weights were obtained from their medical records. The resulting data are listed in Table 2.3.

Notice that the weights range from 98 to 215 pounds. Let's group the weights on stems of 10 units using the hundreds and the tens digits as stems and the units digit as the leaf (see Figure 2.7). The leaves have been arranged in numerical order.

Close inspection of Figure 2.7 suggests that two overlapping distributions may be involved. That is exactly what we have: a distribution of female weights and a distribution of male weights. Figure 2.8 shows a "back-to-back" stem-and-leaf display of this set of data and makes it obvious that two distinct distributions are involved.

TABLE 2.3
Weights of 50 College Students [TA02-03]

Student	1	2	3	4	5	6	7	8	9	10
Male/Female	F	M	F	M	M	F	F	M	M	F
Weight	98	150	108	158	162	112	118	167	170	120
Student	11	12	13	14	15	16	17	18	19	20
Male/Female	M	M	M	F	F	M	F	M	M	F
Weight	177	186	191	128	135	195	137	205	190	120
Student	21	22	23	24	25	26	27	28	29	30
Male/Female	M	M	F	M	F	F	M	M	M	M
Weight	188	176	118	168	115	115	162	157	154	148
Student	31	32	33	34	35	36	37	38	39	40
Male/Female	F	M	M	F	M	F	M	F	M	M
Weight	101	143	145	108	155	110	154	116	161	165
Student	41	42	43	44	45	46	47	48	49	50
Male/Female	F	M	F	M	M	F	F	M	M	M
Weight	142	184	120	170	195	132	129	215	176	183

FIGURE 2.7
Stem-and-Leaf Display

Weights of 50 College Students (lb)

N = 50	Leaf Unit = 1.0
9	8
10	1 8 8
11	0 2 5 5 6 8 8
12	0 0 0 8 9
13	2 5 7
14	2 3 5 8
15	0 4 4 5 7 8
16	1 2 2 5 7 8
17	0 0 6 6 7
18	3 4 6 8
19	0 1 5 5
20	5
21	5

FIGURE 2.8
"Back-to-Back" Stem-and-Leaf Display

Weights of 50 College Students (lb)

Female		Male
8	09	
1 8 8	10	
0 2 5 5 6 8 8	11	
0 0 0 8 9	12	
2 5 7	13	
2	14	3 5 8
	15	0 4 4 5 7 8
	16	1 2 2 5 7 8
	17	0 0 6 6 7
	18	3 4 6 8
	19	0 1 5 5
	20	5
	21	5

Figure 2.9, a "side-by-side" dotplot (same scale) of the same 50 weight data, shows the same distinction between the two subsets.

Based on the information shown in Figures 2.8 and 2.9, and on what we know about people's weight, it seems reasonable to conclude that female college students weigh less than male college students. Situations involving more than one set of data are discussed further in Chapter 3.

FIGURE 2.9
Dotplots with Common Scale

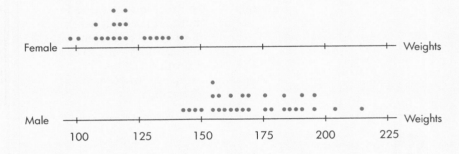

Weights of 50 College Students

T E C H N O L O G Y   I N S T R U C T I O N S :
M U L T I P L E   D O T P L O T S

MINITAB	Input the data into C1 and the corresponding numerical categories into C2; then continue with:

Choose: **Graph > Dotplot . . .**
Select: **One Y, With Groups > OK**
Enter: **Graph variables: C1**
        Categorical variables for grouping: C2 > OK

If the various categories are in separate columns, select Multiple Y's Simple and enter all of the columns under Graph variables.

Excel	Multiple dotplots are not available, but the initial step of ranking the data can be done. Use the commands as shown with the dotplot display on page 37; then finish constructing the dotplots by hand.

TI-83/84	Input the data for the first dotplot into L1 and the data for the second dotplot into L3; then continue with:

Choose: **STAT > EDIT > 2:SortA(**
Enter: **L1 > ENTER**
        In L2, enter counting numbers for each category.
        Ex.   L1   L2
                15   1
                16   1
                16   2
                17   1

Choose:	STAT > EDIT > 2:SortA(
Enter:	L3 > ENTER
	In L4, enter counting numbers (a higher set*) for each category; *for example: use 10,10,11,10,10,11,12, . . . (offsets the two dotplots).
Choose:	2nd > FORMAT > AxesOff (Optional—must return to AxesOn)
Choose:	2nd > STAT PLOT > 1:PLOT1
Choose:	2nd > STAT PLOT > 2:PLOT2
Choose:	Window
Enter:	at most lowest value for both, at least highest value for both, 0 or increment, −2, at least highest counting number, 1, 1
Choose:	Graph > Trace >>>> (gives data values)

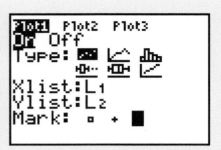

S E C T I O N   2 . 1   E X E R C I S E S

2.1 a. How long do you typically spend on grooming per day?

b. How do you think you compare to the 50,000 college students in "Students, Here's Looking at You" on page 32?

c. How do you think you compare to all college students? What are the similarities? What are the differences?

2.2 [EX02-002] Students in an online statistics course were asked how many different Internet activities they engaged in during a typical week. The following data show the number of activities:

6	7	3	6	9	10	8	9	9	6	4	9	4	9
4	2	3	5	13	12	4	6	4	9	5	6	9	
11	5	6	5	3	7	9	6	5	12	2	6	9	

a. If you were asked to present these data, how would you organize and summarize them?

b. How many different Internet activities did you engage in last week?

c. How do you think you compare to the 40 Internet users in the sample above?

2.3 As a statistical graph the circle graph has limitations. Examine the circle graph in Figure 2.1 and the bar graph in Figure 2.2.

a. What information do they both demonstrate?

b. What information is shown in the circle graph that cannot be shown in the bar graph?

c. "Generally speaking, the bar graph is a better choice for use than the circle graph." Justify this statement.

2.4 Results from a Self.com poll on "What is your top cold-weather beauty concern?" were reported in the December 2008 issue of *Self* magazine: Dry skin—57%; Chapped lips—25%; Dull hair—10%; Rough feet—8%.

a. Construct a pie chart showing the top cold-weather beauty concerns.

b. Construct a bar graph showing the top cold-weather beauty concerns.

c. In your opinion, does the pie chart in part a or the bar graph in part b result in a better representation of the information? Explain.

2.5 American Payroll Association got a big response to this question about company dress code: "The current dress code at my company is . . ."

Final results: a. A little too relaxed—27%
               b. A little too formal—15%
               c. Just right—58%

Most people mentioned the importance of "comfort" in their explanations. The vast majority of respondents were very happy with their company's dress code or policy.

a. Construct a circle graph depicting this information. Label completely.

b. Construct a bar graph depicting this same information. Label completely.

c. Compare the previous two graphs, describing what you see in each one now that the graphs have been drawn and completely labeled. Do you get the same impression about these people's feelings from both graphs? Does one emphasize anything the other one does not?

2.6 How much more young Americans, 17- to 28-year-olds, say they are willing to pay for an environmentally friendly vehicle was reported in a *USA Today* Snapshot on February 5, 2009: A lot more—11%; Somewhat more—36%; Slightly more—33%; Would not pay more—20%.

a. Name the variable of interest.

b. Identify the type of variable.

c. Construct a pie chart showing how young Americans feel about paying for an environmentally friendly vehicle.

d. Construct a bar graph showing how young Americans feel about paying for an environmentally friendly vehicle.

e. In your opinion, which graph is the better representation of the information? Why? Explain.

2.7 The number of points scored by the winning teams on October 28, 2008, the opening night of the 2008–2009 NBA season, are listed below.

Team	Boston	Chicago	LA Lakers
Score	90	108	96

Source: http://www.nba.com/

a. Draw a bar graph of these scores using a vertical scale ranging from 80 to 110.

b. Draw a bar graph of the scores using a vertical scale ranging from 50 to 110.

c. In which bar graph does it appear that the NBA scores vary more? Why?

d. How could you create an accurate representation of the relative size and variation between these scores?

2.8 [EX02-008] The American Community Survey gathers data for population, demographic, and housing unit estimates. The Census Bureau then uses the data to produce and disseminate official estimates of housing units for states and counties. The 2005–2007 housing unit estimates for the town of Webster in New York State estimates are listed here.

Housing Units Webster, NY

Owner-occupied housing units	12,627
Renter-occupied housing units	3,803
Vacant housing units	539
Total	16,969

Source: U.S. Census Bureau

a. Construct a pie chart of this breakdown.

b. Construct a bar graph of this breakdown.

c. Compare the two graphs you constructed in parts a and b; which one seems to be the most informative? Explain why.

2.9 Cleaning behind furniture and washing windows top the list of spring-cleaning chores, according to the Soap and Detergent Association's (SDA) latest National Spring Cleaning Survey. International Communications Research (ICR) completed the independent consumer research study in January/February 2008. The initial survey question was asked of 1013 American adults (507 men and 506 women).

Question asked: Do you regularly engage in spring cleaning?

Results: Yes—77%
          No—23%

More women (86%) than men (68%) spring-clean.

a. Construct and fully label a bar graph showing the results from all adults surveyed.

b. Construct and fully label a bar graph showing the results comparing the women and men separately.

c. Discuss the graphs in parts a and b, being sure to comment on how accurately, or not, the graphs picture the information.

Source: http://www.cleaning101.com/

2.10 [EX02-010] Credit card companies sometimes provide their customers with a year-end summary. The year-end summary provides a comprehensive, easy-to-read report summarizing the transactions into a several categories. Using table on the top of page 44:

a. Explain the meaning of the table entries of $64.02 and $(100.55).

b. Explain the meaning of the totals $1,159.19 and $298.42.

c. Use a pie chart to display the year-end category totals using both dollar amounts and percentages. Be sure to label completely.

d. Use a bar graph to display the monthly totals. Be sure to label completely.

2.11 A shirt inspector at a clothing factory categorized the last 500 defects as: 67—missing button; 153—bad seam; 258—improperly sized; 22—fabric flaw. Construct a Pareto diagram for this information.

Table for Exercise 2.10

Month	Travel	Restaurant	Merchandise	Auto	Services	Utilities	Totals
January	$ –	$ –	$ 87.38	$ –	$ 13.80	$ –	$ 101.18
February	$ –	$ 39.86	$ 9.99	$ 176.90	$ (100.55)	$ –	$ 126.20
March	$ –	$ 24.45	$ –	$ –	$ 60.51	$ –	$ 84.96
April	$ 25.00	$ 135.78	$ –	$ –	$ 260.00	$ –	$ 420.78
May	$ –	$ –	$ –	$ –	$ 175.27	$ –	$ 175.27
June	$ 25.00	$ 19.12	$ 254.30	$ –	$ –	$ –	$ 298.42
July	$ 25.00	$ 46.94	$ 281.12	$ 64.02	$ 30.00	$ –	$ 447.08
August	$ 25.00	$ –	$ 45.54	$ –	$ 21.48	$ 35.40	$ 127.42
September	$ –	$ 22.18	$ –	$ –	$ 55.85	$ –	$ 78.03
October	$ 25.00	$ 38.01	$ –	$ –	$ 61.55	$ –	$ 124.56
November	$ –	$ –	$ 86.51	$ –	$ 15.00	$ –	$ 101.51
December	$ –	$ –	$ 394.35	$ –	$ 22.55	$ –	$ 416.90
Totals	$ 125.00	$ 326.34	$ 1,159.19	$ 240.92	$ 615.46	$ 35.40	$ 2,502.31

2.12 [EX02-012] Definitions of e-mail spam, or junk e-mail, typically include the idea that the e-mail is unsolicited and sent in bulk. Beginning in the early 1990s the amount of e-mail spam grew steadily until recently, with a total volume of over 100 billion e-mails per day in April 2008. The amount being received has begun to decrease due to the use of better filtering software. As unbelievable as it might be, fewer than 200 spammers send about 80% of all spam.

The chart below lists the percentage of e-mail spam relayed by each country in 2007.

Country	Percentage
Brazil	4.1
China	8.4
EU	17.9
France	3.3
Germany	4.2
India	2.5
Italy	2.8
Poland	4.8
Russia	3.1
S Korea	6.5
Turkey	2.9
UK	2.8
US	19.6

Source: http://en.wikipedia.org/

a. Construct a bar graph of this information with the percentages in declining order.

b. Explain why a Pareto diagram of this information cannot be constructed.

2.13 A study completed by International Communications Research for the Soap and Detergent Association (SDA) lists the item Americans say they would be most willing to give up in order to be able to hire someone to do their spring cleaning. The most popular response was $100 (29%), followed by dining out for a month (26%), concert tickets (19%), a weekend trip (9%), and other (17%).

Source: http://www.cleaning101.com/

a. Construct a Pareto diagram displaying this information.

b. Because of the size of the "other" category, the Pareto diagram may not be the best graph to use.

Explain why, and describe what additional information is needed to make the Pareto diagram more appropriate.

2.14 What NOT to get them on Valentine's Day!

Presents Not Wanted

When it comes to Valentine's Day gifts, American adults say they prefer NOT to receive teddy bears.

Flowers 13%
Teddy bears 45%
Chocolate 22%
Jewelery 14%
Don't know 6%

Source: Data from Anne R. Carey and Juan Thomassie, *USA Today*.

a. Draw a bar graph picturing the percentages of "Presents not wanted."

b. Draw a Pareto diagram picturing the "Presents not wanted."

c. If you want to be 80% sure you did not get your Valentine something unwanted, what should you avoid buying? How does the Pareto diagram show this?

d. If 300 adults are to be surveyed, what frequencies would you expect to occur for each unwanted item listed on the graphic?

2.15 The final-inspection defect report for assembly line A12 is reported in a Pareto diagram.

a. What is the total defect count in the report?

b. Verify the 30.0% listed for "Scratch."

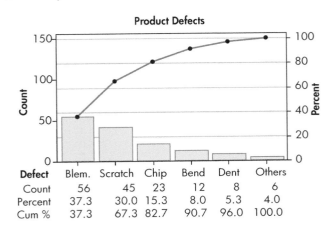

Product Defects

Defect	Blem.	Scratch	Chip	Bend	Dent	Others
Count	56	45	23	12	8	6
Percent	37.3	30.0	15.3	8.0	5.3	4.0
Cum %	37.3	67.3	82.7	90.7	96.0	100.0

c. Explain how the "cum % for bend" value of 90.7% was obtained and what it means.

d. Management has given the production line the goal of reducing their defects by 50%. What two defects would you suggest they give special attention to in working toward this goal? Explain.

2.16 Some cleaning jobs are disliked more than others. According to the July 17, 2009, *USA Today* Snapshot on a survey of women by Consumer Reports National Research Center, the cleaning tasks women dislike the most are presented in the following Pareto diagram.

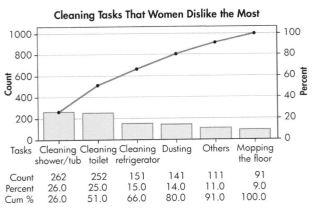

Cleaning Tasks That Women Dislike the Most

Tasks	Cleaning shower/tub	Cleaning toilet	Cleaning refrigerator	Dusting	Others	Mopping the floor
Count	262	252	151	141	111	91
Percent	26.0	25.0	15.0	14.0	11.0	9.0
Cum %	26.0	51.0	66.0	80.0	91.0	100.0

a. How many total women were surveyed?

b. Verify the 15% listed for "Cleaning refrigerator."

c. Explain how the "cum % for dusting" value of 80% was obtained and what it means.

d. What three tasks would make no more than 75% of the women surveyed happy if those tasks were eliminated?

2.17 [EX02-017] The American Time Use Survey presented at the beginning of the chapter outlined the time use of an average weekday for full-time university and college students.

Category	Hours
Sleeping	8.3
Leisure and sports	3.9
Educational activities	3.2
Working and related activities	3.0
Eating and drinking	1.0
Traveling	1.5
Grooming	0.8
Other	2.3
Total	24.0

a. Construct a pareto diagram depicting the average time use for full-time college students.

b. What activities seem to make up 75% of a college student's day?

2.18 [EX02-018] The Office of Aviation Enforcement and Proceedings, U.S. Department of Transportation, posted this table listing the number of consumer complaints against top U.S. airlines by complaint category.

Complaint Category	Number	Complaint Category	Number
Advertising	68	Flight problems	2031
Baggage	1421	Oversales	454
Customer service	1715	Refunds	1106
Disability	477	Reservations/ ticketing/boarding	1159
Fares	523	Other	322

Source: Office of Aviation Enforcement and Proceedings, U.S. Department of Transportation, Air Travel Consumer Report, http://www.infoplease.com/

a. Construct a Pareto diagram depicting this information.

b. What complaints would you recommend the airlines pay the most attention to correcting if they want to have the most effect on the overall number of complaints? Explain how the Pareto diagram from part a demonstrates the validity of your answer.

2.19 [EX02-019] The number of points scored during each game by a high school basketball team last season was as follows: 56, 54, 61, 71, 46, 61, 55, 68, 60, 66, 54, 61, 52, 36, 64, 51. Construct a dotplot of these data.

2.20 [EX02-020] In a July 8, 2009, *USA Today* article titled "Couples saying 'I don't' to expensive weddings," the cutbacks may not be extending to the number of attendants. In a survey of recent weddings, the number of bridesmaids was as follows:

| 7 | 6 | 5 | 2 | 3 | 7 | 6 | 13 | 6 | 3 | 2 | 7 | 8 | 9 |

a. Construct a dotplot of these data.

b. What are the most common numbers of bridesmaids? How does the dotplot show this?

2.21 [EX02-021] Shown below are the heights (in inches) of the basketball players who were the first-round picks by National Basketball Association professional teams for 2009.

82	86	76	77	75	72	75	81	78	74
77	77	81	81	82	80	76	72	74	74
73	82	80	84	74	81	80	77	74	78

Source: http://www.mynbadraft.com/

a.   Construct a dotplot of the heights of these players.

b.   Use the dotplot to uncover the shortest and the tallest players.

c.   What is the most common height and how many players share that height?

d.   What feature of the dotplot illustrates the most common height?

2.22 [EX02-022] The table lists the median house selling prices for the 20 suburbs of Rochester, New York, as listed in the July 18, 2009, *Democrat & Chronicle*.

Median House Prices in $1000s

160	125	122	89	100	110	94	125	108	235
133	121	190	175	218	130	180	113	156	114

Source: Greater Rochester Association of Realtors

a.   Construct a dotplot of these data.

b.   Describe the distribution displayed by the dotplot found in part a.

2.23 [EX02-023] Delco Products, a division of General Motors, produces commutators designed to be 18.810 mm in overall length. (A commutator is a device used in the electrical system of an automobile.) The following sample of 35 commutator lengths was taken while monitoring the manufacturing process:

18.802	18.810	18.780	18.757	18.824	18.827	18.825
18.809	18.794	18.787	18.844	18.824	18.829	18.817
18.785	18.747	18.802	18.826	18.810	18.802	18.780
18.830	18.874	18.836	18.758	18.813	18.844	18.861
18.824	18.835	18.794	18.853	18.823	18.863	18.808

Source: With permission of Delco Products Division, GMC

Use a computer to construct a dotplot of these data values.

2.24 A computer was used to construct this dotplot below.

a.   How many data values are shown?

b.   List the values of the five smallest data.

c.   What is the value of the largest data item?

d.   What value occurred the greatest number of times? How many times did it occur?

2.25 [EX02-025] Construct a stem-and-leaf display of the number of points scored during each basketball game last season:

56	54	61	71	46	61	55	68
60	66	54	61	52	36	64	51

2.26 [EX02-026] Every year *FORTUNE* Magazine publishes a list of The 100 Best Companies to Work For®. The table below shows the top 15 companies and their growth for February 8, 2010.

Company	Growth (%)
1. NetApp	12
2. Edward Jones	9
3. Boston Consulting Group	10
4. Google	40
5. Wegmans Food Markets	6
6. Cisco Systems	7
7. Genentech	5
8. Methodist Hospital System	1
9. Goldman Sachs	2
10. Nugget Market	22
11. Adobe Systems	9
12. Recreational Equipment	11
13. Devon Energy	11
14. Robert W. Baird	4
15. W.L. Gore & Associates	5

Source: http://money.cnn.com/

a.   Construct a stem-and-leaf display of the data.

b.   Based on the stem-and-leaf display, describe the distribution of the percentages of growth.

2.27 [EX02-027] The amounts shown here are the fees charged by Quik Delivery for the 40 small packages it delivered last Thursday afternoon:

4.03	3.56	3.10	6.04	5.62	3.16	2.93	3.82	4.30	3.86
4.57	3.59	4.57	6.16	2.88	5.03	5.46	3.87	6.81	4.91
3.62	3.62	3.80	3.70	4.15	2.07	3.77	5.77	7.86	4.63
4.81	2.86	5.02	5.24	4.02	5.44	4.65	3.89	4.00	2.99

a.   Construct a stem-and-leaf display.

b.   Based on the stem-and-leaf display, describe the distribution of the data.

2.28 [EX02-028] One of the many things the U.S. Census Bureau reports to the public is the increase in population for various geographic areas within the country. The percent of increase in population for the 24 fastest-growing counties in the United States from July 1, 2006, to July 1, 2007, is listed in the accompanying table.

County	State	Percent
St. Bernard Parish	Louisiana	42.9
Orleans Parish	Louisiana	13.8
*** For remainder of data, logon at cengagebrain.com		

Source: http://www.census.gov/

a.   Construct a stem-and-leaf display.

b.   Based on the stem-and-leaf display, describe the distribution of the data.

2.29 Given the following stem-and-leaf display:

Stem-and-Leaf of C1 N = 16
Leaf Unit = 0.010
```
 1      59    7
 4      60    148
(5)     61    02669
 7      62    0247
 3      63    58
 1      64    3
```

a. What is the meaning of "Leaf Unit = 0.010"?

b. How many data are shown on this stem-and-leaf display?

c. List the first four data values.

d. What is the column of numbers down the left-hand side of the figure?

2.30 A term often used in solar energy research is *heating-degree-days*. This concept is related to the difference between an indoor temperature of 65°F and the average outside temperature for a given day. An average outside temperature of 5°F gives 60 heating-degree-days. The annual heating-degree-day normals for several Nebraska locations are shown on the accompanying stem-and-leaf display constructed using MINITAB.

Stem-and-leaf of C1 N = 25
Leaf Unit = 10
```
 2      60    78
 7      61    03699
 9      62    69
11      63    26
(3)     64    233
11      65    48
 9      66    8
 8      67    249
 5      68    18
 3      69    145
```

a. What is the meaning of "Leaf Unit = 10"?

b. List the first four data values.

c. List all the data values that occurred more than once.

2.2 Frequency Distributions and Histograms

Lists of large sets of data do not present much of a picture. Sometimes we want to condense the data into a more manageable form. This can be accomplished with the aid of a *frequency distribution.*

> **Frequency distribution** A listing, often expressed in chart form, that pairs values of a variable with their frequency.

To demonstrate the concept of a frequency distribution, let's use this set of data:

```
3   2   2   3   2   4   4   1   2   2
4   3   2   0   2   2   1   3   3   1
```

If we let x represent the variable, then we can use a frequency distribution to represent this set of data by listing the x values with their frequencies. For example, the value 1 occurs in the sample three times; therefore, the **frequency** for $x = 1$ is 3. The complete set of data is shown in the frequency distribution in Table 2.4.

The frequency, f, is the number of times the value x occurs in the sample. Table 2.4 is an **ungrouped frequency distribution** — "ungrouped" because each value of x in the distribution stands alone. When a large set of data has many different x values instead of a few repeated values, as in the previous example, we can group the values into a set of classes and construct a **grouped frequency distribution**. The stem-and-leaf display in Figure 2.5B (p. 38) shows, in picture form, a grouped frequency distribution. Each stem represents a class. The number of leaves on each stem is the same as the frequency for that same **class** (sometimes called a *bin*). The data represented in Figure 2.5B are listed as a grouped frequency distribution in Table 2.5.

TABLE 2.4 Ungrouped Frequency Distribution

x	f
0	1
1	3
2	8
3	5
4	3

TABLE 2.5 Grouped Frequency Distribution

		Class	Frequency
50 or more to less than 60	⟶	$50 \le x < 60$	1
60 or more to less than 70	⟶	$60 \le x < 70$	3
70 or more to less than 80	⟶	$70 \le x < 80$	8
80 or more to less than 90	⟶	$80 \le x < 90$	5
90 or more to less than 100	⟶	$90 \le x < 100$	2
			19

The stem-and-leaf process can be used to construct a frequency distribution; however, the stem representation is not compatible with all **class widths**. For example, class widths of 3, 4, and 7 are awkward to use. Thus, sometimes it is advantageous to have a separate procedure for constructing a grouped frequency distribution.

E X A M P L E   2 . 6

GROUPING DATA TO FORM A FREQUENCY DISTRIBUTION

To illustrate this grouping (or classifying) procedure, let's use a sample of 50 final exam scores taken from last semester's elementary statistics class. Table 2.6 lists the 50 scores.

Procedure for Constructing a Grouped Frequency Distribution

1. Identify the high score ($H = 98$) and the low score ($L = 39$), and find the range:

$$\text{range} = H - L = 98 - 39 = 59$$

2. Select a number of classes ($m = 7$) and a class width ($c = 10$) so that the product ($mc = 70$) is a bit larger than the range (range $= 59$).

TABLE 2.6
Statistics Exam Scores [TA02-06]

60	47	82	95	88	72	67	66	68	98	90	77	86
58	64	95	74	72	88	74	77	39	90	63	68	97
70	64	70	70	58	78	89	44	55	85	82	83	
72	77	72	86	50	94	92	80	91	75	76	78	

3. Pick a starting point. This starting point should be a little smaller than the lowest score, L. Suppose we start at 35; counting from there by tens (the class width), we get 35, 45, 55, 65, ..., 95, 105. These are called the **class boundaries**. The classes for the data in Table 2.6 are:

35 or more to less than 45	⟶	$35 \le x < 45$
45 or more to less than 55	⟶	$45 \le x < 55$
55 or more to less than 65	⟶	$55 \le x < 65$
65 or more to less than 75	⟶	$65 \le x < 75$
	⋮	$75 \le x < 85$
		$85 \le x < 95$
95 or more to and including 105	⟶	$95 \le x \le 105$

Notes:

1. At a glance you can check the number pattern to determine whether the arithmetic used to form the classes was correct (35, 45, 55, . . . , 105).

2. For the interval $35 \leq x < 45$, 35 is the lower class boundary and 45 is the upper class boundary. Observations that fall on the lower class boundary stay in that interval; observations that fall on the upper class boundary go into the next higher interval, except for the last class.
3. The class width is the difference between the upper and lower class boundaries.
4. Many combinations of class widths, numbers of classes, and starting points are possible when classifying data. There is no one best choice. Try a few different combinations, and use good judgment to decide on the one to use.

Therefore, the following **basic guidelines** are used in constructing a grouped frequency distribution:

1. Each class should be of the same width.
2. Classes (sometimes called *bins*) should be set up so that they do not overlap and so that each data value belongs to exactly one class.
3. For the exercises given in this textbook, 5 to 12 classes are most desirable, because all samples contain fewer than 125 data values. (The square root of n is a reasonable guideline for the number of classes with samples of fewer than 125 data values.)
4. Use a system that takes advantage of a number pattern to guarantee accuracy.
5. When it is convenient, an even class width is often advantageous.

Once the classes are set up, we need to sort the data into those classes. The method used to sort will depend on the current format of the data: If the data are ranked, the frequencies can be counted; if the data are not ranked, we will **tally** the data to find the frequency numbers. When classifying data, it helps to use a standard chart (see Table 2.7).

TABLE 2.7
Standard Chart for Frequency Distribution

Class Number	Class Tallies	Boundaries	Frequency
1	‖	$35 \leq x < 45$	2
2	‖	$45 \leq x < 55$	2
3	‖‖‖ ‖	$55 \leq x < 65$	7
4	‖‖‖ ‖‖‖ ‖‖	$65 \leq x < 75$	13
5	‖‖‖ ‖‖‖ ‖	$75 \leq x < 85$	11
6	‖‖‖ ‖‖‖ ‖	$85 \leq x < 95$	11
7	‖‖‖	$95 \leq x \leq 105$	4
			50

Notes:

1. If the data have been ranked (list form, dotplot, or stem-and-leaf), tallying is unnecessary; just count the data that belong to each class.
2. If the data are not ranked, be careful as you tally.
3. The frequency, f, for each class is the number of pieces of data that belong in that class.
4. The sum of the frequencies should equal the number of pieces of data, n ($n = \Sigma f$). This summation serves as a good check.

Note: See the *Student Solutions Manual* for information about Σ **notation** (read **"summation notation"**).

TABLE 2.8
Frequency Distribution with Class Midpoints

Class Number	Class Boundaries	Frequency, f	Class Midpoints, x
1	$35 \leq x < 45$	2	40
2	$45 \leq x < 55$	2	50
3	$55 \leq x < 65$	7	60
4	$65 \leq x < 75$	13	70
5	$75 \leq x < 85$	11	80
6	$85 \leq x < 95$	11	90
7	$95 \leq x \leq 105$	4	100
		50	

Note: Now you can see why it is helpful to have an even class width. An odd class width would have resulted in a class midpoint with an extra digit. (For example, the class 45–54 is 9 wide and the class midpoint is 49.5.)

APPLIED EXAMPLE 2.7

CLEANING HOUSE

The "Weekly hours devoted to housecleaning" graphic presents a circle graph version of a **relative frequency distribution**. Each sector of the circle represents the amount of time spent cleaning weekly by each person, and the "relative size" of the sector represents the percentage or relative frequency.

Now using statistical terminology, we can say that the *variable* "time spent cleaning" is represented in the graph by sectors of the circle. The *relative frequency* is represented by the size of the angle forming the sector. To form this information into a grouped "relative" frequency distribution, each interval of the variable will be expressed in the form $a \leq x < b$. For example, the 2- to 4-hour category would be expressed as $2 \leq x < 4$. (This way, the lower boundary is part of the interval, but the upper boundary is part of the next larger interval.) The distribution chart for this circle graph would then appear as in the table shown to the left.

Weekly Hours Devoted to Housecleaning

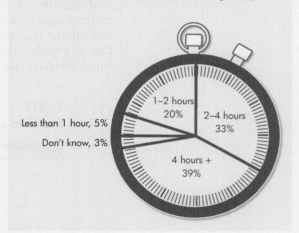

Americans spend an average of 3.4 hours each week cleaning the house. How much time is spent cleaning weekly:

1–2 hours 20%
2–4 hours 33%
Less than 1 hour, 5%
Don't know, 3%
4 hours + 39%

Source: Data from Cindy Hall and Sam Ward, *USA TODAY*; Yankelovich Partners for GCI/ZEP Chemicals.

Class Boundaries	Relative Frequency
$0 \leq x < 1$	0.05
$1 \leq x < 2$	0.20
$2 \leq x < 4$	0.33
$4 \leq x$	0.39
Don't know	0.03

Each class needs a single numerical value to represent all the data values that fall into that class. The **class midpoint** (sometimes called the *class mark*) is the numerical value that is exactly in the middle of each class. It is found by adding the class boundaries and dividing by 2. Table 2.8 shows an additional column for the class midpoint, x. As a check of your arithmetic, successive class midpoints should be a class width apart, which is 10 in this illustration $(40, 50, 60, \ldots, 100$ is a recognizable pattern).

When we classify data into classes, we lose some information. Only when we have all of the raw data do we know the exact values that were actually observed for each class. For example, we put a 47 and a 50 into class 2, with class boundaries of 45 and 55. Once they are placed in the class, their values are lost to us and we use the class midpoint, 50, as their representative value.

> **Histogram** A bar graph that represents a frequency distribution of a quantitative variable. A histogram is made up of the following components:
>
> 1. A title, which identifies the population or sample of concern.
> 2. A vertical scale, which identifies the frequencies in the various classes.
> 3. A horizontal scale, which identifies the variable x. Values for the class boundaries or class midpoints may be labeled along the x-axis. Use whichever method of labeling the axis best presents the variable.

The frequency distribution from Table 2.8 appears in histogram form in Figure 2.10.

FYI Note that the frequency histogram and the relative frequency histogram have the same shape (assuming the same classes are used for both); only the vertical axis labels change.

FYI Be sure to identify both scales so that the histogram tells the complete story.

FIGURE 2.10

Frequency Histogram

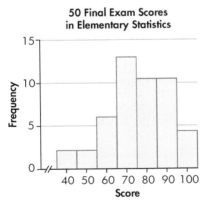

FIGURE 2.11

Relative Frequency Histogram

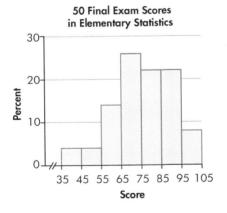

Sometimes the **relative frequency** of a value is important. The relative frequency is a proportional measure of the frequency for an occurrence. It is found by dividing the class frequency by the total number of observations. Relative frequency can be expressed as a common fraction, in decimal form, or as a percentage. For example, in Example 2.6 the frequency associated with the third class (55–65) is 7. The relative frequency for the third class is $\frac{7}{50}$, or 0.14, or 14%. Relative frequencies are often useful in a presentation because most people understand fractional parts when expressed as percents. Relative frequencies are particularly useful when comparing the frequency distributions of two different size sets of data. Figure 2.11 is a **relative frequency histogram** of the sample of the 50 final exam scores from Table 2.8.

A stem-and-leaf display contains all the information needed to create a histogram. Figure 2.5B (p. 38) shows the stem-and-leaf display constructed in

Example 2.4. In Figure 2.12A the stem-and-leaf display has been rotated 90° and labels have been added to show its relationship to a histogram. Figure 2.12B shows the same set of data as a completed histogram.

FIGURE 2.12A
Modified Stem-and-Leaf Display

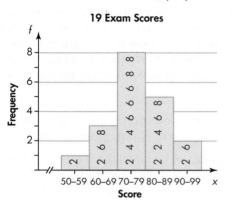

FIGURE 2.12B
Histogram

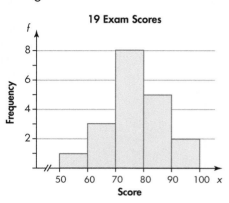

T E C H N O L O G Y   I N S T R U C T I O N S :
H I S T O G R A M

MINITAB

Input the data into C1; then continue with:

Choose: **Graph > Histogram > Simple > OK**
Enter: Graph variables: **C1**
Choose: **Labels > Titles / Footnote**
Enter: **Your title and / or footnote > OK**
Choose: **Scale > Y-Scale Type**
Select: Y scale Type: **Frequency or Percent or Density > OK > OK**

To adjust histogram: Double click anywhere on bars of histogram.

Select: **Binning**
Select: Interval Type: **Midpoint or Cutpoint**
           Interval Definitions:
           Automatic or, **Number of intervals;** Enter: **N** or, **Midpt/cutpt positions;** Enter: **A:B/C > OK**

Notes:
1. Midpoints are the class midpoints, and cutpoints are the class boundaries.
2. Percent is relative frequency.
3. Automatic means MINITAB will make all the choices; N = number of intervals, that is, the number of classes you want used.
4. A = smallest class midpoint or boundary, B = largest class midpoint or boundary, C = class width you want to specify.

The following commands will draw the histogram of a frequency distribution. The end classes can be made full width by adding an extra class with frequency zero to each end of the frequency distribution. Input the class midpoints into C1 and the corresponding frequencies into C2.

Choose: **Graph > Scatterplot > With Connect Line > OK**
Enter: Y variables: **C2**   X variables: **C1**
Select: Data View: Data Display: **Symbols Connect > OK > OK**
Double click on a connect line.
Select: **Options**
           Connection Function: **Step > OK**

Excel

Input the data into column A and the upper class limits* into column B (optional) and (column headings are optional); then continue with:

Choose: Data > Data Analysis† > Histogram > OK
Enter: Input Range: **Data (A1:A6 or select cells)**
             Bin Range: **upper class limits (B1:B6 or select cells)**
             [leave blank if Excel determines the intervals]
Select: **Labels** (if column headings are used)
             Output Range
Enter: **area for freq. distr. & graph (C1 or select cell)**
Select: **Chart Output > OK**

To remove gaps between bars:

Click on: **Any bar on graph**
Click on: **Right mouse button**
Choose: **Format Data Series**
Enter: **Gap Width: 0 % > Close**

To edit the histogram:

Click on: **Anywhere clear on the chart**—use handles to size
             Any title or axis name to change
             Any upper class limit§ or frequency in the frequency distribution
             to change value > **Enter**
             Delete "Frequency" box on right

*If boundary = 50, then limit = 49.9 (depending on the number of decimal places in the data).
†If Data Analysis does not show on the Data menu:

  Choose: Office Button > Excel Options (bottom) > Add-Ins
  Select: Analysis ToolPak
               Analysis ToolPak-VBA

§Note that the upper class limits appear in the center of the bars. Replace with class midpoints. The "More" cell in the frequency distribution may also be deleted.

For tabled data, input the classes into column A (ex. 30–40) and the frequencies into column B; activate both columns; then continue with:

Choose: **Insert > Column > 1st picture** (usually)
Choose: **Chart Layouts > Layout 8**
Enter: **Chart title: your title**
             Category (x) axis: title for **x-axis**
             Value (y) axis: title for **y-axis**

Do as just described to remove gaps and adjust.

TI-83/84 Plus

Input the data into L1; then continue with:

Choose: **2nd > STAT PLOT > 1:Plot1**

Calculator selects classes:

Choose: **Zoom > 9:ZoomStat >**
             Trace > > >

Individual selects classes:

Choose: **Window**
Enter: **at most lowest value, at least highest value, class width, −1, at least highest frequency, 1 (depends on frequency numbers), 1**
Choose: **Graph > Trace** (use values to construct frequency distribution)

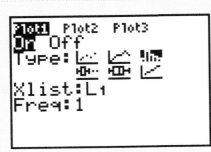

(TI-83/84 Plus continued)

For tabled data, input the class midpoints into L1 and the frequencies into L2; then continue with:

Choose: **2nd > STAT PLOT > 1:Plot1**
Choose: **Window**
Enter: smallest lower class boundary, largest upper class boundary, class width, $-ymax/4$, highest frequency, 0 (for no tick marks), 1
Choose: **Graph > Trace >>>**

To obtain a relative frequency histogram of tabled data instead:

Choose: **STAT > EDIT > 1:EDIT ...**
Highlight: **L3**
Enter: **L3 = L2 / SUM(L2)** [SUM - 2nd LIST > MATH > 5:sum]
Choose: **2nd > STAT PLOT > 1:Plot1**
Choose: **Window**
Enter: smallest lower class boundary, largest upper class boundary, class width, $-ymax/4$, highest rel. frequency, 0 (for no tick marks), 1
Choose: **Graph > Trace >>>**

Histograms are valuable tools. For example, the histogram of a sample should have a distribution shape very similar to that of the population from which the sample was drawn. If the reader of a histogram is at all familiar with the variable involved, he or she will usually be able to interpret several important facts. Figure 2.13 presents histograms with specific shapes that suggest descriptive labels. Possible descriptive labels are listed under each histogram.

FIGURE 2.13
Shapes of Histograms

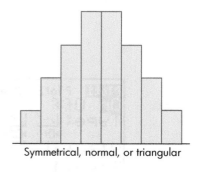
Symmetrical, normal, or triangular

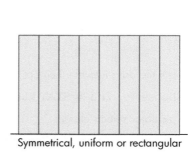
Symmetrical, uniform or rectangular

Skewed to right

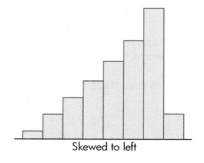
Skewed to left

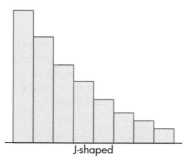
J-shaped

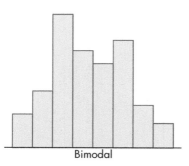
Bimodal

Briefly, the terms used to describe histograms are as follows:

Symmetrical Both sides of this distribution are identical (halves are mirror images).

Normal A symmetrical distribution is mounded up about the mean and becomes sparse at the extremes. (Additional properties are discussed later.)

Uniform (rectangular) Every value appears with equal frequency.

Skewed One tail is stretched out longer than the other. The direction of skewness is on the side of the longer tail.

J-shaped There is no tail on the side of the class with the highest frequency.

Bimodal The two most populous classes are separated by one or more classes. This situation often implies that two populations are being sampled. (See Figure 2.7, p. 40.)

Notes:

1. The **mode** is the value of the data that occurs with the greatest frequency. (Mode will be discussed in Section 2.3, p. 66.)
2. The **modal class** is the class with the highest frequency.
3. A **bimodal distribution** has two high-frequency classes separated by classes with lower frequencies. It is not necessary for the two high frequencies to be the same.

Another way to express a frequency distribution is to use a *cumulative frequency distribution*.

Cumulative frequency distribution A frequency distribution that pairs cumulative frequencies with values of the variable.

The **cumulative frequency** for any given class is the sum of the frequency for that class and the frequencies of all classes of smaller values. Table 2.9 shows the cumulative frequency distribution from Table 2.8 (p. 50).

TABLE 2.9
Using Frequency Distribution to Form a Cumulative
Frequency Distribution

Class Number	Class Boundaries	Frequency, f	Cumulative Frequency
1	$35 \leq x < 45$	2	2 (2)
2	$45 \leq x < 55$	2	4 (2 + 2)
3	$55 \leq x < 65$	7	11 (7 + 4)
4	$65 \leq x < 75$	13	24 (13 + 11)
5	$75 \leq x < 85$	11	35 (11 + 24)
6	$85 \leq x < 95$	11	46 (11 + 35)
7	$95 \leq x \leq 105$	4	50 (4 + 46)
		50	

The same information can be presented by using a *cumulative relative frequency distribution* (see Table 2.10). This combines the cumulative frequency and the relative frequency ideas.

TABLE 2.10 Cumulative Relative Frequency Distribution

Class Number	Class Boundaries	Cumulative Relative Frequency	Cumulative frequencies are for the interval 35 up to the upper boundary of that class.
1	$35 \leq x < 45$	2/50, or 0.04	⟵ from 35 up to less than 45
2	$45 \leq x < 55$	4/50, or 0.08	⟵ from 35 up to less than 55
3	$55 \leq x < 65$	11/50, or 0.22	⟵ from 35 up to less than 65
4	$65 \leq x < 75$	24/50, or 0.48	
5	$75 \leq x < 85$	35/50, or 0.70	⋮
6	$85 \leq x < 95$	46/50, or 0.92	
7	$95 \leq x \leq 105$	50/50, or 1.00	⟵ from 35 up to and including 105

Cumulative distributions can be displayed graphically.

Ogive (pronounced ō′jĭv) A line graph of a cumulative frequency or cumulative relative frequency distribution. An ogive has the following components:

1. A title, which identifies the population or sample.
2. A vertical scale, which identifies either the cumulative frequencies or the cumulative relative frequencies. (Figure 2.14 shows an ogive with cumulative relative frequencies.)
3. A horizontal scale, which identifies the upper class boundaries. (Until the upper boundary of a class has been reached, you cannot be sure you have accumulated all the data in that class. Therefore, the horizontal scale for an ogive is always based on the upper class boundaries.)

FIGURE 2.14

Ogive

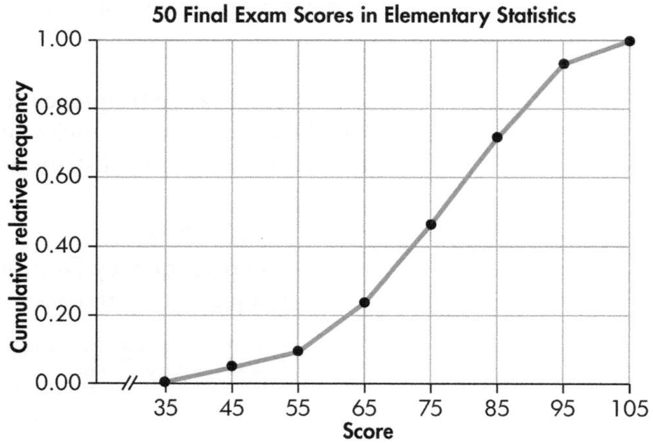

50 Final Exam Scores in Elementary Statistics

FYI Every ogive starts on the left with a relative frequency of zero at the lower class boundary of the first class and ends on the right with a cumulative relative frequency of 1.00 (or 100%) at the upper class boundary of the last class.

The ogive can be used to make percentage statements about numerical data much like a Pareto diagram does for attribute data. For example, suppose we want to know what percent of the final exam scores were not passing if scores of 65 or greater are considered passing. Following vertically from 65 on the horizontal scale to the ogive line and reading from the vertical scale, we could say that approximately 22% of the final exam scores were not passing grades.

TECHNOLOGY INSTRUCTIONS: OGIVE

MINITAB

Input the class boundaries into C1 and the cumulative percentages into C2 (enter 0 [zero] for the percentage paired with the lower boundary of the first class and pair each cumulative percentage with the class upper boundary). Use percentages; that is, use 25% in place of 0.25.

Choose: **Graph > Scatterplot > With Connect Line > OK**
Enter: Y variables: **C2**   X variables: **C1**
Select: Data View: Data Display: **Symbols Connect > OK**
Select: **Labels > Titles/Footnotes**
Enter: **your title or footnotes > OK > OK**

Excel

Input the data into column A and the upper class limits* into column B (include an additional class at the beginning).

Choose: **Data > Data Analysis** ** **> Histogram > OK**
Enter: Input Range: **data (A1:A6 or select cells)**
             Bin Range: **upper class limits (B1:B6 or select cells)**
Select: **Labels** (if column headings were used)
             Output Range
             Enter: area for freq. distr. & graph: **(C1 or select cell)**
             Cumulative Percentage
             Chart Output > OK

To close gaps and edit, see the histogram commands on page 53.

For tabled data, input the upper class boundaries into column A and the cumulative relative frequencies into column B (include an additional class boundary at the beginning with a cumulative relative frequency equal to 0 [zero]); activate column B; then continue with:

Choose: **Insert > Line > 1st picture (usually)**
Right click on chart area
Choose: **Select Data >**
             Horizontal (Category) Axis Labels Edit
Enter: **(A2:A8 or select cells) > OK > OK**
Choose: **Chart Tools > Layout > Labels**
Enter: Chart title: **your title**
             Axis titles: **title for x-axis; title for y-axis**

For editing, see the histogram commands on page 53.

*If the boundary = 50, then the limit = 49.9 (depending on the number of decimal places in the data).

**If Data Analysis does not show on the Data menu, see page 53.

TI-83/84 Plus

Input the class boundaries into L1 and the frequencies into L2 (include an extra class boundary at the beginning with a frequency of zero); then continue with:

Choose: **STAT > EDIT > 1:EDIT . . .**
Highlight: **L3**
Enter: **L3 = 2nd > LIST > OPS >**
             6:cum sum(L2)
Highlight: **L4**
Enter: **L4 = L3 / 2nd > LIST > Math**
             > 5:sum (L2)
Choose: **2nd > STAT PLOT > 1:Plot**
Choose: **Zoom > 9:ZoomStat >**
             Trace > > >

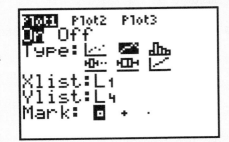

Adjust window if needed for better readability.

⊙ SECTION 2.2 EXERCISES

2.31 a. Form an ungrouped frequency distribution of the following data:

$$1, 2, 1, 0, 4, 2, 1, 1, 0, 1, 2, 4$$

Referring to the preceding distribution:

b. Explain what $f = 5$ represents.

c. What is the sum of the frequency column?

d. What does this sum represent?

2.32 Bar graphs and histograms are not the same thing. Explain their similarities and differences.

2.33 [EX02-033]The players on the Women's National Soccer Team scored 84 points during the 2008 season. The number of goals for those players who scored were:

Player	1	2	3	4	5	6	7	8	9	10	11	12	13	14	15
Goals	1	2	2	1	2	8	15	9	1	10	1	6	12	13	1

Source: U.S. Soccer

a. If you want to show the number of goals scored by each player, would it be more appropriate to display this information on a bar graph or a histogram? Explain.

b. Construct the appropriate graph for part a.

c. If you want to show (emphasize) the distribution of scoring by the team, would it be more appropriate to display this information on a bar graph or a histogram? Explain.

d. Construct the appropriate graph for part c.

2.34 [EX02-034]The California Department of Education gives an annual report on the Advanced Placement test results for each year. In the 2007–2008 school year, Hughson Unified in Stanislaus County had students with the following scores:

AP Scores

3	4	1	4	1	2	4	5	1	3	4	3	2	3	1	3	4	1	1	2	5
2	5	3	2	1	2	4	2	3	3	3	3	2	1	3	3	3	1	2	2	2

Source: http://data1.cde.ca.gov/

a. Construct an ungrouped frequency distribution for the test scores.

b. Construct a frequency histogram of this distribution.

c. Prepare a relative frequency distribution for this same data.

d. If AP scores of at least 3 are often required for college transferability, what percentage of Hughson AP scores will receive college credit?

2.35 [EX02-035] The U.S. Women's Olympic Soccer team had a great year in 2008. One way to describe the players on that team is by their individual heights.

Height (inches)

70	68	65	64	68	66	66	67	68
68	67	65	65	66	64	69	66	65

Source: www.ussoccer.com

a. Construct an ungrouped frequency distribution for the heights.

b. Construct a frequency histogram of this distribution.

c. Prepare a relative frequency distribution for this same data.

d. What percentage of the team is at least 5 ft 6 in. tall?

2.36 [EX02-036] The U.S. Census Bureau posted the following 2006 Report on America's Families and Living Arrangements for all races.

No. in Household	Percentage
1	27%
2	33%
3	17%
4	14%
5	6%
6	2%
7+	1%

Source: http://www.infoplease.com/

a. Draw a relative frequency histogram for the number of people per household.

b. What shape distribution does the histogram suggest?

c. Based on the graph, what do you know about the households in the United States?

2.37 [EX02-037]The 2006 American Community Survey universe is limited to household population and excludes the population living in institutions, college dormitories, and other group quarters. The accompanying table lists the number of rooms in each of the 48,522 housing units in Ellis County, Texas.

Rooms	Housing Units
1 room	403
2 rooms	485
3 rooms	2,171
4 rooms	8,108
5 rooms	12,177
6 rooms	11,251
7 rooms	6,250
8 rooms	4,320
9+ rooms	3,357

Source: U.S. Census Bureau, American Community Survey Office

a. Draw a relative frequency histogram for the number of rooms per household.

b. What shape distribution does the histogram suggest?

c. Based on the graph, what do you know about the number of rooms per household in Ellis County, Texas?

2.38 [EX02-038] Here are the ages of 50 dancers who responded to a call to audition for a musical comedy:

21	19	22	19	18	20	23	19	19	20
19	20	21	22	21	20	22	20	21	20
21	19	21	21	19	19	20	19	19	19
20	20	19	21	21	22	19	19	21	19
18	21	19	18	22	21	24	20	24	17

a. Prepare an ungrouped frequency distribution of these ages.

b. Prepare an ungrouped relative frequency distribution of the same data.

c. Prepare a relative frequency histogram of these data.

d. Prepare a cumulative relative frequency distribution of the same data.

e. Prepare an ogive of these data.

2.39 [EX02-039] The opening-round scores for the Ladies' Professional Golf Association tournament at Locust Hill Country Club were posted as follows:

69	73	72	74	77	80	75	74	72	83	68	73	75	78
76	74	73	68	71	72	75	79	74	75	74	74	68	79
75	76	75	77	74	74	75	75	72	73	73	72	72	71
71	70	82	77	76	73	72	72	72	75	75	74	74	74
76	76	74	73	74	73	72	72	74	71	72	73	72	72
74	74	67	69	71	70	72	74	76	75	75	74	73	74
74	78	77	81	73	73	74	68	71	74	78	70	68	71
72	72	75	74	76	77	74	74	73	73	70	68	69	71
77	78	68	72	73	78	77	79	79	77	75	75	74	73
73	72	71	68	70	71	78	78	76	74	75	72	72	72
75	74	76	77	78	78								

a. Form an ungrouped frequency distribution of these scores.

b. Draw a histogram of the first-round golf scores. Use the frequency distribution from part a.

2.40 Figuring *where* lightning strikes will occur is a nearly impossible task. *When* lightning strikes occur, however, has become more predictable based on research. For a small area in Colorado, data were collected and the results are displayed in the histogram that follows.

Based on the histogram:

a. Data were collected for what variable?

b. What does each bar (interval) represent?

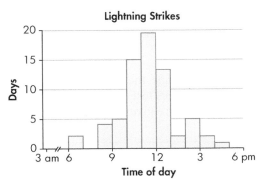

Lightning Strikes

c. What conclusion can be reached about the "when" of lightning strikes in this small area of Colorado?

d. What characteristics of the graph support the conclusion?

2.41 [EX02-041] A survey of 100 resort club managers on their annual salaries resulted in the following frequency distribution:

Annual Salary ($1000s)	15–25	25–35	35–45	45–55	55–65
No. of Managers	12	37	26	19	6

a. The data value "35" belongs to which class?

b. Explain the meaning of "35–45."

c. Explain what "class width" is, give its value, and describe three ways that it can be determined.

d. Draw a frequency histogram of the annual salaries for resort club managers. Label class boundaries.

(Retain these solutions to use in Exercise 2.53 on p. 61.)

2.42 Skillbuilder Applet Exercise demonstrates the procedure of transforming a stem-and-leaf display into a histogram. Type the leaves for the number of stories into the stem-and-leaf display. Click OK to see the corresponding histogram. Comment on the similarities and differences.

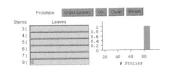

2.43 [EX02-043] The KSW computer science aptitude test was given to 50 students. The following frequency distribution resulted from their scores:

KSW Test Score	0–4	4–8	8–12	12–16	16–20	20–24	24–28
Frequency	4	8	8	20	6	3	1

a. What are the class boundaries for the class with the largest frequency?

b. Give all the class midpoints associated with this frequency distribution.

c. What is the class width?

(continue on page 60)

d. Give the relative frequencies for the classes.

e. Draw a relative frequency histogram of the test scores.

2.44 [EX02-044] During the Spring 2009 semester, 200 students took a statistics test from a particular instructor. The resulting grades are given in the following table.

Test grades	Number
50–60	13
60–70	44
70–80	74
80–90	59
90–100	9
100–110	1
Total	200

a. What is the class width?

b. Draw and completely label a frequency histogram of the statistics test grades.

c. Draw and completely label a relative frequency histogram of the statistics test grades.

d. Carefully examine the two histograms in parts b and c and explain why one of them might be more useful to a student and to the instructor.

FYI Use the computer or calculator commands on page 52–54 to construct a histogram of a frequency distribution.

2.45 [EX02-045] The speeds of 55 cars were measured by a radar device on a city street:

27	23	22	38	43	24	35	26	28	18	20
25	23	22	52	31	30	41	45	29	27	43
29	28	27	25	29	28	24	37	28	29	18
26	33	25	27	25	34	32	36	22	32	33
21	23	24	18	48	23	16	38	26	21	23

a. Classify these data into a grouped frequency distribution by using class boundaries 12–18, 18–24, . . . , 48–54.

b. Find the class width.

c. For the class 24–30, find the class midpoint, the lower class boundary, and the upper class boundary.

d. Construct a frequency histogram of these data.

FYI Use the computer or calculator commands on pages 52–54 to construct a histogram for a given set of data.

2.46 [EX02-046] The hemoglobin A_{1c} test, a blood test given to diabetic patients during their periodic checkups, indicates the level of control of blood sugar during the past 2 to 3 months. The following data values were obtained for 40 different diabetic patients at a university clinic:

6.5	5.0	5.6	7.6	4.8	8.0	7.5	7.9	8.0	9.2
6.4	6.0	5.6	6.0	5.7	9.2	8.1	8.0	6.5	6.6
5.0	8.0	6.5	6.1	6.4	6.6	7.2	5.9	4.0	5.7
7.9	6.0	5.6	6.0	6.2	7.7	6.7	7.7	8.2	9.0

a. Classify these A_{1c} values into a grouped frequency distribution using the classes 3.7–4.7, 4.7–5.7, and so on.

b. What are the class midpoints for these classes?

c. Construct a frequency histogram of these data.

2.47 [EX02-047] All of the third graders at Roth Elementary School were given a physical-fitness strength test. The following data resulted:

12	22	6	9	2	9	5	9	3	5	16	1	22
18	6	12	21	23	9	10	24	21	17	11	18	19
17	5	14	16	19	19	18	3	4	21	16	20	15
14	17	4	5	22	12	15	18	20	8	10	13	20
6	9	2	17	15	9	4	15	14	19	3	24	

a. Construct a dotplot.

b. Prepare a grouped frequency distribution using classes 1–4, 4–7, and so on, and draw a histogram of the distribution. (Retain the solution for use in answering Exercise 2.83, p. 71.)

c. Prepare a grouped frequency distribution using classes 0–3, 3–6, 6–9, and so on, and draw a histogram of the distribution.

d. Prepare a grouped frequency distribution using class boundaries −2.5, 2.5, 7.5, 12.5, and so on, and draw a histogram of the distribution.

e. Prepare a grouped frequency distribution using classes of your choice, and draw a histogram of the distribution.

f. Describe the shape of the histograms found in parts b–e separately. Relate the distribution seen in the histogram to the distribution seen in the dotplot.

g. Discuss how the number of classes used and the choice of class boundaries used affect the appearance of the resulting histogram.

2.48 [EX02-048] People have marveled for years at the continuing eruptions of the geyser Old Faithful in Yellowstone National Park. The times of duration, in minutes, for a sample of 50 eruptions of Old Faithful are listed here.

4.00	3.75	2.25	1.67	4.25	3.92
4.53	1.85	4.63	2.00	1.80	4.00
4.33	3.77	3.67	3.68	1.88	1.97
4.00	4.50	4.43	3.87	3.43	4.13
4.13	2.33	4.08	4.35	2.03	4.57
4.62	4.25	1.82	4.65	4.50	4.10
4.28	4.25	1.68	3.43	4.63	2.50
4.58	4.00	4.60	4.05	4.70	3.20
4.60	4.73				

Source: http://www.stat.sc.edu/

a. Draw a dotplot displaying the eruption-length data.

b. Draw a histogram of the eruption-length data using class boundaries 1.6–2.0–2.4–···–4.8.

c. Draw another histogram of the data using different class boundaries and widths.

d. Repeat part c.

e. Repeat parts a and b using the larger set of 107 eruptions available on [EX02-048].

f. Which graph, in your opinion, does the best job of displaying the distribution? Why?

g. Write a short paragraph describing the distribution.

2.49 [EX02-049] The Office of Coal, Nuclear, Electric and Alternate Fuels reported the following data as the costs (in cents) of the average revenue per kilowatt-hour for sectors in Arkansas:

6.61	7.61	6.99	7.48	5.10	7.56	6.65	5.93	7.92
5.52	7.47	6.79	8.27	7.50	7.44	6.36	5.20	5.48
7.69	8.74	5.75	6.94	7.70	6.67	4.59	5.96	7.26
5.38	8.88	7.49	6.89	7.25	6.89	6.41	5.86	8.04

a. Prepare a grouped frequency distribution for the average revenue per kilowatt-hour using class boundaries 4, 5, 6, 7, 8, 9.

b. Find the class width.

c. List the class midpoints.

d. Construct a relative frequency histogram of these data.

2.50 [EX02-050] Education has long been considered the ticket for upward mobility in the United States. In today's information age, a college education has become the minimum level of educational attainment needed to enter an increasingly competitive market for jobs with more than subsistence wages. A report based on information from the American Fact Finder and the 2007 American Community Survey gave the following percentages of population that have attained a bachelor's degree or higher by state.

> 21.4 26.0 25.3 19.3 29.5 ···
> *** For remainder of data, logon at cengagebrain.com

a. Prepare a grouped frequency distribution for the percentage of population by state that has attained a bachelor's degree or higher using class midpoints 15, 20, 25, ... 50.

b. List the class boundaries.

c. Construct a relative frequency histogram of these data.

2.51 Can you think of variables whose distribution might yield the following different shapes? (See Figure 2.13 on p. 54 if necessary.)

a. A symmetrical, or normal, shape

b. A uniform shape

c. A skewed-to-the-right shape

d. A skewed-to-the-left shape

e. A bimodal shape

2.52 Skillbuilder Applet Exercise demonstrates the effect that the number of classes or bins has on the shape of a histogram.

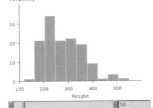

a. What shape distribution does using one class or bin produce?

b. What shape distribution does using two classes or bins produce?

c. What shape distribution does using 10 or 20 bins produce?

2.53 [EX02-041] A survey of 100 resort club managers on their annual salaries resulted in the following frequency distribution.

Ann. Sal. ($1000)	15–25	25–35	35–45	45–55	55–65
No. Mgrs.	12	37	26	19	6

a. Prepare a cumulative frequency distribution for the annual salaries.

b. Prepare a cumulative relative frequency distribution for the annual salaries.

c. Construct an ogive for the cumulative relative frequency distribution found above.

d. What value bounds the cumulative relative frequency of 0.75?

e. 75% of the annual salaries are below what value? Explain the relationship between parts d and e.

2.54 [EX02-034] a. Prepare a cumulative relative frequency distribution for the variable "AP score" in Exercise 2.34.

b. Construct an ogive of the distribution.

c. Using the ogive, find the cumulative relative frequency for the score of 2. Describe its meaning.

d. Using the answer in part c, approximately what percent of the AP scores will receive college credit if a score of at least 3 is required for college transferability? Describe the relationship between answers c and d.

e. Compare your answer to the answer found in 2.34 (d).

2.55 [EX02-043] a. Prepare a cumulative relative frequency distribution for the variable "KSW test score" in Exercise 2.43.

b. Construct an ogive of the distribution.

c. Using the ogive, approximately what percent of the students scored no more than 16 on the KSW computer science aptitude test?

2.56 [EX02-056] Undergraduates who use loans to pay for college average $16,500 in debt. The relative frequency distribution of their monthly debt after graduation is:

Monthly debt, $	Less than 100	100–149	150–199	200–249	250–299	300 or more
Percent	0.17	0.17	0.17	0.19	0.10	0.20

Source: *USA Today* Snapshot, December 23, 2004

a. Prepare a cumulative relative frequency distribution for the monthly debt.

b. Construct an ogive for the cumulative relative frequency distribution found in part a.

c. Based on the ogive, 60% of the monthly debts after graduation are below what approximate amount?

2.57 [EX02-057] American adults spend a large part of their weekdays at work. Commute times can make for an additionally long day. The size and location of the city along with the method of transportation can all make a difference in a commute time. The 2007 American Community Survey reported the following average commute times for each state.

31.5	31.1	30.1	29.8	28.2 . . .

*** For remainder of data, logon at cengagebrain.com

Source: Census Bureau; 2007 American Community Survey

a. Prepare a grouped frequency distribution of the average commute time data using class midpoints of $16, 18, 20, \ldots, 32$.

b. Prepare a grouped relative frequency distribution of these data.

c. Draw a relative frequency histogram of these data.

d. Prepare a cumulative relative frequency distribution of these same data.

e. Draw an ogive of these data.

f. Using the ogive, find the value that exceeds 70% of the data. Describe the relationship among the answer, 70%, the data, and the idea of the cumulative relative frequency distribution.

2.58 The levels of various compounds resulted in the distribution graphs that follow. They all seem to be quite symmetrical about their centers, but they differ in their spreads.

a. For which histogram, A, B, C, or D, would you anticipate the numerical measure of spread to be the largest? The smallest?

b. Which two of the four histograms would you anticipate to have about the same difference between their smallest values and their largest values?

Histograms for Exercise 2.58

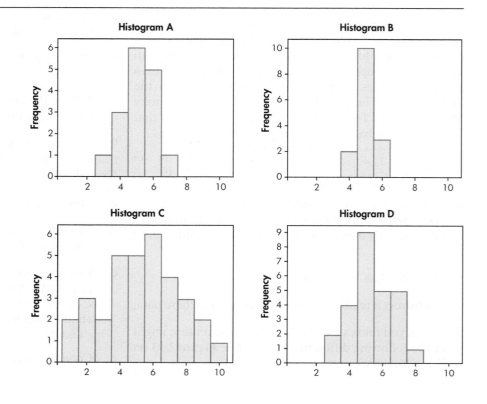

2.3 Measures of Central Tendency

Measures of central tendency are numerical values that locate, in some sense, the center of a set of data. The term *average* is often associated with all measures of central tendency.

> **Mean (arithmetic mean)** The average with which you are probably most familiar. The sample mean is represented by $\bar{x}$ (read "x-bar" or "sample mean"). The mean is found by adding all the values of the variable x (this sum of x values is symbolized Σx) and dividing the sum by the number of these values, n (the "sample size"). We express this in formula form as
>
> $$\text{Sample mean:}\quad \text{x-bar} = \frac{\text{sum of all } x}{\text{number of } x}$$
>
> $$\bar{x} = \frac{\Sigma x}{n} \qquad (2.1)$$

FYI The population mean, μ (lowercase mu, Greek alphabet), is the mean of all x values for the entire population.

Note: See the *Student Solutions Manual* for information about Σ notation ("summation notation").

E X A M P L E   2 . 8

FINDING THE MEAN

A set of data consists of the five values 6, 3, 8, 6, and 4. Find the mean.

Solution

Using formula (2.1), we find

$$\bar{x} = \frac{\Sigma x}{n} = \frac{6 + 3 + 8 + 6 + 4}{5} = \frac{27}{5} = 5.4$$

Therefore, the mean of this sample is **5.4.**

A physical representation of the mean can be constructed by thinking of a number line balanced on a fulcrum. A weight is placed on the number line at the number corresponding to each data value in the sample of Example 2.8. In Figure 2.15 there is one weight each on the 3, 8, and 4 and two weights on the 6, since there are two 6s in the sample. The mean is the value that balances the weights on the number line—in this case, 5.4.

FIGURE 2.15

Physical Representation of the Mean

FYI The mean is the middle point by weight.

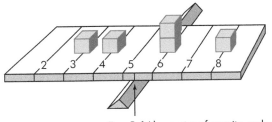

$\bar{x} = \textbf{5.4}$ (the center of gravity, or balance point)

Animated tutorial available—logon and learn more at cengagebrain.com

TECHNOLOGY INSTRUCTIONS: MEAN

MINITAB	Input the data into C1; then continue with: Choose: **Calc > Column Statistics** Select: **Mean** Enter: Input variable: **C1 > OK**

Excel	Input the data into column A and activate a cell for the answer; then continue with: Choose: **Insert Function, f$_x$ > Statistical > AVERAGE > OK** Enter: **Number 1: (A2:A6 or select cells) > OK** [Start at A1 if no header row (column title) is used.]

TI-83/84 Plus	Input the data into L1; then continue with: Choose: **2nd > LIST > Math > 3:mean(** Enter: **L1**

> **Median** The value of the data that occupies the middle position when the data are ranked in order according to size. The sample median is represented by $\tilde{x}$ (read "x-tilde" or "sample median").

Procedure for Finding the Median

Step 1:   Rank the data.

Step 2:   **Determine the depth of the median.** The **depth,** or position (number of positions from either end), of the median is determined by the formula

$$\text{depth of median:}\quad depth\ of\ median = \frac{sample\ size + 1}{2}$$

$$d(\tilde{x}) = \frac{n + 1}{2} \tag{2.2}$$

The median's depth (or position) is found by adding the position numbers of the smallest data (1) and the largest data (n) and dividing the sum by 2 (n is the number of pieces of data).

Step 3:   **Determine the value of the median.** Count the ranked data, locating the data in the $d(\tilde{x})$th position. The median will be the same regardless of which end of the ranked data (high or low) you count from. In fact, counting from both ends will serve as an excellent check.

The following two examples demonstrate this procedure as it applies to both odd-numbered and even-numbered sets of data.

EXAMPLE 2.9

MEDIAN FOR ODD *n*

Find the median for the set of data {6, 3, 8, 5, 3}.

FYI The value of $d(\tilde{x})$ is the depth of the median, NOT the value of the median, $\tilde{x}$.

Solution

Step 1 The data, ranked in order of size, are 3, 3, 5, 6, and 8.

Step 2 Depth of the median: $d(\tilde{x}) = \frac{n+1}{2} = \frac{5+1}{2} = 3$ (the "3rd" position).

Step 3 The median is the third number from either end in the ranked data, or $\tilde{x} = 5$.

Notice that the median essentially separates the ranked set of data into two subsets of equal size (see Figure 2.16).

FIGURE 2.16
Median of {3, 3, 5, 6, 8}

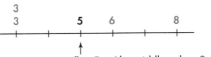

3
3
5
6
8
x̃ = 5 (the middle value; 2 data values are smaller, 2 are larger)

As in Example 2.9, when n is odd, the depth of the median, $d(\tilde{x})$, will always be an integer. When n is even, however, the depth of the median, $d(\tilde{x})$, will always be a half-number, as shown in Example 2.10.

EXAMPLE 2.10

MEDIAN FOR EVEN *n*

Find the median of the sample 9, 6, 7, 9, 10, 8.

FYI The median is the middle point by count.

Solution

Step 1 The data, ranked in order of size, are 6, 7, 8, 9, 9, and 10.

Step 2 Depth of the median: $d(\tilde{x}) = \frac{n+1}{2} = \frac{6+1}{2} = 3.5$ (the "3.5th" position).

Step 3 The median is halfway between the third and fourth data values. To find the number halfway between any two values, add the two values together and divide the sum by 2. In this case, add the third value (8) and the fourth value (9) and then divide the sum (17) by 2. The median is $\tilde{x} = \frac{8+9}{2} = 8.5$, a number halfway between the "middle" two numbers (see Figure 2.17). Notice that the median again separates the ranked set of data into two subsets of equal size.

Example 2.10 (Continued)

FYI The population median, *M* (uppercase mu in the Greek alphabet), is the data value in the middle position of the entire ranked population.

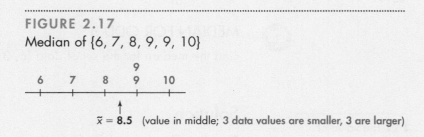

FIGURE 2.17
Median of {6, 7, 8, 9, 9, 10}

$\tilde{x} = 8.5$ (value in middle; 3 data values are smaller, 3 are larger)

TECHNOLOGY INSTRUCTIONS: MEDIAN

MINITAB

Input the data into C1; then continue with:

Choose: **Calc > Column Statistics**
Select: **Median**
Enter: **Input variable: C1 > OK**

Excel

Input the data into column A and activate a cell for the answer; then continue with:

Choose: **Insert Function, f_x > Statistical > MEDIAN > OK**
Enter: **Number 1: (A2:A6 or select cells) > OK**

TI-83/84 Plus

Input the data into L1; then continue with:

Choose: **2nd > LIST > Math > 4:median(**
Enter: **L1**

Mode The mode is the value of *x* that occurs most frequently.

In the set of data from Example 2.9, {3, 3, 5, 6, 8}, the mode is 3 (see Figure 2.18).

In the sample 6, 7, 8, 9, 9, 10, the mode is 9. In this sample, only the 9 occurs more than once; in the data from Example 2.9, only the 3 occurs more than once. If two or more values in a sample are tied for the highest frequency (number of occurrences), we say there is **no mode**. For example, in the sample 3, 3, 4, 5, 5, 7, the 3 and the 5 appear an equal number of times. There is no one value that appears most often; thus, this sample has no mode.

FIGURE 2.18

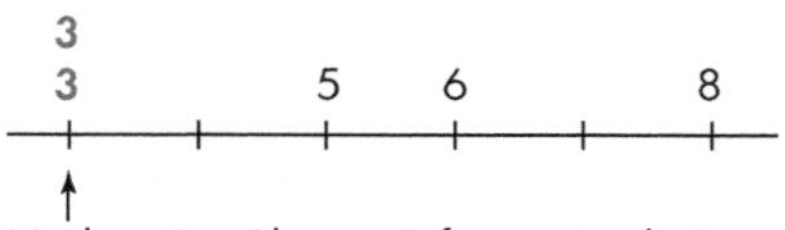

Mode = 3   (the most frequent value)

Midrange The number exactly midway between a lowest-valued data, *L*, and a highest-valued data, *H*. It is found by averaging the low and the high values:

$$midrange = \frac{low\ value + high\ value}{2}$$

$$midrange = \frac{L + H}{2}$$ (2.3)

For the set of data from Example 2.9, {3, 3, 5, 6, 8}, $L = 3$ and $H = 8$ (see Figure 2.19).
Therefore,

$$midrange = \frac{L + H}{2}$$

$$= \frac{3 + 8}{2} = 5.5$$

FIGURE 2.19
Midrange of {3, 3, 5, 6, 8}

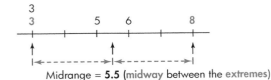

Midrange = **5.5** (**midway** between the **extremes**)

The four measures of central tendency represent four different methods of describing the middle. These four values may be the same, but more likely they will be different.

For the sample data from Example 2.10, the mean, $\bar{x}$, is 8.2; the median, $\tilde{x}$, is 8.5; the mode is 9; and the midrange is 8. Their relationship to one another and to the data is shown in Figure 2.20.

FIGURE 2.20
Measures of Central Tendency for {6, 7, 8, 9, 9, 10}

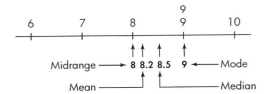

APPLIED EXAMPLE 2.11

"AVERAGE" MEANS DIFFERENT THINGS

When it comes to convenience, few things can match that wonderful mathematical device called averaging. With an average, you can take a fistful of figures on any subject and compute one figure that will represent the whole fistful.

But there is one thing to remember. There are several kinds of measures ordinarily known as *averages*, and each gives a different picture of the figures it is called on to represent.

Take an example. Table 2.11 shows the annual incomes of 10 families.

What would this group's "typical" income be? Averaging would provide the answer, so let's compute the typical income by the simpler and more frequently used kinds of averaging.

TABLE 2.11
Annual Incomes of 10 Families [TA02-11]

| $54,000 | $39,000 | $37,500 | $36,750 | $35,250 | $31,500 | $31,500 | $31,500 | $31,500 | $25,500 |

- *The arithmetic mean.* This is the most common form of average, obtained by adding items in the data set, then dividing by the number of items; for these data, the arithmetic mean is $35,400. The mean is representative of the data set in the sense that the sum of the amounts by which the higher figures exceed the mean is exactly the same as the sum of the amounts by which the lower figures fall short of the mean.

    The higher incomes exceed the mean by a total of $25,650. The lower incomes fall short of the mean by a total of $25,650.

- *The median.* As you may have observed, six families earn less than the mean and four families earn more. You might wish to represent this varied group by the income of the family that is smack dab in the middle of the whole bunch. The median works out to $33,375.

- *The midrange.* Another number that might be used to represent the average is the midrange, computed by calculating the figure that lies halfway between the highest and lowest incomes: $39,750.

- *The mode.* So, three kinds of averages, and not one family actually has an income matching any of them. Say you want to represent the group by stating the income that occurs most frequently. That is called a mode. The modal income would be $31,500.

Four different averages are available, each valid, correct, and informative in its way. But how they differ!

arithmetic mean	*median*	*midrange*	*mode*
$35,400	$33,375	$39,750	$31,500

And they would differ still more if just one family in the group were millionaires—or one were jobless! The large value of $54,000 (extremely different from the other values) is skewing the data out toward larger data values. This skewing causes the mean and midrange to become much larger in value.

So there are three lessons. First, when you see or hear an average, find out which average it is. Then you will know what kind of picture you are being given. Second, think about the figures being averaged so that you can judge whether the average used is appropriate. Third, do not assume that a literal mathematical quantification is intended every time somebody says "average." It isn't. All of us often say "the average person" with no thought of implying a mean, median, or mode. All we intend to convey is the idea of other people who are in many ways a great deal like the rest of us.

Now that we have learned how to calculate several sample statistics, the next question becomes, "How do I express my final answer?"

Round-off rule When rounding off an answer, let's agree to keep one more decimal place in our answer than was present in the original information. To avoid round-off buildup, round off only the final answer, not the intermediate steps. That is, avoid using a rounded value to do further calculations. In our previous examples, the data were composed of whole numbers; therefore, those answers that have decimal values should be rounded to the nearest tenth. See the *Student Solutions Manual* for specific instructions on how to perform the rounding off.

SECTION 2.3 EXERCISES

2.59 Explain why it is possible to find the mean for the data of a quantitative variable, but not for a qualitative variable.

2.60 The number of children, x, belonging to each of eight families registering for swimming was 1, 2, 1, 3, 2, 1, 5, 3. Find the mean, $\bar{x}$.

2.61 [EX02-061] The cost of taking your pet aboard the air flight with you in the continental United States varies according to the airlines. The prices for 14 of the major U.S. airlines in June 2009 were (in dollars):

69 100 100 100 125 150 100 60 100 125 75 100 125 100

Find the mean cost for flying your pet with you.

2.62 Skillbuilder Applet Exercise demonstrates the balancing effect of the mean. A plot is given with one data point at 10. Add more blocks by pointing and clicking on the desired location on the plot until a mean of 1 is achieved.

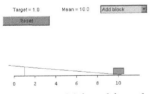

a.   How many blocks were required to balance for a mean of 1?

b.   At what value are these blocks located?

2.63 U.S. Interstate 64 runs between St. Louis, MO, at I-270 on the western end to Portsmouth, VA, at I-264 on the eastern end while passing through six states. The number of miles in each state is: Missouri—16 miles; Illinois—132 miles; Indiana—124 miles; Kentucky—191 miles; West Virginia—183 miles; Virginia—299 miles.

Source: http://www.ihoz.com/

a.   Find the mean number of miles in each state along I-64.

I-64 intersects with nine other interstate highways, besides I-264 and I-270 at its end points.

b.   Find the mean distance between interchanges with other interstate highways along I-64.

(**Hint**: For example, if there were 5 interchanges for a length of highway, there are only 4 sections of highway between those intersections.)

2.64 Interstate 29 intersects many other highways as it crosses four states in Mid-America stretching from the southern end in Kansas City, MO, at I-35 to the northern end in Pembina, ND, at the Canadian border.

U.S. Interstate 29

State	Miles	Number of Intersections
Missouri	123	37
Iowa	161	32
South Dakota	252	44
North Dakota	217	40

Source: Rand McNally and http://www.ihoz.com/

Consider the variable "distance between intersections."

a.   Find the mean distance between interchanges in Missouri.

b.   Find the mean distance between interchanges in Iowa.

c.   Find the mean distance between interchanges in North Dakota.

d.   Find the mean distance between interchanges in South Dakota.

e.   Find the mean distance between interchanges along U.S. I-29.

f.   Find the mean of the four means found in answering parts a through d.

g.   Compare the answers found in parts e and f. Did you expect them to be the same? Explain why they are different.

(**Hint**: For example, if there were 5 interchanges for a length of highway, there are only 4 sections of highway between those intersections.)

2.65 What is the mean weekly pay if 5 employees earn $425 per week, 3 earn $750 per week, and 1 earns $1340?

2.66 Is it possible for eight employees to earn between $300 and $350 while a ninth earns $1250 per week, and the mean to be $430? Verify your answer.

2.67 Find the median height of a basketball team: 73, 76, 72, 70, and 74 inches.

2.68 Find the median rate paid at Jim's Burgers if the workers' hourly rates are $4.25, $4.15, $4.90, $4.25, $4.60, $4.50, $4.60, $4.75.

2.69 For those seventh graders with cell phones, the number of programmed numbers in their phones are:

100	37	12	20	53	10	20	50	35	30

a. Find the mean number of programmed numbers on a seventh grader's cell phone.

b. Find the median number of programmed numbers on a seventh grader's cell phone.

c. Explain the difference in values of the mean and median.

d. Remove the most extreme value and answer parts a through c again.

e. Did removing the extreme value have more effect on the mean or the median? Explain why.

2.70 Skillbuilder Applet Exercise demonstrates the effect one data value can have on the mean and on the median.

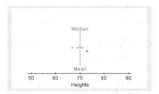

a. Move the red dot to the far right. What happens to the mean? What happens to the median?

b. Move the red dot to the far left. What happens to the mean? What happens to the median?

c. Which measure of central tendency, the mean or the median, gives a better sense of the center when a maverick (or outlier) is present in the data?

2.71 The number of cars per apartment, owned by a sample of tenants in a large complex, is 1, 2, 1, 2, 2, 2, 1, 2, 3, 2. What is the mode?

2.72 Each year around 160 colleges compete in the American Society of Civil Engineer's National Concrete Canoe Competition. Each team must design a seaworthy canoe from concrete, a substance not known for its capacity to float. The canoes must weigh between 100 and 350 pounds. When last year's entries weighed in, the weights ranged from 138 to 349 pounds.

a. Find the midrange.

b. The information given contains 4 weight values; explain why you used two of them in part a and did not use the other two.

2.73 a. Find the mean, median, mode, and midrange for the sample data 9, 6, 7, 9, 10, 8.

b. Verify and discuss the relationship between the answers in part a, as shown in Figure 2.20 on page 67.

2.74 Consider the sample 2, 4, 7, 8, 9. Find the following:

a. mean, $\bar{x}$

b. median, $\tilde{x}$

c. mode

d. midrange

2.75 Consider the sample 6, 8, 7, 5, 3, 7. Find the following:

a. mean, $\bar{x}$

b. median, $\tilde{x}$

c. mode

d. midrange

2.76 Fifteen randomly selected college students were asked to state the number of hours they slept the previous night. The resulting data values are 5, 6, 6, 8, 7, 7, 9, 5, 4, 8, 11, 6, 7, 8, 7. Find the following:

a. mean, $\bar{x}$

b. median, $\tilde{x}$

c. mode

d. midrange

2.77 [EX02-077]

A random sample of 10 of the 2007 NASCAR drivers produced the following ages:

36	26	48	28	45	21	21	38	27	32

a. Find the mean age for the ten 2007 NASCAR drivers.

b. Find the median age for the ten 2007 NASCAR drivers.

c. Find the midrange of age for the ten 2007 NASCAR drivers.

d. Find the mode, if one exists, for age for the ten 2007 NASCAR drivers.

2.78 In January 2009 the unemployment rate in all of New York City was 7.3. The unemployment rates for the five counties forming New York City were: 9.7, 7.7, 6.7, 6.6, 6.5.

a. Do you think the unemployment rate for the whole city and the mean unemployment rate for the five counties are the same? Explain in detail.

b. Find the mean of the unemployment rates for the five counties of New York City.

c. Explain in detail why the mean of the five counties is not the same as the rate for the whole city.

d. What conditions would need to exist in order for the mean of the five counties to be equal to the value for the whole city?

2.79 [EX02-079] A constant objective in the manufacture of contact lenses is to improve those features that affect lens power and visual acuity. One such feature involves the tooling from which lenses are ultimately manufactured. The results of initial process development runs were examined for critical feature X. The resulting data are listed here:

0.026	0.027	0.024	0.023	0.034	0.035	0.035	0.033	0.034
0.033	0.032	0.038	0.041	0.041	0.021	0.022	0.027	0.032
0.023	0.023	0.024	0.017	0.023	0.019	0.027		

Source: Courtesy of Bausch & Lomb (variable not named and data coded at B&L's request)

a. Draw both a dotplot and a histogram of the critical feature X data.

b. Find the mean for critical feature X.

c. Find the median for critical feature X.

d. Find the midrange for critical feature X.

e. Find the mode, if one exists, for critical feature X.

f. What feature of the distribution, as shown by the graphs found in part a, seems unusual? Where do the answers found in parts b, c, and d fall relative to the distribution? Explain.

g. Identify at least one possible cause for this seemingly unusual situation.

2.80 Buick and Jaguar tied for first place in the 2009 Vehicle Dependability Study for J.D. Power & Associates. It is an annual survey of three-year-old cars where consumers check off all the problems they have had with their 2006 model-year vehicles.

Source: J.D. Power & Assoc. 2009 Vehicle Dependability Study

A random sample of the J.D. Power data produced the following numbers of problems:

    148   263   147   159   222   150

a. Calculate the mean.

b. Based on the information given, explain what the mean is telling you. Does this make sense?

c. Reading further into the study, you find out that the "number of problems" data are a total for 100 cars of that vehicle brand. Divide each of the data values in part a by 100 and recalculate the mean.

d. Is there a quicker way you could have calculated the mean for part c?

e. Explain what this new mean is telling you.

f. Which mean would be more useful to both the manufacturer and the consumer? Why?

2.81 [EX02-081] The Rochester Raging Rhinos Professional Soccer Team is hoping for a good 2010 season. The blend of experienced and young, energetic players should make for a solid team. The current ages for the team are:

23	24	25	32	30	20	31	24	30	24
33	36	30	20	25	26	30	31	23	24

a. Construct a grouped frequency histogram using classes 19–21, 21–23, and so on.

b. Describe the distribution shown in the histogram.

c. Based on the histogram and its shape, what would you predict for the mean and the median? Which would be higher? Why?

d. Compute the mean and median. Compare answers to your predicted values in part c.

e. Which measure of central tendency provides the best measure of the center? Why?

2.82 The "average" is a commonly reported statistic. This single bit of information can be very informative or very misleading, with the mean and median being the two most commonly reported.

a. The mean is a useful measure, but it can be misleading. Describe a circumstance when the mean is very useful as the average and a circumstance when the mean is very misleading as the average.

b. The median is a useful measure, but it can be misleading. Describe a circumstance when the median is very useful as the average and a circumstance when the median is very misleading as the average.

2.83 [EX02-047] All the third graders at Roth Elementary School were given a physical-fitness strength test. These data resulted:

12	22	6	9	2	9	5	9	3	5	16	1	22
18	6	12	21	23	9	10	24	21	17	11	18	19
17	5	14	16	19	19	18	3	4	21	16	20	15
14	17	4	5	22	12	15	18	20	8	10	13	20
6	9	2	17	15	9	4	15	14	19	3	24	

a. Construct a dotplot.

b. Find the mode.

(continue on page 72)

c. Prepare a grouped frequency distribution using classes 1–4, 4–7, and so on, and draw a histogram of the distribution.

d. Describe the distribution; specifically, is the distribution bimodal (about what values)?

e. Compare your answers in parts a and c, and comment on the relationship between the mode and the modal values in these data.

f. Could the discrepancy found in the comparison in part e occur when using an ungrouped frequency distribution? Explain.

g. Explain why, in general, the mode of a set of data does not necessarily give us the same information as the modal values do.

2.84 [EX02-084] Consumers are frequently cautioned against eating too much food that is high in calories, fat, and sodium for numerous health and fitness reasons. *Nutrition in Action* published a list of popular low-fat brands of hot dogs commonly labeled "fat-free," "reduced fat," "low-fat," "light," and so on, together with their calories, fat content, and sodium. All measured quantities are for one dog:

Hot Dog Brand	Calories	Fat (g)	Sodium (mg)
Ball Park Fat Free Beef Franks	50	0	460
Butterball Fat Free Franks	40	0	490
*** For remainder of data, logon at cengagebrain.com			

Source: *Nutrition Action HealthLetter, "On the Links," July/August 1998*

a. Find the mean, median, mode, and midrange of the calories, fat, and sodium content of all the frankfurters listed. Use a table to summarize your results.

b. Construct a dotplot of the fat content. Locate the mean, median, mode, and midrange on the plot.

c. In the summer of 2009, the winner of Nathan's Famous Fourth of July Hot Dog Eating Contest consumed 68 hot dogs in 10 minutes. If he had been served the median hot dog, how many calories, grams of fat, and milligrams of sodium would he have consumed in that single sitting? If the recommended daily allowance for sodium intake is 2400 mg, would he likely have exceeded it? Explain.

2.85 [EX02-085]The number of runs scored by Major League Baseball teams is likely influenced by whether the game is played at home or at the opponent's ballpark. In an attempt to measure differences between playing at home or away, the average number of runs scored per game by each MLB team while playing on their home field and while playing away (at the opponents' fields) was calculated. The following table summarizes the data:

Team	Aver. Runs, Home	Aver. Runs, Away
Angels	4.73	4.72
Astros	4.59	4.26
*** For remainder of data, logon at cengagebrain.com		

Source: *MajorLeagueBaseball.com*

a. Find the mean, median, maximum, minimum, and midrange of the runs scored by the teams while playing at home.

b. Find the mean, median, maximum, minimum, and midrange of the runs scored by the teams while playing away.

c. Compare each of the measures you found in parts a and b. What can you conclude?

2.86 [EX02-086] Does everything increase, every year? Sometimes it seems like it! The 2007 annual percentage rate of increase in all motor-fuel consumption by U.S. states was reported in the Highway Statistics, November 2008, and is listed in the table. Notice that the consumption did not increase in every state.

Percent change in all motor-fuel consumption from 2006 to 2007 by state

2.4	−0.4	0.3	6.8	0.5	1.3	0.3	1.5	1.3
−2.1	−0.7	−1.5	4.9	−0.4	−0.8	0.1	3.3	−0.3
−10.2	−1.6	1.0	3.0	−2.7	−0.3	−2.9	0.9	0.6
3.4	−0.5	1.9	2.2	−1.3	−0.4	1.2	4.3	0.1
4.4	0.3	−0.4	−0.9	0.1	3.4	1.4	2.6	3.9
−1.1	2.1	0.6	−0.6	2.1	5.2			

Source: U.S. Department of Transportation—Federal Highway Administration

a. Explain the meaning of: negative and positive values, large and small values, values near zero, values not near zero.

b. Examine the data in the table. What do you anticipate the distribution of "percent change" to look like? What do you think the mean "percent change" will be? Justify your estimate, without any preliminary work or calculations.

c. If you expect very little or no change, what value will the mean have? Explain.

d. Construct a histogram of the percent of change.

e. Calculate the mean percent of changes in consumption from 2006 to 2007.

f. The Federal Highway Administration reported the percentage increase for the entire United States as 0.4 of 1%. The value calculated for the mean in part e is not the same. Explain how this is possible.

2.87 [EX02-087] Students like to engage in the "Battle of the Sexes" when it comes to who's the better driver. But which gender outnumbers the other on the road? The numbers may surprise you. Listed below is the number of licensed male and female drivers in each of 18 randomly selected states.

Number of licensed drivers by gender and state

State	Male	Female
KY	1,451,596	1,481,670
DE	304,455	320,017

*** For remainder of data, logon at cengagebrain.com

Source: Federal Highway Administration, U.S. Dept. of Transportation

a. Do the female drivers outnumber the male drivers? Study the table and see if the data seem to support your thoughts. Explain your initial answer.

b. Define the variable "ratio M/F" as the number of licensed male drivers divided by the number of licensed female drivers in each state. Calculate "ratio M/F" for the states in the sample.

c. If a value of the ratio M/F is near 1.0, what does that mean? Greater than 1.0? Less than 1.0? Explain.

d. Construct a histogram.

e. Describe the distribution shown in the histogram found in part d.

f. Calculate the mean value of the ratio M/F.

g. Explain the meaning of values in each of the tails of the histogram.

h. List two states, not on the table above, that you expect to find near each tail of the distribution of M/F. Explain why you believe these states will have high or low ratios.

i. Answer questions d and f using all 51 data values.

j. Compare the results found in part i to those found in parts d and f.

k. How did you do with your answer to part h? Explain.

2.88 You are responsible for planning the parking needed for a new 256-unit apartment complex and you're told to base the needs on the statistic "average number of vehicles per household is 1.9."

a. Which average (mean, median, mode, midrange) will be helpful to you? Explain.

b. Explain why "1.9" cannot be the median, the mode, or the midrange for the variable "number of vehicles."

c. If the owner wants parking that will accommodate 90% of all the tenants who own vehicles, how many spaces must you plan for?

2.89 In what states do the residents pay the most taxes? Pay the least? Perhaps it depends on the variable used to measure the amount of taxes paid. In 2008 the Tax Policy Center reported the following statistics about the average annual 2006 taxes and percent of personal income paid per person by state.

Tax Revenue 2006

	Taxes Per Capita	Rank	Percent of Personal Income	Rank
DISTRICT OF COLUMBIA	$7,764	1	14.4	4
ALABAMA	$2,782	51	9.6	48
WYOMING	$6,116	3	16.6	1
SOUTH DAKOTA	$2,842	48	9.1	51

Source: Federation of Tax Administrators (2007) and U.S. Bureau of the Census and Bureau of Economic Anaylsis. http://taxpolicycenter.org/

a. Compare and contrast the variables "taxes per capita" and "percent of personal income." How do you account for the differences in ranks for District of Columbia and Wyoming?

b. Based on this information, using the highest and lowest per-state amount of taxes paid per person, what was the "average" amount paid per person?

c. Based on this information, using the highest and lowest percent of income per state paid per person, what was the "average" percent paid per person?

d. Explain why your answers in parts b and c are the only average value you can determine from the given information. What is its name?

2.90 Your instructor and your class have made a deal on the exam just taken and currently being graded. If the class attains a mean score of 74 or better, there will be no homework on the coming weekend. If the class mean is 72 or below, then not only will there be homework as usual but all of the class members will have to show up on Saturday and do 2 hours of general cleanup around the school grounds as a community service project. There are 15 students in your class. Your instructor has graded the first 14 exams, and their mean score is 73.5. Your exam is the only one left to grade.

a. What score must you get in order for the class to win the deal?

b. What score must you get in order that the class will not have to do the community service work?

2.91 Starting with the data values 70 and 100, add three data values to the sample so that the sample has the following: (Justify your answer in each case.)

a. Mean of 100

b. Median of 70

(continue on page 74)

c.   Mode of 87

d.   Midrange of 70

e.   Mean of 100 and a median of 70

f.   Mean of 100 and a mode of 87

g.   Mean of 100 and a midrange of 70

h.   Mean of 100, a median of 70, and a mode of 87

2.92 Skillbuilder Applet Exercise matches means with corresponding histograms. After several practice rounds using "New Plots," explain your method of matching.

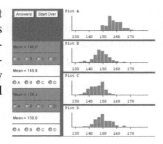

2.4 Measures of Dispersion

Having located the "middle" with the measures of central tendency, our search for information from data sets now turns to the measures of dispersion (spread). The **measures of dispersion** include the *range, variance*, and *standard deviation*. These numerical values describe the amount of spread, or variability, that is found among the data: Closely grouped data have relatively small values, and more widely spread-out data have larger values. The closest possible grouping occurs when the data have no dispersion (all data are the same value); in this situation, the measure of dispersion will be zero. There is no limit to how widely spread out the data can be; therefore, measures of dispersion can be very large. The simplest measure of dispersion is the range.

> **Range** The difference in value between the highest-valued data, *H*, and the lowest-valued data, *L*:
>
> $$range = high value - low value$$
>
> $$range = H - L \qquad (2.4)$$

The sample 3, 3, 5, 6, 8 has a range of $H - L = 8 - 3 = 5$. The range of 5 tells us that these data all fall within a 5-unit interval (see Figure 2.21).

....

FIGURE 2.21
Range of {3, 3, 5, 6, 8}

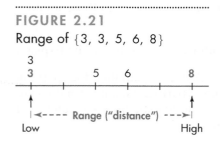

The other measures of dispersion to be studied in this chapter are measures of dispersion about the mean. To develop a measure of dispersion about the mean, let's first answer the question: How far is each *x* from the mean?

> **Deviation from the mean** A deviation from the mean, $x - \bar{x}$, is the difference between the value of *x* and the mean, $\bar{x}$.

Each individual value of *x* deviates from the mean by an amount equal to $(x - \bar{x})$. This deviation $(x - \bar{x})$ is zero when *x* is equal to the mean, $\bar{x}$. The deviation $(x - \bar{x})$ is positive when x is larger than $\bar{x}$ and negative when x is smaller than $\bar{x}$.

Consider the sample 6, 3, 8, 5, 3. Using formula (2.1), $\bar{x} = \frac{\Sigma x}{n}$, we find that the mean is 5. Each deviation, $(x - \bar{x})$, is then found by subtracting 5 from each x value:

Data, x	6	3	8	5	3
Deviation, $x - \bar{x}$	1	−2	3	0	−2

Figure 2.22 shows the four nonzero deviations from the mean.

FIGURE 2.22

Deviations from the Mean

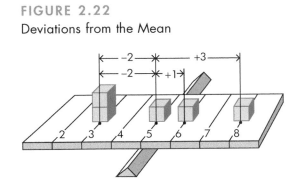

To describe the "average" value of these deviations, we might use the mean deviation, the sum of the deviations divided by n, $\frac{\Sigma(x - \bar{x})}{n}$. However, because the sum of the deviations, $\Sigma(x - \bar{x})$, is exactly zero, the mean deviation will also be zero. In fact, it will always be zero, which means it is not a useful statistic. How does this happen, and why?

The sum of the deviations, $\Sigma(x - \bar{x})$, is always zero because the deviations of x values smaller than the mean (which are negative) cancel out those x values larger than the mean (which are positive). We can remove this neutralizing effect if we do something to make all the deviations positive. This can be accomplished by squaring each of the deviations; squared deviations will all be nonnegative (positive or zero) values. The squared deviations are used to find the *variance*.

Sample variance The sample variance, s^2, is the mean of the squared deviations, calculated using $n - 1$ as the divisor:

$$\text{sample variance:} \quad s \text{ squared} = \frac{\text{sum of (deviations squared)}}{\text{number} - 1}$$

$$s^2 = \frac{\Sigma(x - \bar{x})^2}{n - 1} \tag{2.5}$$

where n is the sample size—that is, the number of data in the sample.

The variance of the sample 6, 3, 8, 5, 3 is calculated in Table 2.12 using formula (2.5).

Notes:

1. The sum of all the x values is used to find $\bar{x}$.
2. The sum of the deviations, $\Sigma(x - \bar{x})$, is always zero, provided the exact value of $\bar{x}$ is used. Use this fact as a check in your calculations, as was done in Table 2.12 (denoted by ⓒⓚ).

3. If a rounded value of $\bar{x}$ is used, then $\Sigma(x - \bar{x})$ will not always be exactly zero. It will, however, be reasonably close to zero.

4. The sum of the squared deviations is found by squaring each deviation and then adding the squared values.

TABLE 2.12 Calculating Variance Using Formula (2.5)

Step 1. Find Σx	Step 2. Find $\bar{x}$	Step 3. Find each $x - \bar{x}$	Step 4. Find $\Sigma(x - \bar{x})^2$	Step 5. Find s^2
6	$\bar{x} = \dfrac{\Sigma x}{n}$	$6 - 5 = 1$	$(1)^2 = 1$	$s^2 = \dfrac{\Sigma(x - \bar{x})}{n - 1}$
3		$3 - 5 = -2$	$(-2)^2 = 4$	
8		$8 - 5 = 3$	$(3)^2 = 9$	
5	$\bar{x} = \dfrac{25}{5}$	$5 - 5 = 0$	$(0)^2 = 0$	$s^2 = \dfrac{18}{4}$
3		$3 - 5 = -2$	$(-2)^2 = 4$	
$\Sigma x = 25$	$\bar{x} = 5$	$\Sigma(x - \bar{x}) = 0$ ⓒⓚ	$\Sigma(x - \bar{x})^2 = 18$	$s^2 = 4.5$

To graphically demonstrate what variances of data sets are telling us, consider a second set of data: $\{1, 3, 5, 6, 10\}$. Note that the data values are more dispersed than the data values in Table 2.12. Accordingly, its calculated variance is larger at $s^2 = 11.5$. An illustrative side-by-side graphical comparison of these two samples and their variances is shown in Figure 2.23.

FIGURE 2.23
Comparison of Data

Sample standard deviation The standard deviation of a sample, s, is the positive square root of the variance:

sample standard deviation: $s = $ square root of sample variance

$$s = \sqrt{s^2} \qquad (2.6)$$

For the samples shown in Figure 2.23, the standard deviations are $\sqrt{4.5}$ or **2.1**, and $\sqrt{11.5}$ or **3.4**.

The numerator for the sample variance, $\Sigma(x - \bar{x})^2$, is often called the *sum of squares for x* and symbolized by $SS(x)$. Thus, formula (2.5) can be expressed as

$$\text{sample variance: } s^2 = \frac{SS(x)}{n - 1} \qquad (2.7)$$

where $SS(x) = \Sigma(x - \bar{x})^2$.

The formulas for variance can be modified into other forms for easier use in various situations. For example, suppose we have the sample 6, 3, 8, 5, 2. The variance for this sample is computed in Table 2.13.

Animated tutorial available—logon and learn more at cengagebrain.com

TABLE 2.13 Calculating Variance Using Formula (2.5)

Step 1. Find Σx	Step 2. Find $\bar{x}$	Step 3. Find each $x - \bar{x}$	Step 4. Find $\Sigma(x - \bar{x})^2$	Step 5. Find s^2
6	$\bar{x} = \dfrac{\Sigma x}{n}$	$6 - 4.8 = 1.2$	$(1.2)^2 = 1.44$	$s^2 = \dfrac{\Sigma(x - \bar{x})^2}{n - 1}$
3		$3 - 4.8 = -1.8$	$(-1.8)^2 = 3.24$	
8		$8 - 4.8 = 3.2$	$(3.2)^2 = 10.24$	
5	$\bar{x} = \dfrac{24}{5}$	$5 - 4.8 = 0.2$	$(0.2)^2 = 0.04$	$s^2 = \dfrac{22.80}{4}$
2		$2 - 4.8 = -2.8$	$(-2.8)^2 = 7.84$	
$\Sigma x = 24$	$\bar{x} = 4.8$	$\Sigma(x - \bar{x}) = 0$ ⓒⓚ	$\Sigma(x - \bar{x})^2 = 22.80$	$s^2 = 5.7$

The arithmetic for this example has become more complicated because the mean contains nonzero digits to the right of the decimal point. However, the "sum of squares for x," the numerator of formula (2.5), can be rewritten so that $\bar{x}$ is not included:

Sum of Squares for x

$$SS(x) = \Sigma x^2 - \frac{(\Sigma x)^2}{n} \tag{2.8}$$

Combining formulas (2.7) and (2.8) yields the "short-cut formula" for sample variance:

Sample Variance, "Short-Cut Formula"

$$s\ squared = \frac{(sum\ of\ x^2) - \left[\dfrac{(sum\ of\ x)^2}{number}\right]}{number - 1}$$

$$\text{sample variance: } s^2 = \frac{\Sigma x^2 - \dfrac{(\Sigma x)^2}{n}}{n - 1} \tag{2.9}$$

Formulas (2.8) and (2.9) are called *shortcuts* because they bypass the calculation of $\bar{x}$. The computations for $SS(x)$, s^2, and s using formulas (2.8), (2.9), and (2.6) are performed as shown in Table 2.14.

TABLE 2.14 Calculating Standard Deviation Using the Shortcut Method

Step 1. Find Σx	Step 2. Find Σx^2	Step 3. Find $SS(x)$	Step 4. Find s^2	Step 5. Find s
6	$6^2 = 36$	$SS(x) = \Sigma x^2 - \dfrac{(\Sigma x)^2}{n}$		$s = \sqrt{s^2}$
3	$3^2 = 9$		$s^2 = \dfrac{\Sigma x^2 - \dfrac{(\Sigma x)^2}{n}}{n - 1}$	$s = \sqrt{5.7}$
8	$8^2 = 64$	$SS(x) = 138 - \dfrac{(24)^2}{5}$		$s = 2.4$
5	$5^2 = 25$		$s^2 = \dfrac{22.8}{4}$	
2	$2^2 = 4$	$SS(x) = 138 - 115.2$		
$\Sigma x = 24$	$\Sigma x^2 = 138$	$SS(x) = 22.8$	$s^2 = 5.7$	

The unit of measure for the standard deviation is the same as the unit of measure for the data. For example, if our data are in pounds, then the standard deviation, s, will also be in pounds. The unit of measure for variance might then be thought of as *units squared*. In our example of pounds, this would be *pounds squared*. As you can see, the unit has very little meaning.

Animated tutorial available—logon and learn more at cengagebrain.com

TECHNOLOGY INSTRUCTIONS: STANDARD DEVIATION

MINITAB	Input the data into C1; then continue with:
	Choose: **Calc > Column Statistics**
	Select: **Standard deviation**
	Enter: **Input variable: C1 > OK**

Excel	Input the data into column A and activate a cell for the answer; then continue with:
	Choose: **Insert Function, f_x > Statistical > STDEV > OK**
	Enter: **Number 1: (A2:A6 or select cells) > OK**

TI-83/84 Plus	Input the data into L1; then continue with:
	Choose: **2nd > LIST > Math > 7:StdDev(**
	Enter: **L1**

TECHNOLOGY INSTRUCTIONS: ADDITIONAL STATISTICS

MINITAB	Input the data into C1; then continue with:
	Choose: **Calc > Column Statistics**
	Then one at a time select the desired statistic
	Select: **N total** — Number of data in column
	Sum — Sum of the data in column
	Minimum — Smallest value in column
	Maximum — Largest value in column
	Range — Range of values in column
	Sum of squares — Sum of squared x-values, Σx^2
	Enter: **Input variable: C1 > OK**

Excel	Input the data into column A and activate a cell for the answer; then continue with:
	Choose: **Insert Function, f_x > Statistical > COUNT**
	> MIN
	> MAX
	OR **> All > SUM**
	> SUMSQ
	Enter: **Number 1: (A2:A6 or select cells)**
	For range, write a formula: **Max( ) − Min( )**

TI-83/84 Plus	Input the data into L1; then continue with:
	Choose: **2nd > LIST > Math > 5:sum(**
	> 1:min(
	> 2:max(
	Enter: **L1**

Standard deviation on your calculator Most calculators have two formulas for finding the standard deviation and mindlessly calculate both, fully expecting the user to decide which one is correct for the given data. How do you decide?

The sample standard deviation is denoted by s and uses the "divide by $n - 1$" formula.

The population standard deviation is denoted by σ and uses the "divide by n" formula.

When you have sample data, always use the s or "divide by $n - 1$" formula. Having the population data is a situation that will probably never occur, other than in a textbook exercise. If you don't know whether you have sample data or population data, it is a "safe bet" that they are sample data—use the s or "divide by $n - 1$" formula!

Multiple formulas Statisticians have multiple formulas for convenience—that is, convenience relative to the situation. The following statements will help you decide which formula to use:

1. When you are working on a computer and using statistical software, you will generally store all the data values first. The computer handles repeated operations easily and can "revisit" the stored data as often as necessary to complete a procedure. The computations for sample variance will be done using formula (2.5), following the process shown in Table 2.12.
2. When you are working on a calculator with built-in statistical functions, the calculator must perform all necessary operations on each data value as the values are entered (most handheld nongraphing calculators do not have the ability to store data). Then after all data have been entered, the computations will be completed using the appropriate summations. The computations for sample variance will be done using formula (2.9), following the procedure shown in Table 2.14.
3. If you are doing the computations either by hand or with the aid of a calculator, but not using statistical functions, the most convenient formula to use will depend on how many data there are and how convenient the numerical values are to work with.

⊡ SECTION 2.4 EXERCISES

2.93 In 2008 the Tax Policy Center reported the following statistics about the average annual 2006 taxes and percent of personal income paid per person by state.

a. Find the range for the amount of taxes paid per person.

b. Find the range for the percentage of personal income paid in taxes per person.

Tax Revenue 2006

	Taxes Per Capita	Rank	Percent of Personal Income	Rank
DISTRICT OF COLUMBIA	$7,764	1	14.4	4
ALABAMA	$2,782	51	9.6	48
WYOMING	$6,116	3	16.6	1
SOUTH DAKOTA	$2,842	48	9.1	51

Source: Federation of Tax Administrators (2007) and U.S. Bureau of the Census and Bureau of Economic Analysis. http://taxpolicycenter.org/

2.94 a. The data value $x = 45$ has a deviation value of 12. Explain the meaning of this.

b. The data value $x = 84$ has a deviation value of -20. Explain the meaning of this.

2.95 The summation $\Sigma(x - \bar{x})$ is always zero. Why? Think back to the definition of the mean (p. 63) and see if you can justify this statement.

2.96 All measures of variation are nonnegative in value for all sets of data.

a. What does it mean for a value to be "nonnegative"?

b. Describe the conditions necessary for a measure of variation to have the value zero.

(continue on page 80)

c. Describe the conditions necessary for a measure of variation to have a positive value.

2.97 A sample contains the data {1, 3, 5, 6, 10}.

a. Use formula (2.5) to find the variance.

b. Use formula (2.9) to find the variance.

c. Compare the results from parts a and b.

2.98 Consider the sample 2, 4, 7, 8, 9. Find the following:

a. Range

b. Variance s^2, using formula (2.5)

c. Standard deviation, s

2.99 Consider the sample 6, 8, 7, 5, 3, 7. Find the following:

a. Range

b. Variance s^2, using formula (2.5)

c. Standard deviation, s

2.100 Given the sample 7, 6, 10, 7, 5, 9, 3, 7, 5, 13, find the following:

a. Variance s^2 using formula (2.5)

b. Variance s^2, using formula (2.9)

c. Standard deviation, s

2.101 Fifteen randomly selected college students were asked to state the number of hours they slept the previous night. The resulting data are 5, 6, 6, 8, 7, 7, 9, 5, 4, 8, 11, 6, 7, 8, 7. Find the following:

a. Variance s^2, using formula (2.5)

b. Variance s^2, using formula (2.9)

c. Standard deviation, s

2.102 [EX02-102] A random sample of 10 of the 2007 NASCAR drivers produced the following ages:

36	26	48	28	45	21	21	38	27	32

a. Find the range.

b. Find the variance.

c. Find the standard deviation

2.103 Adding (or subtracting) the same number from each value in a set of data does not affect the measures of variability for that set of data.

a. Find the variance of this set of annual heating-degree-day data: 6017, 6173, 6275, 6350, 6001, 6300.

b. Find the variance of this set of data (obtained by subtracting 6000 from each value in part a): 17, 173, 275, 350, 1, 300.

2.104 [EX02-104] One aspect of the beauty of scenic landscape is its variability. The elevations (feet above sea level) of 12 randomly selected towns in the Finger Lakes Regions of Upstate New York are recorded here.

559	815	767	668	651	895
1106	1375	861	1559	888	1106

Source: http://www.city-data.com

a. Find the mean.

b. Find the standard deviation.

2.105 [EX02-105] Recruits for a police academy were required to undergo a test that measures their exercise capacity. The exercise capacity (in minutes) was obtained for each of 20 recruits:

25	27	30	33	30	32	30	34	30	27
26	25	29	31	31	32	34	32	33	30

a. Draw a dotplot of the data.

b. Find the mean.

c. Find the range.

d. Find the variance.

e. Find the standard deviation.

f. Using the dotplot from part a, draw a line representing the range. Then draw a line starting at the mean with a length that represents the value of the standard deviation.

g. Describe how the distribution of data, the range, and the standard deviation are related.

2.106 [EX02-106] Jackson wants to buy a new golf club, in particular, a driver. He figures buying online will save time and money. He randomly selects a sample of 10 drivers from the Golflink.com website. Their prices are listed below (in dollars):

Driver Prices

149.99	299.99	49.99	499.99	167.97	299.99	399.99
199.99	99.99	149.99				

Source: http://www.golflink.com/

a. Construct a histogram and determine the shape.

b. Based on the histogram, what do you know about the mean and median?

c.   Calculate the mean and median. Do the results match your answer in part b?

d.   Calculate the range.

e.   Calculate the standard deviation.

f.   Describe what the range and standard deviation tell Jackson about buying a driver online through Golflink.com.

2.107 [EX02-107]*Better Roads* magazine reported the percentage of interstate and state-owned bridges that were structurally deficient or functionally obsolete (%SD/FO) for each U.S. state in 2003. (Percentages are expressed in decimal form [e.g., 0.20 = 20%].)

State	SD/FO*	State	SD/FO*	State	SD/FO*
AK	0.20	AL	0.22	AR	0.20

*** For remainder of data, logon at cengagebrain.com

Source: *Better Roads*, November 2003
*SD/FO = structurally deficient or functionally obsolete.

a.   Construct a histogram.

b.   Does the variable "%SD/FO" appear to have an approximately normal distribution?

c.   Calculate the mean.

d.   Find the median.

e.   Find the range.

f.   Find the standard deviation.

(Retain these solutions to use in Exercise 2.135 on p. 94.)

2.108 [EX02-108] One measure of airline performance is on-time arrival rates. For May 2009 the on-time arrival rates of domestic flights for the 19 largest U.S. airlines were as follows:

Carrier	On-time arrival %
Hawaiian	90.26
SkyWest	86.84
Pinnacle	86.81

*** For remainder of data, logon at cengagebrain.com

Source: U.S. Department of Transportation

a.   Find the range and the standard deviation for the on-time arrival rates.

b.   Describe the relationship among the distribution of the data, the range, and the standard deviation.

2.109 Consider these two sets of data:

Set 1	46	55	50	47	52
Set 2	30	55	65	47	53

Both sets have the same mean, 50. Compare these measures for both sets: $\Sigma(x - \bar{x})$, SS(x), and range. Comment on the meaning of these comparisons.

2.110 Consider the following two sets of data:

Set 1	45	80	50	45	30
Set 2	30	80	35	30	75

Both sets have the same mean, which is 50. Compare these measures for both sets: $\Sigma(x - \bar{x})$, SS(x), and range. Comment on the meaning of these comparisons relative to the distribution.

2.111 Comment on the statement: "The mean loss for customers at First State Bank (which was not insured) was $150. The standard deviation of the losses was $-$ \$125."

2.112 Start with $x = 100$ and add four x values to make a sample of five data such that:

a. $s = 0$

b. $0 < s < 1$

c. $5 < s < 10$

d. $20 < s < 30$

2.113 Each of two samples has a standard deviation of 5. If the two sets of data are made into one set of 10 data values, will the new sample have a standard deviation that is less than, about the same as, or greater than the original standard deviation of 5? Make up two sets of five data values, each with a standard deviation of 5, to justify your answer. Include the calculations.

2.114 Skillbuilder Applet Exercise matches means and standard deviations with corresponding histograms. After several practice rounds using "Start Over," explain your method of matching.

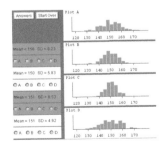

2.5 Measures of Position

Measures of position are used to describe the position a specific data value possesses in relation to the rest of the data when in ranked order. *Quartiles* and *percentiles* are two of the most popular measures of position.

> **Quartiles** Values of the variable that divide the ranked data into quarters; each set of data has three quartiles. The *first quartile*, Q_1, is a number such that at most 25% of the data are smaller in value than Q_1 and at most 75% are larger. The *second quartile* is the median. The *third quartile*, Q_3, is a number such that at most 75% of the data are smaller in value than Q_3 and at most 25% are larger. (See Figure 2.24.)

FIGURE 2.24

Quartiles

The procedure for determining the values of the quartiles is the same as that for percentiles and is shown in the following description of *percentiles*.

Remember that your data must be ranked from low (*L*) to high (*H*).

> **Percentiles** Values of the variable that divide a set of ranked data into 100 equal subsets; each set of data has 99 percentiles (see Figure 2.25). The *k*th percentile, P_k, is a value such that at most *k*% of the data are smaller in value than P_k and at most (100 − *k*)% of the data are larger (see Figure 2.26).

FIGURE 2.25

Percentiles

FIGURE 2.26

*k*th Percentile

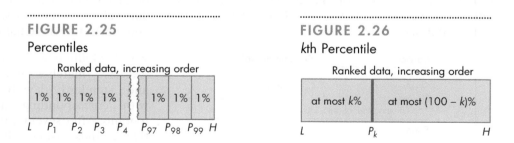

Notes:

1. The first quartile and the 25th percentile are the same; that is, $Q_1 = P_{25}$. Also, $Q_3 = P_{75}$.
2. The median, the second quartile, and the 50th percentile are all the same: $\tilde{x} = Q_2 = P_{50}$. Therefore, when asked to find P_{50} or Q_2, use the procedure for finding the median.

The procedure for determining the value of any *k*th percentile (or quartile) involves four basic steps as outlined on the diagram in Figure 2.27. Example 2.12 demonstrates the procedure.

FIGURE 2.27
Finding P_k Procedure

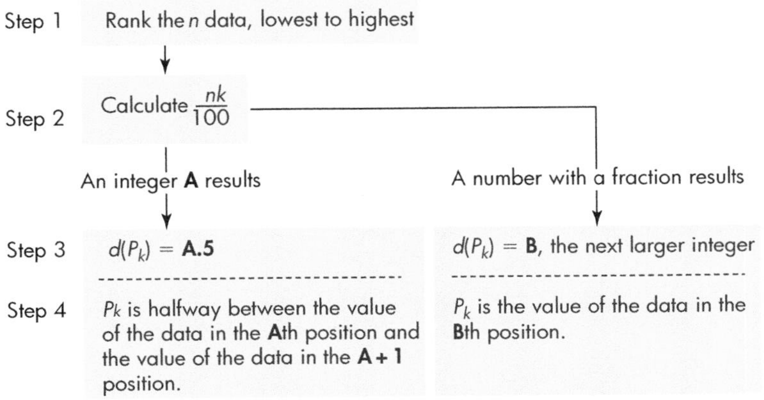

Step 1
Rank the n data, lowest to highest
Step 2
Calculate $\frac{nk}{100}$
An integer **A** results
A number with a fraction results
Step 3
$d(P_k)$ = **A.5**
$d(P_k)$ = **B**, the next larger integer
Step 4
P_k is halfway between the value of the data in the **A**th position and the value of the data in the **A + 1** position.
P_k is the value of the data in the **B**th position.

FYI: $d(P_k)$ = depth or location of the kth percentile

EXAMPLE 2.12

FINDING QUARTILES AND PERCENTILES

Using the sample of 50 elementary statistics final exam scores listed in Table 2.15, find the first quartile, Q_1; the 58th percentile, P_{58}; and the third quartile, Q_3.

TABLE 2.15 [TA02-06]
Raw Scores for Elementary Statistics Exam

60	47	82	95	88	72	67	66	68	98	90	77	86
58	64	95	74	72	88	74	77	39	90	63	68	97
70	64	70	70	58	78	89	44	55	85	82	83	
72	77	72	86	50	94	92	80	91	75	76	78	

Solution

Step 1 Rank the data: A ranked list may be formulated (see Table 2.16), or a graphic display showing the ranked data may be used. The dotplot and the stem-and-leaf display are handy for this purpose. The stem-and-leaf display is especially helpful because it gives depth numbers counted from both extremes when it is computer generated (see Figure 2.28). Step 1 is the same for all three statistics.

Find Q_1:

Step 2 Find $\frac{nk}{100}$: $\frac{nk}{100} = \frac{(50)(25)}{100} = \mathbf{12.5}$

($n = 50$ and $k = 25$, since $Q_1 = P_{25}$.)

Step 3 Find the depth of Q_1: $d(Q_1)$ = **13** (Since 12.5 contains a fraction, **B** is the next larger integer, 13.)

Step 4 Find Q_1: Q_1 is the 13th value, counting from L (see Table 2.16 or Figure 2.28), Q_1 = **67**

Video tutorial available—logon and learn more at cengagebrain.com

TABLE 2.16
Ranked Data: Exam Scores

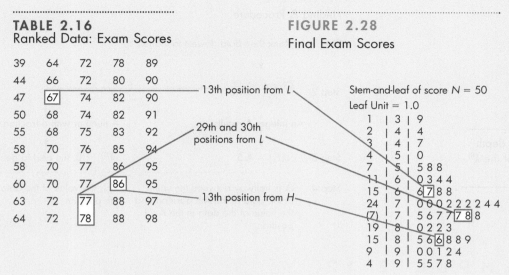

FIGURE 2.28
Final Exam Scores

Stem-and-leaf of score $N = 50$
Leaf Unit = 1.0

```
 1  | 3 | 9
 2  | 4 | 4
 3  | 4 | 7
 4  | 5 | 0
 7  | 5 | 5 8 8
11  | 6 | 0 3 4 4
15  | 6 | 6 7 8 8
24  | 7 | 0 0 0 2 2 2 2 4 4
(7) | 7 | 5 6 7 7 7 8 8
19  | 8 | 0 2 2 3
15  | 8 | 5 6 6 8 8 9
 9  | 9 | 0 0 1 2 4
 4  | 9 | 5 5 7 8
```

Find P_{58}:

Step 2   Find $\dfrac{nk}{100} : \dfrac{nk}{100} = \dfrac{(50)(58)}{100} =$ **29** : ($n = 50$ and $k = 58$ for P_{58}.)

Step 3   Find the depth of P_{58}: $d(P_{58}) =$ **29.5** (Since **A** = 29, an integer, add 0.5 and use 29.5.)

Step 4   Find P_{58}: P_{58} is the value halfway between the values of the 29th and the 30th pieces of data, counting from L (see Table 2.16 or Figure 2.28), so

$$P_{58} = \frac{77 + 78}{2} = 77.5$$

Therefore, it can be stated that "at most, 58% of the exam grades are smaller in value than 77.5." This is also equivalent to stating that "at most, 42% of the exam grades are larger in value than 77.5."

Optional technique: When k is greater than 50, subtract k from 100 and use $(100 - k)$ in place of k in Step 2. The depth is then counted from the highest-value data, H.

FYI An ogive of these exam grades would graphically determine these same percentiles, without the use of formulas.

Find Q_3 using the optional technique:

Step 2   Find $\dfrac{nk}{100} : \dfrac{nk}{100} = \dfrac{(50)(25)}{100} =$ **12.5** ($n = 50$ and $k = 75$, since $Q_3 = P_{75}$, and $k > 50$; use $100 - k = 100 - 75 = 25$.)

Step 3   Find the depth of Q_3 from H: $d(Q_3) = 13$

Step 4   Find Q_3: Q_3 is the 13th value; counting from H (see Table 2.16 or Figure 2.28), $Q_3 =$ **86**

Therefore, it can be stated that "at most, 75% of the exam grades are smaller in value than 86." This is also equivalent to stating that "at most, 25% of the exam grades are larger in value than 86."

An additional measure of central tendency, the *midquartile*, can now be defined.

Animated tutorial available—logon and learn more at cengagebrain.com

Midquartile The numerical value midway between the first quartile and the third quartile.

$$\text{midquartile} = \frac{Q_1 + Q_3}{2} \qquad (2.10)$$

E X A M P L E   2 . 1 3

FINDING THE MIDQUARTILE

Find the midquartile for the set of 50 exam scores given in Example 2.12.

Solution

$Q_1 = 67$ and $Q_3 = 86$, as found in Example 2.12. Thus,

$$\text{midquartile} = \frac{Q_1 + Q_3}{2} = \frac{67 + 86}{2} = 76.5$$

The median, the midrange, and the midquartile are not necessarily the same value. Each is the middle value, but by different definitions of "middle." Figure 2.29 summarizes the relationship of these three statistics as applied to the 50 exam scores from Example 2.12.

FIGURE 2.29
Final Exam Scores

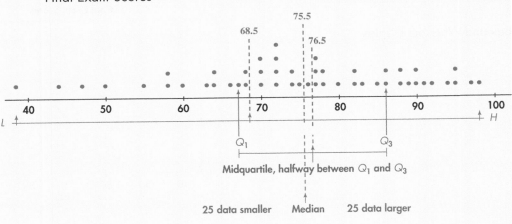

A *5-number summary* is very effective in describing a set of data. It is easy information to obtain and is very informative to the reader.

5-number summary The 5-number summary is composed of the following:

1. L, the smallest value in the data set
2. Q_1, the first quartile (also called P_{25}, the 25th percentile)
3. $\tilde{x}$, the median
4. Q_3, the third quartile (also called P_{75}, the 75th percentile)
5. H, the largest value in the data set

The 5-number summary for the set of 50 exam scores in Example 2.12 is

39	67	75.5	86	98
L	Q_1	$\tilde{x}$	Q_3	H

Notice that these five numerical values divide the set of data into four subsets, with one-quarter of the data in each subset. From the 5-number summary, we can observe how much the data are spread out in each of the quarters. We can now define an additional measure of dispersion.

> **Interquartile range** The difference between the first and third quartiles. It is the range of the middle 50% of the data.

The 5-number summary is even more informative when it is displayed on a diagram drawn to scale. A graphic display that accomplishes this is known as the *box-and-whiskers display*.

> **Box-and-whiskers display** A graphic representation of the 5-number summary. The five numerical values (smallest, first quartile, median, third quartile, and largest) are located on a scale, either vertical or horizontal. The box is used to depict the middle half of the data that lie between the two quartiles. The whiskers are line segments used to depict the other half of the data: One line segment represents the quarter of the data that are smaller in value than the first quartile, and a second line segment represents the quarter of the data that are larger in value than the third quartile.

Figure 2.30 is a box-and-whiskers display of the 50 exam scores.

FIGURE 2.30
Box-and-Whiskers Display

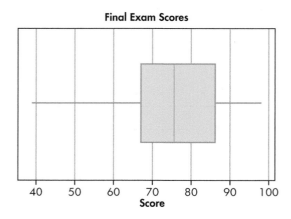

TECHNOLOGY INSTRUCTIONS:
PERCENTILES

MINITAB	Input the data into C1; then continue with:

Choose: **Data > Sort . . .**
Enter: Sort column(s): **C1**   By column: **C1**
Select: Store sorted data in: **Column(s) of current worksheet**
Enter: **C2 > OK**

A ranked list of data will be obtained in C2. Determine the depth position and locate the desired percentile.

Excel	Input the data into column A and activate a cell for the answer; then continue with: Choose: **Formulas > Insert Function, f_x > Statistical > PERCENTILE > OK** Enter: **Array: (A2:A6 or select cells)** **k: K** (desired percentile; ex. .95, .47) **> OK**
TI-83/84 Plus	Input the data into L1; then continue with: Choose: **STAT > EDIT > 2:SortA(** Enter: **L1** Enter: **percentile × sample size** (ex. .25 × 100) Based on product, determine the depth position; then continue with: Enter: **L1(depth position) > Enter**

TECHNOLOGY INSTRUCTIONS: 5-NUMBER SUMMARY

MINITAB	Input the data into C1; then continue with: Choose: **Stat > Basic Statistics > Display Descriptive Statistics . . .** Enter: **Variables: C1 > OK**
Excel	Input the data into column A; then continue with: Choose: **Data > Data Analysis* > Descriptive Statistics > OK** Enter: **Input Range: (A2:A6 or select cells)** Select: **Labels in First Row** (if necessary) **Output Range** Enter: **(B1 or select cell)** Select: **Summary Statistics > OK** To make output readable: Choose: **Home > Cells > Format > AutoFit Column Width** *If Data Analysis does not show on the Data menu, see page 53.
TI-83/84 Plus	Input the data into L1; then continue with: Choose: **STAT > CALC > 1:1-VAR STATS** Enter: **L1**

TECHNOLOGY INSTRUCTIONS: BOX-AND-WHISKERS DISPLAY

MINITAB	Input the data into C1; then continue with: Choose: **Graph > Boxplot . . . > One Y, Simple > OK** Enter: **Graph variables: C1** Optional: Select: **Labels > Titles/Footnoes** Enter: **your title, footnotes > OK** Select: **Scale > Axes and Ticks** Select: **Transpose value and category scales > OK > OK**

For multiple boxplots, enter additional sets of data into C2; then do as just described plus:

Choose: **Graph > Boxplot . . . > Multiple Y's, Simple > OK**
Enter: **Graph variables: C1 C2 > OK**
Optional: See above.

Excel

Input the data into column A; then continue with:
Choose: **Add-Ins > Data Analysis Plus* > BoxPlot > OK**
Enter: **(A2:A6 or select cells)**

To edit the boxplot, review options shown with editing histograms on page 53.

*If Data Analysis Plus does not shown on the Data menu, see page 39.

TI-83/84 Plus

Input the data into L1; then continue with:

Choose: **2nd > STAT PLOT >**
             1:Plot1 . . .
Choose: **ZOOM > 9:ZoomStat >**
             TRACE > > >

If class midpoints are in L1 and frequencies are in L2, do as just described except for:

Enter: **Freq: L2**

For multiple boxplots, enter additional sets of data into L2 or L3; do as just described plus:

Choose: **2nd > STAT PLOT > 2:Plot2 . . .**

The position of a specific value can also be measured in terms of the mean and standard deviation using the *standard score*, commonly called the *z-score*.

Standard score, or z-score The position a particular value of x has relative to the mean, measured in standard deviations. The z-score is found by the formula

$$z = \frac{\text{value} - \text{mean}}{\text{st.dev.}} = \frac{x - \bar{x}}{s} \tag{2.11}$$

E X A M P L E   2 . 1 4

FINDING z-Scores

Find the standard scores for (a) 92 and (b) 72 with respect to a sample of exam grades that have a mean score of 74.92 and a standard deviation of 14.20.

Solution

a. $x = 92$, $\bar{x} = 74.92$, $s = 14.20$. Thus,

$$z = \frac{x - \bar{x}}{s} = \frac{92 - 74.92}{14.20} = \frac{17.08}{14.20} = 1.20.$$

b. $x = 72$, $\bar{x} = 74.92$, $s = 14.20$. Thus,

$$z = \frac{x - \bar{x}}{s} = \frac{72 - 74.92}{14.20} = \frac{-2.92}{14.20} = -0.21.$$

This means that the score 92 is approximately 1.2 standard deviations above the mean and that the score 72 is approximately one-fifth of a standard deviation below the mean.

Notes:

1. Typically, the calculated value of z is rounded to the nearest hundredth.
2. z-scores typically range in value from approximately -3.00 to $+3.00$.

Because the z-score is a measure of relative position with respect to the mean, it can be used to help us compare two raw scores that come from separate populations. For example, suppose you want to compare a grade you received on a test with a friend's grade on a comparable exam in her course. You received a raw score of 45 points; she got 72 points. Is her grade better? We need more information before we can draw a conclusion. Suppose the mean on the exam you took was 38 and the mean on her exam was 65. Your grades are both 7 points above the mean, but we still can't draw a definite conclusion. The standard deviation on the exam you took was 7 points, and it was 14 points on your friend's exam. This means that your score is 1 standard deviation above the mean ($z = 1.0$), whereas your friend's grade is only 0.5 standard deviation above the mean ($z = 0.5$). Your score has the "better" relative position, so you conclude that your score is slightly better than your friend's score. (Again, this is speaking from a relative point of view.)

TECHNOLOGY INSTRUCTIONS: ADDITIONAL COMMANDS

MINITAB

Input the data into C1; then:
To sort the data into ascending order and store them in C2, continue with:

Choose: Data > Sort . . .
Enter: Sort column(s): **C1**   By column: **C1**
Select: Store sorted data in: **Column(s) of current worksheet**
Enter: **C2 > OK**

To form an ungrouped frequency distribution of integer data, continue with:

Choose: Stat > Tables > Tally Individual Variables
Enter: Variables: **C1**
Select: Counts > OK

To print data on the session window, continue with:

Choose: **Data > Display Data**
Enter: **Columns to display: C1 or C1 C2 or C1–C2 > OK**

Excel

Input the data into column A; activate data, then continue with the following to sort the data:

Choose: **Data > AZ↓ (Sort)**

TI-83/84 Plus

Input the data into L1; then continue with the following to sort the data:

Choose: **2nd > STAT > OPS > 1:SortA(**
Enter: **L1**

To form a frequency distribution of the data in L1, continue with:

Choose: **PRGM > EXEC > FREQDIST***
Enter: **L1 > ENTER**
    LW BOUND = first lower class boundary
    UP BOUND = last upper class boundary
    WIDTH = class width (use 1 for ungrouped distribution)

*The program "FREQDIST" is among those available for downloading. See page 35 for details.

T E C H N O L O G Y   I N S T R U C T I O N S :
G E N E R A T E   R A N D O M   S A M P L E S

MINITAB

The data will be put into C1:

Choose: **Calc > Random Data > {Normal, Uniform, Integer, etc.}**
Enter: **Number of rows of data to generate: K**
    Store in column(s): C1
    Population parameters needed: (μ, σ, L, H, A, or B) **> OK**
    (Required parameters will vary depending on the distribution)

Excel

Choose: **Data > Data Analysis* > Random Number Generation > OK**
Enter: **Number of Variables: 1**
    Number of Random Numbers: (desired quantity)
Select: **Distribution: Normal, Discrete, or others**
Enter: **Parameters:** (μ, σ, L, H, A, or B)
    (Required parameters will vary depending on the distribution.)
Select: **Output Range**
Enter: **(A1 or select cell) > OK**

*If Data Analysis does not shown on the Data menu, see page 53.

TI-83/84 Plus

Choose: **STAT > 1:EDIT**
Highlight: **L1**
Choose: **MATH > PRB > 6:randNorm(or 5:randInt(**
Enter: μ, σ, **# of trials or L, H, # of trials**

T E C H N O L O G Y   I N S T R U C T I O N S :
S E L E C T   R A N D O M   S A M P L E S

MINITAB

The existing data to be selected from should be in C1; then continue with:

Choose: **Calc > Random Data > Sample from Columns**
Enter: **Number of rows to sample: K**
    From columns: C1
    Store samples in: C2
Select: **Sample with replacement (optional) > OK**

Excel

The existing data to be selected from should be in column A; then continue with:

Choose: **Data > Data Analysis* > Sampling > OK**
Enter: **Input range: (A2:A10 or select cells)**
Select: **Labels** (optional)
    Random
    Enter: **Number of Samples: K**
    Output range:
    Enter: **(B1 or select cell) > OK**
*If Data Analysis does not show on the Data menu, see page 53.

A P P L I E D   E X A M P L E   2 . 1 5

GROWTH CHART FOR BOYS 2 TO 20 YEARS OLD

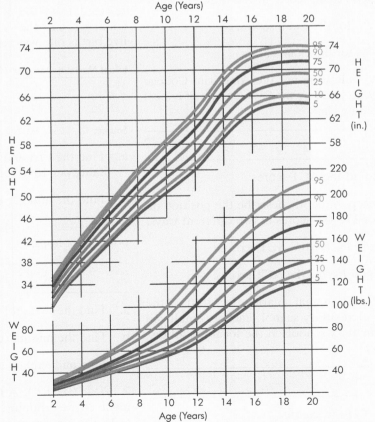

Clinical growth chart showing 5th, 10th, 25th, 50th, 75th, 90th, 95th percentiles, for boys 2 to 20 years.
Source: http://www.cdc.gov/

One very important use for growth charts is to track a child's pattern of growth. If as a youngster, height and weight are approximately at the 40th percentile, the child is larger than approximately 40 percent and smaller than the other 60 percent of those of the same age. The doctor will check this information periodically and if the percentile rating changes dramatically from one year to the next, there might be reason for concern.

Consider this: If you are one of those 5% who are taller than the 95th percentile or one of those 5% who are shorter than the 5th percentile, it is a near certainty that some everyday object is not the right size for you. Height and weight are not the only dimensions that can be compared; it is possible to compare other physical characteristics such as foot size, length of forearm, seated height, and etcetera. Those whose build puts them near one of the extremes are familiar with the problems associated with an extreme size.

SECTION 2.5 EXERCISES

2.115 Refer to the table of exam scores in Table 2.16 on page 84 for the following.

a. Using the concept of depth, describe the position of 91 in the set of 50 exam scores in two different ways.

b. Find P_{20} and P_{35} for the exam scores.

c. Find P_{80} and P_{95} for the exam scores.

2.116 [EX02-116] Following are the American College Test (ACT) scores attained by the 25 members of a local high school graduating class:

21	24	23	17	31	19	19	20	19	25	17	23	16
21	20	28	25	25	21	14	19	17	18	28	20	

a. Draw a dotplot of the ACT scores.

b. Using the concept of depth, describe the position of 24 in the set of 25 ACT scores in two different ways.

c. Find P_5, P_{10}, and P_{20} for the ACT scores.

d. Find P_{99}, P_{90}, and P_{80} for the ACT scores.

2.117 [EX02-117] The annual salaries (in $100) of the kindergarten and elementary school teachers employed at one of the elementary schools in the local school district are listed here:

574	434	455	413	391	471	458	269	501
326	367	433	367	495	376	371	295	317

a. Draw a dotplot of the salaries.

b. Using the concept of depth, describe the position of 295 in the set of 18 salaries in two different ways.

c. Find Q_1 for these salaries.

d. Find Q_3 for these salaries.

2.118 [EX02-118] Fifteen countries were randomly selected from the *World Factbook 2009* list of world countries, and the estimated infant mortality per 1000 live births rate was recorded.

Infant Mortality Rate per 1000 Live Births

151.95	180.21	13.79	15.25	23.07
9.10	17.87	63.34	98.69	18.9
15.96	49.45	12.70	45.36	5.35

Source: *The World Factbook 2009*

a. Find the first and third quartiles for the infant mortality per 1000 rate.

b. Find the midquartile.

2.119 [EX02-119] The following data are the yields (in pounds) of hops:

3.9	3.4	5.1	2.7	4.4	7.0	5.6	2.6	4.8	5.6
7.0	4.8	5.0	6.8	4.8	3.7	5.8	3.6	4.0	5.6

a. Find the first and the third quartiles of the yields.

b. Find the midquartile.

c. Find and explain the percentiles P_{15}, P_{33}, and P_{90}.

2.120 [EX02-120] A research study of manual dexterity involved determining the time required to complete a task. The time required for each of 40 individuals with disabilities is shown here (data are ranked):

7.1	7.2	7.2	7.6	7.6	7.9	8.1	8.1	8.1	8.3	8.3
8.4	8.4	8.9	9.0	9.0	9.1	9.1	9.1	9.1	9.4	9.6
9.9	10.1	10.1	10.1	10.2	10.3	10.5	10.7	11.0	11.1	11.2
11.2	11.2	12.0	13.6	14.7	14.9	15.5				

a. Find Q_1.

b. Find Q_2.

c. Find Q_3.

d. Find P_{95}.

e. Find the 5-number summary.

f. Draw the box-and-whisker display.

2.121 Draw a box-and-whiskers display for the set of data with the 5-number summary 42–62–72–82–97.

2.122 [EX02-122]The U.S. Geological Survey collected atmospheric deposition data in the Rocky Mountains. Part of the sampling process was to determine the concentration of ammonium ions (in percentages). Here are the results from the 52 samples:

2.9	4.1	2.7	3.5	1.4	5.6	13.3	3.9	4.0
2.9	7.0	4.2	4.9	4.6	3.5	3.7	3.3	5.7
3.2	4.2	4.4	6.5	3.1	5.2	2.6	2.4	5.2
4.8	4.8	3.9	3.7	2.8	4.8	2.7	4.2	2.9
2.8	3.4	4.0	4.6	3.0	2.3	4.4	3.1	5.5
4.1	4.5	4.6	4.7	3.6	2.6	4.0		

a. Find Q_1.

b. Find Q_2.

c. Find Q_3.

d. Find the midquartile.

e. Find P_{30}.

f. Find the 5-number summary.

g. Draw the box-and-whiskers display.

2.123 [EX02-123] The NCAA men's basketball "Big Dance" kicks into full gear every March. But if you look at the graduation rate of these athletes, you'll find that many teams don't make the grade, according to a study released in March 2009. Listed are the graduation rates for 63 of the 2009 tournament teams.

Graduation Rates (percentages), 2009 Men's Teams, NCAA Division I Basketball Tournament

63	100	8	89	80	10	53	67	17	37	31	89	100
56	70	34	89	64	55	36	53	77	42	47	53	86
31	91	29	60	40	46	57	55	80	50	46	100	82
20	92	71	100	42	60	45	92	100	57	67	50	
38	30	33	67	100	36	86	69	86	38	100	41	

Source: The Institute for Diversity and Ethics in Sports

a. Draw a dotplot of the graduation rate data.

b. Draw a stem-and-leaf display of these data.

c. Find the 5-number summary and draw a box-and-whiskers display.

d. Find P_5 and P_{95}.

e. Describe the distribution of graduation rates being sure to include information you learned in parts a through d.

f. Are there teams whose graduation rates appear to be quite different from those of the rest? How many? Which ones? Explain.

2.124 [EX02-124] The fatality rate on the nation's highways in 2007 was the lowest since 1994, but these numbers are still mind-boggling. The number of persons killed in motor vehicle traffic crashes, by state including the District of Columbia, in 2007 is listed here.

1110	84	1066	650	3974	554	277	117	44	3214	1641
138	252	1249	898	445	416	864	985	183	614	417
1088	504	884	992	277	256	373	129	724	413	1333
1675	111	1257	754	455	1491	69	1066	146	1210	3363
299	66	1027	568	431	756	150				

Source: http://www-fars.nhtsa.dot.gov/

a. Draw a dotplot of fatality data.

b. Draw a stem-and-leaf display of these data. Describe how the three large-valued data are handled.

c. Find the 5-number summary and draw a box-and-whiskers display.

d. Find P_{10} and P_{90}.

e. Describe the distribution of the number of fatalities per state, being sure to include information learned in parts a through d.

f. Why might it be unfair to draw conclusions about the relative safety level of highways in the 51 states based on these data?

2.125 [EX02-125] Are airline flight arrivals ever on time? The general public thinks they are always late—but are they? The Bureau of Transportation keeps records and periodically reports the findings. Listed here are the percentages of on-time arrivals at the 31 major U.S. airports during the month of April 2009.

ATL 71.2	BOS 77.7	BWI 83.9

*** For remainder of data, logon at cengagebrain.com

Source: U.S. Department of Transportation, Bureau of Transportation Statistics

a. Draw a dotplot of on-time performance data.

b. Draw a stem-and-leaf display of these data.

c. Find the 5-number summary and draw a box-and-whiskers display.

d. Find P_{10} and P_{20}.

e. Describe the distribution of on-time percentage, being sure to include information you learned in parts a through d.

f. Why would you be more likely to talk about the top 80% or 90% of the performance percentages than the middle 80% or 90%?

(continue on page 94)

g. Are there airports whose on-time percentages appear to be quite different from the rest? How many? Which ones? Explain.

2.126 [EX02-126] Major League Baseball stadiums vary in age, style, number of seats, and many other ways. But to the baseball players, the size of the field is of the utmost importance. Suppose we agree to measure the size of the field by using the distance from home plate to the centerfield fence. Below is the distance to center field in the 30 Major League stadiums in 2009.

Distance—Home Plate to the Centerfield Fence

420	400	400	400	400	400	408	400	400	406
434	405	400	415	400	404	407	405	422	404
435	400	400	404	401	396	400	403	408	408

Source: http://mlb.mlb.com

a. Construct a histogram.

b. The interquartile range is described by the bounds of the middle 50% of the data, Q_1 and Q_3. Find the interquartile range.

c. Are there any fields that appear to be considerably smaller or larger than the others?

d. Is there a great deal of difference in the size of these 30 fields as measured by the distance to center field? Justify your answer with statistical evidence.

2.127 What property does the distribution need for the median, the midrange, and the midquartile to all be the same value?

2.128 [EX02-128] Henry Cavendish, an English chemist and physicist (1731–1810), approached many of his experiments using quantitative measurements. He was the first to accurately measure the density of the Earth. Following are 29 measurements (ranked for your convenience) of the density of the Earth done by Cavendish in 1798 using a torsion balance. Density is presented as a multiple of the density of water. (Measurements are in g/cm^3.)

4.88	5.07	5.10	5.26	5.27	5.29	5.29	5.30	5.34	5.34
5.36	5.39	5.42	5.44	5.46	5.47	5.50	5.53	5.55	5.57
5.58	5.61	5.62	5.63	5.65	5.68	5.75	5.79	5.85	

Source: The data and descriptive information are based on material from "Do robust estimators work with real data?" by Stephen M. Stigler, *Annals of Statistics* 5 (1977), 1055–1098.

a. Describe the data set by calculating the mean, median, and standard deviation.

b. Construct a histogram and explain how it demonstrates the values of the descriptive statistics in part a.

c. Find the 5-number summary.

d. Construct a box-and-whiskers display and explain how it demonstrates the values of the descriptive statistics in part c.

e. Based on the two graphs, what "shape" is this distribution of measurements?

f. Assuming that Earth density measurements have an approximately normal distribution, approximately 95% of the data should fall within 2 standard deviations of the mean. Is this true?

2.129 Find the z-score for test scores of 92 and 63 on a test that has a mean of 72 and a standard deviation of 12.

2.130 A sample has a mean of 50 and a standard deviation of 4.0. Find the z-score for each value of x :

a. $x = 54$ b. $x = 50$

c. $x = 59$ d. $x = 45$

2.131 An exam produced grades with a mean score of 74.2 and a standard deviation of 11.5. Find the z-score for each test score x :

a. $x = 54$ b. $x = 68$

c. $x = 79$ d. $x = 93$

2.132 A nationally administered test has a mean of 500 and a standard deviation of 100. If your standard score on this test was 1.8, what was your test score?

2.133 A sample has a mean of 120 and a standard deviation of 20.0. Find the value of x that corresponds to each of these standard scores:

a. $z = 0.0$ b. $z = 1.2$

c. $z = -1.4$ d. $z = 2.05$

2.134 a. What does it mean to say that $x = 152$ has a standard score of +1.5?

b. What does it mean to say that a particular value of x has a z-score of -2.1?

c. In general, the standard score is a measure of what?

2.135 [EX02-107] Consider the percentage of interstate and state-owned bridges that were structurally deficient or functionally obsolete (SD/FO) listed in Exercise 2.107 on page 81.

a. Omit the names of the states and rank the SD/FO values in ascending order, reading horizontally in each row.

b. Construct a 5-number summary table and the corresponding box-and-whiskers display.

c. Find the midquartile percentage and the interquartile range.

d. What are the z-scores for California, Hawaii, Nebraska, Oklahoma, and Rhode Island?

2.136 The ACT Assessment® is designed to assess high school students' general educational development and their ability to complete college-level work. The table lists the mean and standard deviation of scores attained by the 3,908,557 high school students from the 2006 to 2008 graduating classes who took the ACT exams.

2006–2008	English	Mathematics	Reading	Science	Composite
Mean	20.6	21.0	21.4	20.9	21.1
Standard deviation	6.0	5.1	6.1	4.8	4.9

Source: American College Testing

Convert the following ACT test scores to z-scores for both English and math. Compare placement between the two tests.

a. $x = 30$ b. $x = 23$ c. $x = 12$

d. Explain why the relative positions in English and math changed for the ACT scores of 30 and 12.

e. If Jessica had a 26 on one of the ACT exams, on which of the exams would she have the best possible relative score? Explain why.

2.137 Which x value has the higher position relative to the set of data from which it comes?

A: $x = 85$, where mean $= 72$ and standard deviation $= 8$

B: $x = 93$, where mean $= 87$ and standard deviation $= 5$

2.138 Which x value has the lower position relative to the set of data from which it comes?

A: $x = 28.1$, where $\bar{x} = 25.7$ and $s = 1.8$

B: $x = 39.2$, where $\bar{x} = 34.1$ and $s = 4.3$

2.6 Interpreting and Understanding Standard Deviation

Standard deviation is a measure of variation (dispersion) in the data. It has been defined as a value calculated with the use of formulas. Even so, you may be wondering what it really is and how it relates to the data. It is a kind of yardstick by which we can compare the variability of one set of data with that of another. This particular "measure" can be understood further by examining two statements that tell us how the standard deviation relates to the data: the *empirical rule* and *Chebyshev's theorem*.

The Empirical Rule and Testing for Normality

Empirical rule If a variable is normally distributed, then (1) within 1 standard deviation of the mean, there will be approximately 68% of the data; (2) within 2 standard deviations of the mean, there will be approximately 95% of the data; and (3) within 3 standard deviations of the mean, there will be approximately 99.7% of the data. (This rule applies specifically to a normal **[bell-shaped] distribution**, but it is frequently applied as an interpretive guide to any mounded distribution.)

Figure 2.31 shows the intervals of 1, 2, and 3 standard deviations about the mean of an approximately normal distribution. Usually these proportions do not occur exactly in a sample, but your observed values will be close when a large sample is drawn from a normally distributed population.

If a distribution is approximately normal, it will be nearly symmetrical and the mean will divide the distribution in half (the mean and the median are the same in a symmetrical distribution). This allows us to refine the empirical rule, as shown in Figure 2.32.

FIGURE 2.31
Empirical Rule

FIGURE 2.32
Refinement of Empirical Rule

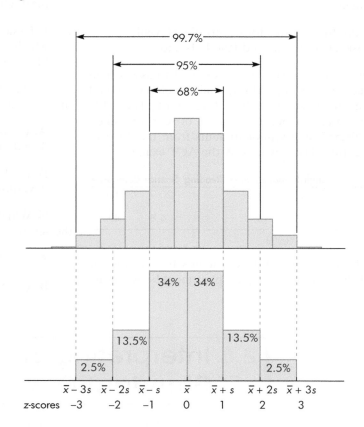

The empirical rule can be used to determine whether a set of data is approximately normally distributed. Let's demonstrate this application by working with the distribution of final exam scores that we have been using throughout this chapter. The mean, $\bar{x}$, was found to be 74.92, and the standard deviation, s, was 14.20. The interval from 1 standard deviation below the mean, $\bar{x} - s$, to 1 standard deviation above the mean, $\bar{x} + s$, is $74.92 - 14.20 = 60.72$ to $74.92 + 14.20 = 89.12$. This interval (60.72 to 89.12) includes 61, 62, 63, . . . , 89. Upon inspection of the ranked data (Table 2.16, p. 84), we see that 34 of the 50 data, or 68%, lie within 1 standard deviation of the mean. Furthermore, $\bar{x} - 2s = 74.92 - (2)(14.20) = 74.92 - 28.40 = 46.52$ to $\bar{x} + 2s = 74.92 + 28.40 = 103.32$ gives the interval from 46.52 to 103.32. Of the 50 data, 48, or 96%, lie within 2 standard deviations of the mean. All 50 data, or 100%, are included within 3 standard deviations of the mean (from 32.32 to 117.52). This information can be placed in a table for comparison with the values given by the empirical rule (see Table 2.17).

TABLE 2.17 Observed Percentages versus
the Empirical Rule

Interval	Empirical Rule Percentage	Percentage Found
$\bar{x} - s$ to $\bar{x} + s$	≈68	68
$\bar{x} - 2s$ to $\bar{x} + 2s$	≈95	96
$\bar{x} - 3s$ to $\bar{x} + 3s$	≈99.7	100

The percentages found are reasonably close to those predicted by the empirical rule. By combining this evidence with the shape of the histogram (see Figure 2.10, p. 51), we can safely say that the final exam data are approximately normally distributed.

There is another way to test for normality—by drawing a probability plot (an ogive drawn on probability paper*) using a computer or graphing calculator. For our illustration, a probability plot of the statistics final exam scores is shown on Figure 2.33. The test for normality, at this point in our study of statistics, is simply to compare the graph of the data (the ogive) with the straight line drawn from the lower left corner to the upper right corner of the graph. If the ogive lies close to this straight line, the distribution is said to be approximately normal. The vertical scale used to construct the probability plot is adjusted so that the ogive for an exactly normal distribution will trace the straight line. The ogive of the exam scores follows the straight line quite closely, suggesting that the distribution of exam scores is approximately normal.

FIGURE 2.33

Probability Plot of Statistics Exam Scores

FYI *On probability paper the vertical scale is not uniform; it has been adjusted to account for the mounded shape of a normal distribution and its cumulative percentages.

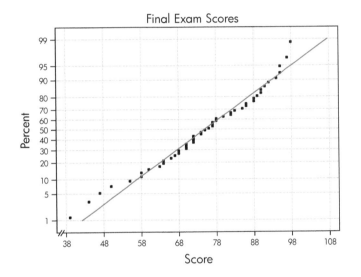

Final Exam Scores

If using a computer, you will obtain an additional piece of information in determining normality. This piece of information comes in the form of a p-value, and if its value is greater than 0.05, you can assume the sample was drawn from an approximately normal distribution (if p-value $\leq$ 0.05, not normal). (p-value will be more fully defined in Chapter 8, Section 8.4.)

TECHNOLOGY INSTRUCTIONS: TESTING FOR NORMALITY

MINITAB

Input the data into C1; then continue with:

Choose: **Stat > Basic Statistics > Normality Test**
Enter: Variable: **C1**
         Title: **your title > OK**

Excel	
	Excel uses a test for normality, not the probability plot.
	Input the data into column A; then continue with:

Choose: **Add-Ins > Data Analysis Plus* > Chi-Squared Test of Normality > OK**
Enter: Input Range: **data** (A1:A6 or select cells)
Select: **Labels** (if column headings were used) **> OK**

Expected values for a normal distribution are given versus the given distribution. If the *p*-value is greater than 0.05, then the given distribution is approximately normal.

*If Data Analysis Plus does not shown on the Data menu, see page 39.

TI-83/84 Plus	
	Input the data into L1; then continue with:

Choose: **Window**
Enter: at most the smallest data value, at least the largest data value, x scale, −5, 5, 1,1
Choose: **2nd > STAT PLOT > 1:Plot**

Chebyshev's Theorem

In the event that the data do not display an approximately normal distribution, Chebyshev's theorem gives us information about how much of the data will fall within intervals centered at the mean for all distributions.

> **Chebyshev's theorem** The proportion of any distribution that lies within k standard deviations of the mean is at least $1 - \frac{1}{k^2}$, where k is any positive number greater than 1. This theorem applies to all distributions of data.

This theorem says that within 2 standard deviations of the mean ($k = 2$), you will always find at least 75% (that is, 75% or more) of the data:

$$1 - \frac{1}{k^2} = 1 - \frac{1}{2^2} = 1 - \frac{1}{4} = \frac{3}{4} = 0.75, \textbf{ at least 75}\%$$

Figure 2.34 shows a mounded distribution that illustrates at least 75%.

If we consider the interval enclosed by 3 standard deviations on either side of the mean ($k = 3$), the theorem says that we will always find at least 89% (that is, 89% or more) of the data:

$$1 - \frac{1}{k^2} = 1 - \frac{1}{3^2} = 1 - \frac{1}{9} = \frac{8}{9} = 0.89, \textbf{ at least 89}\%$$

Figure 2.35 shows a mounded distribution that illustrates at least 89%.

FIGURE 2.34
Chebyshev's Theorem with $k = 2$

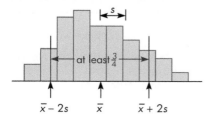

FIGURE 2.35
Chebyshev's Theorem with $k = 3$

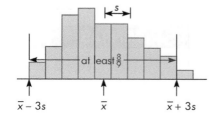

Let's revisit the results of the physical-fitness strength test given to the third-graders in Exercise 2.47 on page 60. Their test results are listed here in rank order and shown on the histogram.

1	2	2	3	3	3	4	4	4	5	5	5	5	6	6	6
8	9	9	9	9	9	9	10	10	11	12	12	12	13	14	14
14	15	15	15	15	16	16	16	17	17	17	17	18	18	18	18
19	19	19	19	20	20	20	21	21	21	22	22	22	23	24	24

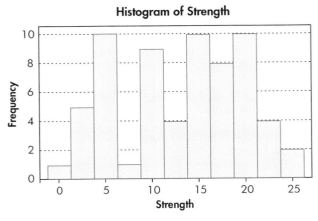

Histogram of Strength

Some questions of interest are: Does this distribution satisfy the empirical rule? Does Chebyshev's theorem hold true? Is this distribution approximately normal?

To answer the first two questions, we need to find the percent of data in each of the three intervals about the mean. The mean is 13.0, and the standard deviation is 6.6.

mean $\pm$ k (Std. Dev.)	Interval	Percentage Found	Empirical	Chebyshev
13.0 ± 1(6.6)	6.4 to 19.6	36/64 = 56.3%	68%	–
13.0 ± 2(6.6)	−0.2 to 26.2	64/64 = 100%	95%	At least 75%
13.0 ± 3(6.6)	−6.8 to 32.8	64/64 = 100%	99.70%	At least 89%

It is left to you to verify the values of the mean, standard deviation, the intervals, and the percentages.

The three percentages found (56.3, 100, and 100) do not approximate the 68, 95, and 99.7 percentages stated in the empirical rule. The two percentages found (100 and 100) do agree with Chebyshev's theorem in that they are greater than 75% and 89%. Remember, Chebyshev's theorem holds for all distributions.

The normality test, introduced on page 97, yields a p-value of 0.009, and along with the distribution seen on the histogram and the three percentages found, it is reasonable to conclude that these test results are not normally distributed.

SECTION 2.6 EXERCISES

2.139 Instructions for an essay assignment include the statement "The length is to be within 25 words of 200." What values of x, number of words, satisfy these instructions?

2.140 The empirical rule indicates that we can expect to find what proportion of the sample included between the following:

a. $\bar{x} - s$ and $\bar{x} + s$ b. $\bar{x} - 2s$ and $\bar{x} + 2s$

c. $\bar{x} - 3s$ and $\bar{x} + 3s$

2.141 Why is it that the z-score for a value that belongs to a normal distribution usually lies between -3 and $+3$?

2.142 The mean lifetime of a certain tire is 30,000 miles and the standard deviation is 2500 miles.

a. If we assume the mileages are normally distributed, approximately what percentage of all such tires will last between 22,500 and 37,500 miles?

(continue on page 100)

[EX00-000] identifies the filename of an exercise's online dataset—available through cengagebrain.com

b. If we assume nothing about the shape of the distribution, approximately what percentage of all such tires will last between 22,500 and 37,500 miles?

2.143 The average cleanup time for a crew of a medium-size firm is 84.0 hours and the standard deviation is 6.8 hours. Assume the empirical rule is appropriate.

a. What proportion of the time will it take the cleanup crew 97.6 hours or more to clean the plant?

b. Within what interval will the total cleanup time fall 95% of the time?

2.144 a. What proportion of a normal distribution is greater than the mean?

b. What proportion is within 1 standard deviation of the mean?

c. What proportion is greater than a value that is 1 standard deviation below the mean?

2.145 Using the empirical rule, determine the approximate percentage of a normal distribution that is expected to fall within the interval described.

a. Less than the mean

b. Greater than 1 standard deviation above the mean

c. Less than 1 standard deviation above the mean

d. Between 1 standard deviation below the mean and 2 standard deviations above the mean

2.146 According to the empirical rule, almost all the data should lie between $(\bar{x} - 3s)$ and $(\bar{x} + 3s)$. The range accounts for all the data.

a. What relationship should hold (approximately) between the standard deviation and the range?

b. How can you use the results of part a to estimate the standard deviation in situations when the range is known?

2.147 Chebyshev's theorem guarantees that what proportion of a distribution will be included between the following:

a. $\bar{x} - 2s$ and $\bar{x} + 2s$ b. $\bar{x} - 3s$ and $\bar{x} + 3s$

2.148 According to Chebyshev's theorem, what proportion of a distribution will be within $k = 4$ standard deviations of the mean?

2.149 Chebyshev's theorem can be stated in an equivalent form to that given on page 98. For example, to say "at least 75% of the data fall within 2 standard deviations of the mean" is equivalent to stating "at most, 25%

will be more than 2 standard deviations away from the mean."

a. At most, what percentage of a distribution will be 3 or more standard deviations from the mean?

b. At most, what percentage of a distribution will be 4 or more standard deviations from the mean?

2.150 The scores achieved by students in America make the news often, and all kinds of conclusions are drawn based on these scores. The ACT Assessment® is designed to assess high school students' general educational development and their ability to complete college-level work. One of the categories tested is Science Reasoning. The mean ACT test score for all high school graduates in 2008 in Science Reasoning was 20.8 with a standard deviation of 4.6.

a. According to Chebyshev's theorem, at least what percent of high school graduates' ACT scores in Science Reasoning were between 11.6 and 30.0?

b. If we know that ACT scores are normally distributed, what percent of ACT Science Reasoning scores were between 11.6 and 30.0?

2.151 According to the U.S. Census Bureau, approximately 45% of the 29 million 18- to 24-year-olds in the United States are enrolled in college. To more accurately gauge these young voters, an Edgewood College, Madison, WI, professor conducted a national college on-campus survey of 945 18- to 24-year-olds at 29 colleges in October 19–21, 2004. The survey analyzed which sources of information most influenced students' votes. Based upon a scale of 0 to 100, with 100 being most influential, students said the top influence was the presidential debates (mean = 69.05, standard deviation = 31.65).

a. According to Chebyshev's theorem, at least what percent of the scores are between 5.75 and 132.35?

b. If it is known that these scores are normally distributed, what percent of these scores are between 5.75 and 132.35?

c. Explain why the relationship between the interval bounds in parts a and b and the mean and standard deviation given in the question suggest that the distribution of scores is not normally distributed. Include specifics.

2.152 [EX02-152] On the first day of class last semester, 50 students were asked for the one-way distance from home to college (to the nearest mile). The resulting data follow:

6	5	3	24	15	15	6	2	1	3
5	10	9	21	8	10	9	14	16	16
10	21	20	15	9	4	12	27	10	10
3	9	17	6	11	10	12	5	7	11
5	8	22	20	13	1	8	13	4	18

a. Construct a grouped frequency distribution of the data by using 1–4 as the first class.

b. Calculate the mean and the standard deviation.

c. Determine the values of $\bar{x} \pm 2s$, and determine the percentage of data within 2 standard deviations of the mean.

2.153 [EX02-153] The Labor Department issued the February 2009 state-by-state unemployment report, and it shows continued declines in the job market. These are the February 2009 unemployment rates for the 50 states and DC.

February 2009 State Unemployment Rates

8.4	8.0	7.4	6.6	10.5

*** For remainder of data, logon at cengagebrain.com

Source: http://blogs.wsj.com/

a. Construct a histogram.

b. Does the histogram suggest an approximately normal distribution?

c. Find the mean and standard deviation.

d. Find the percentage of data falling within the three different intervals about the mean and compare them to the empirical rule. Do the percentages and the empirical rule agree with your answer to part b? Explain.

e. Utilize one of the "testing for normality" Technology Instructions on pages 97–98. Compare the results with your answer to part d.

2.154 [EX02-154] One of the many things the U.S. Census Bureau reports to the public is the increase in population for various geographic areas within the country. The percents of increase in population for the 100 fastest-growing counties with 10,000 or more population in 2008 in the United States from April 1, 2008, to July 1, 2008, are listed in the following table.

Percentage of Population Increase

89.6	83.1	82.1	80.2	71.0	70.8	64.5	63.2	60.5	59.7	58.9
57.7	56.1	55.0	53.7	53.2	52.9	52.3	51.9	50.1	50.0	48.4
48.2	47.7	47.6	47.5	47.4	47.0	47.0	46.4	46.0	44.4	44.1
44.1	44.0	41.4	41.0	41.0	40.5	40.1	40.0	39.9	39.8	39.0
38.7	38.7	38.5	38.5	38.1	38.0	37.9	37.8	37.7	37.6	37.5
37.3	36.9	36.8	36.6	36.4	36.4	36.1	36.0	35.9	35.6	35.6
35.6	35.6	35.5	35.4	35.0	34.8	34.7	34.5	34.4	34.2	34.0
34.0	33.1	33.1	33.0	32.9	32.9	32.8	32.7	32.6	32.6	32.4
32.4	32.1	32.0	31.9	31.8	31.7	31.6	31.3	31.2	31.1	31.0
30.9										

Source: http://www.census.gov/

a. Calculate the mean and standard deviation.

b. Determine the values of $\bar{x} \pm 2s$ and $\bar{x} \pm 3s$, and determine the percentage of data within 2 and 3 standard deviations of the mean.

c. Do the percentages found in part b agree with the empirical rule? What does that mean?

d. Do the percentages found in part b agree with Chebyshev's theorem? What does that mean?

e. Construct a histogram and one other graph of your choice. Does the graph show a distribution that agrees with your answers in parts c and d? Explain.

f. Utilize one of the "testing for normality" Technology Instructions on pages 97–98. Compare the results with your answer to part c.

2.155 [EX02-155] Each year, NCAA college football fans like to learn about the up-and-coming freshman class of players. Following are the heights (in inches) of the nation's top 100 high school football players for 2009.

73	75	71	76	74	77	74	72	73	72	74	72	74	72	72
78	73	76	75	72	77	76	73	72	76	72	73	70	75	72
71	74	77	78	74	75	71	75	71	76	70	76	72	71	74
74	71	72	76	71	75	79	78	79	74	76	76	76	75	73
74	70	74	74	75	75	75	75	76	71	74	75	74	78	72
73	71	72	73	72	74	75	77	73	77	75	77	71	72	70
74	76	71	73	76	76	79	77	74	78					

Source: http://www.takkle.com/

a. Construct a histogram and one other graph of your choice that display the distribution on heights.

b. Calculate the mean and standard deviation.

c. Sort the data into a ranked list.

d. Determine the values of $\bar{x} \pm s$, $\bar{x} \pm 2s$ and $\bar{x} \pm 3s$, and determine the percentage of data within 1, 2, and 3 standard deviations of the mean.

e. Do the percentages found in part d agree with the empirical rule? What does this imply? Explain.

f. Do the percentages found in part d agree with Chebyshev's theorem? What does that mean?

g. Does the graph show a distribution that agrees with your answers in part e? Explain.

h. Utilize one of the "testing for normality" Technology Instructions on pages 97–98. Compare the results with your answer to part e.

2.156 [EX02-156] Each year, NCAA college football fans like to learn about the size of the players in the current year's recruit class. Following are the weights (in pounds) of the nation's top 100 high school football players for 2009.

Weight in Pounds

179	226	210	205	225

*** For remainder of data, logon at cengagebrain.com

Source: http://www.takkle.com/   (continue on page 102)

a. Construct a histogram and one other graph of your choice that displays the distribution of weights.

b. Calculate the mean and standard deviation.

c. Sort the data into a ranked list.

d. Determine the values of $\bar{x} \pm s, \bar{x} \pm 2s$, and $\bar{x} \pm 3s$, and determine the percentage of data within 1, 2, and 3 standard deviations of the mean.

e. Do the percentages found in part d agree with the empirical rule? What does this imply? Explain.

f. Do the percentages found in part d agree with Chebyshev's theorem? What does that mean?

g. Do the graphs show a distribution that agrees with your answers in part e? Explain.

h. Utilize one of the "testing for normality" Technology Instructions on pages 97–98. Compare the results with your answer to part e.

2.157 The empirical rule states that the 1, 2, and 3 standard deviation intervals about the mean will contain 68%, 95%, and 99.7%, respectively.

a. Use the computer or calculator commands on page 90 to randomly generate a sample of 100 data from a normal distribution with mean 50 and standard deviation 10. Construct a histogram using class boundaries that are multiples of the standard deviation 10; that is, use boundaries from 10 to 90 in intervals of 10 (see the commands on pp. 52–54). Calculate the mean and the standard deviation using the commands found on pages 64 and 78; then inspect the histogram to determine the percentage of the data that fell within each of the 1, 2, and 3 standard deviation intervals. How closely do the three percentages compare to the percentages claimed in the empirical rule?

b. Repeat part a. Did you get results similar to those in part a? Explain.

c. Consider repeating part a several more times. Are the results similar each time? If so, in what way?

d. What do you conclude about the truth of the empirical rule?

2.158 Chebyshev's theorem states that "at least $1 - \frac{1}{k^2}$" of the data of a distribution will lie within k standard deviations of the mean.

a. Use the computer commands on page 90 to randomly generate a sample of 100 data from a uniform (nonnormal) distribution that has a low value of 1 and a high value of 10. Construct a histogram using class boundaries of 0 to 11 in increments of 1 (see the commands on pp. 52–54). Calculate the mean and the standard deviation using the commands found on pages 64 and 78; then inspect the histogram to determine the percentage of the data that fell within each of the 1, 2, 3, and 4 standard deviation intervals. How closely do these percentages compare to the percentages claimed in Chebyshev's theorem and in the empirical rule?

b. Repeat part a. Did you get results similar to those in part a? Explain.

c. Consider repeating part a several more times. Are the results similar each time? If so, in what way are they similar?

d. What do you conclude about the truth of Chebyshev's theorem and the empirical rule?

2.7 The Art of Statistical Deception

"There are three kinds of lies—lies, damned lies, and statistics." These remarkable words spoken by Benjamin Disraeli (19th-century British prime minister) represent the cynical view of statistics held by many people. Most people are on the consumer end of statistics and therefore have to "swallow" them.

Good Arithmetic, Bad Statistics

Let's explore an outright statistical lie. Suppose a small business employs eight people who earn between $300 and $350 per week. The owner of the business pays himself $1250 per week. He reports to the general public that the average wage paid to the employees of his firm is $430 per week. That may be an example of good arithmetic, but it is bad statistics. It is a misrepresentation of the situation because only one employee, the owner, receives more than the mean salary. The public will think that most of the employees earn about $430 per week.

Graphic Deception

Graphic representations can be tricky and misleading. The frequency scale (which is usually the vertical axis) should start at zero in order to present a total picture. Usually, graphs that do not start at zero are used to save space. Nevertheless, this can be deceptive. Graphs in which the frequency scale starts at zero tend to emphasize the size of the numbers involved, whereas graphs that are chopped off may tend to emphasize the variation in the numbers without regard to the actual size of the numbers. The labeling of the horizontal scale can be misleading also. You need to inspect graphic presentations very carefully before you draw any conclusions from the "story being told."

Consider the following Applied Example.

Superimposed Misrepresentation

APPLIED EXAMPLE 2.16

CLAIMING WHAT THE READER EXPECTS/ANTICIPATED BAD NEWS

This "clever" graphic overlay, from the *Ithaca Times* (December 7, 2000), has to be the worst graph ever to make a front page. The cover story, "Why does college have to cost so much?" pictures two graphs superimposed on a Cornell University campus scene. The two broken lines represent "Cornell's Tuition" and "Cornell's Ranking," with the tuition steadily increasing and the ranking staggering and falling. A very clear image is created: Students get less, pay more!

Now view the two graphs separately. Notice: (1) The graphs cover two different time periods. (2) The vertical scales differ. (3) The "best" misrepresentation comes from the impression that a "drop in rank" represents a lower quality of education. Wouldn't a rank of 6 be better than a rank of 15?

Courtesy of the *Ithaca Times*

Source: http://www.math.yorku.ca/SCS/Gallery/context.html

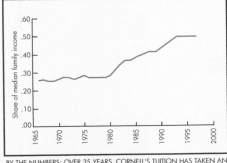

BY THE NUMBERS: OVER 35 YEARS, CORNELL'S TUITION HAS TAKEN AN INCREASINGLY LARGER SHARE OF ITS MEDIAN STUDENT FAMILY INCOME

Source: http://www.math.yorku.ca/SCS/Gallery/context.html

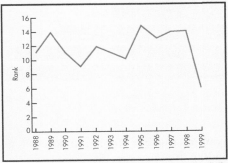

PECKING ORDER: OVER 12 YEARS, CORNELL'S RANKING IN *US NEWS & WORLD REPORT* HAS RISEN AND FALLEN ERRATICALLY.

What it all comes down to is that statistics, like all languages, can be and is abused. In the hands of the careless, the unknowledgeable, or the unscrupulous, statistical information can be as false as "damned lies."

S E C T I O N   2 . 7   E X E R C I S E S

2.159 a. Is the graph below a bar graph or a histogram? Explain how you determined your answer.

Clipping Coupons

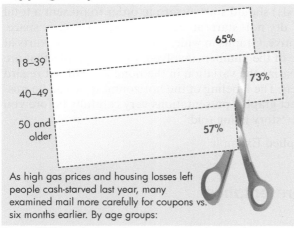

65%

18–39

40–49

73%

50 and older

57%

As high gas prices and housing losses left people cash-starved last year, many examined mail more carefully for coupons vs. six months earlier. By age groups:

Source: DMNews for Pitney Bowes, survey conducted online among 1,003 adults Sept. 9–16, 2008.

b. The age grouping used in the "Clipping Coupons" graphic does not lead to a very informative graph. Describe how the age groups might have been formed and how your suggested grouping would give additional meaning to the graph.

2.160 Did you know... The more you learn the more you earn.

- Don't quit your job.
- Obtain your degree online on your schedule.
- Earn more money.

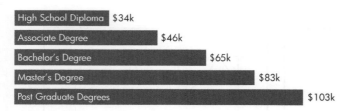

High School Diploma $34k

Associate Degree $46k

Bachelor's Degree $65k

Master's Degree $83k

Post Graduate Degrees $103k

** National Average Income Based on Education Level Source: U.S. Census Bureau Population Survey 2004

a. Examine this bar graph and describe how it is misleading. Be very specific.

b. Redraw the bar graph, correcting the misleading properties.

2.161 "What's wrong with this picture?" That's the question one should be asking when viewing the graphs in Applied Example 2.16 on page 103.

a. Find and describe at least four features of the front-page graph that are incorrectly used.

b. Find and describe at least two features about the "Pecking Order" graph that are misrepresented.

2.162 Did you know that 17.1% is more than double 15.6%? "Ridiculous!" you say.

Death Rate According to Median Household Income:

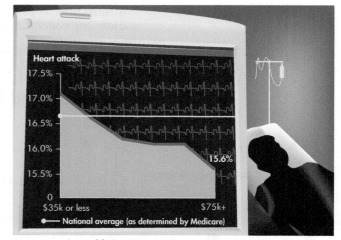

Heart attack

17.5%

17.0%

16.5%

16.0%

15.5%

15.6%

0

$35k or less $75k+

— National average (as determined by Medicare)

Source: Analysis of federal Centers for Medicare & Medicaid Services

a. Explain how the graph suggests such a relationship.

b. How could this misrepresentation be corrected?

c. Redraw this graph to correctly show the relationship between 17.1% and 15.6%.

2.163 The pie chart is drawn correctly, but it gives an incorrect impression.

Working Extra Years for Retirement

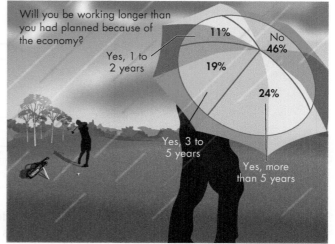

Will you be working longer than you had planned because of the economy?

11%

No 46%

Yes, 1 to 2 years

19%

24%

Yes, 3 to 5 years

Yes, more than 5 years

Source: Sun Life survey of 1,200 adults 30 to 66 years old. Margin of error: ±3 percentage points.

a. The area of each circle segment should be proportional to the percentage it represents. Explain how you can use the ribs of the umbrella to verify that the segments are correctly drawn.

b. Explain why and how the graph is misrepresentative.

2.164 This statistical offering is a rather clever graphic using artistic license along with bills as the bars of a bar graph. An "A for effort," as you have heard before, but the scale aspects of the graph have been compromised.

What do you Plan to Spend your Tax Refund on?

Note: Multiple responses allowed

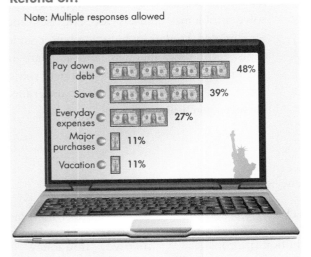

Source: National Retail Federation 2009 Tax Returns Consumer Intentions and Actions survey of 8,426 consumers. Margin of error ±1 percentage point.

a. Identify how and where the scale for percentage is misrepresented.

b. If you were advising the artist, how would you have him or her adjust the bills to correct the problem described in answer to part a?

2.165 What kinds of financial transactions do you do online? Are you worried about your security? According to Consumer Internet Barometer, the source of a March 25, 2009, *USA Today* Snapshot titled "Security of online accounts," the following transactions and percent of people concerned about their online security were reported.

What	Percent
Banking	72
Paying bills	70
Buying stocks, bonds	62
Filing taxes	62

Source: USA Today and Consumer Internet Barometer

Prepare two bar graphs to depict the percentage data. Scale the vertical axis on the first graph from 50 to 80. Scale the second graph from 0 to 100. What is your conclusion concerning how the percentages of the four responses stack up based on the two bar graphs, and what would you recommend, if anything, to improve the presentations?

2.166 Find an article or an advertisement containing a graph that in some way misrepresents the information or statistics.

a. Describe how this graph misrepresents the facts.

b. Redraw the graph in a manner that is more representative of the situation. Describe how your new graph is an improved graphic.

Chapter Review

In Retrospect

You have been introduced to some of the more common techniques of descriptive statistics. There are far too many specific types of statistics used in nearly every specialized field of study for us to review here. We have outlined the uses of only the most universal statistics. Specifically, you have seen several basic graphic techniques (pie charts and bar graphs, Pareto diagrams, dotplots, stem-and-leaf displays, histograms, and box-and-whiskers displays) that are used to present sample data in picture form. You have also been introduced to some of the more common measures of central tendency (mean, median, mode, midrange, and midquartile), measures of dispersion (range, variance, and standard deviation), and measures of position (quartiles, percentiles, and z-scores).

You should now be aware that an average can be any one of five different statistics, and you should understand the distinctions among the different types of averages. The article "'Average' Means Different Things" in Applied Example 2.11 (pp. 67–68) discusses four of the averages studied in this chapter. You might reread it now and find that it has more meaning and is of more interest. It will be time well spent!

You should also have a feeling for, and an understanding of, the concept of a standard deviation. You were introduced to the empirical rule and Chebyshev's theorem for this purpose.

The exercises in this chapter (as in others) are extremely important; they will reinforce the concepts studied before you go on to learn how to use these ideas in later chapters. A good understanding of the descriptive techniques presented in this chapter is fundamental to your success in the later chapters.

CourseMate The **Statistics CourseMate** site for this text brings chapter topics to life with interactive learning, study, and exam preparation tools, including quizzes and flashcards for the Vocabulary and Key Concepts that follow. The site also provides an **eBook** version of the text with highlighting and note taking capabilities. Throughout chapters, the CourseMate icon 🖥 flags concepts and examples that have corresponding interactive resources such as **video** and **animated tutorials** that demonstrate, step by step, how to solve problems; **datasets** for exercises and examples; **Skillbuilder Applets** to help you better understand concepts; **technology manuals**; and software to download including **Data Analysis Plus** (a suite of statistical macros for Excel) and **TI-83/84 Plus** programs—logon at **www.cengagebrain.com**.

Vocabulary and Key Concepts

arithmetic mean (p. 63)
bar graph (p. 33)
bell-shaped distribution (p. 95)
bimodal distribution (p. 55)
box-and-whiskers display (p. 86)
Chebyshev's theorem (p. 98)
circle graph (p. 33)
class (p. 48)
class boundary (p. 48)
class midpoint (class mark) (p. 51)
class width (p. 48)
cumulative frequency distribution (p. 55)
depth (p. 64)
deviation from the mean (p. 74)
distribution (p. 36)
dotplot display (p. 37)
empirical rule (p. 95)
5-number summary (p. 85)
frequency (p. 47)
frequency distribution (p. 47)

grouped frequency distribution (p. 47)
histogram (p. 51)
interquartile range (p. 86)
J-shaped distribution (p. 55)
mean (p. 63)
measure of central tendency (p. 63)
measure of dispersion (p. 74)
measure of position (p. 82)
median (p. 64)
midquartile (p. 85)
midrange (p. 66)
modal class (p. 55)
mode (p. 66)
normal distribution (pp. 55)
ogive (p. 56)
Pareto diagram (p. 35)
percentile (p. 82)
pie chart (p. 33)
qualitative data (p. 33)
quantitative data (p. 36)

quartile (p. 82)
range (p. 74)
rectangular distribution (p. 55)
relative frequency (p. 51)
relative frequency distribution (p. 51)
relative frequency histogram (p. 51)
round-off rule (p. 69)
skewed distribution (p. 55)
standard deviation (pp. 76, 79)
standard score (p. 88)
stem-and-leaf display (p. 38)
summation notation (p. 50)
symmetrical distribution (p. 55)
tally (p. 49)
ungrouped frequency distribution (p. 47)
uniform distribution (p. 55)
variance (pp. 75, 77)
x-bar ($\bar{x}$) (p. 63)
z-score (p. 88)

Learning Outcomes

- Create and interpret graphical displays, including pie charts, bar graphs, Pareto diagrams, dotplots, and stem-and-leaf displays.

  EXP 2.4, Ex. 2.5, 2.13, 2.15, 2.21, 2.27, 2.29

- Understand and be able to describe the difference between grouped and ungrouped frequency distributions, frequency and relative frequency, relative frequency and cumulative relative frequency.

  pp. 47–48, 51, 55–56

- Identify and describe the parts of a frequency distribution: class boundaries, class width, and class midpoint.   EXP 2.6, Ex. 2.43, 2.45

- Create and interpret frequency histograms, relative frequency histograms, and ogives.   pp. 49–52, 55–56, Ex. 2.35, 2.38, 2.40, 2.54

- Identify the shapes of distributions.   pp. 54–55

- Compute, describe, and compare the four measures of central tendency: mean, median, mode, and midrange.   EXP 2.11, Ex. 2.73

- Understand the effect of outliers on each of the four measures of central tendency.   Ex. 2.175, 2.176, 2.205

- Compute, describe, compare, and interpret the two measures of dispersion: range and standard deviation (variance).   pp. 74–76, Ex. 2.99, 2.105

- Compute, describe, and interpret the measures of position: quartiles, percentiles, and z-scores.   EXP 2.12, 2.14, Ex. 2.119, 2.129, 2.192

- Create and interpret boxplots.   Ex. 2.124

- Understand the empirical rule and Chebyshev's theorem and be able to assess a set of data's compliance to these rules.   Ex. 2.140, 2.147, 2.155

- Know when and when not to use certain statistics—graphic and numeric.   pp. 102–103, Ex. 2.159, 2.161

Chapter Exercises

2.167 "Who believes in the 5-second rule?" Most people say food dropped on the floor is not safe to eat.

Who Believes in the 5-second Rule?

When it comes to food that's been dropped on the floor, almost 8 in 10 Americans say it isn't safe to eat it, despite the "rule" of seconds stating otherwise.

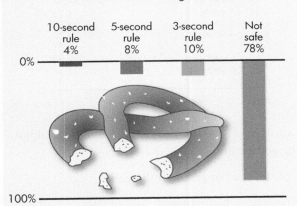

Source: Data from Anne R. Carey and Juan Thomassie, *USA Today.*

a. Draw a circle graph depicting the percentages of adults for each response.

b. If 300 adults are to be surveyed, what frequencies would you expect to occur for each response on the "Do you eat food dropped on the floor" graph?

2.168 The supplies needed for a baby during his or her first year can be expensive—averaging $5000, as shown in this divided bar graph.

Budgeting for Baby

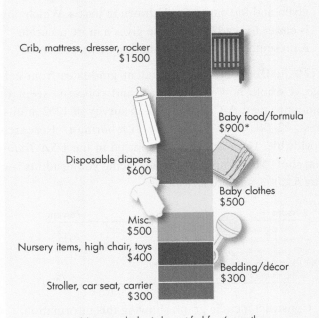

*Assumes baby is breast-fed for 6 months.

Source: Data from Julie Snider, © 2005 *USA Today.*

a. Construct a circle graph showing this same information.

b. Construct a bar graph showing this same information.

c. Compare the appearance of the divided bar graph given with the circle graph drawn in part a and the bar graph drawn in part b. Which one best represents the relationship between various costs of baby supplies?

2.169 There are many types of statistical graphs one can choose from when picturing a set of data. The "divided bar graph" shown next page is an alternative to the circle graph.

If I Win $1 Million...

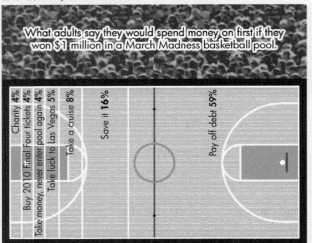

Source: Impulse Research for MSN survey of 1,078 adults in February.

a. Construct a circle graph showing this same information.

b. Compare the appearance of the divided bar graph given and the circle graph drawn in part a. Which one is easier to read? Which one gives a more accurate representation of the information being presented?

2.170 [EX02-170] Once a student graduates from college, a whole new set of issues and concerns seem to come into play. A Charles Schwab survey of 1252 adults, ages 22–28, was done by Lieberman Research Worldwide. The results were reported in the *USA Today* Snapshot "Most important issues facing young adults" on May 5, 2009, and are as follows:

Issues	Percent
Making better money-management choices	52%
Strengthening family relationships	18%
Protecting the environment	11%
Balancing work and personal life	10%
Improving nutrition/health	9%

a. Construct a circle graph showing this information.

b. Construct a bar graph showing this information.

c. Compare the appearance of the circle graph drawn in part a with the bar graph drawn in part b. Which one best represents the relationship between the various issues?

2.171 [EX02-171] The 10 leading causes of death in the United States during 2006 were listed on the Centers for Disease Control and Prevention website.

There were a total of 1,855,610 deaths recorded.

Cause of Death	Number (×10,000)
Alzheimer's	7.2
Chronic Respiratory Disease	12.5
Diabetes	7.2
Heart Disease	63.2
Influenza/Pneumonia	5.6
Malignant Neoplasms	56.0
Accidents	12.2
Nephritis/Nephrosis	4.5
Septicemia	3.4
Stroke	13.7

Source: CDC, Centers for Disease Control and Prevention

a. Construct a Pareto diagram of this information.

b. Write a paragraph describing what the Pareto diagram dramatically shows its reader.

2.172 [EX02-172] The U.S. Census Bureau posted the following 2007 age distribution for the people of Rhode Island. The 2005–2007 American Community Survey universe is limited to the household population and excludes the population living in institutions, college dormitories, and other group quarters.

Gender and Age Distribution	
Male	513,051
Female	549,014
Under 5 years	61,570
5 to 14 years	131,509
15 to 24 years	157,149
25 to 34 years	131,265
35 to 44 years	158,549
45 to 54 years	159,317
55 to 64 years	115,381
65 to 74 years	67,936
75 to 84 years	56,573
85 years and over	22,816

Source: U.S. Census Bureau, American Community Survey

a. Construct a relative frequency distribution of both the gender and the age data.

b. Construct a bar graph of the gender data.

c. Construct a histogram of the age data.

d. Explain why the graph drawn in part b is not a histogram and the graph drawn in part c is a histogram.

2.173 Identify each of the following as examples of (1) attribute (qualitative) or (2) numerical (quantitative) variables.

a. Scores registered by people taking their written state automobile driver's license examination

b. Whether or not a motorcycle operator possesses a valid motorcycle operator's license

c. The number of television sets installed in a house

d. The brand of bar soap being used in a bathroom

e. The value of a cents-off coupon used with the purchase of a box of cereal

2.174 Identify each of the following as examples of (1) attribute (qualitative) or (2) numerical (quantitative) variables.

a. The amount of weight lost in the past month by a person following a strict diet

b. Batting averages of Major League Baseball players

c. Decisions by the jury in felony trials

d. Sunscreen usage before going in the sun (always, often, sometimes, seldom, never)

e. Reason a manager failed to act against an employee's poor performance

2.175 Consider samples A and B. Notice that the two samples are the same except that the 8 in A has been replaced by a 9 in B.

A	2	4	5	5	7	8
B	2	4	5	5	7	9

What effect does changing the 8 to a 9 have on each of the following statistics?

a. Mean   b. Median   c. Mode   d. Midrange

e. Range   f. Variance   g. Std. dev.

2.176 Consider samples C and D. Notice that the two samples are the same except for two values.

C	20	60	60	70	90
D	20	30	70	90	90

What effect does changing the two 60s to 30 and 90 have on each of the following statistics?

a. Mean   b. Median   c. Mode   d. Midrange

e. Range   f. Variance   g. Std. dev.

2.177 [EX02-177] The addition of a new accelerator is claimed to decrease the drying time of latex paint by more than 4%. Several test samples were conducted with the following percentage decreases in drying time:

5.2   6.4   3.8   6.3   4.1   2.8   3.2   4.7

a. Find the sample mean.

b. Find the sample standard deviation.

c. Do you think these percentages average 4 or more? Explain.

(Retain these solutions to use in Exercise 9.29, p. 428.)

2.178 [EX02-178] Gasoline pumped from a supplier's pipeline is supposed to have an octane rating of 87.5. On 13 consecutive days, a sample of octane ratings was taken and analyzed, with the following results:

88.6	86.4	87.2	88.4	87.2	87.6	86.8
86.1	87.4	87.3	86.4	86.6	87.1	

a. Find the sample mean.

b. Find the sample standard deviation.

c. Do you think these readings average 87.5? Explain.

(Retain these solutions to use in Exercise 9.56, p. 431.)

2.179 [EX02-179] These data are the ages of 118 known offenders who committed an auto theft last year in Garden City, Michigan:

11	14	15	15	16	16	17	18	19	21	25	36
12	14	15	15	16	16	17	18	19	21	25	39
13	14	15	15	16	17	17	18	20	22	26	43
13	14	15	15	16	17	17	18	20	22	26	46
13	14	15	16	16	17	17	18	20	22	27	50
13	14	15	16	16	17	17	19	20	23	27	54
13	14	15	16	16	17	18	19	20	23	29	59
13	15	15	16	16	17	18	19	20	23	30	67
14	15	15	16	16	17	18	19	21	24	31	
14	15	15	16	16	17	18	19	21	24	34	

a. Find the mean.   b. Find the median.

c. Find the mode.   d. Find Q_1 and Q_3.

e. Find P_{10} and P_{95}.

2.180 [EX02-180] A survey of 32 workers at building 815 of Eastman Kodak Company was taken last May. Each worker was asked: "How many hours of television did you watch yesterday?" The results were as follows:

0	0	$\frac{1}{2}$	1	2	0	3	$2\frac{1}{2}$	0	0	1
$1\frac{1}{2}$	5	$2\frac{1}{2}$	0	2	$2\frac{1}{2}$	1	0	2	0	$2\frac{1}{2}$
4	0	6	$2\frac{1}{2}$	0	$\frac{1}{2}$	1	$1\frac{1}{2}$	0	2	

a. Construct a stem-and-leaf display.

b. Find the mean.   c. Find the median.

d. Find the mode.   e. Find the midrange.

(continue on page 110)

f. Which measure of central tendency would best represent the average viewer if you were trying to portray the typical television viewer? Explain.

g. Which measure of central tendency would best describe the amount of television watched? Explain.

h. Find the range.   i.   Find the variance.

j. Find the standard deviation.

2.181 [EX02-181] The stopping distance on a wet surface was determined for 25 cars, each traveling at 30 miles per hour. The data (in feet) are shown on the following stem-and-leaf display:

```
 6 | 3  7  6  3  9
 7 | 4  2  0  1  1  2  0  5
 8 | 5  4  5  5  6
 9 | 4  1  0  0  5
10 | 5  4
```

Find the mean and the standard deviation of these stopping distances.

2.182 [EX02-182] Each year, *Sports Illustrated* ranks the highest-earning athletes in the United States. Their earnings include their salary/winnings as well as endorsements. Often the endorsements are more lucrative than their earnings.

The top 20 earnings for 2008 are listed in the following table (in millions of $).

128	62	40	40	35	35	35	31
31	30	27	27	26	26	25	25
25	23	23	23				

a. Find the mean earnings for the top 20 highest-paid athletes.

b. Find the median earnings for the top 20 highest-paid athletes.

c. Find the midrange of the earnings for the top 20 highest-paid athletes.

d. Write a discussion comparing the results from parts a, b, and c.

e. Find the standard deviation of these earnings.

f. Find the percentage of data that is within 1 standard deviation of the mean.

g. Find the percentage of data that is within 2 standard deviations of the mean.

h. Based on these results, discuss why or why not you think that the data are normally distributed.

2.183 [EX02-183] The Office of Aviation Enforcement and Proceedings, U.S. Department of Transportation, reported the number of mishandled baggage reports filed per 1000 airline passengers during October 2007. The industry average was 5.36.

Airline	Reports	Passengers	Reports/1000
JETBLUE AIRWAYS	5345	1641382	3.26
HAWAIIAN AIRLINES	2069	613250	3.37
*** For remainder of data, logon at cengagebrain.com			

Source: Office of Aviation Enforcement and Proceedings, U.S. Department of Transportation

a. Define the terms *population* and *variable* with regard to this information.

b. Are the numbers reported (3.26, 3.37, . . . , 9.57) data or statistics? Explain.

c. Is the average, 5.36, a data value, a statistic, or a parameter value? Explain why.

d. Is the "industry average" the mean of the airline rates of reports per 1000? If not, explain in detail how the 20 airline values are related to the industry average.

2.184 [EX02-184] One of the first scientists to study the density of nitrogen was Lord Raleigh. He noticed that the density of nitrogen produced from the air seemed to be greater than the density of nitrogen produced from chemical compounds. Do his conclusions seem to be justified even though he had so little data?

Lord Raleigh's measurements, which first appeared in *Proceedings, Royal Society* (London, 55, 1894, pp. 340–344) are listed here. The data are the mass of nitrogen filling a certain flask under specified pressure and temperature.

Atmospheric		Chemical	
2.31017	2.31010	2.30143	2.29940
2.30986	2.31028	2.29890	2.29849
2.31010	2.31163	2.29816	2.29889
2.31001	2.30956	2.30182	2.30074
2.31024		2.29869	2.30054

Source: http://exploringdata.cqu.edu.au/datasets/nitrogen.xls

a. Construct side-by-side dotplots of the two sets of data, using a common scale.

b. Calculate mean, median, standard deviation, and first and third quartiles for each set of data.

c. Construct side-by-side boxplots of the two sets of data, using a common scale.

d. Discuss how these two sets of data compare. Do these two very small sets of data show convincing evidence of a difference?

FYI The differences between these sets of data helped lead to the discovery of argon.

2.185 [EX02-185] The lengths (in millimeters) of 100 brown trout in pond 2-B at Happy Acres Fish Hatchery on June 15 of last year were as follows:

15.0	15.3	14.4	10.4

*** For remainder of data, logon at cengagebrain.com

a. Find the mean. b. Find the median.

c. Find the mode. d. Find the midrange.

e. Find the range. f. Find Q_1 and Q_3.

g. Find the midquartile.

h. Find P_{35} and P_{64}.

i. Construct a grouped frequency distribution that uses 10.0–10.5 as the first class.

j. Construct a histogram of the frequency distribution.

k. Construct a cumulative relative frequency distribution.

l. Construct an ogive of the cumulative relative frequency distribution.

2.186 [EX02-186] The national highway system is made up of interstate and noninterstate highways. The Federal Highway Administration reported the number of miles of each type in each state. Listed is a random sample of 20.

Miles of Interstate and Noninterstate Highways by State

State	Interstate	Noninterstate	State	Interstate	Noninterstate
NH	235	589	TN	1,073	2,172
FL	1,471	2,896	NM	1,000	1,935
ME	367	922	LA	904	1,701
HI	55	292	TX	3,233	10,157
MT	1,192	2,683	OH	1,574	2,812
MN	912	3,060	IA	782	2,433
GA	1,245	3,384	NY	1,674	3,476
OK	930	2,431	NE	482	2,496
NC	1,019	2,742	AR	1,167	1,565
RI	71	197	DC	13	70

Source: Federal Highway Administration, U.S. Department of Transportation

Define "ratio I/N" to be number of interstate miles divided by number of noninterstate miles.

a. Inspect the data. What do you estimate the "average" ratio I/N to be?

b. Calculate the "ratio I/N" for each of the 20 states listed.

c. Draw a histogram of the "ratio I/N."

d. Calculate the mean "ratio I/N" for the 20 states listed.

e. Use the total number of 20-state interstate and non-interstate miles to calculate the "ratio I/N" for the combined 20 states.

f. Explain why the answers to parts d and e are not the same.

g. Calculate the standard deviation for the "ratio I/N" for the 20 states listed.

h. Use the dataset indicated to answer the questions in parts b through g using all 51 values.

2.187 [EX02-187] The National Environmental Satellite, Data, and Information Service, U.S. Department of Commerce, posted the area (sq. mi.) and the 2000 population for the 48 contiguous U.S. states.

State	Area (sq. mi.)	Population
AL	51,610	4,447,100
AZ	113,909	5,130,632

*** For remainder of data, logon at cengagebrain.com

Source: U.S. Department of Commerce, http://www5.ncdc.noaa.gov/

When studying how many people live in a country as vast and varied as the United States, perhaps a more interesting variable to study than the population of each state might be the population density of each state since the 48 contiguous states vary so much in area. Define "density" of a state to be the state's population divided by its area.

a. Name three states you believe will be among those with the highest density. Justify your choice.

b. Name three states you believe will be among those with the lowest density. Justify your choice.

c. Describe what you believe the distribution of density will look like. Include ideas of shape of distribution (normal, skewed, etc.).

d. Using the 48 state totals, calculate the overall density for the 48 contiguous U.S. states. Using each state's

(continue on page 112)

population and area, calculate the individual densities for the 48 contiguous U.S. states.

e. Calculate the measures of central tendency.

f. Construct a histogram.

g. Rank the density values. Identify the five states with the highest and the five with the lowest density.

h. Compare the distribution of density information (answers to parts e to g) to your expectation (answers to parts a to c). How did you do?

2.188 [EXO2-188] The volume of Christmas trees sold annually in the United States has declined in recent decades according to a USDA National Agricultural Statistics Service report. All 50 states report contributions to the total U.S. sale of about 25 million Christmas trees annually. Furthermore, each state reports its crop by county. The top 17 producing counties in the United States come from seven states. The number of trees sold by the top 17 counties in 2007 is listed in the following table. This survey is done every 5 years.

Number of Christmas Trees Cut by State (10,000s)

12.0	11.4	11.3	20.0	12.7
157.2	20.2	34.8	309.5	27.3
685.1	118.0	16.7	16.8	31.4
78.5	95.0			

Source: USDA National Agricultural Statistics Service

a. Calculate the mean, median, and midrange for the number of Christmas trees sold annually by the 17 top producing counties.

b. Calculate the standard deviation.

c. What do the answers to parts a and b tell you about the distribution for the number of trees? Explain.

d. Notice that the standard deviation is a larger number than the mean. What does that mean in this situation?

e. Draw a dotplot of the data.

f. Locate the values of the answers to parts a and b to the dotplot drawn for part e.

g. Answer parts c and d again, using the information learned from the dotplot.

2.189 [EXO2-189] Who ate the M&M's? The following table gives the color counts and net weight (in grams) for a sample of 30 bags of M&M's. The advertised net weight is 47.9 grams per bag.

Case	Red	Gr.	Blue	Or.	Yel.	Br.	Weight
1	15	9	3	3	9	19	49.79
2	9	17	19	3	3	8	48.98

*** For remainder of data, logon at cengagebrain.com

Source: http://www.math.uah.edu/stat/
Christine Nickel and Jason York, ST 687 project, Fall 1998

There is something about one case in this dataset that is suspiciously inconsistent with the rest of the data. Find the inconsistency.

a. Construct two different graphs for the weights.

b. Calculate several numerical statistics for the weight data.

c. Did you find any potential inconsistencies in parts a and b? Explain.

d. Find the number of M&M's in each bag.

e. Construct two different graphs for the number of M&M's per bag.

f. Calculate several numerical statistics for the number of M&M's per bag.

g. What inconsistency did you find in parts e and f? Explain.

h. Give a possible explanation as to why the inconsistency does not show up in the weight data but does show up in the number data.

2.190 For a normal (or bell-shaped) distribution, find the percentile rank that corresponds to:

a. $z = 2$ b. $z = -1$

c. Sketch the normal curve, showing the relationship between the z-score and the percentiles for parts a and b.

2.191 For a normal (or bell-shaped) distribution, find the z-score that corresponds to the kth percentile:

a. $k = 20$ b. $k = 95$

c. Sketch the normal curve, showing the relationship between the z-score and the percentiles for parts a and b.

2.192 Bill and Rob are good friends although they attend different high schools in their city. The city school system uses a battery of fitness tests to test all high school students. After completing the fitness tests, Bill and Rob are comparing their scores to see who did better in each event. They need help.

	Sit-ups	Pull-ups	Shuttle Run	50-Yard Dash	Softball Throw
Bill	$z = -1$	$z = -1.3$	$z = 0.0$	$z = 1.0$	$z = 0.5$
Rob	61	17	9.6	6.0	179 ft
Mean	70	8	9.8	6.6	173 ft
Std. Dev.	12	6	0.6	0.3	16 ft

Bill received his test results in z-scores, whereas Rob was given raw scores. Since both boys understand raw scores, convert Bill's z-scores to raw scores in order to make an accurate comparison.

2.193 Twins·Jean and Joan Wong are in fifth grade (different sections), and the class has been given a series of ability tests. If the scores for these ability tests are approximately normally distributed, which girl has the higher relative score on each of the skills listed? Explain your answers.

Skill	Jean: z-Score	Joan: Percentile
Fitness	2.0	99
Posture	1.0	69
Agility	1.0	88
Flexibility	−1.0	35
Strength	0.0	50

2.194 The scores achieved by students in America make the news often, and all kinds of conclusions are drawn based on these scores. The ACT Assessment is designed to assess high school students' general educational development and their ability to complete college-level work. The following table shows the mean and standard deviation for the scores of all high school graduates in 2004 and in 2008 on the four ACT tests and their composite.

	English	Mathematics	Reading	Science Reasoning	Composite
2004					
Mean	20.4	20.7	21.3	20.9	20.9
Std. Dev.	5.9	5.0	6.0	4.6	4.8
2008					
Mean	20.6	21.0	21.4	20.8	21.1
Std. Dev.	6.1	5.2	6.1	4.9	5.0

Source: American College Testing, Digest of Education Statistics

Based on the information in the table:

a. Discuss how the five distributions are similar and different from each other with regard to central value and spread.

b. Discuss any shift in the scores between 2004 and 2008. Include in your answer specifics about how each test distribution has, or has not, changed according to central value and spread.

2.195 [EX02-195] Manufacturing specifications are often based on the results of samples taken from satisfactory pilot runs. The following data resulted from just such a situation, in which eight pilot batches were completed and sampled. The resulting particle sizes are in angstroms (where $1 \text{ Å} = 10^{-8}$ cm):

3923   3807   3786   3710   4010   4230   4226   4133

a. Find the sample mean.

b. Find the sample standard deviation.

c. Assuming that particle size has an approximately normal distribution, determine the manufacturing specification that bounds 95% of the particle sizes (that is, find the 95% interval, $\bar{x} \pm 2s$).

2.196 [EX02-196] Delco Products, a division of General Motors, produces a bracket that is used as part of a power doorlock assembly. The length of this bracket is constantly being monitored. A sample of 30 power door brackets had the following lengths (in millimeters):

11.86	11.88	11.88	11.91	11.88	11.88	11.88	11.88	11.88	11.86
11.88	11.88	11.88	11.88	11.86	11.83	11.86	11.86	11.88	11.88
11.88	11.83	11.86	11.86	11.86	11.88	11.88	11.86	11.88	11.83

Source: With permission of Delco Products Division, GMC

a. Without doing any calculations, what would you estimate for the sample mean?

b. Construct an ungrouped frequency distribution.

c. Draw a histogram of this frequency distribution.

d. Use the frequency distribution and calculate the sample mean and standard deviation.

e. Determine the limits of the $\bar{x} \pm 3s$ interval and mark this interval on the histogram.

f. The product specification limits are 11.7–12.3. Does the sample indicate that production is within these requirements? Justify your answer.

2.197 [EX02-197] The manager of Jerry's Barber Shop recently asked his last 50 customers to punch a time card when they first arrived at the shop and to punch out right

(continue on page 114)

after they paid for their haircut. He then used the data on the cards to measure how long it took Jerry and his barbers to cut hair and used that information to schedule their appointment intervals. The following times (in minutes) were tabulated:

50	21	36	35	35	27	38	51	28	35
32	32	27	25	24	38	43	46	29	45
40	27	36	38	35	31	28	38	33	46
35	31	38	48	23	35	43	31	32	38
43	32	18	43	52	52	49	53	46	19

a.   Construct a stem-and-leaf plot of these data.

b.   Compute the mean, median, mode, range, midrange, variance, and standard deviation of the haircut service times.

c.   Construct a 5-number summary table.

d.   According to Chebyshev's theorem, at least 75% of the haircut service times will fall between what two values? Is this true? Explain why or why not.

e.   How far apart would you recommend that Jerry schedule his appointments to keep his shop operating at a comfortable pace?

2.198   The dotplot below shows the number of attempted passes thrown by the quarterbacks for 22 of the NFL teams that played on one particular Sunday afternoon.

a.   Describe the distribution, including how points A and B relate to the others.

b.   If you remove point A, and maybe point B, would you say the remaining data have an approximately normal distribution? Explain.

c.   Based on the information about distributions that Chebyshev's theorem and the empirical rule give us, how typical an event do you think point A represents? Explain.

2.199   Starting with the data values of 70 and 85, add three data values to your sample so that the sample has the following: (Justify your answer in each case.)

a.   A standard deviation of 5

b.   A standard deviation of 10

c.   A standard deviation of 15

d.   Compare your three samples and the variety of values needed to obtain each of the required standard deviations.

2.200   Make up a set of 18 data (think of them as exam scores) so that the sample meets each of these sets of criteria:

a.   Mean is 75, and standard deviation is 10.

b.   Mean is 75, maximum is 98, minimum is 40, and standard deviation is 10.

c.   Mean is 75, maximum is 98, minimum is 40, and standard deviation is 15.

d.   How are the data in the sample for part b different from those in part c?

2.201   Construct two different graphs of the points (62, 2), (74, 14), (80, 20), and (94, 34).

a.   On the first graph, along the horizontal axis, lay off equal intervals and label them 62, 74, 80, and 94; lay off equal intervals along the vertical axis and label them 0, 10, 20, 30, and 40. Plot the points and connect them with line segments.

b.   On the second graph, along the horizontal axis, lay off equally spaced intervals and label them 60, 65, 70, 75, 80, 85, 90, and 95; mark off the vertical axis in equal intervals and label them 0, 10, 20, 30, and 40. Plot the points and connect them with line segments.

c.   Compare the effect that scale has on the appearance of the graphs in parts a and b. Explain the impression presented by each graph.

Figure for Exercise 2.198

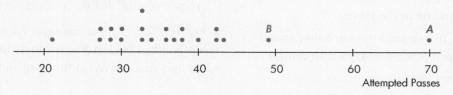

Attempted Passes

2.202 Use a computer to generate a random sample of 500 values of a normally distributed variable x with a mean of 100 and a standard deviation of 20. Construct a histogram of the 500 values.

a. Use the computer commands on page 90 to randomly generate a sample of 500 data from a normal distribution with mean 100 and standard deviation 20. Construct a histogram using class boundaries that are multiples of the standard deviation 20; that is, use boundaries from 20 to 180 in intervals of 20 (see commands on pp. 52–54).

Let's consider the 500 x values found in part a as a population.

b. Use the computer commands on page 91 to randomly select a sample of 30 values from the population found in part a. Construct a histogram of the sample with the same class intervals used in part a.

c. Repeat part b three times.

d. Calculate several values (mean, median, maximum, minimum, standard deviation, etc.) that describe the population and each of the four samples. (See p. 87 for commands.)

e. Do you think a sample of 30 data adequately represents a population? (Compare each of the four samples found in parts b and c to the population.)

2.203 Repeat Exercise 2.202 using a different sample size. You might try a few different sample sizes: $n = 10$, $n = 15, n = 20, n = 40, n = 50, n = 75$. What effect does increasing the sample size have on the effectiveness of the sample in depicting the population? Explain.

2.204 Repeat Exercise 2.202 using populations with different shaped distributions.

a. Use a uniform or rectangular distribution. (Replace the subcommands used in Exercise 2.202; in place of **NORMAL** use: **UNIFORM** with a low of 50 and a high of 150, and use class boundaries of 50 to 150 in increments of 10.)

b. Use a skewed distribution. (Replace the subcommands used in Exercise 2.202; in place of **NORMAL** use: **POISSON** 50 and use class boundaries of 20 to 90 in increments of 5.)

c. Use a J-shaped distribution. (Replace the subcommands used in Exercise 2.202; in place of **NORMAL** use: **EXPONENTIAL** 50 and use class boundaries of 0 to 250 in increments of 10.)

d. Does the shape of the distribution of the population have an effect on how well a sample of size 30 represents the population? Explain.

e. What effect do you think changing the sample size has on the effectiveness of the sample to depict the population? Try a few different sample sizes. Do the results agree with your expectations? Explain.

2.205 [EX02-205] *Outliers!* How often do they occur? What do we do with them? Complete part a to see how often outliers can occur. Then complete part b to decide what to do with outliers.

a. Use the technology of your choice to take samples of various sizes (10, 30, 100, 300 would be good choices) from a normal distribution (mean of 100 and standard deviation of 20 will work nicely) and see how many outliers a randomly generated sample contains. You will probably be surprised. Generate 10 samples of each size for a more representative result. Describe your results—in particular comment on the frequency of outliers in your samples.

MINITAB

Choose:	**Calc > Random Data > Normal**
Enter:	Generate 10 rows of data
	(Use n = 10, 30, 100, 300)
	Store in column(s): **C1–C10**
	Mean: 100
	Stand. Dev. : 20
Choose:	**Graph > Boxplot > Multiple Y's Simple > OK**
Enter:	Graph variables: **C1–C10**
Choose:	**Data View**
Select:	**Interquartile range box**
	Outlier symbols

In practice, we want to do something about the data points that are discovered to be outliers. First, the outlier should be inspected: if there is some obvious reason why it is incorrect, it should be corrected. (For example, a woman's height of 59 inches may well be entered incorrectly as 95 inches, which would be nearly 8 feet tall and is a very unlikely height.) If

(continue on page 116)

the data value can be corrected, fix it! Otherwise, you must weigh the choice between discarding good data (even if they are different) and keeping erroneous data. At this level, it is probably best to make a note about the outlier and continue with using the solution. To help understand the effect of removing an outlier value, let's look at the following set of data, randomly generated from a normal distribution $N(100, 20)$.

74.2	84.5	88.5	110.8	97.6
110.6	93.7	113.3	96.1	86.7
102.8	82.5	107.6	91.1	95.7
100.2	116.4	78.3	154.8	144.7
97.3	102.8	91.8	58.5	120.1
98.0	98.4	81.9	58.5	118.1

b. Construct a boxplot and identify any outliers.

c. Remove the outlier and construct a new boxplot.

d. Describe your findings and comment on why it might be best and less confusing while studying introductory statistics not to discard outliers.

Chapter Practice Test

PART I: Knowing the Definitions

Answer "True" if the statement is always true. If the statement is not always true, replace the words in bold with the words that make the statement always true.

2.1 The **mean** of a sample always divides the data into two halves (half larger and half smaller in value than itself).

2.2 A measure of **central tendency** is a quantitative value that describes how widely the data are dispersed about a central value.

2.3 The sum of the squares of the deviations from the mean, $\Sigma(x - \bar{x})^2$, will **sometimes** be negative.

2.4 For any distribution, the sum of the deviations from the mean equals **zero**.

2.5 The standard deviation for the set of values 2, 2, 2, 2, and 2 is **2**.

2.6 On a test, John scored at the 50th percentile and Jorge scored at the 25th percentile; therefore, John's test score was **twice** Jorge's test score.

2.7 The frequency of a class is the number of pieces of data whose values fall within the **boundaries** of that class.

2.8 **Frequency distributions** are used in statistics to present large quantities of repeating values in a concise form.

2.9 The unit of measure for the standard score is always **standard deviations**.

2.10 For a bell-shaped distribution, the range will be approximately equal to **6 standard deviations**.

PART II: Applying the Concepts

2.11 The results of a consumer study completed at Corner Convenience Store are reported in the accompanying histogram. Answer each question.

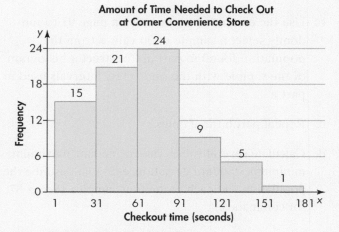

a. What is the class width?

b. What is the class midpoint for the class 31–61?

c. What is the upper boundary for the class 61–91?

d. What is the frequency of the class 1–31?

e. What is the frequency of the class that contains the largest observed value of x ?

f. What is the lower boundary of the class with the largest frequency?

g. How many pieces of data are shown in this histogram?

h. What is the value of the mode?

i. What is the value of the midrange?

j. Estimate the value of the 90th percentile, P_{90}.

2.12 A sample of the purchases of several Corner Convenience Store customers resulted in the following sample data (x = number of items purchased per customer):

x	1	2	3	4	5
f	6	10	9	8	7

a. What does the 2 represent?

b. What does the 9 represent?

c.   How many customers were used to form this sample?

d.   How many items were purchased by the customers in this sample?

e.   What is the largest number of items purchased by one customer?

Find each of the following (show formulas and work):

f.   Mode   g.   Median   h.   Midrange

i.   Mean   j.   Variance   k.   Standard

2.13 Given the set of data 4, 8, 9, 8, 6, 5, 7, 5, 8, find each of the following sample statistics:

a.   Mean   b.   Median   c.   Mode

d.   Midrange   e.   First quartile   f. P_{40}

g.   Variance   h.   Standard deviation

i.   Range

2.14 a.   Find the standard score for the value $x = 452$ relative to its sample, where the sample mean is 500 and the standard deviation is 32.

b.   Find the value of x that corresponds to the standard score of 1.2, where the mean is 135 and the standard deviation is 15.

PART III: Understanding the Concepts

Answer all questions.

2.15 The Corner Convenience Store kept track of the number of paying customers it had during the noon hour each day for 100 days. The resulting statistics are rounded to the nearest integer:

mean = 95   midrange = 93

median = 97   range = 56

mode = 98   standard deviation = 12

first quartile = 85

third quartile = 107

a.   The Corner Convenience Store served what number of paying customers during the noon hour more often than any other number? Explain how you determined your answer.

b.   On how many days were there between 85 and 107 paying customers during the noon hour? Explain how you determined your answer.

c.   What was the greatest number of paying customers during any one noon hour? Explain how you determined your answer.

d.   For how many of the 100 days was the number of paying customers within 3 standard deviations of

the mean ($\bar{x} \pm 3s$)? Explain how you determined your answer.

2.16 Mr. VanCott started his own machine shop several years ago. His business has grown and become very successful in recent years. Currently he employs 14 people, including himself, and pays the following annual salaries:

Owner, President	$80,000	Worker	$25,000
Business Manager	50,000	Worker	25,000
Production Manager	40,000	Worker	25,000
Shop Foreman	35,000	Worker	20,000
Worker	30,000	Worker	20,000
Worker	30,000	Worker	20,000
Worker	28,000	Worker	20,000

a.   Calculate the four "averages": mean, median, mode, and midrange.

b.   Draw a dotplot of the salaries and locate each of the four averages on it.

c.   Suppose you were the feature writer assigned to write this week's feature story, on Mr. VanCott's machine shop, one of a series on local small businesses that are prospering. You plan to interview Mr. VanCott, his business manager, the shop foreman, and one of his newer workers. Which statistical average do you think each will give when asked, "What is the average annual salary paid to the employees here at VanCott's?" Explain why each person interviewed would have a different perspective and why this viewpoint may cause each to cite a different statistical average.

d.   What is there about the distribution of these salaries that causes the four "average values" to be so different?

2.17 Create a set of data containing three or more values in the following cases:

a.   Where the mean is 12 and the standard deviation is 0

b.   Where the mean is 20 and the range is 10

c.   Where the mean, median, and mode are all equal

d.   Where the mean, median, and mode are all different

e.   Where the mean, median, and mode are all different and the median is the largest and the mode is the smallest

f.   Where the mean, median, and mode are all different and the mean is the largest and the median is the smallest

2.18 A set of test papers was machine-scored. Later it was discovered that 2 points should be added to each score. Student A said, "The mean score should also be increased by 2 points." Student B added, "The standard deviation should also be increased by 2 points." Who is right? Justify your answer.

2.19 Student A stated, "Both the standard deviation and the variance preserve the same unit of measurement as the data." Student B disagreed, arguing, "The unit of measurement for variance is a meaningless unit of measurement." Who is right? Justify your answer.

3 Descriptive Analysis and Presentation of Bivariate Data

Image copyright Dec Hogan, 2010. Used under license from Shutterstock.com

3.1 Bivariate Data
Two variables are paired together for analysis.
3.2 Linear Correlation
Does an increase in the value **indicate a change** *in the other?*
3.3 Linear Regression
The **line of best fit** *is a mathematical expression for the relationship between two variables.*

3.1 Bivariate Data

Weighing Your Fish with a Ruler

Ever wanted to know the weight of your catch, but didn't have a scale? Measure the length of a rainbow trout from the snout to the tip of the tail and check this chart. These weights are averages taken from fish collected by DEC (Department of Environmental Conservation) fish management crews from across New York State. [EX03-001]

Length, inch	12	13	14	15	16	17	18	19	20	21	22	23	24	25	26	27	28	29	30
Weight, lb–oz	0–10	0–12	1–0	1–3	1–7	1–12	2–1	2–7	2–14	3–5	3–13	4–6	5–0	5–11	6–6	7–2	8–0	8–14	9–14

Source: NYS DEC 2008–09 *Freshwater Fishing Guide*

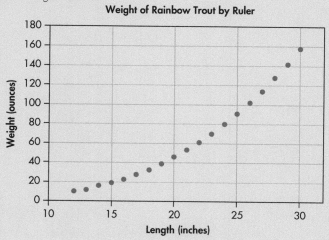

Weight of Rainbow Trout by Ruler

In Chapter 2, we learned how to graphically display and numerically describe sample data for one variable. We will now expand these techniques to cover sample data that involve two paired variables. In particular, the length and weight of rainbow trout, shown on page 120, are two paired quantitative (numerical) variables.

> **Bivariate data** The values of two different variables that are obtained from the same population element.

Each of the two variables may be either *qualitative* or *quantitative*. As a result, three combinations of variable types can form bivariate data:

1. Both variables are qualitative (both attribute).
2. One variable is qualitative (attribute), and the other is quantitative (numerical).
3. Both variables are quantitative (both numerical).

In this section we present tabular and graphic methods for displaying each of these combinations of bivariate data.

Two Qualitative Variables

When bivariate data result from two qualitative (attribute or categorical) variables, the data are often arranged on a **cross-tabulation** or **contingency table**. Let's look at an example.

E X A M P L E   3 . 1

CONSTRUCTING CROSS-TABULATION TABLES

Thirty students from our college were randomly identified and classified according to two variables: gender (M/F) and major (liberal arts, business administration, technology), as shown in Table 3.1. These 30 bivariate data can be summarized on a 2 × 3 cross-tabulation table, where the two rows represent the two genders, male and female, and the three columns represent the three major categories of liberal arts (LA), business administration (BA), and technology (T). The entry in each cell is found by determining how many students fit into each category. Adams is male (M) and liberal arts (LA) and is classified in the cell in the first row, first column. See the red tally mark in Table 3.2. The other 29 students are classified (tallied, shown in black) in a similar fashion.

FYI $m = n$ (rows) $n = n$ (cols) for an $m \times n$ contingency table.

TABLE 3.1 Genders and Majors of 30 College Students [TA03-01]

Name	Gender	Major	Name	Gender	Major	Name	Gender	Major
Adams	M	LA	Feeney	M	T	McGowan	M	BA
Argento	F	BA	Flanigan	M	LA	Mowers	F	BA
Baker	M	LA	Hodge	F	LA	Ornt	M	T
Bennett	F	LA	Holmes	M	T	Palmer	F	LA
Brand	M	T	Jopson	F	T	Pullen	M	T
Brock	M	BA	Kee	M	BA	Rattan	M	BA
Chun	F	LA	Kleeberg	M	LA	Sherman	F	LA
Crain	M	T	Light	M	BA	Small	F	T
Cross	F	BA	Linton	F	LA	Tate	M	BA
Ellis	F	BA	Lopez	M	T	Yamamoto	M	LA

 Video tutorial available—logon and learn more at cengagebrain.com

The resulting 2 × 3 cross-tabulation (contingency) table, Table 3.3, shows the frequency for each cross-category of the two variables along with the row and column totals, called *marginal totals* (or *marginals*). The total of the marginal totals is the *grand total* and is equal to *n*, the *sample size*.

TABLE 3.2 Cross-Tabulation of Gender and Major (tallied)

Gender	Major					
	LA		BA		T	
M	‖‖‖	(5)	‖‖‖ ‖	(6)	‖‖‖ ‖‖	(7)
F	‖‖‖ ‖	(6)	‖‖‖‖	(4)	‖‖	(2)

TABLE 3.3 Cross-Tabulation of Gender and Major (frequencies)

Gender	Major			Row Total
	LA	BA	T	
M	5	6	7	18
F	6	4	2	12
Col. Total	11	10	9	30

Contingency tables often show percentages (relative frequencies). These percentages can be based on the entire sample or on the subsample (row or column) classifications.

Percentages Based on the Grand Total (Entire Sample)

The frequencies in the contingency table shown in Table 3.3 can easily be converted to percentages of the grand total by dividing each frequency by the grand total and multiplying the result by 100. For example, 6 becomes 20% $\left[\left(\dfrac{6}{30}\right) \times 100 = 20\right]$. See Table 3.4.

From the table of percentages of the grand total, we can easily see that 60% of the sample are male, 40% are female, 30% are technology majors, and so on. These same statistics (numerical values describing sample results) can be shown in a bar graph (see Figure 3.1).

Table 3.4 and Figure 3.1 show the distribution of male liberal arts students, female liberal arts students, male business administration students, and so on, relative to the entire sample.

TABLE 3.4 Cross-Tabulation of Gender and Major (relative frequencies; % of grand total)

Gender	Major			Row Total
	LA	BA	T	
M	17%	20%	23%	60%
F	20%	13%	7%	40%
Col. Total	37%	33%	30%	100%

FIGURE 3.1
Bar Graph

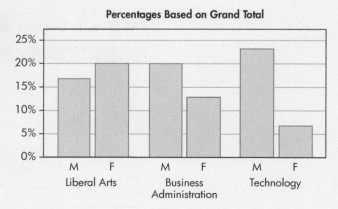

Percentages Based on Row Totals

The frequencies in the same contingency table, Table 3.3, can be expressed as percentages of the row totals (or gender) by dividing each row entry by that row's total and multiplying the results by 100. Table 3.5 is based on row totals.

From Table 3.5 we see that 28% of the male students are majoring in liberal arts, whereas 50% of the female students are majoring in liberal arts. These same statistics are shown in the bar graph in Figure 3.2.

TABLE 3.5 Cross-Tabulation of Gender and Major (% of row totals)

| Gender | Major | | | |
	LA	BA	T	Row Total
M	28%	33%	39%	100%
F	50%	33%	17%	100%
Col. Total	37%	33%	30%	100%

FIGURE 3.2
Bar Graph

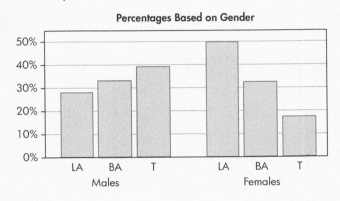

Table 3.5 and Figure 3.2 show the distribution of the three majors for male and female students separately.

Percentages Based on Column Totals

The frequencies in the contingency table, Table 3.3, can be expressed as percentages of the column totals (or major) by dividing each column entry by that column's total and multiplying the result by 100. Table 3.6 is based on column totals.

From Table 3.6 we see that 45% of the liberal arts students are male, whereas 55% of the liberal arts students are female. These same statistics are shown in the bar graph in Figure 3.3.

TABLE 3.6 Cross-Tabulation of Gender and Major (% of column totals)

| Gender | Major | | | |
	LA	BA	T	Row Total
M	45%	60%	78%	60%
F	55%	40%	22%	40%
Col. Total	100%	100%	100%	100%

FIGURE 3.3
Bar Graph

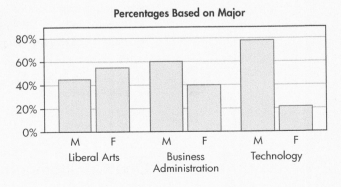

Table 3.6 and Figure 3.3 show the distribution of male and female students for each major separately.

TECHNOLOGY INSTRUCTIONS: CROSS-TABULATION TABLES

MINITAB

Input the single row-variable categorical values into C1 and the corresponding single column-variable categorical values into C2; then continue with:

Choose: **Stat > Tables > Cross** Tabulation and Chi-Square
Enter: Categorical variables: For rows: **C1** For columns: **C2**
Select: **Counts**
            Row Percents
            Column Percents
            Total Percents > OK

Suggestion: The four subcommands that are available for "Display" can be used together; however, the resulting table will be much easier to read if one subcommand at a time is used.

Excel

Using column headings or titles, input the row-variable categorical values into column A and the corresponding column-variable categorical values into column B; then continue with:

Choose: **Insert > Tables > Pivot Table pulldown > Pivot Chart**
Select: **Select a table or range**
Enter: **Range: (A1:B5 or select cells)**
Select: **Existing Worksheet**
Enter: **(C1 or select cell) > OK**
Drag: **Headings to row or column** (depends on preference) **into chart box formed**
            One heading into data area[*]

[*]For other summations, double-click "Count of" in data area box; then continue with:

Choose: Summarize by: **Count**
            Show values as: **% of row** or **% of column** or **% of total > OK**

TI-83/84 Plus

The categorical data must be numerically coded first; use 1, 2, 3, . . . for the various row variables and 1, 2, 3, . . . for the various column variables. Input the numeric row-variable values into L1 and the corresponding numeric column-variable values into L2; then continue with:

Choose: **PRGM > EXEC > CROSSTAB**[*]
Enter: **ROWS: L1 > ENTER**
            COLS: L2 > ENTER

The cross-tabulation table showing frequencies is stored in matrix [A], the cross-tabulation table showing row percentages is in matrix [B], column percentages in matrix [C], and percentages based on the grand total in matrix [D]. All matrices contain marginal totals. To view the matrices, continue with:

Choose: **MATRX > NAMES**
Enter: **1:[A]** or **2:[B]** or **3:[C]** or **4:[D] > ENTER**

[*]Program "CROSSTAB" is one of many programs that are available for downloading. See page 35 for specific instructions.

One Qualitative and One Quantitative Variable

When bivariate data result from one qualitative and one quantitative variable, the quantitative values are viewed as separate samples, each set identified by levels of the qualitative variable. Each sample is described using the techniques from Chapter 2, and the results are displayed side by side for easy comparison.

E X A M P L E   3 . 2

CONSTRUCTING SIDE-BY-SIDE COMPARISONS

The distance required to stop a 3000-pound automobile on wet pavement was measured to compare the stopping capabilities of three tire tread designs (see Table 3.7). Tires of each design were tested repeatedly on the same automobile on a controlled wet pavement.

TABLE 3.7 Stopping Distances (in feet) for Three Tread Designs [TA03-07]

Design A ($n = 6$)			Design B ($n = 6$)			Design C ($n = 6$)		
37	36	38	33	35	38	40	39	40
34	40	32	34	42	34	41	41	43

The design of the tread is a qualitative variable with three levels of response, and the stopping distance is a quantitative variable. The distribution of the stopping distances for tread design A is to be compared with the distribution of stopping distances for each of the other tread designs. This comparison may be made with both numerical and graphic techniques. Some of the available options are shown in Figure 3.4, Table 3.8, and Table 3.9.

FIGURE 3.4

Dotplot and Box-and-Whiskers Display Using a Common Scale

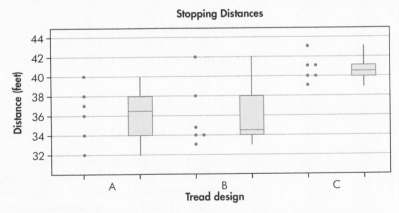

TABLE 3.8 5-Number Summary for Each Design

	Design A	Design B	Design C
High	40	42	43
Q_3	38	38	41
Median	36.5	34.5	40.5
Q_1	34	34	40
Low	32	33	39

TABLE 3.9 Mean and Standard Deviation for Each Design

	Design A	Design B	Design C
Mean	36.2	36.0	40.7
Standard deviation	2.9	3.4	1.4

Video tutorial available—logon and learn more at cengagebrain.com

T E C H N O L O G Y   I N S T R U C T I O N S :
S I D E - B Y - S I D E   B O X P L O T S
A N D   D O T P L O T S

MINITAB	Input the numerical values into C1 and the corresponding categories into C2; then continue with:

Choose: **Graph > Boxplot … > One Y, With Groups > OK**
Enter: Graph variables: C1 Categorical variables: **C2 > OK**

MINITAB commands to construct side-by-side dotplots for data in this form are located on page 41.

If the data for the various categories are in separate columns, use the MINITAB commands for multiple boxplots on page 88. If side-by-side dotplots are needed for data in this form, continue with:

Choose: **Graph > Dotplots**
Select: **Multiple Y's, Simple > OK**
Enter: Graph variables: **C1 C2 > OK**

Excel	Excel commands to construct a single boxplot are on page 88.

TI-83/84 Plus	TI-83/84 commands to construct multiple boxplots are on page 88. TI-83/84 commands to construct multiple dotplots are on page 42.

Much of the information presented here can also be demonstrated using many other statistical techniques, such as stem-and-leaf displays or histograms.

We will restrict our discussion in this chapter to descriptive techniques for the most basic form of correlation and regression analysis—the bivariate linear case.

Two Quantitative Variables

When the bivariate data are the result of two quantitative variables, it is customary to express the data mathematically as **ordered pairs** (x, y), where x is the **input variable** (sometimes called the **independent variable**) and y is the **output variable** (sometimes called the **dependent variable**). The data are said to be *ordered* because one value, x, is always written first. They are called *paired* because for each x value, there is a corresponding y value from the same source. For example, if x is height and y is weight, then a height value and a corresponding weight value are recorded for each person. The input variable, x, is measured or controlled in order to predict the output variable, y. Suppose some research doctors are testing a new drug by prescribing different dosages and observing the lengths of the recovery times of their patients. The researcher can control the amount of drug prescribed, so the amount of drug is referred to as x. In the case of height and weight, either variable could be treated as input and the other as output, depending on the question being asked. However, different results will be obtained from the regression analysis, depending on the choice made.

In problems that deal with two quantitative variables, we present the sample data pictorially on a *scatter diagram*.

> **Scatter diagram** A plot of all the ordered pairs of bivariate data on a coordinate axis system. The input variable, x, is plotted on the horizontal axis, and the output variable, y, is plotted on the vertical axis.

Note: When you construct a scatter diagram, it is convenient to construct scales so that the range of the y values along the vertical axis is equal to or slightly shorter than the range of the x values along the horizontal axis. This creates a "window of data" that is approximately square.

E X A M P L E   3 . 3

CONSTRUCTING A SCATTER DIAGRAM

In Mr. Chamberlain's physical-fitness course, several fitness scores were taken. The following sample is the numbers of push-ups and sit-ups done by 10 randomly selected students:

$$(27, 30) \quad (22, 26) \quad (15, 25) \quad (35, 42) \quad (30, 38)$$
$$(52, 40) \quad (35, 32) \quad (55, 54) \quad (40, 50) \quad (40, 43)$$

Table 3.10 shows these sample data, and Figure 3.5 shows a scatter diagram of the data.

TABLE 3.10 Data for Push-ups and Sit-ups [TA03-10]

Student	1	2	3	4	5	6	7	8	9	10
Push-ups, x	27	22	15	35	30	52	35	55	40	40
Sit-ups, y	30	26	25	42	38	40	32	54	50	43

The scatter diagram from Mr. Chamberlain's physical-fitness course shows a definite pattern. Note that as the number of push-ups increased, so did the number of sit-ups.

FIGURE 3.5

Scatter Diagram

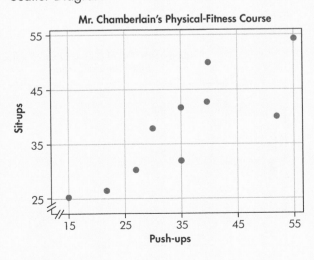

A P P L I E D   E X A M P L E   3 . 4

AMERICANS LOVE THEIR AUTOMOBILES

America's love affair with the sport-utility vehicle (SUV) began in the late 1990s and early 2000s, but may be declining a bit recently due to their gasoline consumption, cost, and poor safety records. The SUV conjures up images of a high-performance, rugged, 4-wheel-drive car built on a truck chassis that: can go off-road; has good towing abilities; can carry more than four passengers; is a safer vehicle than a car because it's larger and heavier built; and gets around better in the snow. However, if the truth were told, most people buy an SUV because they can.

This chart lists 16 of the 4-wheel-drive, 6-cylinder SUVs offered by auto manufacturers in 2009 and the values of four variables for each vehicle.

© iStockphoto.com/Luis Sandoval Mandujano

Variables:

Manu	Vehicle Manufacturer
Model	Vehicle Model
Gas	Regular- or Premium-Grade Gasoline
Cost	Cost of Gasoline to Drive 25 Miles
Fill-up	Cost of a Fill-up
Tank	Capacity of Gasoline Tank in Gallons

TABLE 3.11   2009 4WD, 6-Cyl SUVs [EX03-022]

Manu	Model	Gas	Cost	Fill-up	Tank
Buick	Enclave	Reg.	2.51	37.82	22
Chevrolet	Trailblazer	Reg.	2.98	37.82	22
Chrysler	Aspen	Reg.	3.18	46.41	27
Dodge	Durango	Reg.	3.18	46.41	27
Ford	Escape	Reg.	2.39	28.36	16.5
GMC	Envoy	Reg.	2.98	37.82	22
Honda	Pilot	Reg.	2.65	36.1	21
Jeep	Grd Cherokee	Reg.	2.81	36.27	21.1
Kia	Sportage	Reg.	2.39	29.57	17.2
Lexus	RX 350	Prem.	2.83	37.15	19.2
Lincoln	MKX	Reg.	2.51	32.66	19
Mazda	CX-7	Prem.	2.99	35.22	18.2
Mercury	Mountaineer	Reg.	3.18	38.68	22.5
Mitsubishi	Outlander	Reg.	2.15	27.16	15.8
Nissan	Murano	Prem.	2.69	41.99	21.7
Toyota	RAV4	Reg.	2.27	27.33	15.9

http://www.fueleconomy.gov/

In addition to displaying this information in chart form, let's display the data using some of the techniques from this section in combination with some from Chapter 2.

FIGURE 3.6

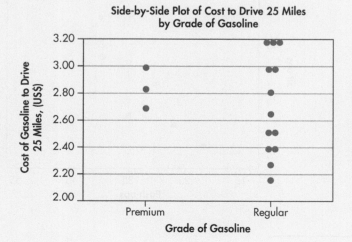

Side-by-Side Plot of Cost to Drive 25 Miles by Grade of Gasoline

Figure 3.6 shows that the cost of gasoline to drive 25 miles is more for three of the SUVs using regular gas than for SUVs using premium. Several of the SUVs using regular gasoline cost less.

FIGURE 3.7

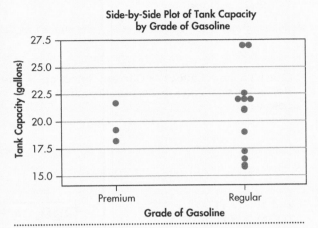

Figure 3.7 shows that six of the SUVs that use regular gas have tanks with larger capacities than the three SUVs using premium. Why would some vehicles need 27-gallon gasoline tanks? See Exercise 3.43 for a possible answer.

FIGURE 3.8

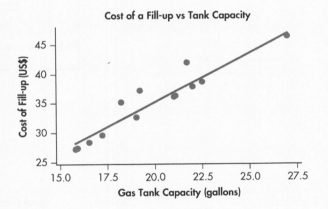

Figure 3.8 probably shows information you already knew: the larger the gas tank, the more it costs to fill it. How could it be any other way? Do you see our three premium-using SUVs? How do the distributions shown in Figure 3.7 appear in Figure 3.8 See Exercise 3.16 for more on this topic.

TECHNOLOGY INSTRUCTIONS: SCATTER DIAGRAM

MINITAB

Input the *x*-variable values into C1 and the corresponding *y*-variable values into C2; then continue with:

Choose: **Graph > Scatter Plot ... > Simple > OK**
Enter: Y variables: **C2** X variables: **C1**
Select: **Labels > Titles/Footnotes**
Enter: Title: **your title > OK > OK**

Excel

Input the *x*-variable values into column A and the corresponding *y*-variable values into column B; activate columns of data; then continue with:

Choose: **Insert > Scatter > 1st picture** (usually)
Choose: **Chart Layouts > Layout 1**
Enter: Chart title: **your title**; (*x*) axis title: **title for x axis**; (*y*) axis title: **title for y axis**[*]

*To remove gridlines:

Choose: **Chart Tools > Layout > Gridlines > Primary Horizontal Gridlines > None**

To edit the scatter diagram, follow the basic editing commands shown for a histogram on page 53.
To change the scale and/or show tick marks, doubleclick on the axis; then continue with:

Choose: **Chart Tools > Layout > Current Selection > Plot A Horiz/ Vertical Axes > Format Selection**
Enter: **new values**
Select: Major tick mark type: **Cross > OK**

TI-83/84 Plus

Input the x-variable values into L1 and the corresponding y-variable values into L2; then continue with:

Choose: **2nd > STATPLOT > 1:Plot1**
Choose: **ZOOM > 9:ZoomStat**
    > TRACE > > >
    or
    WINDOW
    Enter: at most lowest x value, at least highest x value, x-scale, −y-scale, at least highest y value, y-scale, 1
    TRACE > > >

SECTION 3.1 EXERCISES

3.1 [EX03-001] Refer to "Weighing Your Fish with a Ruler" on page 120 to answer the following questions:

a. Is there a relationship (pattern) between the two variables length of a rainbow trout and weight of a rainbow trout? Explain why or why not.

b. Do you think it is reasonable (or possible) to predict the weight of a rainbow trout based on the length of the rainbow trout? Explain why or why not.

3.2 a. Is there a relationship between a person's height and shoe size as he or she grows from an infant to age 16? As one variable gets larger, does the other also get larger? Explain your answers.

b. Is there a relationship between height and shoe size for people who are older than 16 years of age? Do taller people wear larger shoes? Explain your answers.

3.3 [EX03-003] In a national survey of 500 business and 500 leisure travelers, each was asked where he or she would most like "more space."

	On Airplane	Hotel Room	All Other
Business	355	95	50
Leisure	250	165	85

a. Express the table as percentages of the total.

b. Express the table as percentages of the row totals. Why might one prefer the table to be expressed this way?

c. Express the table as percentages of the column totals. Why might one prefer the table to be expressed this way?

3.4 The "In the eye of the beholder" graphic shows two circle graphs, each with four sections. This same information could be represented in the form of a 2 × 4 contingency table of two qualitative variables.

a. Identify the population and name the two variables.

b. Construct the contingency table using entries of percentages based on row totals.

Figure for Exercise 3.4

In the eye of the beholder

Source: Energizer online survey of 1,051 married adults, ages 44–62.

3.5 "The perfect age" graphic shows the results from a 9×2 contingency table for one qualitative and one quantitative variable.

"The Perfect Age"

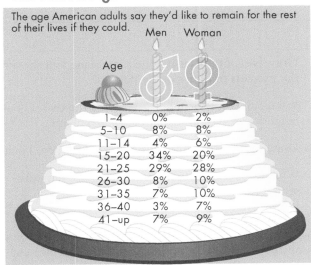

Source: Data from Cindy Hall and Genevieve Lynn, *USA TODAY*; IRC Research for Walt Disney. © 1998 *USA TODAY*, reprinted by permission.

a. Identify the population and name the qualitative and quantitative variables.

b. Construct a bar graph showing the two distributions side by side.

c. Does there seem to be a big difference between the genders on this subject?

3.6 [EX03-006] The National Highway System Designation Act of 1995 allows states to set their own highway speed limits. Most of the states have raised the limits. The November 2008 maximum speed limits on interstate highways (rural) for cars and trucks by each state are given in the following table (in miles per hour):

State	Cars	Trucks	State	Cars	Trucks
Alabama	70	70	Montana	75	65
Alaska	65	65	Nebraska	75	75
Arizona	75	75	Nevada	75	75
Arkansas	70	65	New Hampshire	65	65
California	70	55	New Jersey	65	65
Colorado	75	75	New Mexico	75	75
Connecticut	65	65	New York	65	65
Delaware	65	65	North Carolina	70	70
Florida	70	70	North Dakota	75	75
Georgia	70	70	Ohio	65	55
Hawaii	60	60	Oklahoma	75	75
Idaho	75	65	Oregon	65	55
Illinois	65	55	Pennsylvania	65	65
Indiana	70	65	Rhode Island	65	65
Iowa	70	70	South Carolina	70	70
Kansas	70	70	South Dakota	75	75
Kentucky	65	65	Tennessee	70	70
Louisiana	70	70	Texas	75	65
Maine	65	65	Utah	75	75
Maryland	65	65	Virginia	65	65
Massachusetts	65	65	Vermont	65	65
Michigan	70	60	Washington	70	60
Minnesota	70	70	West Virginia	70	70
Mississippi	70	70	Wisconsin	65	65
Missouri	70	70	Wyoming	75	75

Source: American Trucking Association

a. Build a cross-tabulation of the two variables vehicle type and maximum speed limit on interstate highways. Express the results in frequencies, showing marginal totals.

b. Express the contingency table you derived in part a in percentages based on the grand total.

c. Draw a bar graph showing the results from part b.

d. Express the contingency table you derived in part a in percentages based on the marginal total for speed limit.

e. Draw a bar graph showing the results from part d.

   If you are using a computer or a calculator, try the cross-tabulation table commands on page 124.

3.7 [EX03-007] A statewide survey was conducted to investigate the relationship between viewers' preferences for ABC, CBS, NBC, PBS, or FOX for news information and their political party affiliation. The results are shown in tabular form:

Television Station

Political Affiliation	ABC	CBS	NBC	PBS	FOX
Democrat	203	218	257	156	226
Republican	421	350	428	197	174
Other	156	312	105	57	90

a. How many viewers were surveyed?

(continue on page 132)

b. Why are these bivariate data? Name the two variables. What type of variable is each one?

c. How many viewers preferred to watch CBS?

d. What percentage of the survey was Republican?

e. What percentage of the Democrats preferred ABC?

f. What percentage of the viewers were Republican and preferred PBS?

3.8 [EX03-008] Consider the accompanying contingency table, which presents the results of an advertising survey about the use of credit by Martan Oil Company customers.

	Number of Purchases at Gasoline Station Last Year					
Preferred Method of Payment	0–4	5–9	10–14	15–19	≥20	Sum
Cash	150	100	25	0	0	275
Oil-Company Card	50	35	115	80	70	350
Bank Credit Card	50	60	65	45	5	225
Sum	250	195	205	125	75	850

a. How many customers were surveyed?

b. Why are these bivariate data? What type of variable is each one?

c. How many customers preferred to use an oil-company card?

d. How many customers made 20 or more purchases last year?

e. How many customers preferred to use an oil-company card and made between five and nine purchases last year?

f. What does the 80 in the fourth cell in the second row mean?

3.9 [EX03-009] The June 2009 unemployment rates for eastern and western U.S. states were as follows:

State Unemployment Rates, June 2009

Eastern	8.0	10.6	10.1	7.3	9.2	11.0	12.1	7.2
Western	8.7	11.6	8.4	6.4	12.0	12.2	5.7	9.3

Source: U.S. Bureau of Labor Statistics

Display these rates as two dotplots using the same scale; compare means and medians.

3.10 [EX03-010] What effect does the minimum amount have on the interest rate being offered on 3-month certificates of deposit (CDs)? The following are advertised rates of return, y, for a minimum deposit of $500, $1000, $2500, $5000, or $10,000, x. (Note that x is in $100 and y is annual percentage rate of return.)

Min Deposit	Rate	Min Deposit	Rate	Min Deposit	Rate
100	0.95	25	1.00	25	0.75
100	1.24	50	1.00	10	0.75
10	1.24	100	1.00	100	0.70
10	1.15	5	1.00	5	0.64
100	1.10	10	1.00	10	0.50
50	1.09	10	0.80	100	0.35
100	1.07	10	0.75	25	0.35
5	1.00	10	0.75	5	0.99
25	0.75				

Source: Bankrate.com, July 28, 2009

a. Prepare a dotplot of the five sets of data using a common scale.

b. Prepare a 5-number summary and a boxplot of the five sets of data. Use the same scale for the boxplots.

c. Describe any differences you see among the three sets of data.

If you are using a computer or calculator for Ex. 3.10, try the commands on pages 126.

3.11 [EX03-011] Can a woman's height be predicted from her mother's height? The heights of some mother–daughter pairs are listed; x is the mother's height and y is the daughter's height.

x	63	63	67	65	61	63	61	64	62	63
y	63	65	65	65	64	64	63	62	63	64

x	64	63	64	64	63	67	61	65	64	65	66
y	64	64	65	65	62	66	62	63	66	66	65

a. Draw two dotplots using the same scale and showing the two sets of data side by side.

b. What can you conclude from seeing the two sets of heights as separate sets in part a? Explain.

c. Draw a scatter diagram of these data as ordered pairs.

d. What can you conclude from seeing the data presented as ordered pairs? Explain.

3.12 [EX03-012] The following tables list the ages, heights (in inches), and weights (in pounds) of the players on the 2009 roster for the National Hockey League teams Boston Bruins and Edmonton Oilers.

Boston Bruins			Edmonton Oilers		
Age	Height	Weight	Age	Height	Weight
31	72	193	22	70	180
24	72	186	22	69	178
23	71	176	19	70	191
32	70	195	24	71	183
22	71	194	24	71	190
32	71	209	24	72	190
35	74	186	24	73	195
34	73	175	23	71	200
21	76	220	30	73	202
30	72	195	24	76	217
25	77	215	28	78	265
25	72	192	33	74	220
21	72	189	26	76	243

Boston Bruins			Edmonton Oilers		
Age	Height	Weight	Age	Height	Weight
27	72	195	23	71	180
24	75	188	25	72	191
22	75	196	32	73	203
41	70	195	23	75	217
29	72	192	22	72	196
32	73	209	26	75	210
22	75	185	25	73	195
25	74	225	21	74	223
32	81	261	23	75	204
26	73	211	33	76	227
30	70	189	36	73	200
24	70	187	34	76	225
30	72	220	32	70	188
23	73	185	25	76	189
25	74	218	36	73	208
26	72	200			
34	72	207			
22	74	171			
25	73	190			
28	74	200			
35	71	182			

Source: http://sports.espn.go.com/

a.   Compare each of the three variables—height, weight, and age—using either a dotplot or a histogram (use the same scale).

b.   Based on what you see in the graphs in part a, can you detect a substantial difference between the two teams in regard to these three variables? Explain.

c.   Explain why the data, as used in part a, are not bivariate data.

3.13 Consider the two variables of a person's height and weight. Which variable, height or weight, would you use as the input variable when studying their relationship? Explain why.

3.14 Draw a coordinate axis and plot the points (0, 6), (3, 5), (3, 2), and (5, 0) to form a scatter diagram. Describe the pattern that the data show in this display.

3.15 Does studying for an exam pay off?

a.   Draw a scatter diagram of the number of hours studied, *x*, compared with the exam grade received, *y*.

x	2	5	1	4	2
y	80	80	70	90	60

b.   Explain what you can conclude based on the pattern of data shown on the scatter diagram drawn in part a. (Retain these solutions to use in Exercise 3.55, p. 157.)

3.16 Refer to Figure 3.8 in "Americans Love Their Automobiles" (Applied Example 3.4 on p. 128) to answer the following questions:

a.   Name the two variables used.

b.   Does the scatter diagram suggest a relationship between the two variables? Explain.

c.   What conclusion, if any, can you draw from the appearance of the scatter diagram?

3.17 Growth charts are commonly used by a child's pediatrician to monitor a child's growth. Consider the growth chart that follows.

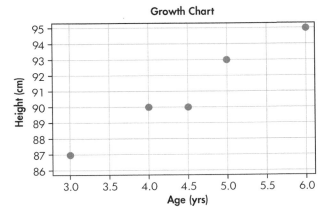

a.   What are the two variables shown in the graph?

b.   What information does the ordered pair (3, 87) represent?

c.   Describe how the pediatrician might use this chart and what types of conclusions might be based on the information displayed by it.

3.18 [EX03-012] a.   Draw a scatter diagram showing height, *x*, and weight, *y*, for the Boston Bruins hockey team, using the data in Exercise 3.12.

b.   Draw a scatter diagram showing height, *x*, and weight, *y*, for the Edmonton Oilers hockey team using the data in Exercise 3.12.

c.   Explain why the data, as used in parts a and b, are bivariate data.

FYI If you are using a computer or calculator, try the commands on page 129–130.

3.19 [EX03-019] The accompanying data show the number of hours, *x*, studied for an exam and the grade received, *y* (*y* is measured in tens; that is, *y* = 8 means that the grade, rounded to the nearest 10 points, is 80). Draw the scatter diagram. (Retain this solution to use in Exercise 3.37, p. 143.)

x	2	3	3	4	4	5	5	6	6	6	7	7	7	8	8
y	5	5	7	5	7	7	8	6	9	8	7	9	10	8	9

3.20 [EX03-020] An experimental psychologist asserts that the older a child is, the fewer irrelevant answers he or she will give during a controlled experiment. To investigate this claim, the following data were collected. Draw

(continue on page 134)

a scatter diagram. (Retain this solution to use in Exercise 3.38, p. 143.)

Age, x	2	4	5	6	6	7	9	9	10	12
Irr Answers, y	12	13	9	7	12	8	6	9	7	5

3.21 [EX03-021] A sample of 15 upper-class students who commute to classes was selected at registration. They were asked to estimate the distance (x) and the time (y) required to commute each day to class (see the following table).

Distance, x (nearest mile)	Time, y (nearest 5 minutes)	Distance, x (nearest mile)	Time, y (nearest 5 minutes)
18	20	2	5
8	15	15	25
20	25	16	30
5	20	9	20
5	15	21	30
11	25	5	10
9	20	15	20
10	25		

a. Do you expect to find a linear relationship between the two variables commute distance and commute time? If so, explain what relationship you expect.

b. Construct a scatter diagram depicting these data.

c. Does the scatter diagram in part b reinforce what you expected in part a?

3.22 [EX03-022] Refer to the 2009 4-wheel-drive, 6-cylinder SUVs chart in Applied Example 3.4 on page 128 and the two variables gas tank capacity, x, and the cost to fill it, y.

a. If you were to draw scatter diagrams of these two variables, on the same graph but separate, for the SUVs that use regular and premium gasoline, do you think the two sets of data would be distinguishable? Explain what you anticipate seeing.

b. Construct a scatter diagram of tank capacity, x, and fill-up cost, y, for the SUVs using regular gasoline.

c. Construct a scatter diagram of tank capacity, x, and fill-up cost, y, for the SUVs using premium gasoline on the scatter diagram for part b.

d. Are the two sets of data distinguishable?

e. How does your answer in part a compare to your answer in part d? Explain any difference.

3.23 [EX03-023] Baseball stadiums vary in age, style and size, and many other ways. Fans might think of the size of a stadium in terms of the number of seats, while players might measure the size of a stadium in terms of the distance from home plate to the centerfield fence.

Seats	CF	Seats	CF	Seats	CF
38,805	420	36,331	434	40,950	435
41,118	400	43,405	405	38,496	400
56,000	400	48,911	400	41,900	400
45,030	400	50,449	415	42,271	404
34,077	400	50,091	400	43,647	401
40,793	400	43,772	404	42,600	396
56,144	408	49,033	407	46,200	400
50,516	400	47,447	405	41,222	403
40,615	400	40,120	422	52,355	408
48,190	406	41,503	404	45,000	408

CF = distance from home plate to centerfield fence
Source: http://mlb.mlb.com

Is there a relationship between these two measurements of the "size" of the 30 Major League Baseball stadiums?

a. What do you think you will find? Bigger fields have more seats? Smaller fields have more seats? No relationship between field size and number of seats? A strong relationship between field size and number of seats? Explain.

b. Construct a scatter diagram.

c. Describe what the scatter diagram tells you, including a reaction to your answer in part a.

3.24 [EX03-024] Most adult Americans drive. But do you have any idea how many licensed drivers there are in each U.S. state? The table here lists the number of male and female drivers licensed in each of 15 randomly selected U.S. states during 2007.

Licensed Drivers per State (× 100,000)

Male	Female	Male	Female
17.92	17.10	59.07	54.62
5.18	5.10	2.38	2.33
21.24	21.85	15.01	16.26
10.03	10.15	75.98	75.86
14.52	14.82	8.32	8.20
15.91	15.59	25.26	23.53
3.74	3.62	2.05	1.93
6.77	6.89		

Source: Federal Highway Admin., U.S. Dept. of Transportation

a. Do you expect to find a linear (straight-line) relationship between number of male and number of female licensed drivers per state? How strong do you anticipate this relationship to be? Describe.

b. Construct a scatter diagram using x for the number of male drivers and y for the number of female drivers.

c. Compare the scatter diagram to your expectations in part a. How did you do? Explain.

d. Are there data points that look like they are separate from the pattern created by the rest of ordered pairs? If they were removed from the dataset, would the results change? What caused these point(s) to be separate from the others, but yet still be part of the extended pattern? Explain.

e. Use the dataset for all 51 states to construct a scatter diagram. Compare the pattern of the sample of 15 to the pattern shown by all 51. Describe in detail.

f. Did the sample provide enough information for you to understand the relationship between the two variables in this situation? Explain.

3.25 [EX03-025] Ronald Fisher, an English statistician (1890–1962), collected measurements for a sample of 150 irises. Of concern were five variables: species, petal width (PW), petal length (PL), sepal width (SW), and sepal length (SL) (all in mm). Sepals are the outermost leaves that encase the flower before it opens. The goal of Fisher's experiment was to produce a simple function that could be used to classify flowers correctly. A random sample of his complete dataset is given in the accompanying table.

Type	PW	PL	SW	SL	Type	PW	PL	SW	SL
0	2	15	35	52	1	24	51	28	58
2	18	48	32	59	1	19	50	25	63
1	19	51	27	58	0	1	15	31	49
0	3	13	35	50	1	23	59	32	68
0	3	15	38	51	2	13	44	23	63
2	12	44	26	55	2	15	42	30	59
1	20	64	38	79	1	25	57	33	67
2	15	49	31	69	1	21	57	33	67
2	15	45	29	60	0	2	15	37	54
2	12	39	27	58	1	18	49	27	63
1	22	56	28	64	1	17	45	25	49
1	13	52	30	67	1	24	56	34	63
0	2	14	29	44	0	2	14	36	50
2	16	51	27	60	2	10	50	22	60
0	5	17	33	51	0	2	12	32	50

a. Construct a scatter diagram of petal length, x, and petal width, y. Use different symbols to represent the three species.*

b. Construct a scatter diagram of sepal length, x, and sepal width, y. Use different symbols to represent the three species.

c. Explain what the scatter diagrams in parts a and b portray.

*In addition to using the commands on pages 129–130, use:
For MINITAB: Select: Scatterplot With Group
              Enter: Categorical variables for grouping: Type
For TI-83/84: Enter different groups into separate x, y columns. Use a separate Stat Plot and "Mark" for each group.

Let's see how well a random sample represents the data from which it was selected.

d. Repeat parts a and b using the dataset containing all 150 of Fisher's data on [EX03-025].

e. Aside from the fact that the scatter diagrams in parts a and b have fewer data, comment on the similarities and differences between the distributions shown for 150 data and for the 30 randomly selected data.

3.26 [EX03-026] Total solar eclipses actually take place nearly as often as total lunar eclipses, but the former are visible over a much narrower path. Both the path width and the duration vary substantially from one eclipse to the next. The table below shows the duration (in seconds) and path width (in miles) of 44 total solar eclipses measured in the past and those projected to the year 2010:

Date	Duration (s)	Width (mi)	Date	Duration (s)	Width (mi)
1950	73	83	1983	310	123
1952	189	85	1984	119	53
1954	155	95	1985	118	430
1955	427	157	1986	1	1
1956	284	266	1987	7	3
1958	310	129	1988	216	104
1959	181	75	1990	152	125
1961	165	160	1991	413	160
1962	248	91	1992	320	182
1963	99	63	1994	263	117
1965	315	123	1995	129	48
1966	117	52	1997	170	221
1968	39	64	1998	248	94
1970	207	95	1999	142	69
1972	155	109	2001	296	125
1973	423	159	2002	124	54
1974	308	214	2003	117	338
1976	286	123	2005	42	17
1977	157	61	2006	247	114
1979	169	185	2008	147	144
1980	248	92	2009	399	160
1981	122	67	2010	320	160

Source: *The World Almanac and Book of Facts 1998.*

a. Draw a scatter diagram showing duration, y, and path width, x, for the total solar eclipses.

b. How would you describe this diagram?

c. The durations and path widths for the years 2006–2009 were projections. The recorded values were:

Year	Path Width	Duration
2006	65 miles	247 sec
2008	147 miles	147 sec
2009	160 miles	399 sec

Compare the recorded values to the projections. Comment on accuracy.

3.2 Linear Correlation

The primary purpose of **linear correlation analysis** is to measure the strength of a linear relationship between two variables. Let's examine some scatter diagrams that demonstrate different relationships between input, or independent variables, x, and output, or dependent variables, y. If as x increases there is no definite shift in the values of y, we say there is **no correlation**, or no relationship between x and y. If as x increases there is a shift in the values of y, then there is a **correlation**. The correlation is **positive** when y tends to increase and **negative** when y tends to decrease. If the ordered pairs (x, y) tend to follow a straight-line path, there is a linear correlation. The preciseness of the shift in y as x increases determines the strength of the **linear correlation**. The scatter diagrams in Figure 3.9 demonstrate these ideas.

FIGURE 3.9

Scatter Diagrams and Correlation

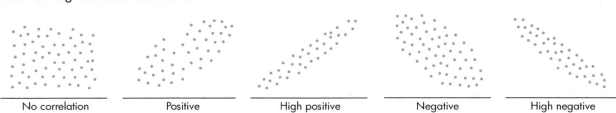

| No correlation | Positive | High positive | Negative | High negative |

Perfect linear correlation occurs when all the points fall exactly along a straight line, as shown in Figure 3.10. The correlation can be either positive or negative, depending on whether y increases or decreases as x increases. If the data form a straight horizontal or vertical line, there is no correlation, because one variable has no effect on the other, as also shown in Figure 3.7.

FIGURE 3.10

Ordered Pairs Forming a Straight Line

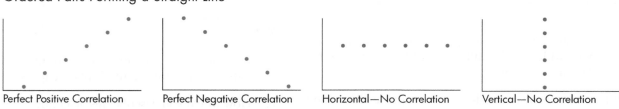

| Perfect Positive Correlation | Perfect Negative Correlation | Horizontal—No Correlation | Vertical—No Correlation |

FIGURE 3.11

No Linear Correlation

Scatter diagrams do not always appear in one of the forms shown in Figures 3.9 and 3.10. Sometimes they suggest relationships other than linear, as in Figure 3.11. There appears to be a definite pattern; however, the two variables are not related linearly, and therefore there is no linear correlation.

The **coefficient of linear correlation**, *r*, is the numerical measure of the strength of the linear relationship between two variables. The coefficient reflects the consistency of the effect that a change in one variable has on the other. The value of the linear correlation coefficient helps us answer the question: Is there a linear correlation between the two variables under consideration? The linear correlation coefficient, r, always has a value between -1 and $+1$. A value of $+1$ signifies a perfect positive correlation, and a value of -1 signifies a perfect negative correlation. If as x increases there is a general increase in the value of y, then r will be positive in value. For example, a positive value of r would be expected for the age and height of children because as children grow older, they grow taller. Also, consider the age, x, and resale value, y, of an automobile. As the car ages, its resale value decreases. Since as x increases, y decreases, the relationship results in a negative value for r.

The value of r is defined by **Pearson's product moment formula:**

Definition Formula

$$r = \frac{\sum (x - \bar{x})(y - \bar{y})}{(n - 1)s_x s_y}$$ (3.1)

Notes:

1. s_x and s_y are the standard deviations of the x- and y-variables.
2. The development of this formula is discussed in Chapter 13.

To calculate r, we will use an alternative formula, formula (3.2), that is equivalent to formula (3.1). As preliminary calculations, we will separately calculate three sums of squares and then substitute them into formula (3.2) to obtain r.

Computational Formula

$$\text{linear correlation coefficient} = \frac{\text{sum of squares for } xy}{\sqrt{(\text{sum of squares for } x)(\text{sum of squares for } y)}}$$

$$r = \frac{SS(xy)}{\sqrt{SS(x)SS(y)}}$$ (3.2)

FYI SS(x) is the numerator of the variance.

Recall the SS(x) calculation from formula (2.8) for sample variance (p. 77):

$$\text{sum of squares for } x = \text{sum of } x^2 - \frac{(\text{sum of } x)^2}{n}$$

$$SS(x) = \sum x^2 - \frac{\left(\sum x\right)^2}{n}$$ (2.8)

We can also calculate:

$$\text{sum of squares for } y = \text{sum of } y^2 - \frac{(\text{sum of } y)^2}{n}$$

$$SS(y) = \sum y^2 - \frac{\left(\sum y\right)^2}{n}$$ (3.3)

$$\text{sum of squares for } xy = \text{sum of } xy - \frac{(\text{sum of } x)(\text{sum of } y)}{n}$$

$$SS(xy) = \sum xy - \frac{\sum x \sum y}{n}$$ (3.4)

EXAMPLE 3.5

CALCULATING THE LINEAR CORRELATION COEFFICIENT, *r*

Find the linear correlation coefficient for the push-up/sit-up data in Example 3.3 (p. 127).

Solution

First, we construct an extensions table (Table 3.12) listing all the pairs of values (*x*, *y*) to aid us in finding x^2, *xy*, and y^2 for each pair and the five column totals.

TABLE 3.12 Extensions Table for Finding Five Summations [TA03-10]

Student	Push-ups, *x*	x^2	Sit-ups, *y*	y^2	*xy*
1	27	729	30	900	810
2	22	484	26	676	572
3	15	225	25	625	375
4	35	1,225	42	1,764	1,470
5	30	900	38	1,444	1,140
6	52	2,704	40	1,600	2,080
7	35	1,225	32	1,024	1,120
8	55	3,025	54	2,916	2,970
9	40	1,600	50	2,500	2,000
10	40	1,600	43	1,849	1,720
	$\Sigma x = 351$	$\Sigma x^2 = 13{,}717$	$\Sigma y = 380$	$\Sigma y^2 = 15{,}298$	$\Sigma xy = 14{,}257$
	sum of *x*	sum of x^2	sum of *y*	sum of y^2	sum of *xy*

Second, to complete the preliminary calculations, we substitute the five summations (the five column totals) from the extensions table into formulas (2.8), (3.3), and (3.4), and calculate the three sums of squares:

$$SS(x) = \Sigma x^2 - \frac{(\Sigma x)^2}{n} = 13{,}717 - \frac{(351)^2}{10} = 1396.9$$

FYI The Σ and SS values will be needed for regression in Section 3.3. Be sure to save them!

$$SS(y) = \Sigma y^2 - \frac{(\Sigma y)^2}{n} = 15{,}298 - \frac{(380)^2}{10} = 858.0$$

$$SS(xy) = \Sigma xy - \frac{\Sigma x \Sigma y}{n} = 14{,}257 - \frac{(351)(380)}{10} = 919.00$$

Third, we substitute the three sums of squares into formula (3.2) to find the value of the correlation coefficient:

$$r = \frac{SS(xy)}{\sqrt{SS(x)SS(y)}} = \frac{919.0}{\sqrt{(1396.9)(858.0)}} = 0.8394 = \mathbf{0.84}$$

Note: Typically, *r* is rounded to the nearest hundredth.

The value of the linear correlation coefficient helps us answer the question: Is there a linear correlation between the two variables under consideration? When the calculated value of *r* is close to zero, we conclude that there is little or no linear correlation. As the calculated value of *r* changes from 0.0 toward either +1.0 or −1.0, it indicates an increasing linear correlation between the two variables. From a graphic viewpoint, when we calculate *r*, we are measuring how well

FYI See this in action with Exercise 3.27 on page 142.

Animated tutorials available—logon and learn more at cengagebrain.com a straight line describes the scatter diagram of ordered pairs. As the value of r changes from 0.0 toward $+1.0$ or -1.0, the data points create a pattern that moves closer to a straight line.

TECHNOLOGY INSTRUCTIONS: CORRELATION COEFFICIENT

MINITAB	Input the x-variable data into C1 and the corresponding y-variable data into C2; then continue with: Choose: **Stat > Basic Statistics > Correlation . . .** Enter: **Variables: C1 C2 > OK**
Excel	Input the x-variable data into column A and the corresponding y-variable data into column B; activate a cell for the answer; then continue with: Choose: **Insert function, f_x > Statistical > CORREL > OK** Enter: **Array 1: x data range** **Array 2: y data range > OK**
TI-83/84 Plus	Input the x-variable data into L1 and the corresponding y-variable data into L2; then continue with: Choose: **2nd > CATALOG > DiagnosticOn* > ENTER > ENTER** Choose: **STAT > CALC > 8: LinReg(a + bx)** Enter: **L1, L2** *DiagnosticOn must be selected for r and r^2 to show. Once set, omit this step.

FIGURE 3.12
The Data Window

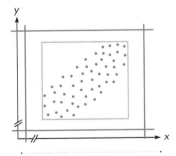

Understanding the Linear Correlation Coefficient

The following method will create (1) a visual meaning for correlation, (2) a visual meaning for what the linear coefficient is measuring, and (3) an estimate for r. The method is quick and generally yields a reasonable estimate when the "window of data" is approximately square.

Note: This estimation technique does not replace the calculation of r. It is very sensitive to the "spread" of the diagram. However, if the "window of data" is approximately square, this approximation will be useful as a mental estimate or check.

Procedure

1. Construct a scatter diagram of your data, being sure to scale the axes so that the resulting graph has an approximately square "window of data," as demonstrated in Figure 3.12 by the light green frame. The window may not be the same region as determined by the bounds of the two scales, shown as a green rectangle on Figure 3.12.

2. Lay two pencils on your scatter diagram. Keeping them parallel, move them to a position so that they are as close together as possible while having all the points on the scatter diagram between them. (See Figure 3.13.)

3. Visualize a rectangular region that is bounded by the two pencils and that ends just beyond the points on the scatter diagram. (See the shaded portion of Figure 3.13.)

FIGURE 3.13
Focusing on Pattern

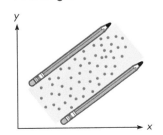

FIGURE 3.14
Finding *k*

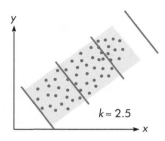

$k \approx 2.5$

4. Estimate the number of times longer the rectangle is than it is wide. An easy way to do this is to mentally mark off squares in the rectangle. (See Figure 3.14.) Call this number of multiples *k*.

5. The value of *r* may be estimated as $\pm \left(1 - \frac{1}{k}\right)$.

6. The sign assigned to *r* is determined by the general position of the length of the rectangular region. If it lies in an increasing position, *r* will be positive; if it lies in a decreasing position, *r* will be negative (see Figure 3.15). If the rectangle is in either a horizontal or a vertical position, then *r* will be zero, regardless of the length–width ratio.

FIGURE 3.15
(a) Increasing Position;   (b) Decreasing Position

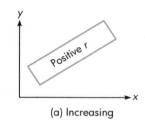   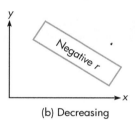

(a) Increasing   (b) Decreasing

Let's use this method to estimate the value of the linear correlation coefficient for the relationship between the number of push-ups and sit-ups. As shown in Figure 3.16, we find that the rectangle is approximately 3.5 times longer than it is wide—that is, $k \approx 3.5$—and the rectangle lies in an increasing position. Therefore, our estimate for *r* is

$$r \approx +\left(1 - \frac{1}{3.5}\right) \approx +0.70$$

FIGURE 3.16
Push-ups versus Sit-ups for
10 Students

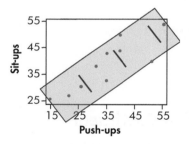

Causation and Lurking Variables

As we try to explain the past, understand the present, and estimate the future, judgments about cause and effect are necessary because of our desire to impose order on our environment.

The **cause-and-effect relationship** is fairly straightforward. You may focus on a situation, the *effect* (e.g., a disease or social problem), and try to determine its *cause(s)*, or you may begin with a *cause* (unsanitary conditions or poverty) and discuss its *effect(s)*. To determine the cause of something, ask yourself why it happened. To determine the effect, ask yourself **what** happened.

> **Lurking variable** A variable that is not included in a study but has an effect on the variables of the study and makes it appear that those variables are related.

A good example is the strong positive relationship shown between the amount of damage caused by a fire and the number of firefighters who work the fire. The "size" of the fire is the lurking variable; it "causes" both the "amount" of damage and the "number" of firefighters.

If there is a strong linear correlation between two variables, then one of the following situations may be true about the relationship between the two variables:

1. There is a direct cause-and-effect relationship between the two variables.
2. There is a reverse cause-and-effect relationship between the two variables.

3. Their relationship may be caused by a third variable.
4. Their relationship may be caused by the interactions of several other variables.
5. The apparent relationship may be strictly a coincidence.

Remember that a strong correlation does not necessarily imply causation.
Here are some pitfalls to avoid:

1. In a direct cause-and-effect relationship, an increase (or decrease) in one variable causes an increase (or decrease) in another. Suppose there is a strong positive correlation between weight and height. Does an increase in weight *cause* an increase in height? Not necessarily. Or to put it another way, does a decrease in weight *cause* a decrease in height? Many other possible variables are involved, such as gender, age, and body type. These other variables are called *lurking variables*.
2. In Applied Example 3.4 (p. 128), a positive correlation existed between the gas tank capacity and the cost of the fill-up. If we had one of the SUVs with a smaller gas tank that cost less to fill, would this save us money on gas?
3. Don't reason from *correlation* to *cause:* Just because all people who move to the city get old doesn't mean that the city *causes* aging. The city may be a factor, but you can't base your argument on the correlation.

APPLIED EXAMPLE 3.6

LIFE INSURANCE RATES

Does a high linear correlation coefficient, *r*, imply that the data are linear in nature? The issue age of the insured and the monthly life insurance rate for nontobacco users appears highly correlated looking at the chart presented here. As the issue age increases, the monthly rate for insurance increases for each of the genders.

© iStockphoto.com

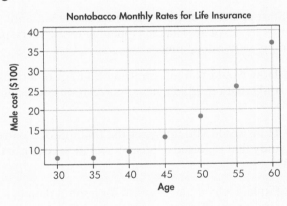

Nontobacco Monthly Rates for Life Insurance

TABLE 3.13 Nontobacco Monthly Rates for Life Insurance [TA03-13]

	$100,000		$250,000		$500,000	
Issue Age	Male ($)	Female ($)	Male ($)	Female ($)	Male ($)	Female ($)
30	7.96	6.59	11.96	9.13	19.25	12.46
35	8.05	6.56	11.96	9.13	19.57	12.46
40	9.63	7.79	15.22	10.89	23.19	16.47
45	13.14	9.80	22.40	15.44	35.87	24.03
50	18.44	12.42	33.69	21.10	53.81	33.38
55	26.01	15.75	49.22	29.37	87.59	48.06
60	37.10	20.83	74.59	42.05	137.38	69.87

Source: http://www.reliaquote.com/
All of the rates listed are for each carrier's best nontobacco classifications.

Let's consider the issue age of the insured and the male monthly rate for a $100,000 policy. The calculated correlation coefficient for this specific class of insurance results in a value of $r = 0.932$. Typically, a value of r this close to 1.0 would indicate a fairly strong straight-line relationship; but wait. Do we have a linear relationship? Only a scatter diagram can tell us that.

The scatter diagram clearly shows a non-straight-line pattern. Yet, the correlation coefficient was so high. It is the elongated pattern in the data that produces a calculated r so large. The lesson from this example is that one should always begin with a scatter diagram when considering linear correlation. The correlation coefficient tells only one side of the story!

SECTION 3.2 EXERCISES

3.27 Skillbuilder Applet Exercise provides scatter diagrams for various correlation coefficients.

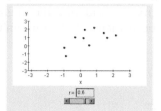

a. Starting at $r = 0$, move the slider to the right until $r = 1$. Explain what is happening to the corresponding scatter diagrams.

b. Starting at $r = 0$, move the slider to the left until $r = -1$. Explain what is happening to the corresponding scatter diagrams.

3.28 How would you interpret the findings of a correlation study that reported a linear correlation coefficient of -1.34?

3.29 How would you interpret the findings of a correlation study that reported a linear correlation coefficient of $+0.37$?

3.30 Explain why it makes sense for a set of data to have a correlation coefficient of zero when the data show a very definite pattern, as in Figure 3.11 (p. 136).

3.31 Does studying for an exam pay off? The number of hours studied, x, is compared with the exam grade received, y:

x	2	5	1	4	2
y	80	80	70	90	60

a. Complete the preliminary calculations: extensions, five sums, SS(x), SS(y), and SS(xy).

b. Find r.

3.32 [EX03-032] Cell phones and iPods are necessities for the current generation. Does the use of one indicate the use of the other? Seven junior high students who own both a cell phone and an iPod were randomly selected, resulting in the following data:

Cell, n (phone #s)	42	7	75	78	126	22	23
iPod, n (songs saved)	303	212	401	500	536	200	278

a. Complete the preliminary calculations: extensions, five sums, and SS(x), SS(y), and SS(xy).

b. Find r.

3.33 [EX03-033] Many organizations offer "special" magazine rates to their members. The American Federation of Teachers is no different, and here are a few of the rates they offer their members.

Magazine	Usual Rate	Your Price
Cosmopolitan	29.97	18.00
Sports Illustrated	89.04	39.95
Time	59.95	29.95
Rolling Stone	25.94	14.95
Martha Stewart Living	28.00	24.00

Source: AFT, Feb. 2009

a. Construct a scatter diagram with "your price" as the dependent variable, y, and "usual rate" as the independent variable, x.

Find:

b. SS(x)

c. SS(y)

d. SS(xy)

e. Pearson's product moment, r

3.34 [EX03-034] A random sample of 10 seventh-grade students gave the following data on x = number of minutes watching television on an average weeknight versus of minutes spent on homework on an average weeknight.

Row	Television	Homework	Row	Television	Homework
1	15	50	6	90	35
2	120	30	7	120	20
3	50	30	8	20	60
4	40	60	9	10	45
5	60	40	10	60	25

a.   Construct a scatter diagram with "homework minutes" as the dependent variable, y, and "television minutes" as the independent variable, x.

Find:

b. $SS(x)$

c. $SS(y)$

d. $SS(xy)$

e.   Pearson's product moment, r

3.35 Manatees swim near the surface of the water. They often run into trouble with the many powerboats in Florida. Consider the graph that follows.

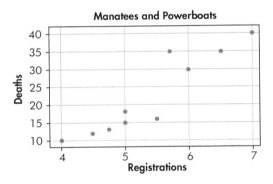

Manatees and Powerboats

a.   What two groups of subjects are being compared?

b.   What two variables are being used to make the comparison?

c.   What conclusion can one make based on this scatterplot?

d.   What might you do if you were a wildlife official in Florida?

3.36 Estimate the correlation coefficient for each of the following:

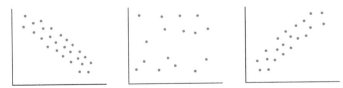

3.37 a.   Use the scatter diagram you drew in Exercise 3.19 (p. 133) to estimate r for the sample data on the number of hours studied and the exam grade.

b.   Calculate r.

3.38 a.   Use the scatter diagram you drew in Exercise 3.20 (p. 133–134) to estimate r for the sample data on the number of irrelevant answers and the child's age.

b.   Calculate r.

FYI Have you tried to use the correlation commands on your computer or calculator?

3.39 [EX03-039] A marketing firm wished to determine whether the number of television commercials broadcast was linearly correlated with the sales of its product. The data, obtained from each of several cities, are shown in the following table.

City	A	B	C	D	E	F	G	H	I	J
Commercials, x	12	6	9	15	11	15	8	16	12	6
Sales Units, y	7	5	10	14	12	9	6	11	11	8

a.   Draw a scatter diagram.   b.   Estimate r.

c.   Calculate r.

3.40 [EX03-040] Movie production companies spend millions of dollars to produce movies, with the great hope of attracting millions of people to the theater. The success of a movie can be measured in many ways, two of which are box office receipts and the number of Oscar nominations received. Below is a list of ten 2008 movies with their "report card." Each movie is measured by its budget cost (in milllions of dollars), its box office receipts (in millions of dollars), and the number of Oscar nominations it received.

Movie	Budget	Box Office	Nominations
The Curious Case of Benjamin Button	150	127.5	13
Slumdog Millionaire	15	141.3	10
Milk	20	31.8	8
The Dark Knight	185	533.3	8
WALL-E	180	223.8	6
Frost/Nixon	25	18.6	5
The Reader	32	34.2	5
Doubt	20	33.4	5
Changeling	55	35.7	3
The Wrestler	6	26.2	2

Source: http://www.boxofficemojo.com/

a.   Draw a scatter diagram using x = budget and y = box office.

b.   Does there appear to be a linear relationship?

c.   Calculate the linear correlation coefficient, r.

d.   What does this value of correlation seem to be telling us? Explain.

e.   Repeat questions a through d using x = box office and y = nominations.

3.41 Skillbuilder Applet Exercise matches correlation coefficients with their scatter diagrams. After several practice rounds using "New Plots," explain your method of matching.

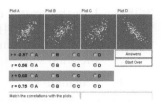

3.42 Skillbuilder Applet Exercise provide practice in constructing scatter diagrams to match given correlation coefficients.

a. After placing just 2 points, what is the calculated r value for each scatter diagram? Why?

b. Which scatter diagram did you find easier to construct?

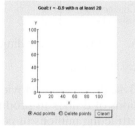

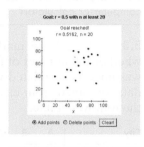

3.43 [EX03-043] Consider the following 2009 4WD, 6-Cyl SUV data.

2009 4WD, 6-Cyl SUVs

Manufacturer	Model	Petro	Tons
Buick	Enclave	18.0	9.6
Chevrolet	Trailblazer	21.4	11.4
Chrysler	Aspen	22.8	12.2
Dodge	Durango	22.8	12.2
Ford	Escape	17.1	9.2
GMC	Envoy	21.4	11.4
Honda	Pilot	19.0	10.2
Jeep	Grd Cherokee	20.1	10.8
Kia	Sportage	17.1	9.2
Lexus	RX 350	18.0	9.6
Lincoln	MKX	18.0	9.6
Mazda	CX-7	19.0	10.2
Mercury	Mountaineer	22.8	12.2
Mitsubishi	Outlander	18.0	9.6
Nissan	Murano	17.1	9.2
Toyota	RAV4	16.3	8.7

a. What value do you anticipate for a correlation coefficient of the two variables annual petroleum consumption in barrels, x, and annual tons of CO_2 emitted, y? Explain.

b. Calculate the linear correlation coefficient for the two variables annual petroleum consumption in barrels, x, and annual tons of CO_2 emitted, y.

c. Is the value found in part b approximately what you anticipated in part a? Explain why or why not.

d. Does it make sense for the data to demonstrate such a high correlation? If the amount of

consumption doubles, what do you think will happen to the tons of CO_2 emitted? Be specific in your explanation.

3.44 [EX03-044] The Children's Bureau of the U.S. Department of Health and Human Services has a monumental job. In 2006, 510,000 children were in foster care. Of those, approximately 51,000 were adopted. Are more males or more females typically adopted? Is there a difference? The table lists the number of males and females adopted in each of 16 randomly identified states.

State	Males	Females	State	Males	Females
Delaware	50	44	Wyoming	27	30
Nevada	231	213	New Jersey	689	636
Alabama	190	197	Arkansas	178	217
Michigan	1296	1296	Idaho	580	603
South Carolina	203	220	Hawaii	202	195
Iowa	512	472	Washington	586	610
Georgia	660	586	Tennessee	497	497
Vermont	90	74	Alaska	112	100

Source: Children's Bureau, Administration for Children and Families, U.S. Department of Health and Human Services, 2006

Is there a linear relationship between the number of males and females adopted from foster care during 2006? Use graphic and numerical statistics to support your answer.

3.45 [EX03-045] Sports drinks are very popular in today's culture around the world. The following table lists 10 different products you can buy in England and the values for three variables: cost per serving (in pence), energy per serving (in kilocalories), and carbohydrates per serving (in grams).

Sports Drink	Cost	Energy	Carbs
Lucozade Sport RTD 330ml pouch/can	72	92	21.1
Lucozade Sport RTD 500ml bot.	79	140	32
Lucozade Sport RTD 650ml sports bot.	119	182	41.6
POWERade 500ml bot.	119	120	30
Gatorade Sports 750ml	89	188	45
Science in Sport Go Electrolyte (500ml)	99	160	40
High Five Isotonic electrolyte (750ml)	99	220	55
Isostar powder (per litre) 5l tub	126	320	77
Isostar RTD 500ml bot.	99	150	35
Maxim Electrolyte (per litre) 2kg bag	66	296	75

Note: Cost is in pence (p), 0.01 of a British pound, worth $0.0187 on March 28, 2005
Source: http://www.simplyrunning.net

a. Draw a scatter diagram using x = carbs/serving and y = energy/serving.

b. Does there appear to be a linear relationship?

c. Calculate the linear correlation coefficient, r.

d. What does this value of correlation seem to be telling us? Explain.

e. Repeat parts a through d using x = cost/serving and y = energy/serving. (Retain these solutions to use in Exercise 3.59, p. 157.)

3.46 [EX03-046] During the 2008 MLB All Star Game Home Run Contest, Josh Hamilton put on an amazing show with his 35 home runs. The recorded Apex and Distance of each home run he hit is listed here:

Apex—The highest point reached by the ball in flight above field level, in feet.

StdDist, Standard Distance—The estimated distance in feet the home run would have traveled if it had flown uninterrupted all the way down to field level. Standard distance factors out the influences of wind, temperature, and altitude, and is thus the best way of comparing home runs hit under a variety of different conditions

Apex	100	114	145	45	98	130	105	94	59
StdDist	459	474	404	378	479	443	393	410	356

Apex	112	50	144	154	153	132	126	123	118
StdDist	430	390	411	418	423	455	421	464	440

Apex	70	152	95	48	162	117	54	110	88
StdDist	432	435	447	386	364	447	379	423	442

Apex	125	47	119	111	84	155	153	116
StdDist	428	387	453	401	387	445	426	463

Source: http://www.hittrackeronline.com

a. Construct a scatter diagram using apex as x and standard distance as y.

b. Do the points seem to suggest a linear pattern? Explain.

c. Does it appear that the apex for the flight of a home run will be useful in predicting the home run's length? Explain, giving at least one reason that is nonstatistical and at least one reason that is statistical.

d. What other factor about the flight of a home run might cause the pattern of points to be so varied?

e. Estimate the value of the linear correlation coefficient.

f. Calculate the correlation coefficient.

3.47 [EX03-047] NBA players, teams, and fans are interested in seeing their leading scorers score lots of points, yet at the same time the number of personal fouls they commit tends to limit their playing time. For the leading scorer on each team, the table lists the number of minutes per game, MPG, and the number of personal fouls committed per game, PFPG, during the 2008/2009 NBA season.

Team	MPG	PFPG	Team	MPG	PFPG
Hawks	39.6	2.23	Bucks	36.4	1.36
Celtics	37.5	2.65	Timberwolves	36.7	2.82
Hornets	37.6	2.96	Nets	36.1	2.38
Bulls	36.6	2.24	Hornets	38.5	2.72
Cavaliers	37.7	1.72	Knicks	29.8	2.78
Mavericks	37.7	2.17	Thunder	39.0	1.81
Nuggets	34.5	2.95	Magic	35.7	3.42
Pistons	34.0	2.63	76ers	39.9	1.85
Warriors	39.6	2.59	Suns	36.8	3.08
Rockets	33.6	3.34	Blazers	37.2	1.63
Pacers	36.2	3.09	Kings	38.2	2.27
Clippers	37.4	3.20	Spurs	34.1	1.53
Lakers	36.1	2.30	Raptors	38.0	2.45
Grizzlies	37.3	2.80	Jazz	36.8	1.97
Heat	38.6	2.25	Wizards	38.2	2.60

Source: NBA.com

a. Construct a scatter diagram.

b. Describe the pattern displayed. Are there any unusual characteristics displayed?

c. Calculate the correlation coefficient.

d. Does the value of the correlation coefficient seem reasonable?

3.48 [EX03-048] Ever wanted to know the weight of your catch, but didn't have a scale? Measure a Muskellunge from the snout to the tip of the tail. The following weights are averages taken from fish collected by DEC fish management crews from across New York State.

Length in.	Muskie lb	Muskie oz	Length in.	Muskie lb	Muskie oz
30	7	4	41	20	7
31	8	1	42	22	2
32	8	15	43	23	15
33	9	15	44	25	14
34	11	0	45	27	14
35	12	1	46	30	0
36	13	4	47	32	3
37	14	8	48	34	8
38	15	14	49	37	0
39	17	5	50	39	9
40	18	13	51	40	4

Source: New York Freshwater Fishing, 2008–2009 *Official Regulations Guide*

© iStockphoto.com/Andrew Hyslop

a. Examine the data and find an approximate pattern for weight gain by inch length for each kind of fish.

b. Explain why the weights cannot be used as given, specifically why 9 lb 4 oz is not 9.4 lbs. Fix the weights so that they are expressed in terms of one unit of measure.

c. Construct a scatter diagram for Muskellunge lengths and weights.

d. Do the points seem to follow a straight line? Explain.

e. What is it about the fish being longer that causes the path of points to be concave up?

f. Calculate the linear correlation coefficient.

(continue on page 146)

g.   Explain why the value of r is so close to 1.0 and yet, graphically, the data do not appear to be linear.

3.49 In many communities there is a strong positive correlation between the amount of ice cream sold in a given month and the number of drownings that occur in that month. Does this mean that ice cream causes drowning? If not, can you think of an alternative explanation for the

strong association? Write a few sentences addressing these questions.

3.50 Explain why one would expect to find a positive correlation between the number of fire engines that respond to a fire and the amount of damage done in the fire. Does this mean that the damage would be less extensive if fewer fire engines were dispatched? Explain.

3.3 Linear Regression

Although the correlation coefficient measures the strength of a linear relationship, it does not tell us about the mathematical relationship between the two variables. In Section 3.2, the correlation coefficient for the push-up/sit-up data was found to be 0.84 (see p. 138). This along with the pattern on the scatter diagram imply that there is a linear relationship between the number of push-ups and the number of sit-ups a student does. However, the correlation coefficient does not help us predict the number of sit-ups a person can do based on knowing that he or she can do 28 push-ups. **Regression analysis** finds the equation of the line that best describes the relationship between two variables. One use of this equation is to make predictions. We make use of these predictions regularly—for example, predicting the success a student will have in college based on high school results and predicting the distance required to stop a car based on its speed. Generally, the exact value of y is not predictable, and we are usually satisfied if the predictions are reasonably close.

The relationship between two variables will be an algebraic expression describing the mathematical relationship between x and y. Here are some examples of various possible relationships, called *models* or **prediction equations:**

Linear (straight-line): $\hat{y} = b_0 + b_1 x$

Quadratic: $\hat{y} = a + bx + cx^2$

Exponential: $\hat{y} = a(b^x)$

Logarithmic: $\hat{y} = a \log_b x$

Figures 3.17, 3.18, and 3.19 show patterns of bivariate data that appear to have a relationship, whereas in Figure 3.20 the variables do not seem to be related.

If a straight-line model seems appropriate, the best-fitting straight line is found by using the **method of least squares.** Suppose that $\hat{y} = b_0 + b_1 x$ is the equation of a straight line, where $\hat{y}$ (read "y-hat") represents the **predicted value of y** that corresponds to a particular value of x. The **least squares criterion** requires that we find the constants b_0 and b_1 such that $\sum(y - \hat{y})^2$ is as small as possible.

FIGURE 3.17

Linear Regression with
Positive Slope

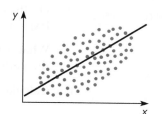

FIGURE 3.18

Linear Regression with
Negative Slope

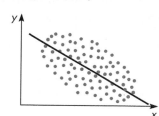

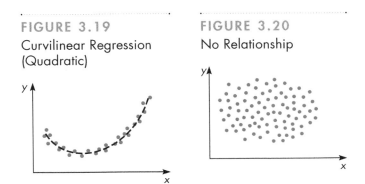

FIGURE 3.19

Curvilinear Regression
(Quadratic)

FIGURE 3.20

No Relationship

Figure 3.21 shows the distance of an observed value of y from a **predicted value of $\hat{y}$.** The length of this distance represents the value $(y - \hat{y})$ (shown as the red line segment in Figure 3.21). Note that $(y - \hat{y})$ is positive when the point (x, y) is above the line and negative when (x, y) is below the line (as shown on figure).

Figure 3.22 shows a scatter diagram with what appears to be the **line of best fit**, along with 10 individual $(y - \hat{y})$ values. (Positive values are shown in red; negative, in green.) The sum of the squares of these differences is minimized (made as small as possible) if the line is indeed the line of best fit.

Figure 3.23 shows the same data points as Figure 3.22. The 10 individual values of $(y - \hat{y})$ are plotted with a line that is definitely not the line of best fit. [The value of $\Sigma(y - \hat{y})^2$ is 149, much larger than the 23 from Figure 3.22.] Every different line drawn through this set of 10 points will result in a different value for $\Sigma(y - \hat{y})^2$. Our job is to find the one line that will make $\Sigma(y - \hat{y})^2$ the smallest possible value.

FIGURE 3.21

Observed and Predicted
Values of y

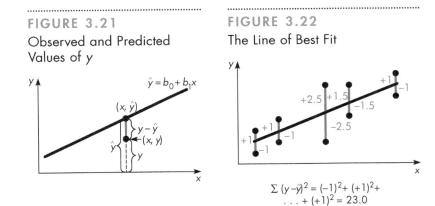

FIGURE 3.22

The Line of Best Fit

$$\Sigma (y - \hat{y})^2 = (-1)^2 + (+1)^2 + \ldots + (+1)^2 = 23.0$$

FIGURE 3.23

Not the Line of Best Fit

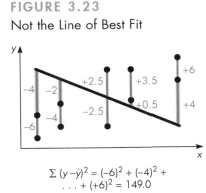

$$\Sigma (y - \hat{y})^2 = (-6)^2 + (-4)^2 + \ldots + (+6)^2 = 149.0$$

The equation of the line of best fit is determined by its **slope** (b_1) and its **y-intercept (b_0)**. (See the *Student Solutions Manual* for a review of the concepts of slope and intercept of a straight line.) The values of the constants—slope and y-intercept—that satisfy the least squares criterion are found by using the formulas presented next:

Definition Formula

$$\text{slope:} \quad b_1 = \frac{\Sigma(x - \bar{x})(y - \bar{y})}{\Sigma(x - \bar{x})^2} \tag{3.5}$$

We will use a mathematical equivalent of formula (3.5) for the slope, b_1, that uses the sums of squares found in the preliminary calculations for correlation:

Computational Formula

$$\text{slope:} \quad b_1 = \frac{SS(xy)}{SS(x)} \qquad (3.6)$$

Notice that the numerator of formula (3.6) is the $SS(xy)$ formula (3.4) (p. 137) and the denominator is formula (2.8) (p. 77) from the correlation coefficient calculations. Thus, if you have previously calculated the linear correlation coefficient using the procedure outlined on page 138 you can easily find the slope of the line of best fit. If you did not previously calculate r, set up a table similar to Table 3.12 (p. 138) and complete the necessary preliminary calculations.

For the y-intercept, we have:

Computational Formula

$$y\text{-intercept} = \frac{(sum\,of\,y) - [(slope)(sum\,of\,x)]}{number}$$

$$b_0 = \frac{\Sigma y - (b_1 \cdot \Sigma x)}{n} \qquad (3.7)$$

Alternative Computational Formula

$$y\text{-intercept} = y\text{-bar} - (slope \cdot x\text{-bar})$$

$$b_0 = \bar{y} - (b_1 \cdot \bar{x}) \qquad (3.7a)$$

Now let's consider the data in Example 3.3 (p. 127) and the question of predicting a student's number of sit-ups based on the number of push-ups. We want to find the line of best fit, $\hat{y} = b_0 + b_1x$. The preliminary calculations have already been completed in Table 3.12 (p. 138). To calculate the slope, b_1, using formula (3.6), recall that $SS(xy) = 919.0$ and $SS(x) = 1396.9$. Therefore,

$$\text{slope:} \quad b_1 = \frac{SS(xy)}{SS(x)} = \frac{919.0}{1396.9} = 0.6579 = \mathbf{0.66}$$

To calculate the y-intercept, b_0, using formula (3.7), recall that $\Sigma x = 351$ and $\Sigma y = 380$ from the extensions table. We have

$$y\text{-intercept:} \quad b_0 = \frac{\Sigma y - (b_1 \cdot \Sigma x)}{n} = \frac{380 - (0.6579)(351)}{10}$$

$$= \frac{380 - 230.9229}{10} = 14.9077 = \mathbf{14.9}$$

By placing the two values just found into the model $\hat{y} = b_0 + b_1x$, we get the equation of the line of best fit:

$$\hat{y} = \mathbf{14.9 + 0.66}x$$

Animated tutorial available—logon and learn more at cengagebrain.com

Notes:

1. Remember to keep at least three extra decimal places while doing the calculations to ensure an accurate answer.
2. When rounding off the calculated values of b_0 and b_1, always keep at least two significant digits in the final answer.

Now that we know the equation for the line of best fit, let's draw the line on the scatter diagram so that we can see the relationship between the line and the data. We need two points in order to draw the line on the diagram. Select two convenient x values, one near each extreme of the domain ($x = 10$ and $x = 60$ are good choices for this illustration), and find their corresponding y values.

$$\text{For } x = 10: \hat{y} = 14.9 + 0.66x = 14.9 + 0.66(10) = 21.5; \quad \textbf{(10, 21.5)}$$
$$\text{For } x = 60: \hat{y} = 14.9 + 0.66x = 14.9 + 0.66(60) = 54.5; \quad \textbf{(60, 54.5)}$$

These two points, $(10, 21.5)$ and $(60, 54.5)$, are then located on the scatter diagram (we use a purple $+$ to distinguish them from data points) and the line of best fit is drawn (shown in red in Figure 3.24).

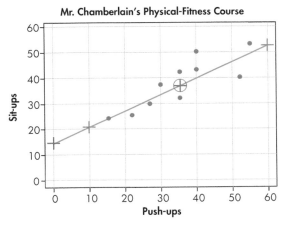

FIGURE 3.24

Line of Best Fit for Push-ups versus Sit-ups

There are some additional facts about the least squares method that we need to discuss.

1. The slope, b_1, represents the predicted change in y per unit increase in x. In our example, where $b_1 = 0.66$, if a student can do an additional 10 push-ups (x), we predict that he or she would be able to do approximately 7 (0.66×10) additional sit-ups (y).
2. The y-intercept is the value of y where the line of best fit intersects the y-axis. (When the vertical scale is located above $x = 0$, the y-intercept is easily seen on the scatter diagram, shown as a blue $+$ in Figure 3.24.) First, however, in interpreting b_0, you must consider whether $x = 0$ is a realistic x value before you can conclude that you would predict $\hat{y} = b_0$ if $x = 0$. To predict that if a student did no push-ups, he or she would still do approximately 15 sit-ups ($b_0 = 14.9$) is probably incorrect. Second, the x value of zero may be outside the domain of the data on which the regression line is based. In predicting y based on an x value, check to be sure that the x value is within the domain of the x values observed.
3. The line of best fit will always pass through the *centroid*, the point $(\bar{x}, \bar{y})$. When drawing the line of best fit on your scatter diagram, use this point as a check. For our illustration,

$$\bar{x} = \frac{\Sigma x}{n} = \frac{351}{10} = 35.1, \quad \bar{y} = \frac{\Sigma y}{n} = \frac{380}{10} = 38.0$$

We see that the line of best fit does pass through $(\bar{x}, \bar{y}) = (35.1, 38.0)$, as shown in green $\oplus$ in Figure 3.24.

Let's work through another example to clarify the steps involved in regression analysis.

EXAMPLE 3.7

CALCULATING THE LINE OF BEST FIT EQUATION

In a random sample of eight college women, each woman was asked her height (to the nearest inch) and her weight (to the nearest 5 pounds). The data obtained are shown in Table 3.14. Find an equation to predict the weight of a college woman based on her height (the equation of the line of best fit), and draw it on the scatter diagram in Figure 3.25.

TABLE 3.14 College Women's Heights and Weights [TA03-14]

	1	2	3	4	5	6	7	8
Height, x	65	65	62	67	69	65	61	67
Weight, y	105	125	110	120	140	135	95	130

Solution

Before we start to find the equation for the line of best fit, it is often helpful to draw the scatter diagram, which provides visual insight into the relationship between the two variables. The scatter diagram for the data on the heights and weights of college women, shown in Figure 3.25, indicates that the linear model is appropriate.

FIGURE 3.25
Scatter Diagram

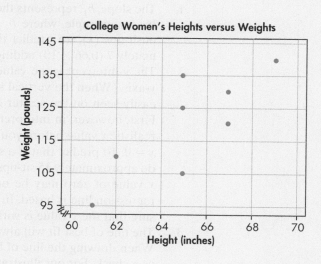

To find the equation for the line of best fit, we first need to complete the preliminary calculations, as shown in Table 3.15 The other preliminary calculations include finding SS(x) from formula (2.8) and SS(xy) from formula (3.4):

TABLE 3.15 Preliminary Calculations Needed to Find b_1 and b_0

Student	Height, x	x^2	Weight, y	xy
1	65	4225	105	6825
2	65	4225	125	8125
3	62	3844	110	6820
4	67	4489	120	8040
5	69	4761	140	9660
6	65	4225	135	8775
7	61	3721	95	5795
8	67	4489	130	8710
	$\sum x = 521$	$\sum x^2 = 33{,}979$	$\sum y = 960$	$\sum xy = 62{,}750$

$$SS(x) = \sum x^2 - \frac{(\sum x)^2}{n} = 33{,}979 - \frac{(521)^2}{8} = 48.875$$

$$SS(xy) = \sum xy - \frac{\sum x \sum y}{n} = 62{,}750 - \frac{(521)(960)}{8} = 230.0$$

Second, we need to find the slope and the y-intercept using formulas (3.6) and (3.7):

slope: $b_1 = \dfrac{SS(xy)}{SS(x)} = \dfrac{230.0}{48.875} = 4.706 = \mathbf{4.71}$

y-intercept: $b_0 = \dfrac{\sum y - (b_1 \cdot \sum x)}{n} = \dfrac{960 - (4.706)(521)}{8} = -186.478 = \mathbf{-186.5}$

Thus, the equation of the line of best fit is $\hat{y} = \mathbf{-186.5 + 4.71x}$.

To draw the line of best fit on the scatter diagram, we need to locate two points. Substitute two values for x—for example, 60 and 70—into the equation for the line of best fit to obtain two corresponding values for $\hat{y}$:

$$\hat{y} = -186.5 + 4.71x = -186.5 + (4.71)(60) = -186.5 + 282.6 = 96.1 \approx 96$$
$$\hat{y} = -186.5 + 4.71x = -186.5 + (4.71)(70) = -186.5 + 329.7 = 143.2 \approx 143$$

The values (60, 96) and (70, 143) represent two points (designated by a red + in Figure 3.26) that enable us to draw the line of best fit.

FIGURE 3.26

Scatter Diagram with Line of Best Fit

Note: In Figure 3.26, $(\bar{x}, \bar{y}) = (65.1, 120)$ is also on the line of best fit. It is the green $\oplus$. Use $(\bar{x}, \bar{y})$ as a check on your work.

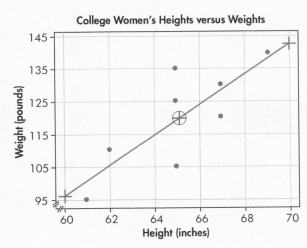

College Women's Heights versus Weights

Making Predictions

One of the main reasons for finding a regression equation is to make predictions. Once a linear relationship has been established and the value of the input variable x is known, we can predict a value of y, $\hat{y}$. Consider the equation $\hat{y} = -186.5 + 4.71x$ relating the height and weight of college women. If a particular female college student is 66 inches tall, what do you predict her weight to be? The predicted value is

$$\hat{y} = -186.5 + 4.71x = -186.5 + (4.71)(66) = -186.5 + 310.86$$
$$= 124.36 \approx 124 \text{ lb}$$

You should not expect this predicted value to occur exactly; rather, it is the average weight you would expect for all female college students who are 66 inches tall.

When you make predictions based on the line of best fit, observe the following restrictions:

1. The equation should be used to make predictions only about the population from which the sample was drawn. For example, using our relationship between the height and the weight of college women to predict the weight of professional athletes given their height would be questionable.
2. The equation should be used only within the sample domain of the input variable. We know that the data demonstrate a linear trend within the domain of the x data, but we do not know what the trend is outside this interval. Hence, predictions can be very dangerous outside the domain of the x data. For instance, in Example 3.7 it is nonsense to predict that a college woman of height zero will weigh -186.5 pounds. Do not use a height outside the sample domain of 61 to 69 inches to predict weight. On occasion you might wish to use the line of best fit to estimate values outside the domain interval of the sample. This can be done, but you should do it with caution and only for values close to the domain interval.
3. If the sample was taken in 2010, do not expect the results to have been valid in 1929 or to hold in 2020. The women of today may be different from the women of 1929 and the women in 2020.

TECHNOLOGY INSTRUCTIONS: LINE OF BEST FIT

MINITAB

Input the x values into C1 and the corresponding y values into C2; then to obtain the equation for the line of best fit, continue with:

Method 1—
Choose: **Stat > Regression > Regression …**
Enter: Response (y): **C2**
                 Predictors (x): **C1 > OK**

To draw the scatter diagram with the line of best fit superimposed on the data points, continue with:

Choose:	**Graph > Scatterplot**
Select:	**With Regression > OK**
Enter:	Y variable: C2 X variable: C1
Select:	**Labels > Titles/ Footnotes**
Enter:	Title: **your title > OK > OK**

OR
Method 2—

Choose:	**Stat > Regression > Fitted Line Plot**
Enter:	Response (Y): C2
	Response (X): C1
Select:	**Linear**
Select:	**Options**
Enter:	Title: **your title > OK > OK**

Excel

Input the x-variable data into column A and the corresponding y-variable data into column B; then continue with:

Choose:	**Data > Data Analysis* > Regression > OK**
Enter:	Input Y Range: (B1:B10 or select cells)
	Input X Range: (A1:A10 or select cells)
Select:	**Labels** (if necessary)
	Output Range
	Enter: (C1 or select cell)
	Line Fits Plots > OK

To make the output readable, continue with:

Choose:	**Home > Cells > Format > AutoFit Column Width**

To form the regression equation, the y-intercept is located at the intersection of the intercept and coefficients columns, whereas the slope is located at the intersection of the x-variable and the coefficients columns.

To draw the line of best fit on the scatter diagram, activate the chart; then continue with:

Choose:	**Chart Tools > Layout > Analysis – Trendline > Linear Trendline**

OR

Choose:	**Chart Tools > Design > Chart Layouts – Layout 9**

(This command also works with the scatter diagram Excel commands on p.129–130

*If Data Analysis does not show on the Data menu, see page 53.

TI-83/84 Plus

Input the x-variable data into L1 and the corresponding y-variable data into L2; then continue with:
   If just the equation is desired:

Choose:	**STAT > CALC > 8: LinReg(a + bx)**
Enter:	L1, L2*

*If the equation and graph on the scatter diagram are desired, use:

Enter:	L1, L2, Y1†

then continue with the same commands for a scatter diagram as shown on page 130.

†To enter Y1, use:

Choose:	**VARS > Y- VARS > 1: Function > 1: Y1 > ENTER**

Understanding the Line of Best Fit

The following method will create (1) a visual meaning for the line of best fit, (2) a visual meaning for what the line of best fit is describing, and (3) an estimate for the slope and y-intercept of the line of best fit. As with the approximation of r, estimations of the slope and y-intercept of the line of best fit should be used only as a mental estimate or check.

Note: This estimation technique *does not* replace the calculations for b_1 and b_0.

Procedure

1. On the scatter diagram of the data, draw the straight line that appears to be the line of best fit. (*Hint:* If you draw a line parallel to and halfway between the two pencils described in Section 3.2 on page 139 [Figure 3.13], you will have a reasonable estimate for the line of best fit.) The two pencils border the "path" demonstrated by the ordered pairs, and the line down the center of this path approximates the line of best fit. Figure 3.27 shows the pencils and the resulting estimated line for Example 3.7.

FIGURE 3.27

Estimate the Line of Best Fit for the College Women Data

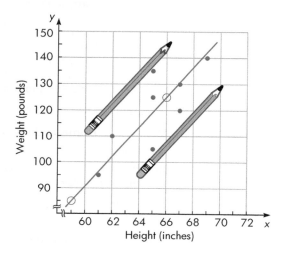

2. This line can now be used to approximate the equation. First, locate any two points (x_1, y_1) and (x_2, y_2) along the line and determine their coordinates. Two such points, circled in Figure 3.27, have the coordinates $(59, 85)$ and $(66, 125)$. These two pairs of coordinates can now be used in the following formula to estimate the slope b_1:

$$\text{estimate of the slope, } b_1: \quad b_1 \approx \frac{y_2 - y_1}{x_2 - x_1} = \frac{125 - 85}{66 - 59} = \frac{40}{7} = 5.7$$

3. Using this result, the coordinates of one of the points, and the following formula, we can determine an estimate for the y-intercept, b_0:

estimate of the y-intercept, b_0:
$$b_0 \approx y - b_1 \cdot x = 85 - (5.7)(59) = 85 - 336.3 = -251.3$$

    Thus, b_0 is approximately -250.

4. We now can write the estimated equation for the line of best fit:

$$\hat{y} = -250 + 5.7x$$

This should serve as a crude estimate. The actual equation calculated using all of the ordered pairs was $\hat{y} = -186.5 + 4.71x$.

APPLIED EXAMPLE 3.8

VIEWING AN OLD FAITHFUL ERUPTION

Old Faithful very faithfully erupts for a short period of time (1.5 to 5 minutes) periodically throughout every day (every 35 to 120 minutes) and has been doing so since 1870, when such records began being kept; thus its name. It is not the most regular, nor the biggest, but is the biggest regular geyser in Yellowstone.

If your luck is like that of many and you traveled to see one of these famous eruptions, you probably arrived minutes after an eruption had stopped. "When will it erupt again?" and "How long will it last?" are typical questions. What you are really asking is, "How long do I have to wait for the next show?" and "Will it be worth the wait?" Since Old Faithful is one of the most studied geysers, the Park

© iStockphoto.com/Sascha Burkard

Rangers are able to predict the next eruption with reasonable accuracy (± 10 minutes). They are able to predict only the next eruption, so you had better stick around for it.

The time until the next eruption, the interval, is predicted based on the length of the previous eruption, the duration. It is not possible to predict the time of occurrence for more than one eruption in advance. Here is a chart summarizing the predicted interval based on the previous duration.

TABLE 3.16

Duration	1.5 min	2.0 min	2.5 min	3.0 min	3.5 min	4.0 min	4.5 min	5.0 min
Interval	50 min	57 min	65 min	71 min	76 min	82 min	89 min	95 min

By looking at the chart, it appears that the time to the next show interval increases 5 to 7 minutes for each extra half minute of eruption. The information on the chart can also be seen on the scatter diagram with the line of best fit. The slope for the line of best fit is 12.64, implying that every extra

FIGURE 3.28

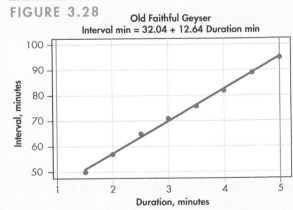

minute of eruption results in an additional 12.6 minutes in wait time for the next eruption, or about 6.3 minutes for every half minute of eruption, as in the information given.

The ordered pairs from Table 3.16 and on the scatter diagram, Figure 3.28, are not data values; they are the result of an averaging effect as hundreds of recorded values were summarized. Old Faithful data will not result in points being exactly distributed along the line of best fit like those shown in Figure 3.28, they will instead show a substantial amount of variability.

Table 3.17 contains data collected by a visitor over a weekend. It is arranged in sequential order.

DID YOU KNOW?

TABLE 3.17

Duration, min	1.7	1.9	2.0	2.3	3.1	3.4	3.5	4.0	4.3	4.5	4.7	4.9
Interval, min	55	49	51	53	57	75	80	76	84	76	93	76

The 12 duration and interval times listed in Table 3.17 and shown on Figure 3.29 offer a different impression than the eight points listed in Table 3.16 on previous page. These data look more realistic, with points scattered above and below the line of best fit. A comparison of the two lines of best fit shows very similar results.

FIGURE 3.29

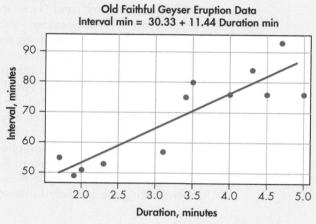

Old Faithful Geyser Eruption Data
Interval min = 30.33 + 11.44 Duration min

SECTION 3.3 EXERCISES

3.51 Draw a scatter diagram for these data:

x	1	2.5	3	4	5	1.5
y	1.5	2.2	3.5	3	4	2.5

Would you be justified in using the techniques of linear regression on these data to find the line of best fit? Explain.

3.52 [EX03-052] Draw a scatter diagram for these data:

x	2	12	4	6	9	4	11	3	10	11	3	1	13	12	14	7	2	8
y	4	8	10	9	10	8	8	5	10	9	8	3	9	8	8	11	6	9

Would you be justified in using the techniques of linear regression on these data to find the line of best fit? Explain.

3.53 [EX03-053] The Children's Bureau of the U.S. Department of Health and Human Services has a monu-

mental job. In 2006, 510,000 children were in foster care. Of those, approximately 303,000 entered during the 2006 year (10/1/05–9/20/06). The following table lists the ages of children who entered foster care during the 2006 year and the number in each age group.

Age	Number	Age	Number	Age	Number
0	47,536	7	12,380	14	18,981
1	20,646	8	11,312	15	22,729
2	18,234	9	10,649	16	21,062
3	16,145	10	10,136	17	12,829
4	14,919	11	10,316	18	702
5	14,159	12	11,910	19	154
6	13,196	13	14,944	20	62

Source: U.S. Department of Health and Human Services

a. Construct a scatter diagram of the ages when children entered foster care, x, and the number of children in each age group, y.

b. What do you think causes the unusual pattern shown on the scatter diagram?

c. Does it appear that these two variables are correlated?

d. Are we justified in using the techniques of linear regression on these data? Explain.

e. Are there any particular age groups where techniques of linear regression may be justified?

3.54 The formulas for finding the slope and the y-intercept of the line of best fit use both summations, Σ's, and sums of squares, SS( )'s. It is important to know the difference. In reference to Example 3.5 (p. 138):

a. Find three pairs of values: Σx^2, SS(x); Σy^2, SS(y); and Σxy, SS(xy).

b. Explain the difference between the numbers for each pair of numbers.

3.55 Does it pay to study for an exam? The number of hours studied, x, is compared to the exam grade received, y:

x	2	5	1	4	2
y	80	80	70	90	60

a. Find the equation for the line of best fit.

b. Draw the line of best fit on the scatter diagram of the data drawn in Exercise 3.15 (p. 133).

c. Based on what you see in your answers to parts a and b, does it pay to study for an exam? Explain.

3.56 [EX03-056] How old is my bass? Ever wondered about the age of the bass you just caught? Measure your bass from the snout to the tip of the tail. The following ages are average ages by length of largemouth and smallmouth bass in New York State.

Length (in.)	SmM Bass Age (yr)	LgM Bass Age (yr)
8	2	2
9	2	2
10	3	3
11	4	4
12	4	4
13	5	5
14	5	6
15	6	6
16	7	7
17	7	8
18	8	8
19	8	9
20	9	10
21	10	10
22	10	11

Source: New York Freshwater Fishing, 2008–2009 *Official Regulations Guide*

a. Examine the data and find an approximate pattern for age increase by inch length for each kind of fish.

b. Construct a scatter diagram for both smallmouth and largemouth bass on the same graph.

c. Do the points for both fish seem to follow a straight line? Explain.

d. Do the points for both fish follow the same line? Explain.

e. Calculate both lines of best fit.

3.57 The values of x used to find points for graphing the line $\hat{y} = 14.9 + 0.66x$ in Figure 3.24 (p. 149) are arbitrary. Suppose you choose to use $x = 20$ and $x = 50$.

a. What are the corresponding $\hat{y}$ values?

b. Locate these two points on Figure 3.24. Are these points on the line of best fit? Explain why or why not.

3.58 If all students from Mr. Chamberlain's physical-fitness course on page 127 who can do 40 push-ups are asked to do as many sit-ups as possible:

a. How many sit-ups do you expect that each can do?

b Will they all be able to do the same number?

c. Explain the meaning of the answer to part a.

3.59 [EX03-045] What is the relationship between carbohydrates consumed and energy released in a sports drink? Let's use the sports drink data listed in Exercise 3.45 on page 144 to investigate the relationship.

a. In Exercise 3.45 a scatter diagram was drawn using x = carbs/serving and y = energy/serving. Review the scatter diagram (if you did not draw it before, do so now), and describe why you believe there is or is not a linear relationship.

b. Find the equation for the line of best fit.

c. Using the equation found in part b, estimate the amount of energy that one can expect to gain from consuming 40 grams of carbohydrates.

d. Using the equation found in part b, estimate the amount of energy that one can expect to gain from consuming 65 grams of carbohydrates.

3.60 Referring to Applied Example 3.8 (p. 155):

a. Explain (in 25 or more words) what you believe the statement, "The ordered pairs from Table 3.16 and on the scatter diagram, Figure 3.28, are not data values; they are the result of an averaging effect as hundreds of recorded values were summarized" is saying.

b. Using the equation shown in Figure 3.28, what is the anticipated interval until the next eruption following a 4.0-minute eruption?

(continue on page 158)

c. Using the equation shown on Figure 3.29 what is the anticipated interval until the next eruption following a 4.0-minute eruption?

d. The two equations result in approximately the same anticipated waiting time for the next eruption. True or false? Explain you answer.

3.61 A.J. used linear regression to help him understand his monthly telephone bill. The line of best fit was $\hat{y} = 23.65 + 1.28x$, where x is the number of long-distance calls made during a month, and y is the total telephone cost for a month. In terms of number of long-distance calls and cost:

a. Explain the meaning of the y-intercept, 23.65.

b. Explain the meaning of the slope, 1.28.

3.62 Geoff is interested in purchasing a moderately priced SUV. He realizes that a car or truck loses its value as soon as it is driven off the dealer's lot. Geoff uses linear regression to get a better sense of how this decline works. The regression line is $\hat{y} = 34.03 - 3.04x$, where x is the age of the car in years, and y is the value of the car ($\times$ $1000). In terms of age and value:

a. Explain the meaning of the y-intercept, 34.03.

b. Explain the meaning of the slope, -3.04.

3.63 For Example 3.7 (p. 150) and the scatter diagram in Figure 3.26 on page 151:

a. Explain how the slope of 4.71 can be seen.

b. Explain why the y-intercept of -186.5 cannot be seen.

3.64 For any basketball player, the number of points scored per game and the number of personal fouls committed are of interest. Data taken for a team last season resulted in the equation $\hat{y} = 1.122 + 3.394x$, where x is the number of personal fouls committed per game and y is the number of points scored per game.

a. If one of the players committed two fouls in a game, how many points would he or she be expected to have made?

b. What is the average number of points a player might expect if he or she commits three fouls in a game?

3.65 A study was conducted to investigate the relationship between the cost, y (in tens of thousands of dollars), per unit of equipment manufactured and the number of units produced per run, x. The resulting equation for the line of best fit was $\hat{y} = 7.31 - 0.01x$, with x being observed for values between 10 and 200. If a production run was scheduled to produce 50 units, what would you predict the cost per unit to be?

3.66 A study was conducted to investigate the relationship between the resale price, y (in hundreds of dollars),

and the age, x (in years), of midsize luxury American automobiles. The equation of the line of best fit was determined to be $\hat{y} = 185.7 - 21.52x$.

a. Find the resale value of such a car when it is 3 years old.

b. Find the resale value of such a car when it is 6 years old.

c. What is the average annual decrease in the resale price of these cars?

3.67 The Federal Highway Administration annually reports on state motor-fuel taxes. Based on the latest report, the amount of receipts, in thousands of dollars, can be estimated using the equation: Receipts $= -5359 + 0.9956$ Collections.

a. If a state collected $500,000, what would you estimate the receipts to be?

b. If a state collected $1,000,000, what would you estimate the receipts to be?

c. If a state collected $1,500,000, what would you estimate the receipts to be?

3.68 A study of the tipping habits of restaurant-goers was completed. The data for two of the variables—x, the amount of the restaurant check, and y, the amount left as a tip for the servers—were used to construct a scatter diagram.

a. Do you expect the two variables to show a linear relationship? Explain.

b. What will the scatter diagram suggest about linear correlation? Explain.

c. What value do you expect for the slope of the line of best fit? Explain.

d. What value do you expect for the y-intercept of the line of best fit? Explain.

The data are used to determine the equation for the line of best fit: $\hat{y} = 0.02 + 0.177x$.

e. What does the slope of this line represent as applied to the actual situation? Does the value 0.177 make sense? Explain.

f. What does the y-intercept of this line represent as applied to the actual situation? Does the value 0.02 make sense? Explain.

g. If the next restaurant check was for $30, what would the line of best fit predict for the tip?

h. Using the line of best fit, predict the tip for a check of $31. What is the difference between this amount and the amount in part g for a $30 check? Does this difference make sense? Where do you see it in the equation for the line of best fit?

3.69 Consider Figure 3.27 on page 154. The graph's y-intercept is -250, not approximately 80, as might be read from the figure. Explain why.

3.70 Consider the College Women's data presented in Example 3.7 and the line of best fit. When estimating the line of best fit from a scatter diagram, the choice for the two points (x_1, y_1) and (x_2, y_2) to be used is somewhat arbitrary. When different points are used, slightly different values for b_0 and b_1 will result, but they should be approximately the same.

a. What points on the scatter diagram (Figure 3.27, p. 154) were used to estimate the slope and y-intercept in the example on page 150? What were the resulting estimates?

b. Use points $(61, 95)$ and $(67, 130)$ and find the approximate slope and y-intercept values.

c. Compare the values found in part b with those described in part a. How similar are they?

d. Compare both sets of estimates with the actual values of slope and y-intercept found in Example 3.7 on pages 150–151. Draw both estimated lines of best fit on the scatter diagram shown in Figure 3.26. How useful do you think estimated values might be? Explain.

3.71 Phi ($\Phi = 1.618033988749895\ldots$), most often pronounced "fī" (like "fly"), is simply an irrational number like pi ($\pi = 3.14159265358979\ldots$), but one with many unusual mathematical properties. Phi is the basis for the Golden Ratio. (Visit http://goldennumber.net/ to learn other interesting things about phi.)

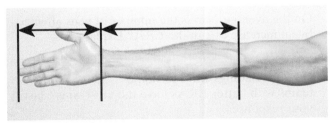

Imagestate/PhotoLibrary

a. If every person's arm displayed the exact Golden Ratio, describe the appearance of a scatter diagram where the length of the forearm, y, and the length of the hand, x, have been plotted.

b. Since body proportions vary from person to person, describe the appearance of a scatter diagram where the length of the forearm, y, and the length of the hand, x, have been plotted for 25 people whose two lengths had been measured.

3.72 Another interesting ratio that uses the length of a person's forearm (as shown in Exercise 3.71) is the ratio of the length of the forearm to the length of a person's foot (in inches). This ratio is 1 to 1.

a. Describe the appearance of a scatter diagram where the length of the foot, y, and length of the forearm, x, have been plotted.

b. What value would you expect the slope of the regression line to be?

3.73 Collect the lengths of the forearm (y) and the hand (x) from 15 or more people, following the picture in Exercise 3.71.

a. Plot the collected data as a scatter diagram; be sure to label completely.

b. Find the equation for the line of best fit.

c. What is the slope? How does its value compare to phi? Explain the similarities or differences found.

3.74 Collect the lengths of the foot (y) and the forearm (x) from 15 or more people.

a. Plot the collected data as a scatter diagram; be sure to label completely.

b. Find the equation for the line of best fit.

c. What is the slope? How does its value compare to your answer in part b of Exercise 3.72? Explain the similarities or differences found.

3.75 [EX03-075] "Now more than ever, a degree matters," according to an upstate New York college advertisement in the May 31, 2009, *Democrat and Chronicle*. The following statistics from the U.S. Bureau of Labor Statistics were presented on median usual weekly earnings.

Amount of Schooling	Median Usual Weekly Earnings	Years of Schooling
Less than a high school diploma	$453	10
High school graduate, no college	$618	12
Bachelor's degree	$1115	16
Advanced degree	$1287	18

a. Construct a scatter diagram with the years of schooling as the independent variable, x, and the median usual weekly earnings as the dependent variable, y.

b. Does there seem to be a linear relationship? Why?

c. Calculate the linear correlation coefficient.

d. Does the value of r seem reasonable compared with the pattern demonstrated in the scatter diagram? Explain.

e. Find the equation of the line of best fit.

(continue on page 160)

f.   Interpret the slope of the equation.

g.   Plot the line of best fit on the scatter diagram.

h.   What is the y-intercept for the equation? Interpret its meaning in this application.

3.76 [EX03-076] The U.S. per capita consumption of bottled water has grown continuously since 1997, by more than 1 gallon annually.

a.   Inspect the data in the chart below and explain how the data show growth of more than 1 gallon annually.

b.   Construct a scatter diagram using years after 1997, x, and consumption, y.

c.   Find the equation for the line of best fit.
d.   Explain how the equation in part c shows that the annual consumption has grown steadily for 10 years at a rate of more than 1 gallon per year. Be specific.

3.77 [EX03-077] Bottled water is big business in the United States, and around the world too. Listed here are annual numbers indicating just how big the U.S. bottled water market is (volume is in gallons and producer revenues are in U.S. dollars).

2000–2008 (Projected)

Year	Millions of Gallons	Millions of Dollars
2000	4,725.10	$6,113.00
2001	5,185.30	$6,880.60
2002	5,795.70	$7,901.40
2003	6,269.80	$8,526.40
2004	6,806.70	$9,169.50
2005	7,538.90	$10,007.40
2006	8,253.50	$10,857.80
2007	8,823.00	$11,705.90
2008	9,418.00	$12,573.50

Source: Beverage Marketing Corporation

a.   Inspect the data in the chart and explain how the numbers show steady and big growth annually.

b.   Construct a scatter diagram using gallons, x, and dollars, y.

c.   Does the scatter diagram show the same steady growth you discussed in part a? Explain any differences.

d.   Find the equation for the line of best fit.

e.   What does the slope found in part d represent?

3.78 [EX03-078] Baseball teams win and lose games. Many fans believe that a team's earned run average (ERA) has a major effect on that team's winning. During the 2008 season, the 30 Major League Baseball teams recorded the following numbers of wins while generating these earned run averages:

Wins	ERA	Wins	ERA	Wins	ERA
89	4.07	92	3.88	89	4.28
88	4.16	84	3.68	72	4.46
63	4.41	86	3.49	81	4.45
89	4.06	95	4.01	84	4.43
97	3.82	74	4.77	61	4.73
90	3.85	75	4.48	75	4.01
67	5.08	74	4.90	82	3.98
86	4.19	74	4.55	59	4.66
100	3.99	72	4.38	86	4.36
97	3.87	79	5.37	68	5.13

Source: http://mlb.mlb.com

a.   What do you think—do the teams with the better ERAs have the most wins? (The lower the ERA, the fewer earned runs the other team scored.)

b.   If this is true, what will the pattern on the scatter diagram look like? Be specific.

c.   Construct a scatter diagram of these data.

d.   Does the scatter diagram suggest that teams tend to win more games when their team ERA is lower? Explain how or how not.

e.   Calculate the equation for the line of best fit using ERA for x and number of wins for y.

f.   On the average, how is the number of wins affected by an increase of 1 in the ERA? Explain how you determined this number.

g.   Do your findings seem to support the idea that teams with better ERAs have the most wins? Justify your response.

3.79 [EX03-079] Consider the saying, "Build it and they will come." This notable saying from a movie may very well apply to shopping malls. Just be sure when you build that there is room for not only the mall but also

Table for Exercise 3.76

Year	1997	1998	1999	2000	2001	2002	2003	2004	2005	2006	2007
Years after 1997	0	1	2	3	4	5	6	7	8	9	10
Gallons per Capita	13.5	14.7	16.2	16.7	18.2	20.1	21.6	23.2	25.4	27.6	29.3

Source: Beverage Marketing Corporation

for those that will come—and thus include enough space for parking. Consider the random sample of major malls in Irvine, California.

Square Feet	Parking Spaces	Number of Stores
270,987	3128	65
258,761	1500	43
1,600,350	8572	120
210,743	793	59
880,000	7100	95
2,700,000	15,000	300

a. Draw a scatter diagram with "parking spaces" as the dependent variable, y, and "square feet" as the independent variable, x. (Suggestion: Use 1000s of square feet.)

b. Does the scatter diagram in part a suggest that a linear regression will be useful? Explain.

c. Calculate the equation for the line of best fit.

d. Draw the line of best fit on the scatter diagram you obtained in part a. Explain the role of a positive slope for this pair of variables.

e. Do you see a potential lurking variable? Explain its possible role.

f. Draw a scatter diagram with "parking spaces" as the dependent variable, y, and "number of stores" as the independent variable, x.

g. Does the scatter diagram in part e suggest that a linear regression will be useful? Explain.

h. Calculate the equation for the line of best fit.

i. Draw the line of best fit on the scatter diagram you obtained in part e.

j. Do you see a potential lurking variable? Explain its possible role.

k. Draw a scatter diagram with "number of stores" as the dependent variable, y, and "square feet" as the predictor variable, x.

l. Does the scatter diagram in part k suggest that a linear regression will be useful? Explain.

m. Calculate the equation for the line of best fit.

n. Draw the line of best fit on the scatter diagram you obtained in part k.

3.80 [EX03-080] The given rule of thumb is that pets age seven times faster than people. The most common pets are dogs and cats.

a. Consider the data below on cat ages versus human ages. Is there a relationship between cat ages and human ages? Comment on the strength. Calculate the equation for the line of best fit. What is the average rate of change for cats?

b. Consider the data below on dog ages versus human ages. Is there a relationship between dog ages and human ages? Comment on the strength. Calculate the equation for the line of best fit. What is the average rate of change for dogs?

c. By age 7, most dogs, particularly larger breeds, are entering their senior years. Do the data support this statement? Why and how?

Human Age	Cat Age	Human Age	Dog Age
23	1	14	1
35	2	23	2
40	3	29	3
45	4	34	4
47	5	38	5
50	6	41	6
53	7	47	7
56	8	50	8
59	9	55	9
61	10	60	10
65	11	64	11
69	12	68	12
72	13	74	13
75	14	78	14
78	15	84	15

3.81 The graph on page 162 shows the relationship between three variables: number of licensed drivers, number of registered vehicles, and the size of the resident population in the United States from 1960 to 2007. Study the graph and answer the questions.

a. Does it seem reasonable for the population line and the drivers line to basically parallel each other, with the population line above the drivers line? Explain what it means for them to be parallel. What would it mean if they were not parallel?

b. What does it mean that the drivers and motor vehicles lines cross? When and what does the point of intersection represent?

c. Explain the relationship between motor vehicles and drivers before 1973.

d. Explain the relationship between motor vehicles and drivers after 1973.

e. Do you predict drivers ever surpassing motor vehicles after 2007? Why or why not?

(continue on page 162)

f. Using the years 1990 and 2002, estimate the slopes of the motor vehicles line and the drivers line. Compare and contrast the slopes found.

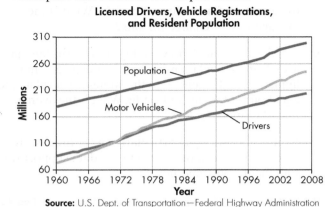

Licensed Drivers, Vehicle Registrations, and Resident Population

Source: U.S. Dept. of Transportation—Federal Highway Administration

3.82 The correlation coefficient and the slope of the line of best fit are related by definition.

a. Verify this statement.

b. Describe how the relationship between correlation coefficient and slope can be seen in the statistics that describe a particular set of data.

c. Show that $b_1 = r(s_y/s_x)$. Comment on this relationship.

Chapter Review

In Retrospect

To sum up what we have just learned: There is a distinct difference between the purpose of regression analysis and the purpose of correlation. In regression analysis, we seek a relationship between the variables. The equation that represents this relationship may be the answer that is desired, or it may be the means to the prediction that is desired. In correlation analysis, we measure the strength of the linear relationship between the two variables.

The Applied Examples in the text show a variety of uses for the techniques of correlation and regression. These examples are worth reading again. When bivariate data appear to fall along a straight line on the scatter diagram, they suggest a linear relationship. But this is not proof of cause and effect. Clearly, if a basketball player commits too many fouls, he or she will not be scoring

more points. Players in foul trouble are "riding the pine" with no chance to score. It also seems reasonable that the more game time they have, the more points they will score and the more fouls they will commit. Thus, a positive correlation and a positive regression relationship will exist between these two variables. Time is a lurking variable here.

The bivariate linear methods we have studied thus far have been presented as a first, descriptive look. More details must, by necessity, wait until additional developmental work has been done. After completing this chapter, you should have a basic understanding of bivariate data, how they are different from just two sets of data, how to present them, what correlation and regression analysis are, and how each is used.

CourseMate The **Statistics CourseMate** site for this text brings chapter topics to life with interactive learning, study, and exam preparation tools, including quizzes and flashcards for the Vocabulary and Key Concepts that follow. The site also provides an **eBook** version of the text with highlighting and note taking capabilities. Throughout chapters, the CourseMate icon ⬚ flags concepts and examples that have

corresponding interactive resources such as **video** and **animated tutorials** that demonstrate, step by step, how to solve problems; **datasets** for exercises and examples; **Skillbuilder Applets** to help you better understand concepts; **technology manuals**; and software to download including **Data Analysis Plus** (a suite of statistical macros for Excel) and **TI-83/84 Plus** programs—logon at **www.cengagebrain.com**.

Vocabulary and Key Concepts

bivariate data (p. 121)
cause-and-effect relationship
(p. 140)
coefficient of linear correlation
(p. 136)
contingency table (p. 121)
correlation (p. 136)
cross-tabulation (p. 121)
dependent variable (p. 126)
independent variable (p. 126)
input variable (p. 126)

least squares criterion (p. 146)
line of best fit (p. 147)
linear correlation (p. 136)
linear correlation analysis (p. 136)
linear regression (p. 146)
lurking variable (p. 140)
method of least squares
(p. 146)
negative correlation (p. 136)
no correlation (p. 136)
ordered pair (p. 126)

output variable (p. 126)
Pearson's product moment, r
(p. 137)
positive correlation (p. 136)
predicted value (p. 146)
prediction equation (p. 146)
regression (p. 146)
regression analysis (p. 146)
scatter diagram (p. 127)
slope, b_1 (p. 147)
y-intercept, b_0 (p. 147)

Learning Outcomes

- Understand and be able to present and describe data in the form of two qualitative variables, both in contingency table format and appropriate graphs. — EXP 3.1, pp. 121–123, Ex. 3.85

- Understand and be able to present and describe data in the form of one qualitative variable and one quantitative variable, in both table format and appropriate graphs. — EXP 3.2, p. 125, Ex. 3.9, 3.10

- Understand and be able to present and describe the relationship between two quantitative variables using a scatter diagram. — EXP 3.3, APP EXP 3.4, pp. 127–129, Ex. 3.15

- Understand and be able to explain a linear relationship. — p. 136

- Compute, describe, and interpret a correlation coefficient. — pp. 136–138, EXP 3.5, Ex. 3.31

- Compute, describe, and interpret a line of best fit. — EXP 3.7

- Define and understand the difference between correlation and causation. — pp. 140–141, Ex. 3.49, 3.50

- Determine and explain possible lurking variables and their effects on a linear relationship. — pp. 140–141, Ex. 3.49, 3.50

- Understand and be able to explain the slope of the line of best fit with respect to the context it is presented in. — Ex. 3.61, 3.68

- Understand and be able to explain the y-intercept of the line of best fit with respect to the context it is presented in. — Ex. 3.61, 3.68

- Create a scatter diagram with the line of best fit drawn on it. — Ex. 3.55

- Compute prediction values based on the line of best fit. — p. 152, Ex. 3.66

- Understand and be able to explain what prediction values are. — pp. 146, 152

- Understand that predictions should be made only for values within the sample domain and that caution must be exercised for values outside that domain. — p. 152

Chapter Exercises

3.83 [EX03-083] Fear of the dentist (or the dentist's chair) is an emotion felt by many people of all ages. A survey of 100 individuals in five age groups was conducted about this fear, and these were the results:

	Elementary	Jr. High	Sr. High	College	Adult
Fear	37	28	25	27	21
Do Not Fear	63	72	75	73	79

a. Find the marginal totals.

b. Express the frequencies as percentages of the grand total.

c. Express the frequencies as percentages of each age group's marginal totals.

(continue on page 164)

[EX00-000] identifies the filename of an exercise's online dataset—available through cengagebrain.com

d. Express the frequencies as percentages of those who fear and those who do not fear.

e. Draw a bar graph based on age groups.

3.84 As summer heats up, Americans turn to ice cream as a way to cool off. One of the questions asked as part of a July 2009 Harris Poll was, "What is your favorite way to eat ice cream?" The study included 2177 U.S. adults.

FAVORITE WAY TO EAT ICE CREAM

Base: All adults who eat ice cream

Favorite Way	Male, %	Female, %
Cup	50	41
Cone	24	34
Sundae	17	18
Sandwich	2	2
Other	8	5
Total	101	100

Image copyright © M. Unal Ozmen. Used under license from Shutterstock.com

The accompanying "Favorite Way to Eat Ice Cream" graphic lists in percentages the distributions of the ways both genders prefer to eat their ice cream.

a. Identify the population, the variables, and the type of variables.

b. Construct a bar graph showing the two distributions side by side.

c. Do the distributions seem to be different for the genders? Explain.

3.85 [EX03-085] Six breeds of dogs have been rather popular in the United States over the past few years. The table below lists the breeds coupled with the number of registrations of each filed with the American Kennel Club in 2004 and 2005:

Breeds	2004	2005
Retrievers (Labrador)	146,692	137,867
Retrievers (Golden)	52,550	48,509
German Shepherd	46,046	45,014
Beagles	44,555	42,592
Dachshunds	40,770	38,566

Source: American Kennel Club

a. A cross-tabulation of the two variables, year (columns) and dog breed (rows), is given. Determine the marginal totals.

b. Express the contingency table in part a in percentages based on the grand total.

c. Draw a bar graph showing the results from part b.

d. Express the contingency table in part a in percentages based on the marginal totals for each year.

e. Draw a bar graph showing the results from part d.

3.86 [EX03-086] When was the last time you saw your doctor? That question was asked for the survey summarized in the following table.

		Time Since Last Consultation		
		Less Than 6 Months	6 Months to Less Than 1 Year	1 Year or More
Age	Younger than 28 years	413	192	295
	28–40	574	208	218
	Older than 40	653	288	259

a. Find the marginal totals.

b. Express the frequencies as percentages of the grand total.

c. Express the frequencies as percentages of each age group's marginal totals.

d. Express the frequencies as percentages of each time period.

e. Draw a bar graph based on the grand total.

3.87 [EX03-087] Part of quality control is keeping track of what is occurring. The following contingency table shows the number of rejected castings last month categorized by their cause and the work shift during which they occurred.

	1st Shift	2nd Shift	3rd Shift
Sand	87	110	72
Shift	16	17	4
Drop	12	17	16
Corebreak	18	16	33
Broken	17	12	20
Other	8	18	22

a. Find the marginal totals.

b. Express the numbers as percentages of the grand total.

c. Express the numbers as percentages of each shift's marginal total.

d. Express the numbers as percentages of each type of rejection.

e. Draw a bar graph based on the shifts.

3.88 Determine whether each of the following questions requires correlation analysis or regression analysis to obtain an answer.

a. Is there a correlation between the grades a student attained in high school and the grades he or she attained in college?

b. What is the relationship between the weight of a package and the cost of mailing it first class?

c. Is there a linear relationship between a person's height and shoe size?

d. What is the relationship between the number of worker-hours and the number of units of production completed?

e. Is the score obtained on a certain aptitude test linearly related to a person's ability to perform a certain job?

3.89 An automobile owner records the number of gallons of gasoline, x, required to fill the gasoline tank and the number of miles traveled, y, between fill-ups.

a. If she does a correlation analysis on the data, what would be her purpose and what would be the nature of her results?

b. If she does a regression analysis on the data, what would be her purpose and what would be the nature of her results?

3.90 These data were generated using the equation $y = 2x + 1$.

x	0	1	2	3	4
y	1	3	5	7	9

A scatter diagram of the data results in five points that fall perfectly on a straight line. Find the correlation coefficient and the equation of the line of best fit.

3.91 Consider this set of bivariate data:

x	1	1	3	3
y	1	3	1	3

a. Draw a scatter diagram.

b. Calculate the correlation coefficient.

c. Calculate the line of best fit.

3.92 Start with the point (5, 5) and add at least four ordered pairs, (x, y), to make a set of ordered pairs that display the following properties. Show that your sample satisfies the requirements.

a. The correlation of x and y is 0.0.

b. The correlation of x and y is $+1.0$.

c. The correlation of x and y is -1.0.

d. The correlation of x and y is between -0.2 and 0.0.

e. The correlation of x and y is between $+0.5$ and $+0.7$.

3.93 A scatter diagram is drawn showing the data for x and y, two normally distributed variables. The data fall within the intervals $20 \le x \le 40$ and $60 \le y \le 100$. Where would you expect to find the data on the scatter diagram, if:

a. The correlation coefficient is 0.0

b. The correlation coefficient is 0.3

c. The correlation coefficient is 0.8

d. The correlation coefficient is -0.3

e. The correlation coefficient is -0.8

3.94 Start with the point (5, 5) and add at least four ordered pairs, (x, y), to make a set of ordered pairs that display the following properties. Show that your sample satisfies the requirements.

a. The correlation of x and y is between $+0.9$ and $+1.0$, and the slope of the line of best fit is 0.5.

b. The correlation of x and y is between $+0.5$ and $+0.7$, and the slope of the line of best fit is 0.5.

c. The correlation of x and y is between -0.7 and -0.9, and the slope of the line of best fit is -0.5.

d. The correlation of x and y is between $+0.5$ and $+0.7$, and the slope of the line of best fit is -1.0.

3.95 [EX03-095] A biological study of a minnow called the blacknose dace* was conducted. The length, y (in millimeters), and the age, x (to the nearest year), were recorded.

*Visit: http://www.dnr.state.oh.us/

x	0	3	2	2	1	3	2	4	1	1
y	25	80	45	40	36	75	50	95	30	15

a. Draw a scatter diagram of these data.

b. Calculate the correlation coefficient.

c. Find the equation of the line of best fit.

d. Explain the meaning of the answers to parts a through c.

3.96 [EX03-096] The following data are a sample of the ages and the asking prices for used Honda Accords that were listed on AutoTrader.com on March 10, 2005:

Age x (years)	Price y (× $1000)	Age x (years)	Price y (× $1000)
3	24.9	2	26.9
7	9.0	4	23.8
5	17.8	5	19.3
4	29.2	4	21.9
6	15.7	6	16.4
3	24.9	4	21.2
2	25.7	3	24.9
7	11.9	5	20.0
6	15.2	7	13.6
2	25.9	5	18.8

Source: http://autotrader.com/

a. Draw a scatter diagram.

b. Calculate the equation of the line of best fit.

c. Graph the line of best fit on the scatter diagram.

d. Predict the average asking price for all Honda Accords that are 5 years old. Obtain this answer in two ways: Use the equation from part b and use the line drawn in part c.

e. Can you think of any potential lurking variables for this situation? Explain any possible role they might play.

3.97 [EX03-097] The sound of crickets chirping is a welcome summer night's sound. In fact, those chirping crickets may be telling you the temperature. In the book *The Song of Insects*, George W. Pierce, a Harvard physics professor, presented real data relating the number of chirps per second, x, for striped ground crickets to the temperature in °F, y. The following table gives real cricket and temperature data. It appears that the number of chirps represents an average, because it is given to the nearest tenth.

x	y	x	y	x	y
20.0	88.6	15.5	75.2	15.0	79.6
16.0	71.6	14.7	69.7	17.2	82.6
19.8	93.3	17.1	82.0	16.0	80.6
18.4	84.3	15.4	69.4	17.0	83.5
17.1	80.6	16.2	83.3	14.4	76.3

Source: George W. Pierce, *The Song of Insects*, Harvard University Press, 1948

a. Draw a scatter diagram of the number of chirps per second, x, and the air temperature, y.

b. Describe the pattern displayed.

c. Find the equation for the line of best fit.

d. Using the equation from part c, find the temperatures that correspond to 14 and 20 chirps,

the approximate bounds for the domain of the study.

e. Does the range of temperature values bounded by the temperature values found in part d seem reasonable for this study? Explain.

f. The next time you are out where crickets chirp on a summer night and you find yourself without a thermometer, just count the chirps and you will be able to tell the temperature. If the count is 16, what temperature would you suspect it is?

3.98 [EX03-098] Lakes are bodies of water surrounded by land and may include seas. The accompanying table lists the areas and maximum depths of 32 lakes throughout the world.

a. Draw a scatter diagram showing area, x, and maximum depth, y, for the lakes.

b. Find the linear correlation coefficient between area and maximum depth. What does the value of this linear correlation imply?

Lake	Area (sq mi)	Max. Depth (ft)
Caspian Sea	143,244	3,363
Superior	31,700	1,330

*** For remainder of data, logon at cengagebrain.com

Source: Geological Survey, U.S. Department of the Interior

3.99 [EX03-099] Wildlife populations are monitored with aerial photographs. The number of animals and their locations relative to areas inhabited by the human population are useful information. Sometimes it is possible to monitor the physical characteristics of the animals. The length of an alligator can be estimated quite accurately from aerial photographs, but its weight cannot. The following data are the lengths, x (in inches), and weights, y (in pounds), of alligators captured in central Florida and can be used to predict the weight of an alligator based on its length.

Weight	Length	Weight	Length	Weight	Length
130	94	38	72	44	61
51	74	366	128	106	90
640	147	84	85	84	89
28	58	80	82	39	68
80	86	83	86	42	76
110	94	70	88	197	114
33	63	61	72	102	90
90	86	54	74	57	78
36	69				

Source: http://exploringdata.cqu.edu.au/

a. Construct a scatter diagram for length, x, and weight, y.

b. Does it appear that the weight of an alligator is predictable from its length? Explain.

c. Is the relationship linear?

d. Explain why the line of best fit, as described in this chapter, is not adequate for estimating weight based on length.

e. Find the value of the linear correlation coefficient.

f. Explain why the value of r can be so high for a set of data that is so obviously not linear in nature.

3.100 [EX03-100] Sugar cane growers are interested in the relationship between the total acres of sugar cane harvested and the total sugar cane production (tons) of those acres. The data listed here are for the 2007 crop from 14 randomly selected sugar cane–producing counties in Louisiana.

Acres	Production	Acres	Production
2,600	70,000	10,100	300,000
28,900	825,000	12,300	375,000
13,600	470,000	25,100	730,000
9,600	295,000	51,000	1,530,000
26,400	800,000	11,100	335,000
39,400	1,220,000	26,500	770,000
30,000	910,000	1,700	55,000

Source: http://www.nass.usda.gov/

a. These data values have many zeros that will be in the way. Change acres harvested to 100s of acres and production to 1000s of tons of production before continuing.

b. Construct a scatter diagram of acres harvested, x, and tons of production, y.

c. Does the relationship between the variables appear to be linear? Explain.

d. Find the equation for the line of best fit.

e. What is the slope for the line of best fit? What does the slope represent? Explain what it means to the sugar cane grower.

3.101 [EX03-101] Relatively few business travelers use mass transit systems when visiting large cities. The payoff could be substantial—in both time and money—if they learned how to use the systems, as noted in the December 28, 2004, *USA Today* article "Mass transit could save business travelers big bucks." *USA Today*

gathered the following information on the busiest U.S. rail systems.

City	Stations	Vehicles	Track (miles)
Atlanta	38	252	193
Baltimore	14	100	34
Boston	53	408	108
Chicago	144	1190	288
Cleveland	18	60	42
Los Angeles	16	102	34
Miami	22	136	57
New York	468	6333	835
Philadelphia	53	371	102
San Francisco	43	669	246
Washington	86	950	226

Source: *USA Today*, December 28, 2004

Suppose a mass transit system is being proposed for a city and you have been put in charge of preparing statistical information (both graphic and numerical) about the relationship between the following three variables: the number of stations, the number of vehicles, and the number of miles of rail. You were provided with the preceding data.

a. Start by inspecting the data given. Do you notice anything unusual about the data? Are there any values that seem quite different from the rest? Explain.

b. Your supervisor suggests that you remove the data for New York. Make a case for that being acceptable. Include some preliminary graphs and calculated statistics to justify removing these values.

Using the data from the other 10 cities:

c. Construct a scatter diagram using miles of track as the independent variable, x, and the number of stations as the dependent variable, y.

d. Is there evidence of a linear relationship between these two variables? Justify your answer.

e. Find the equation of the line of best fit for part c.

f. Interpret the meaning of the equation for the line of best fit. What does it tell you?

g. Construct a scatter diagram using miles of track as the independent variable, x, and the number of vehicles as the dependent variable, y.

h. Is there evidence of a linear relationship between these two variables? Justify your answer.

i. Find the equation of the line of best fit for part g.

(continue on page 168)

j. Interpret the meaning of the equation for the line of best fit. What does it tell you?

k. Construct a scatter diagram using number of stations as the independent variable, x, and the number of vehicles as the dependent variable, y.

l. Is there evidence of a linear relationship between these two variables? Justify your answer.

m. Find the equation of the line of best fit for part k.

n. Interpret the meaning of the equation for the line of best fit. What does it tell you?

o. The city is entertaining initial proposals for a mass transit system of 50 miles of track. Based on the answers found in parts c through n, how many stations and how many vehicles will be needed for the system? Justify your answers.

p. If someone wants an estimate for the number of stations and vehicles needed for a 100-mile system, he or she should not just double the results found in part o. Explain why not.

q. Based on the answers found in parts c through n, how many stations and how many vehicles will be needed for a 100-mile system? Justify your answers.

3.102 [EX03-102] Cicadas are flying, plant-eating insects. One particular species, 13-year cicadas (*Magicicada*), spends five juvenile stages in underground burrows. During the 13 years underground, the cicadas grow from approximately the size of a small ant to nearly the size of an adult cicada. Every 13 years, the animals then emerge from their burrows as adults. The following table presents three different species of these 13-year cicadas and their corresponding adult body weight (BW), in grams, and wing length (WL), in millimeters.

Species	BW	WL	Species	BW	WL
tredecula	0.15	28	tredecula	0.18	29
tredecim	0.29	32	tredecassini	0.21	27
tredecim	0.17	27	tredecula	0.15	30
tredecula	0.18	30	tredecula	0.17	27
tredecim	0.39	35	tredecassini	0.13	27
tredecim	0.26	31	tredecassini	0.17	29
tredecassini	0.17	29	tredecassini	0.23	30
tredecassini	0.16	28	tredecim	0.12	22
tredecassini	0.14	25	tredecula	0.26	30
tredecassini	0.14	28	tredecula	0.19	30
tredecassini	0.28	25	tredecassini	0.20	30
tredecim	0.12	28	tredecula	0.14	23

Source: http://insects/ummz.lsa.umich.edu

a. Construct a scatter diagram of the body weights, x, and the corresponding wing lengths, y. Use a different symbol to represent the ordered pairs for each species.

b. Describe what the scatter diagram displays with respect to relationship and species.

c. Calculate the correlation coefficient, r.

d. Find the equation for the line of best fit.

e. Suppose the body weight of a cicada is 0.20 gram. What wing length would you predict? Which species do you think this cicada might be?

3.103 [EX03-103] Yellowstone National Park's Old Faithful has been a major tourist attraction for a long time. Understanding the duration of eruptions and the time between eruptions is necessary to predict the timing of the next eruption. The Old Faithful dataset variables are as follows: date: an index of the date the observation was taken; duration: the duration of an eruption of the geyser, in minutes; and interruption: the time until the next eruption, in minutes.

Day 1		Day 2		Day 3	
Duration	Interruption	Duration	Interruption	Duration	Interruption
4.4	78	4.3	80	4.5	76
3.9	74	1.7	56	3.9	82
4.0	68	3.9	80	4.3	84
4.0	76	3.7	69	2.3	53
3.5	80	3.1	57	3.8	86
4.1	84	4.0	90	1.9	51
2.3	50	1.8	42	4.6	85
4.7	93	4.1	91	1.8	45
1.7	55	1.8	51	4.7	88
4.9	76	3.2	79	1.8	51
1.7	58	1.9	53	4.6	80
4.6	74	4.6	82	1.9	49
3.4	75	2.0	51	3.5	82

Source: http://comp.uark.edu/

a. Construct a scatter diagram of the 39 durations, x, and interruption, y. Use a different symbol to represent the ordered pairs for each day.

b. Describe the pattern displayed by all 39 ordered pairs.

c. Do the data for the individual days show the same pattern as one another and as the total dataset?

d. Based on information in the scatter diagram, if Old Faithful's last eruption lasted 4 minutes, how long would you predict we will need to wait until the next eruption starts?

e. Find the line of best fit for the data listed in the table.

f. Based on the line of best fit, if Old Faithful's last eruption lasted 4 minutes, how long would you predict we will need to wait until the next eruption starts?

g. What effect do you think the distinctive pattern shown on the scatter diagram has on the line of best fit? Explain.

h. Repeat parts a through g using the dataset for 16 days of observations..

i. Compare the results found in part h to the results in parts a through g. Discuss your conclusions.

3.104 a. Verify, algebraically, that formula (3.2) for calculating r is equivalent to the definition formula (3.1).

b. Verify, algebraically, that formula (3.6) is equivalent to formula (3.5).

3.105 This equation gives a relationship that exists between b_1 and r:

$$r = b_1 \sqrt{\frac{SS(x)}{SS(y)}}$$

a. Verify the equation for these data:

x	4	3	2	3	0
y	11	8	6	7	4

b. Verify this equation using formulas (3.2) and (3.6).

3.106 Show that formula (3.7a) is equivalent to formula (3.7) (p. 148).

Chapter Practice Test

PART I: Knowing the Definitions

Answer "True" if the statement is always true. If the statement is not always true, replace the words shown in bold with words that make the statement always true.

3.1 **Correlation** analysis is a method of obtaining the equation that represents the relationship between two variables.

3.2 The linear correlation coefficient is used to determine the **equation that represents** the relationship between two variables.

3.3 A correlation coefficient of **zero** means that the two variables are perfectly correlated.

3.4 Whenever the slope of the regression line is zero, the **correlation coefficient** will also be zero.

3.5 When r is positive, b_1 will always be **negative.**

3.6 The **slope** of the regression line represents the amount of change expected to take place in y when x increases by one unit.

3.7 When the calculated value of r is positive, the calculated value of b_1 will be **negative.**

3.8 Correlation coefficients range between **0 and +1.**

3.9 The value being predicted is called the **input variable.**

3.10 The line of best fit is used to predict the **average value of y** that can be expected to occur at a given value of x.

PART II: Applying the Concepts

3.11 Refer to the scatter diagram that follows.

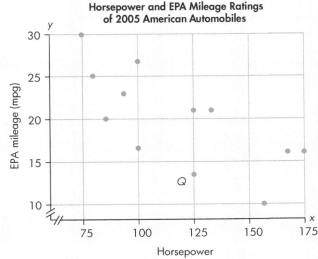

a. Match the descriptions in column 2 with the terms in column 1.

___ population   (a) the horsepower rating for an automobile

___ sample   (b) all 2005 American-made automobiles

___ input variable   (c) the EPA mileage rating for an automobile

___ output variable   (d) the 2005 automobiles with ratings shown on the scatter diagram (continue on page 170)

b. Find the sample size.

c. What is the smallest value reported for the output variable?

d. What is the largest value reported for the input variable?

e. Does the scatter diagram suggest a positive, negative, or zero linear correlation coefficient?

f. What are the coordinates of point Q?

g. Will the slope for the line of best fit be positive, negative, or zero?

h. Will the intercept for the line of best fit be positive, negative, or zero?

3.12 A research group reports a 2.3 correlation coefficient for two variables. What can you conclude from this information?

3.13 For the bivariate data, the extensions, and the totals shown on the table, find the following:

a. $SS(x)$

b. $SS(y)$

c. $SS(xy)$

d. The linear correlation coefficient, r

e. The slope, b_1

f. The y-intercept, b_0

g. The equation of the line of best fit

x	y	x^2	xy	y^2
2	6	4	12	36
3	5	9	15	25
3	7	9	21	49
4	7	16	28	49
5	7	25	35	49
5	9	25	45	81
6	8	36	48	64
28	49	124	204	353

PART III: Understanding the Concepts

3.14 A test was administered to measure the mathematics ability of the people in a certain town. Some of the townspeople were surprised to find out that their test results and shoe sizes correlated strongly. Explain why a strong positive correlation should not have been a surprise.

3.15 Student A collected a set of bivariate data and calculated r, the linear correlation coefficient. Its value was −1.78. Student A proclaimed that there was no correlation between the two variables because the value of r was not between −1.0 and +1.0. Student B argued that −1.78 was impossible and that only values of r near zero imply no correlation. Who is correct? Justify your answer.

3.16 The linear correlation coefficient, r, is a numerical value that ranges from −1.0 to +1.0. Write a sentence or two describing the meaning of r for each of these values:

a. −0.93   d. +0.08

b. +0.89   e. −2.3

c. −0.03

3.17 Make up a set of three or more ordered pairs such that:

a. $r = 0.0$ c. $r = -1.0$

b. $r = +1.0$ d. $b_1 = 0.0$

4 Probability

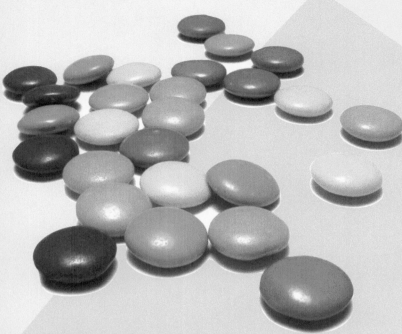

4.1 Probability of Events

Sweet Statistics

Does this "sweet" picture suddenly make you hungry for some candy?   It is pretty hard to resist an M&M. Bet you have a favorite color. Let's see if your favorite color may change—depending on how hungry you are!

Suppose a large bag of M&M's is opened and the resulting distribution of color counts is as shown in Table 4.1

If you were told that you could have all the M&M's of **one color** from this bag, which color would you choose? Remember you are very hungry!

Looks like "blue" is the way to go! It has the largest count for this bag of 692 M&M's. But, how does it compare to the rest of the colors? A convenient way to make a comparison is to use percentages. If we divide 151/692, we get 0.218 ≈ 0.22 or 22%. Thus, 22% of the M&M's in this bag are "blue." Another way to consider this event is that if you were to select one M&M from a thoroughly mixed container without looking, there is a 22% chance of picking a blue M&M.

TABLE 4.1
M&M Colors by Count

Color	Count
Brown	91
Yellow	112
Red	102
Blue	151
Orange	137
Green	99
	692

TABLE 4.2
M&M Colors by Percentage

Color	Percent
Brown	13.2
Yellow	16.2
Red	14.7
Blue	21.8
Orange	19.8
Green	14.3
	100.0

FYI The idea for M&M's® Plain Chocolate Candies was born in the backdrop of the Spanish Civil War. Legend has it that on a trip to Spain, Forrest Mars Sr. encountered soldiers who were eating pellets of chocolate encased in a hard, sugary coating to prevent them from melting. Inspired by this idea, Mr. Mars went back to his kitchen and invented the recipe for M&M's® Plain Chocolate Candies.

You have just completed your first probability experiment! (Granted, actually doing the experiment, and eating the M&M's, would have been more fun!)

We are now ready to define what is meant by probability. Specifically, we talk about "the probability that a certain event will occur."

> **Probability of an event** The relative frequency with which that event can be expected to occur.

The probability of an event may be obtained in three different ways: (1) *empirically*, (2) *theoretically*, or (3) *subjectively*.

The **empirical** method was just illustrated by the M&M's and their percentages and might be called **experimental** or **empirical probability**. This probability is the **observed relative frequency** with which an event occurs. In the M&M example, we observed that 137 of the 692 M&M's were orange. The observed empirical probability for the occurrence of orange was 137/692, or 0.198.

The value assigned to the probability of event A as a result of experimentation can be found by means of the formula:

Empirical (Observed) Probability $P'(A)$

In words: *empirical probability of A* $= \dfrac{number\ of\ times\ A\ occurred}{number\ of\ trials}$

In algebra: $P'(A) = \dfrac{n(A)}{n}$ (4.1)

Notation for empirical probability: When the value assigned to the probability of an event results from experimental or empirical data, we will identify the probability of the event with the symbol $P'(\ )$.

The **theoretical** method for obtaining the probability of an event uses a *sample space*. A **sample space** is a listing of all possible outcomes from the experiment being considered (denoted by the capital letter S). When this method is used, the sample space must contain **equally likely** sample points. For example, the sample space for the rolling of one die is $S = \{1, 2, 3, 4, 5, 6\}$.

The six possible outcomes from one roll

Each **outcome** (i.e., number) is equally likely. An **event** is a subset of the sample space (denoted by a capital letter other than S; A is commonly used for the first event). Therefore, the *probability of an event* A, $P(A)$, is the ratio of the number of points that satisfy the definition of event A, $n(A)$, to the number of **sample points** in the entire sample space, $n(S)$.

Theoretical (Expected) Probability *P*(A)

In words:

$$\text{theoretical probability of A} = \frac{\text{number of times A occurs in sample space}}{\text{number of elements in sample space}}$$

In algebra: $P(A) = \dfrac{n(A)}{n(S)}$, when the elements of S are equally likely   **(4.2)**

Notes:

1. When the value assigned to the probability of an event results from a theoretical source, we will identify the probability of the event with the symbol $P(\ )$.
2. The prime symbol is *not used* with theoretical probabilities; it is used only for empirical probabilities.

E X A M P L E   4 . 1

ONE DIE

Consider one rolling of one die. Define event A as the occurrence of a number "greater than 4." In a single roll of a die, there are six possible outcomes, making $n(S) = 6$. The event "greater than 4" is satisfied by the occurrence of either a 5 or a 6; thus, $n(A) = 2$. Assuming that the die is symmetrical and that each number has an equal likelihood of occurring, the probability of A is $\frac{2}{6}$, or $\frac{1}{3}$.

E X A M P L E   4 . 2

GOLF BALLS

At a golf trade show as each visitor enters, he or she is allowed to reach into a large barrel and select, without looking, one golf ball as a door prize. The barrel contains a mixture of three brands, Titleist, Callaway, and Bridgestone, in the ratio of 2 to 1 to 1. The sample space for this simple probability experiment is S = {Titleist, Callaway, Bridgestone}. However, the sample space expressed this way is not made up of equally likely elements and therefore is not usable for assigning probabilities to the three events of the ball selected as a Titleist (T), Callaway (C), or Bridgestone (B). In order to use the sample space to assign probabilities, it must be modified to have equally likely sample points. This is easily accomplished by listing some of the elements repeatedly, as necessary, to establish the correct ratio of elements. Since there are two Titleists for one Callaway and one Bridgestone, the sample space can be thought of as where the elements are now equally likely.

$$S = \left\{ \text{T1} \quad \text{T2} \quad \text{C} \quad \text{B} \right\}$$

The probability that the ball selected is a Titleist, Callaway, or Bridgestone can now be found using the sample space and formula (4.2): $P(T) = 2/4 = 1/2 = 0.5$, $P(C) = 1/4 = 0.25$, and $P(B) = 1/4 = 0.25$.

E X A M P L E 4 . 3

A PAIR OF DICE

A pair of dice (one white, one black) is rolled one time, and the number of dots showing on each die is observed. The sample space is shown in a chart format:

Chart Representation

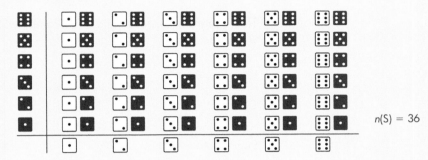

$n(S) = 36$

The sum of their dots is to be considered. A listing of the possible "sums" forms a sample space, $S = \{2, 3, 4, 5, 6, 7, 8, 9, 10, 11, 12\}$ and $n(S) = 11$. However, the elements of this sample space are not equally likely; therefore, this sample space cannot be used to find theoretical probabilities—we must use the 36-point sample space shown in the preceding chart. By using the 36-point sample space, the sample space is entirely made up of equally likely sample points, and the probabilities for the sums of 2, 3, 4, and so on, can be found quite easily. The sum of 2 represents $\{(1, 1)\}$, where the first element of the **ordered pair** is the outcome for the white die and the second element of the ordered pair is the outcome for the black die. The sum of 3 represents $\{(2, 1), (1, 2)\}$; and the sum of 4 represents $\{(1, 3), (3, 1), (2, 2)\}$; and so on. Thus, we can use formula (4.2) and the 36-point sample space to obtain the probabilities for the 11 sums.

$$P(2) = \frac{n(2)}{n(S)} = \frac{1}{36}, \; P(3) = \frac{n(3)}{n(S)} = \frac{2}{36}, \; P(4) = \frac{n(4)}{n(S)} = \frac{3}{36}$$

and so forth.

When a probability experiment can be thought of as a sequence of events, a **tree diagram** often is a very helpful way to picture the sample space.

E X A M P L E 4 . 4

USING TREE DIAGRAMS

A family with two children is to be selected at random, and we want to find the probability that the family selected has one child of each gender. Because there will always be a firstborn and a second-born child, we will use a tree diagram to show the possible arrangements of gender, thus making it possible for us to determine the probability. Start by determining the sequence of events involved—firstborn and second-born in this case. Use the tree to show the possible outcomes of the first event (shown in brown on Figure 4.1) and then add branch segments to show the possible outcomes for the second event (shown in orange in Figure 4.1).

FIGURE 4.1

Tree Diagram Representation* of Family with Two Children

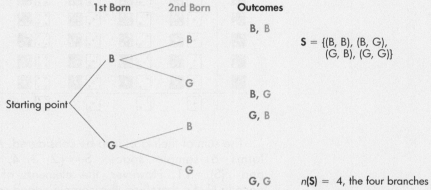

*See the *Student Solutions Manual* for information about tree diagrams.

Notes:

1. The two branch segments representing B and G for the second-born child must be drawn from each outcome for the firstborn child, thus creating the "tree" appearance.

2. There are four branches; each branch starts at the "tree root" and continues to an "end" (made up of two branch segments each), showing a possible outcome.

Because the branch segments are equally likely, assuming equal likeliness of gender, the four branches are then equally likely. This means we need only the count of branches to use formula (4.2) to find the probability of the family having one child of each gender. The two middle branches, (B, G) and (G, B), represent the event of interest, so $n(A) = n(\text{one of each}) = 2$, whereas $n(S) = 4$ because there are a total of four branches. Thus,

$$P(\text{one of each gender in family of two children}) = \frac{2}{4} = \frac{1}{2} = 0.5$$

Now let's consider selecting a family of three children and finding the probability of "at least one boy" in that family. Again the family can be thought of as a sequence of three events—firstborn, second-born, and third-born. To create a tree diagram of this family, we need to add a third set of branch segments to our two-child family tree diagram. The green branch segments represent the third child (see Figure 4.2).

Again, because the branch segments are equally likely, assuming equal likeliness of gender, the eight branches are then equally likely. This means we need only the count of branches to use formula (4.2) to find the probability

FIGURE 4.2

Tree Diagram Representation* of Family with Three Children

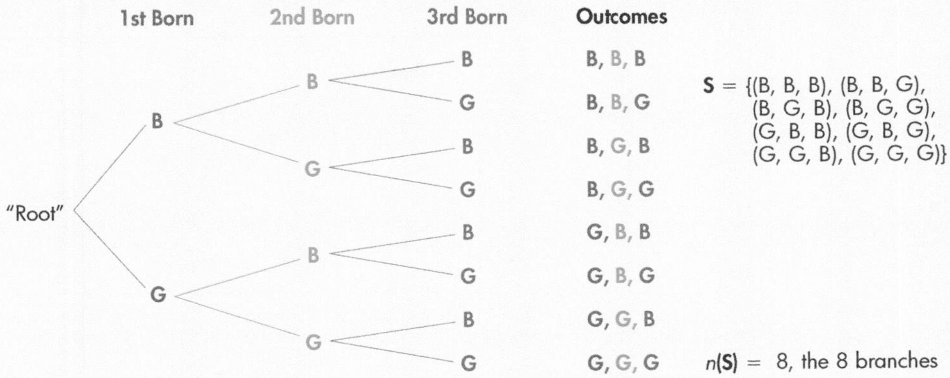

1st Born	2nd Born	3rd Born	Outcomes

S = {(B, B, B), (B, B, G), (B, G, B), (B, G, G), (G, B, B), (G, B, G), (G, G, B), (G, G, G)}

n(S) = 8, the 8 branches

*See the *Student Solutions Manual* for information about tree diagrams.

of the family having at least one boy. The top seven branches all have one or more boys, the equivalent of "at least one."

$$P(\text{at least one boy in a family of three children}) = \frac{7}{8} = 0.875$$

Let's consider one other question before we leave this example. What is the probability that the third child in this family of three children is a girl? The question is actually an easy one; the answer is 0.5, because we have assumed equal likelihood of either gender. However, if we look at the tree diagram in Figure 4.2, there are two ways to view the answer. First, if you look at only the branch segments for the third-born child, you see one of two is for a girl in each set, thus $\frac{1}{2}$, or 0.5. Also, if you look at the entire tree diagram, the last child is a girl on four of the eight branches; thus $\frac{4}{8}$, or 0.5.

When a probability question provides information about the events in the form of the probability of the various events, the number of items per set, or the percentage of each set, a **Venn diagram** often is a very helpful way to display the sample space or the information. Venn diagrams can be used to find both theoretical and empirical probabilities.

EXAMPLE 4.5

USING VENN DIAGRAMS

A lucky customer at Used Car Charlie's lot will get to randomly select one key from a barrel of keys. The barrel contains the keys to all of the cars on Charlie's lot. Charlie's inventory lists 80 cars, of which 38 are foreign models, 50 are compact models, and 22 are foreign compact models. The Venn diagram shown in Figure 4.3 summarizes Charlie's inventory. Notice that some of the 38 foreign models are compact and some are not. The same is true of the compact models; some are foreign, and some are not. Therefore,

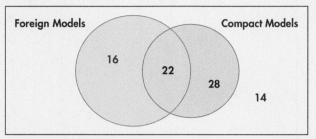

*See the *Student Solutions Manual* for information about Venn diagrams.

when breaking down this kind of information, you must start with the most specific. In this case, 22 cars are foreign and compact; they are represented by the center region of the Venn diagram. From there, you can determine how many cars are foreign but not compact and how many are compact but not foreign. Refer to Figure 4.3.

You are the lucky customer who has won the chance to draw for a free car at Used Car Charlie's, and you are about to draw 1 of the 80 keys. What is the probability that you win a nonforeign compact car? Looking at the Venn diagram, we see that foreign cars are inside the blue circle; therefore, nonforeign cars are outside the blue circle. The event of interest along with nonforeign is compact (inside the red circle), which, based on Figure 4.3, we can determine to be 28 of these cars. Using formula (4.2), we find

$$P(\text{nonforeign compact}) = \frac{28}{80} = 0.35$$

Conveniently, the Venn diagram would work equally well if the information had been given in percentages or probabilities. The diagram would look the same except that the values would be either probabilities or percentages. To be sure that the entire sample space has been covered, the sum for all regions must be exactly 1.0 in order for the labeling to be correct.

Note: Sometimes it is helpful to place a coin on the circle representing an event when you are looking at an event that did "not" occur. In the Venn diagram shown in Figure 4.3, a quarter placed on the "Foreign Models" circle would leave all nonforeign models visible.

Special attention should always be given to the sample space. Like the statistical population, the sample space must be well defined. Once the sample space is defined, you will find the remaining work much easier.

A **subjective probability** generally results from personal judgment. Your local weather forecaster often assigns a probability to the event "precipitation." For example, "There is a 20% chance of rain today," or "There is a 70% chance of snow tomorrow." In such cases, the only method available for assigning probabilities is personal judgment. These probability assignments are called *subjective probabilities*. The accuracy of subjective probabilities depends on an individual's ability to correctly assess the situation.

Properties of Probability Numbers

Whether the probability is *empirical*, *theoretical*, or *subjective*, the following properties must hold.

Property 1

In words: "A probability is always a numerical value between zero and one."

In algebra: $0 \leq$ each $P(A) \leq 1$ or $0 \leq$ each $P'(A) \leq 1$

Notes about Property 1:

1. The probability is 0 if the event cannot occur.
2. The probability is 1 if the event occurs every time.
3. Otherwise, the probability is a fractional number between 0 and 1.

Property 2

In words: "The sum of the probabilities for all outcomes of an experiment is equal to exactly one."

In algebra: $\displaystyle\sum_{\text{all outcomes}} P(A) = 1$ or $\displaystyle\sum_{\text{all outcomes}} P'(A) = 1$

Note about Property 2: The list of "all outcomes" must be a nonoverlapping set of events that includes all the possibilities (**all-inclusive**).

Notes about probability numbers:

1. Probability represents a relative frequency, whether from a sample space or a sample.
2. $P(A)$ is the ratio of the number of times an event can be expected to occur divided by the number of possibilities. $P'(A)$ is the ratio of the number of times an event did occur divided by the number of data.
3. The numerator of the probability ratio must be a positive number or zero.
4. The denominator of the probability ratio must be a positive number (greater than zero).
5. As a result of Notes 1 through 4 above, the probability of an event, whether it be empirical, theoretical, or subjective, will always be a numerical value between zero and one, inclusive.
6. The rules for probability are the same for all three types of probability: empirical, theoretical, and subjective.

How Are Empirical and Theoretical Probabilities Related?

Consider the rolling of one die and define event A as the occurrence of a "1." An ordinary die has six equally likely sides, so the theoretical probability of event A is $P(A) = \frac{1}{6}$.

What does this mean?

Do you expect to see one "1" in each trial of six rolls? Explain. If not, what results do you expect? If we were to roll the die several times and keep track of the proportion of the time event A occurs, we would observe an empirical probability for event A. What value would you expect to observe for $P'(A)$? Explain. How are the two probabilities $P(A)$ and $P'(A)$ related? Explain.

To gain some insight into this relationship, let's perform an experiment.

E X A M P L E   4 . 6

DEMONSTRATION–LAW OF LARGE NUMBERS

The experiment will consist of 20 trials. Each trial of the experiment will consist of rolling a die six times and recording the number of times the "1" occurs. Perform 20 trials.

Each row of Table 4.3 shows the results of one trial; we conduct 20 trials, so there are 20 rows. Column 1 lists the number of 1s observed in each trial (set of six rolls); column 2 lists the observed relative frequency for each trial; and column 3 lists the cumulative relative frequency as each trial was completed.

TABLE 4.3 Experimental Results of Rolling a Die Six Times in Each Trial

Trial	Column 1: Number of 1s Observed	Column 2: Relative Frequency	Column 3: Cumulative Relative Frequency	Trial	Column 1: Number of 1s Observed	Column 2: Relative Frequency	Column 3: Cumulative Relative Frequency
1	1	1/6	1/6 = 0.17	11	1	1/6	10/66 = 0.15
2	2	2/6	3/12 = 0.25	12	0	0/6	10/72 = 0.14
3	0	0/6	3/18 = 0.17	13	2	2/6	12/78 = 0.15
4	1	1/6	4/24 = 0.17	14	1	1/6	13/84 = 0.15
5	0	0/6	4/30 = 0.13	15	1	1/6	14/90 = 0.16
6	1	1/6	5/36 = 0.14	16	3	3/6	17/96 = 0.18
7	2	2/6	7/42 = 0.17	17	0	0/6	17/102 = 0.17
8	2	2/6	9/48 = 0.19	18	1	1/6	18/108 = 0.17
9	0	0/6	9/54 = 0.17	19	0	0/6	18/114 = 0.16
10	0	0/6	9/60 = 0.15	20	1	1/6	19/120 = 0.16

Figure 4.4a shows the fluctuation (above and below) of the observed probability, $P'(A)$ (Table 4.3, column 2), about the theoretical probability, $P(A) = \frac{1}{6}$, whereas Figure 4.4b shows the fluctuation of the cumulative relative frequency (Table 4.3, column 3) and how it becomes more stable. In fact, the cumulative relative frequency becomes relatively close to the theoretical or expected probability $\frac{1}{6}$, or $0.166\overline{6} = 0.167$.

FIGURE 4.4

Fluctuations Found in the Die-Tossing Experiment (a) Relative Frequency

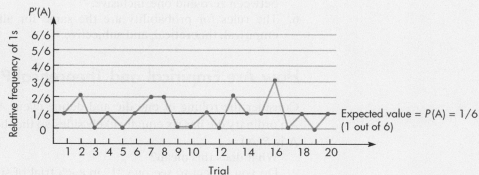

(b) Cumulative
Relative Frequency

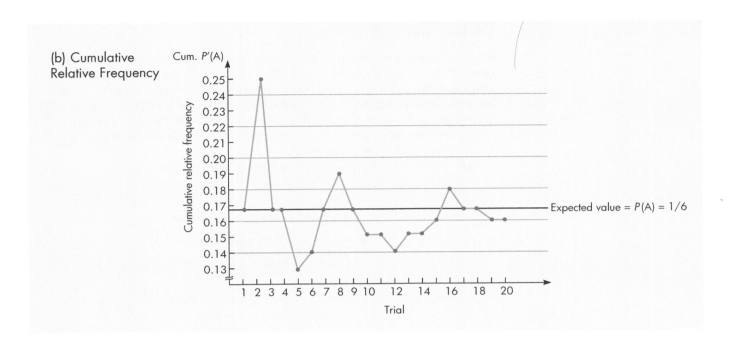

A cumulative graph such as that shown in Figure 4.4b demonstrates the idea of a **long-term average** and is often referred to as the *law of large numbers*.

> **Law of large numbers** As the number of times an experiment is repeated increases, the ratio of the number of successful occurrences to the number of trials will tend to approach the theoretical probability of the outcome for an individual trial.

The law of large numbers is telling us that the larger the number of experimental trials, n, the closer the empirical probability, $P'(A)$, is expected to be to the true or theoretical probability, $P(A)$. This concept has many applications. The preceding die-tossing experiment is an example in which we can easily compare actual results against what we expected to happen; it gave us a chance to verify the claim of the law of large numbers.

Example 4.7 is an illustration in which we live with the results obtained from large sets of data when the theoretical expectation is unknown.

E X A M P L E   4 . 7

USES OF EMPIRICAL PROBABILITIES

The key to establishing proper life insurance rates is using the probability that the insureds will live 1, 2, or 3 years, and so forth, from the time they purchase their policies. These probabilities are derived from actual life and death statistics and hence are empirical probabilities. They are published by the government and are extremely important to the life insurance industry.

Probabilities as Odds

Probabilities can be and are expressed in many ways; we see and hear many of them in the news nearly every day (most of the time, they are subjective probabilities). **Odds** are a way of expressing probabilities by expressing the number of ways an event can happen compared to the number of ways it can't happen. The statement "It is four times more likely to rain tomorrow (R) than not rain (NR)" is a probability statement that can be expressed as odds: "The odds are 4 to 1 in favor of rain tomorrow" (also written 4:1).

The relationship between odds and probability is shown here.

> If the odds in favor of an event A are **a to b** (or **a:b**), then
>
> 1. The odds against event A are **b to a** (or **b:a**).
>
> 2. The probability of event A is $P(A) = \dfrac{a}{a+b}$.
>
> 3. The probability that event A will not occur is $P(\text{not } A) = \dfrac{b}{a+b}$.

To illustrate this relationship, consider the statement "The odds favoring rain tomorrow are 4 to 1." Using the preceding notation, $a = 4$ and $b = 1$. Therefore, the probability of rain tomorrow is $\frac{4}{4+1}$, or $\frac{4}{5} = 0.8$. The odds against rain tomorrow are 1 to 4 (or 1:4), and the probability that there will be no rain tomorrow is $\frac{1}{4+1}$, or $\frac{1}{5} = 0.2$.

APPLIED EXAMPLE 4.8

Making It to the Next Level

250 NCAA Student-Athletes Drafted by Pros

13,600 NCAA Senior Student-Athletes

17,500 NCAA Freshman Roster Positions

306,200 High School Senior Student-Athletes

Key: 1 small football = 500 players

MAKING IT TO THE NEXT LEVEL

Many young men aspire to become professional athletes. Only a few make it to the big time, as indicated in the following graph. For every 13,600 college senior football players, only 250 are drafted by a professional team; that translates to a probability of only 0.018 (250/13,600).

Student-Athletes	Football
High School Senior Student-Athletes	306,200
NCAA Freshman Roster Positions	17,500
NCAA Senior Student-Athletes	13,600
NCAA Student-Athletes Drafted	250

Source: http://www.ncaa.org/

There are many other interesting specifics hidden in this information. For example, many high school boys dream of being a pro football player, but according to these numbers, the probability of a high school senior even being drafted by the pros is only 0.000816 (250/306,200).

Once a player has made a college football team, he might be very interested in the odds that he will play as a senior. Of the 17,500 players making a college freshman team, 13,600 play as seniors, while 3,900 do not. Thus, if a player has made a college team, the odds he will play as a senior are 13,600 to 3,900, which reduces to 136 to 39. The college senior who is playing is interested in his chances of making the next level. We see that of the 13,600 college seniors, only 250 are drafted by the pros, while 13,350 are not; thus the odds against him making the next level are 13,350 to 250, which reduces to 267 to 5. Odds are strongly against him being drafted, and the odds against him making the team are somewhat stronger.

Comparison of Probability and Statistics

Probability and **statistics** are two separate but related fields of mathematics. It has been said that "probability is the vehicle of statistics." That is, if it were not for the laws of probability, the theory of statistics would not be possible.

Let's illustrate the relationship and the difference between these two branches of mathematics by looking at two boxes. We know that the probability box contains five blue, five red, and five white poker chips. Probability tries to answer questions such as, "If one chip is randomly drawn from this box, what is the chance that it will be blue?" On the other hand, in the statistics box we don't know what the combination of chips is. We draw a sample and, based on the findings in the sample, make conjectures about what we believe to be in the box. Note the difference: Probability asks you about the chance that something specific, like drawing a blue chip, will happen when you know the possibilities (that is, you know the population). Statistics, on the other hand, asks you to draw a sample, describe the sample (descriptive statistics), and then make inferences about the population based on the information found in the sample (inferential statistics).

SECTION 4.1 EXERCISES

4.1 a. If you bought a bag of M&M's, what color would you expect to see the most? What color the least? Why?

b. If you bought a bag of M&M's, would you expect to find the percentages listed previously in Table 4.2 (p. 172)? If not, why and what would you expect?

4.2 a. Construct a bar graph showing the Table 4.2 percentages (p. 172) obtained from the 692 M&M's.

b. Based on your graph, which color M&M occurred most often? How does this show on your graph?

c. Based on your graph, which color M&M occurred the least? How does this show on your graph?

4.3 If you were given a small bag of M&M's with 40 candies in it, using the percentages in Table 4.2 (p. 172), how many of each color would you "expect" to find?

4.4 Bad Charts? Just as there are bad graphs (as seen in Section 2.7), there are bad charts—charts that are misleading and hard to read. MADD reported the following data about the 6764 holiday traffic fatalities that occurred in 2002. What is wrong with the numbers on this chart?

Holiday 2002	Traffic Fatalities	Alcohol-Related Fatalities
New Year's Eve (2001)	118	45
New Year's Day	165	94
New Year's holiday	575	301
Super Bowl Sunday	147	86
St. Patrick's Day	158	72
Memorial Day	491	237
Fourth of July	683	330
Labor Day weekend	541	300
Halloween	268	109
Thanksgiving	543	255
Thanksgiving–New Year's	4019	1561
Christmas	130	68
New Year's Eve (2002)	123	57

Source: Mothers Against Drunk Driving (MADD). http://www.infoplease.com/

a. The column totals are not included because they would be meaningless values. Examine the table and explain why.

b. What is the total number of holiday alcohol-related traffic fatalities for 2002?

c. Describe how you would organize this chart to make it more meaningful.

4.5 If you roll a die 40 times and 9 of the rolls result in a "5," what empirical probability do you observe for the event?

4.6 Explain why an empirical probability, an observed proportion, and a relative frequency are actually three different names for the same thing.

4.7 "Is my class watching too much television on school nights?" This was a question that Mrs. Gordon wondered with respect to her seventh graders. She did a quick poll in class and found the following results:

Hours	Number
0	2
1	3
2	2
3	0
4	3
5	2
6	1

a. What percentage of the class is not watching television on school nights?

b. What percentage of the class is watching at most 2 hours of television on school nights?

c. What percentage of the class is watching at least 4 hours of television on school nights?

4.8 Webster Aquatic Center offers various levels of swim lessons year-round. The September 2008 Monday and Wednesday evening lessons included instructions for Water Babies through Adults. The number in each classification is given in the table that follows.

Swim Lesson Types	No. of Participants
Water Babies	9
Tiny Tots	7
Tadpoles	6
Level 2	11
Level 3	10
Level 4	9
Level 5	3
Adults	2
Total	57

If one participant is selected at random, find the probability of the following:

a. The participant is in Tiny Tots.

b. The participant is in the Adults lesson.

c. The participant is in a Level 2 to a Level 5 lesson.

4.9 The table here shows the average number of births per day in the United States as reported by the CDC (Centers for Disease Control).

Based on this information, what is the probability that one baby identified at random was:

a. Born on a Monday?

b. Born on a weekend?

c. Born on a Tuesday or Wednesday?

d. Born on a Wednesday, Thursday, or Friday?

Day	Number
Sunday	7,563
Monday	11,733
Tuesday	13,001
Wednesday	12,598
Thursday	12,514
Friday	12,396
Saturday	8,605
Total	78,410

4.10 The 2007 Current Population Survey reported the following results on U.S. annual household income (in thousands). The survey is a joint effort between the Census Bureau and the Bureau of Labor Statistics.

Annual Household Income	Number
Less than $15,000	15,506
$15,000–$29,999	19,842
$30,000–$49,999	22,739
$50,000–$74,999	21,268
$75,000–$99,999	13,841
$100,000 or more	23,586
Total	116,782

Source: http://www.census.gov/

Suppose one household is to be selected at random for a follow-up interview.
Find the probability of the following events:

a. The annual household income is $49,999 or less.

b.   The annual household income is $75,000 or more.

c.   The annual household income is between $30,000 and $99,999.

d.   The annual household income is at least $100,000.

4.11 There is a large variation in price for private four-year colleges in the United States. The average 2007–2008 tuition and fees ranged from $3,000 to over $36,000 annually, according to CollegeBoard (www. collegeboard.com/). The distribution of full-time undergraduates at private four-year institutions is:

Tuition and Fees	Percent of Full-Time Undergraduates
$36,000 and over	5
$33,000 to $35,999	14
$30,000 to $32,999	8
$27,000 to $29,999	8
$24,000 to $26,999	17
$21,000 to $23,999	12
$18,000 to $20,999	11
$15,000 to $17,999	9
$12,000 to $14,999	6
$9,000 to $11,999	2
$6,000 to $8,999	2
$3,000 to $5,999	6
	100%

Supposing a college student at a private four-year institution is randomly selected to participate in a survey, what is the probability that the student:

a.   Attends a college that costs less than $12,000 annually?

b.   Attends a college that costs $30,000 or more annually?

c.   Attends a college that costs between $15,000 and $26,999 annually?

d.   Attends a college that costs less than $3000 annually?

4.12 A box contains one each of $1, $5, $10, and $20 bills.

a.   One bill is selected at random; list the sample space.

b.   Two bills are drawn at random (without replacement); list the sample space as a tree diagram.

4.13 One single-digit number is selected randomly.

a.   List the sample space.

b.   What is the probability of each single digit?

c.   What is the probability of an even number?

4.14 A single die is rolled. What is the probability that the number on top is the following?

a.   A 3

b.   An odd number

c.   A number less than 5

d.   A number no greater than 3

4.15 A bowl contains two kinds of identical-looking, foil-wrapped, chocolate egg-shaped candies. All but 42 of them are milk chocolate and all but 35 are dark chocolate.

a.   How many of each kind are in the bowl?

b.   How many candies are there in the bowl?

c.   If one chocolate is selected at random, what is the probability that it is milk chocolate?

d.   If one chocolate is selected at random, what is the probability that it is milk or dark chocolate?

e.   If one chocolate is selected at random, what is the probability that it is milk and dark chocolate?

4.16 A bowl contains three kinds of identical-looking, foil-wrapped, chocolate egg-shaped candies. All but 50 of them are milk chocolate, all but 50 are dark chocolate, and all but 50 are semi-sweet chocolate.

a.   How many candies are there in the bowl?

b.   How many of each kind are in the bowl?

c.   If one chocolate is selected at random, what is the probability that it is milk chocolate?

d.   If one chocolate is selected at random, what is the probability that it is milk or dark chocolate?

e.   If one chocolate is selected at random, what is the probability that it is milk and dark chocolate?

4.17 Use either the random number table (Appendix B), a calculator, or a computer (see p. 90) to simulate the following:

a.   The rolling of a die 50 times; express your results as relative frequencies.

b.   The tossing of a coin 100 times; express your results as relative frequencies.

4.18 Use either the random number table (Appendix B), a calculator, or a computer (see p. 90) to simulate the random selection of 100 single-digit numbers, 0 through 9.

a.   List the 100 digits.

b.   Prepare a relative frequency distribution of the 100 digits.

c.   Prepare a relative frequency histogram of the distribution in part b.

4.19 A pair of dice is to be rolled. In Example 4.3, the probability for each of the possible sums was discussed and three of the probabilities, $P(2)$, $P(3)$, and $P(4)$, were found. Find the probability for each of the remaining sums of two dice: $P(5)$, $P(6)$, $P(7)$, $P(8)$, $P(9)$, $P(10)$, $P(11)$, and $P(12)$.

4.20 Two dice are rolled. Find the probabilities in parts b–e. Use the sample space given in Example 4.3 (p. 175).

a. Why is the set $\{2, 3, 4, \ldots, 12\}$ not a useful sample space?

b. P(white die is an odd number)

c. P(sum is 6)

d. P(both dice show odd numbers)

e. P(number on black die is larger than number on white die)

4.21 Take two dice (one white and one colored) and roll them 50 times, recording the results as ordered pairs [(white, color); for example, (3, 5) represents 3 on the white die and 5 on the colored die]. (You could simulate these 50 rolls using a random number table or a computer.) Then calculate each observed probability:

a. P'(white die is an odd number)

b. P'(sum is 6)

c. P'(both dice show odd numbers)

d. P'(number on colored die is larger than number on white die)

e. Explain why these answers and the answers found in Exercise 4.20 are not exactly the same.

4.22 Use a random number table or a computer to simulate rolling a pair of dice 100 times.

a. List the results of each roll as an ordered pair and a sum.

b. Prepare an ungrouped frequency distribution and a histogram of the sums.

c. Describe how these results compare with what you expect to occur when two dice are rolled.

MINITAB

Choose: Calc > Random Data > Integer
Enter: Number of rows to generate: **100**
           Store in column(s): **C1 C2**
           Minimum value: **1**
           Maximum value: **6 > OK**
Choose: Calc > Calculator
Enter: Store result in variable: **C3**
           Expression: **C1 + C2 > OK**

Choose: Stat > Tables > Tally Individual Variables
Enter: Variable: **C3**
Select: **Counts > OK**

Use the MINITAB commands on page 52 to construct a frequency histogram of the data in C3. (Use Binning > midpoint and midpoint positions 2:12/1 if necessary.)

Excel

Enter 1, 2, 3, 4, 5, 6 into column A, label C1: **Die1**; D1: **Die2**; E1: **Dice**, and activate B1.

Choose: Home > Number pulldown >
           Number > Category: Number
Enter: Decimal places: **8 > OK**
Enter: **1/6** in B1
Drag: Bottom right corner of B1 down for 6
           entries
Choose: Data > Data Analysis > Random
           Number Generation > OK
Enter: Number of Variables: **2**
           Number of Random Numbers: **100**
           Distribution: **Discrete**
           Value and Probability Input Range:
           (A1:B6 or select cells)
Select: **Output Range**
Enter: **(C2** or select cells) > OK

Activate the **E2** cell.

Enter: **= C2 + D2 > Enter**
Drag: Bottom right corner of E2 down for 100
           entries
Choose: Insert > Tables > Pivot Table
           pulldown > Pivot Chart
Select: Select a table or range
Enter: Range: **(E1:E101** or select cells) > Next
Select: **Existing Worksheet**
Enter: **(F1** or select cell) > OK

In Pivot Table

Drag: "Dice" heading into both Axis Fields and Σ
           values area
Select: Defer Layout Update > Update

Double-click the "sum of dice" in data area box; then continue with:

Select: Summarize by: **Count**

Label column J "sums" and input the numbers 2, 3, 4, . . . , 12 into it. Use the Excel histogram commands on page 53 with column E as the input range and column J as the bin range, or use chart given.

TI-83/84 Plus

Choose: MATH > PRB >
           5:randInt(
Enter: **1,6,100)**
Choose: STO→ > 2nd L1

Repeat preceding for L2.

Choose: STAT > EDIT >
            1:Edit
Highlight: L3
Enter: L3 = L1 + L2
Choose: 2nd > STAT PLOT >
            1:Plot1
Choose: WINDOW
Enter: −.5, 12.5, 1, −10, 40, 10,1
Choose: TRACE > > >

4.23 Let x be the success rating of a new television show. The following table lists the subjective probabilities assigned to each x for a particular new show by three different media critics. Which of these sets of probabilities are inappropriate because they violate a basic rule of probability? Explain.

	Critic		
	A	B	C
Highly Successful	0.5	0.6	0.3
Successful	0.4	0.5	0.3
Not Successful	0.3	−0.1	0.3

4.24 a. A balanced coin is tossed twice. List a sample space showing the possible outcomes.

  b. A biased coin (it favors heads in a ratio of 3 to 1) is tossed twice. List a sample space showing the possible outcomes.

4.25 A group of files in a medical clinic classifies the patients by gender and by type of diabetes (type 1 or type 2). The groupings may be shown as follows. The table gives the number in each classification.

	Type of Diabetes	
Gender	1	2
Male	30	15
Female	35	20

a. Display the information on this 2 × 2 table as a Venn diagram using "type 1" and "male" as the two events displayed as circles. Explain how the Venn diagram and the given 2 × 2 table display the same information.

If one file is selected at random, find the probability of the following:

b. The selected individual is female.

c. The selected individual has type 2 diabetes.

4.26 Researchers have for a long time been interested in the relationship between cigarette smoking and lung cancer. The following table shows the percentages of adult females observed in a recent study.

	Smokes	Does Not Smoke
Gets Cancer	0.06	0.03
Does Not Get Cancer	0.15	0.76

a. Display the information on this 2 × 2 table as a Venn diagram using "smokes" and "gets cancer" as the two events displayed as circles. Explain how the Venn diagram and the given 2 × 2 table display the same information.

Suppose an adult female is randomly selected from this particular population. What is the probability of the following?

b. She smokes and gets cancer.

c. She smokes.

d. She does not get cancer.

e. She does not smoke or does not get cancer.

f. She gets cancer if she smokes.

g. She does not get cancer, knowing she does not smoke.

4.27 A parts store sells both new and used parts. Sixty percent of the parts in stock are used. Sixty-one percent are used or defective. If 5% of the store's parts are defective, what percentage is both used and defective? Solve using a Venn diagram.

4.28 Union officials report that 60% of the workers at a large factory belong to the union, 90% make more than $12 per hour, and 40% belong to the union and make more than $12 per hour. Do you believe these percentages? Explain. Solve using a Venn diagram.

4.29 a. Explain what is meant by the statement: "When a single die is rolled, the probability of a 1 is $\frac{1}{6}$."

  b. Explain what is meant by the statement: "When one coin is tossed one time, there is a 50-50 chance of getting a tail."

4.30 Skillbuilder Applet Exercise demonstrates the law of large numbers and also allows you to see if you have psychic powers. Repeat the simulations at least 50 times, guessing between picking either a red card or a black card from a deck of cards.

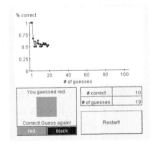

a. What proportion of the time did you guess correctly?

b. As you made more and more guesses, were your proportions starting to stabilize? If so, at what value? Does this value make sense for the experiment? Why?

c. How might you know if you have ESP?

4.31 An experiment consists of two trials. The first is tossing a penny and observing whether it lands with heads or tails facing up; the second is rolling a die and observing a 1, 2, 3, 4, 5, or 6.

a. Construct the sample space using a tree diagram.

b. List your outcomes as ordered pairs, with the first element representing the coin and the second, the die.

4.32 Use a computer (or a random number table) to simulate 200 trials of the experiment described in Exercise 4.31: the tossing of a penny and the rolling of a die. Let 1 = H and 2 = T for the penny, and 1, 2, 3, 4, 5, 6 for the die. Report your results using a cross-tabulated table showing the frequency of each outcome.

a. Find the relative frequency for heads.

b. Find the relative frequency for 3.

c. Find the relative frequency for (H, 3).

4.33 Using a coin, perform the experiment discussed on pages 180–181. Toss a coin 10 times, observe the number of heads (or put 10 coins in a cup, shake and dump them into a box, and use each toss for a block of 10), and record the results. Repeat until you have 200 tosses. Chart and graph the data as individual sets of 10 and as cumulative relative frequencies. Do your data tend to support the claim that $P(\text{head}) = \frac{1}{2}$? Explain.

4.34 A chocolate kiss is to be tossed into the air and will be landing on a smooth hard surface (similar to tossing a coin or rolling dice).

a. What proportion of the time do you believe the kiss will land "point up" 🧪 (as opposed to "point down" 🍼)?

b. Let's estimate the probability that a chocolate kiss lands "point up" when it lands on a smooth hard surface after being tossed. Using a chocolate kiss, with the wrapper still on, perform the die experiment discussed on pages 180–181. Toss the kiss 10 times, record the number of "point up" landings (or put 10 kisses in a cup, shake and dump them onto a hard smooth surface, and use each toss for a block of 10), and record the results. Repeat until you have 200 tosses. Chart and graph the data as individual sets of 10 and as cumulative relative frequencies.

c. What is your best estimate for the true $P($🧪$)$? Explain.

d. If unwrapped kisses were to be tossed, what do you think the probability of "point up" landings would be? Would it be different? Explain.

e. Unwrap the chocolate kisses used in part b and repeat the experiment.

f. Are the results in part e what you anticipated? Explain.

4.35 A box contains marbles of five different colors: red, green, blue, yellow, and purple. There is an equal number of each color. Assign probabilities to each color in the sample space

4.36 Suppose a box of marbles contains equal numbers of red marbles and yellow marbles but twice as many green marbles as red marbles. Draw one marble from the box and observe its color. Assign probabilities to the elements in the sample space.

4.37 If four times as many students pass a statistics course as those who fail and one statistics student is selected at random, what is the probability that the student will pass statistics?

4.38 Events A, B, and C are defined on sample space S. Their corresponding sets of sample points do not intersect, and their union is S. Furthermore, event B is twice as likely to occur as event A, and event C is twice as likely to occur as event B. Determine the probability of each of the three events.

4.39 The odds for the Saints winning next year's Super Bowl are 1 to 6.

a. What is the probability that the Saints will win next year's Super Bowl?

b. What are the odds against the Saints winning next year's Super Bowl?

4.40 The NCAA men's basketball season starts with 327 college teams all dreaming of making it to "the big dance" and attaining the National Championship. Sixty-five teams are selected for the tournament and only one wins it all.

a. What are the odds against a team being selected for the tournament?

b. What are the odds of a team that is in the tournament winning the National Championship?

c. Now wait a minute! What assumption did you make in order to answer the above questions? Does this seem realistic?

4.41 Alan Garole was a jockey at Saratoga Springs Raceway during the 7/23/08 to 9/1/08 season. He had 195 starts, with 39 first places, 17 second places, and 28 third places. If all the 2008 racing season conditions had held for Alan Garole at the beginning of the 2009 season, what would have been:

a.   the odds in favor of Alan Garole coming in first place during the 2009 racing season at Saratoga?

b.   the probability of Alan Garole coming in first place during the 2009 racing season at Saratoga?

c.   the odds in favor of Alan Garole placing (coming in first, second, or third) during the 2009 racing season at Saratoga?

d.   the probability of Alan Garole placing during the 2009 racing season at Saratoga?

e.   Based on the above statistics, should you bet that Alan Garole came in first or placed? Why?

4.42 Applied Example 4.8, "Making It to the Next Level," on page 182, uses two large footballs at the bottom of the graphic. If the scale used for the upper part of the graph were to be used for the high school seniors, how many small footballs would be needed?

4.43 Many young men aspire to become professional athletes. Only a few make it to the big time, as indicated in the table.

Student-Athletes	Baseball
High School Student-Athletes	470,671
High School Senior Student-Athletes	134,477
NCAA Student-Athletes	28,767
NCAA Freshman Roster Positions	8,219
NCAA Senior Student-Athletes	6,393
NCAA Student-Athletes Drafted	600

a.   What are the odds in favor of a high school male athlete playing as an NCAA senior?

b.   What are the odds against a male baseball player who made an NCAA freshman roster being drafted by a pro team?

c.   What is the probability of a high school senior student-athlete being drafted by a pro baseball team?

d.   What is the probability of a high school senior student-athlete still playing baseball as an NCAA senior student-athlete?

4.44 Many young women aspire to become professional athletes. Only a few make it to the big time, as indicated in the table.

Student-Athletes	Women's Basketball
High School Student-Athletes	452,929
High School Senior Student-Athletes	129,408
NCAA Student-Athletes	15,096
NCAA Freshman Roster Positions	4,313
NCAA Senior Student-Athletes	3,355
NCAA Student-Athletes Drafted	32

a.   What are the odds in favor of a high school female athlete being drafted by a pro basketball team?

b.   What are the odds against a female basketball player who makes a freshman college roster playing as a senior?

c.   What is the probability of a high school female student-athlete being drafted by a pro basketball team?

d.   What is the probability of an NCAA senior student-athlete being drafted by a pro basketball team?

4.45 A bowl contains four kinds of identical-looking, foil-wrapped, chocolate egg-shaped candies. All but 50 of them are milk chocolate, all but 50 are dark chocolate, all but 50 are semi-sweet chocolate, and all but 60 are white chocolate.

a.   How many candies are there in the bowl?

b.   How many of each kind of chocolate are in the bowl?

c.   If one chocolate is selected at random, what is the probability that it is white chocolate?

d.   If one chocolate is selected at random, what is the probability that it is white or milk chocolate?

e.   If one chocolate is selected at random, what is the probability that it is milk and dark chocolate?

f.   If two chocolates are selected at random, what is the probability that both are white chocolate?

g.   If two chocolates are selected at random, what is the probability that one is dark and one is semi-sweet chocolate?

h.   If two chocolates are selected at random, what is the probability that neither is milk chocolate?

4.46 A bowl contains 100 identical-looking, foil-wrapped, chocolate egg-shaped candies of four kinds. The candies are either milk or dark chocolate with either a nut or a raisin filling. All but 40 of them are milk chocolate, all but 56 are nut, and all but 29 are nut-filled or milk chocolate.

a.   How many of each kind of chocolate are in the bowl?

(continue on page 190)

b. If one chocolate is selected at random, what is the probability that it is milk chocolate?

c. If one chocolate is selected at random, what is the probability that it is dark or raisin?

d. If one chocolate is selected at random, what is the probability that it is dark and raisin?

e. If one chocolate is selected at random, what is the probability that it is neither dark nor raisin?

f. If one chocolate is selected at random, what is the probability that it is not dark but is nut?

g. If one chocolate is selected at random, what is the probability that it is milk or nut?

4.47 Which of the following illustrates probability? statistics?

a. Determining how likely it is that a "6" will result when a die is rolled

b. Studying the weights of 35 babies to estimate weight gain in the first month after birth

4.48 Which of the following illustrates probability? statistics?

a. Collecting the number of credit hours from 100 students to estimate the average number of credit hours per student at a particular community college

b. Determining how likely it is to win the New York Lottery

4.49 Classify each of the following as a probability or a statistics problem:

a. Determining whether a new drug shortens the recovery time from a certain illness

b. Determining the chance that heads will result when a coin is tossed

c. Determining the amount of waiting time required to check out at a certain grocery store

d. Determining the chance that you will be dealt a "blackjack"

4.50 Classify each of the following as a probability or a statistics problem:

a. Determining how long it takes to handle a typical telephone inquiry at a real estate office

b. Determining the length of life for the 100-watt light bulbs a company produces

c. Determining the chance that a blue ball will be drawn from a bowl that contains 15 balls, of which 5 are blue

d. Determining the shearing strength of the rivets that your company just purchased for building airplanes

e. Determining the chance of getting "doubles" when you roll a pair of dice

4.2 Conditional Probability of Events

Many of the probabilities that we see or hear being used on a daily basis are the result of conditions existing at the time. In this section, we will learn about *conditional probabilities*.

Conditional probability an event will occur A conditional probability is the relative frequency with which an event can be expected to occur under the condition that additional, preexisting information is known about some other event. $P(A|B)$ is used to symbolize the probability of event A occurring under the condition that event B is known to already exist.

Some ways to say or express the conditional probability, $P(A \mid B)$, are:

1st: The "*probability of* A, *given* B"
2nd: The "*probability of* A, *knowing* B"
3rd: The "*probability of* A *happening, knowing* B *has already occurred*"

The concept of conditional probability is actually very familiar and occurs very frequently without our even being aware of it. The news media often report many conditional probability values. However, they don't make the point that it is a conditional probability and it simply passes for everyday arithmetic, as illustrated in the following example.

E X A M P L E   4 . 9

FINDING PROBABILITIES FROM A TABLE OF PERCENTAGES

From a National Election Pool exit poll of 13,660 voters in 250 precincts across the country during the 2008 presidential election, we have the following:

Gender	Percentage of Voters	Percent for Obama	Percent for McCain	Percent for Others
Men	48	44	54	2
Women	52	56	46	1
Age				
18–29	14	63	36	1
30–44	27	44	55	1
45–64	39	45	44	1
65 and over	20	52	48	0

All of the percentages on the table above are to the nearest integer.

One person is to be selected at random from the sample of 13,660 voters. Using the table, find the answers to the following probability questions.

1. What is the probability that the person selected is a man?
   Answer: **0.48**.
   Expressed in equation form:
   P(Voter selected is a man) = 0.48

2. What is the probability that the person selected is of age 18 to 29?
   Answer: **0.14**.
   Expressed in equation form:
   P(Voter selected is of age 18 to 29) = 0.14

3. What is the probability that the person selected voted for McCain, knowing the voter was a woman?   Answer: **0.46**.
   Expressed in equation form: P(McCain | woman) = 0.46

4. What is the probability that the person selected voted for Obama, if the voter was 65 or over?   Answer: **0.52**.
   Expressed in equation form: P(Obama | 65 or over) = 0.52

Note: The first two are simple probabilities, while the last two are conditional probabilities.

EXAMPLE 4.10

FINDING PROBABILITIES FROM A TABLE OF COUNT DATA

From a nationwide exit poll of 1000 voters in 25 precincts across the country during the 2008 presidential election, we have the following:

Education	Number for Obama	Number for McCain	Number for Others	Number of Voters
No high school	19	20	1	40
HS graduate	114	103	3	220
Some college	172	147	1	320
College grad	135	119	6	260
Postgraduate	70	88	2	160
	510	477	13	1000

One person is to be selected at random from the above sample of 1000 voters. Using the table, find the answers to the following probability questions.

1. What is the probability that the person selected voted for McCain, knowing that the voter is a high school graduate?
   Answer: **103/220** = 0.46818 = 0.47.
   Expressed in equation form:
   $P(\text{McCain} \mid \text{HS graduate}) = 103/220 = 0.46818 = \underline{0.47}$

2. What is the probability that the person selected voted for Obama, given that the voter has some college education?
   Answer: **172/320** = 0.5375 = 0.54.
   Expressed in equation form:
   $P(\text{Obama} \mid \text{some college}) = 172/320 = 0.5375 = \underline{0.54}$

3. Knowing the selected person voted for McCain, what is the probability that the voter has a postgraduate education?
   Answer: **88/477** = 0.1844 = 0.18.
   Expressed in equation form:
   $P(\text{postgraduate} \mid \text{McCain}) = 88/477 = 0.1844 = \underline{0.18}$

4. Given that the selected person voted for Obama, what is the probability that the voter does not have a high school education?
   Answer: **19/510** = 0.0372 = 0.04.
   Expressed in equation form:
   $P(\text{no high school} \mid \text{Obama}) = 19/510 = 0.0372 = \underline{0.04}$

Notes:

1. The conditional probability notation is very informative and useful. When you express a conditional probability in equation form, it is to your advantage to use the most complete notation—that way, when you read the information back, all the information is there.
2. When finding a conditional probability, some of the possibilities will be eliminated as soon as the condition is known. Consider question 4 in Example 4.10. As soon as the conditional "given that the selected person voted for Obama" is stated, the 477 who voted for McCain and the 13 voting for Others are eliminated, leaving the 510 possible outcomes.

S E C T I O N   4 . 2   E X E R C I S E S

4.51 Three hundred viewers were asked if they were satisfied with the TV coverage of a recent disaster.

	Female	Male
Satisfied	80	55
Not satisfied	120	45

One viewer is to be randomly selected from those surveyed.

a. Find P(satisfied).

b. Find P(S | female).

c. Find P(S male).

4.52 Saturday mornings are busy times at the Webster Aquatic Center. Swim lessons ranging from Red Cross Level 2, Fundamental Aquatic Skills, through Red Cross Level 6, Swimming and Skill Proficiency, are offered during two sessions.

Level	Number of People in 10 a.m. Class	Number of People in 11 a.m. Class
2	12	12
3	15	10
4	8	8
5	2	0
6	2	0

Lauren, the program coordinator, is going to randomly select one swimmer to be interviewed for a local television spot on the center and its swim program. What is the probability that the selected swimmer is in the following?

a. A Level 3 class

b. The 10 a.m. class

c. A Level 2 class, given that it is the 10 a.m. session

d. The 11 a.m. session, given that it is the Level 6 class

4.53 *The World Factbook*, 2008, reports that U.S. airports have the following numbers of meters of runways that are either paved or unpaved.

Total Runway (meters)	Number of Airports	
	Paved	Unpaved
Over 3047 m	190	0
2438 to 3047 m	227	6
1524 to 2437 m	1464	156
914 to 1523 m	2307	1734
Under 914 m	958	7909
Total	5146	9805

Source: *The World Factbook*, January 2008. https://www.cia.gov/

If one of these airports is selected at random for inspection, what is the probability that it will have

a. paved runways?

b. 914 to 2437 meters of runway?

c. less than 1524 meters of runway and be unpaved?

d. more than 2437 meters of runway and be paved?

e. paved runway, given that it has more than 1523 meters of runway?

f. unpaved runway, knowing that it has less than 1524 meters of runway?

g. less than 1524 meters of runway, given that it is unpaved?

4.54 During the Spring 2009 semester at Monroe Community College, a random sample of students was questioned on their knowledge of the meaning of "sustainability." The primary motivation for the survey was to investigate how interested students might be in a Sustainability Certificate and to discover the best means of informing them of this option. The following table lists how much 224 students agreed with the statement "Sustainability is important to me."

Level of Agreement with Statement "Sustainability is important to me"

Generation (ages)	Strongly Agree	Agree	Disagree	Strongly Disagree	Total
Millennium Y (18–29)	74	109	11	1	195
Generation X (30–44)	14	8	1	0	23
Baby Boomers (45+)	2	3	0	1	6
All Respondents	90	120	12	2	224

Source: Monroe Community College, Sustainability Certificate Survey

Find the probability that a randomly selected student:

a. "strongly agrees" that sustainability is important to her.

b. is a member of Generation X.

c. "disagrees" with the importance of sustainability to her, given that she is a member of the Millennium Y generation.

d. is a member of the Baby Boomers, given that she "agrees" with the importance of sustainability.

4.55 A *USA Today* article, "Yum Brands builds dynasty in China" (February 7, 2005), reports on how Yum Brands, the world's largest restaurant company, is bringing the fast food industry to China, India, and other big countries. Yum Brands, a spin-off of PepsiCo, has been delivering growth of double-digit earnings in the past year.

Location and Number of Yum Brands Fast Food Stores

Store	USA	Abroad	Total
KFC	5450	7676	13,126
Pizza Hut	6306	4680	10,986
Taco Bell	5030	193	5223
Long John Silver's	1200	33	1233
A&W All-American	485	209	694
Total	18,471	12,791	31,262

Source: *USA Today*, 2/7/2005, and Yum Brands

(continue on page 194)

Suppose that when the CEO of Yum Brands was interviewed for this article, he was asked the following questions. How might he answer based on the chart?

a. What percent of your locations are in the United States?

b. What percent of your locations are abroad?

c. What percent of your stores are Pizza Huts?

d. What percent of your stores are Taco Bells, given that the location is the United States?

e. What percent of your locations are abroad, given that the store is an A&W All-American?

f. What percent of your stores are KFCs, given that the location is abroad?

g. What percent of your abroad locations are KFCs?

h. What do you notice about your answers to parts f and g? Why is this happening?

4.56 In 2007, data from two Youth Risk Behavior Surveys were analyzed to investigate seatbelt use among high school students ages 16 or older. The results were published in the September 2008 issue of *American Journal of Preventive Medicine*. Results (in percents) include the table that follows:

If one student is selected at random from this population, what is the probability that the student selected:

a. Always uses a seatbelt when driving and always uses a seatbelt when a passenger?

b. Always uses a seatbelt when driving but not always when a passenger, given that he or she is 18 or older?

c. Does not always use a seatbelt when driving but always does when a passenger, knowing he or she is 16?

d. Always wears a seatbelt when driving?

e. Does not always wear a seatbelt when driving and is 17 years old?

4.57 The *American Community Survey* reported its findings about workers' principal means of transportation to work during 2007.

Means of Transportation	Number (thousands)
All workers	139,260
Automobile	120,442
Drives self	105,955
Carpool	14,487
2-person	11,139
3-person	1,963
4+ person	1,385
Public transportation[1]	6,801
Taxicab	179
Bicycle or motorcycle	949
Walks only	3,954
Other means[2]	1,258
Works at home	5,677

NOTE: Principal means of transportation refers to the mode used most often by individual.
1. Public transportation refers to bus, streetcar, subway, or elevated trains.
2. Other means include ferryboats, surface trains, and van service.
Source: U.S. Census Bureau, Bureau of Transportation Statistics, 2007 *American Community Survey.* http://factfinder.census.gov/

a. The column total is not included because it would be a meaningless value. Examine the table and explain why.

One person is to be selected and asked additional questions as part of this survey. If that person is selected at random, find the probability for each of the following events.

b. Person selected is a member of a carpool.

c. Person selected is a member of a 2-person carpool, given that he or she carpools.

d. Person selected does not arrive by car.

e. Person selected uses public transportation, knowing that he or she doesn't use an automobile.

4.58 The five most popular colors for sport/compact cars manufactured during the 2006 model year in North America are reported here in percentages.

Sport/Compact	Percentage
1. Silver	18
2. Black	15
3. Gray	15
4. Red	15
5. Blue	13

Source: DuPont Herberts Automotive Systems, Troy, Mich. 2006 DuPont Automotive Color Popularity Survey Results. http://www.infoplease.com/

Table for Exercise 4.56

Characteristic	Always use when driving		Do not always use when driving	
	Always use when passenger	Do not always use when passenger	Always use when passenger	Do not always use when passenger
Total	38.4	20.6	3.4	37.6
Age (years)				
16	38.2	22.5	3.2	36.1
17	38.1	19.9	3.6	38.4
≥18	39.4	18.4	3.6	38.6

Source: http://www.ajpm-online.net/

a. Why does the column of percentages not total 100%?

b. Why are all probabilities based on this table conditional? What is that condition?

c. Did your favorite color make the list?

If one 2006 sport/compact car was picked at random from all sport/compact cars manufactured in the United States in 2006, what is the probability that its color is

d. black, silver, gray, red, or blue?

e. not silver?

f. black, knowing that the sport/compact car has one of the five most popular colors?

g. black, knowing that the sport/compact car has one of the five most popular colors, but not red?

4.3 Rules of Probability

Often, one wants to know the probability of a **compound event** but the only data available are the probabilities of the related simple events. (Compound events are combinations of more than one simple event.) In the next few paragraphs, the relationship between these probabilities is summarized.

Finding the Probability of "Not A"

The concept of complementary events is fundamental to finding the probability of "not A."

> **Complementary events** The *complement of an event A, $\overline{A}$,* is the set of all sample points in the sample space that do not belong to event A.

Note: The complement of event A is denoted by $\overline{A}$ (read "A complement").

A few examples of complementary events are (1) the complement of the event "success" is "failure," (2) the complement of "selected voter is Republican" is "selected voter is not Republican," and (3) the complement of "no heads" on 10 tosses of a coin is "at least one head."

By combining the information in the definition of complement with Property 2 (p. 179), we can say that

$$P(A) + P(\overline{A}) = 1.0 \text{ for any event A}$$

As a result of this relationship, we have the complement rule:

> **Complement Rule**
>
> In words: *probability of A complement = one − probability of A*
>
> In algebra: $P(\overline{A}) = 1 − P(A)$ (4.3)

Note: Every event A has a complementary event $\overline{A}$. Complementary probabilities are very useful when the question asks for the probability of "at least one." Generally, this represents a combination of several events, but the complementary event "none" is a single outcome. It is easier to solve for the complementary event and get the answer by using formula (4.3).

E X A M P L E   4 . 1 1

USING COMPLEMENTS TO FIND PROBABILITIES

Two dice are rolled. What is the probability that the sum is at least 3 (that is, 3, 4, 5, . . . , 12)?

Solution

Suppose one of the dice is black and the other is white. (See the chart in Example 4.3 on page 175; it shows all 36 possible pairs of results when rolling a pair of dice.)

Rather than finding the probability for each of the sums 3, 4, 5, . . . , 12 separately and adding, it is much simpler to find the probability that the sum is 2 ("less than 3") and then use formula (4.3) to find the probability of "at least 3," because "less than 3" and "at least 3" are complementary events.

$$P(\text{sum of 2}) = P(A) = \frac{1}{36} \text{ ("2" occurs only once in the 36-point sample space)}$$

$$P(\text{sum is at least 3}) = P(\overline{A}) = 1 - P(A) = 1 - \frac{1}{36} = \frac{35}{36} \text{ [using formula (4.3)]}$$

Finding the Probability of "A or B"

An hourly wage earner wants to estimate the chances of "receiving a promotion or getting a pay raise." The worker would be happy with either outcome. Historical information is available that will allow the worker to estimate the probability of "receiving a promotion" and "getting a pay raise" separately. In this section we will learn how to apply the **addition rule** to find the compound probability of interest.

General Addition Rule

Let A and B be two events defined in a sample space, S.

In words: *probability of A or B =*
            probability of A + probability of B − probability of A and B

In algebra: $P(A \text{ or } B) = P(A) + P(B) - P(A \text{ and } B)$ **(4.4)**

To see if the relationship expressed by the general addition rule works, let's look at Example 4.12.

E X A M P L E   4 . 1 2

UNDERSTANDING THE ADDITION RULE

A statewide poll of 800 registered voters in 25 precincts from across New York State was taken. Each voter was identified as being registered as Republican, Democrat, or other and then asked, "Are you in favor of or against the current budget proposal awaiting the governor's signature?" The resulting tallies are shown here.

	Number in Favor	Number Against	Number of Voters
Republican	136	88	224
Democrat	314	212	526
Other	14	36	50
Totals	464	336	800

Suppose one voter is to be selected at random from the 800 voters summarized in the preceding table. Let's consider the two events "The voter selected is in favor" and "The voter is a Republican." Find the four probabilities: P(in favor), P(Republican), P(in favor or Republican), and P(in favor and Republican). Then use the results to check the truth of the addition rule.

Solution

Probability the voter selected is "in favor" = P(in favor) = **464**/800 = <u>0.58.</u>

Probability the voter selected is "Republican" =
P(Republican) = **224**/800 = <u>0.28.</u>

Probability the voter selected is "in favor or Republican" = P(in favor or Republican) = (**136 + 314 + 14 + 88**)/800 = 552/800 = <u>0.69.</u>

Probability the voter selected is "in favor" and "Republican" = P(in favor and Republican) = **136**/800 = <u>0.17.</u>

Notes about finding the preceding probabilities:

1. The connective "or" means "one or the other or both"; thus, "in favor or Republican" means all voters who satisfy either event.
2. The connective "and" means "both" or "in common"; thus, "in favor and Republican" means all voters who satisfy both events.

Now let's use the preceding probabilities to demonstrate the truth of the addition rule.

Let A = "in favor" and B = "Republican." The general addition rule then becomes:

P(in favor or Republican) = P(in favor) + P(Republican) − P(in favor and Republican)

Remember: Previously we found: P(in favor or Republican) = <u>0.69</u>.

Using the other three probabilities, we see:

P(in favor) + P(Republican) − P(in favor and Republican)
= 0.58 + 0.28 − 0.17 = <u>0.69</u>

Thus, we obtain identical answers by applying the addition rule and by referring to the relevant cells in the table. You typically do not have the option of finding P(A or B) two ways, as we did here. You will instead be asked to find P(A or B) starting with P(A) and P(B). However, you will need a third piece of information. In the previous situation, we needed P(A and B). We will either need to know P(A and B) or some information that allows us to find it.

Finding the Probability of "A and B"

Suppose a criminal justice professor wants his class to determine the likeliness of the event "a driver is ticketed for a speeding violation and the driver had previously attended a defensive driving class." The students are confident they can find the probabilities of "a driver being ticketed for speeding" and "a driver who has attended a defensive driving class" separately. In this section we will learn how to apply the **multiplication rule** to find the compound probability of interest.

General Multiplication Rule

Let A and B be two events defined in sample space S.

In words: *probability of A and B = probability of A × probability of B, knowing A*

In algebra: $P(A \text{ and } B) = P(A) \cdot P(B|A)$ **(4.5)**

Note: When two events are involved, either event can be identified as A, with the other identified as B. The general multiplication rule could also be written as $P(B \text{ and } A) = P(B) \cdot P(A|B)$.

EXAMPLE 4.13

UNDERSTANDING THE MULTIPLICATION RULE

A statewide poll of 800 registered voters in 25 precincts from across New York State was taken. Each voter was identified as being registered as Republican, Democrat, or other and then asked, "Are you in favor of or against the current budget proposal awaiting the governor's signature?" The resulting tallies are shown here.

	Number in Favor	Number Against	Number of Voters
Republican	136	88	224
Democrat	314	212	526
Other	14	36	50
Totals	464	336	800

Suppose one voter is to be selected at random from the 800 voters summarized in the preceding table. Let's consider the two events: "The voter selected is in favor" and "The voter is a Republican." Find the three probabilities: P(in favor), P(Republican | in favor), and P(in favor and Republican). Then use the results to check the truth of the multiplication rule.

Solution

Probability the voter selected is "in favor" = P(in favor) = **464/800** = 0.58.
Probability the voter selected is "Republican, given in favor"

= P(Republican|in favor) = 136/464 = 0.29.

Probability the voter selected is in "favor" and "Republican"

$$= P(\text{in favor and Republican}) = \mathbf{136/800} = \frac{136}{800} = \underline{0.17}.$$

Notes about finding the preceding probabilities:

1. The conditional "given" means there is a restriction; thus, "Republican | in favor" means we start with only those voters who are "in favor." In this case, this means we are looking only at 464 voters when determining this probability.
2. The connective "and" means "both" or "in common"; thus, "in favor and Republican" means all voters who satisfy both events.

Now let's use the previous probabilities to demonstrate the truth of the multiplication rule.

Let A = "in favor" and B = "Republican." The general multiplication rule then becomes:

$$P(\text{in favor and Republican}) = P(\text{in favor}) \cdot P(\text{Republican}|\text{in favor})$$

Previously we found: $P(\text{in favor and Republican}) = \dfrac{136}{800} = \underline{0.17}$.

Using the other two probabilities, we see:

$$P(\text{in favor}) \cdot P(\text{Republican}|\text{in favor}) = \frac{464}{800} \cdot \frac{136}{464} = \frac{136}{800} = \underline{0.17}.$$

You typically do not have the option of finding $P(A \text{ and } B)$ two ways, as we did here. When you are asked to find $P(A \text{ and } B)$, you will often be given $P(A)$ and $P(B)$. However, you will not always get the correct answer by just multiplying those two probabilities together. You will need a third piece of information: the conditional probability of one of the two events or information that will allow you to find it.

E X A M P L E   4 . 1 4

DRAWING WITHOUT REPLACEMENT

At a carnival game, the player blindly draws one colored marble at a time from a box containing two red and four blue marbles. The chosen marble is not returned to the box after being selected; that is, each drawing is done without replacement. The marbles are mixed before each drawing. It costs $1 to play, and if the first two marbles drawn are red, the player receives a $2 prize. If the first four marbles drawn are all blue, the player receives a $5 prize. Otherwise, no prize is awarded. To find the probability of winning a prize, let's look first at the probability of drawing red or blue on consecutive drawings and organize the information on a tree diagram.

On the first draw (represented by the purple branch segments in Figure 4.5), the probability of red is two chances out of six, 2/6 or 1/3, whereas the probability of blue is 4/6, or 2/3. Because the marble is not replaced, only five marbles are left in the box; the number of each color remaining depends on the color of the first marble drawn. If the first marble was red, then the probabilities are 1/5 and 4/5, as shown on the tree diagram (green branch segments in Figure 4.5). If the first marble was blue, then the probabilities are 2/5 and 3/5, as shown on the tree diagram (orange branch segments in Figure 4.5). The probabilities change with each drawing, because the number of marbles available keeps decreasing as each drawing takes place. The tree diagram is a marvelous pictorial aid in following the progression.

FIGURE 4.5
Tree Diagram—First Two Drawings, Carnival Game

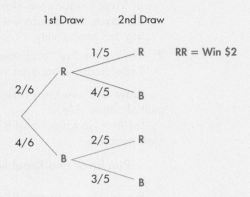

The probability of winning the $2 prize can now be found using formula (4.5):

$$P(A \text{ and } B) = P(A) \cdot P(B|A)$$

$$P(\text{winning } \$\,2) = P(R_1 \text{ and } R_2) = P(R_1) \cdot P(R_2|R_1) = \frac{2}{6} \cdot \frac{1}{5} = \frac{1}{15} = 0.067$$

(Winning the $5 prize is left as Exercise 4.79.)

Note: The tree diagram, when labeled, has the probabilities needed for multiplying listed along the branch representing the winning effort.

SECTION 4.3 EXERCISES

4.59 a. If the probability that event A occurs during an experiment is 0.7, what is the probability that event A does not occur during that experiment?

b. If the results of a probability experiment can be any integer from 16 to 28 and the probability that the integer is less than 20 is 0.78, what is the probability that the integer will be 20 or more?

4.60 a. If the probability that you will pass the next exam in statistics is accurately assessed at 0.75, what is the probability that you will not pass the next statistics exam?

b. The weather forecaster predicts that there is a "70 percent" chance of less than 1 inch of rain during the next 30-day period. What is the probability of at least 1 inch of rain in the next 30 days?

4.61 According to the American Pet Products Manufacturers Association 2007–2008 National Pet Owners Survey, about 63% of all American dog owners—some 60 million—are owners of one dog. Based on this information, find the probability that an American dog owner owns more than one dog.

4.62 According to the Sleep Channel (http://www .sleepdisorderchannel.com/ July 2009), sleep apnea affects 18 million individuals in the United States. The sleep disorder interrupts breathing and can awaken its sufferers as often as five times an hour. Many people do not recognize the condition even though it causes loud snoring. Assuming there are 304 million people in the

United States, what is the probability that an individual chosen at random will not be affected by sleep apnea?

4.63 If $P(A) = 0.4$, $P(B) = 0.5$, and $P(A \text{ and } B) = 0.1$, find $P(A \text{ or } B)$.

4.64 If $P(A) = 0.5$, $P(B) = 0.3$, and $P(A \text{ and } B) = 0.2$, find $P(A \text{ or } B)$.

4.65 If $P(A) = 0.4$, $P(B) = 0.5$, and $P(A \text{ or } B) = 0.7$, find $P(A \text{ and } B)$.

4.66 If $P(A) = 0.4$, $P(A \text{ or } B) = 0.9$, and $P(A \text{ and } B) = 0.1$, find $P(B)$.

4.67 The entertainment sports industry employs athletes, coaches, referees, and related workers. Of these, 0.37 work part time and 0.50 earn more than $20,540 annually. If 0.32 of these employees work full time and earn more than $20,540, what proportion of the industry's employees are full time or earn more than $20,540?

4.68 Jason attends his high school reunion. Of the attendees, 50% are female. Common knowledge has it that 88% of people are right-handed. Being a left-handed male, Jason knows that of a given crowd, only approximately 6% are left-handed males. If Jason talks to the first person he meets at the reunion, what is the probability that the person is a male or left-handed?

4.69 A parts store sells both new and used parts. Sixty percent of the parts in stock are used. Sixty-one percent are used or defective. If 5% of the store's parts are defective, what percentage is both used and defective? Solve using formulas. Compare your solution to your answer to Exercise 4.27.

4.70 Union officials report that 60% of the workers at a large factory belong to the union, 90% make more than $12 per hour, and 40% belong to the union and make more than $12 per hour. Do you believe these percentages? Explain. Solve using formulas. Compare your solution to your answer to Exercise 4.28.

4.71 A and B are events defined on a sample space, with $P(A) = 0.7$ and $P(B|A) = 0.4$. Find $P(A \text{ and } B)$

4.72 A and B are events defined on a sample space, with $P(A|B) = 0.5$ and $P(B) = 0.8$. Find $P(A \text{ and } B)$

4.73 A and B are events defined on a sample space, with $P(A) = 0.6$ and $P(A \text{ and } B) = 0.3$. Find $P(B|A)$.

4.74 A and B are events defined on a sample space, with $P(B) = 0.5$ and $P(A \text{ and } B) = 0.4$. Find $P(A|B)$.

4.75 It is known that steroids give users an advantage in athletic contests, but it is also known that steroid use is banned in athletes. As a result, a steroid testing program has been instituted and athletes are randomly tested. The test procedures are believed to be equally effective on both users and nonusers and claim to be 98% accurate. If 90% of the athletes affected by this testing program are clean, what is the probability that the next athlete tested will be a user and fail the test?

4.76 Juan lives in a large city and commutes to work daily by subway or by taxi. He takes the subway 80% of the time because it costs less, and he takes a taxi the other 20% of the time. When taking the subway, he arrives at work on time 70% of the time, whereas he makes it on time 90% of the time when traveling by taxi.

a. What is the probability that Juan took the subway and is at work on time on any given day?

b. What is the probability that Juan took a taxi and is at work on time on any given day?

4.77 Nobody likes paying taxes, but cheating is not the way to get out of it! It is believed that 10% of all taxpayers intentionally claim some deductions to which they are not entitled. If 9% of all taxpayers both intentionally claim extra deductions and deny doing so when audited, find the probability that a taxpayer who does take extra deductions intentionally will deny it.

4.78 Casey loves his mid-morning coffee and always stops by one of his favorite coffeehouses for a cup. When he gets take-out, there is a 0.6 chance that he will also get a pastry. He gets both coffee and a pastry as take-out with a probability of 0.48. What is the probability that he does take-out?

4.79 Find the probability of winning $5 if you play the carnival game described in Example 4.14.

a. Complete the branches of the tree diagram started in Figure 4.5, listing the probabilities for all possible drawings.

b. What is the probability of drawing a red marble on the second drawing? What additional information is needed to find the probability? What "conditions" could exist?

c. Calculate the probability of winning the $5 prize.

d. Is the $2 prize or the $5 prize harder to win? Which is more likely? Justify your answer.

4.80 Suppose the rules for the carnival game in Example 4.14 are modified so that the marble drawn each time is returned to the box before the next drawing.

a. Redraw the tree diagram drawn for Exercise 4.79, listing the probabilities for the game when played "with replacement."

b. What is the probability of drawing a red marble on the second drawing? What additional information is needed to find the probability? What effect does this have on $P(\text{red on 2nd drawing})$?

(continue on page 202)

c.   Calculate the probability of winning the $2 prize.

d.   Calculate the probability of winning the $5 prize.

e.   When the game is played with replacement, is the $2 or the $5 prize harder to win? Which is more likely? Justify your answer.

4.81 Suppose that A and B are events defined on a common sample space and that the following probabilities are known: $P(A) = 0.3$, $P(B) = 0.4$, and $P(A|B) = 0.2$. Find $P(A \text{ or } B)$.

4.82 Suppose that A and B are events defined on a common sample space and that the following probabilities are known: $P(A \text{ or } B) = 0.7$, $P(B) = 0.5$, and $P(A|B) = 0.2$. Find $P(A)$.

4.83 Suppose that A and B are events defined on a common sample space and that the following probabilities are known: $P(A) = 0.4$, $P(B) = 0.3$, and $P(A \text{ or } B) = 0.66$. Find $P(A|B)$.

4.84 Suppose that A and B are events defined on a common sample space and that the following probabilities are known: $P(A) = 0.5$, $P(A \text{ and } B) = 0.24$, and $P(A|B) = 0.4$. Find $P(A \text{ or } B)$.

4.85 Given $P(A \text{ or } B) = 1.0$, $P(\overline{A \text{ and } B}) = 0.7$, and $P(\overline{B}) = 0.4$, find:

a. $P(B)$ b. $P(A)$ c. $P(A|B)$

4.86 Given $P(A \text{ or } B) = 1.0$, $P(\overline{A \text{ and } B}) = 0.3$, and $P(\overline{B}) = 0.4$, find:

a. $P(B)$ b. $P(A)$ c. $P(A|B)$

4.87 The probability of A is 0.5. The conditional probability that A occurs given that B occurs is 0.25. The conditional probability that B occurs given that A occurs is 0.2.

a.   What is the probability that B occurs?

b.   What is the conditional probability that B does not occur given that A does not occur?

4.88 The probability of C is 0.4. The conditional probability that C occurs given that D occurs is 0.5. The conditional probability that C occurs given that D does not occur is 0.25.

a.   What is the probability that D occurs?

b.   What is the conditional probability that D occurs given that C occurs?

4.4 Mutually Exclusive Events

To further our discussion of compound events, the concept of "mutually exclusive" must be introduced.

Mutually exclusive events Nonempty events defined on the same sample space with each event excluding the occurrence of the other. In other words, they are events that share no common elements.

In algebra: $P(A \text{ and } B) = 0$
In words: There are several equivalent ways to express the concept of mutually exclusive:

1. If you know that either one of the events has occurred, then the other event is excluded or cannot have occurred.
2. If you are looking at the lists of the elements making up each event, none of the elements listed for either event will appear on the other event's list; there are "no shared elements."
3. If you are looking at a Venn diagram, the closed areas representing each event "do not intersect"—that is, there are "no shared elements," or as another way to say it, "they are disjoint."
4. The equation says, "the **intersection** of the two events has a probability of zero," meaning "the intersection is an empty set" or "there is no intersection."

Note: The concept of mutually exclusive events is based on the relationship between the sets of elements that satisfy the events. Mutually exclusive is not a probability concept by definition; it just happens to be easy to express the concept using a probability statement.

Let's look at some examples.

EXAMPLE 4.15

UNDERSTANDING MUTUALLY EXCLUSIVE EVENTS

From a nationwide exit poll of 1000 voters in 25 precincts across the country on November 4, 2008, we have the following.

Education	Number for McCain	Number for Obama	Number for Other	Number of Voters
No high school	19	20	1	40
High school graduate	114	103	3	220
Some college	172	147	1	320
College graduate	135	119	6	260
Postgraduate	70	88	2	160
Total	510	477	13	1000

Consider the two events "the selected voter voted for McCain" and "the selected voter voted for Obama." Suppose one voter is selected at random from the 1000 voters summarized in the table. In order for the event "the selected voter voted for McCain" to occur, the selected voter must be 1 of the 510 voters listed in "Number for McCain" column. In order for the event "the selected voter voted for Obama" to occur, the selected voter must be 1 of the 477 voters listed in "Number for Obama" column. Because no voter listed in the McCain column is also listed in the Obama column and because no voter listed in the Obama column is also listed in the McCain column, these two events are mutually exclusive.

In equation form: P(voted for McCain and voted for Obama) = 0.

EXAMPLE 4.16

UNDERSTANDING NOT MUTUALLY EXCLUSIVE EVENTS

From a nationwide exit poll of 1000 voters in 25 precincts across the country on November 4, 2008 we have the following.

Education	Number for McCain	Number for Obama	Number for Other	Number of Voters
No high school	19	20	1	40
High school graduate	114	103	3	220
Some college	172	147	1	320
College graduate	135	119	6	260
Postgraduate	70	88	2	160
Total	510	477	13	1000

Consider the two events "the selected voter voted for McCain" and "the selected voter had some college education." Suppose one voter is selected at random from the 1000 voters summarized in the table. In order for the event "the selected voter voted for McCain" to occur, the selected voter must be 1 of the 510 voters listed in "Number for McCain" column. In order for the event "the selected voter had some college education" to occur, the selected voter must be 1 of the 320 voters listed in the "Some college" row. Because the 172 voters shown in the intersection of the "Number for McCain" column and the "Some college" row belong to both of the events ("the selected voter voted for McCain" and "the selected voter had some college education"), these two events are NOT mutually exclusive.

In equation form: P(voted for McCain and some college education) = **172**/1000 = 0.172, which is not equal to zero.

E X A M P L E   4 . 1 7

MUTUALLY EXCLUSIVE CARD EVENTS

Consider a regular deck of playing cards and the two events "card drawn is a queen" and "card drawn is an ace." The deck is to be shuffled and one card randomly drawn. In order for the event "card drawn is a queen" to occur, the card drawn must be one of the four queens: queen of hearts, queen of diamonds, queen of spades, or queen of clubs. In order for the event "card drawn is an ace" to occur, the card drawn must be one of the four aces: ace of hearts, ace of diamonds, ace of spades, or ace of clubs. Notice that there is no card that is both a queen and an ace. Therefore, these two events, "card drawn is a queen" and "card drawn is an ace," are mutually exclusive events.

In equation form: P(queen and ace) = 0.

E X A M P L E   4 . 1 8

NOT MUTUALLY EXCLUSIVE CARD EVENTS

Consider a regular deck of playing cards and the two events "card drawn is a queen" and "card drawn is a heart." The deck is to be shuffled and one card randomly drawn. Are the events "queen" and "heart" mutually exclusive? The event "card drawn is a queen" is made up of the four queens: queen of hearts, queen of diamonds, queen of spades, and queen of clubs. The event "card drawn is a heart" is made up of the 13 hearts: ace of hearts, king of hearts, queen of hearts, jack of hearts, and the other nine hearts. Notice that the "queen of hearts" is on both lists, thereby making it possible for both events "card drawn is a queen" and "card drawn is a heart" to occur simultaneously. That means, when one of these two events occurs, it does not exclude the possibility of the other's occurrence. These events are not mutually exclusive events.

In equation form: P(queen and heart) = 1/52, which is not equal to zero.

EXAMPLE 4.19

VISUAL DISPLAY AND UNDERSTANDING OF MUTUALLY EXCLUSIVE EVENTS

Consider an experiment in which two dice are rolled. Three events are defined as follows:

A: The sum of the numbers on the two dice is 7.
B: The sum of the numbers on the two dice is 10.
C: Each of the two dice shows the same number.

Let's determine whether these three events are mutually exclusive.

We can show that three events are mutually exclusive by showing that each pair of events is mutually exclusive. Are events A and B mutually exclusive? Yes, they are, because the sum on the two dice cannot be both 7 and 10 at the same time. If a sum of 7 occurs, it is impossible for the sum to be 10.

Figure 4.6 presents the sample space for this experiment. This is the same sample space shown in Example 4.3, except that ordered pairs are used in place of the pictures. The ovals, diamonds, and rectangles show the ordered pairs that are in events A, B, and C, respectively. We can see that events A and B do not intersect. Therefore, they are mutually exclusive. Point (5, 5) in Figure 4.6 satisfies both events B and C. Therefore, B and C are not mutually exclusive. Two dice can each show a 5, which satisfies C, and the total satisfies B. Since we found one pair of events that are not mutually exclusive, events A, B, and C are not mutually exclusive.

FIGURE 4.6
Sample Space for the Roll of Two Dice

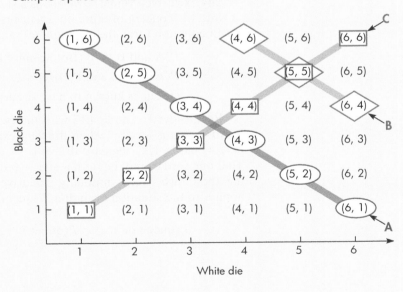

Special Addition Rule

The addition rule simplifies when the events involved are mutually exclusive.

If we know two events are mutually exclusive, then by applying $P(A \text{ and } B) = 0$ to the addition rule for probabilities, it follows that

$P(A \text{ or } B) = P(A) + P(B) - P(A \text{ and } B)$ becomes
$P(A \text{ or } B) = P(A) + P(B)$.

Special Addition Rule

Let A and B be two mutually exclusive events defined in a sample space S.

In words: *Probability of A or B = probability of A + probability of B*

In algebra: $P(A \text{ or } B) = P(A) + P(B)$ (4.6)

This formula can be expanded to consider more than two mutually exclusive events:

$$P(A \text{ or } B \text{ or } C \text{ or } \ldots \text{ or } E) = P(A) + P(B) + P(C) + \ldots + P(E)$$

This equation is often convenient for calculating probabilities, but it does not help us understand the relationship between the events A and B. It is the *definition* that tells us how we should think about mutually exclusive events. Students who understand mutual exclusiveness this way gain insight into what mutual exclusiveness is all about. This should lead you to think more clearly about situations dealing with mutually exclusive events, thereby making you less likely to confuse the concept of mutually exclusive events with independent events (to be defined in Section 4.5) or to make other common mistakes regarding the concept of mutually exclusive.

Notes:

1. Define mutually exclusive events in terms of the sets of elements satisfying the events and test for mutual exclusiveness in that manner.
2. Do not use $P(A \text{ and } B) = 0$ as the definition of mutually exclusive events. It is a property that results from the definition. It can be used as a test for mutually exclusive events; however, as a statement, it shows no meaning or insight into the concept of mutually exclusive events.
3. In equation form, the *definition* of mutually exclusive events states:

   $P(A \text{ and } B) = 0$ (Both cannot happen at same time.)

   $P(A|B) = 0$ and $P(B|A) = 0$

   (If one is known to have occurred, then the other has not.)

Reconsider Example 4.17, with the two events "card drawn is a queen" and "card drawn is an ace" when drawing exactly one card from a deck of regular playing cards. The one card drawn is a queen, or the one card drawn is an ace. That one card cannot be both a queen and an ace at the same time, thereby making these two events mutually exclusive. The special addition rule therefore applies to the situation of finding $P(\text{queen or ace})$.

$$P(\text{queen or ace}) = P(\text{queen}) + P(\text{ace}) = \frac{4}{52} + \frac{4}{52} = \frac{8}{52} = \frac{2}{13}$$

SECTION 4.4 EXERCISES

4.89 Determine whether each of the following pairs of events is mutually exclusive.

a. Five coins are tossed: "one head is observed," "at least one head is observed."

b. A salesperson calls on a client and makes a sale: "the sale exceeds $100," "the sale exceeds $1000."

c. One student is selected at random from a student body: the person selected is "male," the person selected is "older than 21 years of age."

d. Two dice are rolled: the total showing is "less than 7," the total showing is "more than 9."

4.90 Determine whether each of the following sets of events is mutually exclusive.

a. Five coins are tossed: "no more than one head is observed," "two heads are observed," "three or more heads are observed."

b. A salesperson calls on a client and makes a sale: the amount of the sale is "less than $100," is "between $100 and $1000," is "more than $500."

c. One student is selected at random from the student body: the person selected is "female," is "male," is "older than 21."

d. Two dice are rolled: the numbers of dots showing on the dice are "both odd," "both even," "total 7," "total 11."

4.91 Explain why $P(A \text{ and } B) = 0$ when events A and B are mutually exclusive.

4.92 Explain why $P(A \text{ occurring when B has occurred}) = 0$ when events A and B are mutually exclusive.

4.93 If $P(A) = 0.3$ and $P(B) = 0.4$, and if A and B are mutually exclusive events, find:

a. $P(\overline{A})$ c. $P(A \text{ or } B)$

b. $P(\overline{B})$ d. $P(A \text{ and } B)$

4.94 If $P(A) = 0.4$ and $P(B) = 0.5$, and if A and B are mutually exclusive events, find $P(A \text{ or } B)$.

4.95 One student is selected from the student body of your college. Define the following events: M—the student selected is male, F—the student selected is female, S—the student selected is registered for statistics.

a. Are events M and F mutually exclusive? Explain.

b. Are events M and S mutually exclusive? Explain.

c. Are events F and S mutually exclusive? Explain.

d. Are events M and F complementary? Explain.

e. Are events M and S complementary? Explain.

f. Are complementary events also mutually exclusive events? Explain.

g. Are mutually exclusive events also complementary events? Explain.

4.96 One student is selected at random from a student body. Suppose the probability that this student is female is 0.5 and the probability that this student works part time is 0.6. Are the two events "female" and "working part time" mutually exclusive? Explain.

4.97 Two dice are rolled. Define events as follows: A—sum of 7, C—doubles, E—sum of 8.

a. Which pairs of events, A and C, A and E, or C and E, are mutually exclusive? Explain.

b. Find the probabilities $P(A \text{ or } C)$, $P(A \text{ or } E)$, and $P(C \text{ or } E)$.

4.98 An aquarium at a pet store contains 40 orange swordfish (22 females and 18 males) and 28 green swordtails (12 females and 16 males). You randomly net one of the fish.

a. What is the probability that it is an orange swordfish?

b. What is the probability that it is a male fish?

c. What is the probability that it is an orange female swordfish?

d. What is the probability that it is a female or a green swordtail?

e. Are the events "male" and "female" mutually exclusive? Explain.

f. Are the events "male" and "swordfish" mutually exclusive? Explain.

4.99 Do people take indoor swimming lessons in the middle of the hot summer? They sure do at the Webster Aquatic Center. During the month of July 2009 alone, 283 people participated in various forms of lessons.

Swim Categories	Daytime	Evenings
Preschool	66	80
Levels	69	56
Adult and diving	10	2
Total	145	138

If one swimmer was selected at random from the July participants:

a. Are the events the selected participant is "daytime" and "evening" mutually exclusive? Explain.

b. Are the events the selected participant is "preschool" and "levels" mutually exclusive? Explain.

c. Are the events the selected participant is "daytime" and "preschool" mutually exclusive? Explain.

d. Find $P(\text{preschool})$.

e. Find $P(\text{daytime})$.

f. Find $P(\text{not levels})$.

g. Find $P(\text{preschool or evening})$.

h. Find $P(\text{preschool and daytime})$.

i. Find $P(\text{daytime} \mid \text{levels})$.

j. Find $P(\text{adult and diving} \mid \text{evening})$.

4.100 Injuries are unfortunately part of every sport. High school basketball is no exception, as the following table shows. The percentages listed are the percent of reported injuries that occur to high school male and female

(continue on page 208)

basketball players and the location on their body that was injured.

Injury Location	Males	Females
Ankle/foot	38.3%	36.0%
Hip/thigh/leg	14.7%	16.6%
Knee	10.3%	13.0%
Forearm/wrist/hand	11.5%	11.2%
Face/scalp	12.2%	8.8%
Other	13.0%	14.4%
Total	100.0%	100.0%

If one player is selected at random from those included in the table:

a. Are the events the selected player was "male" and "female" mutually exclusive? Explain.

b. Are the events the selected player's injury was "ankle/foot" and "knee" mutually exclusive? Explain.

c. Are the events "female" and "face/scalp" mutually exclusive? Explain.

d. Find P(ankle/foot | male).

e. Find P(ankle/foot | female).

f. Find P(not leg related | male).

g. Find P(knee or face/scalp | male).

h. Find P(knee or face/scalp | female).

i. Explain why P(knee) for all high school basketball players cannot be found using the information in the table. What additional information is needed?

4.101 Most Americans, 70% in fact, say frequent hand washing is the best way to fend off the flu. Despite that, when using public restrooms, women wash their hands only 62% of the time and men wash only 43% of the time. Of the adults using the public restroom at a large grocery chain store, 58% are women. What is the probability that the next person to enter the restroom in this store washes his or her hands?

4.102 He is the last guy you want to see in your rearview mirror when you are speeding down the highway, but research shows that a traffic ticket reduces a driver's chance of being involved in a fatal accident, at least for a few weeks. By age group, 13.3% of all drivers are younger than age 25, 58.6% are between ages 25 and 54, and 28.1% are 55 or older. Statistics show that 1.6% of the drivers younger than age 25, 2.2% of those ages 25 to 54, and 0.5% of those 55 or older will have an accident in the next month. What is the probability that a randomly identified driver will have an accident in the next month?

4.5 Independent Events

The concept of independent events is necessary to continue our discussion on compound events.

Independent events Two events are *independent* if the occurrence (or nonoccurrence) of one gives us no information about the likeliness of occurrence of the other. In other words, if the probability of A remains unchanged after we know that B has happened (or has not happened), the events are independent.

In algebra: $P(A) = P(A|B) = P(A|\text{not } B)$

In words: There are several equivalent ways to express the concept of independence:

1. The probability of event A is unaffected by knowledge that a second event, B, has occurred, knowledge that B has not occurred, or no knowledge about event B whatsoever.
2. The probability of event A is unaffected by knowledge, or no knowledge, about a second event, B, having occurred or not occurred.
3. The probability of event A (with no knowledge about event B) is the same as the probability of event A, knowing event B has occurred, and both are the same as the probability of event A, knowing event B has not occurred.

Not all events are independent.

Dependent events Events that are not independent. That is, the occurrence of one event does have an effect on the probability for occurrence of the other event.

Let's look at some examples.

E X A M P L E   4 . 2 0

UNDERSTANDING INDEPENDENT EVENTS

A statewide poll of 750 registered Republicans and Democrats in 25 precincts from across New York State was taken. Each voter was identified as being registered as a Republican or a Democrat and then asked, "Are you are in favor of or against the current budget proposal awaiting the governor's signature?" The resulting tallies are shown here.

	Number in Favor	Number Against	Number of Voters
Republican	135	90	225
Democrat	315	210	525
Totals	450	300	750

Suppose one voter is to be selected at random from the 750 voters summarized in the preceding table. Let's consider the two events "The selected voter is in favor" and "The voter is a Republican." Are these two events independent?

To answer this, consider the following three probabilities: (1) probability the selected voter is in favor; (2) probability the selected voter is in favor, knowing the voter is a Republican; and (3) probability the selected voter is in favor, knowing the voter is not a Republican.

Probability the selected voter is in favor = $P(\text{in favor}) = 450/750 = \underline{0.60}$.

Probability the selected voter is in favor, knowing voter is a Republican
= $P(\text{in favor}|\text{Republican}) = 135/225 = \underline{0.60}$.

Probability the selected voter is in favor, knowing voter is not a Republican
= Probability the selected voter is in favor, knowing voter is a Democrat

$P(\text{in favor}|\text{not Republican}) = P(\text{in favor}|\text{Democrat}) = 315/525 = \underline{0.60}$.

Does knowing the voter's political affiliation have an influencing effect on the probability that the voter is in favor of the budget proposal? With no information about political affiliation, the probability of being in favor is 0.60. Information about the event "Republican" does not alter the probability of "in favor." They are all the value 0.60. Therefore, these two events are said to be *independent events*.

When checking the three probabilities, $P(A)$, $P(A|B)$, and $P(A|\text{not } B)$, we need to compare only two of them. If any two of the three probabilities are equal, the third will be the same value. Furthermore, if any two of the three probabilities are unequal, then all three will be different in value.

Note: Determine all three values, using the third as a check. All will be the same, or all will be different—there is no other possible outcome.

EXAMPLE 4.21

UNDERSTANDING NOT INDEPENDENT EVENTS

From a National Election Pool exit poll of 13,660 voters in 250 precincts across the country on November 4, 2008, we have the following:

	Percentage of Voters	Percentage for Obama	Percentage for McCain	Percentage for Others
Men	48	44	54	2
Women	52	56	43	1

Suppose one voter is selected at random from the 13,660 voters summarized on the table. Let's consider the two events "The voter is a woman" and "The voter voted for Obama." Are these two events independent? To answer this, consider this question, "Does knowing the voter is a woman have an influencing effect on the probability that the voter voted for Obama?" What is the probability of voting for Obama if the voter is a woman? You say, "0.56." Now compare this to the probability of voting for Obama if the voter is not a woman. You say that the probability is **0.44.** So I ask you, "Did knowing the voter was a woman influence the probability of voting for Obama?" Yes, it did; it is **0.56** when the voter is a woman and 0.44 when the voter is not a woman. Information about event "woman" does alter the probability of "voted for Obama." Therefore, these two events are *not independent* and are said to be *dependent events*.

In equation form:

P(voted for Obama|knowing voter is a woman) = $P(O|W)$ = 0.56 and
P(voted for Obama|knowing voter is not a woman) = $P(O|\overline{W})$ = 0.44.

Therefore: $P(O|W) \neq P(O|\overline{W})$, and the two events are not independent.

EXAMPLE 4.22

INDEPENDENT CARD EVENTS

Cengage Learning

Consider a regular deck of playing cards and the two events "card drawn is a queen" and "card drawn is a heart." Suppose that I shuffle the deck, randomly draw one card, and, before looking at the card, ask you the probability that it is a "queen." You say 4/52, or 1/13. Then I peek at the card and tell you that it is a "heart." Now, what is the probability that the card is a "queen"? You say it is 1/13, the same as before knowing the card was a "heart."

The hint that the card was a heart provided you with additional information, but that information did not change the probability that it was a queen. Therefore, "queen" and "heart" are independent. Furthermore, suppose that after I drew the card and looked at it, I had told you the card was "not a heart." What would be the probability the card is a "queen"? You say 3/39, or 1/13. Again, notice that knowing the card was "not a heart" did provide additional information, but that information did not change the probability that it was a "queen." This is what it means for the two events "card is a queen" and "card is a heart" to be independent.

In equation form:

$$P(\text{queen}|\text{card is heart}) = P(Q|H) = P(Q)$$

$$P(\text{queen}|\text{card is not heart}) = P(Q|\text{not } H) = P(Q)$$

Therefore, $P(Q) = P(Q|H) = P(Q|\text{not } H)$, and the two events are independent.

E X A M P L E   4 . 2 3

NOT INDEPENDENT CARD EVENTS

Now, let's consider the two events "card drawn is a heart" and "card drawn is red." Are the events "heart" and "red" independent? Following the same scenario as in Example 4.22, I shuffle the deck of 52 cards, randomly draw one card, and before looking at it, you say the probability that the unknown card is "red" is $26/52 = 1/2$. However, when told the additional information that the card is a "heart," you change your probability that the card is "red" to $13/13$, or 1. This additional information results in a different probability of "red."

$P(\text{red}|\text{card is heart}) = P(R|H) = 13/13 = 1$, and $P(\text{red}) = P(\text{red}|\text{having}$ no additional information$) = 26/52 = 1/2$. Therefore, the additional information did change the probability of the event "red." These two events are not independent and are therefore said to be dependent events.

In equation form, the definition states:

A and B are independent if and only if $P(A|B) = P(A)$

Note: Define **independence** in terms of conditional probability, and test for independence in that manner.

Special Multiplication Rule

The multiplication rule simplifies when the events involved are independent.

If we know two events are independent, then by applying the definition of independence, $P(B|A) = P(B)$, to the multiplication rule, it follows that:

$$P(A \text{ and } B) = P(A) \cdot P(B|A) \text{ becomes } P(A \text{ and } B) = P(A) \cdot P(B)$$

Special Multiplication Rule

Let A and B be two independent events defined in a sample space S.

In words: *probability of A and B = probability of A × probability of B*

In algebra: $P(A \text{ and } B) = P(A) \cdot P(B)$ **(4.7)**

This formula can be expanded to consider more than two independent events:

$$P(A \text{ and } B \text{ and } C \text{ and } \ldots \text{ and } E) = P(A) \cdot P(B) \cdot P(C) \cdot \ldots \cdot P(E)$$

This equation is often convenient for calculating probabilities, but it does not help us understand the independence relationship between the events A and B. It is *the definition* that tells us how we should think about independent events. Students who understand independence this way gain insight into what independence is all about. This should lead you to think more clearly about situations dealing with independent events, thereby making you less likely to confuse the concept of independent events with mutually exclusive events or to make other common mistakes regarding independence.

Note: Do not use $P(A \text{ and } B) = P(A) \cdot P(B)$ as the definition of independence. It is a property that results from the definition. It can be used as a test for independence, but as a statement, it shows no meaning or insight into the concept of independent events.

SECTION 4.5 EXERCISES

4.103 Determine whether each of the following pairs of events is independent:

a. Rolling a pair of dice and observing a "1" on the first die and a "1" on the second die

b. Drawing a "spade" from a regular deck of playing cards and then drawing another "spade" from the same deck without replacing the first card

c. Same as part b except the first card is returned to the deck before the second drawing

d. Owning a red automobile and having blonde hair

e. Owning a red automobile and having a flat tire today

f. Studying for an exam and passing the exam

4.104 Determine whether each of the following pairs of events is independent:

a. Rolling a pair of dice and observing a "2" on one of the dice and having a "total of 10"

b. Drawing one card from a regular deck of playing cards and having a "red" card and having an "ace"

c. Raining today and passing today's exam

d. Raining today and playing golf today

e. Completing today's homework assignment and being on time for class

4.105 A and B are independent events, and $P(A) = 0.7$ and $P(B) = 0.4$. Find $P(A \text{ and } B)$.

4.106 A and B are independent events, and $P(A) = 0.5$ and $P(B) = 0.8$. Find $P(A \text{ and } B)$.

4.107 A and B are independent events, and $P(A) = 0.6$ and $P(A \text{ and } B) = 0.3$. Find $P(B)$.

4.108 A and B are independent events, and $P(A) = 0.4$ and $P(A \text{ and } B) = 0.5$. Find $P(B)$.

4.109 If $P(A) = 0.3$ and $P(B) = 0.4$ and A and B are independent events, what is the probability of each of the following?

a. $P(A \text{ and } B)$ b. $P(B|A)$ c. $P(A|B)$

4.110 Suppose that $P(A) = 0.3$, $P(B) = 0.4$, and $P(A \text{ and } B) = 0.12$.

a. What is $P(A|B)$?

b. What is $P(B|A)$?

c. Are A and B independent?

4.111 Suppose that $P(A) = 0.3$, $P(B) = 0.4$, and $P(A \text{ and } B) = 0.20$.

a. What is $P(A|B)$?

b. What is $P(B|A)$?

c. Are A and B independent?

4.112 One student is selected at random from a group of 200 students known to consist of 140 full-time (80 female and 60 male) students and 60 part-time (40 female and 20 male) students. Event A is "the student selected is full time," and event C is "the student selected is female."

a. Are events A and C independent? Justify your answer.

b. Find the probability $P(A \text{ and } C)$.

4.113 A single card is drawn from a standard deck. Let A be the event that "the card is a face card" (a jack, a queen, or a king), B is a "red card," and C is "the card is a heart." Determine whether the following pairs of events are independent or dependent:

a. A and B   b. A and C   c. B and C

4.114 A box contains four red and three blue poker chips. Three poker chips are to be randomly selected, one at a time.

a. What is the probability that all three chips will be red if the selection is done with replacement?

b. What is the probability that all three chips will be red if the selection is done without replacement?

c. Are the drawings independent in either part a or b? Justify your answer.

4.115 Excluding job benefit coverage, approximately 49% of adults have purchased life insurance. The likelihood that those aged 18 to 24 without life insurance will purchase life insurance in the next year is 15%, and for those aged 25 to 34, it is 26%. (Opinion Research)

a. Find the probability that a randomly selected adult has not purchased life insurance.

b. What is the probability that an adult aged 18 to 24 will purchase life insurance within the next year?

c. Find the probability that a randomly selected adult will be 25 to 34 years old, does not currently have life insurance, and will purchase it within the next year.

4.116 The U.S. space program has a history of many successes and some failures. Space flight reliability is of the utmost importance in the launching of space shuttles. The reliability of the complete mission is based on the reliability of all its components. Each of the six joints in the space shuttle *Challenger*'s booster rocket had a 0.977 reliability. The six joints worked independently.

a. What does it mean to say that the six joints worked independently?

b. What was the reliability (probability) of all six of the joints working together?

4.117 In a 2008 study by Experian Automotive, it was found that the average number of vehicles per household in the United States is 2.28 vehicles. The results also showed that nearly 35% of households have three or more vehicles. http://www.autospies.com/

a. If two U.S. households are randomly selected, find the probability that both will have three or more vehicles.

b. If two U.S. households are randomly selected, find the probability that neither of the two has three or more vehicles.

c. If four U.S. households are randomly selected, find the probability that all four will have three or more vehicles.

4.118 A *USA Today* Snapshot titled "Weighing heavily" (February 5, 2009) provided the results from the National College Health Assessment 2007 Web Summary, in which 34% of the students said that "stress" was the health and mental health issue that most often hampered their academic performance. If five college students are randomly selected, what is the probability that all five will say that "stress" is the health and mental health issue that most often hampers their academic performance?

4.119 The June 16, 2009, issue of *Democrat and Chronicle* presented the article "For the most part, the kids are all right." According to information from the CDC (Centers for Disease Control) and Safe Kids USA, a nonprofit advocacy group, 77% of children ages 19 to 35 months receive all recommended vaccinations. If three children ages 19 to 35 months are randomly selected, what is the probability that all three will have received all recommended vaccinations?

4.120 You have applied for two scholarships: a merit scholarship (M) and an athletic scholarship (A). Assume the probability that you receive the athletic scholarship is 0.25, the probability that you receive both scholarships is 0.15, and the probability that you get at least one of the scholarships is 0.37. Use a Venn diagram to answer these questions:

a. What is the probability that you receive the merit scholarship?

b. What is the probability that you do not receive either of the two scholarships?

c. What is the probability that you receive the merit scholarship, given that you have been awarded the athletic scholarship?

d. What is the probability that you receive the athletic scholarship, given that you have been awarded the merit scholarship?

e. Are the events "receiving an athletic scholarship" and "receiving a merit scholarship" independent events? Explain.

4.121 The owners of a two-person business make their decisions independently of each other and then compare their decisions. If they agree, the decision is made; if they do not agree, then further consideration is necessary before a decision is reached. If each person has a history of making the right decision 60% of the time, what is the probability that together they:

a. Make the right decision on the first try?

b. Make the wrong decision on the first try?

c. Delay the decision for further study?

4.122 The odds against throwing a pair of dice and getting a total of 5 are 8 to 1. The odds against throwing a pair of dice and getting a total of 10 are 11 to 1. What is the probability of throwing the dice twice and getting a total of 5 on the first throw and 10 on the second throw?

4.123 Consider the set of integers 1, 2, 3, 4, and 5.

a. One integer is selected at random. What is the probability that it is odd?

b. Two integers are selected at random (one at a time with replacement so that each of the five is available for a second selection). Find the probability that neither is odd; exactly one of them is odd; both are odd.

4.124 A box contains 25 parts, of which 3 are defective and 22 are nondefective. If 2 parts are selected without replacement, find the following probabilities:

a. P(both are defective)

b. P(exactly one is defective)

c. P(neither is defective)

4.125 According to the U.S. Department of Education, the percentage of college students who graduate within 4 years from a private institution is 79%. That percent drops to 49% for public institutions. One of the reasons for this might be that 38% of college students attend only part time.

Source: http://www.naicu.edu/

What additional information do you need to determine the probability that a student selected at random is part time and will graduate within 4 years?

4.126 From a survey of adults, 48% plan to buy candy this year at Easter. The types of candy they will buy are described in the following table.

What additional information do you need to determine the probability that a customer selected at random will buy candy and it will be chocolate?

Table for Exercise 4.126

Chocolate	Nonchocolate	Jellybeans	Cream-Filled	Marshmallow	Malted	Don't Know
30%	25%	13%	11%	8%	7%	6%

Source: International Mass Retail Association

4.6 Are Mutual Exclusiveness and Independence Related?

Mutually exclusive events and independent events are two very different concepts based on definitions that start from very different orientations. The two concepts can easily be confused because they interact with each other and are intertwined by the probability statements we use in describing these concepts.

To describe these two concepts and eventually understand the distinction between them as well as the relationship between them, we need to agree that the events being considered are two nonempty events defined on the same sample space and therefore each has nonzero probabilities.

Note: Students often have a hard time realizing that when we say, "Event A is a nonempty event" and write, "$P(A) > 0$," we are describing the same situation. The words and the algebra often do not seem to have the same meaning. In this case the words and the probability statement both tell us that event A exists within the sample space.

Mutually Exclusive

Mutually exclusive events are two nonempty events defined on the same sample space and share no common elements.

This means:

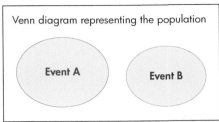

Venn diagram representing the population

Event A   Event B

1.   In words: In this Venn diagram, the closed areas representing each event "do not intersect"; in other words, they are disjoint sets, or no intersection occurs between their respective sets.
2.   In algebra: $P(A \text{ and } B) = 0$, which says, "The intersection of the two events is an empty set"; in other words, there is no intersection between their respective sets.

Notice that the concept of mutually exclusive is based on the relationship of the elements that satisfy the events. Mutually exclusive is not a probability concept by definition—it just happens to be easy to express the concept using a probability statement.

Independence

Independent events are two nonempty events defined on the same sample space that are related in such a way that the occurrence of either event does not affect the probability of the other event.
This means that:

1.   In words: If event A has (or is known to have) already occurred, the probability of event B is unaffected (that is, the probability of B after knowing event A had occurred remains the same as it was before knowing event A had occurred).

      In addition, it is also the case when A and B interchange roles that if event B has (or is known to have) already occurred, the probability of event A is unaffected (i.e., the probability of A is still the same after knowing event B had occurred as it was before).
      This is a "mutual relationship"; it works both ways.

2.   In algebra: $P(B|A) = P(B|\text{not } A) = P(B)$ and
            $P(A|B) = P(A|\text{not } B) = P(A)$
      Or with a few words to help interpret the algebra, $P(B$, knowing A has occurred$) = P(B$, knowing A has not occurred$) = P(B)$ and $P(A$, knowing B has occurred$) = P(A$, knowing B has not occurred$) = P(A)$.

      Notice that the concept of independence is based on the effect one event (in this case, the lack of effect) has on the probability of the other event.

Let's look at the following four demonstrations relating to mutually exclusive and independent events:

Demonstration I
Given: $P(A) = 0.4$, $P(B) = 0.5$, and A and B are mutually exclusive; are they independent?

Answer: If A and B are mutually exclusive events, $P(A|B) = 0.0$, and because we are given $P(A) = 0.4$, we see that the occurrence of B has an effect on the probability of A. Therefore, A and B are not independent events.

Conclusion I: If the events are mutually exclusive, they are NOT independent.

Demonstration II
Given: $P(A) = 0.4$, $P(B) = 0.5$, and A and B are independent; are events A and B mutually exclusive?

Answer: If A and B are independent events, then the $P(A \text{ and } B)$ $= P(A) \cdot P(B) = 0.4 \cdot 0.5 = 0.20$, and because $P(A \text{ and } B)$ is greater than zero, events A and B must intersect, meaning the events are not mutually exclusive.

Conclusion II: If the events are independent, they are NOT mutually exclusive.

Demonstration III

Given: $P(A) = 0.4$, $P(B) = 0.5$, and A and B are not mutually exclusive; are events A and B independent?

Answer: Because A and B are not mutually exclusive events, it must be that $P(A \text{ and } B)$ is greater than zero. Now, if $P(A \text{ and } B)$ happens to be exactly 0.20, then events A and B are independent $[P(A) \cdot P(B) = 0.4 \cdot 0.5 = 0.20]$, but if $P(A \text{ and } B)$ is any other positive value, say 0.1, then events A and B are not independent. Therefore, events A and B could be either independent or dependent; some other information is needed to make that determination.

Conclusion III: If the events are not mutually exclusive, they MAY be either independent or dependent; additional information is needed to determine which.

Demonstration IV

Given: $P(A) = 0.4$, $P(B) = 0.5$, and A and B are not independent; are events A and B mutually exclusive?

Answer: Because A and B are not independent events, it must be that $P(A \text{ and } B)$ is different from 0.20, the value it would be if they were independent $[P(A) \cdot P(B) = 0.4 \cdot 0.5 = 0.20]$. Now, if $P(A \text{ and } B)$ happens to be exactly 0.00, then events A and B are mutually exclusive, but if $P(A \text{ and } B)$ is any other positive value, say 0.1, then events A and B are not mutually exclusive. Therefore, events A and B could be either mutually exclusive or not; some other information is needed to make that determination.

Conclusion IV: If the events are NOT independent, they MAY be either mutually exclusive or not mutually exclusive; additional information is needed to determine which.

Advice

Work very carefully, starting with the information you are given and the definitions of the concepts involved.

What not to do

Do not rely on the first "off-the-top" example you can think of to lead you to the correct answer. It typically will not!

The following examples give further practice with these probability concepts.

E X A M P L E   4 . 2 4

CALCULATING PROBABILITIES AND THE ADDITION RULE

A pair of dice is rolled. Event T is defined as the occurrence of a "total of 10 or 11," and event D is the occurrence of "doubles." Find the probability $P(T \text{ or } D)$.

 Video tutorial available—logon and learn more at cengagebrain.com

Solution

Look at the sample space of 36 ordered pairs for the rolling of two dice in Figure 4.6 (p. 205). Event T occurs if any one of 5 ordered pairs occurs: (4, 6), (5, 5), (6, 4), (5, 6), (6, 5). Therefore, $P(T) = \frac{5}{36}$. Event D occurs if any one of 6 ordered pairs occurs: (1, 1), (2, 2), (3, 3), (4, 4), (5, 5), (6, 6). Therefore, $P(D) = \frac{6}{36}$. Notice, however, that these two events are not mutually exclusive.

The two events "share" the ordered pair (5, 5). Thus, the probability $P(T \text{ and } D) = \frac{1}{36}$. As a result, the probability $P(T \text{ or } D)$ will be found using formula (4.4).

$$P(T \text{ or } D) = P(T) + P(D) - P(T \text{ and } D)$$
$$= \frac{5}{36} + \frac{6}{36} - \frac{1}{36} = \frac{10}{36} = \frac{5}{18}$$

Look at the sample space in Figure 4.6 and verify $P(T \text{ or } D) = \frac{5}{18}$.

EXAMPLE 4.25

USING CONDITIONAL PROBABILITIES TO DETERMINE INDEPENDENCE

In a sample of 150 residents, each person was asked if he or she favored the concept of having a single, countywide police agency. The county is composed of one large city and many suburban townships. The residence (city or outside the city) and the responses of the residents are summarized in Table 4.4. If one of these residents was to be selected at random, what is the probability that the person will (a) favor the concept? (b) favor the concept if the person selected is a city resident? or (c) favor the concept if the person selected is a resident from outside the city? (d) Are the events F (favor the concept) and C (reside in city) independent?

TABLE 4.4
Sample Results for Example 4.25

Residence	Favor (F)	Oppose (F̄)	Total
In city (C)	80	40	120
Outside of city (C̄)	20	10	30
Total	100	50	150

Solution

(a) $P(F)$ is the proportion of the total sample that favors the concept. Therefore,

$$P(F) = \frac{n(F)}{n(S)} = \frac{100}{150} = \frac{2}{3}$$

(continue on page 218)

 Video tutorial available—logon and learn more at cengagebrain.com

(b) $P(F|C)$ is the probability that the person selected favors the concept, given that he or she lives in the city. The condition, "is city resident," reduces the sample space to the 120 city residents in the sample. Of these, 80 favored the concept; therefore,

$$P(F|C) = \frac{n(F \text{ and } C)}{n(C)} = \frac{80}{120} = \frac{2}{3}$$

(c) $P(F|\overline{C})$ is the probability that the person selected favors the concept, knowing that the person lives outside the city. The condition, "lives outside the city," reduces the sample space to the 30 noncity residents; therefore,

$$P(F|\overline{C}) = \frac{n(F \text{ and } \overline{C})}{n(\overline{C})} = \frac{20}{30} = \frac{2}{3}$$

(d) All three probabilities have the same value, $\frac{2}{3}$. Therefore, we can say that the events F (favor) and C (reside in city) are independent. The location of residence did not affect $P(F)$.

EXAMPLE 4.26

DETERMINING INDEPENDENCE AND USING THE MULTIPLICATION RULE

One student is selected at random from a group of 200 known to consist of 140 full-time (80 female and 60 male) students and 60 part-time (40 female and 20 male) students. Event A is "the student selected is full time," and event C is "the student selected is female."

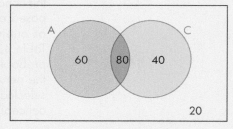

(a) Are events A and C independent?

(b) Find the probability $P(A \text{ and } C)$ using the multiplication rule.

Solution 1

(a) First find the probabilities $P(A)$, $P(C)$, and $P(A|C)$:

$$P(A) = \frac{n(A)}{n(S)} = \frac{140}{200} = 0.7$$

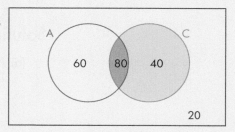

$$P(C) = \frac{n(C)}{n(S)} = \frac{120}{200} = 0.6$$

$$P(A|C) = \frac{n(A \text{ and } C)}{n(C)} = \frac{80}{120} = 0.67$$

A and C are dependent events because $P(A) \neq P(A|C)$.

(b) $P(A \text{ and } C) = P(C) \cdot P(A|C) = \frac{120}{200} \cdot \frac{80}{120} = \frac{80}{200} = \mathbf{0.4}$

Solution 2

(a) First find the probabilities $P(A)$, $P(C)$, and $P(C|A)$:

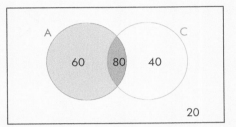

$$P(A) = \frac{n(A)}{n(S)} = \frac{140}{200} = 0.7$$

$$P(C) = \frac{n(C)}{n(S)} = \frac{120}{200} = 0.6$$

$$P(C|A) = \frac{n(C \text{ and } A)}{n(A)} = \frac{80}{140} = 0.57$$

A and C are dependent events because $P(C) \neq P(C|A)$.

(b) $P(C \text{ and } A) = P(A) \cdot P(C|A) = \frac{140}{200} \cdot \frac{80}{140} = \frac{80}{200} = \mathbf{0.4}$

EXAMPLE 4.27

FYI Misclassification can happen two ways!

USING SEVERAL PROBABILITY RULES

A production process produces thousands of items. On the average, 20% of all items produced are defective. Each item is inspected before it is shipped. The inspector misclassifies an item 10% of the time; that is,

$P(\text{classified good}|\text{defective item}) = P(\text{classified defective}|\text{good item})$

$$= 0.10$$

What proportion of the items will be "classified good"?

Solution

What do we mean by the event "classified good"?

    G: The item is good.
    D: The item is defective.
    CG: The item is classified good by the inspector.
    CD: The item is classified defective by the inspector.

(continue on page 220)

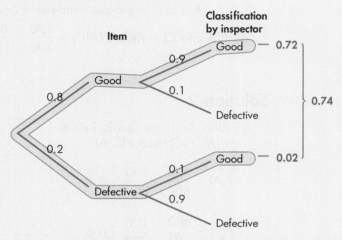

FIGURE 4.7
Using Several Probability Rules

CG consists of two possibilities: "the item is good and is correctly classified good" and "the item is defective and is misclassified good." Thus,

$$P(CG) = P[(CG \text{ and } G) \text{ or } (CG \text{ and } D)]$$

Since the two possibilities are mutually exclusive, we can start by using the addition rule, formula (4.6):

$$P(CG) = P(CG \text{ and } G) + P(CG \text{ and } D)$$

The condition of an item and its classification by the inspector are not independent. The multiplication rule for dependent events must be used. Therefore,

$$P(CG) = [P(G) \cdot P(CG \mid G)] + [P(D) \cdot P(CG \mid D)]$$

Substituting the known probabilities in Figure 4.7, we get

$$P(CG) = [(0.8)(0.9)] + [(0.2)(0.1)]$$
$$= 0.72 + 0.02$$
$$= \mathbf{0.74}$$

That is, 74% of the items are classified good.

S E C T I O N   4 . 6   E X E R C I S E S

4.127 a. Describe in your own words what it means for two events to be mutually exclusive.

   b. Describe in your own words what it means for two events to be independent.

   c. Explain how mutually exclusive and independent are two very different properties.

4.128 a. Describe in your own words why two events cannot be independent if they are already known to be mutually exclusive.

   b. Describe in your own words why two events cannot be mutually exclusive if they are already known to be independent.

4.129 $P(G) = 0.5$, $P(H) = 0.4$, and $P(G \text{ and } H) = 0.1$ (see the diagram).

a. Find $P(G|H)$.

b. Find $P(H|G)$.

c. Find $P(\overline{H})$.

d. Find $P(G \text{ or } H)$.

e. Find $P(G \text{ or } \overline{H})$.

f. Are events G and H mutually exclusive? Explain.

g. Are events G and H independent? Explain.

4.130 $P(R) = 0.5$, $P(S) = 0.3$, and events R and S are independent.

a. Find $P(R \text{ and } S)$.

b. Find $P(R \text{ or } S)$.

c. Find $P(\overline{S})$.

d. Find $P(R|S)$.

e. Find $P(S|R)$.

f. Are events R and S mutually exclusive? Explain.

4.131 $P(M) = 0.3$, $P(N) = 0.4$, and events M and N are mutually exclusive.

a. Find $P(M \text{ and } N)$.

b. Find $P(M \text{ or } N)$.

c. Find $P(M \text{ or } \overline{N})$.

d. Find $P(M|N)$.

e. Find $P(M|\overline{N})$.

f. Are events M and N independent? Explain.

4.132 Two flower seeds are randomly selected from a package that contains five seeds for red flowers and three seeds for white flowers.

a. What is the probability that both seeds will result in red flowers?

b. What is the probability that one of each color is selected?

c. What is the probability that both seeds are for white flowers?

FYI Draw a tree diagram.

4.133 One thousand employees at the Russell Microprocessor Company were polled about worker satisfaction. One employee is selected at random.

	Male		Female		
	Skilled	Unskilled	Skilled	Unskilled	Total
Satisfied	350	150	25	100	625
Unsatisfied	150	100	75	50	375
Total	500	250	100	150	1000

a. Find the probability that an unskilled worker is satisfied with work.

b. Find the probability that a skilled female employee is satisfied with work.

c. Is satisfaction for female employees independent of their being skilled or unskilled?

4.134 A company that manufactures shoes has three factories. Factory 1 produces 25% of the company's shoes, Factory 2 produces 60%, and Factory 3 produces 15%. One percent of the shoes produced by Factory 1 are mislabeled, 0.5% of those produced by Factory 2 are mislabeled, and 2% of those produced by Factory 3 are mislabeled. If you purchase one pair of shoes manufactured by this company, what is the probability that the shoes are mislabeled?

Chapter Review

In Retrospect

You have been studying the basic concepts of probability. These fundamentals need to be mastered before we continue with our study of statistics. Probability is the vehicle of statistics, and we have begun to see how probabilistic events occur. We have explored theoretical and experimental probabilities for the same event. Does the experimental probability turn out to have the same value as the theoretical? Not exactly, but we have seen that over the long run, it does have approximately the same value.

Upon completion of this chapter, you should understand the properties of mutual exclusiveness and independence and be able to apply the multiplication and addition rules to "and" and "or" compound events. You should also be able to calculate conditional probabilities.

In the next three chapters we will look at distributions associated with probabilistic events. This will prepare us for the statistics that follows. We must be able to predict the variability that the sample will show with respect to the population before we can be successful at "inferential statistics," in which we describe the population based on the sample statistics available.

CourseMate The **Statistics CourseMate** site for this text brings chapter topics to life with interactive learning, study, and exam preparation tools, including quizzes and flashcards for the Vocabulary and Key Concepts that follow. The site also provides an **eBook** version of the text with highlighting and note taking capabilities. Throughout chapters, the CourseMate icon 🖥 flags concepts and examples that have corresponding interactive resources such as **video** and **animated tutorials** that demonstrate, step by step, how to solve problems; **datasets** for exercises and examples; **Skillbuilder Applets** to help you better understand concepts; **technology manuals**; and software to download including **Data Analysis Plus** (a suite of statistical macros for Excel) and **TI-83/84 Plus** programs—logon at **www.cengagebrain.com**.

Vocabulary and Key Concepts

addition rule (p. 196)
all-inclusive events (p. 179)
complement rule (p. 195)
complementary event (p. 195)
compound event (p. 195)
conditional probability (p. 190)
dependent events (p. 209)
empirical probability (p. 173)
equally likely events (p. 173)
event (p. 173)
experimental probability (p. 173)
general addition rule (p. 196)

general multiplication rule (p. 198)
independence (p. 211)
independent events (p. 208)
intersection (p. 202)
law of large numbers (p. 181)
long-term average (p. 181)
multiplication rule (p. 198)
mutually exclusive events (p. 202)
observed relative frequency (p. 173)
odds (p. 182)
ordered pair (p. 175)
outcome (p. 173)

probability of an event (p. 173)
sample points (p. 173)
sample space (p. 173)
special addition rule (p. 206)
special multiplication rule (p. 211)
statistics (p. 183)
subjective probability (p. 178)
theoretical probability (p. 173)
tree diagram (p. 175)
Venn diagram (p. 177)

Learning Outcomes

- Understand and be able to describe the basic concept of probability. pp. 172–174
- Understand and describe a simple event. EXP. 4.1
- Understand and be able to describe the differences between empirical, theoretical, and subjective probabilities. pp. 173–175, 178
- Compute and interpret relative frequencies. Ex. 4.5, 4.8, 4.9, 4.135
- Identify and describe a sample space for an experiment. pp. 173–174; Ex. 4.13, 4.31
- Construct tables, tree diagrams, and/or Venn diagrams to aid in computing and interpreting probabilities. EXP. 4.3, 4.4, 4.5, Ex. 4.12, 4.25
- Understand the properties of probability numbers: p. 179; Ex. 4.23, 4.37
  1. $0 \leq$ Each $P(A) \leq 1$
  2. $\sum_{\text{all outcomes}} P(A) = 1$
- Understand, describe, and use the law of large numbers to determine probabilities. EXP. 4.6, p. 180; Ex. 4.30, 4.32
- Understand, compute, and interpret odds of an event. EXP. 4.8; Ex. 4.39, 4.41, 4.122
- Understand that compound events involve the occurrence of more than one event. Ex. 4.31, 4.53

- Construct, describe, compute, and interpret a conditional probability.
- Understand and be able to utilize the complement rule.
- Compute probabilities of compound events using the addition rule.
- Compute probabilities of compound events using the multiplication rule.
- Understand, describe, and determine mutually exclusive events.
- Compute probabilities of compound events using the addition rule for mutually exclusive events.
- Understand, describe, and determine independent events.
- Compute probabilities of compound events using the multiplication rule for independent events.
- Recognize and compare the differences between mutually exclusive events and independent events.

EXP. 4.10, Ex. 4.51, 4.55, 4.152

EXP. 4.11, Ex. 4.61, 4.62

EXP. 4.12, Ex. 4.67, EXP. 4.24

EXP. 4.13, Ex. 4.76

p. 202, EXP. 4.15, 4.16, Ex. 4.89, 4.95

EXP. 4.19, Ex. 4.99

p. 208, EXP 4.20, 4.21, Ex. 4.103

EXP. 4.25, 4.26, Ex. 4.116, 4.117

pp. 214–216, Ex. 4.129, 4.149, 4.157

Chapter Exercises

4.135 The U.S. Department of Transportation and the Federal Motor Carrier Safety Administration produce an annual report on various traffic violations. There were 2092 "moving violations" in the state of New York in 2008, as described in the following table.

Moving Violations	2008 Numbers
Failure to obey traffic control device	1050
Following too close	37
Improper lane change	67
Improper passing	9
Reckless driving	4
Speeding	857
Improper turns	33
Failure to yield right of way	13
Operating a CMV while ill or fatigued	22
Total	2092

If one violation is selected at random for review, what is the probability that the moving violation is due to:

a. Speeding?

b. Reckless driving?

c. Improper passing or improper turns?

d. If two violations are selected for review, would this be an example of sampling with or without replacement? Explain why.

4.136 [EX04-136] The number of people living in the 50 U.S. states and the District of Columbia in September 2004 is reported by age groups in the following table.

Age Group	Percentage	Number (1000s)
0–17	25%	73,447.7
18–24	10%	28,855.7
25–34	13%	39,892.5
35–49	23%	66,620.3
50	29%	84,119.8

Source: Sales & Marketing Management Survey of Buying Power, September 2004, for the 50 U.S. states and the District of Columbia

a. Verify the percentages reported in the table.

If one person is picked at random from all the people represented in the table, what is the probability of the following events?

b. "Between 18 and 24." How is this related to the 10% listed on the table?

c. "Older than 17"

d. "Between 18 and 24" and "older than 17"

e. "Between 18 and 24" or "older than 17"

f. "At least 25"

g. "No more than 24"

4.137 One thousand persons screened for a certain disease are given a clinical exam. As a result of the exam, the sample of 1000 persons is classified according to height and disease status.

Height	Disease Status				
	None	Mild	Moderate	Severe	Total
Tall	122	78	139	61	400
Medium	74	51	90	35	250
Short	104	71	121	54	350
Total	300	200	350	150	1000

Use the information in the table to estimate the probability of being medium or short and of having moderate or severe disease status.

4.138 [EX04-138] The Federal Highway Administration periodically tracks the number of licensed vehicle drivers by sex and by age. The following table shows the results of the administration's findings in 2007:

(continue on page 224)

[EX00-000] identifies the filename of an exercise's online dataset—datasets available through cengagebrain.com

Age Group (yr)	Male	Female
19 and under	5,077,141	4,843,033
20–24	8,669,114	8,520,482
25–29	9,072,595	9,077,275
30–34	8,852,063	8,766,584
35–39	9,762,966	9,935,291
40–44	10,117,084	10,041,634
45–49	10,583,203	10,641,856
50–54	9,869,590	9,994,330
55–59	8,581,110	8,723,673
60–64	6,891,032	6,976,462
65–69	4,981,745	5,095,436
70–74	3,733,751	3,877,392
75–79	2,933,321	3,187,834
80–84	1,999,765	2,305,836
85 and over	1,340,456	1,589,791
Total	102,464,936	103,576,909

Source: U.S. Department of Transportation, Federal Highway Administration, *Highway Statistics 2007.*

Suppose you encountered a driver of a vehicle at random. Find the probabilities of the following events:

a. The driver is a male and over the age of 59.

b. The driver is a female or under the age of 30.

c. The driver is under the age of 25.

d. The driver is a female.

e. The driver is a male between the ages of 35 and 49.

f. The driver is over the age of 69.

g. The driver is a female, given the driver is between the ages of 25 and 44.

h. The driver is between the ages of 25 and 44, given the driver is female.

4.139 Let's assume there are three traffic lights between your house and a friend's house. As you arrive at each light, it may be red (R) or green (G).

a. List the sample space showing all possible sequences of red and green lights that could occur on a trip from your house to your friend's. (RGG represents red at the first light and green at the other two.) Assume that each element of the sample space is equally likely to occur.

b. What is the probability that on your next trip to your friend's house, you will have to stop for exactly one red light?

c. What is the probability that you will have to stop for at least one red light?

4.140 Assuming that a woman is equally likely to bear a boy or a girl, use a tree diagram to compute the probability that a four-child family consists of one boy and three girls.

4.141 Skillbuilder Applet Exercise simulates generating a family. The "family" will stop having children when it has a boy or three girls, whichever comes first. Assuming that a woman is equally likely to bear 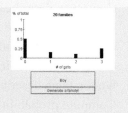 a boy or a girl, perform the simulation 24 times. What is the probability that the family will have a boy?

4.142 A coin is flipped three times.

a. Draw a tree diagram that represents all possible outcomes.

b. Identify all branches that represent the event "exactly one head occurred."

c. Find the probability of "exactly one head occurred."

4.143 A recent survey of New York State families asked about their vacation habits. The accompanying two-way table shows the number of families according to where they live (rural, suburban, urban) and the length of their last vacation (1–7 days, 8 days or more).

	Rural	Suburban	Urban	Total
1–7 Days	90	57	52	199
8 Days or More	74	38	21	133
Total	164	95	73	332

If one family is selected at random from these 332 families, what is the probability of the following?

a. They spent 8 days or more on vacation.

b. They were a rural family.

c. They were an urban family and spent 8 days or more on vacation.

d. They were a rural family or spent 1 to 7 days on vacation.

e. They spent 8 days or more on vacation, given they were a suburban family.

f. They were a rural family, given they spent 1 to 7 days on vacation.

4.144 The age and gender demographics for the Fall 2008 Monroe Community College full-time students are outlined in the table below.

	19 and under	20–24	25–29	30 and over
Female	2928	1658	420	649
Male	2883	1705	377	438
Total	5811	3363	797	1087

If one of these students is selected at random, what is the probability that the student is

a. a male?

b. between 20 and 24 years of age?

c. a female and 30 and over?

d. a male or 19 years old and under?

e. between 25 and 29 years of age, given the student is female?

f. a male student, given the student is 20 or older?

4.145 This bar graph shows the number of registered automobiles in each of several countries.
a. Name at least two countries not included in the information.

b. Why are all probabilities resulting from this information conditional probabilities?

Based on the information in the accompanying graph:

c. What percentage of all cars in these countries is registered in the United States?

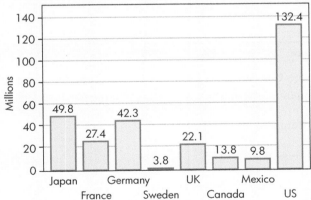

Number of Automobiles

d. If one registered car was selected at random from all of these cars, what is the probability that it is registered in the United States?

e. Explain the relationship between your answers to parts c and d.

4.146 Probabilities for events A, B, and C are distributed as shown in the figure. Find:

a. $P(A \text{ and } B)$

b. $P(A \text{ or } C)$

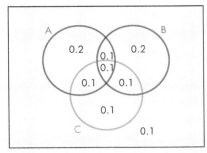

c. $P(A|C)$

4.147 Show that if event A is a subset of event B, then $P(A \text{ or } B) = P(B)$.

4.148 Explain why these probabilities cannot be legitimate: $P(A) = 0.6, P(B) = 0.4, P(A \text{ and } B) = 0.7$.

4.149 A shipment of grapefruit arrived containing the following proportions of types: 10% pink seedless, 20% white seedless, 30% pink with seeds, and 40% white with seeds. A grapefruit is selected at random from the shipment. Find the probability of these events:

a. It is seedless.

b. It is white.

c. It is pink and seedless.

d. It is pink or seedless.

e. It is pink, given that it is seedless.

f. It is seedless, given that it is pink.

4.150 A traffic analysis at a busy traffic circle in Washington, DC, showed that 0.8 of the autos using the circle entered from Connecticut Avenue. Of those entering the traffic circle from Connecticut Avenue, 0.7 continued on Connecticut Avenue at the opposite side of the circle. What is the probability that a randomly selected auto observed in the traffic circle entered from Connecticut and will continue on Connecticut?

4.151 Suppose that when a job candidate interviews for a job at RJB Enterprises, the probability that he or she will want the job (A) after the interview is 0.68. Also, the probability that RJB will want the candidate (B) is 0.36. The probability $P(A|B)$ is 0.88.

a. Find $P(A \text{ and } B)$.

b. Find $P(B|A)$.

c. Are events A and B independent? Explain.

(continue on page 226)

d. Are events A and B mutually exclusive? Explain.

e. What would it mean to say that A and B are mutually exclusive events in this exercise?

4.152 The probability that thunderstorms are in the vicinity of a particular Midwestern airport on an August day is 0.70. When thunderstorms are in the vicinity, the probability that an airplane lands on time is 0.80. Find the probability that thunderstorms are in the vicinity and that the plane lands on time.

4.153 Tires salvaged from a train wreck are on sale at the Getrich Tire Company. Of the 15 tires offered in the sale, 5 have suffered internal damage and the remaining 10 are damage free. You randomly selected and purchased two of these tires.

a. What is the probability that the tires you purchased are both damage free?

b. What is the probability that exactly one of the tires you purchased is damage free?

c. What is the probability that at least one of the tires you purchased is damage free?

4.154 According to automobile accident statistics, one out of every six accidents results in an insurance claim of $100 or less in property damage. Three cars insured by an insurance company are involved in different accidents. Consider these two events:

A: The majority of claims exceed $100.
B: Exactly two claims are $100 or less.

a. List the sample points for this experiment.

b. Are the sample points equally likely?

c. Find $P(A)$ and $P(B)$.

d. Are events A and B independent? Justify your answer.

4.155 A testing organization wishes to rate a particular brand of television. Six TVs are selected at random from stock. If nothing is found wrong with any of the six, the brand is judged satisfactory.

a. What is the probability that the brand will be rated satisfactory if 10% of the TVs actually are defective?

b. What is the probability that the brand will be rated satisfactory if 20% of the TVs actually are defective?

c. What is the probability that the brand will be rated satisfactory if 40% of the TVs actually are defective?

4.156 Suppose a certain ophthalmic trait is associated with eye color. Three hundred randomly selected individuals are studied, with results given in the following table.

Trait	Eye Color			Total
	Blue	Brown	Other	
Yes	70	30	20	120
No	20	110	50	180
Total	90	140	70	300

a. What is the probability that a person selected at random has blue eyes?

b. What is the probability that a person selected at random has the trait?

c. Are events A (has blue eyes) and B (has the trait) independent? Justify your answer.

d. How are the two events A (has blue eyes) and C (has brown eyes) related—independent, mutually exclusive, complementary, or all-inclusive? Explain why or why not each term applies.

4.157 As listed in *The World Factbook*, 2009, the age structure of the U.S. population is shown in the table.

	Male	Female
0–14 years	31,639,127	30,305,704
15–64 years	102,665,043	103,129,321
65 years and over	16,901,232	22,571,696

Source: *The World Factbook*, July 2009. https://www.cia.gov/

If one U.S. citizen were to be selected at random from this population, what is the probability that the person selected is

a. female?

b. 0–14 years old?

c. male and 15–64 years old?

d. female or 65 years and over?

e. under 15 years of age, knowing the person is female?

f. male, given that the person is 15 to 64 years old?

g. Are the events "person selected is male" and "person selected is female" independent events? Justify your answer. What is the relationship between female and male in this situation?

4.158 The following table shows the sentiments of 2500 wage-earning employees at the Spruce Company on a proposal to emphasize fringe benefits rather than wage increases during their impending contract discussions.

Employee	Opinion			Total
	Favor	Neutral	Opposed	
Male	800	200	500	1500
Female	400	100	500	1000
Total	1200	300	1000	2500

a. Calculate the probability that an employee selected at random from this group will be opposed.

b. Calculate the probability that an employee selected at random from this group will be female.

c. Calculate the probability that an employee selected at random from this group will be opposed, given that the person is male.

d. Are the events "opposed" and "female" independent? Explain.

4.159 Events R and S are defined on a sample space. If $P(R) = 0.2$ and $P(S) = 0.5$, explain why each of the following statements is either true or false:

a. If R and S are mutually exclusive, then $P(R \text{ or } S) = 0.10$.

b. If R and S are independent, then $P(R \text{ or } S) = 0.6$.

c. If R and S are mutually exclusive, then $P(R \text{ and } S) = 0.7$.

d. If R and S are mutually exclusive, then $P(R \text{ or } S) = 0.6$.

4.160 It is believed that 3% of a clinic's patients have cancer. A particular blood test yields a positive result for 98% of patients with cancer, but it also shows positive for 4% of patients who do not have cancer. One patient is chosen at random from the clinic's patient list and is tested. What is the probability that if the test result is positive, the person actually has cancer?

4.161 Box 1 contains two red balls and three green balls, and Box 2 contains four red balls and one green ball. One ball is randomly selected from Box 1 and placed in Box 2. Then one ball is randomly selected from Box 2. What is the probability that the ball selected from Box 2 is green?

4.162 Salespersons Adams and Jones call on three and four customers, respectively, on a given day. Adams could make 0, 1, 2, or 3 sales, whereas Jones could make 0, 1, 2, 3, or 4 sales. The sample space listing the number of possible sales for each person on a given day is shown in the table. (3, 1 stands for 3 sales by Jones and 1 sale by Adams.)

Adams	Jones				
	0	1	2	3	4
0	0, 0	1, 0	2, 0	3, 0	4, 0
1	0, 1	1, 1	2, 1	3, 1	4, 1
2	0, 2	1, 2	2, 2	3, 2	4, 2
3	0, 3	1, 3	2, 3	3, 3	4, 3

Assume that each sample point is equally likely. Consider these events:

A: At least one of the salespersons made no sales.
B: Together they made exactly three sales.
C: Each made the same number of sales.
D: Adams made exactly one sale.

Find the probabilities by counting sample points:

a. $P(A)$ b. $P(B)$ c. $P(C)$

d. $P(D)$ e. $P(A \text{ and } B)$ f. $P(B \text{ and } C)$

g. $P(A \text{ or } B)$ h. $P(B \text{ or } C)$ i. $P(A|B)$

j. $P(B|D)$ k. $P(C|B)$ l. $P(B|\overline{A})$

m. $P(C|\overline{A})$ n. $P(A \text{ or } B \text{ or } C)$

Are the following pairs of events mutually exclusive? Explain.

o. A and B p. B and C q. B and D

Are the following pairs of events independent? Explain.xx

r. A and B s. B and C t. B and D

4.163 Alex, Bill, and Chen each, in turn, toss a balanced coin. The first one to throw a head wins.

a. What are their respective chances of winning if each tosses only one time?

b. What are their respective chances of winning if they continue, given a maximum of two tosses each?

FYI Draw a tree diagram.

4.164 Coin A is loaded in such a way that $P(\text{heads})$ is 0.6. Coin B is a balanced coin. Both coins are tossed. Find:

a. The sample space that represents this experiment; assign a probability measure to each outcome

b. $P(\text{both show heads})$

(continue on page 228)

c. P(exactly one head shows)

d. P(neither coin shows a head)

e. P(both show heads| coin A shows a head)

f. P(both show heads| coin B shows a head)

g. P(heads on coin A| exactly one head shows)

4.165 Professor French forgets to set his alarm with a probability of 0.3. If he sets the alarm, it rings with a probability of 0.8. If the alarm rings, it wakes him on time to make his first class with a probability of 0.9. If the alarm does not ring, he wakes in time for his first class with a probability of 0.2. What is the probability that Professor French will wake in time to make his first class tomorrow?

4.166 The probability that a certain door is locked is 0.6. The key to the door is one of five unidentified keys hanging on a key rack. You randomly select two keys before approaching the door. What is the probability that you can open the door without returning for another key?

4.167 Your local art museum has planned next year's 52-week calendar by scheduling a mixture of 1-week and 2-week shows that feature the works of 22 painters and 20 sculptors. There is a showing scheduled for every week of the year, and only one artist is featured at a time. There are 42 different shows scheduled for next year. You have randomly selected one week to attend and have been told the probability of it being a 2-week show of sculpture is 3/13.

a. What is the probability that the show you have selected is a painter's showing?

b. What is the probability that the show you have selected is a sculptor's showing?

c. What is the probability that the show you have selected is a 1-week show?

d. What is the probability that the show you have selected is a 2-week show?

4.168 A two-page typed report contains an error on one of the pages. Two proofreaders review the copy. Each has an 80% chance of catching the error. What is the probability that the error will be identified in the following cases?

a. Each reads a different page.

b. They each read both pages.

c. The first proofreader randomly selects a page to read and then the second proofreader randomly selects a page, unaware of which page the first selected.

4.169 In sports, championships are often decided by two teams playing in a championship series. Often the fans of the losing team claim they were unlucky and their team is actually the better team. Suppose Team A is the better team, and the probability it will defeat Team B in any one game is 0.6.

a. What is the probability that the better team, Team A, will win the series if it is a one-game series?

b. What is the probability that the better team, Team A, will win the series if it is a best out of three series?

c. What is the probability that the better team, Team A, will win the series if it is a best out of seven series?

d. Suppose the probability that A would beat B in any given game were actually 0.7. Recompute parts a–c.

e. Suppose the probability that A would beat B in any given game were actually 0.9. Recompute parts a–c.

f. What is the relationship between the "best" team winning and the number of games played? The best team winning and the probabilities that each will win?

4.170 A woman and a man (unrelated) each has two children. At least one of the woman's children is a boy, and the man's older child is a boy. Is the probability that the woman has two boys greater than, equal to, or less than the probability that the man has two boys?

a. Demonstrate the truth of your answer by using a simple sample to represent each family.

b. Demonstrate the truth of your answer by taking two samples, one from men with two-children families and one from women with two-children families.

c. Demonstrate the truth of your answer using computer simulation. Using the Bernoulli probability function with $p = 0.5$ (let 0 = girl and 1 = boy), generate 500 "families of two children" for the man and the woman. Determine which of the 500 satisfy the condition for each and determine the observed proportion with two boys.

d. Demonstrate the truth of your answer by repeating the computer simulation several times. Repeat the simulation in part c several times.

e. Do the preceding procedures seem to yield the same results? Explain.

Chapter Practice Test

PART I: Knowing the Definitions

Answer "True" if the statement is always true. If the statement is not always true, replace the words shown in bold with words that make the statement always true.

4.1 The probability of an event is a **whole number**.

4.2 The concepts of probability and relative frequency as related to an event are **very similar**.

4.3 The **sample space** is the theoretical population for probability problems.

4.4 The sample points of a sample space are **equally likely** events.

4.5 The value found for experimental probability will **always be** exactly equal to the theoretical probability assigned to the same event.

4.6 The probabilities of complementary events always are **equal**.

4.7 If two events are mutually exclusive, they are also **independent**.

4.8 If events A and B are **mutually exclusive**, the sum of their probabilities must be exactly 1.

4.9 If the sets of sample points that belong to two different events do not intersect, the events are **independent**.

4.10 A compound event formed with the word "and" requires the use of the **addition rule**.

PART II: Applying the Concepts

4.11 A computer is programmed to generate the eight single-digit integers 1, 2, 3, 4, 5, 6, 7, and 8 with equal frequency. Consider the experiment "the next integer generated" and these events:

A: odd number, {1, 3, 5, 7}
B: number greater than 4, {5, 6, 7, 8}
C: 1 or 2, {1, 2}

a. Find $P(A)$.
b. Find $P(B)$.
c. Find $P(C)$.
d. Find $P(\overline{C})$.
e. Find $P(A \text{ and } B)$.
f. Find $P(A \text{ or } B)$.
g. Find $P(B \text{ and } C)$.
h. Find $P(B \text{ or } C)$.
i. Find $P(A \text{ and } C)$.
j. Find $P(A \text{ or } C)$.
k. Find $P(A|B)$.
l. Find $P(B|C)$.

m. Find $P(A|C)$.
n. Are events A and B mutually exclusive? Explain.
o. Are events B and C mutually exclusive? Explain.
p. Are events A and C mutually exclusive? Explain.
q. Are events A and B independent? Explain.
r. Are events B and C independent? Explain.
s. Are events A and C independent? Explain.

4.12 Events A and B are mutually exclusive and $P(A) = 0.4$ and $P(B) = 0.3$.
a. Find $P(A \text{ and } B)$.
b. Find $P(A \text{ or } B)$.
c. Find $P(A|B)$.
d. Are events A and B independent? Explain.

4.13 Events E and F have probabilities $P(E) = 0.5$, $P(F) = 0.4$, and $P(E \text{ and } F) = 0.2$.
a. Find $P(E \text{ or } F)$.
b. Find $P(E|F)$.
c. Are E and F mutually exclusive? Explain.
d. Are E and F independent? Explain.
e. Are G and H independent? Explain.

4.14 Janice wants to become a police officer. She must pass a physical exam and then a written exam. Records show that the probability of passing the physical exam is 0.85 and that once the physical is passed, the probability of passing the written exam is 0.60. What is the probability that Janice passes both exams?

PART III: Understanding the Concepts

4.15 Student A says that independence and mutually exclusive are basically the same thing; namely, both mean neither event has anything to do with the other one. Student B argues that although Student A's statement has some truth in it, Student A has missed the point of these two properties. Student B is correct. Carefully explain why.

4.16 Using complete sentences, describe the following in your own words:
a. Mutually exclusive events
b. Independent events
c. The probability of an event
d. A conditional probability

4.171 Three balanced coins are tossed simultaneously. Find the probability of obtaining three heads, given that at least one of the coins shows heads.
a. Solve using an equally likely sample space.
b. Solve using the formula for conditional probability.

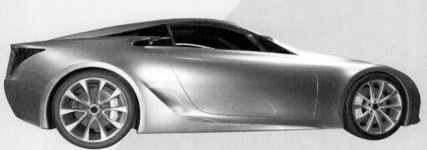

5 Probability Distributions (Discrete Variables)

5.1 Random Variables
*A **numerical value** assigned to each outcome*

5.2 Probability Distributions of a Discrete Random Variable
*The probability for each value of the random variable is listed in a **probability distribution**.*

5.3 The Binomial Probability Distribution
*Binomial situations occur when each trial has **two possible outcomes**.*

Image copyright Michael Shake, 2012. Used under license from Shutterstock.com

---

5.1 Random Variables

USA and Its Automobiles

Americans are very much in love with the automobile, and many have more than one available to them. The national average is 2.28 vehicles per household, with nearly 34% being single-vehicle and 31% being two-vehicle households. However, nearly 35% of all households have three or more vehicles.

Vehicles, x	1	2	3	4	5	6	7	8
$P(x)$	0.34	0.31	0.22	0.06	0.03	0.02	0.01	0.01

By pairing the number of vehicles per household as the variable x with the probability for each value of x, a probability distribution is created. This is much like the relative frequency distribution that we studied in Chapter 2.

If each outcome of a probability **experiment** is assigned a numerical value, then as we observe the results of the experiment, we are observing the values of a random variable. This numerical value is the *random variable value*.

> **Random variable** A variable that assumes a unique numerical value for each of the outcomes in the sample space of a probability experiment.

In other words, a random variable is used to denote the outcomes of a probability experiment. The random variable can take on any numerical value that belongs to the set of all possible outcomes of the experiment. (It is called "random" because the value it assumes is the result of a chance, or random, event.) Each event in a probability experiment must also be defined in such a way that only one value of the random variable is assigned to it **(mutually exclusive events)**, and every event must have a value assigned to it **(all-inclusive events)**.

The following example demonstrates random variables.

E X A M P L E   5 . 1

RANDOM VARIABLES

a. We toss five coins and observe the "number of heads" visible. The random variable x is the number of heads observed and may take on integer values from 0 to 5.

b. Let the "number of phone calls received" per day by a company be the random variable. Integer values ranging from zero to some very large number are possible values.

c. Let the "length of the cord" on an electrical appliance be a random variable. The random variable is a numerical value between 12 and 72 inches for most appliances.

d. Let the "qualifying speed" for race cars trying to qualify for the Indianapolis 500 be a random variable. Depending on how fast the driver can go, the speeds are approximately 220 and faster and are measured in miles per hour (to the nearest thousandth).

Numerical random variables can be subdivided into two classifications: *discrete random variables* and *continuous random variables*.

FYI Discrete and continuous variables were defined on page 8.

Discrete random variable A quantitative random variable that can assume a countable number of values.

Continuous random variable A quantitative random variable that can assume an uncountable number of values.

The random variables "number of heads" and "number of phone calls received" in Example 5.1 parts a and b are discrete. They each represent a count, and therefore there is a countable number of possible values. The random variables "length of the cord" and "qualifying speed" in Example 5.1 parts c and d are continuous. They each represent measurements that can assume any value along an interval, and therefore there is an infinite number of possible values.

⬛ S E C T I O N   5 . 1   E X E R C I S E S

5.1 Refer to the chart accompanying "USA and Its Automobiles" on page 230.

a. What percentage of households have three vehicles?

b. What number of vehicles per household has the highest likelihood?

c. What variable could be used to describe all eight of the events shown on the chart?

d. Are the events mutually exclusive? Explain.

5.2 Based on the information depicted in "USA and Its Automobiles" on page 230,

a. what statistical graph could be used to picture or display this information? Draw it.

b. what other statistical methods could be used to describe this information?

5.3 Survey your classmates about the number of siblings they have and the length of the last conversation they had with their mother. Identify the two random variables of interest and list their possible values.

5.4 a. Explain why the variable "number of saved telephone numbers on a person's cell phone" is discrete.

b. Explain why the variable "weight of a statistics textbook" is continuous.

5.5 a. The variables in Exercise 5.3 are either discrete or continuous. Which are they and why?

b. Explain why the variable "number of dinner guests for Thanksgiving dinner" is discrete.

c. Explain why the variable "number of miles to your grandmother's house" is continuous.

5.6 A social worker is involved in a study about family structure. She obtains information regarding the number of children per family in a certain community from the census data. Identify the random variable of interest, determine whether it is discrete or continuous, and list its possible values.

5.7 *FORTUNE* Magazine released its annual list of The 100 Best Companies to Work For® on February 8, 2010. Many of the companies on the list plan to hire during 2010. Those adding the most employees include:

Company	New Jobs
51. Ernst + Young	2800
5. Wegmans	2000
2. Edward Jones	1040

Source: http://money.cnn.com

a. What is the random variable involved in this study?

b. Is the random variable discrete or continuous? Explain.

5.8 Above-average hot weather extended over the northwest on August 3, 2009. The day's forecasted high temperatures in four cities in the affected area were:

City	Temperature
Boise, ID	100°
Spokane, WA	95°
Portland, OR	91°
Helena, MT	91°

a. What is the random variable involved in this study?

b. Is the random variable discrete or continuous? Explain.

5.9 An archer shoots arrows at the bull's-eye of a target and measures the distance from the center of the target to the arrow. Identify the random variable of interest, determine whether it is discrete or continuous, and list its possible values.

5.10 A *USA Today* Snapshot titled "What women 'splurge' on" (July 21, 2009) reported that 34% of women said "shoes"; 22% said "handbags"; 15% said "work clothing"; 12% said "formal wear"; and 10% said "jewelry."

a. What is the variable involved, and what are the possible values?

b. Why is this variable not a random variable?

5.11 A March 11, 2009, *USA Today* article titled "College freshmen study booze more than books" presents the following chart depicting average hours per week spent on various activities by college freshmen. The study's sponsor, Outside the Classroom, surveyed more than 30,000 first-year students on 76 campuses.

Activity	Average Amount of Time/Week
Partying	10.2 hours
Studying	8.4 hours
Exercising	5.0 hours
Online social networking or playing video games	4.1 hours
Social networking	2.5 hours
Working for pay	2.2 hours

a. What is the random variable involved in this study?

b. Is the random variable discrete or continuous? Explain.

5.12 [EX05-012] If you could stop time and live forever in good health, what age would you pick? Answers to this question were reported in a *USA Today* Snapshot. The average ideal age for each age group is listed in the following table; the average ideal age for all adults was found to be 41. Interestingly, those younger than 30 years want to be older, whereas those older than 30 years want to be younger.

Age Group	18–24	25–29	30–39	40–49	50–64	65+
Ideal Age	27	31	37	40	44	59

Age is used as a variable twice in this application.

a. The age of the person being interviewed is not the random variable in this situation. Explain why and describe how "age" is used with regard to age group.

b. What is the random variable involved in this study? Describe its role in this situation.

c. Is the random variable discrete or continuous? Explain.

5.2 Probability Distributions of a Discrete Random Variable

Consider a coin-tossing experiment where two coins are tossed and no heads, one head, or two heads are observed. If we define the random variable x to be the number of heads observed when two coins are tossed, x can take on the value 0, 1, or 2. The probability of each of these three events can be calculated using techniques from Chapter 4:

$$P(x = 0) = P(0\text{H}) = P(\text{TT}) = \frac{1}{2} \cdot \frac{1}{2} = \frac{1}{4} = 0.25$$

$$P(x = 1) = P(1\text{H}) = P(\text{HT or TH}) = \frac{1}{2} \cdot \frac{1}{2} + \frac{1}{2} \cdot \frac{1}{2} = \frac{1}{2} = 0.50$$

$$P(x = 2) = P(2\text{H}) = P(\text{HH}) = \frac{1}{2} \cdot \frac{1}{2} = \frac{1}{4} = 0.25$$

These probabilities can be listed in any number of ways. One of the most convenient is a table format known as a *probability distribution* (see Table 5.1).

TABLE 5.1
Probability Distribution: Tossing Two Coins

x	P(x)
0	0.25
1	0.50
2	0.25

> **Probability distribution** A distribution of the probabilities associated with each of the values of a random variable. The probability distribution is a theoretical distribution; it is used to represent populations.

In an experiment in which a single die is rolled and the number of dots on the top surface is observed, the random variable is the number observed. The probability distribution for this random variable is shown in Table 5.2.

FYI Can you see why the name "probability distribution" is used?

TABLE 5.2
Probability Distribution: Rolling a Die

x	1	2	3	4	5	6
P(x)	$\frac{1}{6}$	$\frac{1}{6}$	$\frac{1}{6}$	$\frac{1}{6}$	$\frac{1}{6}$	$\frac{1}{6}$

Sometimes it is convenient to write a rule that algebraically expresses the probability of an event in terms of the value of the random variable. This expression is typically written in formula form and is called a *probability function*.

> **Probability function** A rule that assigns probabilities to the values of the random variables.

A probability function can be as simple as a list that pairs the values of a random variable with their probabilities. Tables 5.1 and 5.2 show two such listings. However, a probability function is most often expressed in formula form.

Consider a die that has been modified so that it has one face with one dot, two faces with two dots, and three faces with three dots. Let x be the number of dots observed when this die is rolled. The probability distribution for this experiment is presented in Table 5.3.

TABLE 5.3
Probability Distribution: Rolling the Modified Die

x	P(x)
1	$\frac{1}{6}$
2	$\frac{2}{6}$
3	$\frac{3}{6}$

FYI These properties were presented in Chapter 4.

Each of the probabilities can be represented by the value of x divided by 6; that is, each $P(x)$ is equal to the value of x divided by 6, where $x = 1, 2,$ or 3. Thus,

$$P(x) = \frac{x}{6} \qquad \text{for} \qquad x = 1, 2, 3$$

is the formula for the probability function of this experiment.

The probability function for the experiment of rolling one ordinary die is

$$P(x) = \frac{1}{6} \qquad \text{for} \qquad x = 1, 2, 3, 4, 5, 6$$

This particular function is called a **constant function** because the value of $P(x)$ does not change as x changes.

Every probability function must display the two basic properties of probability (see p. 179). These two properties are (1) the probability assigned to each value of the random variable must be between zero and one, inclusive, and (2) the sum of the probabilities assigned to all the values of the random variable must equal one — that is,

Property 1 $0 \leq$ each $P(x) \leq 1$

Property 2 $\displaystyle\sum_{\text{all } x} P(x) = 1$

E X A M P L E   5 . 2

TABLE 5.4
Probability Distribution for $P(x) = \frac{x}{10}$ for $x = 1, 2, 3, 4$

x	P(x)
1	$\frac{1}{10} = 0.1$ ✓
2	$\frac{2}{10} = 0.2$ ✓
3	$\frac{3}{10} = 0.3$ ✓
4	$\frac{4}{10} = 0.4$ ✓
	$\frac{10}{10} = 1.0$ (ck)

DETERMINING A PROBABILITY FUNCTION

Is $P(x) = \frac{x}{10}$ for $x = 1, 2, 3, 4$ a probability function?

Solution

To answer this question we need only test the function in terms of the two basic properties. The probability distribution is shown in Table 5.4.

Property 1 is satisfied because 0.1, 0.2, 0.3, and 0.4 are all numerical values between zero and one. (See the ✓ indicating each value was checked.) Property 2 is also satisfied because the sum of all four probabilities is exactly one. (See the (ck) indicating the sum was checked.) Since both properties are satisfied, we can conclude that $P(x) = \frac{x}{10}$ for $x = 1, 2, 3, 4$ is a probability function.

What about $P(x = 5)$ (or any value other than $x = 1, 2, 3,$ or 4) for the function $P(x) = \frac{x}{10}$ for $x = 1, 2, 3, 4$? $P(x = 5)$ is considered to be zero. That is, the probability function provides a probability of zero for all values of x other than the values specified as part of the domain.

Probability distributions can be presented graphically. Regardless of the specific graphic representation used, the values of the random variable are plotted on the horizontal scale, and the probability associated with each value of the random variable is plotted on the vertical scale. The probability distribution of a discrete random variable could be presented by a set of line segments drawn at the values of x with lengths that represent the probability of each x. Figure 5.1 shows the probability distribution of $P(x) = \frac{x}{10}$ for $x = 1, 2, 3, 4$.

A regular histogram is used more frequently to present probability distributions. Figure 5.2 shows the probability distribution of Figure 5.1 as a

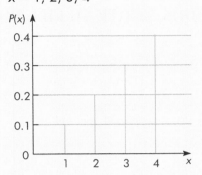

FIGURE 5.1

Line Representation: Probability Distribution for $P(x) = \frac{x}{10}$ for $x = 1, 2, 3, 4$

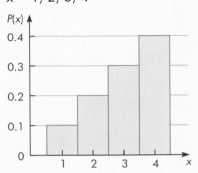

FIGURE 5.2

Histogram: Probability Distribution for $P(x) = \frac{x}{10}$ for $x = 1, 2, 3, 4$

FYI The graph in Figure 5.1 is sometimes called a needle graph.

probability histogram. The histogram of a probability distribution uses the physical area of each bar to represent its assigned probability. The bar for $x = 2$ is 1 unit wide (from 1.5 to 2.5) and 0.2 unit high. Therefore, its area (length × width) is $(0.2)(1) = 0.2$, the probability assigned to $x = 2$. The areas of the other bars can be determined in similar fashion. This area representation will be an important concept in Chapter 6 when we begin to work with continuous random variables.

TECHNOLOGY INSTRUCTIONS: GENERATE RANDOM DATA

MINITAB

Input the possible values of the random variable into C1 and the corresponding probabilities into C2; then continue with:

Choose: **Calc > Random Data > Discrete**
Enter: Number of rows of data to generate: **25** (number wanted)
         Store in column(s): **C3**
         Values (of x) in: **C1**
         Probabilities in: **C2 > OK**

Excel

Input the possible values of the random variable into column A and the corresponding probabilities into column B; then continue with:

Choose: **Data > Data Analysis > Random Number Generation > OK**
Enter: Number of Variables: **1**
         Number of Random Numbers: **25** (# wanted)
         Distribution: **Discrete**
         Value & Prob. Input Range: **(A2:B5 select data cells, not labels)**
Select: **Output Range**
Enter: **(C1 or select cell)**

APPLIED EXAMPLE 5.3

APPLYING FOR ADMISSION

Students hedge their bets

Most students apply to more than one school, making it difficult for colleges to predict how many will actually enroll. Last fall's freshman class was asked:

To how many colleges, other than the one where you enrolled, did you apply for admission this year?

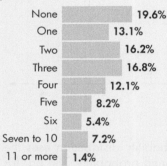

None	19.6%
One	13.1%
Two	16.2%
Three	16.8%
Four	12.1%
Five	8.2%
Six	5.4%
Seven to 10	7.2%
11 or more	1.4%

Source: The American Freshman: National Norms for Fall 2001, survey of 281,064 freshmen entering 421 four-year colleges and universities.

Data from Julie Snider, © 2002 USA Today.

COLLEGES STRIVE TO FILL DORMS

By Mary Beth Marklein, *USA Today*

Colleges and universities will mail their last batch of admission offers in the next few days, but the process is far from over.

Now, students have until May 1 to decide where they'll go this fall. And with lingering concerns about the economy and residual fears about travel and security since Sept. 11, many admissions officials are less able this year to predict how students will respond.

Note the distribution depicted on the bar graph. It has the makings of a discrete probability distribution. The random variable, "number of colleges applied to," is a discrete random variable with values from zero to 11 or more. Each of the values has a corresponding probability, and the sum of the probabilities is equal to 1.

Mean and Variance of a Discrete Probability Distribution

Recall that in Chapter 2 we calculated several numerical sample statistics (mean, variance, standard deviation, and others) to describe empirical sets of data. Probability distributions may be used to represent theoretical populations, the counterpart to samples. We use **population parameters** (mean, variance, and standard deviation) to describe these probability distributions just as we use **sample statistics** to describe samples.

Notes:

1. $\bar{x}$ is the mean of the sample.
2. s^2 and s are the variance and standard deviation of the sample, respectively.
3. $\bar{x}, s^2$, and s are called *sample statistics*.
4. μ (lowercase Greek letter mu) is the mean of the population.
5. σ^2 (sigma squared) is the variance of the population.
6. σ (lowercase Greek letter sigma) is the standard deviation of the population.
7. μ, σ^2, and σ are called *population parameters*. (A parameter is a constant; μ, σ^2, and σ are typically unknown values in real statistics problems. About the only time they are known is in a textbook problem setting for the purposes of learning and understanding.)

The *mean of the probability distribution* of a discrete random variable, or the *mean of a discrete random variable*, is found in a manner that takes full advantage of the table format of a discrete probability distribution. The mean of a discrete random variable is often referred to as its *expected value*.

Mean of a discrete random variable (expected value) The mean, μ, of a discrete random variable x is found by multiplying each possible value of x by its own probability and then adding all the products together:

mean of x: mu = sum of (each x multiplied by its own probability)

$$\mu = \Sigma[xP(x)] \tag{5.1}$$

The variance of a discrete random variable is defined in much the same way as the variance of sample data, the mean of the squared deviations from the mean.

Variance of a discrete random variable The variance, σ^2, of a discrete random variable x is found by multiplying each possible value of the squared deviation from the mean, $(x - \mu)^2$, by its own probability and then adding all the products together:

variance: sigma squared = sum of (squared deviation times probability)

$$\sigma^2 = \Sigma[(x - \mu)^2 P(x)] \tag{5.2}$$

Formula (5.2) is often inconvenient to use; it can be reworked into the following form(s):

variance: sigma squared = sum of (x^2 times probability)
$$- [\text{sum of } (x \text{ times probability})]^2$$

$$\sigma^2 = \Sigma[x^2 P(x)] - \{\Sigma[xP(x)]\}^2 \tag{5.3a}$$

or

$$\sigma^2 = \Sigma[x^2 P(x)] - \mu^2 \tag{5.3b}$$

Likewise, standard deviation of a random variable is calculated in the same manner as is the standard deviation of sample data.

Standard deviation of a discrete random variable The positive square root of variance.

standard deviation: $\sigma = \sqrt{\sigma^2}$ (5.4)

E X A M P L E   5 . 4

STATISTICS FOR A PROBABILITY FUNCTION (DISTRIBUTION)

Find the mean, variance, and standard deviation of the probability function

$$P(x) = \frac{x}{10} \text{ for } x = 1, 2, 3, 4$$

Solution

We will find the mean using formula (5.1), the variance using formula (5.3a), and the standard deviation using formula (5.4). The most convenient way to organize the products and find the totals we need is to expand the probability distribution into an extensions table (see Table 5.5).

TABLE 5.5

Extensions Table: Probability Distribution, $P(x) = \dfrac{x}{10}$ for x = 1, 2, 3, 4

x	P(x)	xP(x)	x^2	$x^2P(x)$
1	$\frac{1}{10}$ = 0.1 ✓	0.1	1	0.1
2	$\frac{2}{10}$ = 0.2 ✓	0.4	4	0.8
3	$\frac{3}{10}$ = 0.3 ✓	0.9	9	2.7
4	$\frac{4}{10}$ = 0.4 ✓	1.6	16	6.4
	$\frac{10}{10}$ = 1.0 (ck)	$\Sigma[xP(x)]$ = 3.0		$\Sigma[x^2P(x)]$ = 10.0

Find the mean of x: The xP(x) column contains each value of x multiplied by its corresponding probability, and the sum at the bottom is the value needed in formula (5.1):

$$\mu = \Sigma[xP(x)] = \mathbf{3.0}$$

Find the variance of x: The totals at the bottom of the xP(x) and $x^2P(x)$ columns are substituted into formula (5.3a):

$$\sigma^2 = \Sigma[x^2P(x)] - \{\Sigma[xP(x)]\}^2$$
$$= 10.0 - \{3.0\}^2 = \mathbf{1.0}$$

Find the standard deviation of x: Use formula (5.4):

$$\sigma = \sqrt{\sigma^2} = \sqrt{1.0} = \mathbf{1.0}$$

Notes:

1. The purpose of the extensions table is to organize the process of finding the three column totals: $\Sigma[P(x)]$, $\Sigma[xP(x)]$, and $\Sigma[x^2P(x)]$.
2. The other columns, x and x^2, should not be totaled; they are not used.
3. $\Sigma[P(x)]$ will always be 1.0; use this only as a check.
4. $\Sigma[xP(x)]$ and $\Sigma[x^2P(x)]$ are used to find the mean and variance of x.

E X A M P L E   5 . 5

MEAN, VARIANCE, AND STANDARD DEVIATION OF A DISCRETE RANDOM VARIABLE

A coin is tossed three times. Let the "number of heads" that occur in those three tosses be the random variable, x. Find the mean, variance, and standard deviation of x.

Video tutorial available—logon and learn more at cengagebrain.com

Solution

There are eight possible outcomes (all equally likely) to this experiment: {HHH, HHT, HTH, HTT, THH, THT, TTH, TTT}. One outcome results in $x = 0$, three in $x = 1$, three in $x = 2$, and one in $x = 3$. Therefore, the probabilities for this random variable are $\frac{1}{8}$, $\frac{3}{8}$, $\frac{3}{8}$, and $\frac{1}{8}$. The probability distribution associated with this experiment is shown in Figure 5.3 and in Table 5.6. The necessary extensions and summations for the calculation of the mean, variance, and standard deviation are also shown in Table 5.6.

FIGURE 5.3

Probability Distribution: Number of Heads in Three Tosses of Coin

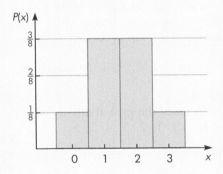

TABLE 5.6 Extensions Table of Probability Distribution of Number of Heads in Three Coin Tosses

x	$P(x)$	$xP(x)$	x^2	$x^2P(x)$
0	$\frac{1}{8}$ ✓	$\frac{0}{8}$	0	$\frac{0}{8}$
1	$\frac{3}{8}$ ✓	$\frac{3}{8}$	1	$\frac{3}{8}$
2	$\frac{3}{8}$ ✓	$\frac{6}{8}$	4	$\frac{12}{8}$
3	$\frac{1}{8}$ ✓	$\frac{3}{8}$	9	$\frac{9}{8}$

$$\Sigma[P(x)] = \frac{8}{8} = 1.0 \, \text{(ck)} \quad \Sigma[xP(x)] = \frac{12}{8} = 1.5 \quad\quad\quad \Sigma[x^2P(x)] = \frac{24}{8} = 3.0$$

The mean is found using formula (5.1):

$$\mu = \Sigma[xP(x)] = \mathbf{1.5}$$

This result, 1.5, is the mean of the theoretical distribution for the random variable "number of heads" observed per set of three coin tosses. It is expected that the mean for many observed values of the random variable will also be approximately equal to this value.

The variance is found using formula (5.3a):

$$\sigma^2 = \Sigma[x^2P(x)] - \{\Sigma[xP(x)]\}^2$$
$$= 3.0 - (1.5)^2 = 3.0 - 2.25 = \mathbf{0.75}$$

The standard deviation is found using formula (5.4):

$$\sigma = \sqrt{\sigma^2} = \sqrt{0.75} = 0.866 = \mathbf{0.87}$$

That is, 0.87 is the standard deviation of the theoretical distribution for the random variable "number of heads" observed per set of three coin tosses. It is expected that the standard deviation for many observed values of the random variable will also be approximately equal to this value.

⌨ S E C T I O N   5 . 2   E X E R C I S E S

5.13 Express the tossing of one coin as a probability distribution of x, the number of heads occurring (that is, $x = 1$ if a head occurs and $x = 0$ if a tail occurs).

5.14 a. Express $P(x) = \frac{1}{6}$, for $x = 1, 2, 3, 4, 5, 6$, in distribution form.

   b. Construct a histogram of the probability distribution $P(x) = \frac{1}{6}$, for $x = 1, 2, 3, 4, 5, 6$.

   c. Describe the shape of the histogram in part b.

5.15 a. Explain how the various values of x in a probability distribution form a set of mutually exclusive events.

   b. Explain how the various values of x in a probability distribution form a set of "all-inclusive" events.

5.16 Test the following function to determine whether it is a probability function. If it is not, try to make it into a probability function.

$$R(x) = 0.2, \text{ for } x = 0, 1, 2, 3, 4$$

a. List the distribution of probabilities.

b. Sketch a histogram.

5.17 Test the following function to determine whether it is a probability function.

$$P(x) = \frac{x^2 + 5}{50}, \text{ for } x = 1, 2, 3, 4$$

a. List the probability distribution.

b. Sketch a histogram.

5.18 Test the following function to determine whether it is a probability function. If it is not, try to make it into a probability function.

$$S(x) = \frac{6 - |x - 7|}{36}, \text{ for } x = 2, 3, 4, 5, 6, 7, \ldots, 11, 12$$

a. List the distribution of probabilities and sketch a histogram.

b. Do you recognize $S(x)$? If so, identify it.

5.19 Census data are often used to obtain probability distributions for various random variables. Census data for families in a particular state with a combined income of $50,000 or more show that 20% of these families have no children, 30% have one child, 40% have two children, and 10% have three children. From this information, construct the probability distribution for x, where x represents the number of children per family for this income group.

5.20 In a *USA Today* Snapshot (June 1, 2009), the following statistics were reported on the number of hours of sleep that adults get.

Number of Hours	Percentage
5 or less	12%
6	29%
7	37%
8 or more	24%

Source: StrategyOne survey for Tempur-Pedic of 1004 adults in April

a. Are there any other values that the number of hours can attain?

b. Explain why the total of the percentages is not 100%.

c. Is this a discrete probability distribution? Is it a probability distribution? Explain.

5.21 Verify that formulas (5.3a) and (5.3b) are equivalent to formula (5.2).

5.22 a. Form the probability distribution table for $P(x) = \frac{x}{6}$, for $x = 1, 2, 3$.

b. Find the extensions $xP(x)$ and $x^2P(x)$ for each x.

c. Find $\Sigma[xP(x)]$ and $\Sigma[x^2P(x)]$.

d. Find the mean for $P(x) = \frac{x}{6}$, for $x = 1, 2, 3$.

e. Find the variance for $P(x) = \frac{x}{6}$, for $x = 1, 2, 3$.

f. Find the standard deviation for $P(x) = \frac{x}{6}$, for $x = 1, 2, 3$.

5.23 If you find the sum of the x and the x^2 columns on the extensions table, exactly what have you found?

5.24 Given the probability function $P(x) = \frac{5-x}{10}$, for $x = 1, 2, 3, 4$, find the mean and standard deviation.

5.25 Given the probability function $R(x) = 0.2$, for $x = 0, 1, 2, 3, 4$, find the mean and standard deviation.

5.26 The number of ships to arrive at a harbor on any given day is a random variable represented by x. The probability distribution for x is as follows:

x	10	11	12	13	14
P(x)	0.4	0.2	0.2	0.1	0.1

Find the mean and standard deviation of the number of ships that arrive at a harbor on a given day.

5.27 The College Board website provides much information for students, parents, and professionals with respect to the many aspects involved in Advanced Placement (AP) courses and exams. One particular annual report provides the percent of students who obtain each of the possible AP grades (1 through 5). The 2008 grade distribution for all subjects was as follows:

AP Grade	Percent
1	20.9
2	21.3
3	24.1
4	19.4
5	14.3

a. Express this distribution as a discrete probability distribution.

b. Find the mean and standard deviation of the AP exam scores for 2008.

5.28 The number of children per household, x, in the United States in 2008 is expressed as a probability distribution here.

x	0	1	2	3	4	5+
P(x)	0.209	0.384	0.249	0.106	0.032	0.020

Source: U.S. Census Bureau

a. Is this a discrete probability distribution? Explain.

b. Draw a histogram for the distribution of x, the number of children per household.

c. Replacing "5+" with exactly "5," find the mean and standard deviation.

5.29 Is a dog "man's best friend"? One would think so, with 60 million pet dogs nationwide. But how many friends are needed? In the American Pet Products Association's 2007–2008 National Pet Owners Survey, the following statistics were reported.

Number of Pet Dogs	Percentage
One	63
Two	25
Three or more	12

Source: APPMA 2007–2008 National Pet Owners Survey

a. Is this a discrete probability distribution? Explain.

b. Draw a relative frequency histogram to depict the results shown in the table.

c. Replacing the category "Three or more" with exactly "Three," find the mean and standard deviation of the number of pet dogs per household.

d. How do you interpret the mean?

e. Explain the effect that replacing the category "Three or more" with "Three" has on the mean and standard deviation.

5.30 As reported in the chapter opener "USA and Its Automobiles," Americans are in love with the automobile—the majority have more than one vehicle per household. In fact, the national average is 2.28 vehicles per household. The number of vehicles per household in the United States can be described as follows:

Vehicles, x	P(x)
1	0.34
2	0.31
3	0.22
4	0.06
5	0.03
6	0.02
7	0.01
8 or more	0.01

a. Replacing the category "8 or more" with exactly "8," find the mean and standard deviation of the number of vehicles per household in the United States.

b. How does the mean calculated in part a correspond to the national average of 2.28?

(continue on page 242)

c. Explain the effect that replacing the category "8 or more" with "8" has on the mean and standard deviation.

5.31 The random variable A has the following probability distribution:

A	1	2	3	4	5
$P(A)$	0.6	0.1	0.1	0.1	0.1

a. Find the mean and standard deviation of A.

b. How much of the probability distribution is within 2 standard deviations of the mean?

c. What is the probability that A is between $\mu - 2\sigma$ and $\mu + 2\sigma$?

5.32 The random variable $\bar{x}$ has the following probability distribution:

$\bar{x}$	1	2	3	4	5
$P(\bar{x})$	0.6	0.1	0.1	0.1	0.1

a. Find the mean and standard deviation of $(\bar{x})$.

b. What is the probability that $\bar{x}$ is between $\mu - \sigma$ and $\mu + \sigma$?

5.33 a. Draw a histogram of the probability distribution for the single-digit random numbers 0, 1, 2, ..., 9.

b. Calculate the mean and standard deviation associated with the population of single-digit random numbers.

c. Represent (1) the location of the mean on the histogram with a vertical line and (2) the magnitude of the standard deviation with a line segment.

d. How much of this probability distribution is within 2 standard deviations of the mean?

5.34 Skillbuilder Applet Exercise simulates playing a game where a player has a 0.2 probability of winning $3 and a 0.8 probability of losing $1. Repeat the simulations for several sets of 100 plays using the "Play 25 times" button.

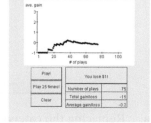

a. What would you estimate for your expected value (average gain or loss) from the results?

b. Using the following probability distribution, calculate the mean.

x	$P(x)$
$3	0.2
-$1	0.8

c. How do your answers to parts a and b compare? Would you consider this a fair game? Why?

5.35 A *USA Today* Snapshot (March 4, 2009) presented a pie chart depicting how workers damage their laptops. Statistics were derived from a survey conducted by Ponemon Institute for Dell of 714 IT managers. Is this a probability distribution? Explain.

Reason for Damage to Laptop	Percentage (%)
Spilled food or liquids	34
Dropping them	28
Not protecting during travel	25
Worker anger	13

5.36 a. Use a computer (or random number table) to generate a random sample of 25 observations drawn from the following discrete probability distribution.

x	1	2	3	4	5
$P(x)$	0.2	0.3	0.3	0.1	0.1

Compare the resulting data to your expectations.

b. Form a relative frequency distribution of the random data.

c. Construct a probability histogram of the given distribution and a relative frequency histogram of the observed data using class midpoints of 1, 2, 3, 4, and 5.

d. Compare the observed data with the theoretical distribution. Describe your conclusions.

e. Repeat parts a through d several times with $n = 25$. Describe the variability you observe between samples.

f. Repeat parts a through d several times with $n = 250$. Describe the variability you see between samples of this much larger size.

MINITAB

a. Input the x values of the random variable into C1 and their corresponding probabilities, $P(x)$, into C2; then continue with the generating random data MINITAB commands on page 235.

b. To obtain the frequency distribution, continue with:

Choose: **Stat > Tables > Cross Tabulation**
Enter: Categorical variables: For rows: **C3**
Select: Display: **Total percents > OK**

c. To construct the histogram of the generated data in C3, continue with the histogram MINITAB commands on page 53, selecting scale > Y-Scale Type > Percent. (Use Binning followed by midpoint and midpoint positions 1:5/1 if necessary.)

To construct a bar graph of the given distribution:

Choose:	**Graph > Bar Chart > Bars represent: Values from a table > One Column of values: Simple > OK**
Enter:	**Graph variables: C2** Categorical variables: **C1**
Select:	**Labels > Data Labels > Label Type: Use y-value labels > OK**
Select:	**Data View > Data Display: Bars > OK > OK**

Excel

a. Input the x values of the random variable in column A and their corresponding probabilities, P(x), in column B; then continue with the generating random data Excel commands on page 235 for n = 25.

b.& c. The frequency distribution is given with the histogram of the generated data. Use the histogram Excel commands on page 53 using the data in column C and the bin range in column A.

To construct a histogram of the given distribution, activate A1:B6 or select cells and continue with:

Choose:	**Insert > Column > 1ˢᵗ picture**(usually) **Chart Layouts > Layout 9**
Choose:	**Select Data > Series 1 > Remove > OK**
Enter:	**Chart and axes titles** (Edit as needed)

5.37 a. Use a computer (or random number table) and generate a random sample of 100 observations drawn from the discrete probability population $P(x) = \frac{5-x}{10}$, for $x = 1, 2, 3, 4$. List the resulting sample. (Use the computer commands in Exercise 5.36; just change the arguments.)

b. Form a relative frequency distribution of the random data.

c. Form a probability distribution of the expected probability distribution. Compare the resulting data with your expectations.

d. Construct a probability histogram of the given distribution and a relative frequency histogram of the observed data using class midpoints of 1, 2, 3, and 4.

e. Compare the observed data with the theoretical distribution. Describe your conclusions.

f. Repeat parts a–d several times with $n = 100$. Describe the variability you observe between samples.

5.38 Every Tuesday, Jason's Video has "roll-the-dice" day. A customer may roll two fair dice and rent a second movie for an amount (in cents) determined by the numbers showing on the dice, the larger number first. For example, if the customer rolls a one and a five, a second movie may be rented for $0.51. Let x represent the amount paid for a second movie on roll-the-dice Tuesday.

a. Use the sample space for the rolling of a pair of dice and express the rental cost of the second movie, x, as a probability distribution.

b. What is the expected mean rental cost (mean of x) of the second movie on roll-the-dice Tuesday?

c. What is the standard deviation of x?

d. Using a computer and the probability distribution found in part a, generate a random sample of 30 values for x and determine the total cost of renting the second movie for 30 rentals.

e. Using a computer, obtain an estimate for the probability that the total amount paid for 30 second movies will exceed $15.00 by repeating part d 500 times and using the 500 results.

5.3 The Binomial Probability Distribution

Consider the following probability experiment. Your instructor gives the class a surprise four-question multiple-choice quiz. You have not studied the material, and therefore you decide to answer the four questions by randomly guessing the answers without reading the questions or the answers.

Answer Page to Quiz

Directions: Circle the best answer to each question.

1. a   b   c
2. a   b   c
3. a   b   c
4. a   b   c

FYI That's right, guess!

Circle your answers before continuing.

Before we look at the correct answers to the quiz and find out how you did, let's think about some of the things that might happen if you answered a quiz this way.

1. How many of the four questions are you likely to have answered correctly?
2. How likely are you to have more than half of the answers correct?
3. What is the probability that you selected the correct answers to all four questions?
4. What is the probability that you selected wrong answers for all four questions?
5. If an entire class answers the quiz by guessing, what do you think the class "average" number of correct answers will be?

To find the answers to these questions, let's start with a tree diagram of the sample space, showing all 16 possible ways to answer the four-question quiz. Each of the four questions is answered with the correct answer (C) or with a wrong answer (W). See Figure 5.4.

FIGURE 5.4

Tree Diagram: Possible Answers to a Four-Question Quiz

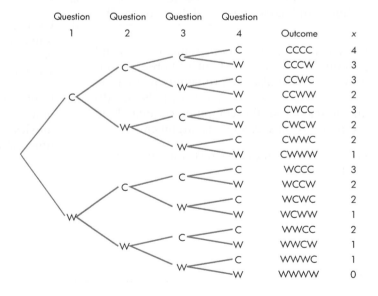

FYI WWWW? represents wrong on 1 and wrong on 2 and wrong on 3 and wrong on 4; therefore, its probability is found using the multiplication rule, formula (4.7).

We can convert the information on the tree diagram into a probability distribution. Let x be the "number of correct answers" on one person's quiz when the quiz was taken by randomly guessing. The random variable x may take on any one of the values 0, 1, 2, 3, or 4 for each quiz. Figure 5.4 shows 16 branches representing five different values of x. Notice that the event $x = 4$, "four correct answers," is represented by the top branch of the tree diagram, and the event

$x = 0$, "zero correct answers," is shown on the bottom branch. The other events, "one correct answer," "two correct answers," and "three correct answers," are each represented by several branches of the tree. We find that the event $x = 1$ occurs on four different branches, event $x = 2$ occurs on six branches, and event $x = 3$ occurs on four branches.

Each individual question has only one correct answer among the three possible answers, so the probability of selecting the correct answer to an individual question is $\frac{1}{3}$. The probability that a wrong answer is selected on an individual question is $\frac{2}{3}$. The probability of each value of x can be found by calculating the probabilities of all the branches and then combining the probabilities for branches that have the same x values. The calculations follow, and the resulting probability distribution appears in Table 5.7.

$P(x = 0)$ is the probability that the correct answers are given for zero questions and the wrong answers are given for four questions (there is only one branch on Figure 5.4 where all four are wrong—WWWW):

$$P(x = 0) = \frac{2}{3} \times \frac{2}{3} \times \frac{2}{3} \times \frac{2}{3} = \left(\frac{2}{3}\right)^4 = \frac{16}{81} = \mathbf{0.198}$$

Note: Answering each individual question is a separate and independent event, thereby allowing us to use formula (4.7), which states that we should multiply the probabilities.

$P(x = 1)$ is the probability that the correct answer is given for exactly one question and wrong answers are given for the other three (there are four branches on Figure 5.4 where this occurs—namely, CWWW, WCWW, WWCW, WWWC—and each has the same probability):

$$P(x = 1) = (4) \times \frac{1}{3} \times \frac{2}{3} \times \frac{2}{3} \times \frac{2}{3} = (4)\left(\frac{1}{3}\right)^1\left(\frac{2}{3}\right)^3 = \mathbf{0.395}$$

$P(x = 2)$ is the probability that correct answers are given for exactly two questions and wrong answers are given for the other two (there are six branches on Figure 5.4 where this occurs—CCWW, CWCW, CWWC, WCCW, WCWC, WWCC—and each has the same probability):

$$P(x = 2) = (6) \times \frac{1}{3} \times \frac{1}{3} \times \frac{2}{3} \times \frac{2}{3} = (6)\left(\frac{1}{3}\right)^2\left(\frac{2}{3}\right)^2 = \mathbf{0.296}$$

$P(x = 3)$ is the probability that correct answers are given for exactly three questions and a wrong answer is given for the other question (there are four branches on Figure 5.4 where this occurs—CCCW, CCWC, CWCC, WCCC—and each has the same probability):

$$P(x = 3) = (4) \times \frac{1}{3} \times \frac{1}{3} \times \frac{1}{3} \times \frac{2}{3} = (4)\left(\frac{1}{3}\right)^3\left(\frac{2}{3}\right)^1 = \mathbf{0.099}$$

$P(x = 4)$ is the probability that correct answers are given for all four questions (there is only one branch on Figure 5.4 where all four are correct—CCCC):

$$P(x = 4) = \frac{1}{3} \times \frac{1}{3} \times \frac{1}{3} \times \frac{1}{3} = \left(\frac{1}{3}\right)^4 = \frac{1}{81} = \mathbf{0.012}$$

Now we can answer the five questions that were asked about the four-question quiz (p. 244).

Answer 1: The most likely occurrence would be to get one answer correct; it has a probability of 0.395. Zero, one, or two correct answers are expected to result approximately 89% of the time $(0.198 + 0.395 + 0.296 = 0.889)$.

TABLE 5.7
Probability Distribution for the Four-Question Quiz

x	P(x)
0	0.198
1	0.395
2	0.296
3	0.099
4	0.012
	1.000 (ck)

Answer 2: Having more than half correct is represented by $x = 3$ or 4; their total probability is $0.099 + 0.012 = 0.111$. (You will pass this quiz only 11% of the time by random guessing.)

Answer 3: $P(\text{all four correct}) = P(x = 4) = 0.012$. (All correct occurs only 1% of the time.)

Answer 4: $P(\text{all four wrong}) = P(x = 0) = 0.198$. (That's almost 20% of the time.)

Answer 5: The class average is expected to be $\frac{1}{3}$ of 4, or 1.33 correct answers.

The correct answers to the quiz are b, c, b, a. How many correct answers did you have? Which branch of the tree in Figure 5.4 represents your quiz results? You might ask several people to answer this same quiz by guessing the answers. Then construct an observed relative frequency distribution and compare it with the distribution shown in Table 5.7.

Many experiments are composed of repeated trials whose outcomes can be classified into one of two categories: **success** or **failure**. Examples of such experiments are coin tosses, right/wrong quiz answers, and other, more practical experiments such as determining whether a product did or did not do its prescribed job and whether a candidate gets elected or not. There are experiments in which the trials have many outcomes that, under the right conditions, may fit this general description of being classified in one of two categories. For example, when we roll a single die, we usually consider six possible outcomes. However, if we are interested only in knowing whether a "one" shows or not, there are really only two outcomes: the "one" shows or "something else" shows. The experiments just described are called *binomial probability experiments*.

Binomial probability experiment An experiment that is made up of repeated trials that possess the following properties:
1. There are n repeated identical independent trials.
2. Each trial has two possible outcomes (success or failure).
3. $P(\text{success}) = p$, $P(\text{failure}) = q$, and $p + q = 1$.
4. The **binomial random variable** x is the count of the number of successful trials that occur; x may take on any integer value from zero to n.

Notes:
1. Properties 1 and 2 describe the two basic characteristics of any binomial experiment.
2. **Independent trials** mean that the result of one trial does not affect the probability of success on any other trial in the experiment. In other words, the probability of success remains constant throughout the entire experiment.
3. Property 3 gives the algebraic notation for each trial.
4. Property 4 concerns the algebraic notation for the complete experiment.
5. It is of utmost importance that both x and p be associated with "success."

The four-question quiz qualifies as a binomial experiment made up of four trials when all four of the answers are obtained by random guessing.

Property 1: A **trial** is the **answering of one question**, and it is repeated $n = 4$ times. The trials are **independent** because the probability of a correct answer on any one question is not affected by the answers on other questions.

Property 2: The two possible outcomes on each trial are **success = C**, correct answer, and **failure = W**, wrong answer.

Property 3: For each trial (each question): $p = P(correct) = \frac{1}{3}$ and $q = P(incorrect) = \frac{2}{3}$. $[p + q = 1$ ⓒⓚ$]$

Property 4: For the total experiment (the quiz): $x =$ **number of correct answers** and can be any integer value from zero to $n = 4$.

E X A M P L E   5 . 6

DEMONSTRATING THE PROPERTIES OF A BINOMIAL PROBABILITY EXPERIMENT

Consider the experiment of rolling a die 12 times and observing a "one" or "something else." At the end of all 12 rolls, the number of "ones" is reported. The random variable x is the number of times that a "one" is observed in the $n = 12$ trials. Since "one" is the outcome of concern, it is considered "success"; therefore, $p = P(one) = \frac{1}{6}$ and $q = P(not\ one) = \frac{5}{6}$. This experiment is binomial.

E X A M P L E   5 . 7

DEMONSTRATING THE PROPERTIES OF A BINOMIAL PROBABILITY EXPERIMENT

If you were an inspector on a production line in a plant where television sets are manufactured, you would be concerned with identifying the number of defective television sets. You probably would define "success" as the occurrence of a defective television. This is not what we normally think of as success, but if we count "defective" sets in a binomial experiment, we must define "success" as a "defective." The random variable x indicates the number of defective sets found per lot of n sets; $p = P(\text{television is defective})$ and $q = P(\text{television is good})$.

The key to working with any probability experiment is its probability distribution. All binomial probability experiments have the same properties, and therefore the same organization scheme can be used to represent all of them. The *binomial probability function* allows us to find the probability for each possible value of x.

Binomial probability function For a binomial experiment, let p represent the probability of a "success" and q represent the probability of a "failure" on a single trial. Then $P(x)$, the probability that there will be exactly x successes in n trials, is

$$P(x) = \binom{n}{x}(p^x)(q^{n-x}) \text{ for } x = 0, 1, 2, \ldots, n \qquad (5.5)$$

When you look at the probability function, you notice that it is the product of three basic factors:

1. The number of ways that exactly x successes can occur in n trials, $\binom{n}{x}$
2. The probability of exactly x successes, p^x
3. The probability that failure will occur on the remaining $(n - x)$ trials, q^{n-x}

The number of ways that exactly x successes can occur in a set of n trials is represented by the symbol $\binom{n}{x}$, which must always be a positive integer. This term is called the **binomial coefficient** and is found by using the formula

$$\binom{n}{x} = \frac{n!}{x!(n - x)!} \tag{5.6}$$

Notes:
1. $n!$ (*"n factorial"*) is an abbreviation for the product of the sequence of integers starting with n and ending with one. For example, $3! = 3 \cdot 2 \cdot 1 = 6$ and $5! = 5 \cdot 4 \cdot 3 \cdot 2 \cdot 1 = 120$. There is one special case, $0!$, that is defined to be 1. For more information about **factorial notation**, see the *Student Solutions Manual*.
2. The values for $n!$ and $\binom{n}{x}$ can be readily found using most scientific calculators.
3. The binomial coefficient $\binom{n}{x}$ is equivalent to the number of combinations $_nC_x$, the symbol most likely on your calculator.
4. See the *Student Solutions Manual* for general information on the binomial coefficient.

Let's reconsider Example 5.5 (pp. 238–240): A coin is tossed three times and we observe the number of heads that occur in the three tosses. This is a binomial experiment because it displays all the properties of a binomial experiment:

1. There are $n = 3$ repeated **independent** trials (each coin toss is a separate trial, and the outcome of any one trial has no effect on the probability of another trial).
2. Each trial (each toss of the coin) results in one of two possible outcomes: success = **heads** (what we are counting) or failuare = **tails**.
3. The probability of success is $p = P(\text{H}) = \mathbf{0.5}$, and the probability of failure is $q = P(\text{T}) = \mathbf{0.5}$. $[p + q = 0.5 + 0.5 = 1 \text{ ck}]$
4. The random variable x is the **number of heads** that occur in the three trials. x will assume exactly one of the values **0, 1, 2, or 3** when the experiment is complete.

The binomial probability function for the tossing of three coins is

$$P(x) = \binom{n}{x}(p^x)\ (q^{n-x}) = \binom{3}{x}(0.5)^x(0.5)^{3-x} \quad \text{for} \quad x = 0, 1, 2, 3$$

Let's find the probability of $x = 1$ using the preceding binomial probability function:

FYI In Table 5.6 (p. 239), $P(1) = \frac{3}{8}$. Here, $P(1) = 0.375$ and $\frac{3}{8} = 0.375$.

$$P(x = 1) = \binom{3}{1}(0.5)^1(0.5)^2 = 3(0.5)(0.25) = \mathbf{0.375}$$

Note that this is the same value found in Example 5.5 (p. 238).

EXAMPLE 5.8

DETERMINING A BINOMIAL EXPERIMENT AND ITS PROBABILITIES

Consider an experiment that calls for drawing five cards, one at a time with replacement, from a well-shuffled deck of playing cards. The drawn card is identified as a spade or not a spade, it is returned to the deck, the deck is reshuffled, and so on. The random variable x is the number of spades observed in the set of five drawings. Is this a binomial experiment? Let's identify the four properties.

1. There are **five repeated drawings**; $n = 5$. These individual trials are **independent** because the drawn card is returned to the deck and the deck is reshuffled before the next drawing.
2. Each drawing is a trial, and each drawing has two outcomes: **spade** or **not spade**.
3. $p = P(\text{spade}) = \frac{13}{52}$ and $q = P(\text{not spade}) = \frac{39}{52}$
4. x is the **number of spades** recorded upon completion of the five trials; the possible values are **0, 1, 2, . . . , 5**.

The binomial probability function is

$$P(x) = \binom{5}{x}\left(\frac{13}{52}\right)^{x}\left(\frac{39}{52}\right)^{5-x} = \binom{5}{x}\left(\frac{1}{4}\right)^{x}\left(\frac{3}{4}\right)^{5-x} = \binom{5}{x}(0.25)^{x}(0.75)^{5-x}$$
$$\text{for } x = 0, 1, ..., 5$$

$$P(0) = \binom{5}{0}(0.25)^{0}(0.75)^{5} = (1)(1)(0.2373) = \mathbf{0.2373}$$

$$P(1) = \binom{5}{1}(0.25)^{1}(0.75)^{4} = (5)(0.25)(0.3164) = \mathbf{0.3955}$$

$$P(2) = \binom{5}{2}(0.25)^{2}(0.75)^{3} = (10)(0.0625)(0.421875) = \mathbf{0.2637}$$

$$P(3) = \binom{5}{3}(0.25)^{3}(0.75)^{2} = (10)(0.15625)(0.5625) = \mathbf{0.0879}$$

The two remaining probabilities are left for you to compute in Exercise 5.52.

FYI Answer: five

The preceding distribution of probabilities indicates that the single most likely value of x is one, the event of observing exactly one spade in a hand of five cards. What is the least likely number of spades that would be observed?

EXAMPLE 5.9

BINOMIAL PROBABILITY OF "BAD EGGS"

The manager of Steve's Food Market guarantees that none of his cartons of a dozen eggs will contain more than one bad egg. If a carton contains more than one bad egg, he will replace the whole dozen and allow the customer to keep the original eggs. If the probability that an individual egg is bad is 0.05, what is the probability that the manager will have to replace a given carton of eggs?

Solution

At first glance, the manager's situation appears to fit the properties of a binomial experiment if we let x be the number of bad eggs found in a carton of a dozen eggs, let $p = P(bad) = 0.05$, and let the inspection of each egg be a trial that results in finding a "bad" or "not bad" egg. There will be $n = 12$ trials to account for the 12 eggs in a carton. However, trials of a binomial experiment must be independent; therefore, we will assume that the quality of one egg in a carton is independent of the quality of any of the other eggs. (This may be a big assumption! But with this assumption, we will be able to use the binomial probability distribution as our model.) Now, based on this assumption, we will be able to find/estimate the probability that the manager will have to make good on his guarantee. The probability function associated with this experiment will be:

$$P(x) = \binom{12}{x}(0.05)^x(0.95)^{12-x} \qquad \text{for } x = 0, 1, 2, \ldots, 12$$

The probability that the manager will replace a dozen eggs is the probability that $x = 2, 3, 4, \ldots, 12$. Recall that $\Sigma P(x) = 1$; that is,

$$P(0) + P(1) + P(2) + \cdots + P(12) = 1$$

$$P(\text{replacement}) = P(2) + P(3) + \cdots + P(12) = 1 - [P(0) + P(1)]$$

It is easier to find the probability of replacement by finding $P(x = 0)$ and $P(x = 1)$ and subtracting their total from 1 than by finding all of the other probabilities. We have

$$P(x) = \binom{12}{x}(0.05)^x(0.95)^{12-x}$$

$$P(0) = \binom{12}{0}(0.05)^0(0.95)^{12} = \mathbf{0.540}$$

$$P(1) = \binom{12}{1}(0.05)^1(0.95)^{11} = \mathbf{0.341}$$

$$P(\text{replacement}) = 1 - (0.540 + 0.341) = \mathbf{0.119}$$

If $p = 0.05$ is correct, then the manager will be busy replacing cartons of eggs. If he replaces 11.9% of all the cartons of eggs he sells, he certainly will be giving away a substantial proportion of his eggs. This suggests that he should adjust his guarantee (or market better eggs). For example, if he were to replace a carton of eggs only when four or more were found to be bad, he would expect to replace only 3 out of 1000 cartons $[1.0 - (0.540 + 0.341 + 0.099 + 0.017)]$, or 0.3% of the cartons sold. Notice that the manager will be able to control his "risk" (probability of replacement) if he adjusts the value of the random variable stated in his guarantee.

Note: The value of many binomial probabilities for values of $n \leq 15$ and common values of p are found in Table 2 of Appendix B. In this example, we have $n = 12$ and $p = 0.05$, and we want the probabilities for $x = 0$ and 1. We need to locate the section of Table 2 where $n = 12$, find the column headed $p = 0.05$, and read the numbers across from $x = 0$ and $x = 1$. We find .540 and .341, as shown in Table 5.8. (Look up these values in Table 2 in Appendix B.)

TABLE 5.8
Excerpt of Table 2 in Appendix B, Binomial Probabilities

									p							
n	x	0.01	0.05	0.10	0.20	0.30	0.40	0.50	0.60	0.70	0.80	0.90	0.95	0.99	x
	⋮														
12	0	.886	.540	.282	.069	.014	.002	0+	0+	0+	0+	0+	0+	0+	0
	1	.107	.341	.377	.206	.071	.017	.003	0+	0+	0+	0+	0+	0+	1
	2	.006	.099	.230	.283	.168	.064	.016	.002	0+	0+	0+	0+	0+	2
	3	0+	.017	.085	.236	.240	.142	.054	.012	.001	0+	0+	0+	0+	3
	4	0+	.002	.021	.133	.231	.213	.121	.042	.008	.001	0+	0+	0+	4
	⋮														

Note: A convenient notation to identify the binomial probability distribution for a binomial experiment with $n = 12$ and $p = 0.05$ is $B(12, 0.05)$. $B(12, 0.05)$, read *"binomial distribution for $n = 12$ and $p = 0.05$,"* represents the entire distribution or "block" of probabilities shown in purple in Table 5.8. When used in combination with the $P(x)$ notation, $P(x = 1 | B(12, 0.05))$ indicates the probability of $x = 1$ from this distribution, or 0.341, as shown on Table 5.8.

TECHNOLOGY INSTRUCTIONS: BINOMIAL AND CUMULATIVE BINOMIAL PROBABILITIES

MINITAB

For binomial probabilities, input x values into C1; then continue with:

Choose: **Calc > Probability Distributions > Binomial**
Select: **Probability***
Enter: **Number of trials: n**
             Event probability: p
Select: **Input column**
Enter: **C1**
             Optional Storage: C2 (not necessary) > OK

Or
Select: **Input constant**
Enter: **One single x value > OK**

*For cumulative binomial probabilities, repeat the preceding commands but replace the probability selection with:
Select: **Cumulative Probability**

Excel

For binomial probabilities, input x values into column A and activate the column B cell across from the first x value; then continue with:

Choose: **Insert function, f_x > Statistical > BINOMDIST > OK**
Enter: **Number_s: (A1:A4 or select "x value" cells)**
             Trials: n
             Probability_s: p
             Cumulative: false* (gives individual probabilities) > OK

Drag: **Bottom-right corner of probability value cell in column B down to give other probabilities**

*For cumulative binomial probabilities, repeat the preceding commands but replace the false cumulative with:
Cumulative: **true (gives cumulative probabilities) > OK**

TI-83/84 Plus

To obtain a complete list of probabilities for a particular n and p, continue with:

Choose: **2nd > DISTR > 0:binompdf(**
Enter: **n, p)**

Use the right arrow key to scroll through the probabilities.
To scroll through a vertical list in L1:

Choose: **STO → > L1 > ENTER**
               STAT > EDIT > 1:Edit

To obtain individual probabilities for a particular n, p, and x, continue with:

Choose: **2nd > DISTR > 0:binompdf(**
Enter: **n, p, x)**

To obtain cumulative probabilities for $x = 0$ to $x = n$ for a particular n and p, continue with:

Choose: **2nd > DISTR > A:binomcdf(**
Enter: **n, p)*** (see previous for scrolling through probabilities)

*To obtain individual cumulative probabilities for a particular n, p, and x, repeat the preceding commands but replace the enter with:
Enter: **n, p, x)**

APPLIED EXAMPLE 5.10

LIVING WITH THE LAW

WHAT IS AN AFFIRMATIVE ACTION PROGRAM (AAP)?

As a condition of doing business with the federal government, federal contractors meeting certain contract and employee population levels agree to prepare, in accordance with federal regulations at 41 CFR 60-1, 60-2, etc., an Affirmative Action Program (AAP). A contractor's AAP is a combination of numerical reports, commitments of action and description of policies. A quick overview of an AAP based on the federal regulations (41 CFR 60-2.10), is as follows:

AAPs must be developed for
*   Minorities and women (41 CFR 60-1 and 60-2)
*   Special disabled veterans, Vietnam era veterans, and other covered veterans (41 CFR 60-250)
*   Individuals with disabilities (41 CFR 60-741)

Source: http://eeosource.peopleclick.com/aap/

The AAP regulations do not endorse the use of a specific test for determining whether the percentage of minorities or women is less than would be reasonably expected. However, several tests are commonly used. One of the tests is called the *exact binomial test* as defined here.

EXACT BINOMIAL TEST

The variables used are:
  T = The total number of employees in the job group
  M = The number of females or minorities in the job group
  A = The availability percentage of females or minorities for the job group
  This test involves the calculation of a probability, denoted as P, and the comparison of that probability to 0.05. If P is less than or equal to 0.05, the percentage of minorities or women is considered to be "less than would be reasonably expected." The formula for calculating P is as follows:

1. Calculate probability, Q, the cumulative binomial probability for the binomial probability distribution with $n = \text{T}$, $x = \text{M}$, and $p = \text{A}/100$.
2. If Q is less than or equal to 0.5, then $P = 2Q$; otherwise, $P = Q$.

For example, if T = 50 employees and M = 2 females, A = 6% female availability.
  Using a computer, the value Q is found: Q = 0.41625. Since Q is less than 0.5, P = 2Q = 0.8325. P, 0.8325, is greater than 0.05, so the percentage of women is found to be "**not** less than would be reasonably expected."

Mean and Standard Deviation of the Binomial Distribution

The mean and standard deviation of a theoretical binomial probability distribution can be found by using these two formulas:

Mean of Binomial Distribution

$$\mu = np \tag{5.7}$$

and

Standard Deviation of Binomial Distribution

$$\sigma = \sqrt{npq} \tag{5.8}$$

The formula for the mean, μ, seems appropriate: the number of trials multiplied by the probability of "success." [Recall that the mean number of correct answers on the binomial quiz (Answer 5, p. 246) was expected to be $\frac{1}{3}$ of 4, $4(\frac{1}{3})$, or np.] The formula for the standard deviation, σ, is not as easily understood. Thus, at this point it is appropriate to look at an example that demonstrates that formulas (5.7) and (5.8) yield the same results as formulas (5.1), (5.3a), and (5.4).

  In Example 5.5 (pp. 236–238), x is the number of heads in three coin tosses, $n = 3$, and $p = \frac{1}{2} = 0.5$. Using formula (5.7), we find the mean of x to be

$$\mu = np = (3)(0.5) = 1.5$$

Using formula (5.8), we find the standard deviation of x to be

$$\sigma = \sqrt{npq} = \sqrt{(3)(0.5)(0.5)} = \sqrt{0.75} = 0.866 = 0.87$$

Now look back at the solution for Example 5.5 (p. 237). Note that the results are the same, regardless of which formula you use. However, formulas (5.7) and (5.8) are much easier to use when x is a binomial random variable.

EXAMPLE 5.11

CALCULATING THE MEAN AND STANDARD DEVIATION OF A BINOMIAL DISTRIBUTION

Find the mean and standard deviation of the binomial distribution when $n = 20$ and $p = \frac{1}{5}$ (or 0.2, in decimal form). Recall that the "binomial distribution where $n = 20$ and $p = 0.2$" has the probability function

$$P(x) = \binom{20}{x}(0.2)^x(0.8)^{20-x} \quad \text{for} \quad x = 0, 1, 2, \ldots, 20$$

and a corresponding distribution with 21 x values and 21 probabilities, as shown in the distribution chart, Table 5.9, and on the histogram in Figure 5.5.

TABLE 5.9
Binomial Distribution: $n = 20$, $p = 0.2$

x	$P(x)$
0	0.012
1	0.058
2	0.137
3	0.205
4	0.218
5	0.175
6	0.109
7	0.055
8	0.022
9	0.007
10	0.002
11	0+
12	0+
13	0+
.	.
:	:
20	0+

FIGURE 5.5
Histogram of Binomial Distribution $B(20, 0.2)$

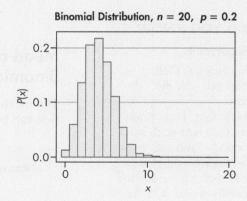

Let's find the mean and the standard deviation of this distribution of x using formulas (5.7) and (5.8):

$$\mu = np = (20)(0.2) = \mathbf{4.0}$$
$$\sigma = \sqrt{npq} = \sqrt{(20)(0.2)(0.8)} = \sqrt{3.2} = \mathbf{1.79}$$

FIGURE 5.6
Histogram of Binomial Distribution $B(20, 0.2)$

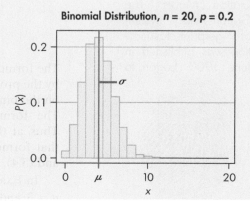

Figure 5.6 shows the mean, $\mu = 4$ (shown by the location of the vertical blue line along the x-axis), relative to the variable x. This 4.0 is the mean value expected for x, the number of successes in each random sample of size 20 drawn from a population with $p = 0.2$. Figure 5.6 also shows the size of the standard deviation, $\sigma = 1.79$ (as shown by the length of the horizontal red line segment). It is the expected standard deviation for the values of the random variable x that occur in samples of size 20 drawn from this same population.

SECTION 5.3 EXERCISES

5.39 Consider the four-question multiple-choice quiz presented at the beginning of this section (pp. 244–246).

a. Explain why the four questions represent four independent trials.

b. Explain why the number 4 is multiplied into the $P(x = 1)$.

c. In Answer 5 on page 246, where did $\frac{1}{3}$ and 4 come from? Why multiply them to find an expected average?

5.40 Identify the properties that make flipping a coin 50 times and keeping track of heads a binomial experiment.

5.41 State a very practical reason why the defective item in an industrial situation might be defined to be the "success" in a binomial experiment.

5.42 What does it mean for the trials to be independent in a binomial experiment?

5.43 Evaluate each of the following.

a. 4!   b. 7!   c. 0!   d. $\dfrac{6!}{2!}$

e. $\dfrac{5!}{2!3!}$ f. $\dfrac{6!}{4!(6-4)!}$ g. $(0.3)^4$ h. $\binom{7}{3}$

i. $\binom{5}{2}$ j. $\binom{3}{0}$ k. $\binom{4}{1}(0.2)^1(0.8)^3$

l. $\binom{5}{0}(0.3)^0(0.7)^5$

5.44 Show that each of the following is true for any values of n and k. Use two specific sets of values for n and k to show that each is true.

a. $\binom{n}{0} = 1$ and $\binom{n}{n} = 1$

b. $\binom{n}{1} = n$ and $\binom{n}{n-1} = n$ c. $\binom{n}{k} = \binom{n}{n-k}$

5.45 A carton containing 100 T-shirts is inspected. Each T-shirt is rated "first quality" or "irregular." After all 100 T-shirts have been inspected, the number of irregulars is reported as a random variable. Explain why x is a binomial random variable.

5.46 A die is rolled 20 times, and the number of "fives" that occur is reported as being the random variable. Explain why x is a binomial random variable.

5.47 Four cards are selected, one at a time, from a standard deck of 52 playing cards. Let x represent the number of aces drawn in the set of four cards.

a. If this experiment is completed without replacement, explain why x is not a binomial random variable.

b. If this experiment is completed with replacement, explain why x is a binomial random variable.

5.48 The employees at a General Motors assembly plant are polled as they leave work. Each is asked, "What brand of automobile are you riding home in?" The random variable to be reported is the number of each brand mentioned. Is x a binomial random variable? Justify your answer.

5.49 Consider a binomial experiment made up of three trials with outcomes of success, S, and failure, F, where $P(S) = p$ and $P(F) = q$.

a. Complete the accompanying tree diagram. Label all branches completely.

b. In column (b) of the tree diagram, express the probability of each outcome represented by the branches as a product of powers of p and q.

(continue on page 256)

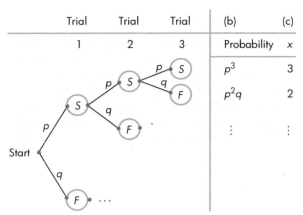

	Trial 1	Trial 2	Trial 3	(b) Probability	(c) x

c. Let x be the random variable, the number of successes observed. In column (c), identify the value of x for each branch of the tree diagram.

d. Notice that all the products in column (b) are made up of three factors and that the value of the random variable is the same as the exponent for the number p. Write the equation for the binomial probability function for this situation.

5.50 Draw a tree diagram picturing a binomial experiment of four trials.

5.51 Use the probability function for three coin tosses as demonstrated on page 239 and verify the probabilities for $x = 0, 2$, and 3.

5.52 a. Calculate $P(4)$ and $P(5)$ for Example 5.8 on page 249.

  b. Verify that the six probabilities $P(0)$, $P(1)$, $P(2), \ldots, P(5)$ form a probability distribution.

5.53 Skillbuilder Applet Exercise demonstrates calculating a binomial probability along with a visual interpretation. Suppose that you buy 20 plants from a nursery and the nursery claims that 95% of its plants survive when planted. Inputting $n = 20$ and $p = 0.95$, compute the following:

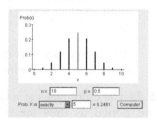

a. The probability that all 20 will survive

b. The probability that at most 16 survive

c. The probability that at least 18 survive

5.54 Skillbuilder Applet Exercise demonstrates calculating a binomial probability along with a visual interpretation. Suppose that you are in a class of 30 students and it is assumed that approximately

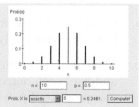

11% of the population is left-handed. Inputting $n = 30$ and $p = 0.11$, compute the following:

a. The probability that exactly five students are left-handed

b. The probability that at most four students are left-handed

c. The probability that at least six students are left-handed

5.55 If x is a binomial random variable, calculate the probability of x for each case.

a. $n = 4, x = 1, p = 0.3$ b. $n = 3, x = 2, p = 0.8$

c. $n = 2, x = 0, p = \dfrac{1}{4}$ d. $n = 5, x = 2, p = \dfrac{1}{3}$

e. $n = 4, x = 2, p = 0.5$ f. $n = 3, x = 3, p = \dfrac{1}{6}$

5.56 If x is a binomial random variable, use Table 2 in Appendix B to determine the probability of x for each of the following:

a. $n = 10, x = 8, p = 0.3$ b. $n = 8, x = 7, p = 0.95$

c. $n = 15, x = 3, p = 0.05$ d. $n = 12, x = 12, p = 0.99$

e. $n = 9, x = 0, p = 0.5$ f. $n = 6, x = 1, p = 0.01$

g. Explain the meaning of the symbol 0+ that appears in Table 2.

5.57 Test the following function to determine whether or not it is a binomial probability function. List the distribution of probabilities and sketch a histogram.

$$T(x) = \binom{5}{x}\left(\frac{1}{2}\right)^{x}\left(\frac{1}{2}\right)^{5-x} \quad \text{for} \quad x = 0, 1, 2, 3, 4, 5$$

5.58 Let x be a random variable with the following probability distribution:

x	0	1	2	3
$P(x)$	0.4	0.3	0.2	0.1

Does x have a binomial distribution? Justify your answer.

5.59 According to a December 2008 *Self* magazine online poll, 66% responded "Yes" to "Do you want to relive your college days?" What is the probability that exactly half of the next 10 randomly selected poll participants also respond "Yes" to this question?

5.60 According to a National Safety Council report, up to 78% of automobile collisions are a result of distractions such as text messaging, phoning a pal, or fumbling with the stereo. Consider a randomly selected group of 18 reported collisions.

Source: *Self* magazine, December 2008, "Cruise Control"

a. What is the probability that all of the collisions will be due to the distractions mentioned?

b. What is the probability that 15 of the collisions will be due to the distractions mentioned?

5.61 According to the article "Season's Cleaning," the U.S. Department of Energy reports that 25% of people with two-car garages don't have room to park any cars inside.

Source: January 1, 2009, Rochester D&C

Assuming this to be true, what is the probability of the following?

a. Exactly 3 two-car-garage households of a random sample of 5 two-car-garage households do not have room to park any cars inside.

b. Exactly 7 two-car-garage households of a random sample of 15 two-car-garage households do not have room to park any cars inside.

c. Exactly 20 two-car-garage households of a random sample of 30 two-car-garage households do not have room to park any cars inside.

5.62 Can playing video games as a child and teenager lead to a gambling or substance addiction? According to the April 11, 2009, *USA Today* article "Kids show addiction symptoms," research published in the Journal *Psychological Science* found that 8.5% of video-game-playing children and teens displayed behavioral signs that may indicate addiction. Suppose a randomly selected group of 30 video-gaming eighth-grade students is selected.

a. What is the probability that exactly 2 will display addiction symptoms?

b. If the study also indicated that 12% of video-gaming boys display addiction symptoms, what is the probability that exactly 2 out of the 17 boys in the group will display addiction symptoms?

c. If the study also indicated that 3% of video-gaming girls display addiction symptoms, what is the probability that exactly 2 out of the 13 girls in the group will display addiction symptoms?

5.63 Of the parts produced by a particular machine, 0.5% are defective. If a random sample of 10 parts produced by this machine contains 2 or more defective parts, the machine is shut down for repairs. Find the probability that the machine will be shut down for repairs based on this sampling plan.

5.64 As a quality-control inspector of toy trucks, you have observed that 3% of the time, the wooden wheels are bored off-center. If six wooden wheels are used on each toy truck, what is the probability that a randomly selected toy truck has no off-center wheels?

5.65 The survival rate during a risky operation for patients with no other hope of survival is 80%. What is the probability that exactly four of the next five patients survive this operation?

5.66 Of all the trees planted by a landscaping firm, 90% survive. What is the probability that 8 or more of the 10 trees they just planted will survive? (Find the answer by using a table.)

5.67 In the biathlon event of the Olympic Games, a participant skis cross-country and on four intermittent occasions stops at a rifle range and shoots a set of five shots. If the center of the target is hit, no penalty points are assessed. If a particular man has a history of hitting the center of the target with 90% of his shots, what is the probability of the following?

a. He will hit the center of the target with all five of his next set of five shots.

b. He will hit the center of the target with at least four of his next set of five shots. (Assume independence.)

5.68 The May 26, 2009, *USA Today* Snapshot "Overcoming identity theft" reported the results from a poll of identity-theft victims. According to the source, Affinion Security Center, 20% of the victims stated that it took "one week to one month" to recover from identity theft. A group of 14 identity-theft victims are randomly selected in your hometown.

a. What is the probability that none of them was able to recover from the theft in one week to one month?

b. What is the probability that exactly 3 were able to recover from the theft in one week to one month?

c. What is the probability that at least 5 were able to recover from the theft in one week to one month?

d. What is the probability that no more than 4 were able to recover from the theft in one week to one month?

5.69 A January 2005 survey of bikers, commissioned by the Progressive Group of Insurance Companies, showed that 40% of bikers have body art, such as tattoos and piercings. A group of 10 bikers are in the process of buying motorcycle insurance.

Source: http://www.syracuse.com/

a. What is the probability that none of the 10 has any body art?

b. What is the probability that exactly 3 have some body art?

c. What is the probability that at least 4 have some body art?

(continue on page 258)

d.   What is the probability that no more than 2 have some body art?

5.70 Consider the manager of Steve's Food Market as illustrated in Example 5.9. What would be the manager's "risk" if he bought "better" eggs, say with $P(\text{bad}) = 0.01$ using the "more than one" guarantee?

5.71 If boys and girls are equally likely to be born, what is the probability that in a randomly selected family of six children, there will be at least one boy? (Find the answer using a formula.)

5.72 One-fourth of a certain breed of rabbits are born with long hair. What is the probability that in a litter of six rabbits, exactly three will have long hair? (Find the answer by using a formula.)

5.73 Find the mean and standard deviation for the binomial random variable x with $n = 30$ and $p = 0.6$, using formulas (5.7) and (5.8).

5.74 Consider the binomial distribution where $n = 11$ and $p = 0.05$.

a.   Find the mean and standard deviation using formulas (5.7) and (5.8).

b.   Using Table 2 in Appendix B, list the probability distribution and draw a histogram.

c.   Locate μ and σ on the histogram.

5.75 Consider the binomial distribution where $n = 11$ and $p = 0.05$ (see Exercise 5.74).

a.   Use the distribution [Exercise 5.74(b) or Table 2] and find the mean and standard deviation using formulas (5.1), (5.3a), and (5.4).

b.   Compare the results of part a with the answers found in Exercise 5.74(a).

5.76 Given the binomial probability function

$$P(x) = \binom{5}{x} \cdot \left(\tfrac{1}{2}\right)^x \cdot \left(\tfrac{1}{2}\right)^{5-x} \quad \text{for} \quad x = 0, 1, 2, 3, 4, 5$$

a.   Calculate the mean and standard deviation of the random variable by using formulas (5.1), (5.3a), and (5.4).

b.   Calculate the mean and standard deviation using formulas (5.7) and (5.8).

c.   Compare the results of parts a and b.

5.77 Find the mean and standard deviation of x for each of the following binomial random variables:

a.   The number of tails seen in 50 tosses of a quarter

b.   The number of left-handed students in a classroom of 40 students (Assume that 11% of the population is left-handed.)

c.   The number of cars found to have unsafe tires among the 400 cars stopped at a roadblock for inspection (Assume that 6% of all cars have one or more unsafe tires.)

d.   The number of melon seeds that germinate when a package of 50 seeds is planted (The package states that the probability of germination is 0.88.)

5.78 Find the mean and standard deviation for each of the following binomial random variables in parts a–c:

a.   The number of sixes seen in 50 rolls of a die

b.   The number of defective televisions in a shipment of 125 (The manufacturer claimed that 98% of the sets were operative.)

c.   The number of operative televisions in a shipment of 125 (The manufacturer claimed that 98% of the sets were operative.)

d.   How are parts b and c related? Explain.

5.79 According to United Mileage Plus Visa (November 22, 2004), 41% of passengers say they "put on the earphones" to avoid being bothered by their seatmates during flights. To show how important, or not important, the earphones are to people, consider the variable x to be the number of people in a sample of 12 who say they "put on the earphones" to avoid their seatmates. Assume the 41% is true for the whole population of airline travelers and that a random sample is selected.

a.   Is x a binomial random variable? Justify your answer.

b.   Find the probability that $x = 4$ or 5.

c.   Find the mean and standard deviation of x.

d.   Draw a histogram of the distribution of x: label it completely, highlight the area representing $x = 4$ and $x = 5$, draw a vertical line at the value of the mean, and mark the location of x that is 1 standard deviation larger than the mean.

5.80 According to the *USA Today* Snapshot "Knowing drug addicts," 45% of Americans know somebody who became addicted to a drug other than alcohol. Assuming this to be true, what is the probability of the following?

a.   Exactly 3 people of a random sample of 5 know someone who became addicted. Calculate the value.

b.   Exactly 7 people of a random sample of 15 know someone who became addicted. Estimate using Table 2 in Appendix B.

c.   At least 7 people of a random sample of 15 know someone who became addicted. Estimate using Table 2.

d.   No more than 7 people of a random sample of 15 know someone who became addicted. Estimate using Table 2.

5.81 a. Use a calculator or computer to find the probability that $x = 3$ in a binomial experiment where $n = 12$ and $p = 0.30$: $P(x = 3|B(12, 0.30))$. (See Note about this notation on p. 251.)

b. Use Table 5.8 to verify the answer in part a.

5.82 If the binomial $(q + p)$ is squared, the result is $(q + p)^2 = q^2 + 2qp + p^2$. For the binomial experiment with $n = 2$, the probability of no successes in two trials is q^2 (the first term in the expansion), the probability of one success in two trials is $2qp$ (the second term in the expansion), and the probability of two successes in two trials is p^2 (the third term in the expansion). Find $(q + p)^3$ and compare its terms to the binomial probabilities for $n = 3$ trials.

5.83 Use a computer to find the probabilities for all possible x values for a binomial experiment where $n = 30$ and $p = 0.35$.

MINITAB

Choose: **Calc > Make Patterned Data > Simple Set of Numbers**

Enter: Store patterned data in: **C1**
          From first value: **0**
          To last value: **30**
          In steps of: **1 > OK**

Continue with the binomial probability MINITAB commands on page 251, using $n = 30$, $p = 0.35$, and C2 for optional storage.

Excel

Enter: **0,1,2, . . . , 30** into column A

Continue with the binomial probability Excel commands on page 251–252, using $n = 30$ and $p = 0.35$.

TI-83/84 Plus

Use the binomial probability TI-83 commands on page 252, using $n = 30$ and $p = 0.35$.

5.84. Use a computer to find the cumulative probabilities for all possible x values for a binomial experiment where $n = 45$ and $p = 0.125$.

a. Explain why there are so many 1.000s listed.

b. Explain what is represented by each number listed.

MINITAB

Choose: **Calc > Make Patterned Data > Simple Set of Numbers . . .**

Enter: Store patterned data in: **C1**
          From first value: **0**
          To last value: **45**
          In steps of: **1 > OK**

Continue with the **cumulative** binomial probability MINITAB commands on page 251, using $n = 45$, $p = 0.125$, and C2 as optional storage.

Excel

Enter: **0,1,2, . . . , 45** into column A

Continue with the **cumulative** binomial probability Excel commands on pages 251–252, using $n = 45$ and $p = 0.125$.

TI-83/84 Plus

Use the **cumulative** binomial probability TI-83 commands on page 252, using $n = 45$ and $p = 0.125$.

5.85 Where does all that Halloween candy go? The October 2004 issue of *Readers' Digest* quoted that "90% of parents admit taking Halloween candy from their children's trick-or-treat bags." The source of information was the National Confectioners Association. Suppose that 25 parents are interviewed. What is the probability that 20 or more took Halloween candy from their children's trick-or-treat bags?

5.86 Harris Interactive conducted a survey for Tylenol PM asking U.S. drivers what they do if they are driving while drowsy. The results were reported in a *USA Today* Snapshot on January 18, 2005, with 40% of the respondents saying they "open the windows" to fight off sleep. Suppose that 35 U.S. drivers are interviewed. What is the probability that between 10 and 20 of the drivers will say they "open the windows" to fight off sleep?

5.87 Of all mortgage foreclosures in the United States, 48% are caused by disability. People who are injured or ill cannot work—they then lose their jobs and thus their incomes. With no income, they cannot make their mortgage payments and the bank forecloses.

Source: http://www.ricedelman.com

Given that 20 mortgage foreclosures are audited by a large lending institution, find the probability of the following:

a. Five or fewer of the foreclosures are due to a disability.

b. At least three foreclosures are due to a disability.

5.88 The increase in Internet usage over the past few years has been phenomenal, as demonstrated by the February 2004 report from the Pew Internet & American Life Project. The survey of Americans 65 or older (about 8 million adults) reported that 22% have access to the Internet. By contrast, 58% of 50 to 64-year-olds, 75% of 30- to 49-year-olds, and 77% of 18- to 29-year-olds currently go online.

Source: http://www.suddenlysenior.com/

Suppose that 50 adults in each age group are to be interviewed.

a.   What is the probability that "have Internet access" is the response of 10 to 20 adults in the 65 or older group?

b.   What is the probability that "have Internet access" is the response of 30 to 40 adults in the 50- to 64-year-old group?

c.   What is the probability that "have Internet access" is the response of 30 to 40 adults in the 30- to 49-year-old group?

d.   What is the probability that "have Internet access" is the response of 30 to 40 adults in the 18- to 29-year-old group?

e.   Why are the answers for parts a and d nearly the same? Explain.

f.   What effect did the various values of p have on the probabilities? Explain.

5.89 A binomial random variable has a mean equal to 200 and a standard deviation of 10. Find the values of n and p.

5.90 The probability of success on a single trial of a binomial experiment is known to be $\frac{1}{4}$. The random variable x, number of successes, has a mean value of 80. Find the number of trials involved in this experiment and the standard deviation of x.

5.91 A binomial random variable x is based on 15 trials with the probability of success equal to 0.4. Find the probability that this variable will take on a value more than 2 standard deviations above the mean.

5.92 A binomial random variable x is based on 15 trials with the probability of success equal to 0.2. Find the probability that this variable will take on a value more than 2 standard deviations from the mean.

5.93 a. When using the exact binomial test (Applied Example 5.10, pp. 252–253), what is the interpretation of the situation when the calculated value of P is less than or equal to 0.05?

b.   When using the exact binomial test, what is the interpretation of the situation when the calculated value of P is larger than 0.05?

c.   An employer has 15 employees in a very specialized job group, of which 2 are minorities. Based on 2000 census information, the proportion of minorities available for this type of work is 5%. Using the binomial test, is the percentage of minorities what would be reasonably expected?

d.   For this same employer and the same job group, there are three female employees. The percentage of female availability for this position is 50%. Does it appear that the percentage of females is what would be reasonably expected?

5.94 Extended to overtime in game 7 on the road in the 2002 NBA play-offs, the two-time defending champion Los Angeles Lakers did what they do best—thrived when the pressure was at its highest. Both of the Lakers' star players had their chance at the foul line late in overtime.

a.   With 1:27 minutes left in overtime and the game tied at 106–106, Shaquille (Shaq) O'Neal was at the line for two free-throw attempts. He has a history of making 0.555 of his free-throw attempts, and during this game, prior to these two shots, he had made 9 of his 13 attempts. Justify the statement "The law of averages was working against him."

b.   With 0:06 seconds left in overtime and the game score standing at 110–106, Kobe Bryant was at the line for two free-throw shots. He has a history of making 0.829 of his free throws, and during this game, prior to these two shots, he had made 6 of his 8 attempts. Justify the statement "The law of averages was working for him."

Both players made both shots, and the series with the Sacramento Kings was over.

5.95 Imprints Galore buys T-shirts (to be imprinted with an item of the customer's choice) from a manufacturer who guarantees that the shirts have been inspected and that no more than 1% are imperfect in any way. The shirts arrive in boxes of 12. Let x be the number of imperfect shirts found in any one box.

a.   List the probability distribution and draw the histogram of x.

b.   What is the probability that any one box has no imperfect shirts?

c.   What is the probability that any one box has no more than one imperfect shirt?

d.   Find the mean and standard deviation of x.

e.   What proportion of the distribution is between $\mu - \sigma$ and $\mu + \sigma$?

f.   What proportion of the distribution is between $\mu - 2\sigma$ and $\mu + 2\sigma$?

g.   How does this information relate to the empirical rule and Chebyshev's theorem? Explain.

h. Use a computer to simulate Imprints Galore's buying 200 boxes of shirts and observing x, the number of imperfect shirts per box of 12. Describe how the information from the simulation compares to what was expected (answers to parts a–g describe the expected results).

i. Repeat part h several times. Describe how these results compare with those of parts a–g and with part h.

MINITAB

a.

Choose:	**Calc > Make Patterned Data > Simple Set of Numbers . . .**
Enter:	Store patterned data in: **C1**
	From first value: **−1** (see note)
	To last value: **12**
	In steps of: **1 > OK**

c. Continue with the binomial probability MINITAB commands on page 251, using $n = 12$, $p = 0.01$, and C2 for optional storage.

Choose:	**Graph > Scatterplot > Simple > OK**
Enter:	Y variables: **C2** X variables: **C1**
Select:	Data view: Data Display: **Area > OK**

The graph is not a histogram, but can be converted to a histogram by double clicking on "area" of graph.

Select:	**Options** Select: **Step > OK > OK**

h. Continue with the **cumulative** binomial probability MINITAB commands on page 251, using $n = 12$, $p = 0.01$, and C3 for optional storage.

Choose:	**Calc > Random Data > Binomial**
Enter:	Number of rows of data to generate:
	200 rows of data
	Store in column **C4**
	Number of trials: **12**
	Probability: **.01 > OK**
Choose:	**Stat > Tables > Cross Tabulation**
Enter:	Categorical variables: For rows: **C4**
Select:	Display: **Total percents > OK**
Choose:	**Calc > Column Statistics**
Select:	Statistic: **Mean**
Enter:	Input variable: **C4 > OK**
Choose:	**Calc > Column Statistics**
Select:	Statistic: **Standard deviation**
Enter:	Input variable: **C4 > OK**

Continue with the histogram MINITAB commands on page 53, using the data in C4 and selecting the options: percent and midpoint with intervals 0:12/1.

Note: The binomial variable x cannot take on the value −1. The use of −1 (the next would-be class midpoint to left of 0) allows MINITAB to draw the histogram of a probability distribution. Without −1, PLOT will draw only half of the bar representing $x = 0$.

Excel

a.

Enter:	0, 1, 2, . . . , 12 into column A

Continue with the binomial probability Excel commands on pages 251–252, using $n = 12$ and $p = 0.01$. Activate columns A and B; then continue with:

Choose:	**Insert > Column > 1st picture**(usually)
Choose:	**Select Data > Series 1 > Remove > OK**
If necessary:	
Click on:	**Anywhere clear on the chart**
	—use handles to size so x values fall under corresponding bars

Continue with the **cumulative** binomial probability Excel commands on pages 251–252, using $n = 12$, $p = 0.01$, and column C for the activated cell.

h.

Choose:	**Data > Data analysis > Random Number Generation > OK**
Enter:	Number of Variables: **1**
	Number of Random Numbers: **200**
	Distribution: **Binomial**
	p Value = **0.01**
	Number of Trials = **12**
Select:	Output Options: **Output Range**
Enter	(D1 or select cell) **> OK**
	Activate the E1 cell, then:
Choose:	**Insert function, f_x > Statistical > AVERAGE > OK**
Enter:	Number 1: **D1:D200 > OK**
	Activate the E2 cell, then:
Choose:	**Insert function, f_x > Statistical > STDEV > OK**
Enter:	Number 1: **D1:D200 > OK**

Continue with the histogram Excel commands on pages 53–54, using the data in column D and the bin range in column A.

TI-83/84 Plus

a.

Choose:	**STAT > EDIT > 1:Edit**
Enter:	L1: 0,1,2,3,4,5,6,7,8,9,10,11,12
Choose:	**2nd QUIT > 2nd DISTR > 0:binompdf(**
Enter:	12, 0.01) **> ENTER**
Choose:	**STO→ > L2 > ENTER**
Choose:	**2nd > STAT PLOT > 1:Plot1**
Choose:	**WINDOW**
Enter:	0, 13, 1, −.1, .9, .1, 1
Choose:	**TRACE > > >**

c.

Choose:	**2nd > DISTR > A:binomcdf(**
Enter:	12, 0.01)
Choose:	**STO→ > L3 > ENTER**
	STAT > EDIT > 1:Edit

h.

Choose:	**MATH > PRB > 7:randBin(**
Enter:	12, .01, 200) (takes a while to process)
Choose:	**STO→ > L4 > ENTER**

(continue on page 262)

Choose: 2nd LIST > Math > 3:mean(
Enter: L4
Choose: 2nd LIST > Math > 7:StdDev(
Enter: L4

Continue with the histogram TI-83/84 commands on page 54, using the data in column L4 and adjusting the window after the initial look using ZoomStat.

5.96 Did you ever buy an incandescent light bulb that failed (either burned out or did not work) the first time you turned the light switch on? When you put a new bulb into a light fixture, you expect it to light, and most of the time it does. Consider 8-packs of 60-watt bulbs and let x be the number of bulbs in a pack that "fail" the first time they are used. If 0.02 of all bulbs of this type fail on their first use and each 8-pack is considered a random sample,

a. List the probability distribution and draw the histogram of x.

b. What is the probability that any one 8-pack has no bulbs that fail on first use?

c. What is the probability that any one 8-pack has no more than one bulb that fails on first use?

d. Find the mean and standard deviation of x.

e. What proportion of the distribution is between $\mu - \sigma$ and $\mu + \sigma$?

f. What proportion of the distribution is between $\mu - 2\sigma$ and $\mu + 2\sigma$?

g. How does this information relate to the empirical rule and Chebyshev's theorem? Explain.

h. Use a computer to simulate testing 100 8-packs of bulbs and observing x, the number of failures per 8-pack. Describe how the information from the simulation compares with what was expected (answers to parts a–g describe the expected results).

i. Repeat part h several times. Describe how these results compare with those of parts a–g and with part h.

Chapter Review

In Retrospect

In this chapter we combined concepts of probability with some of the ideas presented in Chapter 2. We now are able to deal with distributions of probability values and find means, standard deviations, and other statistics.

In Chapter 4 we explored the concepts of mutually exclusive events and independent events. We used the addition and multiplication rules on several occasions in this chapter, but very little was said about mutual exclusiveness or independence. Recall that every time we add probabilities, as we did in each of the probability distributions, we need to know that the associated events are mutually exclusive. If you look back over the chapter, you will notice that the random variable actually requires events to be mutually exclusive; therefore, no real emphasis was placed on this concept. The same basic comment can be made in reference to the multiplication of probabilities and the concept of independent events. Throughout this chapter, probabilities were multiplied, and occasion-

ally independence was mentioned. Independence, of course, is necessary to be able to multiply probabilities.

Now, after completing Chapter 5, if we were to take a close look at some of the sets of data in Chapter 2, we would see that several problems could be reorganized to form probability distributions. Here are some examples: (1) Let x be the number of credit hours for which a student is registered this semester, paired with the percentage of the entire student body reported for each value of x. (2) Let x be the number of correct passageways through which an experimental laboratory animal passes before going down a wrong one, paired with the probability of each x value. (3) Let x be the number of college applications made other than to the college where you enrolled (Applied Example 5.3), paired with the probability of each x value. The list of examples is endless.

We are now ready to extend these concepts to continuous random variables in Chapter 6.

CourseMate The **Statistics CourseMate** site for this text brings chapter topics to life with interactive learning, study, and exam preparation tools, including quizzes and flashcards for the Vocabulary and Key Concepts that follow. The site also provides an **eBook** version of the text with highlighting and note taking capabilities. Throughout chapters, the CourseMate icon 🖥 flags concepts and examples that have corresponding interactive resources such as **video** and **animated tutorials** that demonstrate, step by step, how to solve problems; **datasets** for exercises and examples; **Skillbuilder Applets** to help you better understand concepts; **technology manuals**; and software to download including **Data Analysis Plus** (a suite of statistical macros for Excel) and **TI-83/84 Plus** programs—logon at **www.cengagebrain.com**.

Vocabulary List and Key Concepts

all-inclusive events (p. 230)
binomial coefficient (p. 248)
binomial probability experiment (p. 246)
binomial probability function (p. 247)
binomial random variable (p. 246)
constant function (p. 234)
continuous random variable (p. 231)

discrete random variable (p. 231)
experiment (p. 230)
factorial notation (p. 248)
failure (p. 246)
independent trials (p. 246)
mean of discrete random variable (p. 237)
mutually exclusive events (p. 230)
population parameter (p. 236)
probability distribution (p. 233)

probability function (p. 233)
probability histogram (p. 235)
random variable (p. 230)
sample statistic (p. 236)
standard deviation of discrete random variable (p. 237)
success (p. 246)
trial (p. 246)
variance of discrete random variable (p. 237)

Learning Outcomes

- Understand that a random variable is a numerical quantity whose value depends on the conditions and probabilities associated with an experiment. pp. 230–231, EXP. 5.1

- Understand the difference between a discrete and a continuous random variable. Ex. 5.4, 5.5, 5.9

- Be able to construct a discrete probability distribution based on an experiment or given function. pp. 233–234, Ex. 5.13, 5.19

- Understand the terms *mutually exclusive* and *all-inclusive* as they apply to the variables for probability distributions. p. 231, Ex. 5.15

- Understand the similarities and differences between frequency distributions and probability distributions. p. 231, Ex. 5.98

- Understand and be able to utilize the two main properties of probability distributions to verify compliance. p. 234, EXP. 5.2, Ex. 5.17, 5.97, 5.99

- Understand that a probability distribution is a theoretical probability distribution and that the mean and standard deviation (μ and σ, respectively) are parameters. pp. 236–238, Ex. 5.98

- Compute, describe, and interpret the mean and standard deviation of a probability distribution. EXP. 5.5, Ex. 5.26, 5.27

- Understand the key elements of a binomial experiment and be able to define $x, n, p,$ and q. p. 246, EXP. 5.6, 5.7

- Know and be able to calculate binomial probabilities using the binomial probability function. EXP. 5.8, Ex. 5.55, 5.61

- Understand and be able to use Table 2 in Appendix B, Binomial Probabilities, to determine binomial probabilities. p. 251, Ex. 5.56, 5.111

- Compute, describe, and interpret the mean and standard deviation of a binomial probability distribution. EXP. 5.11, Ex. 5.77, 5.79

Chapter Exercises

5.97 What are the two basic properties of every probability distribution?

5.98 a. Explain the difference and the relationship between a probability distribution and a probability function.

b. Explain the difference and the relationship between a probability distribution and a frequency distribution, and explain how they relate to a population and a sample.

5.99 Verify whether or not each of the following is a probability function. State your conclusion and explain.

a. $f(x) = \dfrac{\frac{3}{4}}{x!(3-x)!}$, for $x = 0, 1, 2, 3$

b. $f(x) = 0.25$, for $x = 9, 10, 11, 12$

c. $f(x) = (3 - x)/2$, for $x = 1, 2, 3, 4$

d. $f(x) = (x^2 + x + 1)/25$, for $x = 0, 1, 2, 3$

5.100 Verify whether or not each of the following is a probability function. State your conclusion and explain.

a. $f(x) = \dfrac{3x}{8x!}$, for $x = 1, 2, 3, 4$

b. $f(x) = 0.125$, for $x = 0, 1, 2, 3$, and $f(x) = 0.25$, for $x = 4, 5$

c. $f(x) = (7 - x)/28$, for $x = 0, 1, 2, 3, 4, 5, 6, 7$

d. $f(x) = (x^2 + 1)/60$, for $x = 0, 1, 2, 3, 4, 5$

5.101 The number of ships to arrive at a harbor on any given day is a random variable represented by x. The probability distribution for x is as follows:

x	10	11	12	13	14
$P(x)$	0.4	0.2	0.2	0.1	0.1

Find the probability of the following for any a given day:

a. Exactly 14 ships arrive.

b. At least 12 ships arrive.

c. At most 11 ships arrive.

5.102 "How many TVs are there in your household?" was one of the questions on a questionnaire sent to 5000 people in Japan. The collected data resulted in the following distribution:

Number of TVs/Household	0	1	2	3	4	5 or more
Percentage	1.9	31.4	23.0	24.4	13.0	6.3

Source: http://www.japan-guide.com/

a. What percentage of the households have at least one television?

b. What percentage of the households have at most three televisions?

c. What percentage of the households have three or more televisions?

d. Is this a binomial probability experiment? Justify your answer.

e. Let x be the number of televisions per household. Is this a probability distribution? Explain.

f. Assign $x = 5$ for "5 or more" and find the mean and standard deviation of x.

5.103 Patients who have hip-replacement surgery experience pain the first day after surgery. Typically, the pain is measured on a subjective scale using values of 1 to 5. Let x represent the random variable, the pain score as determined by a patient. The probability distribution for x is believed to be:

x	1	2	3	4	5
$P(x)$	0.10	0.15	0.25	0.35	0.15

a. Find the mean of x.

b. Find the standard deviation of x.

5.104 The coffee consumption per capita in the United States is approximately 1.9 cups per day for men and 1.4 cups for women. The number of cups consumed per day, x, by women coffee drinkers is expressed as the accompanying distribution.

x	1	2	3	4	5	6	7
$P(x)$	0.20	0.33	0.28	0.10	0.05	0.02	0.02

a. Is this a discrete probability distribution? Explain.

b. Draw a histogram of the distribution.

c. Find the mean and standard deviation of x.

5.105 Imagine that you are in the midst of purchasing a lottery ticket and the person behind the counter prints too many tickets with your numbers. What would you do? The results of an online survey were as follows:

Let them keep the tickets.	30.77%
Trust the person to delete them.	15.38%
Buy the extra ones and hope they win.	30.77%
Other	23.08%

Is this a probability distribution? Explain.

5.106 "Sustainability" is quite the buzzword for environmentalists. When they think about sustainability, the word that typically comes to mind for most Americans is "recycling." A May 2008 Harris Poll of 2602 U.S. adults surveyed online asked the question: "Have you heard the phrase 'environmental sustainability' used?" The percentage of adults who answered "Yes" for each age group was reported as follows:

Age Group	18–31	32–43	44–62	63+
Percent	46%	47%	42%	30%

Is this a probability distribution? Explain.

5.107 A doctor knows from experience that 10% of the patients to whom she gives a certain drug will have undesirable side effects. Find the probabilities that among the 10 patients to whom she gives the drug:

a. At most two will have undesirable side effects.

b. At least two will have undesirable side effects.

5.108 In a recent survey of women, 90% admitted that they had never looked at a copy of *Vogue* magazine. Assuming that this is accurate information, what is the probability that a random sample of three women will show that fewer than two have read the magazine?

5.109 Of those seeking a driver's license, 70% admitted that they would not report someone if he or she copied some answers during the written exam. You have just entered the room and see 10 people waiting to take the written exam. What is the probability that if copying took place, 5 of the 10 would not report what they saw?

5.110 The engines on an airliner operate independently. The probability that an individual engine operates for a given trip is 0.95. A plane will be able to complete a trip successfully if at least one-half of its engines operate for the entire trip. Determine whether a four-engine or a two-engine plane has the higher probability of a successful trip.

5.111 The Pew Internet & American Life Project found that nearly 70% of "wired" senior citizens go online

every day. In a randomly selected group of 15 "wired" senior citizens:

a. What is the probability that more than four will say they go online every day?

b. What is the probability that exactly 10 will say that they go online every day?

c. What is the probability that fewer than 10 will say that they go online every day?

5.112 There are 750 players on the active rosters of the 30 Major League Baseball teams. A random sample of 15 players is to be selected and tested for use of illegal drugs.

a. If 5% of all the players are using illegal drugs at the time of the test, what is the probability that 1 or more players test positive and fail the test?

b. If 10% of all the players are using illegal drugs at the time of the test, what is the probability that 1 or more players test positive and fail the test?

c. If 20% of all the players are using illegal drugs at the time of the test, what is the probability that 1 or more players test positive and fail the test?

5.113 A box contains 10 items, of which 3 are defective and 7 are nondefective. Two items are selected without replacement, and x is the number of defective items in the sample of two. Explain why x is not a binomial random variable.

5.114 A box contains 10 items, of which 3 are defective and 7 are nondefective. Two items are randomly selected, one at a time, with replacement, and x is the number of defectives in the sample of two. Explain why x is a binomial random variable.

5.115 A large shipment of radios is accepted upon delivery if an inspection of 10 randomly selected radios yields no more than 1 defective radio.

a. Find the probability that this shipment is accepted if 5% of the total shipment is defective.

b. Find the probability that this shipment is not accepted if 20% of this shipment is defective.

c. The binomial probability distribution is often used in situations similar to this one, namely, large populations sampled without replacement. Explain why the binomial yields a good estimate.

5.116 The town council has nine members. A proposal to establish a new industry in this town has been tabled, and all proposals must have at least two-thirds of the votes to be accepted. If we know that two members of the town council are opposed and that the others randomly vote "in favor" and "against," what is the probability that the proposal will be accepted?

5.117 The state bridge design engineer has devised a plan to repair North Carolina's 4706 bridges that are currently listed as being in either poor or fair condition. The state has a total of 13,268 bridges. Before the governor will include the cost of this plan in his budget, he has decided to personally visit and inspect five bridges, which are to be randomly selected. What is the probability that in the sample of five bridges, the governor will visit the following?

a. No bridges rated as poor or fair

b. One or two bridges rated as poor or fair

c. Five bridges rated as poor or fair

5.118 A discrete random variable has a standard deviation equal to 10 and a mean equal to 50. Find $\Sigma x^2 P(x)$.

5.119 A binomial random variable is based on $n = 20$ and $p = 0.4$. Find $\Sigma x^2 P(x)$.

5.120 [EX05-120] In a germination trial, 50 seeds were planted in each of 40 rows. The number of seeds germinating in each row was recorded as listed in the following table.

Number Germinated	Number of Rows	Number Germinated	Number of Rows
39	1	45	8
40	2	46	4
41	3	47	3
42	4	48	1
43	6	49	1
44	7		

a. Use the preceding frequency distribution table to determine the observed rate of germination for these seeds.

b. The binomial probability experiment with its corresponding probability distribution can be used with the variable "number of seeds germinating per row" when 50 seeds are planted in every row. Identify the specific binomial function and list its distribution using the germination rate found in part a. Justify your answer.

c. Suppose you are planning to repeat this experiment by planting 40 rows of these seeds, with 50 seeds in each row. Use your probability model from part b to find the frequency distribution for x that you would expect to result from your planned experiment.

d. Compare your answer in part c with the results that were given in the preceding table. Describe any similarities and differences.

5.121 In another germination experiment involving old seed, 50 rows of seeds were planted. The number of seeds germinating in each row were recorded in the following table (each row contained the same number of seeds).

Number Germinating	Number of Rows	Number Germinating	Number of Rows
0	17	3	2
1	20	4	1
2	10	5 or more	0

a. What probability distribution (or function) would be helpful in modeling the variable "number of seeds germinating per row"? Justify your choice.

b. What information is needed in order to apply the probability distribution you chose in part a?

c. Based on the information you do have, what is the highest or lowest rate of germination that you can estimate for these seeds? Explain.

5.122 A business firm is considering two investments. It will choose the one that promises the greater payoff. Which of the investments should it accept? (Let the mean profit measure the payoff.)

Invest in Tool Shop		Invest in Book Store	
Profit	Probability	Profit	Probability
$100,000	0.10	$400,000	0.20
50,000	0.30	90,000	0.10
20,000	0.30	−20,000	0.40
−80,000	0.30	−250,000	0.30
	Total 1.00		Total 1.00

5.123 Bill has completed a 10-question multiple-choice test on which he answered 7 questions correctly. Each question had one correct answer to be chosen from five alternatives. Bill says that he answered the test by randomly guessing the answers without reading the questions or answers.

a. Define the random variable x to be the number of correct answers on this test, and construct the probability distribution if the answers were obtained by random guessing.

b. What is the probability that Bill guessed 7 of the 10 answers correctly?

c. What is the probability that anybody can guess six or more answers correctly?

d. Do you believe that Bill actually randomly guessed as he claims? Explain.

Chapter Practice Test

PART I: Knowing the Definitions

Answer "True" if the statement is always true. If the statement is not always true, replace the words shown in bold with words that make the statement always true.

5.1 The number of hours you waited in line to register this semester is an example of a **discrete** random variable.

5.2 The number of automobile accidents you were involved in as a driver last year is an example of a **discrete** random variable.

5.3 The sum of all the probabilities in any probability distribution is always exactly **two**.

5.4 The various values of a random variable form a list of **mutually exclusive events**.

5.5 A binomial experiment always has **three or more** possible outcomes to each trial.

5.6 The formula $\mu = np$ may be used to compute the mean of a **discrete** population.

5.7 The binomial parameter p is the probability of **one success occurring in n** trials when a binomial experiment is performed.

5.8 A parameter is a statistical measure of some aspect of a **sample**.

5.9 **Sample statistics** are represented by letters from the Greek alphabet.

5.10 The probability of event A or B is equal to the sum of the probability of event A and the probability of event B when A and B are **mutually exclusive events**.

PART II: Applying the Concepts

5.11 a. Show that the following is a probability distribution:

x	1	3	4	5
P(x)	0.2	0.3	0.4	0.1

b. Find $P(x = 1)$.
c. Find $P(x = 2)$.
d. Find $P(x > 2)$.
e. Find the mean of x.
f. Find the standard deviation of x.

5.124 A random variable that can assume any one of the integer values $1, 2, \ldots, n$ with equal probabilities of $\frac{1}{n}$ is said to have a uniform distribution. The probability function is written $P(x) = \frac{1}{n}$, for $x = 1, 2, 3, \cdots, n$. Show that $\mu = \frac{n+1}{2}$.

(*Hint:* $1 + 2 + 3 + \cdots + n = [n(n+1)]/2$.)

5.12 A T-shirt manufacturing company advertises that the probability of an individual T-shirt being irregular is 0.1. A box of 12 such T-shirts is randomly selected and inspected.

a. What is the probability that exactly 2 of these 12 T-shirts are irregular?

b. What is the probability that exactly 9 of these 12 T-shirts are not irregular?

Let x be the number of T-shirts that are irregular in all such boxes of 12 T-shirts.

c. Find the mean of x.
d. Find the standard deviation of x.

PART III: Understanding the Concepts

5.13 What properties must an experiment possess in order for it to be a binomial probability experiment?

5.14 Student A uses a relative frequency distribution for a set of sample data and calculates the mean and standard deviation using formulas from Chapter 5. Student A justifies her choice of formulas by saying that since relative frequencies are empirical probabilities, her sample is represented by a probability distribution and therefore her choice of formulas was correct. Student B argues that since the distribution represents a sample, the mean and standard deviation involved are known as $\bar{x}$ and s and must be calculated using the corresponding frequency distribution and formulas from Chapter 2. Who is correct, A or B? Justify your choice.

5.15 Student A and Student B were discussing one entry in a probability distribution chart:

x	P(x)
−2	0.1

Student B thought this entry was okay because $P(x)$ is a value between 0.0 and 1.0. Student A argued that this entry was impossible for a probability distribution because x is −2 and negatives are not possible. Who is correct, A or B? Justify your choice.

6 Normal Probability Distributions

© 2010/Jupiterimages Corporation

6.1 Normal Probability Distributions
*The domain of **bell-shaped distributions** is the set of all real numbers.*

6.2 The Standard Normal Distribution
*To work with normal distributions, we need the **standard score**.*

6.3 Applications of Normal Distributions
*The normal distribution can help us to determine **probabilities**.*

6.4 Notation
The z notation is critical in the use of normal distributions.

6.5 Normal Approximation of the Binomial
***Binomial probabilities** can be **estimated** by using a normal distribution.*

6.1 Normal Probability Distributions

Intelligence Scores

The normal probability distribution is considered the single most important probability distribution. An unlimited number of **continuous random variables** have either a normal or an approximately **normal distribution**.

We are all familiar with IQ (*intelligence quotient*) scores and/or SAT (*Scholastic Aptitude Test*) scores. IQ scores have a mean of 100 and a standard deviation of 16. SAT scores have a mean of 500 with a standard deviation of 100. But did you know that these continuous random variables also follow a normal distribution?

Figure A, pictures the comparison of several deviation scores and the normal distribution: Standard scores have a mean of zero and a standard deviation of 1.0. Scholastic Aptitude Test scores have a mean of 500 and a standard deviation of 100.

Binet Intelligence Scale scores have a mean of 100 and a standard deviation of 16. In each case there are 34 percent of the scores between the mean and one standard deviation, 14 percent between one and two standard deviations, and 2 percent beyond two standard deviations.

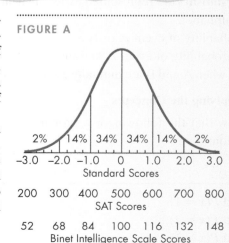

FIGURE A

Source: Beck, *Applying Psychology: Critical and Creative Thinking,* Figure 6.2 "Pictures the Comparison of Several Deviation Scores and the Normal Distribution," © 1992 Prentice-Hall, Inc. Reproduced by permission of Pearson Education, Inc.

FYI The Binet Intelligence Scale. Alfred Binet, who devised the first general aptitude test at the beginning of the 20th century, defined intelligence as *the ability to make adaptations*. The general purpose of the test was to determine which children in Paris could benefit from school. Binet's test, like its subsequent revisions, consists of a series of progressively more difficult tasks that children of different ages can successfully complete. A child who can solve problems typically solved by children at a particular age level is said to have that mental age. For example, if a child can successfully do the same tasks that an average 8-year-old can do, he or she is said to have a mental age of 8. The *intelligence quotient*, or IQ, is defined by the formula:

Intelligence Quotient = 100 ×
(Mental Age/Chronological Age)

There has been a great deal of controversy in recent years over what intelligence tests measure. Many of the test items depend on either language or other specific cultural experiences for correct answers. Nevertheless, such tests can rather effectively predict school success. If school requires language and the tests measure language ability at a particular point of time in a child's life, then the test is a better-than-chance predictor of school performance.

Source: Beck, *Applying Psychology: Critical and Creative Thinking*.

The empirical rule from Chapter 2 of this text (see p. 96) reinforces Figure A in the preceding excerpt and what we already know about a symmetrical shape that is mounded in the middle. The percents within so many deviations were presented and accepted in Chapter 2. But where do they come from?

Recall that in Chapter 5 we learned how to use a probability function to calculate probabilities associated with **discrete random variables**. The normal probability distribution has a **continuous random variable** and uses two functions: one function to determine the ordinates (*y* values) of the graph displaying the distribution, and a second function to determine probabilities.

> **Probability distribution, continuous variable** A formula or a list that provides the probability for a continuous random variable having a value falling within a specified interval. The probability distribution is a theoretical distribution; it is used to represent populations.

FYI Formula (6.1) expresses the ordinate (*y* value) that corresponds to each abscissa (*x* value).
Normal Probability Distribution Function

$$y = f(x) = \frac{e^{-\frac{1}{2}\left(\frac{x-\mu}{\sigma}\right)^2}}{\sigma\sqrt{2\pi}} \text{ for all real } x \qquad (6.1)$$

Note: Each different pair of values for mean (μ) and standard deviation (σ) will result in a different normal probability distribution function.

When a graph of all such points is drawn, the normal (bell-shaped) curve will appear as shown in Figure 6.1.

FIGURE 6.1

The Normal Probability Distribution

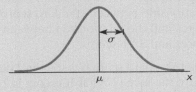

Formula (6.2) yields the probability associated with the interval from $x = a$ to $x = b$. Using calculus to find probability

$$P(a \leq x \leq b) = \int_a^b f(x)\,dx \qquad (6.2)$$

> The probability that x is within the interval from $x = a$ to $x = b$ is shown as the shaded area in Figure 6.2.
>
> **FIGURE 6.2**
>
> Shaded Area: $P(a \leq x \leq b)$
>
>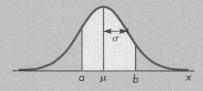

The definite integral of formula (6.2) is a calculus topic and is mathematically beyond what is expected in elementary statistics. Instead of using formulas (6.1) and (6.2), we will use a table to find probabilities for normal distributions. Before we learn to use the table, however, it must be pointed out that the table is expressed in "standardized" form. It is standardized so that this one table can be used to find probabilities for all combinations of mean μ and standard deviation σ values. That is, the normal probability distribution with mean 38 and standard deviation 7 is very similar to the normal probability distribution with mean 123 and standard deviation 32. Recall the empirical rule and the percentage of the distribution that falls within certain intervals of the mean (see p. 96). The same three percentages hold true for all normal distributions.

> **FYI** **Percentage, proportion,** and **probability** are basically the same concepts. Percentage (25%) is usually used when talking about a proportion $\left(\frac{1}{4}\right)$ of a population. Probability is usually used when talking about the chance that the next individual item will possess a certain property. Area is the graphic representation of all three when we draw a picture to illustrate the situation.
>
> The empirical rule is a fairly crude measuring device; with it we are able to find probabilities associated only with whole number multiples of the standard deviation (within 1, 2, or 3 standard deviations of the mean). We will often be interested in the probabilities associated with fractional parts of the standard deviation. For example, we might want to know the probability that x is within 1.37 standard deviations of the mean. Therefore, we must refine the empirical rule so that we can deal with more precise measurements. This refinement is discussed in the next section.

SECTION 6.1 EXERCISES

6.1 a.   Explain why the IQ score is a continuous variable.

b.   What are the mean and the standard deviation for the distribution of: IQ scores? SAT scores? standard scores?

c.   Express, algebraically or as an equation, the relationship between standard scores and IQ scores and the relationship between standard scores and SAT scores.

d.   What standard score is 2 standard deviations above the mean? What IQ score is 2 standard deviations above the mean? What SAT score is 2 standard deviations above the mean?

e.   Compare the information about percentage of distribution, in Figure A on page 268, with the empirical rule studied in Chapter 2. Explain the similarities.

6.2 Examine the intelligence quotient, or IQ, as it is defined by the formula:

Intelligence Quotient $= 100 \times$ (Mental Age/Chronological Age)

Justify why it is reasonable for the mean to be 100.

6.3 Percentage, proportion, or probability—identify which is illustrated by each of the following statements.

a.   One-third of the crowd had a clear view of the event.

b.   Fifteen percent of the voters were polled as they left the voting precinct.

c. The chance of rain during the day tomorrow is 0.2.

6.4 Percentage, proportion, or probability—in your own words, using between 25 and 50 words for each, describe how:

a. percentage is different from the other two.

b. proportion is different from the other two.

c. probability is different from the other two.

d. all three are basically the same thing.

6.2 The Standard Normal Distribution

There are an unlimited number of normal probability distributions, but fortunately they are all related to one distribution, the **standard normal distribution**. The standard normal distribution is the normal distribution of the standard variable z (called "**standard score**" or "**z-score**").

Properties of the Standard Normal Distribution

1. The total area under the **normal curve** is equal to 1.
2. The distribution is mounded and symmetric; it extends indefinitely in both directions, approaching but never touching the horizontal axis.
3. The distribution has a mean of 0 and a standard deviation of 1.
4. The mean divides the area in half, 0.50 on each side.
5. Nearly all the area is between $z = -3.00$ and $z = 3.00$.

Table 3 in Appendix B lists the probabilities associated with the **cumulative area** to the left of a specified value of z. Probabilities associated with other intervals may be found by using the table entries along with the operations of addition and subtraction, in accordance with the preceding properties. Let's look at several examples demonstrating how to use Table 3 to find probabilities of the standard normal score, z.

Recall that we have seen the **standard normal distribution** in earlier chapters, where it appeared as the empirical rule. When using the empirical rule, the values

FYI Ogives are the graphic representation of cumulative relative frequency distributions, as you learned in Chapter 2. Table 3 in Appendix B is a listing of the cumulative standard normal probability distribution. The graph below shows the relationship between the standard normal probability curve (shown in blue) and the cumulative standard normal distribution (shown in red). Even though single vertical scale is used, the units of measure for the two curves are totally different: The vertical scale for the cumulative distribution is probability, while the scale for the normal curve (blue) is probability density.

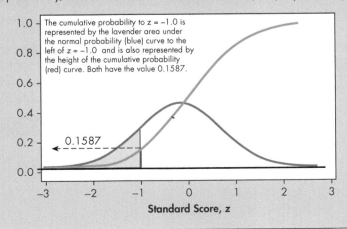

The cumulative probability to $z = -1.0$ is represented by the lavender area under the normal probability (blue) curve to the left of $z = -1.0$ and is also represented by the height of the cumulative probability (red) curve. Both have the value 0.1587.

of z were typically integer values; see Figure 6.3. By using Table 3, the z-score will be measured to the nearest one-hundredth and allow increased accuracy.

FIGURE 6.3
Standard Normal Distribution According to Empirical Rule

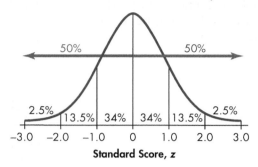

Recall also that one of the basic properties of a probability distribution is that the sum of all probabilities is exactly 1.0. Since the area under the normal curve represents the measure of probability, the total area under the bell-shaped curve is exactly 1. Notice in Figure 6.3 that the distribution is also symmetrical with respect to a vertical line drawn through $z = 0$. That is, the area under the curve to the left of the mean is one-half, 0.5, and the area to the right is also one-half, 0.5. Note $z = 0.00$ on Table 3 in Appendix B. Areas (probabilities, percentages) not given directly by the table can be found with the aid of these properties.

Now let's look at some examples.

EXAMPLE   6 . 1

FINDING AREA TO THE LEFT OF A NEGATIVE z-VALUE

Find the area under the standard normal curve to the left of $z = -1.52$ (see Figure 6.4).

FIGURE 6.4
Area to the Left of $z = -1.52$

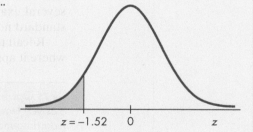

Solution

Table 3 of Appendix B is designed to give the area to the left of -1.52 directly. The z-score is located on the margins, with the units and tenths digits along the left side and the hundredths digit across the top. For $z = -1.52$, locate the row labeled -1.5 and the column labeled 0.02; at their intersection you will find 0.0643, the measure of the cumulative area to the left of $z = -1.52$ (see Table 6.1). Expressed as a probability: $P(z < -1.52) = 0.0643$.

TABLE 6.1 A Portion of Table 3

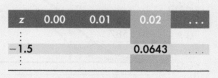

z	0.00	0.01	0.02	...
⋮				
−1.5			0.0643	...
⋮				

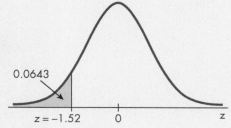

EXAMPLE 6.2

FINDING AREA TO THE LEFT OF A POSITIVE z-VALUE

Find the area under the normal curve to the left of $z = 1.52$: $P(z < 1.52)$.

Solution

Table 3 is designed to also give the area to the left of positive z-values directly. Note the right portion of Table 3, which shows the positive z-values. Likewise, the z-score is located on the margins, with the units and tenths digits along the left side and the hundredths digit across the top. For $z = 1.52$, locate the row labeled 1.5 and the column labeled 0.02; at their intersection you will find 0.9357, the measure of the cumulative area to the left of $z = +1.52$.

TABLE 6.2 A Portion of Table 3

z	0.00	0.01	0.02	...
⋮				
1.5			0.9357	...
⋮				

$P(z < 1.52) = \mathbf{0.9357}$

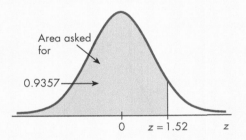

Area asked for

0.9357

0 $z = 1.52$ z

Notes:

1. Probabilities associated with positive z-values are greater than 0.5000 since they include the entire left half of the normal curve.
2. As we have done in Examples 6.1 and 6.2, always draw and label a sketch. It is most helpful.
3. Make it a habit to write the z-score with two decimal places, and the areas (probabilities, percentages) with four decimal places, as in Table 3. This will help with distinguishing between the two concepts.

   "The area under the entire normal distribution curve is equal to 1" is the key factor in determining probabilities associated with the values to the right of a z-value.

EXAMPLE 6.3

FINDING AREA TO THE RIGHT OF A z-VALUE

Find the area under the normal curve to the right of $z = -1.52$: $P(z > -1.52)$.

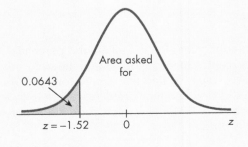

0.0643

Area asked for

$z = -1.52$ 0 z

Solution

The problem asks for the area that is not included in the 0.0643 shaded area. Since the area under the entire normal curve is 1, we subtract 0.0643 from 1:

$$P(z > -1.52) = 1.000 - 0.0643 = 0.9357$$

When finding the area to the right of any z-score, the method is the same as that demonstrated in Example 6.3: look up the area to the left and subtract the table value from 1.0. The total of the area to the left (value from Table 3) and the area to the right will always be 1.0.

Sometimes the area between two z-scores is needed. The following example demonstrates this case.

EXAMPLE 6.4

FINDING AREA BETWEEN ANY TWO z-VALUES

Find the area under the normal curve between $z = -1.36$ and $z = 2.14$: $P(-1.36 < z < 2.14)$.

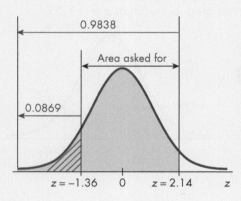

Solution

The area between $z = -1.36$ and $z = 2.14$ is found using subtraction. The cumulative area to the left of the larger z, $z = 2.14$, includes both the area asked for and the area to the left of the smaller z, $z = -1.36$. Therefore, we subtract the area to the left of the smaller z, $z = -1.36$, from the area to the left of the larger z, $z = 2.14$:

$$P(-1.36 < z < 2.14) = 0.9838 - 0.0869 = \mathbf{0.8969}$$

Note: There are several situations that are similar to Example 6.4. As in Example 6.4, one z-score can be negative while the other is positive, or both can be negative, or both can be positive. In all three cases, one of the z-scores is larger (to the right on the figure), the other is smaller (to the left on the figure), and the area in between is found as shown in the example above.

Table 3 can also be used to find the z-score(s) that bound(s) a specified area. By finding the area or probability within the table, the z-score can be read from along the left side and top margins.

EXAMPLE 6.5

FINDING z-SCORE ASSOCIATED WITH A PERCENTILE

What is the z-score associated with the 75th percentile of a normal distribution?

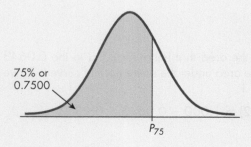

Solution

Table 3's cumulative area matches the definition of a percentile. Recall that the 75th percentile means that 75% of the data are less than the percentile's value. To find the z-score for the 75th percentile, look in Table 3 and find the "area" entry that is closest to 0.7500; this area entry is 0.7486. Now read the z-score that corresponds to this area.

TABLE 6.3 A Portion of Table 3

z	...	0.07		0.08
0.6	...	0.7486	*0.7500*	0.7518

From the table, the z-score is found to be **z = 0.67**. This says that the 75th percentile in a normal distribution is 0.67 (approximately 2/3) standard deviation above the mean.

E X A M P L E 6 . 6

FINDING THE z-SCORE THAT BOUNDS AN AREA

What z-score forms the lower boundary for the upper 14% of a normal distribution?

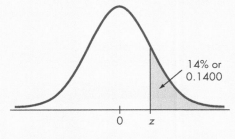

14% or 0.1400

Solution

Table 3 lists the cumulative area. In order to match the table, the area to the left must be determined by subtracting 0.1400 from 1.0, the total area.

$1.0000 - 0.1400 = 0.8600$, the value to look for in Table 3.

In Table 3, the "area" entry that is closest to 0.8600 is 0.8599. Now, read the z-score that corresponds to this area.

From the table, the z-score is found to be **z = 1.08**. This says that z = 1.08 is the lower boundary for the upper 14% of the standard normal distribution.

TABLE 6.4 A Portion of Table 3

z	...	0.08		0.09
1.0	...	0.8599	*0.8600*	0.8621

E X A M P L E 6 . 7

FINDING TWO z-SCORES THAT BOUND AN AREA

What z-scores bound the middle 95% of a normal distribution?

Solution

The 95% is split into two equal parts by the mean, so 0.4750 is the area (percentage) between the z-score at the left boundary and z = 0, the mean (as well as the area between z = 0, the mean, and the right boundary). See Figure 6.5.

FIGURE 6.5

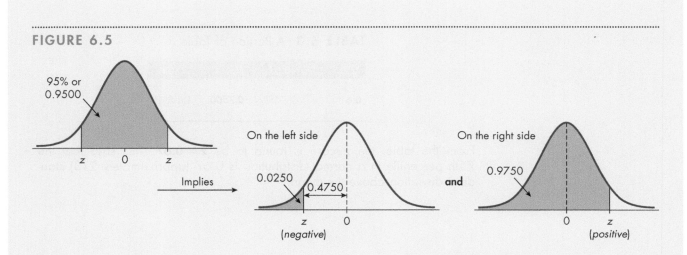

The area that is not included in either tail can be found by recalling that the area for each half of the normal curve is equal to 0.5000 and that the curve is symmetric. Thus on the left side, $0.5000 - 0.4750 = 0.0250$ is needed; and on the right side, $0.5000 + 0.4750 = 0.9750$ is needed. To find the left boundary z-score, use the area 0.0250 in Table 3 and find the "area" entry that is closest to 0.0250; this entry is exactly 0.0250.

TABLE 6.5 A Portion of   and   A Portion of Table 3
Table 3 (negative z-side)   (positive z-side)

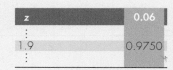

z		0.06
⋮		
−1.9	0.0250	

z		0.06
⋮		
1.9		0.9750

Reading the table, the z-score that corresponds to this area is found to be $z = -1.96$. Likewise, to find the right boundary z-score, use the area 0.9750 in Table 3 and find the "area" entry that is closest to 0.9750; this entry is exactly 0.9750. Reading this z-score gives $z = +1.96$.

Therefore, you can look up either one and utilize the symmetry of the normal distribution. $z = -1.96$ and $z = 1.96$ bound the middle 95% of a normal distribution.

As a check, consider doing it one way, and then check the result using the other way.

S E C T I O N   6 . 2   E X E R C I S E S

6.5 a. Describe the distribution of the standard normal score, z.

  b. Why is this distribution called standard normal?

6.6 Find the area under the standard normal curve to the left of $z = -1.37$.

6.7 Find the area under the normal curve that lies to the left of the following z-values.

a. $z = -1.30$ b. $z = -2.56$

c. $z = -3.20$ d. $z = -0.64$

6.8 Find the area under the standard normal curve to the left of $z = 2.13$.

6.9 Find the probability that a piece of data picked at random from a normal population will have a standard score (z) that lies to the left of the following z-values.

a. $z = 2.10$ b. $z = 1.20$

c. $z = 3.26$ d. $z = 0.71$

6.10 Find the area under the standard normal curve to the right of $z = -2.35$, $P(z > -2.35)$.

6.11 Find the following areas under the standard normal curve.

a.   To the right of $z = -0.47$, $P(z > -0.47)$

b.   To the right of $z = -1.01$, $P(z > -1.01)$

c.   To the right of $z = -3.39$, $P(z > -3.39)$

6.12 Find the area under the standard normal curve to the right of $z = 2.03$, $P(z > 2.03)$.

6.13 Find the following areas under the standard normal curve.

a.   To the right of $z = 3.18$, $P(z > 3.18)$

b.   To the right of $z = 1.84$, $P(z > 1.84)$

c.   To the right of $z = 0.75$, $P(z > 0.75)$

6.14 a.   Find the area under the standard normal curve to the left of $z = 0$, $P(z < 0)$.

   b.   Find the area under the standard normal curve to the right of $z = 0$, $P(z > 0)$.

6.15 Find the area under the standard normal curve between -1.39 and the mean, $P(-1.39 < z < 0.00)$.

6.16 Find the area under the standard normal curve between $z = -3.01$ and the mean, $P(-3.01 < z < 0)$.

6.17 Find the area under the standard normal curve between $z = -1.83$ and $z = 1.23$, $P(-1.83 < z < 1.23)$.

6.18 Find the area under the standard normal curve between $z = -2.46$ and $z = 1.46$, $P(-2.46 < z < 1.46)$.

6.19 Find the area under the standard normal curve that corresponds to the following z-values.

a.   Between 0 and 1.55

b.   To the right of 1.55

c.   To the left of 1.55

d.   Between -1.55 and 1.55

6.20 Find the probability that a piece of data picked at random from a normal population will have a standard score (z) that lies

a.   between 0 and 0.74.

b.   to the right of 0.74.

c.   to the left of 0.74.

d.   between -0.74 and 0.74.

6.21 Find the following areas under the normal curve.

a.   To the right of $z = 0.00$

b.   To the right of $z = 1.05$

c.   To the right of $z = -2.30$

d.   To the left of $z = 1.60$

e.   To the left of $z = -1.60$

6.22 Find the probability that a piece of data picked at random from a normally distributed population will have a standard score that is

a.   less than 3.00.

b.   greater than -1.55.

c.   less than -0.75.

d.   less than 1.24.

e.   greater than -1.24.

6.23 Find the following:

a. $P(0.00 < z < 2.35)$

b. $P(-2.10 < z < 2.34)$

c. $P(z > 0.13)$

d. $P(z < 1.48)$

6.24 Find the following:

a. $P(-2.05 < z < 0.00)$

b. $P(-1.83 < z < 2.07)$

c. $P(z < -1.52)$

d. $P(z < -0.43)$

6.25 Find the following:

a. $P(0.00 < z < 0.74)$

b. $P(-1.17 < z < 1.94)$

c. $P(z > 1.25)$

d. $P(z < 1.75)$

6.26 Find the following:

a. $P(-3.05 < z < 0.00)$

b. $P(-2.43 < z < 1.37)$

c. $P(z < -2.17)$

d. $P(z > 2.43)$

6.27 Find the area under the standard normal curve between $z = 0.75$ and $z = 2.25$, $P(0.75 < z < 2.25)$.

6.28 Find the area under the standard normal curve between $z = -2.75$ and $z = -1.28$, $P(-2.75 < z < -1.28)$.

6.29 Find the area under the normal curve that lies between the following pairs of z-values:

a. $z = -1.20$ to $z = -0.22$

b. $z = -1.75$ to $z = -1.54$

c. $z = 1.30$ to $z = 2.58$

d. $z = 0.35$ to $z = 3.50$

6.30 Find the probability that a piece of data picked at random from a normal population will have a standard score (z) that lies between the following pairs of z-values:

a. $z = -2.75$ to $z = -1.38$

b. $z = 0.67$ to $z = 2.95$

c. $z = -2.95$ to $z = -1.18$

6.31 Find the z-score for the standard normal distribution shown in each of the following diagrams.

a.
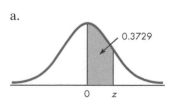

b.
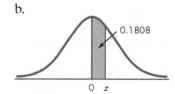

c.
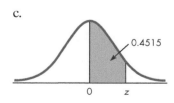

d.
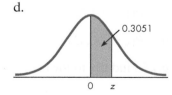

e.
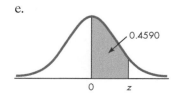

f.
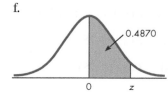

6.32 Find the z-score for the standard normal distribution shown in each of the following diagrams.

a.
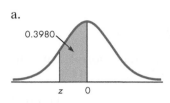

b.
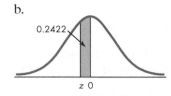

c.
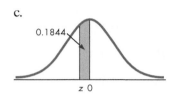

d.
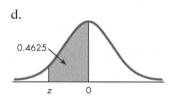

e.
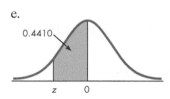

f.
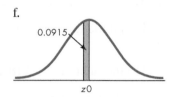

6.33 Find the standard score, z, shown in each of the following diagrams.

a.
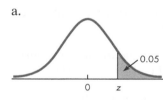

b.
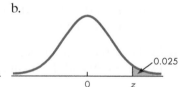

c.
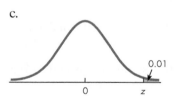

6.34 Find the standard score, z, shown in each of the following diagrams.

a.
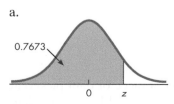

b.
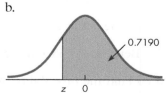

Diagram for Exercise 6.34

c.

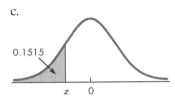

0.1515

z   0

6.35 Assuming a normal distribution, what is the *z*-score associated with the 90th percentile? the 95th percentile? the 99th percentile?

6.36 Assuming a normal distribution, what is the *z*-score associated with the 1st quartile? the 2nd quartile? the 3rd quartile?

6.37 Find the *z*-score that forms the upper boundary for the lower 20% of a normal distribution.

6.38 Find a value of *z* such that 40% of the distribution lies between it and the mean. (There are two possible answers.)

6.39 a.   Find the standard *z*-score such that 80% of the distribution is below (to the left of) this value.

b.   Find the standard *z*-score such that the area to the right of this value is 0.15.

c.   Find the two *z*-scores that bound the middle 50% of a normal distribution.

6.40 Find two standard *z*-scores such that

a.   the middle 90% of a normal distribution is bounded by them.

b.   the middle 98% of a normal distribution is bounded by them.

6.41 a.   Find the *z*-score for the 80th percentile of the standard normal distribution.

b.   Find the *z*-scores that bound the middle 75% of the standard normal distribution.

6.42 a.   Find the *z*-score for the 33rd percentile of the standard normal distribution.

b.   Find the *z*-scores that bound the middle 40% of the standard normal distribution.

6.3 Applications of Normal Distributions

In Section 6.2 we learned how to use Table 3 in Appendix B to convert information about the standard normal variable *z* into probability, or the opposite, to convert probability information about the standard normal distribution into *z*-scores. Now we are ready to apply this methodology to all normal distributions. The key is the standard score, *z*. The information associated with a normal distribution will be in terms of *x* values or probabilities. We will use the *z*-score and Table 3 as the tools to "go between" the given information and the desired answer.

Recall that the standard score, *z*, was defined in Chapter 2.

FYI When $x = \mu$, the standard score $z = 0$.

Standard Score, *z*

$$z = \frac{x - (\text{mean of } x)}{(\text{standard deviation of } x)}$$

$$z = \frac{x - \mu}{\sigma} \qquad (6.3)$$

E X A M P L E   6 . 8

CONVERTING TO A STANDARD NORMAL CURVE TO FIND PROBABILITIES

FYI Remember, when finding the area between two z-scores, subtract the area corresponding to the smaller z from the area corresponding to the larger z.

Consider the intelligence quotient (IQ) scores for people. IQ scores are normally distributed, with a mean of 100 and a standard deviation of 16. If a person is picked at random, what is the probability that his or her IQ is between 100 and 115? That is, what is $P(100 < x < 115)$?

Solution

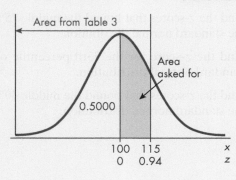

Area from Table 3

0.5000

Area asked for

| 100 | 115 | x |
| 0 | 0.94 | z |

$P(100 < x < 115)$ is represented by the shaded area in the figure.

The variable x must be standardized using formula (6.3).

The z-values are shown on the figure to the left.

$$z = \frac{x - \mu}{\sigma}$$

When $x = 100$: $z = \dfrac{100 - 100}{16} = \mathbf{0.00}$

When $x = 115$: $z = \dfrac{115 - 100}{16} = \mathbf{0.94}$

Therefore,

$$P(100 < x < 115) = P(0.00 < z < 0.94) = 0.8264 - 0.5000 = \mathbf{0.3264}$$

Thus, the probability is 0.3264 that a person picked at random has an IQ between 100 and 115.

E X A M P L E   6 . 9

CALCULATING PROBABILITY UNDER "ANY" NORMAL CURVE

Find the probability that a person selected at random will have an IQ greater than 90 ($\mu = 100$. $\sigma = 16$).

Solution

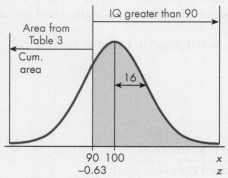

IQ greater than 90

Area from Table 3

Cum. area

16

| 90 | 100 | x |
| -0.63 | | z |

$$z = \frac{x - \mu}{\sigma} = \frac{90 - 100}{16} = \frac{-10}{16} = -0.625 = -0.63$$

$$P(x > 90) = P(z > -0.63)$$
$$= 1.0000 - 0.2643 = \mathbf{0.7357}$$

Thus, the probability is 0.7357 that a person selected at random will have an IQ greater than 90.

The normal table, Table 3, can be used to answer many kinds of questions that involve a normal distribution. Many times a problem will call for the location of a "cutoff point," that is, a particular value of x such that there is exactly a certain percentage in a specified area. The following examples concern some of these problems.

E X A M P L E   6 . 1 0

USING THE NORMAL CURVE AND z TO DETERMINE PERCENTILES

Find the 33rd percentile for IQ scores ($\mu = 100$. $\sigma = 16$; from Example 6.8, p. 280).

Solution

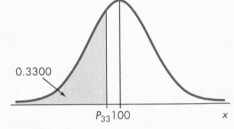

$$P(z < P_{33}) = 0.3300$$

33rd percentile is at $z = -0.44$

Now we convert the 33rd percentile of the z-scores, -0.44, to an x-score:

$$\text{Formula (6.3), } z = \frac{x - \mu}{\sigma}: \quad -0.44 = \frac{x - 100}{16}$$

$$x - 100 = 16(-0.44)$$

$$x = 100 - 7.04 = \mathbf{92.96}$$

Thus, 92.96 is the 33rd percentile for IQ scores.

E X A M P L E   6 . 1 1

USING THE NORMAL CURVE AND z TO DETERMINE DATA VALUES

In a large class, suppose your instructor tells you that you need to obtain a grade in the top 10% of your class to get an A on a particular exam. From past experience she is able to estimate that the mean and standard deviation on this exam will be 72 and 13, respectively. What will be the minimum grade needed to obtain an A? (Assume that the grades will be approximately normally distributed.)

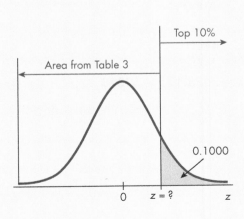

Solution

Start by converting the 10% to information that is compatible with Table 3 by subtracting:

$$10\% = 0.1000; \ 1.0000 - 0.1000 = 0.9000$$

Look in Table 3 to find the value of z associated with the area entry closest to 0.9000; it is $z = 1.28$. Thus, $P(z > 1.28) = 0.10$.

Now find the x-value that corresponds to $z = 1.28$ by using formula (6.3):

$$z = \frac{x - \mu}{\sigma}: \quad 1.28 = \frac{x - 72}{13}$$

$$x - 72 = (13)(1.28)$$

$$x = 72 + (13)(1.28) = 72 + 16.64 = 88.64, \text{ or } \mathbf{89}$$

Thus, if you receive an 89 or higher, you can expect to be in the top 10% (which means you get an A).

Example 6.12 concerns a normal distribution in which you are asked to find the standard deviation σ when given the related information.

E X A M P L E   6 . 1 2

USING THE NORMAL CURVE AND z TO DETERMINE POPULATION PARAMETERS

The incomes of junior executives in a large corporation are approximately normally distributed. A pending cutback will not discharge those junior executives with earnings within $4900 of the mean. If this represents the middle 80% of the incomes, what is the standard deviation for the salaries of this group of junior executives?

FYI Recall that the z-score is the number of multiples of the standard deviation that an x-value is from the mean.

Solution

Table 3 indicates that the middle 80%, or 0.8000, of a normal distribution is bounded by -1.28 and 1.28. Consider point B shown in the figure: 4900 is the difference between the x value at B and the value of the mean, the numerator of formula (6.3): $x - \mu = 4900$.

Using formula (6.3) we can find the value of σ:

$$z = \frac{x - \mu}{\sigma}: \quad 1.28 = \frac{4900}{\sigma}$$

$$\sigma = \frac{4900}{1.28}$$

$$\sigma = 3828.125 = \mathbf{\$3828}$$

That is, the current standard deviation for the salaries of junior executives is $3828.

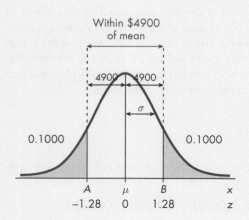

FYI A standard notation used to abbreviate "normal distribution with mean μ and standard deviation σ" is $N(\mu, \sigma)$. That is, $N(58, 7)$ represents "normal distribution, mean = 58 and standard deviation = 7."

Additional Insight

Referring again to the IQ scores, what is the probability that a person picked at random has an IQ of 125, $P(x = 125)$? (IQ scores are normally distributed, with a mean of 100 and a standard deviation of 16.) This situation has two interpretations: (1) theoretical and (2) practical. Let's look at the theoretical interpretation first. Recall that the probability associated with an interval for a continuous random variable is represented by the area under the curve. That is, $P(a \le x \le b)$ is equal to the area between a and b under the curve. $P(x = 125)$ (that is, x is exactly 125) is then $P(125 \le x \le 125)$, or the area of the vertical line segment at $x = 125$. This area is zero. However, this is not the practical meaning of $x = 125$, which generally means 125 to the nearest integer value. Thus, $P(x = 125)$ would most likely be interpreted as

$$P(124.5 < x < 125.5)$$

The interval from 124.5 to 125.5 under the curve has a measurable area and is then nonzero. In situations of this nature, you must be sure of the meaning being used.

TECHNOLOGY INSTRUCTIONS: GENERATE RANDOM DATA FROM A NORMAL DISTRIBUTION

MINITAB

Choose: **Calc > Random Data > Normal**
Enter: Number of rows of data to generate: **n**
          Store in column(s): **C1**
          Mean: μ
          Stand. dev.: σ **> OK**

If multiple samples (say, 12), all of the same size are wanted, modify the above commands:
Store in column(s): **C1–C12**

Note: To find descriptive statistics for each of these samples, use the commands: **Stat > Basic Statistics > Display Descriptive Statistics** for C1–C12.

Excel

Choose: **Data > Data Analysis > Random Number Generation > OK**
Enter: Number of Variables: **1**
          Number of Random Numbers: **n**
          Distribution: **Normal**
          Mean: μ
          Stand. dev.: σ
Select: Output Options: **Output Range**
Enter: (**A1 or select cell**) **> OK**

If multiple samples (say, 12), all of the same size are wanted, modify the above commands:
Number of variables: **12**.

Note: To find descriptive statistics for each of these samples, use the commands: **Data > Data Analysis > Descriptive Statistics** for columns A–L.

TI-83/84 Plus

Choose: **MATH > PRB > 6:randNorm(**
Enter: μ, σ, **# of trials)**
Choose: **STO→ > L1 > ENTER**

If multiple samples (say, 6), all of the same size are wanted, repeat the previous commands six times and store in L1–L6.

Note: To find descriptive statistics for each of these samples, use the commands: **STAT > CALC > 1:1-Var Stats** for L1–L6.

TECHNOLOGY INSTRUCTIONS: CALCULATING ORDINATE VALUES (y's) FOR A NORMAL DISTRIBUTION CURVE

MINITAB

Input the desired abscissas (x's) into C1; then continue with:

Choose: **Calc > Probability Distributions > Normal**
Select: **Probability Density**
Enter: Mean: μ
           Stand. dev.: σ
           Input column: **C1**
           Optional Storage: **C2 > OK**

To draw the graph of a normal probability curve with the x values in C1 and the y values in C2, continue with:

Choose: **Graph > Scatterplot**
Select: **With Connect Line > OK**
Enter: Y variables: **C2**   X variables: **C1 > OK**

Excel

Input the desired abscissas (x's) into column A and activate B1; then continue with:

Choose: **Insert function f$_x$ > Statistical > NORMDIST > OK**
Enter: X: **(A1:A100 or select "x value" cells)**
           Mean: μ
           Stand. dev.: σ
           Cumulative: **False > OK**
Drag: **Bottom right corner of the ordinate value box down to give other ordinates**

To draw the graph of a normal probability curve with the x values in column A and the y values in column B, activate both columns and continue with:

Choose: **Insert > Scatter > 1st picture**

TI-83/84 Plus

The ordinate values can be calculated for individual abscissa values, "x."

Choose: **2nd > DISTR > 1:normalpdf(**
Enter: **x, μ, σ)**

To draw the graph of the normal probability curve for a particular μ and σ, continue with:

Choose: **WINDOW**
Enter: **$\mu - 3\sigma$, $\mu + 3\sigma$, σ, −.05, 1, .1, 0**
Choose: **Y= > 2nd > DISTR > 1:normalpdf(**
Enter: **x, μ, σ)**

After an initial graph, adjust with 0:ZoomFit from the ZOOM menu.

TECHNOLOGY INSTRUCTIONS: CUMULATIVE PROBABILITY FOR NORMAL DISTRIBUTIONS

MINITAB

Input the desired abscissas (x's) into C1; then continue with:

Choose: **Calc > Probability Distributions > Normal**
Select: **Cumulative probability**
Enter: Mean: μ
              Stand. dev.: σ
              Input column: **C1**
              Optional Storage: **C3 > OK**

Notes:
1. To find the probability between two x values, enter the two values in C1, use the above commands, and subtract using the numbers in C3.
2. To draw a graph of the cumulative probability distribution (ogive), use the Scatterplot commands on page 284 with C3 as the y-variable.

Excel

Input the desired abscissas (x's) into column A and activate C1; then continue with:

Choose: **Insert function f$_x$ > Statistical > NORMDIST > OK**
Enter: **X: (A1:A100 or select "x value" cells)**
           Mean: μ
           Stand. dev.: σ
           Cumulative: True > OK
Drag: **Bottom right corner of the cumulative probability box down to give other cumulative probabilities**

Notes:
1. To find the probability between two x values, enter the two values in column A, use the above commands, and subtract using the numbers in column C.
2. To draw a graph of the cumulative probability distribution (ogive), use the Insert commands on page 284, choosing the subcommand Select Data to remove Series 1.

TI-83/84 Plus

The cumulative probabilities can be calculated for individual abscissa values, "x".

Choose: **2nd > DISTR > 2:normalcdf(**
Enter: **-1 EE 99, x, μ, σ)**

Notes:
1. To find the probability between two x values, enter the two values in place of -1 EE 99 and the x.
2. To draw a graph of the cumulative probability distribution (ogive), use either the Scatter command under STATPLOTS, with the x values and their cumulative probabilities in a pair of lists, or normalcdf(-1EE99, x, μ, σ) in the Y = editor.

APPLIED EXAMPLE 6.13

BOTTLE STOPPERS–CORKS

You're probably aware of that seemingly insignificant little cylinder of squeezable woody material called a bottle cork. But are you aware that the process by which raw bark from the Oak cork tree becomes a cork is anything but simple? The cork industry has very high standards, and there are

Image copyright Kenneth V. Pilon, 2012. Used under license from Shutterstock.com

very strict international laws covering everything from the harvesting of the cork to the delivery of the corks to the user.

Corks start as bark of the *Quercus suber* tree, and after being peeled from the tree, the bark goes through a series of storage and cooking processes to stabilize, clean, and increase the elasticity of the cork. Next it is cut into strips, and the corks are punched out. This is followed by a series of washing, bleaching, disinfecting, and coloring processes, all the while being inspected and sorted. The finishing processes include inspections, coatings, printing, cessation of moisture, surface treating, sterilization, packing, and quality-control certification.

The standard no. 9 size cork is 24 mm in diameter by 1.75 inches (45 mm) in length. Some of its characteristics (and specifications) that must pass inspection are as follows:

- Defects/faults (e.g., worm holes, cracks, pores, green wood)
- Length (45.0 + 1.0 mm/ − 0.5 mm)
- Average diameter (24 mm + 0.6 mm/ − 0.4 mm)
- Ovality (out of round, <1.0 mm)
- Weight (grams)
- Specific weight (g/cc)
- Humidity (customer's requirements ±1.5%)
- Residual peroxide (<0.2 ppm)
- Extraction force (300 N + 100 N/ − 150 N)

Length is the one variable that is not very important in evaluating the quality of corks because it has little to do with the effectiveness of a cork in preserving wine. Long corks are chosen over shorter corks largely because of their aesthetic appeal—the loud pop when you uncork the bottle.

Some of the aforementioned variables have normal distributions; others do not. Two of them with normal distributions are the average diameter of the cork and the extraction force. The diameter of each cork is measured in several places, and an average diameter is reported for the cork. It has a normal distribution with a mean of 24.0 mm and standard deviation of 0.13 mm. A sample of 250 corks produced the following summary.

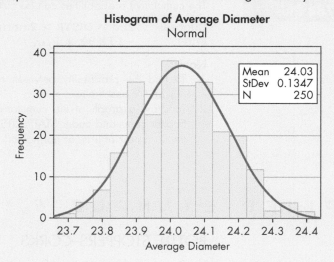

To obtain the extraction force, each bottle is filled, corked, and allowed to sit for 24 hours. It is then placed on a machine that removes the cork and records the force required to extract it from the bottle. This force has a normal distribution with a mean of 310 Newtons and a standard deviation of 36 Newtons. (A Newton is a unit of force; 1 N = 1 Newton = 1 kilogram meter/sec².) A sample of 400 corks produced this summary.

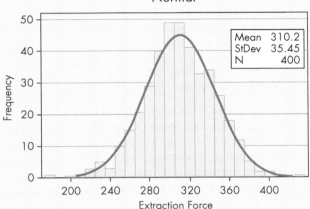

Histogram of Extraction Force
Normal

Mean 310.2
StDev 35.45
N 400

Ovality (the measure of out of round) is the difference between a cork's maximum diameter and minimum diameter. As you might expect, ovality does not have a normal distribution. Its lowest possible value is 0, and it increases from there. It does have a mounded but skewed right distribution.

What kind of distribution do you anticipate for the variables length, weight, and specific weight?

Source: Courtesy of Gültig GmbH

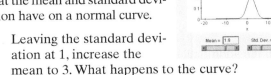

SECTION 6.3 EXERCISES

6.43 Skillbuilder Applet Exercise demonstrates that probability is equal to area under a curve. Given that college students sleep an average of 7 hours per night, with a standard deviation equal to 1.7 hours, use the scroll bar in the applet to find:

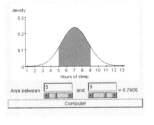

a. P(a student sleeps between 5 and 9 hours)

b. P(a student sleeps between 2 and 4 hours)

c. P(a student sleeps between 8 and 11 hours)

6.44 Skillbuilder Applet Exercise demonstrates the effects that the mean and standard deviation have on a normal curve.

a. Leaving the standard deviation at 1, increase the mean to 3. What happens to the curve?

b. Reset the mean to 0 and increase the standard deviation to 2. What happens to the curve?

c. If you could decrease the standard deviation to 0.5, what do you think would happen to the normal curve?

6.45 Given $x = 58$, $\mu = 43$ and $\sigma = 5.2$, find z.

6.46 Given $x = 237$, $\mu = 220$ and $\sigma = 12.3$, find z.

6.47 Given that x is a normally distributed random variable with a mean of 60 and a standard deviation of 10, find the following probabilities.

a. $P(x > 60)$ b. $P(60 < x < 72)$

c. $P(57 < x < 83)$ d. $P(65 < x < 82)$

e. $P(38 < x < 78)$ f. $P(x < 38)$

6.48 Given that x is a normally distributed random variable with a mean of 28 and a standard deviation of 7, find the following probabilities.

(continue on page 288)

[EX00-000] identifies the filename of an exercise's online dataset—datasets and Skillbuilder Applets available through cengagebrain.com

a. $P(x < 28)$

b. $P(28 < x < 38)$

c. $P(24 < x < 40)$

d. $P(30 < x < 45)$

e. $P(19 < x < 35)$

f. $P(x < 48)$

6.49 As shown in Example 6.8, IQ scores are considered normally distributed, with a mean of 100 and a standard deviation of 16.

a.   Find the probability that a randomly selected person will have an IQ score between 100 and 120.

b.   Find the probability that a randomly selected person will have an IQ score above 80.

6.50 Based on a survey conducted by Greenfield Online, 25 to 34-year-olds spend the most each week on fast food. The average weekly amount of $44 was reported in a May 2009 *USA Today* Snapshot. Assuming that weekly fast food expenditures are normally distributed with a standard deviation of $14.50, what is the probability that a 25- to 34-year-old will spend:

a.   less than $25 a week on fast food?

b.   between $30 and $50 a week on fast food?

c.   more than $75 a week on fast food?

6.51 Depending on where you live and on the quality of the day care, costs of day care can range from $3000 to $15,000 a year (or $250 to $1250 a month) for one child, according to the Baby Center. Day care centers in large cities such as New York and San Francisco are notoriously expensive. Suppose that day care costs are normally distributed with a mean equal to $9000 and a standard deviation equal to $1800. *Source*: http://www.babycenter.com/

a.   What percentage of day care centers cost between $7200 and $10,800?

b.   What percentage of day care centers cost between $5400 and $12,600?

c.   What percentage of day care centers cost between $3600 and $14,400?

d.   Compare the results in a through c with the empirical rule. Explain the relationship.

6.52 According to Collegeboard.com [http://www.collegeboard.com/] the national average salary for a plumber as of 2007 is $47,350. If we assume that the annual salaries for plumbers are normally distributed with a standard deviation of $5250, find the following:

a.   What percentage earn below $30,000?

b.   What percentage earn above $63,000?

6.53 According to the Federal Highway Administration's 2006 highway statistics, the distribution of ages for licensed drivers has a mean of 47.5 years and a standard deviation of 16.6 years [www.fhwa.dot.gov]. Assuming the distribution of ages is normally distributed, what percentage of the drivers are:

a.   between the ages of 17 and 22?

b.   younger than 25 years of age?

c.   older than 21 years of age?

d.   between the ages of 48 and 68?

e.   older than 75 years of age?

6.54 There is a new working class with money to burn, according to the *USA Today* March 1, 2005, article "New 'gold-collar' young workers gain clout." "Gold-collar" is a subset of blue-collar workers, defined by researchers as those working in fast food and retail jobs, or as security guards, office workers, or hairdressers. These 18- to 25-year-old "gold-collar" workers are spending an average of $729 a month on themselves (versus $267 for college students and $609 for blue-collar workers). Assuming this spending is normally distributed with a standard deviation of $92.00, what percentage of gold-collar workers spend:

a.   between $600 and $900 a month on themselves?

b.   between $400 and $1000 a month on themselves?

c.   more than $1050 a month on themselves?

d.   less than $500 a month on themselves?

6.55 Findings from a survey of American adults conducted by Yankelovich Partners for the International Bottled Water Association indicate that Americans on the average drink 6.1 8-ounce servings of water a day [http://www.pangaeawater.com/]. Assuming that the number of 8-ounce servings of water is approximately normally distributed with a standard deviation of 1.4 servings, what proportion of Americans drink

a.   more than the recommended 8 servings?

b.   less than half the recommended 8 servings?

6.56 As shown in Example 6.12, incomes of junior executives are normally distributed with a standard deviation of $3828.

a.   What is the mean for the salaries of junior executives, if a salary of $62,900 is at the top end of the middle 80% of incomes?

b.   With the additional information learned in part a, what is the probability that a randomly selected junior executive earns less than $50,000?

6.57 According to ACT, results from the 2008 ACT testing found that students had a mean reading score of 21.4 with a standard deviation of 6.0. Assuming that the scores are normally distributed,

a. find the probability that a randomly selected student had a reading ACT score less than 20.

b. find the probability that a randomly selected student had a reading ACT score between 18 and 24.

c. find the probability that a randomly selected student had a reading ACT score greater than 30.

d. find the value of the 75th percentile for ACT scores.

6.58 On a given day, the number of square feet of office space available for lease in a small city is a normally distributed random variable with a mean of 750,000 square feet and a standard deviation of 60,000 square feet. The number of square feet available in a second small city is normally distributed with a mean of 800,000 square feet and a standard deviation of 60,000 square feet.

a. Sketch the distribution of leasable office space for both cities on the same graph.

b. What is the probability that the number of square feet available in the first city is less than 800,000?

c. What is the probability that the number of square feet available in the second city is more than 750,000?

6.59 A brewery's filling machine is adjusted to fill quart bottles with a mean of 32.0 oz of ale and a variance of 0.003. Periodically, a bottle is checked and the amount of ale is noted.

a. Assuming the amount of fill is normally distributed, what is the probability that the next randomly checked bottle contains more than 32.02 oz?

b. Let's say you buy 100 quart bottles of this ale for a party; how many bottles would you expect to find containing more than 32.02 oz of ale?

6.60 Using the standard normal curve and z:

a. Find the minimum score needed to receive an A if the instructor in Example 6.11 said the top 15% were to get A's.

b. Find the 25th percentile for IQ scores in Example 6.10.

c. If SAT scores are normally distributed with a mean of 500 and a standard deviation of 100, what score does a student need to at least be considered by a college that takes only students with scores within the top 7%?

6.61 Final averages are typically approximately normally distributed with a mean of 72 and a standard deviation of 12.5. Your professor says that the top 8% of the class will receive an A; the next 20%, a B; the next 42%, a C; the next 18%, a D; and the bottom 12%, an F.

a. What average must you exceed to obtain an A?

b. What average must you exceed to receive a grade better than a C?

c. What average must you obtain to pass the course? (You'll need a D or better.)

6.62 A radar unit is used to measure the speed of automobiles on an expressway during rush-hour traffic. The speeds of individual automobiles are normally distributed with a mean of 62 mph.

a. Find the standard deviation of all speeds if 3% of the automobiles travel faster than 72 mph.

b. Using the standard deviation found in part a, find the percentage of these cars that are traveling less than 55 mph.

c. Using the standard deviation found in part a, find the 95th percentile for the variable "speed."

6.63 The weights of ripe watermelons grown at Mr. Smith's farm are normally distributed with a standard deviation of 2.8 lb. Find the mean weight of Mr. Smith's ripe watermelons if only 3% weigh less than 15 lb.

6.64 A machine fills containers with a mean weight per container of 16.0 oz. If no more than 5% of the containers are to weigh less than 15.8 oz, what must the standard deviation of the weights equal? (Assume normality.)

6.65 "On-hold" times for callers to a local cable television company are known to be normally distributed with a standard deviation of 1.3 minutes. Find the average caller "on-hold" time if the company maintains that no more than 10% of callers have to wait more than 6 minutes.

6.66 [EX06-066] The data below are the net weights (in grams) for a sample of 30 bags of M&M's. The advertised net weight is 47.9 grams per bag.

46.22	46.72	46.94	47.61	47.67	47.70
47.98	48.28	48.33	48.45	48.49	48.72
48.74	48.95	48.98	49.16	49.40	49.69
49.79	49.80	49.80	50.01	50.23	50.40
50.43	50.97	51.53	51.68	51.71	52.06

Source: http://www.math.uah.edu/

The FDA requires that (nearly) every bag contain the advertised weight; otherwise, violations (less than 47.9 grams per bag) will bring about mandated fines. (M&M's are manufactured and distributed by Mars Inc.)

(continue on page 290)

a. What percentage of the bags in the sample are in violation?

b. If the weight of all filled bags is normally distributed with a mean weight of 47.9 g, what percentage of the bags will be in violation?

c. Assuming the bag weights are normally distributed with a standard deviation of 1.5 g, what mean value would leave 5% of the weights below 47.9 g?

d. Assuming the bag weights are normally distributed with a standard deviation of 1.0 g, what mean value would leave 5% of the weights below 47.9 g?

e. Assuming the bag weights are normally distributed with a standard deviation of 1.5 g, what mean value would leave 1% of the weights below 47.9 g?

f. Why is it important for Mars to keep the percentage of violations low?

g. It is important for Mars to keep the standard deviation as small as possible so that in turn the mean can be as small as possible to maintain net weight. Explain the relationship between the standard deviation and the mean. Explain why this is important to Mars.

6.67 The extraction force required to remove a cork from a bottle of wine has a normal distribution with a mean of 310 Newtons and a standard deviation of 36 Newtons.

a. The specs for this variable, given in Applied Example 6.13, were "300 N + 100 N/−150 N" Express these specs as an interval.

b. What percentage of the corks is expected to fall within the specs?

c. What percentage of the tested corks will have an extraction force of more than 250 Newtons?

d. What percentage of the tested corks will have an extraction force within 50 Newtons of 310?

6.68 The diameter of each cork, as described in Applied Example 6.13, is measured in several places and an average diameter is reported for the cork. The average diameter has a normal distribution with a mean of 24.0 mm and standard deviation of 0.13 mm.

a. The specs for this variable, given in Applied Example 6.13, were "24 mm + 0.6 mm/−0.4 mm" Express these specs as an interval.

b. What percentage of the corks is expected to fall within the specs?

c. What percentage of the tested corks will have an average diameter of more than 24.5 millimeters?

d. What percentage of the tested corks will have an average diameter within 0.35 millimeter of 24?

6.69 a. Generate a random sample of 100 data from a normal distribution with mean 50 and standard deviation 12.

b. Using the random sample of 100 data found in part a and the technology commands for calculating ordinate values on page 284, find the 100 corresponding y values for the normal distribution curve with mean 50 and standard deviation 12.

c. Use the 100 ordered pairs found in part b to draw the curve for the normal distribution with mean 50 and standard deviation 12. (Technology commands are included with part b commands on p. 284.)

d. Using the technology commands for cumulative probability on page 285, find the probability that a randomly selected value from a normal distribution with mean 50 and standard deviation 12 will be between 55 and 65. Verify your results by using Table 3.

6.70 Use a computer or calculator to find the probability that one randomly selected value of x from a normal distribution, with mean 584.2 and standard deviation 37.3, will have a value

a. less than 525.

b. between 525 and 590.

c. of at least 590.

d. Verify the result using Table 3.

e. Explain any differences you may find.

MINITAB
Input 525 and 590 into C1; then continue with the cumulative probability commands on page 285, using 584.2 as μ, 37.3 as σ, and C2 as optional storage.

Excel
Input 525 and 590 into column A and activate the B1 cell; then continue with the cumulative probability commands on page 285, using 584.2 as μ and 37.3 as σ.

TI-83/84 Plus
Input 525 and 590 into L1; then continue with the cumulative probability commands on page 285 in L2, using 584.2 as μ, and 37.3 as σ.

6.71 Use a computer to compare a random sample to the population from which the sample was drawn. Consider the normal population with mean 100 and standard deviation 16.

a. List values of x from $\mu - 4\sigma$ to $\mu + 4\sigma$ in increments of half standard deviations and store them in a column.

b. Find the ordinate (y value) corresponding to each abscissa (x value) for the normal distribution curve for $N(100, 16)$ and store them in a column.

c. Graph the normal probability distribution curve for $N(100, 16)$.

d. Generate 100 random data values from the $N(100, 16)$ distribution and store them in a column.

e. Graph the histogram of the 100 data obtained in part d using the numbers listed in part a as class boundaries.

f. Calculate other helpful descriptive statistics of the 100 data values and compare the data to the expected distribution. Comment on the similarities and the differences you see.

MINITAB

a. Choose: **Calc > Make Patterned Data > Simple Set of Numbers**
   Enter: Store patterned data in: **C1**
            From first value: **36**
            To last value: **164**
            In steps of: **8 > OK**

b. Choose: **Calc > Prob. Dist. > Normal**
   Select: **Probability density**
   Enter: Mean: 100
             Stand. dev.:   16
             Input column: C1
             Optional Storage: **C2 > OK**

c. Use the Scatterplot commands on page 284 for the data in C1 and C2.

d. Use the Calculate RANDOM DATA commands on page 283, replacing n with 100, store in with C3, mean with 100, and standard deviation with 16.

e. Use the HISTOGRAM With Fits commands on page 53 for the data in C3. To adjust the histogram, select Binning with cutpoint and cutpoint positions 36:148/8

f. Use the MEAN and STANDARD DEVIATION commands on pages 65 and 79 for the data in C3.

Excel

a. Choose: **Data > Data Analysis > Random Number Generation > OK**
   Enter: Number of variables: 1
             Distribution: **Patterned**
             From: **36** to **172** in steps of **8**
             Repeat each number: **1 time**
   Select: **Output Range**
   Enter: **(A1 or select cell)**

b. Activate B1; then continue with:

   Choose: **Insert function f_x > Statistical > NORMDIST > OK**
   Enter: X: (A1:A? or select "x value" cells)
             Mean: 100
             Stand. dev.: 16

Cumulative: **False > OK**

Drag: **Bottom right corner of the ordinate value box down to give other ordinates**

c. Use the Insert > Scatter commands on page 284 for the data in columns A and B.

d. Activate cell C1; then use the Normal RANDOM NUMBER GENERATION commands on page 283, replacing number of random numbers with 100, mean with 100, and standard deviation with 16.

e. Use the HISTOGRAM commands on pages 53–54 with column C as the input range and column A as the bin range.

f. Use the MEAN and STANDARD DEVIATION commands on pages 65 and 79 for the data in column C.

6.72 Use a computer to compare a random sample to the population from which the sample was drawn. Consider the normal population with mean 75 and standard deviation 14. Answer questions a through f of Exercise 6.71 using $N(75, 14)$.

6.73 Suppose you were to generate several random samples, all the same size, all from the same normal probability distribution. Will they all be the same? How will they differ? By how much will they differ?

a. Use a computer or calculator to generate 10 different samples, all of size 100, all from the normal probability distribution of mean 200 and standard deviation 25.

b. Draw histograms of all 10 samples using the same class boundaries.

c. Calculate several descriptive statistics for all 10 samples, separately.

d. Comment on the similarities and the differences you see.

MINITAB

a. Use the Generate RANDOM DATA commands on page 283, replacing n with 100 store in with C1–C10, mean with 200, and standard deviation with 25.

b. Use the HISTOGRAM commands on page 53 for the data in C1–C10. To adjust the histogram, select Binning with cutpoint and cutpoint positions 36:148/8.

c. Use the DISPLAY DESCRIPTIVE STATISTICS command on page 88 for the data in C1–C10.

Excel

a. Use the Normal RANDOM NUMBER GENERATION commands on page 283, replacing number of variables with 10, number of random numbers with 100, mean with 200, and standard deviation with 25.

b. Use the RANDOM NUMBER GENERATION Patterned Distribution commands in Exercise 6.71, replacing the first value with 100, the last value with 300, the steps with 25, and the output range with K1. Use the HISTOGRAM commands on pages 53–54 for each of the columns A through J (input range), with column K as the bin range

(continue on page 292)

c. Use the DESCRIPTIVE STATISTICS commands on page 88 for the data in columns A through J.

TI-83/84 Plus

a. Use the 6:randNorm commands on pages 283–284, replacing the mean with 200, the standard deviation with 25, and the number of trials with 100. Repeat 6 times using L1–L6 for storage.

b. Use the HISTOGRAM commands on page 54 for the data in L1–L6, entering WINDOW values: 100, 300, 25, −10, 60, 10, 1. Adjust with ZoomStat.

c. Use the 1-Var Stats command on page 88 for the data in L1–L6.

6.74 Generate 10 random samples, each of size 25, from a normal distribution with mean 75 and standard deviation 14. Answer questions b through d of Exercise 6.73.

6.4 Notation

The z-score is used throughout statistics in a variety of ways; however, the relationship between the numerical value of z and the area under the standard normal distribution curve does not change. Since z is used with great frequency, we want a convenient notation to identify the necessary information. The convention that we will use as an "algebraic name" for a specific z-score is $z(\alpha)$, where α represents the "area to the right" of the z being named.

EXAMPLE 6.14

VISUAL INTERPRETATION OF $z(\alpha)$

$z(0.05)$ (read "z of 0.05") is the algebraic name for the z such that the area to the right and under the standard normal curve is exactly 0.05, as shown in Figure 6.6.

FIGURE 6.6
Area Associated with $z(0.05)$

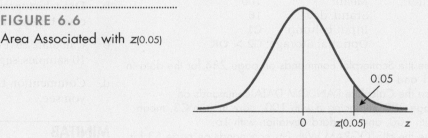

EXAMPLE 6.15

VISUAL INTERPRETATION OF $z(\alpha)$

$z(0.90)$ (read "z of 0.90") is that value of z such that 0.90 of the area lies to its right, as shown in Figure 6.7.

FIGURE 6.7
Area Associated with $z(0.90)$

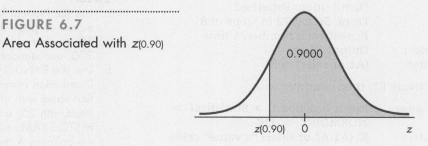

Now let's find the numerical values of $z(0.05)$ and $z(0.90)$.

EXAMPLE 6 . 1 6

DETERMINING CORRESPONDING z-VALUES FOR $z(\alpha)$

Find the numerical value of $z(0.05)$.

Solution

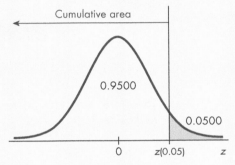

FIGURE 6.8
Find the Value of $z(0.05)$

Remember that the area under the entire normal curve is 1. Therefore, subtracting 0.05 from 1 gives 0.95, the area to the left of the $z(0.05)$. The 0.9500 area is the area we can use with Table 3 in Appendix B, or with the cumulative function on a calculator or computer; see the areas shown in Figure 6.8.

When we look in Table 3, we look for an area as close as possible to 0.9500.

z	. . .	0.04		0.05	. . .
⋮					
1.6	. . .	0.9495	0.9500	0.9505	. . .
⋮					

We use the z that corresponds to the area closest in value. When the value happens to be exactly halfway between the table entries as above, always use the larger value of z.

Therefore, $z(0.05) = $ **1.65**.

FYI It is customary to round up to the next larger value due to the typical uses of critical values, as will be seen in Chapter 8.

EXAMPLE 6 . 1 7

DETERMINING CORRESPONDING z-VALUES FOR $z(\alpha)$

Find the value of $z(0.90)$.

Solution

As in Example 6.16, the 0.90 area needs to be subtracted from 1, thereby giving an area of 0.10 to the left of $z(0.90)$. The 0.1000 area is the area we can use with Table 3 in Appendix B, as shown in the following diagram.

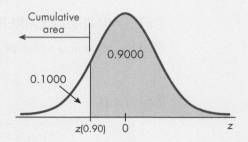

The closest values in Table 3 are 0.1003 and 0.0985, with 0.1003 being closer to 0.1000.

z	. . .	0.08		0.09
⋮				
−1.2	. . .	0.1003	0.1000	0.0985
⋮				

Therefore, $z(0.90)$ is related to -1.28. Since $z(0.90)$ is below the mean, it makes sense that $z(0.90) = \mathbf{-1.28}$.

The $z(\alpha)$ notation is used regularly in connection with inferential situations involving the area of a tail (extreme ends of a distribution curve—either left or right) region. In later chapters this notation will be used on a regular basis. The values of z that will be used regularly come from one of the following situations: (1) the z-score such that there is a specified area in one tail of the normal distribution, or (2) the z-scores that bound a specified middle proportion of the normal distribution.

Example 6.16 showed a commonly used one-tail situation; $z(0.05) = 1.65$ is located so that 0.05 of the area under the normal distribution curve is in the tail to the right. Example 6.17 also showed a commonly used one-tail situation; $z(0.90) = -1.28$ is located so that 0.10 of the area under the normal distribution curve is in the tail to the left.

Because of the symmetrical nature of the normal distribution, $z(\alpha)$ and $z(1 - \alpha)$ are closely related, the only difference being that one is positive and the other is negative. Let's look at an example that demonstrates this.

EXAMPLE 6.18

DEMONSTRATING THE RELATIONSHIP BETWEEN $z(\alpha)$ AND $z(1 - \alpha)$

The value of $z(0.05)$ was found to be 1.65 in Example 6.16 (p. 293). Find $z(0.95)$.

Solution

$z(0.95)$ is located on the left-hand side of the normal distribution since the area to the right is 0.95. The area in the tail to the left then contains the other 0.05, as shown in Figure 6.9.

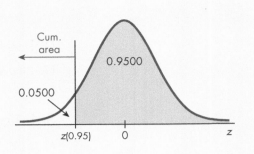

Using Table 3, $z(0.95) = -\mathbf{1.65}$.

Because of the symmetrical nature of the normal distribution, $z(0.95) = -1.65$ and $z(0.05) = 1.65$ differ only in sign and the side of the distribution to which they belong.

Thus, $z(0.95) = -z(0.05) = -\mathbf{1.65}$.

In many situations, it will be more convenient to refer to the area of the tail than to either the cumulative area or the area to the right; thus we introduce an alternative algebraic name for the z-values bounding a left-side-tail situation. For example, since $z(0.95)$ and $z(0.05)$ have the same numerical value and differ only in sign, we saw that we can identify $z(0.95)$ as $-z(0.05)$.

In general, when $1 - \alpha$ is larger than 0.5000, the notation convention we will use is $z(1 - \alpha) = -z(\alpha)$.

E X A M P L E   6 . 1 9

USING TABLE 4 TO DETERMINE $z(\alpha)$ AND $z(1 - \alpha)$

Find the values of $z(0.05)$ and $z(0.95)$ using Table 4 in Appendix B.

Solution

Table 4, Critical Values of Standard Normal Distribution, was designed to provide only the most commonly used values of z when the area(s) of the tail regions are given. Part A, One-Tailed Situations, is used when the area of a tail is given.

A Portion of Table 4A, One-Tailed Situations

		Amount of α in One Tail			
α	. . .	0.10	0.05	0.025	. . .
$z(\alpha)$	. . .	1.28	**1.65**	1.96	. . .

$z(0.05) = \mathbf{1.65}$, and since the standard normal distribution is symmetrical, the value of $z(0.95) = -z(0.05) = -\mathbf{1.65}$.

When the middle proportion of a normal distribution is specified, we can also use the "area-to-the-right" notation to identify the specific z-score involved.

E X A M P L E   6 . 2 0

DETERMINING z-SCORES FOR BOUNDED AREAS

Find the z-scores that bound the middle 0.95 of the normal distribution.

Solution 1—Using One Tail

Given 0.95 as the area in the middle (Figure 6.10), the two tails must contain a total of 0.05. Therefore, each tail contains $\frac{1}{2}$ of 0.05, or 0.025, as shown in Figure 6.11.

FIGURE 6.10
Area Associated with Middle 0.95

FIGURE 6.11
Finding z-Scores for Middle 0.95

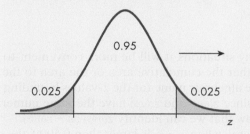

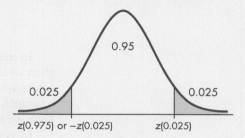

The right tail value, z(0.025), is found using Table 4, Part A, One-Tailed Situations, as demonstrated in Example 6.19.

A Portion of Table 4A, One-Tailed Situations

		Amount of α in One Tail			
α	...	0.05	0.025	0.02	...
$z(\alpha)$	...	1.65	**1.96**	2.05	...

$z(0.025) = 1.96$, and since the standard normal distribution is symmetrical, the value of $z(0.975) = -z(0.025) = -1.96$.

Solution 2—Using Two Tails

Given 0.95 as the area in the middle (Figure 6.12), the two tails must contain a total of 0.05. Table 4, Part B, Two-Tailed Situations, can be used when the combined area of both tails (or the area in the center) is given. Locate the column that corresponds to $\alpha = 0.05$ or $(1 - \alpha) = 0.95$.

FIGURE 6.12
Finding z-Scores for Middle 0.95

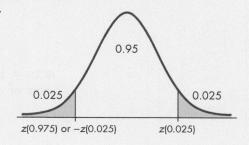

FYI There is another option for finding values of $z(\alpha)$: Use the inverse cumulative function on your calculator or computer. For specific instructions, see page 285.

A Portion of Table 4B, Two-Tailed Situations
Amount of α in Two Tails

α	...	0.10	0.05	0.02	...
$z(\alpha/2)$	...	1.65	**1.96**	2.33	...
$1-\alpha$	...	0.90	0.95	0.98	...

Area in the "center"

From Table 4B we find $z(0.05/2) = z(0.025) = 1.96$. Using the symmetry property of the distribution, we find $z(0.975) = -z(0.025) = -1.96$.
The middle 0.95 of the normal distribution is bounded by -1.96 and 1.96.

SECTION 6.4 EXERCISES

6.75 Using the $z(\alpha)$ notation (identify the value of α used within the parentheses), name each of the standard normal variable z's shown in the following diagrams.

a.  0.03

b.  0.14

c. 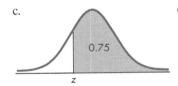 0.75

d. 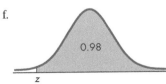 0.22

e. 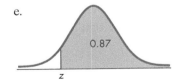 0.87

f. 0.98

6.76 Using the $z(\alpha)$ notation (identify the value of α used within the parentheses), name each of the standard normal variable z's shown in the following diagrams.

a. 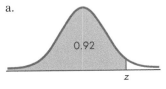 0.92

b. 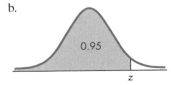 0.95

c.  0.05

d. 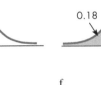 0.18

e. 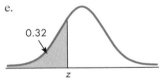 0.32

f. 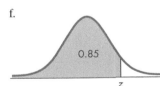 0.85

6.77 Using the $z(\alpha)$ notation (identify the value of α used within the parentheses), name each of the standard normal variable z's shown in the following diagrams.

a.  0.01

b. 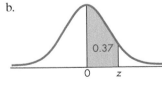 0.37

c. 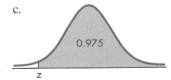 0.975

d. 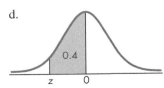 0.4

6.78 Using the $z(\alpha)$ notation (identify the value of α used within the parentheses), name each of the standard normal variable z's shown in the following diagrams.

(continue on page 298)

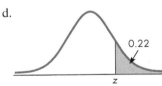

a.
b.

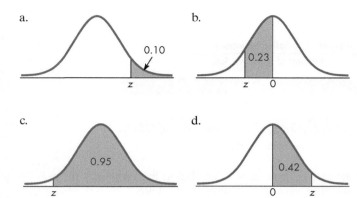

c.
d.

6.79 Draw a figure of the standard normal curve showing:
a. $z(0.15)$ b. $z(0.82)$

6.80 Draw a figure of the standard normal curve showing:
a. $z(0.04)$ b. $z(0.94)$

6.81 We are often interested in finding the value of z that bounds a given area in the right-hand tail of the normal distribution, as shown in the accompanying figure. The notation $z(\alpha)$ represents the value of z such that $P(z > z(\alpha)) = \alpha$.

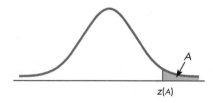

Find the following:

a. $z(0.025)$ b. $z(0.05)$ c. $z(0.01)$

6.82 We are often interested in finding the value of z that bounds a given area in the left-hand tail of the normal distribution, as shown in the accompanying figure. The notation $z(\alpha)$ represents the value of z such that $P(z > z(\alpha)) = \alpha$.

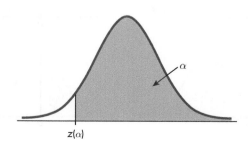

Find the following:

a. $z(0.98)$ b. $z(0.80)$ c. $z(0.70)$

6.83 Use Table 4A, Appendix B, and the symmetry property of normal distributions to find the following values of z.

a. $z(0.05)$ b. $z(0.01)$ c. $z(0.025)$

d. $z(0.975)$ e. $z(0.98)$

6.84 Using Table 4A and the symmetry property of the normal distribution, complete the following charts of z-scores. The area given in the tables is the area to the right under the normal distribution in the figures.

a. z-scores associated with the right-hand tail: Given the area A, find $z(A)$.

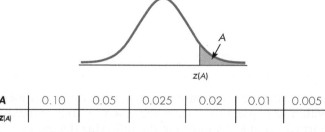

A	0.10	0.05	0.025	0.02	0.01	0.005
z(A)						

b. z-scores associated with the left-hand tail: Given the area B, find $z(B)$.

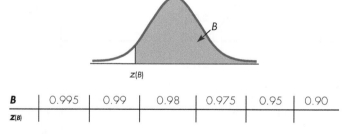

B	0.995	0.99	0.98	0.975	0.95	0.90
z(B)						

6.85 Using Table 4B, find the z-scores that bound the middle 0.80 of the normal distribution.

Verify the z-scores using Table 4A.

6.86 Using Table 4B, find the z-scores that bound the middle 0.98 of the normal distribution.

Verify the z-scores using Table 4A.

6.87 Using Table 4B, complete the following chart of z-scores that bound a middle area of a normal distribution.

Middle	0.75	0.90	0.95	0.99
±z				

6.88 a.   Find the area under the normal curve for z between $z(0.95)$ and $z(0.025)$.

b.   Find $z(0.025) - z(0.95)$.

6.89 The z notation, $z(\alpha)$, combines two related concepts, the z-score and the area to the right, into a mathematical symbol. Identify the letter in each of the following as being a z-score or an area, and then with the aid of a diagram explain what both the given number and the letter represent on the standard curve.

a. $z(A) = 0.10$

b. $z(0.10) = B$

c. $z(C) = -0.05$

d. $-z(0.05) = D$

6.90 Understanding the z notation, $z(\alpha)$, requires us to know whether we have a z-score or an area. Each of the following expressions uses the z notation in a variety of ways, some typical and some not so typical. Find the value asked for in each of the following, and then with the aid of a diagram explain what your answer represents.

a. $z(0.08)$

b.   the area between $z(0.98)$ and $z(0.02)$

c. $z(1.00 - 0.01)$

d. $z(0.025) - z(0.975)$

6.5 Normal Approximation of the Binomial

In Chapter 5 we introduced the **binomial distribution**. Recall that the binomial distribution is a probability distribution of the discrete random variable x, the number of successes observed in n repeated independent trials. We will now see how **binomial probabilities**—that is, probabilities associated with a binomial distribution—can be reasonably approximated by using the normal probability distribution.

Let's look first at a few specific binomial distributions. Figure 6.13 shows the probabilities of x for 0 to n in three situations: $n = 4$, $n = 8$, and $n = 24$. For each of these distributions, the probability of success for one trial is 0.5. Notice that as n becomes larger, the distribution appears more and more like the normal distribution.

FIGURE 6.13

Binomial Distributions

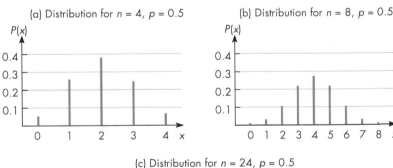

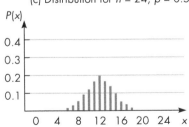

To make the desired approximation, we need to take into account one major difference between the binomial and the normal probability distribution. The binomial random variable is **discrete**, whereas the normal random variable is **continuous**. Recall that Chapter 5 demonstrated that the probability assigned to a particular value of x should be shown on a diagram by means of a straight-line segment whose length represents the probability (as in Figure 6.13). Chapter 5 suggested, however, that we can also use a histogram in which the area of each bar is equal to the probability of x.

Let's look at the distribution of the binomial variable x, when $n = 14$ and $p = 0.5$. The probabilities for each x value can be obtained from Table 2 in Appendix B. This distribution of x is shown in Figure 6.14. We see the very same distribution in Figure 6.15 in histogram form.

FIGURE 6.14

The Distribution of x when $n = 14$, $p = 0.5$

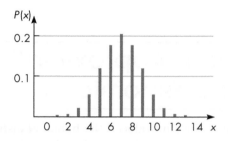

FIGURE 6.15

Histogram for the Distribution of x when $n = 14$, $p = 0.5$

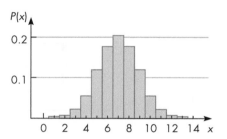

Let's examine $P(x = 4)$ for $n = 14$ and $p = 0.5$ to study the approximation technique. $P(x = 4)$ is equal to 0.061 (see Table 2 in Appendix B), the area of the bar (rectangle) above $x = 4$ in Figure 6.16.

FIGURE 6.16

Area of Bar above $x = 4$ is 0.061, for $B(n = 14, p = 0.5)$

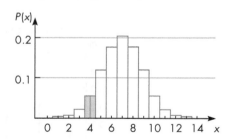

The area of a rectangle is the product of its width and height. In this case the height is 0.061 and the width is 1.0, so the area is 0.061. Let's take a closer look at the width. For $x = 4$, the bar starts at 3.5 and ends at 4.5, so we are looking at an area bounded by $x = 3.5$ and $x = 4.5$. The addition and subtraction of 0.5 to the x value is commonly called the **continuity correction factor**. It is our method of converting a discrete variable into a continuous variable.

Now let's look at the normal distribution related to this situation. We will first need a normal distribution with a mean and a standard deviation equal to those of the binomial distribution we are discussing. Formulas (5.7) and (5.8) give us these values:

$$\mu = np = (14)(0.5) = \textbf{7.0}$$

$$\sigma = \sqrt{npq} = \sqrt{(14)(0.5)(0.5)} = \sqrt{3.5} = \textbf{1.87}$$

The probability that $x = 4$ is approximated by the area under the normal curve between $x = 3.5$ and $x = 4.5$ is shown in Figure 6.17. Figure 6.18 shows the entire distribution of the binomial variable x with a normal distribution of the same mean and standard deviation superimposed. Notice that the bars and the interval areas under the curve cover nearly the same area.

FIGURE 6.17
FIGURE 6.17

Probability that $x = 4$ is Approximated by Shaded Area

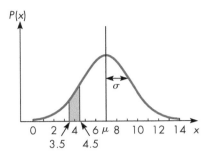

FIGURE 6.18

Normal Distribution Superimposed over Distribution for Binomial Variable x

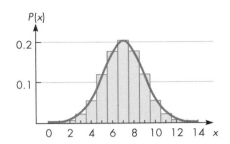

The probability that x is between 3.5 and 4.5 under this normal curve is found by using formula (6.3), Table 3, and the methods outlined in Section 6.3:

$$z = \frac{x - \mu}{\sigma}: \quad P(3.5 < x < 4.5) = P\left(\frac{3.5 - 7.0}{1.87} < z < \frac{4.5 - 7.0}{1.87}\right)$$

$$= P(-1.87 < z < -1.34)$$

$$= 0.0901 - 0.0307 = \mathbf{0.0594}$$

Since the binomial probability of 0.061 and the normal probability of 0.0594 are reasonably close, the normal probability distribution seems to be a reasonable approximation of the binomial distribution.

The normal approximation of the binomial distribution is also useful for values of p that are not close to 0.5. The binomial probability distributions shown in Figures 6.19 and 6.20 suggest that binomial probabilities can be approximated

FIGURE 6.19   Binomial Distributions

(a) Distribution for $n = 4$, $p = 0.3$

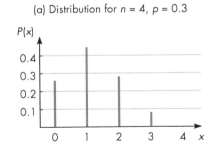

(b) Distribution for $n = 8$, $p = 0.3$

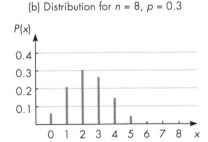

(c) Distribution for $n = 24$, $p = 0.3$

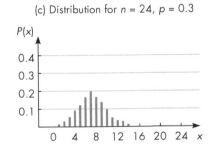

FIGURE 6.20   Binomial Distributions

(a) Distribution for $n = 4$, $p = 0.1$

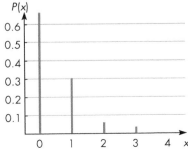

(b) Distribution for $n = 8$, $p = 0.1$

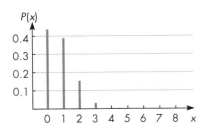

(c) Distribution for $n = 50$, $p = 0.1$

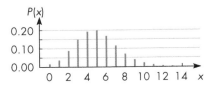

using the normal distribution. Notice that as *n* increases, the binomial distribution begins to look like the normal distribution. As the value of *p* moves away from 0.5, a larger *n* is needed in order for the normal approximation to be reasonable. The following *rule of thumb* is generally used as a guideline:

> **Rule** The normal distribution provides a reasonable approximation to a binomial probability distribution whenever the values of np and $n(1-p)$ both equal or exceed 5.

By now you may be thinking, "So what? I will just use the binomial table and find the probabilities directly and avoid all the extra work." But consider for a moment the situation presented in Example 6.21.

EXAMPLE 6.21

SOLVING A BINOMIAL PROBABILITY PROBLEM WITH THE NORMAL DISTRIBUTION

An unnoticed mechanical failure has caused $\frac{1}{3}$ of a machine shop's production of 5000 rifle firing pins to be defective. What is the probability that an inspector will find no more than 3 defective firing pins in a random sample of 25?

Solution

In this example of a binomial experiment, *x* is the number of defectives found in the sample, $n = 25$, and $p = P(\text{defective}) = \frac{1}{3}$. To answer the question using the binomial distribution, we will need to use the binomial probability function, formula (5.5):

$$P(x) = \binom{25}{x}\left(\frac{1}{3}\right)^{x}\left(\frac{2}{3}\right)^{25-x} \text{ for } x = 0, 1, 2, \ldots, 25$$

We must calculate the values for $P(0)$, $P(1)$, $P(2)$, and $P(3)$, because they do not appear in Table 2. This is a very tedious job because of the size of the exponent. In situations such as this, we can use the normal approximation method.

Now let's find $P(x \le 3)$ by using the normal approximation method. We first need to find the mean and standard deviation of *x*, formulas (5.7) and (5.8):

$$\mu = np = (25)\left(\frac{1}{3}\right) = \textbf{8.333}$$

$$\sigma = \sqrt{npq} = \sqrt{(25)\left(\frac{1}{3}\right)\left(\frac{2}{3}\right)} = \sqrt{5.55556} = \textbf{2.357}$$

These values are shown in the figure. The area of the shaded region $(x < 3.5)$ represents the probability of $x = 0$, 1, 2, or 3. Remember that $x = 3$, the discrete binomial variable, covers the continuous interval from 2.5 to 3.5.

$$P(x \text{ is no more than } 3) = P(x \le 3) \text{ (for a discrete variable } x\text{)}$$
$$= P(x < 3.5) \text{ (for a continuous variable } x\text{)}$$

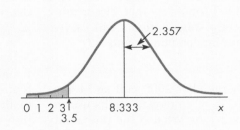

$$z = \frac{x - \mu}{\sigma}: \quad P(x < 3.5) = P\left(z < \frac{3.5 - 8.333}{2.357}\right) = P(z < -2.05)$$

$$= \mathbf{0.0202}$$

Thus, P(no more than three defectives) is approximately 0.02.

S E C T I O N   6 . 5   E X E R C I S E S

6.91 Find the values np and nq (recall: $q = 1 - p$) for a binomial experiment with $n = 100$ and $p = 0.02$. Does this binomial distribution satisfy the rule for normal approximation? Explain.

6.92 In which of the following binomial distributions does the normal distribution provide a reasonable approximation? Use computer commands to generate a graph of the distribution and compare the results to the "rule of thumb." State your conclusions.

a. $n = 10, p = 0.3$ b. $n = 100, p = 0.005$

c. $n = 500, p = 0.1$ d. $n = 50, p = 0.2$

MINITAB
Insert the specific n and p as needed in the procedure below. Use the **Make Patterned Data** commands in Exercise 6.71, replacing the first value with 0, the last value with n, and the steps with 1.
Use the **Binomial Probability Distribution** commands on page 251, using C2 as optional storage.
Use the **Scatterplot Simple** commands for the data in C1 and C2. Select **Data View, Data Display, Project Lines** to complete the graph.

Excel
Insert the specific n and p as needed in the procedure below. Use the **RANDOM NUMBER GENERATION Patterned Distribution** commands in Exercise 6.71, replacing the first value with 0, the last value with n, the steps with 1, and the output range with A1. Activate cell B1; then use the **Binomial Probability Distribution** commands on pages 251–252.
Use the **Insert >** commands for the data in columns A and B Choosing the **Select Data** subcommand remove Series 1.

6.93 In order to see what happens when the normal approximation is improperly used, consider the binomial distribution with $n = 15$ and $p = 0.05$. Since $np = 0.75$, the rule of thumb ($np > 5$ and $nq > 5$) is not satisfied. Using the binomial tables, find the probability of one or fewer successes and compare this with the normal approximation.

6.94 Find the normal approximation for the binomial probability $P(x = 6)$, where $n = 12$ and $p = 0.6$. Compare this to the value of $P(x = 6)$ obtained from Table 2.

6.95 Find the normal approximation for the binomial probability $P(x = 4, 5)$, where $n = 14$ and $p = 0.5$. Compare this to the value of $P(x = 4, 5)$ obtained from Table 2.

6.96 Find the normal approximation for the binomial probability $P(x \le 8)$, where $n = 14$ and $p = 0.4$. Compare this to the value of $P(x \le 8)$ obtained from Table 2.

6.97 Find the normal approximation for the binomial probability $P(x \ge 9)$, where $n = 13$ and $p = 0.7$. Compare this to the value of $P(x \ge 9)$ obtained from Table 2.

6.98 Referring to Example 6.21 (p. 302):

a.   Calculate $P(x \le 3 | B(25, \frac{1}{3}))$.

b.   How good was the normal approximation? Explain. (*Hint:* If you use a computer or calculator, use the commands on pp. 251–252.)

6.99 Melanoma is the most serious form of skin cancer and is increasing at a rate higher than any other cancer in the United States. If it is caught in its early stage, the

(continue on page 304)

5-year survival rate for patients on average is 98% in the United States. What is the probability that 235 or more of some group of 250 early stage patients will survive 5 years or more after their melanoma diagnosis?

Source: http://www.health.com/

6.100 According to a September 2008 poll and report done by Pew/Internet, 62% of employed adults use the Internet or e-mail on their jobs. What is the probability that more than 180 out of 250 employed adults use the Internet or e-mail on their jobs?

Source: http://www.pewinternet.org

6.101 According to Federal Highway Administration 2007 statistics, the percent of licensed female drivers has just surpassed the percent of licensed male drivers. Of the drivers in the United States, 50.2% are females. If a random sample of 50 drivers is to be selected for a survey,

a.   what is the probability that no more than half (25) of the drivers are female?

b.   what is the probability that at least three-fourths (38) of the drivers are female?

6.102 According to a November 2008 survey completed by the Pew Internet & American Life Project [http://www.pewinternet.org], about 74% of all adult Internet users say they went online for news and information about the 2008 election or to communicate with others about the election race. Assuming the percentage is correct, use the normal approximation to the binomial to find the probability that in a survey of 2000 American adult Internet users,

a.   at least 1400 used the Internet for information about the 2008 election.

b.   at least 1565 used the Internet for information about the 2008 election.

c.   at most 1500 used the Internet for information about the 2008 election.

d.   at most 1425 used the Internet for information about the 2008 election.

6.103 Not all NBA coaches who enjoy lengthy careers consistently put together winning seasons with the teams they coach. For example, Bill Fitch, who coached 25 seasons of professional basketball after starting his coaching career at the University of Minnesota, won 944 games but lost 1106 while working with the Cavaliers, Celtics, Rockets, Nets, and Clippers. If you were to randomly select 60 box scores from the historical records of games in which Bill Fitch coached one of the teams, what is the probability that fewer than half of them show his team winning? To obtain your answer, use the normal approximation to the binomial distribution.

Source: basketball-reference.com

6.104 Eighty-eight percent of voters would vote for a female presidential candidate if she was qualified, a poll found. The poll was conducted in February 2007 by Gallup and reported by the Pew Research Center [http://pewresearch.org]. Only 53% of voters felt this way in 1969. Assuming 88% is the true current proportion, what is the probability that another poll of 1125 registered voters conducted randomly will result in

a.   more than eight-ninths saying they would vote for a female presidential candidate, if she was qualified?

b.   less than 85% saying they would vote for a female presidential candidate, if she was qualified?

6.105 According to a December 2008 report from the Join Together website of the Boston University School of Public Health, approximately half (42%) of U.S. children are exposed to secondhand smoke on a weekly basis, with more than 25% of parents reporting that their child has been exposed to smoke in their homes. This statistic was one of many results from the *Social Climate Survey of Tobacco Control* [http://www.socialclimate.org/]. Use the normal approximation to the binomial distribution to find the probability that in a poll of 1200 randomly selected parents, between 450 and 500 inclusive will report that their child has been exposed to smoke on a weekly basis.

Source: http://www.jointogether.org

a.   Solve using normal approximation and Table 3 in Appendix B.

b.   Solve using a computer or calculator and the normal approximation method.

c.   Solve using a computer or calculator and the binomial probability function.

6.106 You are not alone if your garage is so cluttered that you cannot fit your car inside. According to the *Democrat & Chronicle* article "Season's cleaning" (January 1, 2009), the U.S. Department of Energy reports that 25% of people with two-car garages don't have room to park any cars inside it. Use the normal approximation to the binomial distribution to find the probability that in a poll of 1200 homeowners with two-car garages, between 250 and 340 inclusive cannot fit their cars inside their garage.

a.   Solve using normal approximation and Table 3.

b.   Solve using a computer or calculator and the normal approximation method.

6.107 Technology is the key to our future. Apparently, students believe this also. According to an April 2009 poll of high school students by Ridgid, the top career choice for high school students was information technology, selected by 25% of the surveyed students. Suppose you randomly select 200 students from your local high school. Use the normal approximation to the binomial distribution to find the probability that within your sample:

a. more than 65 of the students picked information technology as their career choice.

b. fewer than 27 of the students picked information technology as their career choice.

c. between 45 and 60 of the students picked information technology as their career choice.

© 2010/Jupiterimages Corporation

Chapter Review

In Retrospect

We have learned about the standard normal probability distribution, the most important family of continuous random variables. We have learned to apply it to all other normal probability distributions and how to use it to estimate probabilities of binomial distributions. We have seen a wide variety of variables that have this normal distribution or are reasonably well approximated by it.

In the next chapter we will examine sampling distributions and learn how to use the standard normal probability to solve additional applications.

CourseMate

The **Statistics CourseMate** site for this text brings chapter topics to life with interactive learning, study, and exam preparation tools, including quizzes and flashcards for the Vocabulary and Key Concepts that follow. The site also provides an **eBook** version of the text with highlighting and note taking capabilities. Throughout chapters, the CourseMate icon 🖥 flags concepts and examples that have corresponding interactive resources such as **video** and **animated tutorials** that demonstrate, step by step, how to solve problems; **datasets** for exercises and examples; **Skillbuilder Applets** to help you better understand concepts; **technology manuals**; and software to download including **Data Analysis Plus** (a suite of statistical macros for Excel) and **TI-83/84 Plus** programs—logon at **www.cengagebrain.com**.

Vocabulary and Key Concepts

area representation for probability (p. 270)
bell-shaped distribution (p. 268)
binomial distribution (p. 299)
binomial probability (p. 299)
continuity correction factor (p. 300)
continuous random variable (pp. 268, 299)

cumulative area (p. 271)
discrete random variable (p. 269)
normal approximation of binomial (p. 301)
normal curve (p. 271)
normal distribution (p. 268)
percentage (p. 270)
probability (p. 270)

probability distribution, continuous variable (p. 271)
proportion (p. 270)
random variable (p. 269)
standard normal distribution (p. 271)
standard score (p. 271)
z notation (p. 292)
z-score (p. 271)

Learning Outcomes

- Understand the difference between a discrete and a continuous random variable. p. 269
- Understand the relationship between the empirical rule and the normal curve. pp. 268–269, Ex. 6.1
- Understand that a normal curve is a bell-shaped curve, with total area under the curve equal to 1. pp. 269–271 Ex. 6.44
- Understand that the normal curve is symmetrical about the mean, with an area of 0.5000 on each side of the mean. pp. 268–271, Ex. 6.39
- Be able to draw a normal curve, labeling the mean and various z-scores. p. 268
- Understand and be able to use Table 3, Areas of the Standard Normal Distribution, in Appendix B. EXP. 6.1–6.4
- Calculate probabilities for intervals defined on the standard normal distribution. Ex. 6.7, 6.13, 6.23, 6.29
- Determine z-values for corresponding intervals on the standard normal distribution. EXP. 6.5–6.7, Ex. 6.31, 6.34, 6.35, 6.110
- Compute, describe, and interpret a z-value for a data value from a normal distribution. EXP. 6.8, 6.9, Ex. 6.47
- Compute z-scores and probabilities for applications of the normal distribution. Ex. 6.47, 6.49, 6.59
- Draw, compute, and interpret z of alpha notation, $z(\alpha)$. EXP. 6.16, 6.17, Ex. 6.79, 6.83, 6.84
- Understand the key elements of a binomial experiment: x, n, p, q. Know its mean and standard deviation formulas. pp. 299–300
- Understand that the normal distribution can be used to calculate binomial probabilities, provided certain conditions are met. pp. 300–301, Ex. 6.91
- Understand and be able to use the continuity correction factor when calculating z-scores. p. 301, Ex. 6.95, 6.96
- Compute z-scores and probabilities for normal approximations to the binomial. EXP. 6.21, Ex. 6.99, 6.117

Chapter Exercises

6.108 According to Chebyshev's theorem, at least how much area is there under the standard normal distribution between $z = -2$ and $z = +2$? What is the actual area under the standard normal distribution between $z = -2$ and $z = +2$?

6.109 The middle 60% of a normally distributed population lies between what two standard scores?

6.110 Find the standard score z such that the area above the mean and below z under the normal curve is

a. 0.3962. b. 0.4846. c. 0.3712.

6.111 Find the standard score z such that the area below the mean and above z under the normal curve is

a. 0.3212. b. 0.4788. c. 0.2700.

6.112 Given that z is the standard normal variable, find the value of k such that

a. $P(|z| > 1.68) = k$. b. $P(|z| < 2.15) = k$.

6.113 Given that z is the standard normal variable, find the value of c such that:

a. $P(|z| > c) = 0.0384$. b. $P(|z| < c) = 0.8740$.

6.114 Find the following values of z:

a. $z(0.12)$ b. $z(0.28)$ c. $z(0.85)$ d. $z(0.99)$

6.115 Find the area under the normal curve that lies between the following pairs of z-values:

a. $z = -3.00$ and $z = 3.00$

b. $z(0.975)$ and $z(0.025)$

c. $z(0.10)$ and $z(0.01)$

6.116 Based on data from ACT in 2008, the average science reasoning test score was 20.8, with a standard deviation of 4.6. Assuming that the scores are normally distributed,

a. find the probability that a randomly selected student has a science reasoning ACT score of least 25.

b. find the probability that a randomly selected student has a science reasoning ACT score between 20 and 26.

c. find the probability that a randomly selected student has a science reasoning ACT score less than 16.

6.117 The 70-year long-term record for weather shows that for New York State, the annual precipitation has a mean of 39.67 inches and a standard deviation of 4.38 inches [Department of Commerce; State, Regional and National Monthly Precipitation Report]. If the annual precipitation amount has a normal distribution, what is the probability that next year the total precipitation will be:

a. more than 50.0 inches?

b. between 42.0 and 48.0 inches?

c. between 30.0 and 37.5 inches?

d. more than 35.0 inches?

e. less than 45.0 inches?

f. less than 32.0 inches?

6.118 A company that manufactures rivets used by commercial aircraft manufacturers knows that the shearing strength of (force required to break) its rivets is of major concern. They believe the shearing strength of their rivets is normally distributed, with a mean of 925 pounds and a standard deviation of 18 pounds.

a. If they are correct, what percentage of their rivets has a shearing strength greater than 900 pounds?

b. What is the upper bound for the shearing strength of the weakest 1% of the rivets?

c. If one rivet is randomly selected from all of the rivets, what is the probability that it will require a force of at least 920 pounds to break it?

d. Using the probability found in part c, what is the probability, rounded to the nearest tenth, that 3 rivets in a random sample of 10 will break at a force less than 920 pounds?

6.119 In a study of the length of time it takes to play Major League Baseball games during the early 2008 season, the variable "time of game" appeared to be normally distributed, with a mean of 2 hours 49 minutes and a standard deviation of 21 minutes.

Source: http://mlb.com/

a. Some fans describe a game as "unmanageably long" if it takes more than 3 hours. What is the probability that a randomly identified game was unmanageably long?

b. Many fans describe a game lasting less than 2 hours, 30 minutes as "quick." What is the probability that a randomly selected game was quick?

c. What are the bounds of the interquartile range for the variable time of game?

d. What are the bounds for the middle 90% of the variable time of game?

6.120 The length of the life of a certain type of refrigerator is approximately normally distributed with a mean of 4.8 years and a standard deviation of 1.3 years.

a. If this machine is guaranteed for 2 years, what is the probability that the machine you purchased will require replacement under the guarantee?

b. What period of time should the manufacturer give as a guarantee if it is willing to replace only 0.5% of the machines?

6.121 A machine is programmed to fill 10-oz containers with a cleanser. However, the variability inherent in any machine causes the actual amounts of fill to vary. The distribution is normal with a standard deviation of 0.02 oz. What must the mean amount μ be so that only 5% of the containers receive less than 10 oz?

6.122 In a large industrial complex, the maintenance department has been instructed to replace light bulbs before they burn out. It is known that the life of light bulbs is normally distributed with a mean life of 900 hours of use and a standard deviation of 75 hours. When should the light bulbs be replaced so that no more than 10% of them burn out while in use?

6.123 The grades on an examination whose mean is 525 and whose standard deviation is 80 are normally distributed.

a. Anyone who scores below 350 will be retested. What percentage does this represent?

b. The top 12% are to receive a special commendation. What score must be surpassed to receive this special commendation?

c. The interquartile range of a distribution is the difference between Q_1 and Q_3, $Q_3 - Q_1$. Find the interquartile range for the grades on this examination.

d. Find the grade such that only 1 out of 500 will score above it.

6.124 A soft drink vending machine can be regulated so that it dispenses an average of μ oz of soft drink per cup.

a. If the ounces dispensed per cup are normally distributed with a standard deviation of 0.2 oz, find the setting for μ that will allow a 6-oz glass to hold (without overflowing) the amount dispensed 99% of the time.

b. Use a computer or calculator to simulate drawing a sample of 40 cups of soft drink from the machine (set using your answer to part a).

MINITAB
Use the Calculate RANDOM DATA commands on page 283, replacing n with 40, store in with C1, mean with the value calculated in part a, and standard deviation with 0.2.
Use the HISTOGRAM commands on page 53 for the data in C1. To adjust the histogram, select Binning with cutpoint and cutpoint positions 5:6.2/0.05.

Excel
Use the Normal RANDOM NUMBER GENERATION commands on page 283, replacing n with 40, the mean with the value calculated in part a, the standard deviation with 0.2, and the output range with A1.
Use the RANDOM NUMBER GENERATION Patterned Distribution on page 291, replacing the first value with 5, the last value with 6.2, the steps with 0.05, and the output range with B1.
Use the histogram commands on pages 53–54 with column A as the input range and column B as the bin range.

TI-83/84 Plus
Use the 6:randNorm commands on page 283, replacing the mean with the value calculated in part a, the standard deviation

with 0.2, and the number of trials with 40. Store in with L1. Use the HISTOGRAM commands on page 54 for the data in L1, entering WINDOW VALUES: 5, 6.2, 0.05, −1, 10, 1, 1.

c. What percentage of your sample would have over-flowed the cup?

d. Does your sample seem to indicate the setting for μ is going to work? Explain.

FYI Repeat part b a few times. Try a different value for part a and repeat part b. Observe how many cups would overflow in each set of 40.

6.125 Suppose that x has a binomial distribution with $n = 25$ and $p = 0.3$.

a. Explain why the normal approximation is reasonable.

b. Find the mean and standard deviation of the normal distribution that is used in the approximation.

6.126 Let x be a binomial random variable for $n = 30$ and $p = 0.1$.

a. Explain why the normal approximation is not reasonable.

b. Find the function used to calculate the probability of any x from $x = 0$ to $x = 30$.

c. Use a computer or calculator to list the probability distribution.

6.127 a. Use a computer or calculator to list the bino-mial probabilities for the distribution where $n = 50$ and $p = 0.1$.

b. Use the results from part a to find $P(x \le 6)$.

c. Find the normal approximation for $P(x \le 6)$, and compare the results with those in part b.

6.128 a. Use a computer or calculator to list both the probability distribution and the cumu-lative probability distribution for the binomial probability experiment with $n = 40$ and $p = 0.4$.

b. Explain the relationship between the two dis-tributions found in part a.

c. If you could use only one of these lists when solving problems, which one would you use and why?

6.129 Consider the binomial experiment with $n = 300$ and $p = 0.2$.

a. Set up, but do not evaluate, the probability expression for 75 or fewer successes in the 300 trials.

b. Use a computer or calculator to find $P(x \leq 75)$ using the binomial probability function.

c. Use a computer or calculator to find $P(x \leq 75)$ using the normal approximation.

d. Compare the answers in parts b and c.

FYI Use the cumulative probability commands.

6.130 A test-scoring machine is known to record an incorrect grade on 5% of the exams it grades. Find, by the appropriate method, the probability that the machine records

a. exactly 3 wrong grades in a set of 5 exams.

b. no more than 3 wrong grades in a set of 5 exams.

c. no more than 3 wrong grades in a set of 15 exams.

d. no more than 3 wrong grades in a set of 150 exams.

6.131 A company asserts that 80% of the customers who purchase its special lawn mower will have no repairs during the first two years of ownership. Your personal study has shown that only 70 of the 100 lawn mowers in your sample lasted the two years without repair expenses. What is the probability of your sample outcome or less if the actual expenses-free percentage is 80%?

6.132 It is believed that 58% of married couples with children agree on methods of disciplining their children. Assuming this to be the case, what is the probability that in a random survey of 200 married couples, we would find

a. exactly 110 couples who agree?

b. fewer than 110 couples who agree?

c. more than 100 couples who agree?

6.133 It turns out that making a lot of money doesn't necessarily make you sexy. In a poll conducted by Salary.com, firefighters hosed down the competition and won the title of "sexiest job" with 16% of the votes.

Source: http://salary.com/

Suppose you randomly select 50 adults. Use the normal approximation to the binomial distribution to find the probability that within your selection,

a. more than 12 of the adults pick firefighter as the sexiest job.

b. fewer than 8 of the adults pick firefighter as the sexiest job.

c. from 7 to 14 of the adults pick firefighter as the sexiest job.

6.134 *National Coffee Drinking Trends* is "the publication" in the coffee industry. Each year it tracks consumption patterns in a wide variety of situations and categories and has done so for over five decades. A recent edition says that 39% of the total coffee drinkers age 18 years and over have purchased shade-grown coffee in the last year.

If this percentage is true for coffee drinkers at Crimson Light's coffeehouse, what is the probability that of the next 50 customers purchasing coffee at Crimson Light's,

a. more than 20 will ask for a shade-grown variety?

b. fewer than 15 will ask for a shade-grown variety?

6.135 Apparently playing video games, watching TV, and instant messaging friends isn't relaxing enough. In a poll from Yesawich, Pepperdine, Brown and Russell found that the vast majority of children say they "need" a vacation. One-third of the children polled said they helped research some aspect of their family's vacation on the Internet. If a follow-up survey of 100 of these children is taken, what is the probability that

a. less than 25% of the new sample will say they help research the family vacation on the Internet.

b. more than 40% of the new sample will say they help research the family vacation on the Internet.

6.136 [EX06-136] Infant-mortality rates are often used to assess quality of life and adequacy of health care. The rate is based on the number of deaths of infants under 1 year old in a given year per 1000 live births in the same year. Listed here are the infant-mortality rates, to the nearest integer, in eight nations throughout the world, as found in *The World Factbook*.

(continue on page 310)

[EX00-000] identifies the filename of an exercise's online dataset—available through cengagebrain.com

Nation	Infant Mortality
China	25
Germany	4
India	58
Japan	3
Mexico	22
Russia	17
S. Africa	62
United States	7

Source: http://www.cia.gov

Suppose the next 2000 births within each nation are tracked for the occurrence of infant deaths.

a. Construct a table showing the mean and standard deviation of the associated binomial distributions.

b. In the final column of the table, find the probability that at least 70 infants from the samples within each nation will become casualties that contribute to the nation's mortality rate. Show all work.

6.137 [EX06-137] A large sample of lenses was randomly selected and evaluated for a particular lens dimension. It was then compared to its specification range of nominal (0.000) ± 0.030 units. One hundred ten lenses were evaluated. Data were coded in two ways and are shown below:

−0.020	−0.043	−0.002	0.002	−0.018
−0.016	−0.051	0.024	−0.024	−0.032
−0.002	0.003	−0.014	0.022	0.000
−0.004	0.035	−0.006	−0.004	0.000
−0.014	−0.017	0.014	−0.008	−0.002
−0.006	0.032	0.034	−0.004	−0.012
−0.006	0.034	−0.032	0.012	−0.016
0.004	0.029	−0.030	0.026	−0.028
0.024	−0.016	−0.014	−0.040	−0.010
0.000	−0.020	−0.016	0.008	0.026
−0.008	−0.019	−0.018	0.012	0.014
−0.014	−0.026	−0.028	−0.032	0.010
0.010	−0.065	0.016	0.010	0.010
0.010	−0.011	0.008	0.000	0.006
0.004	−0.018	0.026	0.044	−0.006
0.014	−0.036	0.002	0.001	−0.008
0.004	−0.022	−0.012	0.014	−0.024
0.078	−0.005	0.000	0.006	−0.016
0.012	0.000	−0.010	−0.002	−0.018
0.006	0.029	−0.020	−0.024	−0.002
0.006	0.018	−0.022	−0.018	−0.014
−0.010	0.010	−0.016	−0.018	−0.016

Source: Courtesy of Bausch & Lomb [Variable not named & data coded at B&L's request]

a. Calculate the mean and standard deviation of the data.

b. Create a histogram, and comment on the pattern of variability of the data.

c. Use tests for normality and/or the empirical rule as confirmation of the normal appearance. Explain your findings.

d. Determine the observed percentage of conformance to specification. That is, what percentage of the measurements fall within the specification range of 0.000 ± 0.030 units?

6.138 Assume that the distribution of data in Exercise 6.137 was exactly normally distributed, with a mean of 0.00 and a standard deviation of 0.020.

a. Find the bounds of the middle 95% of the distribution.

b. What percent of the data is actually within the interval found in part a?

c. Using z-scores, determine the percentage of estimated conformance to specification. That is, what percentage of the measurements would be expected to fall within the specification range of 0.000 ± 0.030 units?

Chapter Practice Test

PART I: Knowing the Definitions

Answer "True" if the statement is always true. If the statement is not always true, replace the words shown in bold with words that make the statement always true.

6.1 The normal probability distribution is symmetric about **zero**.

6.2 The total area under the curve of any normal distribution is **1.0**.

6.3 The theoretical probability that a particular value of a **continuous** random variable will occur is exactly zero.

6.4 The unit of measure for the standard score is the **same as the unit of measure of the data**.

6.5 All **normal** distributions have the same general probability function and distribution.

6.6 In the notation $z(0.05)$, the number in parentheses is the measure of the area to the **left** of the z-score.

6.7 Standard normal scores have a mean of **one** and a standard deviation of **zero**.

6.8 Probability distributions of **all** continuous random variables are normally distributed.

6.9 We are able to add and subtract the areas under the curve of a continuous distribution because these areas represent probabilities of **independent** events.

6.10 The most common distribution of a continuous random variable is the **binomial** probability.

PART II: Applying the Concepts

6.11 Find the following probabilities for z, the standard normal score:
  a. $P(0 < z < 2.42)$ b. $P(z < 1.38)$
  c. $P(z < -1.27)$ d. $P(-1.35 < z < 2.72)$

6.12 Find the value of each z-score:
  a. $P(z > ?) = 0.2643$ b. $P(z < ?) = 0.17$
  c. $z(0.04)$

6.13 Use the symbolic notation $z(\alpha)$ to give the symbolic name for each z-score shown in the figure below.

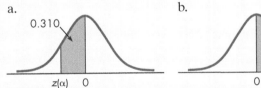

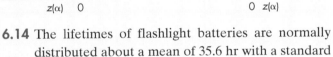

6.14 The lifetimes of flashlight batteries are normally distributed about a mean of 35.6 hr with a standard deviation of 5.4 hr. Kevin selected one of these batteries at random and tested it. What is the probability that this one battery will last less than 40.0 hr?

6.15 The lengths of time, x, spent commuting daily, one way, to college by students are believed to have a mean of 22 min with a standard deviation of 9 min. If the lengths of time spent commuting are approximately normally distributed, find the time, x, that separates the 25% who spend the most time commuting from the rest of the commuters.

6.16 Thousands of high school students take the SAT each year. The scores attained by the students in a certain city are approximately normally distributed with a mean of 490 and a standard deviation of 70. Find:
  a. the percentage of students who score between 600 and 700
  b. the percentage of students who score less than 650
  c. the third quartile
  d. the 15th percentile, P_{15}
  e. the 95th percentile, P_{95}

PART III: Understanding the Concepts

6.17 In 50 words, describe the standard normal distribution.

6.18 Describe the meaning of the symbol $z(\alpha)$.

6.19 Explain why the standard normal distribution, as computed in Table 3 in Appendix B, can be used to find probabilities for all normal distributions.

7 Sample Variability

Image copyright cosma, 2012. Used under license from Shutterstock.com

7.1 Sampling Distributions
*A distribution of repeated values for a **sample statistic***

7.2 The Sampling Distribution of Sample Means
*The theorem describing the distribution of **sample means***

7.3 Application of the Sampling Distribution of Sample Means
*The behavior of sample means is **predictable**.*

7.1 Sampling Distributions

Everyday Sampling

Samples are taken every day for many reasons. Industries monitor their products continually to be sure of their quality, agencies monitor our environment, medical professionals monitor our health; the list is limitless. Many of these samples are one-time samples, while many are samples that are repeated for ongoing monitoring.

Population Sampling

A census, a 100% survey or sampling, in the United States is done only every 10 years. It is an enormous and overwhelming job, but the information that is obtained is vital to our country's organization and structure. Issues come up and times change; information is needed and a census is impractical. This is where representative and everyday samples come in.

Census enumerator checking data on
a hand-held computer complete with
GPS capabilities to record data

Census worker doing follow-up

Thus to make inferences about a population, we need to discuss sample results a little more. A sample mean, $\bar{x}$, is obtained from a sample. Do you expect this value, $\bar{x}$, to be exactly equal to the value of the population mean, μ? Your answer should be no. We do not expect the means to be identical, but we will be satisfied with our sample results if the sample mean is "close" to the value of the population mean. Let's consider a second question: If a second sample is taken, will the second sample have a mean equal to the population mean? Equal to the first sample mean? Again, no, we do not expect the sample mean to be equal to the population mean, nor do we expect the second sample mean to be a repeat of the first one. We do, however, again expect the values to be "close." (This argument should hold for any other sample statistic and its corresponding population value.)

The next questions should already have come to mind: What is "close"? How do we determine (and measure) this closeness? Just how will **repeated sample statistics** be distributed? To answer these questions we must look at a *sampling distribution*.

> **Sampling distribution of a sample statistic** The distribution of values for a sample statistic obtained from repeated samples, all of the same size and all drawn from the same population.

THE SAMPLING ISSUE

The fundamental goal of a survey is to come up with the same results that would have been obtained had every single member of a population been interviewed. For national Gallup polls the objective is to present the opinions of a sample of people that are exactly the same opinions that would have been obtained had it been possible to interview all adult Americans in the country.

The key to reaching this goal is a fundamental principle called *equal probability of selection*, which states that if every member of a population has an equal probability of being selected in a sample, then that sample will be representative of the population. It's that straightforward.

Thus, it is Gallup's goal in selecting samples to allow every adult American an equal chance of falling into the sample. How that is done, of course, is the key to the success or failure of the process.

Source: Reprinted by permission of the Gallup Organization, http://www.gallup.com/

Let's start by investigating two different small theoretical sampling distributions.

EXAMPLE 7.1

FORMING A SAMPLING DISTRIBUTION OF MEANS AND RANGES

Consider as a population the set of single-digit even integers {0, 2, 4, 6, 8}. In addition, consider all possible samples of size 2. We will look at two different sampling distributions that might be formed: the sampling distribution of sample means and the sampling distribution of sample ranges.

First we need to list all possible samples of size 2; there are 25 possible samples:

{0, 0}	{2, 0}	{4, 0}	{6, 0}	{8, 0}
{0, 2}	{2, 2}	{4, 2}	{6, 2}	{8, 2}
{0, 4}	{2, 4}	{4, 4}	{6, 4}	{8, 4}
{0, 6}	{2, 6}	{4, 6}	{6, 6}	{8, 6}
{0, 8}	{2, 8}	{4, 8}	{6, 8}	{8, 8}

FYI Samples are drawn with replacement.

Each of these samples has a mean $\bar{x}$. These means are, respectively:

0	1	2	3	4
1	2	3	4	5
2	3	4	5	6
3	4	5	6	7
4	5	6	7	8

TABLE 7.1
Probability Distribution: Sampling Distribution of Sample Means

$\bar{x}$	$P(\bar{x})$
0	0.04
1	0.08
2	0.12
3	0.16
4	0.20
5	0.16
6	0.12
7	0.08
8	0.04

Each of these samples is equally likely, and thus each of the 25 sample means can be assigned a probability of $\frac{1}{25} = 0.04$. The **sampling distribution of sample means** is shown in Table 7.1 as a **probability distribution** and shown in Figure 7.1 as a histogram.

For the same set of all possible samples of size 2, let's find the sampling distribution of sample ranges. Each sample has a range R. The ranges are:

0	2	4	6	8
2	0	2	4	6
4	2	0	2	4
6	4	2	0	2
8	6	4	2	0

FIGURE 7.1
Histogram: Sampling Distribution of Sample Means

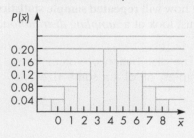

Again, each of these 25 sample ranges has a probability of 0.04. Table 7.2 shows the sampling distribution of sample ranges as a probability distribution, and Figure 7.2 shows the sampling distribution as a histogram.

TABLE 7.2
Probability Distribution: Sampling Distribution of Sample Ranges

R	$P(R)$
0	0.20
2	0.32
4	0.24
6	0.16
8	0.08

FIGURE 7.2
Histogram: Sampling Distribution of Sample Ranges

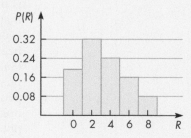

Example 7.1 is theoretical in nature and therefore expressed in probabilities. Since this population is small, it is easy to list all 25 possible samples of size 2 (a sample space) and assign probabilities. However, it is not always possible to do this.

Now, let's empirically (that is, by experimentation) investigate another sampling distribution.

E X A M P L E   7 . 2

CREATING A SAMPLING DISTRIBUTION OF SAMPLE MEANS

Let's consider a population that consists of five equally likely integers: 1, 2, 3, 4, and 5. Figure 7.3 shows a histogram representation of the population. We can observe a portion of the sampling distribution of sample means when 30 samples of size 5 are randomly selected.

Table 7.3 shows 30 samples and their means. The resulting sampling distribution, a **frequency distribution**, of sample means is shown in Figure 7.4. Notice that this distribution of sample means does not look like the population. Rather, it seems to display the characteristics of a normal distribution: it is mounded and nearly symmetrical about its mean (approximately 3.0).

FIGURE 7.3

The Population: Theoretical Probability Distribution

TABLE 7.3
30 Samples of Size 5 [TA07-03]

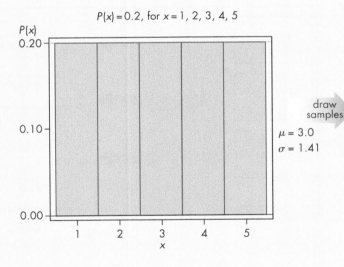

$P(x) = 0.2$, for $x = 1, 2, 3, 4, 5$

$\mu = 3.0$
$\sigma = 1.41$

No.	Sample	$\bar{x}$	No.	Sample	$\bar{x}$
1	4,5,1,4,5	3.8	16	4,5,5,3,5	4.4
2	1,1,3,5,1	2.2	17	3,3,1,2,1	2.0
3	2,5,1,5,1	2.8	18	2,1,3,2,2	2.0
4	4,3,3,1,1	2.4	19	4,3,4,2,1	2.8
5	1,2,5,2,4	2.8	20	5,3,1,4,2	3.0
6	4,2,2,5,4	3.4	21	4,4,2,2,5	3.4
7	1,4,5,5,2	3.4	22	3,3,5,3,5	3.8
8	4,5,3,1,2	3.0	23	3,4,4,2,2	3.0
9	5,3,3,3,5	3.8	24	3,3,4,5,3	3.6
10	5,2,1,1,2	2.2	25	5,1,5,2,3	3.2
11	2,1,4,1,3	2.2	26	3,3,3,5,2	3.2
12	5,4,3,1,1	2.8	27	3,4,4,4,4	3.8
13	1,3,1,5,5	3.0	28	2,3,2,4,1	2.4
14	3,4,5,1,1	2.8	29	2,1,1,2,4	2.0
15	3,1,5,3,1	2.6	30	5,3,3,2,5	3.6

draw samples →

using the 30 means →

FIGURE 7.4

Empirical Distribution of Sample Means

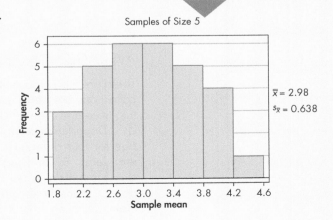

Samples of Size 5

$\bar{x} = 2.98$
$s_{\bar{x}} = 0.638$

Note: The variable for the sampling distribution is $\bar{x}$; therefore, the mean of the $\bar{x}$'s is $\bar{\bar{x}}$ and the standard deviation of $\bar{x}$ is $s_{\bar{x}}$.

The theory involved with sampling distributions that will be described in the remainder of this chapter requires *random sampling*.

> **Random sample** A sample obtained in such a way that each possible sample of fixed size *n* has an equal probability of being selected (see p.20).

Figure 7.5 shows how the sampling distribution of sample means is formed.

FIGURE 7.5 The Sampling Distribution of Sample Means

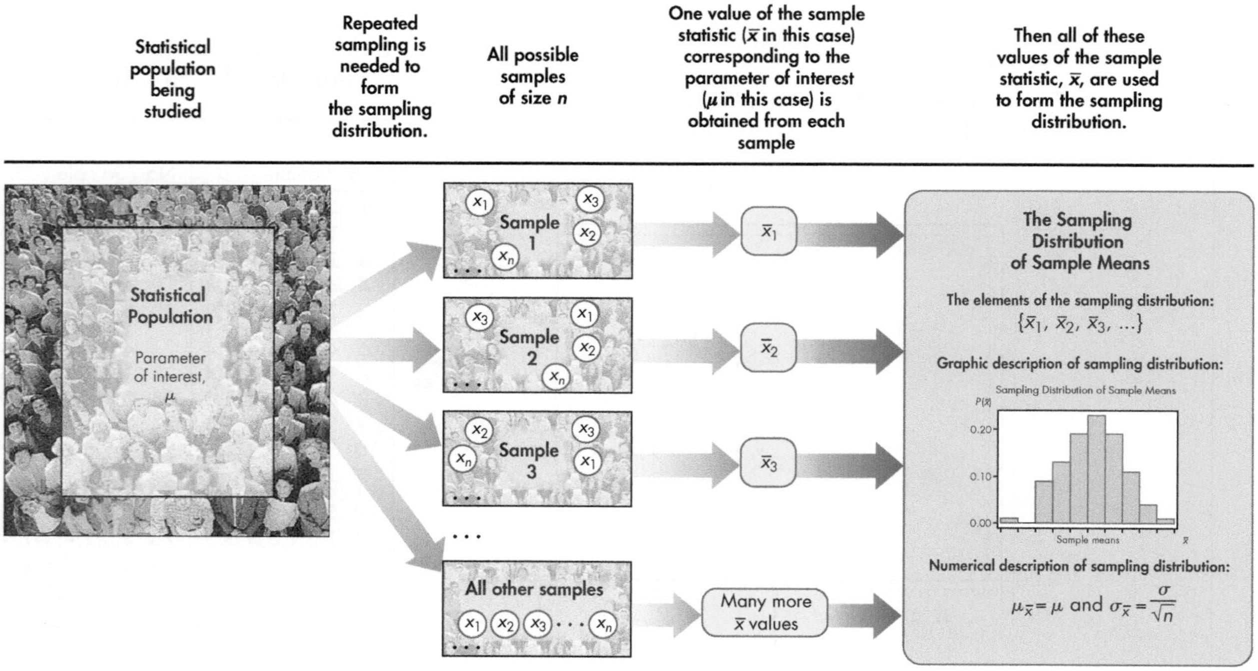

APPLIED EXAMPLE 7.3

AVERAGE AGE OF URBAN TRANSIT RAIL VEHICLES

There are many reasons for collecting data repeatedly. Not all repeated data collections are performed in order to form a sampling distribution. Consider the "Average Age of Urban Transit Rail Vehicles (Years)" statistics from the U.S. Department of Transportation that follow. The table shows the average age for four different classifications of transit rail vehicles tracked over several years. By studying the pattern of change in the average age for each class of vehicle, a person can draw conclusions about what has been happening to the fleet over several years. Chances are the people involved in maintaining each fleet can also detect when a change in policies regarding replacement of older vehicles is needed. However useful this information is, there is no sampling distribution involved here.

Average Age of Urban Transit Rail Vehicles (Years)

	1985	1990	1995	2000	2003	2007
Transit rail						
Commuter rail locomotives[a]	16.3	15.7	15.9	13.4	16.6	18.4
Commuter rail passenger coaches	19.1	17.6	21.4	16.9	20.5	18.9
Heavy-rail passenger cars	17.1	16.2	19.3	22.9	19.0	21.6
Light-rail vehicles (streetcars)	20.6	15.2	16.8	16.1	15.6	16.1

[a]Locomotives used in Amtrak intercity passenger services are not included.
Source: U.S. Department of Transportation, Federal Transit Administration

SECTION 7.1 EXERCISES

7.1 [EX07-01] Suppose a random sample of 100 ages was taken from the 2000 census distribution.

```
45  78  55  15  47  85  93  46  13  41
87  78   7   7  94  48  11  41  81  32
59   8  15  20  49  66  11  61  16  19
39  74  34   6  46   8  46  21  44  41
52  84  27  53  33  48  80   6  62  21
47  11  17   3  31  43  46  23  52  20
35  24  30  37  54  90  26  55  89   2
58  44  30  45  15  25  47  13  28  10
80  41  30  57  63  79  75   7  26   4
 2  10  21  19   5  62  32  59  40  16
```

a. How would you describe the "ages" sample data above graphically? Construct the graph.

b. Using the graph that you constructed in part a, describe the shape of the distribution of the sample data.

c. If another sample were to be collected, would you expect the same results? Explain.

7.2 a. What numerical statistics would you use to describe the "ages" sample data in Exercise 7.1? Calculate those statistics.

According to the 2000 census (2010 census is not complete), 275 million Americans have a mean age of 36.5 years and a standard deviation of 22.5 years.

b. How well do the statistics calculated in part a compare to the parameters from the 2000 census? Be specific.

c. If another sample were to be collected, would you expect the same results? Explain.

7.3 Manufacturers use random samples to test whether or not their product is meeting specifications.

These samples could be people, manufactured parts, or even samples during the manufacturing of potato chips.

a. Do you think that all random samples taken from the same population will lead to the same result?

b. What characteristic (or property) of random samples could be observed during the sampling process?

7.4 Refer to Table 7.1 in Example 7.1 (p. 314) and explain why the samples are equally likely; that is, why $P(0) = 0.04$, and why $P(2) = 0.12$.

7.5 a. What is a sampling distribution of sample means?

b. A sample of size 3 is taken from a population, and the sample mean is found. Describe how this sample mean is related to the sampling distribution of sample means.

7.6 Consider the set of odd single-digit integers {1, 3, 5, 7, 9}.

a. Make a list of all samples of size 2 that can be drawn from this set of integers. (Sample with replacement; that is, the first number is drawn, observed, and then replaced [returned to the sample set] before the next drawing.)

b. Construct the sampling distribution of sample means for samples of size 2 selected from this set.

c. Construct the sampling distributions of sample ranges for samples of size 2.

7.7 Consider the set of even single-digit integers {0, 2, 4, 6, 8}.

a. Make a list of all the possible samples of size 3 that can be drawn from this set of integers. (Sample with replacement; that is, the first number is drawn, observed, and then replaced [returned to the sample set] before the next drawing.)

b. Construct the sampling distribution of the sample medians for samples of size 3.

c. Construct the sampling distribution of the sample means for samples of size 3.

7.8 Using the phone numbers listed in your telephone directory as your population, randomly obtain 20 samples of size 3. From each phone number identified as a source, take the fourth, fifth, and sixth digits. (For example, for 245-8268, you would take the 8, the 2, and the 6 as your sample of size 3.)

a. Calculate the mean of the 20 samples.

b. Draw a histogram showing the 20 sample means. (Use classes −0.5 to 0.5, 0.5 to 1.5, 1.5 to 2.5, and so on.)

c. Describe the distribution of $\bar{x}$'s that you see in part b (shape of distribution, center, and amount of dispersion).

d. Draw 20 more samples and add the 20 new $\bar{x}$'s to the histogram in part b. Describe the distribution that seems to be developing.

7.9 Using a set of five dice, roll the dice and determine the mean number of dots showing on the five dice. Repeat the experiment until you have 25 sample means.

a. Draw a dotplot showing the distribution of the 25 sample means. (See Example 7.2, p. 315.)

b. Describe the distribution of $\bar{x}$'s in part a.

c. Repeat the experiment to obtain 25 more sample means and add these 25 $\bar{x}$'s to your dotplot. Describe the distribution of 50 means.

7.10 Considering the population of five equally likely integers in Example 7.2:

a. Verify μ and σ for the population in Example 7.2.

b. Table 7.3 lists 30 $\bar{x}$ values. Construct a grouped frequency distribution to verify the frequency distribution shown in Figure 7.4.

c. Find the mean and standard deviation of the 30 $\bar{x}$ values in Table 7.3 to verify the values for $\bar{\bar{x}}$ and $s_{\bar{x}}$. Explain the meaning of the two symbols $\bar{\bar{x}}$ and $s_{\bar{x}}$.

7.11 In reference to Applied Example 7.3 on page 316:

a. Explain why the numerical values on this table do not form a sampling distribution.

b. Explain how this repeated gathering of data differs from the idea of repeated sampling to gather information about a sampling distribution.

7.12 From the table of random numbers in Table 1 in Appendix B, construct another table showing 20 sets of 5 randomly selected single-digit integers. Find the mean of each set (the grand mean) and compare this value with the theoretical population mean, μ, using the absolute difference and the % error. Show all work.

7.13 a. Using a computer or a random number table, simulate the drawing of 100 samples, each of size 5, from the uniform probability distribution of single-digit integers, 0 to 9.

b. Find the mean for each sample.

c. Construct a histogram of the sample means. (Use integer values as class midpoints.)

d. Describe the sampling distribution shown in the histogram in part c.

MINITAB

a. Use the Integer RANDOM DATA commands on page 91, replacing generate with 100, store in with C1–C5, minimum value with 0, and maximum value with 9.

b. Choose: **Calc > Row Statistics**
    Select: **Mean**
    Enter: Input variables: **C1–C5**
                    Store result in: **C6 > OK**

c. Use the HISTOGRAM commands on page 53 for the data in C6. To adjust the histogram, select Binning with midpoint and midpoint positions 0:9/1.

Excel

a. Input 0 through 9 into column A and corresponding 0.1's into column B; then continue with:

Choose: **Data > Data Analysis >**
                Random Number Generation > OK
Enter: Number of Variables: **5**
                Number of Random Numbers: **100**
                Distribution: **Discrete**
                Value and Probability Input Range:
                (A1:B10 or select cells)
Select: **Output Range:**
Enter: **(C1 or select cell) > OK**

b. Activate cell H1.

Choose: **Insert function, f_x > Statistical >**
                AVERAGE > OK
Enter: Number1: **(C1:G1 or select cells)**
Drag: **Bottom right corner of average value box**
                down to give other averages

c. Use the HISTOGRAM commands on pages 53–54 with column H as the input range and column A as the bin range.

TI-83/84 Plus

a. Use the Integer RANDOM DATA and STO commands on page 91, replacing the Enter with 0,9,100). Repeat preceding commands four more times, storing data in L2, L3, L4, and L5, respectively.

b. Choose: **STAT > EDIT > 1: Edit**
   Highlight: **L6 (column heading)**
   Enter: **(L1 + L2 + L3 + L4 + L5)/5**

c. Choose: **2nd > STAT PLOT > 1: Plot1**
   Choose: **Window**
   Enter: **0, 9, 1, 0, 30, 5, 1**
   Choose: **Trace > > >**

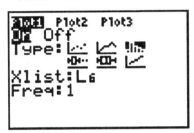

7.14 a. Using a computer or a random number table, simulate the drawing of 250 samples, each of size 18, from the uniform probability distribution of single-digit integers, 0 to 9.

   b. Find the mean for each sample.

   c. Construct a histogram of the sample means.

   d. Describe the sampling distribution shown in the histogram in part c.

7.15 a. Use a computer to draw 200 random samples, each of size 10, from the normal probability distribution with mean 100 and standard deviation 20.

   b. Find the mean for each sample.

   c. Construct a frequency histogram of the 200 sample means.

   d. Describe the sampling distribution shown in the histogram in part c.

MINITAB

a. Use the Normal RANDOM DATA commands on page 91, replacing generate with 200, store in with C1–C10, mean with 100, and standard deviation with 20.

b. Choose: **Calc > Row Statistics**
   Select: **Mean**
   Enter: Input variables: **C1–C10**
                Store result in: **C11 > OK**

c. Use the HISTOGRAM commands on page 53 for the data in C11. To adjust the histogram, select Binning with midpoint and midpoint positions 74.8:125.2/6.3.

Excel

a. Use the Normal RANDOM NUMBER GENERATION commands on page 91, replacing number of variables with 10, number of random numbers with 200, mean with 100, and standard deviation with 20.

b. Activate cell K1.

   Choose: **Insert function, f$_x$ > Statistical > AVERAGE > OK**
   Enter: Number1: **(A1:J1 or select cells)**
   Drag: **Bottom right corner of average value box down to give other averages**

c. Use the RANDOM NUMBER GENERATION Patterned Distribution commands in Exercise 6.71(a) on page 290, replacing the first value with 74.8, the last value with 125.2, the steps with 6.3, and the output range with L1. Use the HISTOGRAM commands on pages 53–54 with column K as the input range and column L as the bin range.

7.16 a. Use a computer to draw 500 random samples, each of size 20, from the normal probability distribution with mean 80 and standard deviation 15.

   b. Find the mean for each sample.

   c. Construct a frequency histogram of the 500 sample means.

   d. Describe the sampling distribution shown in the histogram in part c, including the mean and standard deviation.

7.2 The Sampling Distribution of Sample Means

On the preceding pages we discussed the sampling distributions of two statistics: sample means and sample ranges. Many others could be discussed; however, the only sampling distribution of concern to us at this time is the sampling distribution of sample means.

Sampling distribution of sample means (SDSM) If all possible random samples, each of size n, are taken from any population with mean μ and standard deviation σ, then the sampling distribution of sample means will have the following:

1. A mean $\mu_{\bar{x}}$ equal to μ
2. A standard deviation $\sigma_{\bar{x}}$ equal to $\frac{\sigma}{\sqrt{n}}$

Furthermore, if the sampled population has a normal distribution, then the sampling distribution of $\bar{x}$ will also be normal for samples of all sizes.

This is a very interesting two-part statement. The first part tells us about the relationship between the population mean and standard deviation, and the sampling distribution mean and standard deviation for all sampling distributions of sample means. The standard deviation of the sampling distribution is denoted by $\sigma_{\bar{x}}$ and given a specific name to avoid confusion with the population standard deviation, σ.

Standard error of the mean $(\sigma_{\bar{x}})$ The standard deviation of the sampling distribution of sample means.

The second part indicates that this information is not always useful. Stated differently, it says that the mean value of only a few observations will be normally distributed when samples are drawn from a normally distributed population, but it will not be normally distributed when the sampled population is uniform, skewed, or otherwise not normal. However, the *central limit theorem* gives us some additional and very important information about the sampling distribution of sample means.

Central limit theorem (CLT) The sampling distribution of sample means will more closely resemble the normal distribution as the sample size increases.

If the sampled distribution is normal, then the sampling distribution of sample means (SDSM) is normal, as stated previously, and the central limit theorem (CLT) is not needed. But, if the sampled population is not normal, the CLT tells us that the sampling distribution will still be approximately normally distributed under the right conditions. If the sampled population distribution is nearly normal, the $\bar{x}$ distribution is approximately normal for fairly small n (possibly as small as 15). When the sampled population distribution lacks symmetry, n may have to be quite large (maybe 50 or more) before the normal distribution provides a satisfactory approximation.

By combining the preceding information, we can describe the sampling distribution of $\bar{x}$ completely: (1) the location of the center (mean), (2) a measure of spread indicating how widely the distribution is dispersed (standard error of the mean), and (3) an indication of how it is distributed.

1. $\mu_{\bar{x}} = \mu$; the mean of the sampling distribution $(\mu_{\bar{x}})$ is equal to the mean of the population (μ).
2. $\sigma_{\bar{x}} = \frac{\sigma}{\sqrt{n}}$; the standard error of the mean $(\sigma_{\bar{x}})$ is equal to the standard deviation of the population (σ) divided by the square root of the sample size, n.
3. The distribution of sample means is normal when the parent population is normally distributed, and the CLT tells us that the distribution of sample means becomes approximately normal (regardless of the shape of the parent population) when the sample size is large enough.

Note: The n referred to is the size of each sample in the sampling distribution. (The number of repeated samples used in an empirical situation has no effect on the standard error.)

We do not show the proof for the preceding three facts in this text; however, their validity will be demonstrated by examining two examples. For the first example, let's consider a population for which we can construct the theoretical sampling distribution of all possible samples.

EXAMPLE 7.4

CONSTRUCTING A SAMPLING DISTRIBUTION OF SAMPLE MEANS

Let's consider all possible samples of size 2 that could be drawn from a population that contains the three numbers 2, 4, and 6. First let's look at the population itself. Construct a histogram to picture its distribution, Figure 7.6; calculate the mean, μ, and the standard deviation, σ, Table 7.4. (Remember: We must use the techniques from Chapter 5 for discrete probability distributions.)

FIGURE 7.6
Population

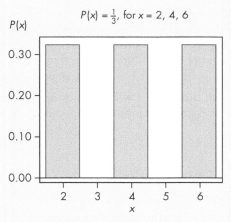

$P(x) = \frac{1}{3}$, for $x = 2, 4, 6$

TABLE 7.4
Extensions Table for x

x	$P(x)$	$xP(x)$	$x^2P(x)$
2	$\frac{1}{3}$	$\frac{2}{3}$	$\frac{4}{3}$
4	$\frac{1}{3}$	$\frac{4}{3}$	$\frac{16}{3}$
6	$\frac{1}{3}$	$\frac{6}{3}$	$\frac{36}{3}$
Σ	$\frac{3}{3}$ ⓒⓚ	$\frac{12}{3}$	$\frac{56}{3}$
	1.0	4.0	18.6$\overline{6}$

$\mu = \mathbf{4.0}$

$\sigma = \sqrt{18.6\overline{6} - (4.0)^2} = \sqrt{2.6\overline{6}} = \mathbf{1.63}$

Table 7.5 lists all the possible samples of size 2 that can be drawn from this population. (One number is drawn, observed, and then returned to the population before the second number is drawn.) Table 7.5 also lists the means of these samples. The sample means are then collected to form the sampling distribution. The distribution for these means and the extensions are given in Table 7.6 (p. 322, along with the calculation of the mean and the standard error of the mean for the sampling distribution. The histogram for the sampling distribution of sample means is shown in Figure 7.7 (p. 322).

TABLE 7.5
All Nine Possible Samples of Size 2

Sample	$\overline{x}$	Sample	$\overline{x}$	Sample	$\overline{x}$
2,2	2	4,2	3	6,2	4
2,4	3	4,4	4	6,4	5
2,6	4	4,6	5	6,6	6

TABLE 7.6
Extensions Table for $\bar{x}$

$\bar{x}$	$P(\bar{x})$	$\bar{x}P(\bar{x})$	$\bar{x}^2P(\bar{x})$
2	$\frac{1}{9}$	$\frac{2}{9}$	$\frac{4}{9}$
3	$\frac{2}{9}$	$\frac{6}{9}$	$\frac{18}{9}$
4	$\frac{3}{9}$	$\frac{12}{9}$	$\frac{48}{9}$
5	$\frac{2}{9}$	$\frac{10}{9}$	$\frac{50}{9}$
6	$\frac{1}{9}$	$\frac{6}{9}$	$\frac{36}{9}$
Σ	$\frac{9}{9}$ (ck)	$\frac{36}{9}$	$\frac{156}{9}$
	1.0	4.0	17.3$\overline{3}$

$\mu_{\bar{x}} = \textbf{4.0}$

$\sigma_{\bar{x}} = \sqrt{17.3\overline{3} - (4.0)^2} = \sqrt{1.3\overline{3}} = \textbf{1.15}$

FIGURE 7.7
Sampling Distribution of Sample Means

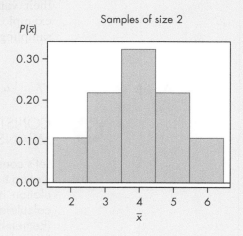

Let's now check the truth of the three facts about the sampling distribution of sample means:

1. The mean $\mu_{\bar{x}}$ of the sampling distribution will equal the mean μ of the population: both μ and $\mu_{\bar{x}}$ have the value **4.0**.

2. The standard error of the mean $\sigma_{\bar{x}}$ for the sampling distribution will equal the standard deviation σ of the population divided by the square root of the sample size, n: $\sigma_{\bar{x}} = \textbf{1.15}$ and $\sigma = 1.63$, $n = 2$, $\frac{\sigma}{\sqrt{n}} = \frac{1.63}{\sqrt{2}} = \textbf{1.15}$; they are equal: $\sigma_{\bar{x}} = \frac{\sigma}{\sqrt{n}}$.

3. The distribution will become approximately normally distributed: the histogram in Figure 7.7 very strongly suggests normality.

Example 7.4, a theoretical situation, suggests that all three facts appear to hold true. Do these three facts hold when actual data are collected? Let's look back at Example 7.2 (p. 315) and see if all three facts are supported by the empirical sampling distribution there.

First, let's look at the population—the theoretical probability distribution from which the samples in Example 7.2 were taken. Figure 7.3 is a histogram showing the probability distribution for randomly selected data from the population of equally likely integers 1, 2, 3, 4, 5. The population mean μ equals 3.0. The population standard deviation σ is $\sqrt{2}$, or 1.41. The population has a uniform distribution.

Now let's look at the empirical distribution of the 30 sample means found in Example 7.2. From the 30 values of $\bar{x}$ in Table 7.3, the observed mean of the $\bar{x}$'s, $\bar{\bar{x}}$, is 2.98 and the observed standard error of the mean, $s_{\bar{x}}$, is 0.638. The histogram of the sampling distribution in Figure 7.4 appears to be mounded, approximately symmetrical, and centered near the value 3.0.

Now let's check the truth of the three specific properties:

1. $\mu_{\bar{x}}$ and μ will be equal. The mean of the population μ is 3.0, and the observed sampling distribution mean $\bar{\bar{x}}$ is 2.98; they are very close in value.

2. $\sigma_{\bar{x}}$ will equal $\frac{\sigma}{\sqrt{n}}$. $\sigma = 1.41$ and $n = 5$; therefore, $\frac{\sigma}{\sqrt{n}} = \frac{1.41}{\sqrt{5}} = \mathbf{0.632}$, and $s_{\bar{x}} = \mathbf{0.638}$; they are very close in value. (Remember that we have taken only 30 samples, not all possible samples, of size 5.)

3. The **sampling distribution** of $\bar{x}$ will be approximately normally distributed. Even though the population has a rectangular distribution, the histogram in Figure 7.4 suggests that the $\bar{x}$ distribution has some of the properties of normality (mounded, symmetrical).

Although Examples 7.2 and 7.4 do not constitute a proof, the evidence seems to strongly suggest that both statements, the sampling distribution of sample means and the CLT, are true.

Having taken a look at these two specific examples, let's now look at four graphic illustrations that present the sampling distribution information and the CLT in a slightly different form. Each of these illustrations has four distributions. The first graph shows the distribution of the parent population, the distribution of the individual x values. Each of the other three graphs shows a sampling distribution of sample means, $\bar{x}$'s, using three different sample sizes.

In Figure 7.8 we have a uniform distribution, much like Figure 7.3 for the integer illustration, and the resulting distributions of sample means for samples of sizes 2, 5, and 30.

(Continued)

trials using the normal distribution. The definition of statistical independence also made its debut along with many dice and other games. de Moivre proved that the central limit theorem holds for numbers resulting from games of chance. With the use of mathematics, he also successfully predicted the date of his own death.

FIGURE 7.8

Uniform Distribution

(a) Population

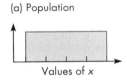

Values of x

(b) Sampling distribution of $\bar{x}$ when $n = 2$

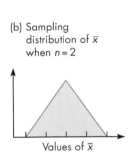

Values of $\bar{x}$

(c) Sampling distribution of $\bar{x}$ when $n = 5$

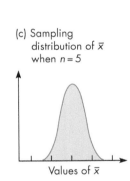

Values of $\bar{x}$

(d) Sampling distribution of $\bar{x}$ when $n = 30$

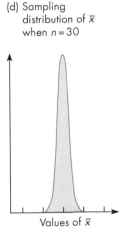

Values of $\bar{x}$

Figure 7.9 shows a U-shaped population and the three sampling distributions.

FIGURE 7.9

U-Shaped Distribution

(a) Population

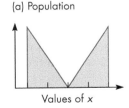

Values of x

(b) Sampling distribution of $\bar{x}$ when $n = 2$

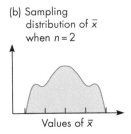

Values of $\bar{x}$

(c) Sampling distribution of $\bar{x}$ when $n = 5$

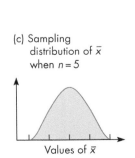

Values of $\bar{x}$

(d) Sampling distribution of $\bar{x}$ when $n = 30$

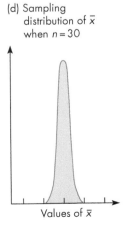

Values of $\bar{x}$

Figure 7.10 shows a J-shaped population and the three sampling distributions.

FIGURE 7.10
J-Shaped Distribution

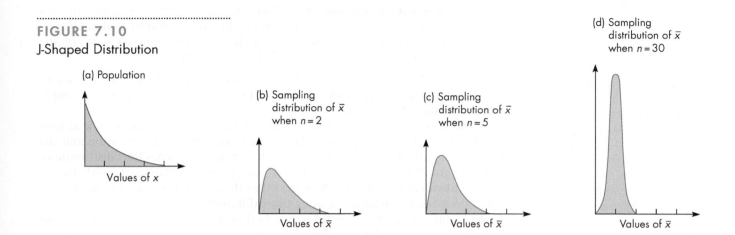

FIGURE 7.10
J-Shaped Distribution

All three nonnormal population distributions seem to verify the CLT; the sampling distributions of sample means appear to be approximately normal for all three when samples of size 30 are used. Now consider Figure 7.11, which shows a normally distributed population and the three sampling distributions. With the normal population, the sampling distributions of the sample means for all sample sizes appear to be normal. Thus, you have seen an amazing phenomenon: No matter what the shape of a population, the sampling distribution of sample means either is normal or becomes approximately normal when n becomes sufficiently large.

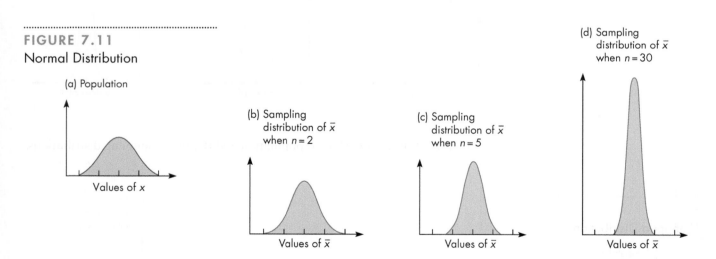

FIGURE 7.11
Normal Distribution

You should notice one other point: The sample mean becomes less variable as the sample size increases. Notice that as n increases from 2 to 30, all the distributions become narrower and taller.

SECTION 7.2 EXERCISES

7.17 Skillbuilder Applet Exercise simulates taking samples of size 4 from an approximately normal population, where $\mu = 65.15$ and $\sigma = 2.754$.

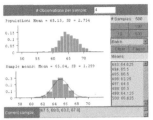

a. Click "1" for "# Samples." Note the four data values and their mean. Change "slow" to "batch" and take at least 1000 samples using the "500" for "# Samples."

b. What is the mean for the 1001 sample means? How close is it to the population mean, μ?

c. Compare the sample standard deviation to the population standard deviation, σ. What is happening to the sample standard deviation? Compare it with $\sigma/\sqrt{n}$, which is $2.754/\sqrt{4}$.

d. Does the histogram of sample means have an approximately normal shape?

e. Relate your findings to the SDSM.

7.18 Skillbuilder Applet Exercise simulates sampling from a skewed population, where $\mu = 6.029$ and $\sigma = 10.79$.

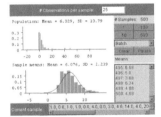
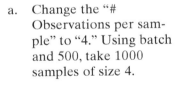

a. Change the "# Observations per sample" to "4." Using batch and 500, take 1000 samples of size 4.

b. Compare the mean and standard deviation for the sample means with μ and σ. Compare the sample standard deviation with $\sigma/\sqrt{n}$, which is $10.79/\sqrt{4}$. Does the histogram have an approximately normal shape? If not, what shape is it?

c. Using the "clear" button each time, repeat the directions in parts a and b for samples of size 25, 100, and 1000. Table your findings for each sample size.

d. Relate your findings to the SDSM and the CLT.

7.19 a. What is the total measure of the area for any probability distribution?

b. Justify the statement "$\bar{x}$ becomes less variable as n increases."

7.20 If a population has a standard deviation σ of 25 units, what is the standard error of the mean if samples of size 16 are selected? Samples of size 36? Samples of size 100?

7.21 A certain population has a mean of 500 and a standard deviation of 30. Many samples of size 36 are randomly selected and the means calculated.

a. What value would you expect to find for the mean of all these sample means?

b. What value would you expect to find for the standard deviation of all these sample means?

c. What shape would you expect the distribution of all these sample means to have?

7.22 According to *Nielsen's Television Audience Report*, in 2009 the average American home had 2.86 television sets (more than the average number of people per household, at 2.5 people). If the standard deviation for the number of televisions in a U.S. household is 1.2 and a random sample of 80 American households is selected, the mean of this sample belongs to a sampling distribution.

a. What is the shape of this sampling distribution?

b. What is the mean of this sampling distribution?

c. What is the standard deviation of this sampling distribution?

7.23 The September 21, 2006, *USA Today* article "Average home has more TVs than people" stated that Americans watch an average of 4.58 hours of television per person per day.

Source: Nielsen Media Research

If the standard deviation for the number of hours of television watched per day is 2.1 and a random sample of 250 Americans is selected, the mean of this sample belongs to a sampling distribution.

a. What is the shape of this sampling distribution?

b. What is the mean of this sampling distribution?

c. What is the standard deviation of this sampling distribution?

7.24 According to *The World Factbook*, 2009, the total fertility rate (estimated mean number of children born per woman) for Uganda is 6.77. Suppose that the standard deviation of the total fertility rate is 2.6. The mean number of children for a sample of 200 randomly

(continue on page 326)

Skillbuilder Applets are available online through cengagebrain.com.

selected women is one value of many that form the sampling distribution of sample means.

a. What is the mean value for this sampling distribution?

b. What is the standard deviation of this sampling distribution?

c. Describe the shape of this sampling distribution.

7.25 The American Meat Institute published the 2007 report "U.S. Meat and Poultry Production & Consumption: An Overview." The 2007 Fact Sheet lists the annual consumption of chicken as 86.5 pounds per person. Suppose the standard deviation for the consumption of chicken per person is 29.3 pounds. The mean weight of chicken consumed for a sample of 150 randomly selected people is one value of many that form the sampling distribution of sample means.

a. What is the mean value for this sampling distribution?

b. What is the standard deviation of this sampling distribution?

c. Describe the shape of this sampling distribution.

7.26 A researcher wants to take a simple random sample of about 5% of the student body at each of two schools. The university has approximately 20,000 students, and the college has about 5000 students. Identify each of the following as true or false and justify your answer.

a. The sampling variability is the same for both schools.

b. The sampling variability for the university is higher than that for the college.

c. The sampling variability for the university is lower than that for the college.

d. No conclusion about the sampling variability can be stated without knowing the results of the study.

7.27 a. Use a computer to randomly select 100 samples of size 6 from a normal population with mean $\mu = 20$ and standard deviation $\sigma = 4.5$.

b. Find mean $\bar{x}$ for each of the 100 samples.

c. Using the 100 sample means, construct a histogram, find mean $\bar{\bar{x}}$, and find the standard deviation $s_{\bar{x}}$.

MINITAB

a. Use the Normal RANDOM DATA commands on page 91, replacing generate with 100, store in with C1–C6, mean with 20, and standard deviation with 4.5.

b. Use the ROW STATISTICS commands on page 318, replacing input variables with C1–C6 and store result in with C7.

c. Use the HISTOGRAM commands on page 53 for the data in C7. To adjust the histogram, select Binning with midpoint and midpoint positions 12.8:27.2/1.8. Use the MEAN and STANDARD DEVIATION commands on pages 65 and 79 for the data in C7.

Excel

a. Use the Normal RANDOM NUMBER GENERATION commands on page 91, replacing number of variables with 6, number of random numbers with 100, mean with 20, and standard deviation with 4.5.

b. Activate cell G1.

Choose:	Insert function, f_x > Statistical > AVERAGE > OK
Enter:	Number1: (A1:F1 or select cells)
Drag:	Bottom right corner of average value box down to give other averages

c. Use the RANDOM NUMBER GENERATION Patterned Distribution commands in Exercise 6.71(a) on page 291 replacing the first value with 12.8, the last value with 27.2, the steps with 1.8, and the output range with H1. Use the HISTOGRAM commands on pages 53–54 with column G as the input range and column H as the bin range. Use the MEAN and STANDARD DEVIATION commands on pages 65 and 79 for the data in column G.

TI-83/84 Plus

a. Use the Normal RANDOM DATA and STO commands on page 91, replacing Enter with 20, 4.5,100). Repeat the preceding commands five more times, storing data in L2, L3, L4, L5, and L6, respectively.

b. Enter:	(L1 + L2 + L3 + L4 + L5 + L6)/6
Choose:	STO → L7 (use ALPHA key for the "L" or use "MEAN")

c. Choose:	2 nd > STAT PLOT > 1: Plot1
Choose:	Window
Enter:	12.8, 27.2, 1.8, 0, 40, 5, 1
Choose:	Trace > > >
Choose:	STAT > CALC > 1:1-VAR STATS > 2nd > LIST
Select:	L7

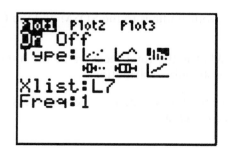

d.   Compare the results of part c with the three statements made in the SDSM.

7.28   a.   Use a computer to randomly select 200 samples of size 24 from a normal population with mean $\mu = 20$ and standard deviation $\sigma = 4.5$.

b.   Find mean $\bar{x}$ for each of the 200 samples.

c.   Using the 200 sample means, construct a histogram, find mean $\bar{\bar{x}}$, and find the standard deviation $s_{\bar{x}}$.

d.   Compare the results of part c with the three statements made for the SDSM and CLT on page 320.

e.   Compare these results with the results obtained in Exercise 7.27. Specifically, what effect did the increase in sample size from 6 to 24 have? What effect did the increase from 100 samples to 200 samples have?

FYI If you use a computer, see Exercise 7.27.

7.3 Application of the Sampling Distribution of Sample Means

When the sampling distribution of sample means is normally distributed, or approximately normally distributed, we will be able to answer probability questions with the aid of the standard normal distribution (Table 3 of Appendix B).

E X A M P L E   7 . 5

CONVERTING $\bar{x}$ INFORMATION INTO z-SCORES

Consider a normal population with $\mu = 100$ and $\sigma = 20$. If a random sample of size 16 is selected, what is the probability that this sample will have a mean value between 90 and 110? That is, what is $P(90 < \bar{x} < 110)$?

Solution

Since the population is normally distributed, the sampling distribution of $\bar{x}$'s is normally distributed. To determine probabilities associated with a normal distribution, we will need to convert the statement $P(90 < \bar{x} < 110)$ to a probability statement involving the **z-score**. This will allow us to use Table 3 in Appendix B, the standard normal distribution table. The sampling distribution is shown in the figure, where the shaded area represents $P(90 < \bar{x} < 110)$.

The formula for finding the z-score corresponding to a known value of $\bar{x}$ is

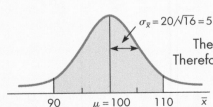

$$z = \frac{\bar{x} - \mu_{\bar{x}}}{\sigma_{\bar{x}}} \qquad (7.1)$$

The mean and standard error of the mean are $\mu_{\bar{x}} = \mu$ and $\sigma_{\bar{x}} = \frac{\sigma}{\sqrt{n}}$. Therefore, we will rewrite formula (7.1) in terms of μ, σ, and n:

$$z = \frac{\bar{x} - \mu}{\sigma/\sqrt{n}} \qquad (7.2)$$

Returning to the example and applying formula (7.2), we find:

$$\text{z-score for } \bar{x} = 90: \quad z = \frac{\bar{x} - \mu}{\sigma/\sqrt{n}} = \frac{90 - 100}{20/\sqrt{16}} = \frac{-10}{5} = \mathbf{-2.00}$$

$$\text{z-score for } \bar{x} = 110: \quad z = \frac{\bar{x} - \mu}{\sigma/\sqrt{n}} = \frac{110 - 100}{20/\sqrt{16}} = \frac{10}{5} = \mathbf{2.00}$$

Therefore,

$$P(90 < \bar{x} < 110) = P(-2.00 < z < 2.00) = 0.9773 - 0.0228 = \mathbf{0.9545}$$

Before we look at additional examples, let's consider what is implied by $\sigma_{\bar{x}} = \frac{\sigma}{\sqrt{n}}$. To demonstrate, let's suppose that $\sigma = 20$ and let's use a sampling distribution of samples of size 4. Now $\sigma_{\bar{x}}$ is $20/\sqrt{4}$, or 10, and approximately 95% (0.9545) of all such sample means should be within the interval from 20 below to 20 above the population mean (within 2 standard deviations of the population mean). However, if the sample size is increased to 16, $\sigma_{\bar{x}}$ becomes $20/\sqrt{16} = 5$ and approximately 95% of the sampling distribution should be within 10 units of the mean, and so on. As the sample size increases, the size of $\sigma_{\bar{x}}$ becomes smaller and the distribution of sample means becomes much narrower. Figure 7.12 illustrates what happens to the distribution of $\bar{x}$'s as the size of the individual samples increases.

FIGURE 7.12
Distributions of Sample Means

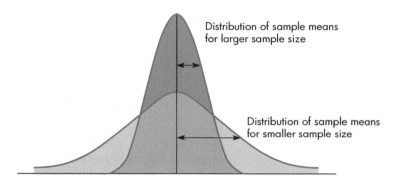

Distribution of sample means for larger sample size

Distribution of sample means for smaller sample size

Recall that the area (probability) under the normal curve is always exactly 1. So as the width of the curve narrows, the height has to increase to maintain this area.

EXAMPLE 7.6

CALCULATING PROBABILITIES FOR THE MEAN HEIGHT OF KINDERGARTEN CHILDREN

Kindergarten children have heights that are approximately normally distributed about a mean of 39 inches and a standard deviation of 2 inches. A random sample of size 25 is taken, and the mean $\bar{x}$ is calculated. What is the probability that this mean value will be between 38.5 and 40.0 inches?

Solution

We want to find $P(38.5 < \bar{x} < 40.0)$. The values of $\bar{x}$, 38.5 and 40.0, must be converted to z-scores (necessary for use of Table 3 in Appendix B) using $z = \frac{\bar{x} - \mu}{\sigma/\sqrt{n}}$:

$$\bar{x} = 38.5: \quad z = \frac{\bar{x} - \mu}{\sigma/\sqrt{n}} = \frac{38.5 - 39.0}{2/\sqrt{25}} = \frac{-0.5}{0.4} = \mathbf{-1.25}$$

Video tutorial available—logon and learn more at cengagebrain.com

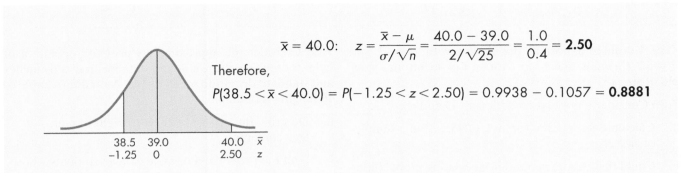

$$\bar{x} = 40.0: \quad z = \frac{\bar{x} - \mu}{\sigma/\sqrt{n}} = \frac{40.0 - 39.0}{2/\sqrt{25}} = \frac{1.0}{0.4} = \mathbf{2.50}$$

Therefore,

$$P(38.5 < \bar{x} < 40.0) = P(-1.25 < z < 2.50) = 0.9938 - 0.1057 = \mathbf{0.8881}$$

E X A M P L E   7 . 7

CALCULATING MEAN HEIGHT LIMITS FOR THE MIDDLE 90% OF KINDERGARTEN CHILDREN

Use the heights of kindergarten children given in Example 7.6. Within what limits does the middle 90% of the sampling distribution of sample means for samples of size 100 fall?

Solution

The two tools we have to work with are formula (7.2) and Table 3 in Appendix B. The formula relates the key values of the population to the key values of the sampling distribution, and Table 3 relates areas to z-scores. First, using Table 3, we find that the middle 0.9000 is bounded by $z = \pm 1.65$.

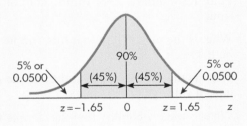

FYI Remember: If value is exactly halfway, use the larger z.

z	...	0.04		0.05	...
$\vdots$					$\vdots$
-1.6	...	0.0505	**0.0500**	0.0495	...
$\vdots$					$\vdots$

Second, we use formula (7.2), $z = \dfrac{\bar{x} - \mu}{\sigma/\sqrt{n}}$:

$z = -1.65: \quad -1.65 = \dfrac{\bar{x} - 39.0}{2/\sqrt{100}}$

$\bar{x} - 39 = (-1.65)(0.2)$

$\bar{x} = 39 - 0.33$

$= \mathbf{38.67}$

$z = 1.65: \quad 1.65 = \dfrac{\bar{x} - 39.0}{2/\sqrt{100}}$

$\bar{x} - 39 = (1.65)(0.2)$

$\bar{x} = 39 + 0.33$

$= \mathbf{39.33}$

Thus,

$$P(38.67 < \bar{x} < 39.33) = 0.90$$

Therefore, 38.67 inches and 39.33 inches are the limits that capture the middle 90% of the sample means.

SECTION 7.3 EXERCISES

7.29 Consider a normal population with $\mu = 43$ and $\sigma = 5.2$. Calculate the z-score for an $\bar{x}$ of 46.5 from a sample of size 16.

7.30 Consider a population with $\mu = 43$ and $\sigma = 5.2$.

a. Calculate the z-score for an $\bar{x}$ of 46.5 from a sample of size 35.

b. Could this z-score be used in calculating probabilities using Table 3 in Appendix B? Why or why not?

7.31 In Example 7.5, explain how the 0.9773 and the 0.0228 were obtained and what they are used for.

7.32 What is the probability that the sample of kindergarten children in Example 7.6 has a mean height of less than 39.75 inches?

7.33 A random sample of size 36 is to be selected from a population that has a mean $\mu = 50$ and a standard deviation σ of 10.

a. This sample of 36 has a mean value of $\bar{x}$, which belongs to a sampling distribution. Find the shape of this sampling distribution.

b. Find the mean of this sampling distribution.

c. Find the standard error of this sampling distribution.

d. What is the probability that this sample mean will be between 45 and 55?

e. What is the probability that the sample mean will have a value greater than 48?

f. What is the probability that the sample mean will be within 3 units of the mean?

7.34 The local bakery bakes more than a thousand 1-pound loaves of bread daily, and the weights of these loaves vary. The mean weight is 1 lb and 1 oz, or 482 grams. Assume that the standard deviation of the weights is 18 grams and that a sample of 40 loaves is to be randomly selected.

a. This sample of 40 has a mean value of $\bar{x}$, which belongs to a sampling distribution. Find the shape of this sampling distribution.

b. Find the mean of this sampling distribution.

c. Find the standard error of this sampling distribution.

d. What is the probability that this sample mean will be between 475 and 495?

e. What is the probability that the sample mean will have a value less than 478?

f. What is the probability that the sample mean will be within 5 grams of the mean?

7.35 Consider the approximately normal population of heights of male college students with mean $\mu = 69$ inches and standard deviation $\sigma = 4$ inches. A random sample of 16 heights is obtained.

a. Describe the distribution of x, height of male college students.

b. Find the proportion of male college students whose height is greater than 70 inches.

c. Describe the distribution of $\bar{x}$, the mean of samples of size 16.

d. Find the mean and standard error of the $\bar{x}$ distribution.

e. Find $P(\bar{x} > 70)$.

f. Find $P(\bar{x} < 67)$.

7.36 The amount of fill (weight of contents) put into a glass jar of spaghetti sauce is normally distributed with mean $\mu = 850$ grams and standard deviation $\sigma = 8$ grams.

a. Describe the distribution of x, the amount of fill per jar.

b. Find the probability that one jar selected at random contains between 848 and 855 grams.

c. Describe the distribution of $\bar{x}$, the mean weight for a sample of 24 such jars of sauce.

d. Find the probability that a random sample of 24 jars has a mean weight between 848 and 855 grams.

7.37 The heights of the kindergarten children mentioned in Example 7.6 (p. 328) are approximately normally distributed with $\mu = 39$ and $\sigma = 2$.

a. If an individual kindergarten child is selected at random, what is the probability that he or she has a height between 38 and 40 inches?

b. A classroom of 30 of these children is used as a sample. What is the probability that the class mean $\bar{x}$ is between 38 and 40 inches?

c. If an individual kindergarten child is selected at random, what is the probability that he or she is taller than 40 inches?

d. A classroom of 30 of these kindergarten children is used as a sample. What is the probability that the class mean $\bar{x}$ is greater than 40 inches?

7.38 Salaries for various positions can vary significantly, depending on whether or not the company is in the public or private sector. The U.S. Department of Labor posted the 2007 average salary for human resource managers employed by the federal government as $76,503. Assume that annual salaries for this type of job are normally distributed and have a standard deviation of $8850.

a. What is the probability that a randomly selected human resource manager received over $100,000 in 2007?

b. A sample of 20 human resource managers is taken and annual salaries are reported. What is the probability that the sample mean annual salary falls between $70,000 and $80,000?

7.39 Based on data from 1996 through 2006 from the Western Regional Climate Center, the average speed of winds in Honolulu, Hawaii, equals 10.6 miles per hour. Assume that wind speeds are approximately normally distributed with a standard deviation of 3.5 miles per hour.

a. Find the probability that the wind speed in any one reading will exceed 13.5 miles per hour.

b. Find the probability that the mean of a random sample of 9 readings exceeds 13.5 miles per hour.

c. Do you think the assumption of normality is reasonable? Explain.

d. What effect do you think the assumption of normality had on the answers to parts a and b? Explain.

7.40 TIMSS 2007 (Trends in International Mathematics and Science Study) focused on the mathematics and science achievement of eighth-grade students throughout the world. A total of 8 countries (including the United States) participated in the study. The mean math exam score for U.S. students was 509 with a standard deviation of 88.

Source: http://nces.ed.gov/

Assuming the scores are normally distributed, find the following for a sample of 150 students.

a. Find the probability that the mean TIMSS score for a randomly selected group of eighth-grade students would be between 495 and 515.

b. Find the probability that the mean TIMSS score for a randomly selected group of eighth-grade students would be less than 520.

c. Do you think the assumption of normality is reasonable? Explain.

7.41 According to the June 2004 *Readers' Digest* article "Only in America," the average amount that a 17-year-old spends on his or her high school prom is $638. Assume that the amounts spent are normally distributed with a standard deviation of $175.

a. Find the probability that the mean cost to attend a high school prom for 36 randomly selected 17-year-olds is between $550 and $700.

b. Find the probability that the mean cost to attend a high school prom for 36 randomly selected 17-year-olds is greater than $750.

c. Do you think the assumption of normality is reasonable? Explain.

7.42 The Bureau of Labor Statistics provides compensation information on and services for various positions. As of May 2008, the national average salary for an RN (registered nurse) was $65,130. Suppose the standard deviation is $9385. Find the following for the mean of a random sample of 100 such nurses.

a. The probability that the mean of the sample is less than $62,500

b. The probability that the sample mean is between $64,000 and $67,500

c. The probability that the sample mean is greater than $66,000

d. Explain why the assumption of normality about the distribution of wages was not involved in the solutions parts to a, b, and c.

7.43 Referring to Example 7.6 (p. 328), what height would bound the lower 25% of all samples of size 25?

7.44 A popular flashlight that uses two D-size batteries was selected, and several of the same models were purchased to test the "continuous-use life" of D batteries. As fresh batteries were installed, each flashlight was turned on and the time noted. When the flashlight no longer produced light, the time was again noted. The resulting "life" data from Rayovac batteries had a mean of 21.0 hours.

Source: http://www.rayovac.com.

Assume these values have a normal distribution with a standard deviation of 1.38 hours.

a. What is the probability that one randomly selected Rayovac battery will have a test life of between 20.5 and 21.5 hours?

b. What is the probability that a randomly selected sample of 4 Rayovac batteries will have a mean test life of between 20.5 and 21.5 hours?

c. What is the probability that a randomly selected sample of 16 Rayovac batteries will have a mean test life of between 20.5 and 21.5 hours?

d. What is the probability that a randomly selected sample of 64 Rayovac batteries will have a mean test life of between 20.5 and 21.5 hours?

e. Describe the effect that the increase in sample size had on the answers for parts b–d.

7.45 a. Find $P(4 < \bar{x} < 6)$ for a random sample of size 4 drawn from a normal population with $\mu = 5$ and $\sigma = 2$.

b. Use a computer to randomly generate 100 samples, each of size 4, from a normal probability distribution with $\mu = 5$ and $\sigma = 2$. Calculate the mean, $\bar{x}$, for each sample.

c. How many of the sample means in part b have values between 4 and 6? What percentage is that?

d. Compare the answers to parts a and c, and explain any differences that occurred.

MINITAB

a. Input the numbers 4 and 6 into C1. Use the CUMULATIVE NORMAL PROBABILITY DISTRIBUTION commands on page 285, replacing the mean with 5, the standard deviation with 1 ($2/\sqrt{4}$), the input column with C1, and the optional storage with C2. Find CDF(6) – CDF(4).

b. Use the Normal RANDOM DATA commands on page 91, replacing generate with 100, store in with C3–C6, mean with 5, and standard deviation with 2. Use the ROW STATISTICS commands on page 318, replacing input variables with C3–C6 and store result in with C7.

c. Use the HISTOGRAM commands on page 53 for the data in C7. Select Labels, Data Labels, Label Type; use *y*-value levels. To adjust the histogram, select Binning with midpoint and midpoint positions 0:10/1.

Excel

a. Input the numbers 4 and 6 into column A. Activate cell B1. Use the CUMULATIVE NORMAL DISTRIBUTION commands on page 285, replacing X with A1:A2. Find CDF(6)–CDF(4).

b. Use the Normal RANDOM NUMBER GENERATION commands on page 91, replacing number of variables with 4, number of random numbers with 100, mean with 5, standard deviation with 2, and output range with C1. Activate cell G1. Use the AVERAGE INSERT FUNCTION

commands in Exercise 7.13(b) on page 318, replacing Number1 with C1:F1.

c. Use the RANDOM NUMBER GENERATION Patterned Distribution commands in Exercise 6.71(a) on page 291, replacing the first value with 0, the last value with 9, the steps with 1, and the output range with H1. Use the HISTOGRAM commands on pages 53–54 with column G as the input range, column H as the bin range, and column I as the output range.

TI-83/84 Plus

a. Use the CUMULATIVE NORMAL PROBABILITY commands on page 285, replacing the Enter with 4,6,5,1). (The standard deviation is 1; from $2/\sqrt{4}$.)

b. Use the Normal RANDOM DATA and STO commands on page 91, replacing the Enter with 5,2,100). Repeat these commands three more times, storing data in L2, L3, and L4, respectively.

Choose: **STAT > EDIT > 1: Edit**
Highlight: **L5** (column heading)
Enter: **(L1 + L2 + L3 + L4)/4**

c. Use the HISTOGRAM and TRACE commands on page 54 to count. Enter 0,9,1,0,45,1 for the Window.

7.46 a. Find $P(46 < \bar{x} < 55)$ for a random sample of size 16 drawn from a normal population with mean $\mu = 50$ and standard deviation $\sigma = 10$.

b. Use a computer to randomly generate 200 samples, each of size 16, from a normal probability distribution with mean $\mu = 50$ and standard deviation $\sigma = 10$. Calculate the mean, $\bar{x}$, for each sample.

c. How many of the sample means in part b have values between 46 and 55? What percentage is that?

d. Compare the answers to parts a and c, and explain any differences that occurred.

FYI If you use a computer, see Exercise 7.45.

Chapter Review

In Retrospect

In Chapters 6 and 7 we have learned to use the standard normal probability distribution. We now have two formulas for calculating a z-score:

$$z = \frac{x - \mu}{\sigma} \quad \text{and} \quad z = \frac{\bar{x} - \mu}{\sigma/\sqrt{n}}$$

You must be careful to distinguish between these two formulas. The first gives the standard score when we have individual values from a normal distribution (x values). The second formula deals with a sample mean ($\bar{x}$ value). The key to distinguishing between the formulas is to decide whether the problem deals with an individual x or a sample mean $\bar{x}$. If it deals with the individual values of x, we use the first formula, as presented in Chapter 6. If the problem deals with a sample mean, $\bar{x}$, we use the second formula and proceed as illustrated in this chapter.

The basic purpose for considering what happens when a population is repeatedly sampled, as discussed in this chapter, is to form sampling distributions. The sampling distribution is then used to describe the variability that occurs from one sample to the next. Once this pattern of variability is known and understood for a specific sample statistic, we are able to make predictions about the corresponding population parameter with a measure of how accurate the prediction is. The SDSM and the central limit theorem help describe the distribution for sample means. We will begin to make inferences about population means in Chapter 8.

There are other reasons for repeated sampling. Repeated samples are commonly used in the field of production control, in which samples are taken to determine whether a product is of the proper size or quantity. When the sample statistic does not fit the standards, a mechanical adjustment of the machinery is necessary. The adjustment is then followed by another sampling to be sure the production process is in control.

The "standard error of the _____" is the name used for the standard deviation of the sampling distribution for whatever statistic is named in the blank. In this chapter we have been concerned with the standard error of the mean. However, we could also work with the standard error of the proportion, median, or any other statistic.

You should now be familiar with the concept of a sampling distribution and, in particular, with the sampling distribution of sample means. In Chapter 8 we will begin to make predictions about the values of population parameters.

CourseMate The **Statistics CourseMate** site for this text brings chapter topics to life with interactive learning, study, and exam preparation tools, including quizzes and flashcards for the Vocabulary and Key Concepts that follow. The site also provides an **eBook** version of the text with highlighting and note taking capabilities. Throughout chapters, the CourseMate icon flags concepts and examples that have corresponding interactive resources such as **video** and **animated tutorials** that demonstrate, step by step, how to solve problems; **datasets** for exercises and examples; **Skillbuilder Applets** to help you better understand concepts; **technology manuals**; and software to download including **Data Analysis Plus** (a suite of statistical macros for Excel) and **TI-83/84 Plus** programs—logon at **www.cengagebrain.com**.

Vocabulary and Key Concepts

central limit theorem (p. 320)
frequency distribution (p. 315)
probability distribution (p. 314)
random sample (p. 316)

repeated sampling (p. 313)
sampling distribution (p. 323)
sampling distribution of sample means (pp. 314, 320)

standard error of the mean (p. 320)
z-score (p. 327)

Learning Outcomes

- Understand what a sampling distribution of a sample statistic is and that the distribution is obtained from repeated samples, all of the same size.

  pp. 313–314, EXP 7.1

- Be able to form a sampling distribution for a mean, median, or range based on a small, finite population.

  EXP 7.1, Ex. 7.6, 7.7

- Understand that a sampling distribution is a probability distribution for a sample statistic.

  EXP 7.2

- Understand and be able to present and describe the sampling distribution of sample means and the central limit theorem.

  pp. 319–321, EXP 7.4

- Understand and be able to explain the relationship between the sampling distribution of sample means and the central limit theorem.

  pp. 319–321, Ex. 7.17, 7.18, 7.21

- Determine and be able to explain the effect of sample size on the standard error of the mean.

  pp. 323–324, Ex. 7.20, 7.26, 7.47

- Understand when and how the normal distribution can be used to find probabilities corresponding to sample means.

  EXP 7.5

- Compute, describe, and interpret z-scores corresponding to known values of $\bar{x}$.

  EXP 7.6, EXP 7.7, Ex. 7.29, 7.30, 7.48

- Compute z-scores and probabilities for applications of the sampling distribution of sample means.

  Ex. 7.33, 7.35

Chapter Exercises

7.47 If a population has a standard deviation σ of 18.2 units, what is the standard error of the mean if samples of size 9 are selected? Samples of size 25? Samples of size 49? Samples of size 100?

7.48 Consider a normal population with $\mu = 24.7$ and $\sigma = 4.5$.

a. Calculate the z-score for an x of 21.5.

b. Calculate the z-score for an $\bar{x}$ of 21.5 from a sample of size 25.

c. Explain how 21.5 can have such different z-scores.

7.49 The dean of nursing tells students being recruited for the incoming class that the school's graduates can expect to earn a mean weekly income of $775 one year after graduation. Assume that the dean's statement is true and that the weekly salaries one year after graduation are normally distributed with a standard deviation of $115.

If one graduate is randomly selected:

a. Describe the distribution of the weekly salary being earned one year after graduation.

b. What is the probability that the selected graduate is making between $625 and $825?

If a random sample of 25 graduates is selected:

c. Describe the mean weekly salary being earned one year after graduation.

d. What is the probability that the sample mean is between $710 and $785?

e. Why is the z-score used in answering parts b and d?

f. Why is the formula for z used in part d different from that used in part b?

7.50 The diameters of Red Delicious apples in a certain orchard are normally distributed with a mean of 2.63 inches and a standard deviation of 0.25 inch.

a. What percentage of the apples in this orchard have diameters less than 2.25 inches?

b. What percentage of the apples in this orchard are larger than 2.56 inches in diameter?

A random sample of 100 apples is gathered, and the mean diameter obtained is $\bar{x} = 2.56$.

c. If another sample of size 100 is taken, what is the probability that its sample mean will be greater than 2.56 inches?

d. Why is the z-score used in answering parts a–c?

e. Why is the formula for the z-score used in part c different from that used in parts a and b?

7.51 a. Find a value for e such that 95% of the apples in Exercise 7.50 are within e units of the mean, 2.63. That is, find e such that $P(2.63 - e < x < 2.63 + e) = 0.95$.

b. Find a value for E such that 95% of the samples of 100 apples taken from the orchard in Exercise 7.50 will have mean values within E units of the mean, 2.63. That is, find E such that $P(2.63 - E < \bar{x} < 2.63 + E) = 0.95$.

7.52 Americans spend billions on veterinary care each year. According to the APPA National Pet Owners Survey, U.S. citizens spent $10.1 billion on pet care in 2007. The health care services offered to animals rival those provided to humans, with the typical surgery costing from $1700 to $3000 or more. On average, a dog owner spent an estimated $670 on veterinary-related expenses in that year.

Source: American Pet Products Manufacturers Association

Assume that annual dog owner expenditure on health care is normally distributed with a mean of $670 and a standard deviation of $290.

a. What is the probability that a dog owner, randomly selected from the population, spent over $1000 in health care in 2007?

b. Suppose a survey of 300 dog owners is conducted, and each is asked to report the total of his or her vet care bills for 2007. What is the probability that the mean annual expenditure of this sample falls between $700 and $750?

c. The assumption of a normal distribution in this situation is likely misguided. Explain why and what effect this had on the answers to parts a and b.

7.53 The statistics-conscious store manager at Marketview records the number of customers who walk through the door each day. Years of records show the mean number of customers per day to be 586 with a standard deviation of 165. Assume the number of customers is normally distributed.

a. What is the probability that on any given day, the number of customers exceeds 1000?

b. If 20 days are randomly selected, what is the probability that the mean of this sample is less than 550?

c. The assumption of normality allowed you to calculate the probabilities; however, this may not be a reasonable assumption. Explain why and how that affects the probabilities found in parts a and b.

7.54 Everyone needs to cut costs, even those planning a wedding, according to the July 8, 2009, *USA Today* article "Today's bride is 'definitely the opposite of bridezilla.'" The article quoted the average wedding gown cost, based on information from The Knot Real Wedding Survey of 2008, as $1032. If we assume that the cost of wedding gowns is normally distributed with a standard deviation of $550, what is the probability that the mean cost of wedding gowns for a sample of 20 randomly selected brides-to-be is between $800 and $1200?

7.55 A shipment of steel bars will be accepted if the mean breaking strength of a random sample of 10 steel bars is greater than 250 pounds per square inch. In the past, the breaking strength of such bars has had a mean of 235 and a variance of 400.

a. Assuming that the breaking strengths are normally distributed, what is the probability that one randomly selected steel bar will have a breaking strength in the range from 245 to 255 pounds per square inch?

b. What is the probability that the shipment will be accepted?

7.56 A report in *The Washington Post* (April 26, 2009) stated that the average age for men to marry in the United States is now 28 years of age. If the standard deviation is assumed to be 3.2 years, find the probability that a random sample of 40 U.S. men would show a mean age less than or equal to 27 years.

7.57 A manufacturer of light bulbs claims that its light bulbs have a mean life of 700 hours and a standard deviation of 120 hours. You purchased 144 of these bulbs and decided that you would purchase more if the mean life of your current sample exceeds 680 hours. What is the probability that you will not buy again from this manufacturer?

7.58 A tire manufacturer claims (based on years of experience with its tires) that the mean mileage of its tires is 35,000 miles and the standard deviation is 5000 miles. A consumer agency randomly selects 100 of these tires and finds a sample mean of 31,000. Should the consumer agency doubt the manufacturer's claim?

7.59 For large samples, the sample sum (Σx) has an approximately normal distribution. The mean of the sample sum is $n \cdot \mu$ and the standard deviation is $\sqrt{n} \cdot \sigma$. The distribution of savings per account for a savings and loan institution has a mean equal to $750 and a standard deviation equal to $25. For a sample of 50 such accounts, find the probability that the sum in the 50 accounts exceeds $38,000.

7.60 The baggage weights for passengers using a particular airline are normally distributed with a mean of 20 lb and a standard deviation of 4 lb. If the limit on total luggage weight is 2125 lb, what is the probability that the limit will be exceeded for 100 passengers?

7.61 A trucking firm delivers appliances for a large retail operation. The packages (or crates) have a mean weight of 300 lb and a variance of 2500.

a. If a truck can carry 4000 lb and 25 appliances need to be picked up, what is the probability that the 25 appliances will have an aggregate weight greater than the truck's capacity? Assume that the 25 appliances represent a random sample.

b. If the truck has a capacity of 8000 lb, what is the probability that it will be able to carry the entire lot of 25 appliances?

7.62 A pop music record firm wants the distribution of lengths of cuts on its records to have an average of 2 minutes and 15 seconds (135 seconds) and a standard deviation of 10 seconds so that disc jockeys will have plenty of time for commercials within each 5-minute period. The population of times for cuts is approximately normally distributed with only a negligible skew to the right. You have just timed the cuts on a new release and have found that the 10 cuts average 140 seconds.

a. What percentage of the time will the average be 140 seconds or longer if the new release is randomly selected?

b. If the music firm wants 10 cuts to average no longer than 140 seconds less than 5% of the time, what must the population mean be, given that the standard deviation remains at 10 seconds?

7.63 Let's simulate the sampling distribution related to the disc jockey's concern for "length of cut" in Exercise 7.62.

a. Use a computer to randomly generate 50 samples, each of size 10, from a normal distribution with mean 135 and standard deviation 10. Find the "sample total" and the sample mean for each sample.

b. Using the 50 sample means, construct a histogram and find their mean and standard deviation.

c. Using the 50 sample "totals," construct a histogram and find their mean and standard deviation.

d. Compare the results obtained in parts b and c. Explain any similarities and any differences observed.

MINITAB

a. Use the Normal RANDOM DATA commands on page 91, replacing generate with 50, store in with C1–C10, mean with 135, and standard deviation with 10. Use the ROW STATISTICS commands on page 318, selecting Sum and replacing input variables with C1–C10 and store result in with C11. Use the ROW STATISTICS commands, again selecting Mean and then replacing input variables with C1–C10 and store result in with C12.

b. Use the HISTOGRAM commands on page 53 for the data in C12. To adjust the histogram, select Binning with midpoint. Use the MEAN and STANDARD DEVIATION commands on pages 65 and 79 for the data in C12.

c. Use the HISTOGRAM commands on page 53 for the data in C11. To adjust the histogram, select Binning with midpoints. Use the MEAN and STANDARD DEVIATION commands on pages 65 and 79 for the data in C11.

d. Use the DISPLAY DESCRIPTIVE STATISTICS commands on page 88 for the data in C11 and C12.

Excel

a. Use the Normal RANDOM NUMBER GENERATION commands on page 91, replacing number of variables with 10, number of random numbers with 50, mean with 135, and standard deviation with 10.

Activate cell K1.

Choose: **Insert function, f_x > All > SUM > OK**
Enter: Number1: (A1:J1 or select cells)
Drag: **Bottom right corner of sum value box down to give other sums**

Activate cell L1. Use the AVERAGE INSERT FUNCTION commands in Exercise 7.13(b) on page 318, replacing Number1 with A1:J1.

b. Use the RANDOM NUMBER GENERATION Patterned Distribution commands in Exercise 6.71(a) on page 291, replacing the first value with 125.4, the last value with 144.6, the steps with 3.2, and the output range with M1. Use the HISTOGRAM commands on pages 53–54 with column L as the input range and column M as the bin range. Use the MEAN and STANDARD DEVIATION commands on pages 65 and 79 for the data in column L.

c. Use the RANDOM NUMBER GENERATION Patterned Distribution commands in Exercise 6.71(a) on page 291, replacing the first value with 1254, the last value with 1446, the steps with 32, and the output range with M20. Use the HISTOGRAM commands on pages 53–54 with column L as the input range and cells M20–? as the bin range. Use the MEAN and STANDARD DEVIATION commands on pages 65 and 79 for the data in column K.

d. Use the DESCRIPTIVE STATISTICS commands on page 88 for the data in columns K and L.

7.64 a. Find the mean and standard deviation of x for a binomial probability distribution with $n = 16$ and $p = 0.5$.

b. Use a computer to construct the probability distribution and histogram for the binomial probability experiment with $n = 16$ and $p = 0.5$.

c. Use a computer to randomly generate 200 samples of size 25 from a binomial probability distribution with $n = 16$ and $p = 0.5$. Calculate the mean of each sample.

d. Construct a histogram and find the mean and standard deviation of the 200 sample means.

e. Compare the probability distribution of x found in part b and the frequency distribution of $\bar{x}$ in part d. Does your information support the CLT? Explain.

MINITAB

a. Use the MAKE PATTERNED DATA commands in Exercise 6.71(a) on page 291, replacing the first value with 0, the last value with 16, and the steps with 1. Use the BINOMIAL PROBABILITY DISTRIBUTIONS commands on page 251, replacing n with 16, p with 0.5, input column with C1, and optional storage with C2. Use the Scatterplot with Connect Line commands on page 129, replacing Y with C2 and X with C1.

b. Use the BINOMIAL RANDOM DATA commands on page 261, replacing generate with 200, store in with C3–C27, number of trials with 16, and probability with 0.5. Use the ROW STATISTICS commands for a mean on page 318 replacing input variables with C3–C27 and store result in with C28. Use the HISTOGRAM commands on page 53 for the data in C28. To adjust the histogram, select Binning with midpoints. Use the MEAN and STANDARD DEVIATION commands on pages 65 and 79 for the data in C28.

Excel

a. Input 0 through 16 into column A. Continue with the binomial probability commands on pages 251–252, using

$n = 16$ and $p = 0.5$. Activate columns A and B; then continue with:

Choose: **Insert > Column > 1st picture > Next > Series**
Choose: **Select Data > Series1 > Remove > OK**

b. Use the Binomial RANDOM NUMBER GENERATION commands from Exercise 5.95 on page 261, replacing number of variables with 25, number of random numbers with 200, p value with 0.5, number of trials with 16, and output range with C1. Activate cell BB1. Use the AVERAGE INSERT FUNCTION commands in Exercise 7.13(b) on page 318, replacing Number1 with C1:AA1.

c. Use the RANDOM NUMBER GENERATION Patterned Distribution commands in Exercise 6.71(a) on page 291, replacing the first value with 6.8, the last value with 9.2, the steps with 0.4, and the output range with CC1. Use the HISTOGRAM commands on pages 53–54 with column BB as the input range and column CC as the bin range. Use the MEAN and STANDARD DEVIATION commands on pages 65 and 79 for the data in column BB.

7.65 a. Find the mean and standard deviation of x for a binomial probability distribution with $n = 200$ and $p = 0.3$.

b. Use a computer to construct the probability distribution and histogram for the random variable x of the binomial probability experiment with $n = 200$ and $p = 0.3$.

c. Use a computer to randomly generate 200 samples of size 25 from a binomial probability distribution with $n = 200$ and $p = 0.3$. Calculate the mean $\bar{x}$ of each sample.

d. Construct a histogram and find the mean and standard deviation of the 200 sample means.

e. Compare the probability distribution of x found in part b and the frequency distribution of $\bar{x}$ found in part d. Does your information support the CLT? Explain.

FYI Use the commands in Exercise 7.64, making the necessary adjustments.

7.66 A sample of 144 values is randomly selected from a population with mean, μ, equal to 45 and standard deviation, σ, equal to 18.

a. Determine the interval (smallest value to largest value) within which you would expect a sample mean to lie.

(continue on page 338)

b. What is the amount of deviation from the mean for a sample mean of 45.3?

c. What is the maximum deviation you have allowed for in your answer to part a?

Chapter Practice Test

PART I: Knowing the Definitions

Answer "True" if the statement is always true. If the statement is not always true, replace the words shown in bold with words that make the statement always true.

7.1 A sampling distribution **is** a distribution listing all the sample statistics that describe a particular sample.

7.2 The histograms of **all** sampling distributions are symmetrical.

7.3 The mean of the sampling distribution of $\bar{x}$'s is equal to the mean of the **sample**.

7.4 The standard error of the mean is the standard deviation of the population **from which the samples have been taken**.

7.5 The standard error of the mean **increases** as the sample size increases.

7.6 The shape of the distribution of sample means is always that of a **normal** distribution.

7.7 A **probability** distribution of a sample statistic is a distribution of all the values of that statistic that were obtained from all possible samples.

7.8 The sampling distribution of sample means provides us with a description of the three characteristics of a sampling distribution of sample **medians**.

7.9 A **frequency** sample is obtained in such a way that all possible samples of a given size have an equal chance of being selected.

7.10 We **do not need** to take repeated samples in order to use the concept of the sampling distribution.

PART II: Applying the Concepts

7.11 The lengths of the lake trout in Conesus Lake are believed to have a normal distribution with a mean of 15.6 inches and a standard deviation of 3.8 inches.

a. Kevin is going fishing at Conesus Lake tomorrow. If he catches one lake trout, what is the probability that it is less than 15.0 inches long?

b. If Captain Brian's fishing boat takes 10 people fishing on Conesus Lake tomorrow and they catch a random sample of 16 lake trout, what is

d. How is this maximum deviation related to the standard error of the mean?

the probability that the mean length of their total catch is less than 15 inches?

7.12 Cigarette lighters manufactured by EasyVice Company are claimed to have a mean lifetime of 20 months with a standard deviation of 6 months. The money-back guarantee allows you to return the lighter if it does not last at least 12 months from the date of purchase.

a. If the lifetimes of these lighters are normally distributed, what percentage of the lighters will be returned to the company?

b. If a random sample of 25 lighters is tested, what is the probability the sample mean lifetime will be more than 18 months?

7.13 Aluminum rivets produced by Rivets Forever, Inc., are believed to have shearing strengths that are distributed about a mean of 13.75 with a standard deviation of 2.4. If this information is true and a sample of 64 such rivets is tested for shear strength, what is the probability that the mean strength will be between 13.6 and 14.2?

PART III: Understanding the Concepts

7.14 "Two heads are better than one." If that's true, then how good would several heads be? To find out, a statistics instructor drew a line across the chalkboard and asked her class to estimate its length to the nearest inch. She collected their estimates, which ranged from 33 to 61 inches, and calculated the mean value. She reported that the mean was 42.25 inches. She then measured the line and found it to be 41.75 inches long. Does this show that "several heads are better than one"? What statistical theory supports this occurrence? Explain how.

7.15 The sampling distribution of sample means is more than just a distribution of the mean values that occur from many repeated samples taken from the same population. Describe what other specific condition must be met in order to have a sampling distribution of sample means.

7.16 Student A states, "A sampling distribution of the standard deviations tells you how the standard deviation varies from sample to sample." Student B argues, "A population distribution tells you that." Who is right? Justify your answer.

7.17 Student A says it is the "size of each sample used" and Student B says it is the "number of samples used" that determines the spread of an empirical sampling distribution. Who is right? Justify your choice.

8 Introduction to Statistical Inferences

© 2010 Image Source/Jupiterimages Corporation

8.1 The Nature of Estimation

Are We Taller or Shorter Today?

The average height for an early 17th-century Englishman was approximately 5 ft 6 in. For 17th-century Englishwomen, it was about 5 ft $\frac{1}{2}$ in. While average heights in England remained virtually unchanged in the 17th and 18th centuries, American colonists grew taller. Averages for modern Americans are just over 5 ft 9 in. for men and about 5 ft $3\frac{3}{4}$ in. for women.

Source: http://www.plimoth.org/

The National Center for Health Statistics (NCHS) provides statistical information that will guide actions and policies to improve the health of the American people. Recent data from NCHS give the average height of females in the United States to be 63.7 inches with a standard deviation of 2.75 inches.

Suppose a sample of heights was gathered from 50 randomly selected American female health professionals. Do you expect the mean of this random sample of 50 female heights to be exactly equal to the population mean of 63.7 inches given by NCHS (an **estimation question**)? If the sample mean is greater than 63.7 inches, does it mean that female health professionals are taller than American females (a **hypothesis-testing question**)? These are inferential questions about "Are We Taller or Shorter Today?"

As you recall, the central limit theorem gave us some very important information about the sampling distribution of sample means (SDSM). Specifically, it stated that in many realistic cases (when the random sample is large enough) a distribution of sample means is normally or approximately normally distributed about the mean of the population. With this information we were able to make probability statements about the likelihood of certain sample mean values occurring when samples are drawn from a population with a known mean and a known standard deviation. We are now ready to turn this situation around to the case in which the population mean is not known. We will draw one sample, calculate its mean value, and then make an inference about the value of the population mean based on the sample's mean value.

The objective of inferential statistics is to use the information contained in the sample data to increase our knowledge of the sampled population. We will learn about making two types of inferences: (1) estimating the value of a population parameter and (2) testing a hypothesis. The sampling distribution of sample means (SDSM) is the key to making these inferences as shown in Figure 8.1.

FIGURE 8.1

Where the Sampling Distribution Fits into the Statistical Process

The Statistical Process

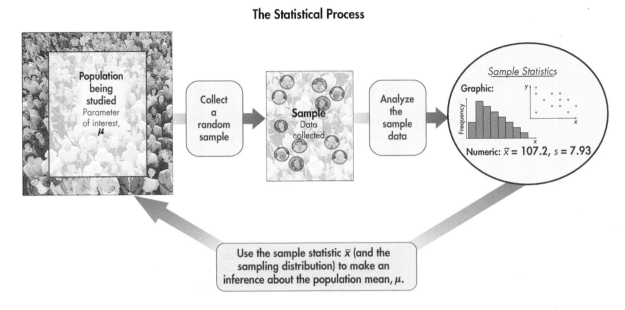

In this chapter, we deal with questions about the population mean using two methods that assume the value of the population standard deviation is a known quantity. This assumption is seldom realized in real-life problems, but it will make our first look at the techniques of inference much simpler.

Starting with the concept of **estimation**, let's consider a company that manufactures rivets for use in building aircraft. One characteristic of extreme importance is the "shearing strength" of each rivet. The company's engineers must monitor production to be certain that the shearing strength of the rivets meets the required specs. To accomplish this, they take a sample and determine the mean shearing strength of the sample. Based on this sample information, the company can estimate the mean shearing strength for all the rivets it is manufacturing.

A random sample of 36 rivets is selected, and each rivet is tested for shearing strength. The resulting sample mean is $\bar{x} = 924.23$ lb. Based on this sample, we say, "We believe the mean shearing strength of all such rivets is 924.23 lb."

Notes:

1. Shearing strength is the force required to break a material in a "cutting" action. Obviously, the manufacturer is not going to test all rivets because the test destroys each rivet tested. Therefore, samples are tested and the information about each sample must be used to make inferences about the population of all such rivets.

2. Throughout Chapter 8 we will treat the standard deviation, σ, as a known, or given, quantity and concentrate on learning the procedures for making statistical inferences about the population mean, μ. Therefore, to continue the explanation of statistical inferences, we will assume $\sigma = 18$ for the specific rivets described in our example.

> **Point estimate for a parameter** A single number designed to estimate a quantitative parameter of a population, usually the value of the corresponding **sample statistic**.

That is, the sample mean, $\bar{x}$, is the point estimate (single-number value) for the mean, μ, of the sampled population. For our rivet example, 924.23 is the point estimate for μ, the mean shearing strength of all rivets.

The quality of this point estimate should be questioned. Is the estimate exact? Is the estimate likely to be high? Or low? Would another sample yield the same result? Would another sample yield an estimate of nearly the same value? Or a value that is very different? How is "nearly the same" or "very different" measured? The quality of an estimation procedure (or method) is greatly enhanced if the sample statistic is both *less variable* and *unbiased*. The variability of a statistic is measured by the standard error of its sampling distribution. The sample mean can be made less variable by reducing its standard error, $\sigma/\sqrt{n}$. That requires using a larger sample because as n increases, the standard error decreases.

> **Unbiased statistic** A sample statistic whose sampling distribution has a mean value equal to the value of the population parameter being estimated. A statistic that is not unbiased is a **biased statistic**.

Figure 8.2 illustrates the concept of being unbiased and the effect of variability on the point estimate. The value A is the parameter being estimated, and the dots represent possible sample statistic values from the sampling distribution of the statistic. If A represents the true population mean, μ, then the dots represent possible sample means from the $\bar{x}$ sampling distribution.

FIGURE 8.2

Effects of Variability and Bias

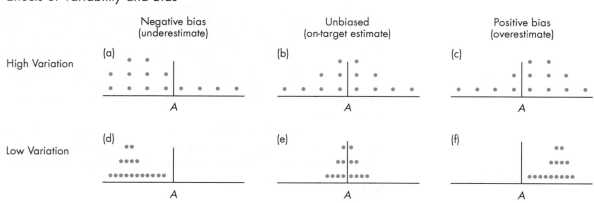

Figure 8.2(a), (c), (d), and (f) show biased statistics; (a) and (d) show sampling distributions whose mean values are less than the value of the parameter, whereas (c) and (f) show sampling distributions whose mean values are greater than the parameter. Figure 8.2(b) and (e) show sampling distributions that appear to have a mean value equal to the value of the parameter; therefore, they are unbiased. Figure 8.2(a), (b), and (c) show more variability, whereas (d), (e), and (f) show less variability in the sampling distributions. Diagram (e) represents the best situation, an estimator that is unbiased (on-target) and has low variability (all values close to the target).

The sample mean, $\bar{x}$, is an unbiased statistic because the mean value of the sampling distribution of sample means, $\mu_{\bar{x}}$, is equal to the population mean, μ. (Recall that the sampling distribution of sample means has a mean $\mu_{\bar{x}} = \mu$.) Therefore, the sample statistic $\bar{x} = 924.23$ is an unbiased point estimate for the mean strength of all rivets being manufactured in our example.

Sample means vary in value and form a sampling distribution in which not all samples result in $\bar{x}$ values equal to the population mean. Therefore, we should not expect this sample of 36 rivets to produce a point estimate (sample mean) that is exactly equal to the mean μ of the sampled population. We should, however, expect the point estimate to be fairly close in value to the population mean. The sampling distribution of sample means (SDSM) and the central limit theorem (CLT) provide the information needed to describe how close the point estimate, $\bar{x}$, is expected to be to the population mean, μ.

Recall that approximately 95% of a normal distribution is within 2 standard deviations of the mean and that the CLT describes the sampling distribution of sample means as being nearly normal when samples are large enough. Samples of size 36 from populations of variables like rivet strengths are generally considered large enough. Therefore, we should anticipate that 95% of all random samples selected from a population with unknown mean μ and standard deviation $\sigma = 18$ will have means $\bar{x}$ between

FYI $\sigma = 18$ was given in Note 2 on page 341.

$$\mu - 2(\sigma_{\bar{x}}) \quad \text{and} \quad \mu + 2(\sigma_{\bar{x}})$$

$$\mu - 2\left(\frac{\sigma}{\sqrt{n}}\right) \quad \text{and} \quad \mu + 2\left(\frac{\sigma}{\sqrt{n}}\right)$$

$$\mu - 2\left(\frac{18}{\sqrt{36}}\right) \quad \text{and} \quad \mu + 2\left(\frac{18}{\sqrt{36}}\right)$$

$$\mu - 6 \quad \text{and} \quad \mu + 6$$

This suggests that 95% of all random samples of size 36 selected from the population of rivets should have a mean $\bar{x}$ between $\mu - 6$ and $\mu + 6$. Figure 8.3 shows the middle 95% of the distribution, the bounds of the interval covering the 95%, and the mean μ.

FIGURE 8.3
Sampling Distribution of $\bar{x}$'s, Unknown μ

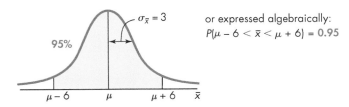

or expressed algebraically:
$P(\mu - 6 < \bar{x} < \mu + 6) = 0.95$

Now let's put all of this information together in the form of a *confidence interval*.

Interval estimate An interval bounded by two values and used to estimate the value of a population parameter. The values that bound this interval are statistics calculated from the sample that is being used as the basis for the estimation.

Level of confidence $1 - \alpha$ The portion of all interval estimates that include the parameter being estimated.

Confidence interval An interval estimate with a specified level of confidence.

To construct the confidence interval, we will use the point estimate $\bar{x}$ as the central value of an interval in much the same way as we used the mean μ as the central value to find the interval that captures the middle 95% of the $\bar{x}$ distribution in Figure 8.3.

For our rivet example, we can find the bounds to an interval centered at $\bar{x}$:

$$\bar{x} - 2(\sigma_{\bar{x}}) \quad \text{to} \quad \bar{x} + 2(\sigma_{\bar{x}})$$
$$924.23 - 6 \quad \text{to} \quad 924.23 + 6$$

The resulting interval is 918.23 to 930.23

The level of confidence assigned to this interval is approximately 95%, or 0.95. The bounds of the interval are two multiples ($z = 2.0$) of the standard error from the sample mean, and by looking at Table 3 in Appendix B, we can more accurately determine the level of confidence as 0.9545. Putting all of this information together, we express the estimate as a confidence interval: **918.23 to 930.23** *is the 95.45% confidence interval for the mean shear strength of the rivets.* Or in an abbreviated form: **918.23 to 930.23**, *the 95.45% confidence interval for μ.*

APPLIED EXAMPLE 8.1

FYI Visit the Old Faithful WebCam. When is the next eruption predicted to occur?

DID YOU KNOW

Yellowstone contains approximately one-half of the world's hydrothermal features. There are over 10,000 hydrothermal features, including over 300 geysers, in the park.

YELLOWSTONE PARK'S OLD FAITHFUL

Welcome to the Old Faithful WebCam.

Predictions for the time of the next eruption of Old Faithful are made by the rangers using a formula that takes into account the length of the previous eruption. The formula used has proved to be accurate, plus or minus 10 minutes, 90% of the time. At 3:05 p.m. on August 14, 2009, the posted prediction time of the next eruption was:

Next Prediction:
3:19 p.m. ±10 min.

Source: http://www.nps.gov/yell/oldfaithfulcam.htm

Note the time at which the picture was recorded: 3:25:19 p.m. Right on time!

Old Faithful is predicted to erupt at 3:19 PM +/- 10 min
Fri Aug 14, 2009 03:25:19 pm
© SCPhotos/Alamy

SECTION 8.1 EXERCISES

8.1 [EX08-001] A random sample of 50 female American health professionals yielded the following height data.

65	66	64	67	59	69	66	69	64	62
63	62	63	64	72	66	65	64	67	68
70	63	63	68	58	60	64	66	64	62
65	69	64	69	62	58	66	68	59	56
64	66	65	69	67	67	68	62	70	62

a. What population was sampled to obtain the height data listed above?

b. Describe the sample data using the mean and standard deviation, plus any other numerical statistic(s) that help describe the sample.

c. Construct a histogram and comment on the shape of the distribution. Construct any other graph(s) that help describe the sample.

d. Using the statistics found in parts b and c, estimate the mean height of all American female health professionals using a single value. Using an interval.

e. What quality of the interval estimate would improve the worth of the interval?

8.2 Referring to Exercise 8.1:

a. How is the distribution of the sample height data on p. 344 related to: (1) The distribution of the population? (2) The sampling distribution of sample means?

b. Using the techniques of Chapter 7, find the limits that would bound the middle 90% of the sampling distribution of sample means for samples of size 50 selected randomly from the population of female heights with a known mean of 63.7 inches and a standard deviation of 2.75 inches.

c. On the histogram drawn in Exercise 8.1: (1) Draw a vertical line at the population mean of 63.7. (2) Draw a horizontal line segment showing the interval found in part b. Does the sample mean found in Exercise 8.1 (b) fall in the interval? Answer yes or no, and explain what this means.

d. Using the techniques of Chapter 7, find $P(\bar{x} \geq 64.7)$ for a random sample of 50 drawn from a population with a known mean of 63.7 inches and a standard deviation of 2.75 inches. Explain the meaning of the resulting value.

e. Does the sample of 50 height data values appear to belong to the population described by the NCHS? Explain.

f. Review the above answers and consider how the SDSM and CLT from Chapter 7 might be used to make an improvement in the interval estimate.

8.3 Explain the difference between a point estimate and an interval estimate.

8.4 Identify each numerical value by "name" (e.g., *mean, variance*) and by symbol (e.g., $\bar{x}$):

a. The mean height of 24 junior high school girls is 4 ft 11 in.

b. The standard deviation for IQ scores is 16.

c. The variance among the test scores on last week's exam was 190.

d. The mean height of all cadets who have ever entered West Point is 69 inches.

8.5 [EX08-005] A random sample of the amount paid (in dollars) for taxi fare from downtown to the airport was obtained:

| 15 | 19 | 17 | 23 | 21 | 17 | 16 | 18 | 12 | 18 | 20 | 22 | 15 | 18 | 20 |

Use the data to find a point estimate for each of the following parameters.

a. Mean

b. Variance

c. Standard deviation

8.6 [EX08-006] The number of engines owned per fire department was obtained from a random sample taken from the profiles of fire departments from across the United States (*Firehouse*/June 2003).

| 29 | 8 | 7 | 33 | 21 | 26 | 6 | 11 | 4 | 54 | 7 | 4 |

Use the data to find a point estimate for each of the following parameters.

a. Mean

b. Variance

c. Standard deviation

8.7 In each diagram below, I and II represent sampling distributions of two statistics that might be used to estimate a parameter. In each case, identify the statistic that you think would be the better estimator, or neither, and describe why it is your choice.

a.
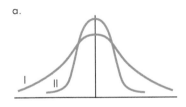

b.
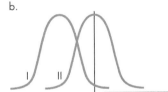

c.
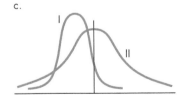

8.8 Suppose that there are two statistics that will serve as an estimator for the same parameter. One of them is biased, and the other is unbiased.

a. Everything else being equal, explain why you usually would prefer an unbiased estimator to a biased estimator.

b. If a statistic is unbiased, does that ensure that it is a good estimator? Why or why not? What other considerations must be taken into account?

c. Describe a situation that might occur in which the biased statistic might be a better choice as an estimator than the unbiased statistic.

8.9 The use of a tremendously large sample does not ensure quality in an estimator. What problems do you anticipate with very large samples?

8.10 Being unbiased and having a small variability are two desirable characteristics of a statistic if it is going to be used as an estimator. Describe how the SDSM addresses both of these properties when estimating the mean of a population.

8.11 The U.S. Census Bureau reports that the estimated mean U.S. married-couple family income is $90,835 ± $101. The Bureau describes the margin of error as providing a 90% probability that the interval defined by the estimate minus the margin of error and the estimate plus the margin of error (the lower and upper confidence bounds) contains the true value.

Source: U.S. Census Bureau, 2005–2007 American Community Survey

a. What is the population and variable of interest?

b. What parameter is being estimated? What is its estimated value?

c. How is the margin of error related to the maximum error of estimate?

d. What value is being reported as the margin of error?

e. What level of confidence is being reported?

f. Find the confidence interval and state exactly what it represents.

8.12 *Consumer Reports* National Research Center reported that 76% of women respond, "Daily or more often" when asked, "How often is your bed made?" As a footnote, this additional information is included: margin of error ±3.2 percentage points.

a. What is the population and variable of interest?

b. What parameter is being estimated? What is its estimated value?

c. What value is being reported as the margin of error?

d. Find the interval and state exactly what it represents.

e. What additional information might you want to know about this confidence interval?

8.13 Explain why the standard error of sample means is 3 for the rivet example on page 343.

8.14 a.   Verify that a 95% level of confidence requires a 1.96-standard-deviation interval.

b.   Verify that the level of confidence for a 2-standard-deviation interval is 95.45%.

8.15 Find the level of confidence assigned to an interval estimate of the mean formed using the following intervals:

a. $\bar{x} - 1.28 \cdot \sigma_{\bar{x}}$ to $\bar{x} + 1.28 \cdot \sigma_{\bar{x}}$

b. $\bar{x} - 1.44 \cdot \sigma_{\bar{x}}$ to $\bar{x} + 1.44 \cdot \sigma_{\bar{x}}$

c. $\bar{x} - 1.96 \cdot \sigma_{\bar{x}}$ to $\bar{x} + 1.96 \cdot \sigma_{\bar{x}}$

d. $\bar{x} - 2.33 \cdot \sigma_{\bar{x}}$ to $\bar{x} + 2.33 \cdot \sigma_{\bar{x}}$

8.16 Find the level of confidence assigned to an interval estimate of the mean formed using the following intervals:

a. $\bar{x} - 1.15 \cdot \sigma_{\bar{x}}$ to $\bar{x} + 1.15 \cdot \sigma_{\bar{x}}$

b. $\bar{x} - 1.65 \cdot \sigma_{\bar{x}}$ to $\bar{x} + 1.65 \cdot \sigma_{\bar{x}}$

c. $\bar{x} - 2.17 \cdot \sigma_{\bar{x}}$ to $\bar{x} + 2.17 \cdot \sigma_{\bar{x}}$

d. $\bar{x} - 2.58 \cdot \sigma_{\bar{x}}$ to $\bar{x} + 2.58 \cdot \sigma_{\bar{x}}$

8.17 "Population requirement for primary hip-replacement surgery: a cross-sectional study" was conducted by the University of Bristol in the United Kingdom. The findings resulted in the following statement: "The prevalence of self-reported hip pain was 107 per 1000 (95% CI 101–113) for men and 173 per 1000 (166–180) for women."

a. Explain the meaning of the confidence interval, 95% CI 101–113.

b. Find the standard error for the men's self-reported hip pain 95% confidence interval.

c. Assuming the women's data was also from a 95% confidence interval, find the standard error.

8.18 A sample of 25 of 174 funded projects revealed that 19 were valued at $17,320 each and 6 were valued at $20,200 each. From the sample data, estimate the total value of the funding for all the projects.

8.19 Using the Old Faithful eruption information in Applied Example 8.1 on page 344:

a. What does "3:19 PM ±10 min." mean? Explain.

b. Did this eruption occur during the predicted time interval?

c. What does "90% of the time" mean? Explain.

8.20 A recruiter estimates that if you are hired to work for her company and you put in a full week at the commissioned sales representative position she is offering, you will make "$525 plus or minus $250, 80% of the time." She adds, "It all depends on you!"

a.   What does "$525 plus or minus $250" mean?

b.   What does "80% of the time" mean?

c.   If you make $300 to the nearest $10 most weeks, will she have told you the truth? Explain.

8.2 Estimation of Mean μ (σ Known)

In Section 8.1 we surveyed the basic ideas of estimation: point estimate, interval estimate, level of confidence, and confidence interval. These basic ideas are inter-related and used throughout statistics when an inference calls for an estimate. In this section we formalize the interval estimation process as it applies to estimating the population mean μ based on a random sample under the restriction that the population standard deviation σ is a known value.

The sampling distribution of sample means and the CLT provide us with the information we need to ensure that the necessary *assumptions* for estimating a population mean are satisfied.

> **The assumption for estimating mean μ using a known σ** The sampling distribution of $\bar{x}$ has a normal distribution.

Note: The word *assumptions* is somewhat of a misnomer. It does not mean that we "assume" something to be the situation and continue but rather that we must be sure that the conditions expressed by the assumptions do exist before we apply a particular statistical method.

The information needed to ensure that this assumption (or condition) is satisfied is contained in the SDSM and in the CLT (see Chapter 7, pp. 319–320):

> The sampling distribution of sample means $\bar{x}$ is distributed about a mean equal to μ with a standard error equal to $\sigma/\sqrt{n}$; and (1) if the randomly sampled population is normally distributed, then $\bar{x}$ is normally distributed for all sample sizes, or (2) if the randomly sampled population is not normally distributed, then $\bar{x}$ is approximately normally distributed for sufficiently large sample sizes.

FYI If the assumptions are not met for estimating the mean μ using a known σ, the level of confidence is most likely lower than stated.

FYI The help of a professional statistician should be sought when treating extremely skewed data.

Therefore, we can satisfy the required **assumption** by either (1) knowing that the sampled population is normally distributed or (2) using a random sample that contains a sufficiently large amount of data. The first possibility is obvious. We either know enough about the population to know that it is normally distributed or we don't. The second way to satisfy the assumption is by applying the CLT. Inspection of various graphic displays of the sample data should yield an indication of the type of distribution the population possesses. The CLT can be applied to smaller samples (say, $n = 15$ or larger) when the data provide a strong indication of a unimodal distribution that is approximately symmetrical. If there is evidence of some skewness in the data, then the sample size needs to be much larger (perhaps $n \geq 50$). If the data provide evidence of an extremely skewed or J-shaped distribution, the CLT will still apply if the sample is large enough. In extreme cases, "large enough" may be unrealistically or impracticably large.

Note: There is no hard-and-fast rule defining "large enough"; the sample size that is "large enough" varies greatly according to the distribution of the population.

The $1 - \alpha$ confidence interval for the estimation of mean μ is found using formula (8.1).

Confidence Interval for Mean

$$\bar{x} - z(\alpha/2)\left(\frac{\sigma}{\sqrt{n}}\right) \quad \text{to} \quad \bar{x} + z(\alpha/2)\left(\frac{\sigma}{\sqrt{n}}\right) \qquad (8.1)$$

Here are the parts of the confidence interval formula:

1. $\bar{x}$ is the point estimate and the center point of the confidence interval.
2. $z(\alpha/2)$ is the **confidence coefficient**. It is the number of multiples of the standard error needed to formulate an interval estimate of the correct width to have a level of confidence of $1 - \alpha$. Figure 8.4 shows the relationship among the level of confidence $1 - \alpha$ (the middle portion of the distribution), $\alpha/2$ (the "area to the right" used with the critical-value notation), and the confidence coefficient $z(\alpha/2)$ (whose value is found using Table 4B of Appendix B). Alpha, α, is the first letter of the Greek alphabet and represents the portion associated with the tails of the distribution.
3. $\sigma/\sqrt{n}$ is the **standard error of the mean**, or the standard deviation of the sampling distribution of sample means.
4. $z(\alpha/2)\left(\frac{\sigma}{\sqrt{n}}\right)$ is one-half the width of the confidence interval (the product of the confidence coefficient and the standard error) and is called the **maximum error of estimate, E.**
5. $\bar{x} - z(\alpha/2)\left(\frac{\sigma}{\sqrt{n}}\right)$ is called the **lower confidence limit** (LCL), and $\bar{x} + z(\alpha/2)\left(\frac{\sigma}{\sqrt{n}}\right)$ is called the **upper confidence limit** (UCL) for the confidence interval.

The estimation procedure is organized into a five-step process that will take into account all of the preceding information and produce both the point estimate and the confidence interval.

FIGURE 8.4

Confidence Coefficient $z(\alpha/2)$

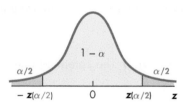

FYI Basically, the confidence interval is "point estimate $\pm$ maximum error."

THE CONFIDENCE INTERVAL: A FIVE-STEP PROCEDURE

Step 1 The Set-Up:
Describe the population parameter of interest.

Step 2 The Confidence Interval Criteria:
  a. Check the assumptions.
  b. Identify the probability distribution and the formula to be used.
  c. State the level of confidence, $1 - \alpha$.

Step 3 The Sample Evidence:
Collect the sample information.

Step 4 The Confidence Interval:
 a. Determine the confidence coefficient.
 b. Find the maximum error of estimate.
 c. Find the lower and upper confidence limits.

Step 5 The Results:
State the confidence interval.

Example 8.2 will illustrate this five-step confidence interval procedure.

E X A M P L E   8 . 2

CONSTRUCTING A CONFIDENCE INTERVAL FOR A MEAN ONE-WAY COMMUTE DISTANCE

The student body at many community colleges is considered a "commuter population." The student activities office wishes to obtain an answer to the question: "How far (one way) does the average community college student commute to college each day?" (Typically the "average student's commute distance" is meant to be the "mean distance" commuted by all students who commute.) A random sample of 100 commuting students was identified, and the one-way distance each commuted was obtained. The resulting sample mean distance was 10.22 miles.

Estimate the mean one-way distance commuted by all commuting students using (a) a point estimate and (b) a 95% confidence interval. (Use $\sigma = 6$ miles.)

Solution

 (a) The point estimate for the mean one-way distance is **10.22** miles (the sample mean).
 (b) We use the five-step procedure to find the 95% confidence interval.

Step 1 The Set-Up:
Describe the population parameter of interest.
 The mean μ of the one-way distances commuted by all commuting community college students is the parameter of interest.

Step 2 The Confidence Interval Criteria:
 a. **Check the assumptions.**
    σ is known. The variable "distance commuted" most likely has a skewed distribution because the vast majority of the students commute between 0 and 25 miles, with fewer commuting more than 25 miles. A sample size of 100 should be large enough for the CLT to satisfy the assumption; the $\bar{x}$ sampling distribution is approximately normal.
 b. **Identify the probability distribution and the formula to be used.**
    The standard normal distribution, z, will be used to determine the confidence coefficient, and formula (8.1) with $\sigma = 6$.
 c. **State the level of confidence, $1 - \alpha$.**
    The question asks for 95% confidence, or $1 - \alpha = 0.95$.

Step 3 The Sample Evidence:

Collect the sample information.

The sample information is given in the statement of the problem: $n = 100$, $\bar{x} = 10.22$.

Step 4 The Confidence Interval:

a. **Determine the confidence coefficient.**
   The confidence coefficient is found using Table 4B:

A Portion of Table 4B

	α	. . .	0.05	
Enter table with Level of confidence: $1 - \alpha = 0.95 \rightarrow$	$z(\alpha/2)$	. . .	**1.96**	$\rightarrow$ Leave table with Confidence coefficient:
	$1 - \alpha$	. . .	0.95	$z(\alpha/2) = \mathbf{1.96}$

b. **Find the maximum error of estimate.**
   Use the maximum error part of formula (8.1):

$$E = z(\alpha/2)\left(\frac{\sigma}{\sqrt{n}}\right) = 1.96\left(\frac{6}{\sqrt{100}}\right) = (1.96)(0.6) = 1.176$$

c. **Find the lower and upper confidence limits.**
   Using the point estimate, $\bar{x}$, from Step 3 and the maximum error, E, from Step 4b, we find the confidence interval limits:

$$\bar{x} - z(\alpha/2)\left(\frac{\sigma}{\sqrt{n}}\right) \quad \text{to} \quad \bar{x} + z(\alpha/2)\left(\frac{\sigma}{\sqrt{n}}\right)$$

$$10.22 - 1.176 \quad \text{to} \quad 10.22 + 1.176$$

$$9.044 \quad \text{to} \quad 11.396$$

$$9.04 \quad \text{to} \quad 11.40$$

Step 5 The Results:

State the confidence interval.

9.04 to 11.40 is the 95% confidence interval for μ. That is, with 95% confidence we can say, "The mean one-way distance is between 9.04 and 11.40 miles."

Let's look at another example of the estimation procedure.

EXAMPLE 8.3

CONSTRUCTING A CONFIDENCE INTERVAL FOR MEAN PARTICLE SIZE

"Particle size" is an important property of latex paint and is monitored during production as part of the quality-control process. Thirteen particle-size measurements were taken using the Dwight P. Joyce Disc, and the sample mean was 3978.1 angstroms (where 1 angstrom $[1\text{Å}] = 10^{-8}$ cm). The particle size, x, is normally distributed with a standard deviation $\sigma = 200$ angstroms. Find the 98% confidence interval for the mean particle size for this batch of paint.

Video tutorial available—logon and learn more at cengagebrain.com

Solution

Step 1 The Set-Up:

Describe the population parameter of interest.
The mean particle size, μ, for the batch of paint from which the sample was drawn

Step 2 The Confidence Interval Criteria:

a. **Check the assumptions.**
σ is known. The variable "particle size" is normally distributed; therefore, the sampling distribution of sample means is normal for all sample sizes.

b. **Identify the probability distribution and the formula to be used.**
The standard normal variable z, and formula (8.1) with $\sigma = 200$

c. **State the level of confidence, $1 - \alpha$.**
98%, or $1 - \alpha = 0.98$

Step 3 The Sample Evidence:

Collect the sample information: $n = 13$ and $\bar{x} = 3978.1$.

Step 4 The Confidence Interval:

a. **Determine the confidence coefficient.**
The confidence coefficient is found using Table 4B:
$z(\alpha/2) = z(0.01) = 2.33$.

A Portion of Table 4B

	α	$\ldots$	0.02	
Enter table with Level of confidence:	$z(\alpha/2)$	$\ldots$	**2.33**	Leave table with Confidence coefficient:
$1 - \alpha = 0.95 \rightarrow$	$1 - \alpha$	$\ldots$	0.98	$z(\alpha/2) = \mathbf{2.33}$

b. **Find the maximum error of estimate.**

$$E = z(\alpha/2)\left(\frac{\sigma}{\sqrt{n}}\right) = 2.33\left(\frac{200}{\sqrt{13}}\right) = (2.33)(55.47) = 129.2$$

c. **Find the lower and upper confidence limits.**
Using the point estimate, $\bar{x}$, from Step 3 and the maximum error, E, from Step 4b, we find the confidence interval limits:

$$\bar{x} - z(\alpha/2)\left(\frac{\sigma}{\sqrt{n}}\right) \quad \text{to} \quad \bar{x} + z(\alpha/2)\left(\frac{\sigma}{\sqrt{n}}\right)$$

$$3978.1 - 129.2 = 3848.9 \quad \text{to} \quad 3978.1 + 129.2 = 4107.3$$

Step 5 The Results:

State the confidence interval.
3848.9 to 4107.3 is the 98% confidence interval for μ. With 98% confidence we can say, "The mean particle size is between 3848.9 and 4107.3 angstroms."

Let's take another look at the concept "level of confidence." It was defined to be the probability that the sample to be selected will produce interval bounds that contain the parameter.

E X A M P L E   8 . 4

DEMONSTRATING THE MEANING OF A CONFIDENCE INTERVAL

Single-digit random numbers, like the ones in Table 1 in Appendix B, have a mean value $\mu = 4.5$ and a standard deviation $\sigma = 2.87$ (see Exercise 5.33, p. 242). Draw a sample of 40 single-digit numbers from Table 1 and construct the 90% confidence interval for the mean. Does the resulting interval contain the expected value of μ, 4.5? If we were to select another sample of 40 single-digit numbers from Table 1, would we get the same result? What might happen if we selected a total of 15 different samples and constructed the 90% confidence interval for each? Would the expected value for μ—namely, 4.5—be contained in all of them? Should we expect all 15 confidence intervals to contain 4.5? Think about the definition of "level of confidence"; it says that in the long run, 90% of the samples will result in bounds that contain μ. In other words, 10% of the samples will not contain μ. Let's see what happens.

First we need to address the assumptions; if the assumptions are not satisfied, we cannot expect the 90% and the 10% to occur. We know: (1) the distribution of single-digit random numbers is rectangular (definitely not normal), (2) the distribution of single-digit random numbers is symmetrical about their mean, (3) the $\bar{x}$ distribution for very small samples ($n = 5$) in Example 7.2 (p. 315) displayed a distribution that appeared to be approximately normal, and (4) there should be no skewness involved. Therefore, it seems reasonable to assume that $n = 40$ is large enough for the CLT to apply.

The first random sample was drawn from Table 1 in Appendix B:

TABLE 8.1   Random Sample of Single-Digit Numbers [TA08-01]

2	8	2	1	5	5	4	0	9	1
0	4	6	1	5	1	1	3	8	0
3	6	8	4	8	6	8	9	5	0
1	4	1	2	1	7	1	7	9	3

The sample statistics are $n = 40$, $\Sigma x = 159$, and $\bar{x} = 3.98$. Here is the resulting 90% confidence interval:

$$\bar{x} \pm z(\alpha/2)\left(\frac{\sigma}{\sqrt{n}}\right): \qquad 3.98 \pm 1.65\left(\frac{2.87}{\sqrt{40}}\right)$$

$$3.98 \pm (1.65)(0.454)$$

$$3.98 \pm 0.75$$

$$3.98 - 0.75 = 3.23 \quad \text{to} \quad 3.98 + 0.75 = 4.73$$

3.23 to 4.73 is the 90% confidence interval for μ.

Figure 8.5 shows this confidence interval, its bounds, and the expected mean μ.

FIGURE 8.5
The 90% Confidence Interval

With 90% confidence, we think μ is somewhere within this interval.

3.23 $\mu = 4.50$ 4.73 $\bar{x}$

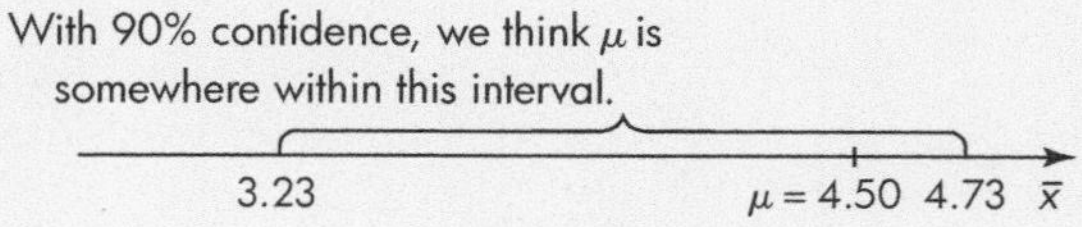

The expected value for the mean, 4.5, does fall within the bounds of the confidence interval for this sample. Let's now select 14 more random samples from Table 1 in Appendix B, each of size 40.

Table 8.2 lists the mean from the first sample and the means obtained from the 14 additional random samples of size 40. The 90% confidence intervals for the estimation of μ based on each of the 15 samples are listed in Table 8.2 and shown in Figure 8.6.

TABLE 8.2 Fifteen Samples of Size 40 [TA08-02]

Sample Number	Sample Mean, $\bar{x}$	90% Confidence Interval Estimate for μ	Sample Number	Sample Mean, $\bar{x}$	90% Confidence Interval Estimate for μ
1	3.98	3.23 to 4.73	9	4.08	3.33 to 4.83
2	4.64	3.89 to 5.39	10	5.20	4.45 to 5.95
3	4.56	3.81 to 5.31	11	4.88	4.13 to 5.63
4	3.96	3.21 to 4.71	12	5.36	4.61 to 6.11
5	5.12	4.37 to 5.87	13	4.18	3.43 to 4.93
6	4.24	3.49 to 4.99	14	4.90	4.15 to 5.65
7	3.44	2.69 to 4.19	15	4.48	3.73 to 5.23
8	4.60	3.85 to 5.35			

FIGURE 8.6

Confidence Intervals from Table 8.2

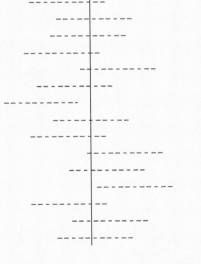

We see that 86.7% (13 of the 15) of the intervals contain μ and 2 of the 15 samples (sample 7 and sample 12) do not contain μ. The results here are "typical"; repeated experimentation might result in any number of intervals that contain 4.5. However, in the long run, we should expect approximately $1 - \alpha = 0.90$ (or 90%) of the samples to result in bounds that contain 4.5 and approximately 10% that do not contain 4.5.

APPLIED EXAMPLE 8.5

MEAN TRAVEL TIME TO WORK

AP Photo/Mary Altaffer

Americans spend more than 100 hours commuting to work each year, according to American Community Survey (ACS) data released by the U.S. Census Bureau. This exceeds the two weeks of vacation time (80 hours) workers frequently take over the course of a year. For the nation as a whole, the average daily commute to work lasted about 24.3 minutes in 2003.

Here are the mean travel times for a few cities and states and the respective 90% confidence intervals:

Rank	Place	Mean	Lower Bound	Upper Bound
	United States	24.4	24.2	24.6
Cities				
1	New York, NY	38.4	37.9	38.9
2	Chicago, IL	32.7	31.9	33.5
4	Riverside, CA	29.8	26.7	32.9
66	Oklahoma City, OK	17.8	17.0	18.6
States				
1	New York	30.8	30.5	31.1
2	Maryland	30.0	29.5	30.5
26	Kentucky	22.7	21.7	23.7
51	North Dakota	14.8	14.0	15.6

Source: U.S. Census Bureau

The table above shows the lower and upper bounds (limits) of the 90% confidence interval, an interval that gives a range of values likely to include the population true value.

Note: <u>U.S. Census Bureau Definition:</u> **<u>Travel time to work</u>** refers to the total number of minutes that it usually took the person to get from home to work each day during the reference week. The elapsed time includes time spent waiting for public transportation, time spent picking up passengers in carpools, and time spent in other activities related to getting to work.

TECHNOLOGY INSTRUCTIONS: CONFIDENCE INTERVAL FOR MEAN μ WITH A GIVEN σ

MINITAB	Input the data into C1; then continue with:

Choose:	**Stat > Basic Statistics > 1-Sample Z**
Enter:	**Samples in columns: C1**
	Standard deviation: σ
Select:	**Options**
Enter:	**Confidence Level: $1 - \alpha$** (ex.: 0.95 or 95.0)
Select:	**Alternative: not equal > OK > OK**

Excel	Input the data into column A; then continue with:

Choose: **Add-Ins > Data Analysis Plus > Z-Estimate: Mean > OK**
Enter: **Input Range: (A1:A20 or select cells) > OK**
            Standard Deviation (SIGMA): σ > OK
            Alpha: α (ex.: 0.05) > OK

TI-83/84 PLUS	Input the data into L1; then continue with the following, entering the appropriate values and highlighting Calculate:

Choose: **STAT > TESTS > 7:ZInterval**

```
ZInterval
 Inpt:DATA Stats
 σ:0
 List:L₁
 Freq:1
 C-Level:.95
 Calculate
```

Sample Size

The confidence interval has two basic characteristics that determine its quality: its level of confidence and its width. It is preferable for the interval to have a high level of confidence and be precise (narrow) at the same time. The higher the level of confidence, the more likely the interval is to contain the parameter, and the narrower the interval, the more precise the estimation. However, these two properties seem to work against each other, because it would seem that a narrower interval would tend to have a lower probability and a wider interval would be less precise. The maximum error part of the confidence interval formula specifies the relationship involved.

Maximum Error of Estimate

$$E = z_{(\alpha/2)}\left(\frac{\sigma}{\sqrt{n}}\right) \tag{8.2}$$

This formula has four components: (1) the maximum error E, half of the width of the confidence interval; (2) the confidence coefficient, $z(\alpha/2)$, which is determined by the level of confidence; (3) the sample size, n; and (4) the standard deviation, σ. The standard deviation σ is not a concern in this discussion because it is a constant (the standard deviation of a population does not change in value). That leaves three factors. Inspection of formula (8.2) indicates the following: increasing the level of confidence will make the confidence coefficient larger and thereby require either the maximum error to increase or the sample size to increase; decreasing the maximum error will require the level of confidence to decrease or the sample size to increase; and decreasing the sample size will force the maximum error to become larger or the level of confidence to decrease. We thus have a "three-way tug-of-war," as pictured in Figure 8.7. An increase or decrease in any one of the three factors has an effect on one or both of the other two factors. The statistician's job is to "balance" the level of confidence, the sample size, and the maximum error so that an acceptable interval results.

FYI When the denominator increases, the value of the fraction decreases.

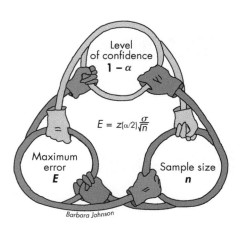

FIGURE 8.7
The "Three-Way Tug-of-War"
between $1 - \alpha$, n, and E

Level
of confidence
$1 - \alpha$

$E = z_{(\alpha/2)}\frac{\sigma}{\sqrt{n}}$

Maximum
error
E

Sample size
n

Barbara Johnson

Let's look at an example of this relationship in action.

EXAMPLE 8.6

DETERMINING THE SAMPLE SIZE FOR A CONFIDENCE INTERVAL

Determine the sample size needed to estimate the mean weight of all second-grade boys if we want to be accurate within 1 lb with 95% confidence. Assume a normal distribution and that the standard deviation of the boys' weights is 3 lb.

Solution

FYI Instructions for using Table 4B are given on page 350.

The desired level of confidence determines the confidence coefficient: the confidence coefficient is found using Table 4B: $z_{(\alpha/2)} = z(0.025) = \mathbf{1.96}$.

The desired maximum error is $E = 1.0$. Now you are ready to use the maximum error formula:

$$E = z_{(\alpha/2)}\left(\frac{\sigma}{\sqrt{n}}\right): \quad 1.0 = 1.96\left(\frac{3}{\sqrt{n}}\right)$$

$$\text{Solve for } n: 1.0 = \frac{5.88}{\sqrt{n}}$$

$$\sqrt{n} = 5.88$$

$$n = (5.88)^2 = 34.57 = \mathbf{35}$$

Therefore, **$n = 35$** is the sample size needed if you want a 95% confidence interval with a maximum error no greater than 1 lb.

Note: When we solve for the sample size n, it is customary to round up to the next larger integer, no matter what fraction (or decimal) results.

Using the maximum error formula (8.2) can be made a little easier by rewriting the formula in a form that expresses n in terms of the other values.

Sample Size

$$n = \left(\frac{z_{(\alpha/2)} \cdot \sigma}{E}\right)^2 \tag{8.3}$$

 Video tutorial available—logon and learn more at cengagebrain.com

If the maximum error is expressed as a multiple of the standard deviation σ, then the actual value of σ is not needed in order to calculate the sample size.

EXAMPLE 8.7

DETERMINING THE SAMPLE SIZE WITHOUT A KNOWN VALUE OF SIGMA (σ)

Find the sample size needed to estimate the population mean to within $\frac{1}{5}$ of a standard deviation with 99% confidence.

Solution

Determine the confidence coefficient (using Table 4B): $1 - \alpha = 0.99$, $z(\alpha/2) = 2.58$. The desired maximum error is $E = \frac{\sigma}{5}$. Now you are ready to use the sample size formula (8.3):

$$n = \left(\frac{z(\alpha/2) \cdot \sigma}{E}\right)^2: \quad n = \left(\frac{(2.58) \cdot \sigma}{\sigma/5}\right)^2 = \left(\frac{(2.58\sigma)(5)}{\sigma}\right)^2 = [(2.58)(5)]^2$$

$$= (12.90)^2 = 166.41 = \mathbf{167}$$

SECTION 8.2 EXERCISES

8.21 Discuss the conditions that must exist before we can estimate the population mean using the interval techniques of formula (8.1).

8.22 Determine the value of the confidence coefficient $z(\alpha/2)$ for each situation described:

a. $1 - \alpha = 0.90$

b. $1 - \alpha = 0.95$

8.23 Determine the value of the confidence coefficient $z(\alpha/2)$ for each situation described:

a.   98% confidence

b.   99% confidence

8.24 Determine the level of the confidence given the confidence coefficient $z(\alpha/2)$ for each situation:

a. $z(\alpha/2) = 1.645$ b. $z(\alpha/2) = 1.96$

c. $z(\alpha/2) = 2.575$ d. $z(\alpha/2) = 2.05$

8.25 Given the information, the sampled population is normally distributed, $n = 16, \bar{x} = 28.7$, and $\sigma = 6$:

a.   Find the 0.95 confidence interval for μ.

b.   Are the assumptions satisfied? Explain.

8.26 Given the information, the sampled population is normally distributed, $n = 55, \bar{x} = 78.2$, and $\sigma = 12$:

a.   Find the 0.98 confidence interval for μ.

b.   Are the assumptions satisfied? Explain.

8.27 Given the information, $n = 86$, $\bar{x} = 128.5$, and $\sigma = 16.4$:

a.   Find the 0.90 confidence interval for μ.

b.   Are the assumptions satisfied? Explain.

8.28 Given the information, $n = 22, \bar{x} = 72.3$, and $\sigma = 6.4$:

a.   Find the 0.99 confidence interval for μ.

b.   Are the assumptions satisfied? Explain.

8.29 Based on the confidence interval formed in Exercise 8.27, give the value for each of the following:

a.   Point estimate

b.   Confidence coefficient

c.   Standard error of the mean

d.   Maximum error of estimate, E

(continue on page 358)

 Video tutorial available—logon and learn more at cengagebrain.com

e.   Lower confidence limit

f.   Upper confidence limit

8.30 Based on the confidence interval formed in Exercise 8.26, give the value for each of the following:

a.   Point estimate

b.   Confidence coefficient

c.   Standard error of the mean

d.   Maximum error of estimate, E

e.   Lower confidence limit

f.   Upper confidence limit

8.31 In your own words, describe the relationship between/among the following:

a.   Sample mean and point estimate

b.   Sample size, sample standard deviation, and standard error

c.   Standard error and maximum error

8.32 "Mean Travel Time to Work" (Applied Example 8.5) provides commuting information for various cities and states in the United States. Considering the city of Chicago and the confidence interval information given:

a.   What confidence interval term is the 32.7?

b.   What confidence interval term is the 31.9? The 33.5?

c.   What confidence interval term is the 90%?

d.   What is the maximum error?

e.   Calculate the standard error.

8.33   Skillbuilder Applet Exercise demonstrates the effect that the level of confidence $(1 - \alpha)$ has on the width of a confidence interval. Consider sampling from a population where $\mu = 300$ and $\sigma = 80$.

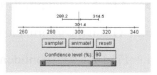

a.   Set the slider for level of confidence to 68%. Click "sample!" to construct one 68% confidence interval. Note the upper and lower confidence limits and calculate the width of the interval. Using "animate!" construct many samples and note the percent of intervals containing the true mean of 300. Click "stop" and "reset."

b.   Set the slider for level of confidence to 95%. Click "sample!" to construct one 95% confidence interval.

Note the upper and lower confidence limits and calculate the width of the interval. Using "animate!" construct many samples and note the percent of intervals containing the true mean of 300. Click "stop" and "reset."

c.   Set the slider for level of confidence to 99%. Click "sample!" to construct one 99% confidence interval. Note the upper and lower confidence limits and calculate the width of the interval. Using "animate!" construct many samples and note the percent of intervals containing the true mean of 300. Click "stop."

d.   Using the information collected in parts a–c, what effect does the level of confidence have on the width of the interval? Why is this happening?

8.34 Discuss the effect that each of the following has on the confidence interval:

a.   Point estimate

b.   Level of confidence

c.   Sample size

d.   Variability of the characteristic being measured

8.35 A machine produces parts with lengths that are normally distributed with $\sigma = 0.5$. A sample of 10 parts has a mean length of 75.92.

a.   Find the point estimate for μ.

b.   Find the 98% confidence maximum error of estimate for μ.

c.   Find the 98% confidence interval for μ.

8.36 A sample of 60 night-school students' ages is obtained in order to estimate the mean age of night-school students. $\bar{x} = 25.3$ years. The population variance is 16.

a.   Give a point estimate for μ.

b.   Find the 95% confidence interval for μ.

c.   Find the 99% confidence interval for μ.

8.37 Two hundred fish caught in Cayuga Lake had a mean length of 14.3 inches. The population standard deviation is 2.5 inches.

a.   Find the 90% confidence interval for the population mean length.

b.   Find the 98% confidence interval for the population mean length.

8.38 Based on a survey conducted by Greenfield Online, 25- to 34-year-olds spend the most each week on fast food. The average weekly amount of $44 (based on 115 respondents) was reported in a May 2009 *USA Today* Snapshot. Assuming that weekly fast food expenditures are normally distributed with a known standard deviation of $14.50, construct a 90% confidence interval for the mean weekly amount that 25- to 34-year-olds spend each week on fast food.

8.39 The Eurostar was Europe's first international train, designed to take advantage of the Channel Tunnel that connects England with Continental Europe. It carries nearly 800 passengers and occasionally reaches a peak speed of more than 190 mph [http://www.o-keating.com/]. Assume the standard deviation of train speed is 19 mph in the course of all the journeys back and forth and that the train's speed is normally distributed. Suppose speed readings are made during the next 20 trips of the Eurostar and the mean speed of these measurements is 184 mph.

a. What is the variable being studied?

b. Find the 90% confidence interval estimate for the mean speed.

c. Find the 95% confidence interval estimate for the mean speed.

8.40 The Trends in International Mathematics and Science Study (TIMSS) in 2007 examined eighth-grade proficiency in math and science. The mean mathematics scale score for the sample of eighth-grade students in the United States was 508.5 with a standard error of 2.83. Construct a 95% confidence interval for the mean mathematics score for all eighth-grade students in the United States.

8.41 [EX08-041] A certain adjustment to a machine will change the length of the parts it makes but will not affect the standard deviation. The length of the parts is normally distributed, and the standard deviation is 0.5 mm. After an adjustment is made, a random sample is taken to determine the mean length of the parts now being produced. The resulting lengths are as follows:

75.3	76.0	75.0	77.0	75.4	76.3	77.0	74.9	76.5	75.8

a. What is the parameter of interest?

b. Find the point estimate for the mean length of all parts now being produced.

c. Find the 0.99 confidence interval for μ.

8.42 "Tweens or teens" refers to the age group of students in seventh grade. Growth spurts are very common at this age. A sample of 12 randomly selected seventh-grade females in a New York City school resulted in the following heights:

67	63	65	64	63	64	63	57	67	68	63	65

Assuming heights of females in the 12- to 13-year-age group are normally distributed with a standard deviation of 2.56 inches:

a. What is the parameter of interest?

b. Find the point estimate for the population mean height of seventh-grade females.

c. Find the 0.95 confidence interval for the population mean height of seventh-grade females.

8.43 [EX08-043] The atomic weight of a reference sample of silver was measured at the National Institute of Standards and Technology (NIST) using two nearly identical mass spectrometers. This project was undertaken in conjunction with the redetermination of the Faraday constant. Following are 48 observations:

107.8681568	107.8681465	107.8681572

*** For remainder of data, logon at cengagebrain.com

Source: StatLib, http://lib.stat.cmu.edu/

Notice that the data differ only in the fifth, sixth, and seventh decimal places. Most computers will round the data and their calculated results; thus, the variation is seemingly lost. The statistics can be computed using just the last three digits of each data value (i.e., 107.8681568 will become 568). Algebraically this coding looks like this:

$$\text{Atom Wt Coded} = (\text{Atomic weight} - 107.8681000) \times 10,000,000$$

The data are listed in both the original and coded formats at cengagebrain.com.

a. Construct a graph of the coded data. How does the coding appear on the graph?

b. Find the mean and standard deviation of the coded data.

c. Convert the answers found in part b to original units.

d. Determine whether the data have an approximately normal distribution. Present your case.

e. Do the SDSM and CLT apply? Explain.

f. Is sigma known?

g. If the goal is to find the 95% confidence interval for the mean value of all observations, what would you do?

h. Find the 95% confidence interval for the mean value of all such observations. Justify your method.

8.44 [EX08-044] The force required to extract a cork from a wine bottle is an important property of the cork. If the force is too little, the cork probably is not a good

(continue on page 360)

protector of the wine inside. If the force is too great, it will be difficult to remove. Neither is desirable. The no. 9 corks in Applied Example 6.13 (p. 285) are believed to have an extraction force that is normally distributed with a standard deviation of 36 Newtons.

a. A sample of 20 randomly chosen bottles is selected for testing.

Extraction Force in Newtons

296	338	341	261	250	347	336	297	279	297
259	334	281	284	279	266	300	305	310	253

Find the 98% confidence interval for the mean extraction force.

b. During a different testing, a sample of eight bottles was randomly selected and tested.

Extraction Force in Newtons

331.9	312.0	289.4	303.6	346.9	308.1	346.9	276.0

Find the 98% confidence interval for the mean extraction force.

c. What effect did the two different sample means have on the answers in parts a and b? Explain.

d. What effect did the two different sample sizes have on the answers in parts a and b? Explain.

e. The mean extraction force was claimed to be 310 Newtons. Does either sample show sufficient reason to doubt the truthfulness of the claim? Explain.

8.45 "College costs rise" (October 29, 2008), an article on the CNN Money website, gave the latest figures from the College Board on annual tuition, fees, and room and board. The average total figures are $34,132 for private colleges and $14,333 for public colleges.

Source: http://money.cnn.com/

In an effort to compare those same costs in New York State, a sample of 32 junior students was randomly selected statewide from the private colleges and 32 more from the public colleges. The private college sample resulted in a mean of $34,020, and the public college sample mean was $14,045.

a. Assuming the annual college fees for private colleges have a mounded distribution and the standard deviation is $2200, find the 95% confidence interval for the mean college costs.

b. Assuming the annual college fees for public colleges have a mounded distribution and the standard deviation is $1500, find the 95% confidence interval for the mean college costs.

c. How do the New York State college costs compare to the College Board's values? Explain.

d. Compare the confidence intervals found in parts a and b and describe the effect the two different sample means had on the resulting answers.

e. Compare the confidence intervals found in parts a and b and describe the effect the two different sample standard deviations had on the resulting answers.

8.46 Using a computer or calculator, randomly select a sample of 40 single-digit numbers and find the 90% confidence interval for μ. Repeat several times, observing whether or not 4.5 is in the interval each time. Refer to Example 8.4, page 352. Describe your results.

FYI Use commands for generating integer data on page 91; then continue with confidence interval commands on pages 354–355.

8.47 Find the sample size needed to estimate μ of a normal population with $\sigma = 3$ to within 1 unit at the 98% level of confidence.

8.48 How large a sample should be taken if the population mean is to be estimated with 99% confidence to within $75? The population has a standard deviation of $900.

8.49 A high-tech company wants to estimate the mean number of years of college education its employees have completed. A good estimate of the standard deviation for the number of years of college is 1.0. How large a sample needs to be taken to estimate μ to within 0.5 of a year with 99% confidence?

8.50 By measuring the amount of time it takes a component of a product to move from one workstation to the next, an engineer has estimated that the standard deviation is 5 seconds.

a. How many measurements should be made to be 95% certain that the maximum error of estimation will not exceed 1 second?

b. What sample size is required for a maximum error of 2 seconds?

8.51 The new mini-laptop computers can deliver as much computing power as machines several times their size, but they weigh in at less than 3 lb. How large a sample would be needed to estimate the population mean weight if the maximum error of estimate is to be 0.4 of 1 standard deviation with 95% confidence?

8.52 According to the *USA Today* (March 11, 2009) article "College freshmen study booze more than book," freshmen spend an average of 8.4 hours studying per week. A large upstate college is interested in estimating this statistic with respect to its freshmen. How large will the sample need to be to estimate the mean within 1/8 of 1 standard deviation with 0.98 confidence?

8.3 The Nature of Hypothesis Testing

We make decisions every day of our lives. Some of these decisions are of major importance; others are seemingly insignificant. All decisions follow the same basic pattern. We weigh the alternatives; then, based on our beliefs and preferences and whatever evidence is available, we arrive at a decision and take the appropriate action. The statistical hypothesis test follows much the same process, except that it involves statistical information. In this section we develop many of the concepts and attitudes of the hypothesis test while looking at several decision-making situations without using any statistics.

A friend is having a party (Super Bowl party, home-from-college party—you know the situation, any reason will do), and you have been invited. You must make a decision: attend or not attend. That's simple—well, except that you want to go only if you can be convinced the party is going to be more fun than your friend's typical parties. Furthermore, you definitely do *not* want to go if the party is going to be just another dud. You have taken the position that "the party will be a dud" and you will not go unless you become convinced otherwise. Your friend assures you, "Guaranteed, the party will be a great time!" Do you go or not?

The decision-making process starts by identifying *something of concern* and then formulating *two hypotheses* about it.

> **Hypothesis** A statement that something is true.

Your friend's statement, "The party will be a great time," is a hypothesis. Your position, "The party will be a dud," is also a hypothesis.

> **Statistical hypothesis test** A process by which a decision is made between two opposing hypotheses. The two opposing hypotheses are formulated so that each hypothesis is the negation of the other. (That way, one of them is always true, and the other one is always false.) Then one hypothesis is tested in hopes that it can be shown to be a very improbable occurrence, thereby implying that the other hypothesis is likely the truth.

The two hypotheses involved in making a decision are known as the *null hypothesis* and the *alternative hypothesis*.

> **Null hypothesis,* H_o** The hypothesis we will test. Generally, this is a statement that a population parameter has a specific value. The null hypothesis is so named because it is the "starting point" for the investigation. (The phrase "there is no difference" is often used in its interpretation.)
>
> **Alternative hypothesis, H_a** A statement about the same population parameter that is used in the null hypothesis. Generally, this is a statement that specifies the population parameter has a value different, in some way, from the value given in the null hypothesis. The rejection of the null hypothesis will imply the likely truth of this alternative hypothesis.

With regard to your friend's party, the two opposing viewpoints or hypotheses are "The party will be a great time" and "The party will be a dud." Which statement becomes the null hypothesis, and which becomes the alternative hypothesis?

---

*We use the notation H_o for the null hypothesis to contrast it with H_a for the alternative hypothesis. Other texts may use H_0 (subscript zero) in place of H_o and H_1 in place of H_a.

Determining the statement of the null hypothesis and the statement of the alternative hypothesis is a very important step. The *basic idea* of the hypothesis test is for the evidence to have a chance to "disprove" the null hypothesis. The null hypothesis is the statement that the evidence might disprove. *Your concern* (belief or desired outcome), as the person doing the testing, is expressed in the alternative hypothesis. As the person making the decision, you believe that the evidence will demonstrate the feasibility of your "theory" by demonstrating the *unlikeliness* of the truth of the null hypothesis. The alternative hypothesis is sometimes referred to as the *research hypothesis* because it represents what the researcher hopes will be found to be "true."

Because the "evidence" (who's going to the party, what is going to be served, and so on) can demonstrate only the unlikeliness of the party being a dud, your initial position, "The party will be a dud," becomes the null hypothesis. Your friend's claim, "The party will be a great time," then becomes the alternative hypothesis.

H_o: "Party will be a dud "   vs. H_a: "Party will be a great time"

The following examples will illustrate the formation of and the relationship between null and alternative hypotheses.

EXAMPLE 8.8

WRITING HYPOTHESES

You are testing a new design for air bags used in automobiles, and you are concerned that they might not open properly. State the null and alternative hypotheses.

Solution

The two opposing possibilities are "Bags open properly" and "Bags do not open properly." Testing could produce evidence that discredits the hypothesis "Bags open properly"; plus your concern is that "Bags do not open properly." Therefore, "Bags do not open properly" would become the alternative hypothesis and "Bags open properly" would be the null hypothesis.

The alternative hypothesis can also be the statement the experimenter wants to show to be true.

EXAMPLE 8.9

WRITING HYPOTHESES

An engineer wishes to show that the new formula that was just developed results in a quicker-drying paint. State the null and alternative hypotheses.

Solution

The two opposing possibilities are "does dry quicker" and "does not dry quicker." Because the engineer wishes to show "does dry quicker," the alternative hypothesis is "Paint made with the new formula does dry quicker" and the null hypothesis is "Paint made with the new formula does not dry quicker."

Occasionally it might be reasonable to hope that the evidence does not lead to a rejection of the null hypothesis. Such is the case in Example 8.10.

EXAMPLE 8.10

WRITING HYPOTHESES

You suspect that a brand-name detergent outperforms the store's brand of detergent, and you wish to test the two detergents because you would prefer to buy the cheaper store brand. State the null and alternative hypotheses.

Solution

Your suspicion, "The brand-name detergent outperforms the store brand," is the reason for the test and therefore becomes the alternative hypothesis.

H_o: "There is no difference in detergent performance."
H_a: "The brand-name detergent performs better than the store brand."

However, as a consumer, you are hoping not to reject the null hypothesis for budgetary reasons.

APPLIED EXAMPLE 8.11

EVALUATION OF TEACHING TECHNIQUES

ABSTRACT: THIS STUDY TESTS THE EFFECT OF HOMEWORK COLLECTION AND QUIZZES ON EXAM SCORES.

The hypothesis for this study is that an instructor can improve a student's performance (exam scores) through influencing the student's perceived effort-reward probability. An instructor accomplishes this by assigning tasks (teaching techniques) which are a part of a student's grade and are perceived by the student as a means of improving his or her grade in the class. The student is motivated to increase effort to complete those tasks which should also improve understanding of course material. The expected final result is improved exam scores. The null hypothesis for this study is:

H_o: Teaching techniques have no significant effect on students' exam scores. . . .

Source: "Evaluation of Teaching Techniques" by David R. Vruwink and Janon R. Otto, published in *The Accounting Review*, Vol. LXII, No. 2, April 1987. Reprinted by permission.

Before returning to our example about the party, we need to look at the four possible outcomes that could result from the null hypothesis being either true or false and the decision being either to "reject H_o" or to "fail to reject H_o." Table 8.3 shows these four possible outcomes.

A **type A correct decision** occurs when the null hypothesis is true and we decide in its favor. A **type B correct decision** occurs when the null hypothesis is false and the decision is in opposition to the null hypothesis. A **type I error** is committed when a true null hypothesis is rejected—that is, when the null hypothesis is true but we decide against it. A **type II error** is committed when we decide in favor of a null hypothesis that is actually false.

TABLE 8.3 Four Possible Outcomes in a Hypothesis Test

	Null Hypothesis	
Decision	True	False
Fail to reject H_o	Type A correct decision	Type II error
Reject H_o	Type I error	Type B correct decision

EXAMPLE 8.12

DESCRIBING THE POSSIBLE OUTCOMES AND RESULTING ACTIONS (ON HYPOTHESIS TESTS)

Describe the four possible outcomes and the resulting actions that would occur for the hypothesis test in Example 8.10.

Solution

Recall: H_o: "There is no difference in detergent performance."
H_a: "The brand-name detergent performs better than the store brand."

	Null Hypothesis Is True	Null Hypothesis Is False
Fail to Reject H_o	**Type A Correct Decision** Truth of situation: There is no difference between the detergents. Conclusion: It was determined that there is no difference. Action: The consumer buys the cheaper detergent, saving money and getting the same results.	**Type II Error** Truth of situation: The brand-name detergent is better. Conclusion: It was determined that there is no difference. Action: The consumer buys the cheaper detergent, saving money but getting inferior results.
Reject H_o	**Type I Error** Truth of situation: There is no difference between the detergents. Conclusion: It was determined that the brand-name detergent is better. Action: The consumer buys the brand-name detergent, spending extra money to attain no better results.	**Type B Correct Decision** Truth of situation: The brand-name detergent is better. Conclusion: It was determined that the brand-name detergent is better. Action: The consumer buys the brand-name detergent, spending more and getting better results.

Notes:

1. The truth of the situation is not known before the decision is made, the conclusion reached, and the resulting actions take place. The truth of H_o may never be known.
2. The type II error often results in what represents a "lost opportunity"; lost in this situation is the chance to use a product that yields better results.

When a decision is made, it would be nice to always make the correct decision. This, however, is not possible in statistics because we make our decisions on the basis of sample information. The best we can hope for is to control the

 Video tutorial available—logon and learn more at cengagebrain.com

probability with which an error occurs. The probability assigned to the type I error is α (called **"alpha"**). The probability of the type II error is β (called **"beta"**; β is the second letter of the Greek alphabet). See Table 8.4.

TABLE 8.4 Probability with which Decisions Occur

Error in Decision	Type	Probability	Correct Decision	Type	Probability
Rejection of a true H_o	I	α	Failure to reject a true H_o	A	$1 - \alpha$
Failure to reject a false H_o	II	β	Rejection of a false H_o	B	$1 - \beta$

To control these errors we assign a small probability to each of them. The most frequently used probability values for α and β are 0.01 and 0.05. The probability assigned to each error depends on its seriousness. The more serious the error, the less willing we are to have it occur; therefore, a smaller probability will be assigned. α and β are probabilities of errors, each under separate conditions, and they cannot be combined. Therefore, we cannot determine a single probability for making an incorrect decision. Likewise, the two correct decisions are distinctly separate, and each has its own probability; $1 - \alpha$ is the probability of a correct decision when the null hypothesis is true, and $1 - \beta$ is the probability of a correct decision when the null hypothesis is false. $1 - \beta$ is called the *power of the statistical test* because it is the measure of the ability of a hypothesis test to reject a false null hypothesis, a very important characteristic.

Note: Regardless of the outcome of a hypothesis test, you can never be certain that a correct decision has been reached.

Let's look back at the two possible errors in decision that could occur in Example 8.10. Most people would become upset if they found out they were spending extra money for a detergent that performed no better than the cheaper brand. Likewise, many people would become upset if they found out they could have been buying a better detergent. Evaluating the relative seriousness of these errors requires knowing whether this is your personal laundry or a professional laundry business, how much extra the brand-name detergent costs, and so on.

There is an interrelationship among the probability of the type I error (α), the probability of the type II error (β), and the sample size (n). This is very much like the interrelationship among level of confidence, maximum error, and sample size discussed on pages 355–356. Figure 8.8 shows the "three-way tug-of-war" among α, β, and n. If any one of the three is increased or decreased, it has an effect on one or both of the others. The statistician's job is thus to "balance" the three values of α, β, and n to achieve an acceptable testing situation.

FIGURE 8.8

The "Three-Way Tug-of-War" between α, β, and n

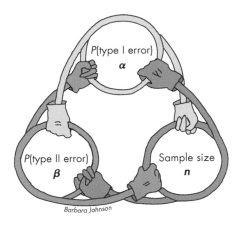

Barbara Johnson

If α is reduced, then either β must increase or n must be increased; if β is decreased, then either α increases or n must be increased; if n is decreased, then either α increases or β increases. The choices for α, β, and n are definitely not arbitrary. At this time in our study of statistics, only the sample size, n, and α, P(type I error), will be given and used to complete a hypothesis test. β, P(type II error), is further investigated in the section exercises but will not be utilized in this introduction to hypothesis testing.

Sample size, n, is self-explanatory, so let's look at the role of α.

> **Level of significance** α The probability of committing a type I error.

Establishing the level of significance can be thought of as a "managerial decision." Typically, someone in charge determines the level of probability with which he or she is willing to risk a type I error.

At this point in the hypothesis test procedure, the evidence is collected and summarized and the value of a *test statistic* is calculated.

> **Test statistic** A random variable whose value is calculated from the sample data and is used in making the decision "reject H_o" or "fail to reject H_o."

The value of the calculated test statistic is used in conjunction with a decision rule to determine either "reject H_o" or "fail to reject H_o." This **decision rule** must be established prior to collecting the data; it specifies how you will reach the decision.

Back to your friend's party: You have to weigh the history of your friend's parties, the time and place, others going, and so on, against your own criteria and then make your decision. As a result of the decision about the null hypothesis ("The party will be a dud"), you will take the appropriate action; you will either go or not go to the party.

To complete a hypothesis test, you will need to write a conclusion that carefully describes the meaning of the decision relative to the intent of the hypothesis test.

> **The Conclusion**
>
> a. If the decision is "reject H_o," then the conclusion should be worded something like, "There is sufficient evidence at the α level of significance to show that . . . [the meaning of the alternative hypothesis]."
> b. If the decision is "fail to reject H_o," then the conclusion should be worded something like, "There is not sufficient evidence at the α level of significance to show that . . . [the meaning of the alternative hypothesis]."

When writing the decision and the conclusion, remember that (1) the decision is about H_o and (2) the conclusion is a statement about whether or not the contention of H_a was upheld. This is consistent with the "attitude" of the whole hypothesis test procedure. The null hypothesis is the statement that is "on trial," and therefore the decision must be about it. The contention of the alternative hypothesis is the thought that brought about the need for a decision. Therefore, the question that led to the alternative hypothesis must be answered when the conclusion is written.

We must always remember that when the decision is made, nothing has been proved. Both decisions can lead to errors: "fail to reject H_o" could be a type II error (the lack of sufficient evidence has led to great parties being missed more than once), and "reject H_o" could be a type I error (more than one person has decided to go to a party that was a dud).

SECTION 8.3 EXERCISES

8.53 "Positively the Most Effective Sciatica & Back Pain Relief System on the Planet...," according to Dr. Craig Mueller. This claim appeared on www.EraseYourBackPain.com.

a. How would you attempt to show that the above statement is true? What evidence would you need to collect?

b. How would you attempt to disprove the above statement? What evidence would you need to collect?

c. Would it make more sense to attempt to prove the above statement true or to disprove it? Explain.

8.54 You have heard that the average American male is 69.7 inches tall. Yet almost all of the adult males in your personal world seem to be between 5 and 6 feet tall, with a few just over 6 feet. So how does the average get to be almost 6 feet? In fact, you are quite convinced that the average height might be something less than 69.7 inches.

a. How would you attempt to show that the above statement of "less than 69.7 inches" is true? What evidence would you collect? What would the collected evidence need to show in order to convince you of the truth of the statement?

b. How would you attempt to disprove the above statement of "less than 69.7 inches"? What evidence would you collect? What would the collected evidence need to show in order to disprove the statement?

c. Would it be easier to prove that "less than 69.7 inches" is true or to disprove it? Explain.

8.55 You are testing a new detonating system for explosives and are concerned that the system is not reliable. State the null and alternative hypotheses.

8.56 Referring to Applied Example 8.11, state the instructor's hypothesis, the alternative hypothesis.

8.57 State the null and alternative hypotheses for each of the following:

a. You are investigating a complaint that "special delivery mail takes too much time" to be delivered.

b. You want to show that people find the new design for a recliner chair more comfortable than the old design.

c. You are trying to show that cigarette smoke affects the quality of a person's life.

d. You are testing a new formula for hair conditioner and hope to show that it is effective on "split ends."

8.58 State the null and alternative hypotheses for each of the following:

a. You want to show an increase in buying and selling of single-family homes this year when compared with last year's rate.

b. You are testing a new recipe for "low-fat" cheesecake and expect to find that its taste is not as good as traditional cheesecake.

c. You are trying to show that music lessons have a positive effect on a child's self-esteem.

d. You are investigating the relationship between a person's gender and the automobile he or she drives—specifically you want to show that males tend to drive truck-type vehicles more than females do.

8.59 Using the example of your friend's party (pp. 361 and 366) with H_o: "Party will be a dud" versus H_a: "The party will be a great time," describe the four possible decisions and the resulting actions as described in Example 8.12.

8.60 When a parachute is inspected, the inspector is looking for anything that might indicate the parachute might not open.

a. State the null and alternative hypotheses.

b. Describe the four possible outcomes that can result depending on the truth of the null hypothesis and the decision reached.

c. Describe the seriousness of the two possible errors.

8.61 When a medic at the scene of a serious accident inspects each victim, she administers the appropriate medical assistance to all victims, unless she is certain the victim is dead.

a. State the null and alternative hypotheses.

b. Describe the four possible outcomes that can result depending on the truth of the null hypothesis and the decision reached.

c. Describe the seriousness of the two possible errors.

8.62 A supplier of highway construction materials claims he can supply an asphalt mixture that will make roads that are paved with his materials less slippery when wet. A general contractor who builds roads wishes to test the supplier's claim. The null hypothesis is "Roads paved with this asphalt mixture are no less slippery than roads

(continue on page 368)

paved with other asphalt." The alternative hypothesis is "Roads paved with this asphalt mixture are less slippery than roads paved with other asphalt."

a. Describe the meaning of the two possible types of errors that can occur in the decision when this hypothesis test is completed.

b. Describe how the null hypothesis, as stated previously, is a "starting point" for the decision to be made about the asphalt.

8.63 Using the information from Exercise 8.59, describe how the type II error in the party example represents a "lost opportunity."

8.64 Describe the actions that would result in a type I error and a type II error if each of the following null hypotheses were tested. (Remember, the alternative hypothesis is the negation of the null hypothesis.)

a. H_o: The majority of Americans favor laws against assault weapons.

b. H_o: The choices on the fast food menu are not low in salt.

c. H_o: This building must not be demolished.

d. H_o: There is no waste in government spending.

8.65 Describe the action that would result in a correct decision type A and a correct decision type B if each of the null hypotheses in Exercise 8.64 were tested.

8.66 Describe the action that would result in a correct decision type A and a correct decision type B if the hypotheses for the new detonating system for explosives of Exercise 8.55 were tested.

8.67 Consider the null hypothesis in Applied Example 8.11, "H_o: Teaching techniques have no significant effect on students' exam scores." Describe the actions that would result in a type I and a type II error if H_o were tested.

8.68 Consider the null hypothesis in Applied Example 8.11, "H_o: Teaching techniques have no significant effect on students' exam scores." Describe the actions that would result in a correct decision type A and correct decision type B if H_o were tested.

8.69 a. If the null hypothesis is true, what decision error could be made?

b. If the null hypothesis is false, what decision error could be made?

c. If the decision "reject H_o" is made, what decision error could have been made?

d. If the decision "fail to reject H_o" is made, what decision error could have been made?

8.70 The director of an advertising agency is concerned with the effectiveness of a television commercial.

a. What null hypothesis is she testing if she commits a type I error when she erroneously says that the commercial is effective?

b. What null hypothesis is she testing if she commits a type II error when she erroneously says that the commercial is effective?

8.71 The director of an advertising agency is concerned with the effectiveness of a television commercial.

a. What null hypothesis is she testing if she makes a correct decision type A when she correctly says that the commercial is not effective?

b. What null hypothesis is she testing if she makes a correct decision type B when she correctly says that the commercial is not effective?

8.72 A politician is concerned with winning an upcoming election.

a. What null hypothesis is he testing if he commits a type I error when he erroneously says that he will win the election?

b. What null hypothesis is he testing if he commits a type II error when he erroneously says that he will win the election?

8.73 a. If α is assigned the value 0.001, what are we saying about the type I error?

b. If α is assigned the value 0.05, what are we saying about the type I error?

c. If α is assigned the value 0.10, what are we saying about the type I error?

8.74 a. If β is assigned the value 0.001, what are we saying about the type II error?

b. If β is assigned the value 0.05, what are we saying about the type II error?

c. If β is assigned the value 0.10, what are we saying about the type II error?

8.75 a. If the null hypothesis is true, the probability of a decision error is identified by what name?

b. If the null hypothesis is false, the probability of a decision error is identified by what name?

8.76 Suppose that a hypothesis test is to be carried out by using $\alpha = 0.05$. What is the probability of committing a type I error?

8.77 Explain why α is not always the probability of rejecting the null hypothesis.

8.78 Explain how assigning a small probability to an error controls the likelihood of its occurrence.

8.79 The conclusion is the part of the hypothesis test that communicates the findings of the test to the reader. As such, it needs special attention so that the reader receives an accurate picture of the findings.

a.   Carefully describe the "attitude" of the statistician and the statement of the conclusion when the decision is "reject H_o."

b.   Carefully describe the "attitude" and the statement of the conclusion when the decision is "fail to reject H_o."

8.80 Find the power of a test when the probability of the type II error is:

a.   0.01   b.   0.05   c.   0.10

8.81 A normally distributed population is known to have a standard deviation of 5, but its mean is in question. It has been argued to be either $\mu = 80$ or $\mu = 90$, and the following hypothesis test has been devised to settle the argument. The null hypothesis, H_o: $\mu = 80$, will be tested by using one randomly selected data value and comparing it with the critical value of 86. If the data value is greater than or equal to 86, the null hypothesis will be rejected.

a.   Find α, the probability of the type I error.

b.   Find β, the probability of the type II error.

8.82 Suppose the argument in Exercise 8.81 was to be settled using a sample of size 4; find α and β.

8.83 [EX08-083] You are a quality-control inspector and are in a position to make the decision as to whether a large shipment of cork stoppers for use in bottling still (versus bubbly) wine passes inspection. Once you inspect the mandatory number in the approved manner, you will make a decision to accept or reject the lot.

   Part 1 of the inspection requires you to randomly select 32 corks and measure three physical dimensions of the cylindrical stopper according to defined procedures.

Specification Limits

Diameter	24 mm $\pm$ 0.5 mm
Ovalization	$\leq$ 0.7 mm
Length	45 mm $\pm$ 0.7 mm

Acceptance Quality Levels (AQL)

The batch is accepted if no more than two corks present an inferior or superior result to the limits of specification.

   The batch may be refused if three or more corks present an inferior or superior result to the limits of specification.

Source: http://www.codiliege.org

The results of inspecting the mandated sample follow. (All measurements are in millimeters.)

a.   Determine the number of corks that pass part 1 of the inspection.

b.   State the decision and explain how it was reached.

c.   Prepare a short written report summarizing the requirements and your findings and decision.

8.84 [EX08-084] As the quality-control inspector in Exercise 8.83, you are ready for the second phase of the inspection.

(continue on page 370)

Table for Exercise 8.83

Cork	1	2	3	4	5	6	7	8
Diameter	24.51	24.13	24.28	24.27	23.79	24.11	24.08	23.66
Ovalization	0.20	0.88	0.38	0.20	0.29	0.14	0.20	0.32
Length	44.89	44.69	45.36	44.94	44.65	45.50	44.86	44.67

Cork	9	10	11	12	13	14	15	16
Diameter	24.41	24.08	24.02	23.94	23.71	24.18	24.13	24.30
Ovalization	0.03	0.43	0.50	0.43	0.51	0.46	0.53	0.14
Length	45.13	44.92	44.88	45.14	44.87	44.67	45.01	44.86

Cork	17	18	19	20	21	22	23	24
Diameter	23.78	24.01	24.03	24.10	23.77	24.28	23.85	24.39
Ovalization	0.07	0.32	0.34	0.23	0.76	0.39	0.47	0.43
Length	45.12	45.21	45.70	44.95	44.27	45.23	45.29	44.98

Cork	25	26	27	28	29	30	31	32
Diameter	24.27	23.92	24.23	24.17	23.77	24.40	24.31	23.85
Ovalization	0.20	0.47	0.23	0.23	0.28	0.34	0.56	0.05
Length	44.80	45.06	45.38	45.11	44.75	45.42	45.04	44.53

Part 2 requires that the humidity percentage of 20 cork stoppers be determined while following the prescribed procedure.

Specification Limits

Nominal value: 6%

Limits of specification: ±2% (i.e., from 4% to 8%)

Acceptance Quality Levels (AQL)

The batch is accepted if no more than two corks present an inferior or superior result to the limits of specification.

The batch may be refused if three or more corks present an inferior or superior result to the limits of specification.

Source: http://www.codiliege.org

Listed are three different samples, each taken from different lots. Review the sample results and answer these questions for each sample separately.

Sample 1	5	5	6	3	7	6	6	7	8	6
	6	7	5	7	6	6	7	6	4	5
Sample 2	1	6	6	8	6	5	7	6	10	6
	7	5	7	6	5	6	6	8	5	9
Sample 3	5	7	3	5	5	5	6	5	9	3
	5	7	7	9	7	8	5	10	8	9

a. Construct a dotplot of the data.

b. Completely label the dotplot and circle the dots representing cork percents inferior or superior to the limits of specification.

c. State the decision and explain how it was reached.

d. Prepare a short written report summarizing the requirements and your findings and decision for each sample.

8.4 Hypothesis Test of Mean μ (σ Known): A Probability-Value Approach

In Section 8.3 we surveyed the concepts and much of the reasoning behind a hypothesis test while looking at nonstatistical examples. In this section we are going to formalize the hypothesis test procedure as it applies to statements concerning the mean μ of a population under the restriction that σ, the population standard deviation, is a known value.

> **The assumption for hypothesis tests about mean μ using a known σ** The sampling distribution of $\bar{x}$ has a normal distribution.

The information we need to ensure that this assumption is satisfied is contained in the sampling distribution of sample means and in the CLT:

FYI If the assumptions are not met for hypothesis tests about the mean μ using a known σ, the calculated p-value could cause a wrong decision about H_o.

The sampling distribution of sample means $\bar{x}$ is distributed about a mean equal to μ with a standard error equal to $\sigma/\sqrt{n}$; and (1) if the randomly sampled population is normally distributed, then $\bar{x}$ is normally distributed for all sample sizes, or (2) if the randomly sampled population is not normally distributed, then $\bar{x}$ is approximately normally distributed for sufficiently large sample sizes.

The hypothesis test is a well-organized, step-by-step procedure used to make a decision. Two different formats are commonly used for hypothesis testing. The *probability-value approach*, or simply *p-value approach*, is the hypothesis test process that has gained popularity in recent years, largely as a result of the convenience and the "number-crunching" ability of the computer. This approach is organized as a five-step procedure.

THE PROBABILITY-VALUE HYPOTHESIS TEST: A FIVE-STEP PROCEDURE

Step 1 The Set-Up:
a. Describe the population parameter of interest.
b. State the null hypothesis (H_o) and the alternative hypothesis (H_a).

Step 2 The Hypothesis Test Criteria:
a. Check the assumptions.
b. Identify the probability distribution and the test statistic to be used.
c. Determine the level of significance, α.

Step 3 The Sample Evidence:
a. Collect the sample information.
b. Calculate the value of the test statistic.

Step 4 The Probability Distribution:
a. Calculate the p-value for the test statistic.
b. Determine whether or not the p-value is smaller than α.

Step 5 The Results:
a. State the decision about H_o.
b. State the conclusion about H_a.

FYI Think about the consequences of using weak rivets.

A commercial aircraft manufacturer buys rivets to use in assembling airliners. Each rivet supplier that wants to sell rivets to the aircraft manufacturer must demonstrate that its rivets meet the required specifications. One of the specs is, "The mean shearing strength of all such rivets, μ, is at least 925 lb." Each time the aircraft manufacturer buys rivets, it is concerned that the mean strength might be less than the 925-lb specification.

Note: Each individual rivet has a shearing strength, which is determined by measuring the force required to shear ("break") the rivet. Clearly, not all the rivets can be tested. Therefore, a sample of rivets will be tested, and a decision about the mean strength of all the untested rivets will be based on the mean from those sampled and tested.

STEP 1 The Set-Up:

a. **Describe the population parameter of interest.**
The population parameter of interest is the mean μ, the mean shearing strength of (or mean force required to shear) the rivets being considered for purchase.

FYI More specific instructions are given on pages 361–363.

b. **State the null hypothesis (H_o) and the alternative hypothesis (H_a).**
The null hypothesis and the alternative hypothesis are formulated by inspecting the problem or statement to be investigated and first formulating two opposing statements about the mean μ. For our example, these two opposing statements are (A) "The mean shearing strength is less than 925" ($\mu < 925$, the aircraft manufacturer's concern), and (B) "The mean shearing strength is at least 925" ($\mu = 925$, the rivet supplier's claim and the aircraft manufacturer's spec).

Note: The trichotomy law from algebra states that two numerical values must be related in exactly one of three possible relationships: $<$, $=$, or $>$. All three of these possibilities must be accounted for in the two opposing hypotheses in order for the two hypotheses to be negations of each other. The three possible combinations of signs and hypotheses are shown in Table 8.5. Recall that the null hypothesis assigns a specific value to the parameter in question, and therefore "equals" will always be part of the null hypothesis.

TABLE 8.5 The Three Possible Statements of Null and Alternative Hypotheses

Null Hypothesis	Alternative Hypothesis
1. Greater than or equal to ($\geq$)	Less than ($<$)
2. Less than or equal to ($\leq$)	Greater than ($>$)
3. Equal to ($=$)	Not equal to ($\neq$)

The parameter of interest, the population mean μ, is related to the value 925. Statement (A) becomes the alternative hypothesis:

$$H_a: \mu < 925 \text{ (the mean is less than 925)}$$

This statement represents the aircraft manufacturer's concern and says, "The rivets do not meet the required specs." Statement (B) becomes the null hypothesis:

$$H_o: \mu = 925 \; (\geq) \text{ (the mean is at least 925)}$$

This hypothesis represents the negation of the aircraft manufacturer's concern and says, "The rivets do meet the required specs."

Note: We will write the null hypothesis with just the equal sign, thereby stating the exact value assigned. When "equal" is paired with "less than" or with "greater than," the combined symbol is written beside the null hypothesis as a reminder that all three signs have been accounted for in these two opposing statements.

Before continuing with our example, let's look at three examples that demonstrate formulating the statistical null and alternative hypotheses involving the population mean μ. Examples 8.13 and 8.14 each demonstrate a "one-tailed" alternative hypothesis.

EXAMPLE 8.13

WRITING NULL AND ALTERNATIVE HYPOTHESES (ONE-TAILED SITUATION)

Suppose the Environmental Protection Agency is suing the city of Rochester for noncompliance with carbon monoxide standards. Specifically, the EPA would want to show that the mean level of carbon monoxide in downtown Rochester's air is dangerously high, higher than 4.9 parts per million. State the null and alternative hypotheses.

Solution

To state the two hypotheses, we first need to identify the population parameter in question: the "mean level of carbon monoxide in Rochester." The parameter μ is being compared with the value 4.9 parts per million, the specific value of interest. The EPA is questioning the value of μ and wishes to show it is higher than 4.9 (i.e., $\mu > 4.9$). The three possible relationships—(1) $\mu < 4.9$, (2) $\mu = 4.9$, and (3) $\mu > 4.9$—must be arranged to form two opposing statements: one states the EPA's position, "The mean level is higher than 4.9 ($\mu > 4.9$)," and the other states the negation, "The mean level is not higher than 4.9 ($\mu \leq 4.9$)." One of these two statements will become the null hypothesis, H_o, and the other will become the alternative hypothesis, H_a.

Recall that there are two rules for forming the hypotheses: (1) the null hypothesis states that the parameter in question has a specified value ("H_o must contain the equal sign"), and (2) the EPA's contention becomes the alternative hypothesis ("higher than"). Both rules indicate:

$$H_o: \mu = 4.9 \ (\leq) \quad \text{and} \quad H_a: \mu > 4.9$$

EXAMPLE 8.14

WRITING NULL AND ALTERNATIVE HYPOTHESES (ONE-TAILED SITUATION)

An engineer wants to show that applications of paint made with the new formula dry and are ready for the next coat in a mean time of less than 30 minutes. State the null and alternative hypotheses for this test situation.

Solution

The parameter of interest is the mean drying time per application, and 30 minutes is the specified value. $\mu < 30$ corresponds to "The mean time is less than 30," whereas $\mu \geq 30$ corresponds to the negation, "The mean time is not less than 30." Therefore, the hypotheses are

$$H_o: \mu = 30 \ (\geq) \quad \text{and} \quad H_a: \mu < 30$$

Example 8.15 demonstrates a "two-tailed" alternative hypothesis.

EXAMPLE 8.15

WRITING NULL AND ALTERNATIVE HYPOTHESES (TWO-TAILED SITUATION)

Job satisfaction is very important to worker productivity. A standard job-satisfaction questionnaire was administered by union officers to a sample of assembly-line workers in a large plant in hopes of showing that the assembly workers' mean score on this questionnaire would be different from the established mean of 68. State the null and alternative hypotheses.

Solution

Either the mean job-satisfaction score is different from 68 ($\mu \neq 68$) or the mean is equal to 68 ($\mu = 68$). Therefore,

$$H_o: \mu = 68 \quad \text{and} \quad H_a: \mu \neq 68$$

 Video tutorial available—logon and learn more at cengagebrain.com

Notes:

1. The alternative hypothesis is referred to as being "two-tailed" when H_a is "not equal."
2. When "less than" is combined with "greater than," they become "not equal to."

The viewpoint of the experimenter greatly affects the way the hypotheses are formed. Generally, the experimenter is trying to show that the parameter value is different from the value specified. Thus, the experimenter is often hoping to be able to reject the null hypothesis so that the experimenter's theory can be substantiated. Examples 8.13, 8.14, and 8.15 also represent the three possible arrangements for the $<$, $=$, and $>$ relationships between the parameter μ and a specified value.

Table 8.6 lists some additional common phrases used in claims and indicates their negations and the hypothesis in which each phrase will be used. Again, notice that "equals" is always in the null hypothesis. Also notice that the negation of "less than" is "greater than or equal to." Think of negation as "all the others" from the set of three signs.

TABLE 8.6 Common Phrases and Their Negations

H_o: ($\geq$) vs. H_a: ($<$)	H_o: ($\leq$) vs. H_a: ($>$)	H_o: ($=$) vs. H_a: ($\neq$)
At least Less than	At most More than	Is Is not
No less than Less than	No more than More than	Not different from Different from
Not less than Less than	Not greater than Greater than	Same as Not same as

After the null and alternative hypotheses are established, we will work under the assumption that the null hypothesis is a true statement until there is sufficient evidence to reject it. This situation might be compared with a courtroom trial, where the accused is assumed to be innocent (H_o: Defendant is innocent versus H_a: Defendant is not innocent) until sufficient evidence has been presented to show that innocence is totally unbelievable ("beyond reasonable doubt"). At the conclusion of the hypothesis test, we will make one of two possible decisions. We will decide in opposition to the null hypothesis and say that we "reject H_o" (this corresponds to "conviction" of the accused in a trial), or we will decide in agreement with the null hypothesis and say that we "fail to reject H_o" (this corresponds to "fail to convict" or an "acquittal" of the accused in a trial).

Let's return to the rivet example we interrupted on page 371 and continue with Step 2. Recall that

$$H_o: \mu = 925 \ (\geq) \ \text{(at least 925)} \qquad H_a: \mu < 925 \ \text{(less than 925)}$$

STEP 2 The Hypothesis Test Criteria:

    a. Check the assumptions.

        Assume that the standard deviation of the shearing strength of rivets is known from past experience to be $\sigma = 18$. Variables like shearing strength typically have a mounded distribution; therefore, a sample of size 50 should be large enough for the CLT to apply and ensure that the SDSM will be normally distributed.

    b. Identify the probability distribution and the test statistic to be used.

        The standard normal probability distribution is used because $\bar{x}$ is expected to have a normal distribution.

For a hypothesis test of μ, we want to compare the value of the sample mean with the value of the population mean as stated in the null hypothesis. This comparison is accomplished using the test statistic in formula (8.4):

Test Statistic for Mean

$$z\star = \frac{\bar{x} - \mu}{\sigma / \sqrt{n}} \qquad\qquad (8.4)$$

The resulting calculated value is identified as $z\star$ ("z star") because it is expected to have a standard normal distribution when the null hypothesis is true and the assumptions have been satisfied. The $\star$ ("star") is to remind us that this is the calculated value of the test statistic.

$$\text{The test statistic to be used is } z\star = \frac{\overline{x} - \mu}{\sigma/\sqrt{n}} \text{ with } \sigma = 18.$$

c. Determine the level of significance, α.

Setting α was described as a managerial decision in Section 8.3. To see what is involved in determining α, the probability of the type I error, for our rivet example, we start by identifying the four possible outcomes, their meanings, and the action related to each.

The type I error occurs when a true null hypothesis is rejected. This would occur when the manufacturer tested rivets that did meet the specs and rejected them. Undoubtedly this would lead to the rivets not being purchased even though they do meet the specs. In order for the manager to set a level of significance, related information is needed—namely, how soon is the new supply of rivets needed? If they are needed tomorrow and this is the only vendor with an available supply, waiting a week to find acceptable rivets could be very expensive; therefore, rejecting good rivets could be considered a serious error. On the other hand, if the rivets are not needed until next month, then this error may not be very serious. Only the manager will know all the ramifications, and therefore the manager's input is important here.

After much consideration, the manager assigns the level of significance: $\alpha = 0.05$.

STEP 3 The Sample Evidence:

a. Collect the sample information.
The sample must be a random sample drawn from the population whose mean μ is being questioned. A random sample of 50 rivets is selected, each rivet is tested, and the sample mean shearing strength is calculated: $\overline{x} = 921.18$ and $n = 50$.

b. Calculate the value of the test statistic.
The sample evidence ($\overline{x}$ and n found in Step 3a) is next converted into the **calculated value of the test statistic, $z\star$**, using formula (8.4). (μ is 925 from H_o, and $\sigma = 18$ is a known quantity.) We have

$$z\star = \frac{\overline{x} - \mu}{\sigma/\sqrt{n}}: \quad z\star = \frac{921.18 - 925.0}{18/\sqrt{50}} = \frac{-3.82}{2.5456} = -1.50$$

STEP 4 The Probability Distribution:

a. Calculate the p-value for the test statistic.

> **Probability value, or p-value** The probability that the test statistic could be the value it is or a more extreme value (in the direction of the alternative hypothesis) when the null hypothesis is true. (*Note:* The symbol **P** will be used to represent the p-value, especially in algebraic situations.)

Draw a sketch of the standard normal distribution and locate $z\star$ (found in Step 3b) on it. To identify the area that represents the p-value, look at the sign in the alternative hypothesis. For this test, the alternative hypothesis indicates that we are interested in that part of the sampling distribution that is "*less than*" $z\star$. Therefore, the p-value is the area that lies to the *left* of $z\star$. Shade this area.

FYI Complete instructions for using Table 3 are given on pages 272–276.

FYI You will use only one of these three equivalent methods.

FYI Instructions for using this computer command are given on page 285. Try it! See if you get the same answer.

To find the p-value, you may use any one of the three methods outlined here. The method you use is not the important thing, because each method is just the tool of choice to help you find the p-value.

Method 1: Use Table 3 in Appendix B to determine the tabled area related to the left of $z = -1.50$:

$$p\text{-value} = P(z < z\star) = P(z < -1.50) = \mathbf{0.0668}$$

Method 2: Use Table 5 in Appendix B and the symmetry property: Table 5 is set up to allow you to read the p-value directly from the table. Since $P(z < -1.50) = P(z > 1.50)$, simply locate $z\star = 1.50$ on Table 5 and read the p-value:

$$P(z < -1.50) = \mathbf{0.0668}$$

Method 3: Use the cumulative probability function on a computer or calculator to find the p-value:

$$P(z < -1.50) = \mathbf{0.0668}$$

b. **Determine whether or not the p-value is smaller than α.**
The p-value (0.0668) is not smaller than α (0.05).

STEP 5 The Results:

a. **State the decision about H_o.**
Is the p-value small enough to indicate that the sample evidence is highly unlikely in the event that the null hypothesis is true? In order to make the decision, we need to know the *decision rule*.

> **Decision rule**
>
> a. If the p-value is *less than or equal to* the level of significance α, then the decision must be **reject H_o**.
> b. If the p-value is *greater than* the level of significance α, then the decision must be **fail to reject H_o**.

Decision about H_o: Fail to reject H_o.

FYI Specific information about writing the conclusion is given on page 366.

b. **State the conclusion about H_a.**
There is not sufficient evidence at the 0.05 level of significance to show that the mean shearing strength of the rivets is less than 925. "We failed to convict" the null hypothesis. In other words, a sample mean as small as 921.18 is likely to occur (as defined by α) when the true population mean value is 925.0 and $\bar{x}$ is normally distributed. The resulting action by the manager would be to buy the rivets.

Note: When the decision reached is "fail to reject H_o," it simply means "For the lack of better information, act as if the null hypothesis is true" (that is, "Accept H_o" is a misnomer).

Before looking at another example, let's look at the procedures for finding the p-value. The p-value is represented by the area under the curve of the probability distribution for the test statistic that is more extreme than the calculated value of the test statistic. There are three separate cases, and the direction (or sign) of the alternative hypothesis is the key. Table 8.7 outlines the procedure for all three cases.

TABLE 8.7 Finding *p*-Values Using the Cumulative Distribution

Case 1 H_a contains ">" "Right tail"	*p*-value is the *area to right of z★* *p*-value = $P(z > z★)$	*p*-Value in Right Tail
Case 2 H_a contains "<" "Left tail"	*p*-value is the *area to left of z★* *p*-value = $P(z < z★)$	*p*-Value in Left Tail
Case 3 H_a contains "≠" "Two-tailed"	*p*-value is the *total area of both tails* *p*-value = $P(z < -\|z★\|) + P(z > \|z★\|)$ $z★$ may be in either tail, and since both areas are equal, find the probability of one tail and double it. Thus, *p*-value = $2 \times P(z < -\|z★\|)$	*p*-Value in Two Tails

Let's look at an example involving the two-tailed procedure.

E X A M P L E   8 . 1 6

TWO-TAILED HYPOTHESIS TEST

For years, many large companies in a certain city have used the Kelley Employment Agency for testing prospective employees. The employment selection test used has historically resulted in scores normally distributed about a mean of 82 with a standard deviation of 8. The Brown Agency has developed a new test that is quicker and easier to administer and therefore less expensive. Brown claims that its test results are the same as those obtained from the Kelley test. Many of the companies are considering a change from the Kelley Agency to the Brown Agency to cut costs. However, they are unwilling to make the change if the Brown test results have a different mean value. An independent testing firm tested 36 prospective employees with the Brown test. A sample mean of 79 resulted. Determine the *p*-value associated with this hypothesis test. (Assume $\sigma = 8$.)

Video tutorial available—logon and learn more at cengagebrain.com

Solution

Step 1 The Set-Up:

a. **Describe the population parameter of interest.**
The population mean μ, the mean of all test scores using the Brown Agency test

b. **State the null hypothesis (H_o) and the alternative hypothesis (H_a).**
The Brown Agency's test results "will be different" (the concern) if the mean test score is not equal to 82. They "will be the same" if the mean is equal to 82. Therefore,

$H_o: \mu = 82$ (test results have the same mean)

$H_a: \mu \neq 82$ (test results have a different mean)

Step 2 The Hypothesis Test Criteria:

a. **Check the assumptions.**
σ is known. If the Brown test scores are distributed the same as the Kelley test scores, they will be normally distributed and the sampling distribution will be normal for all sample sizes.

b. **Identify the probability distribution and the test statistic to be used.**
The standard normal probability distribution and the test statistic

$$z\star = \frac{\bar{x} - \mu}{\sigma/\sqrt{n}} \text{ will be used with } \sigma = 8.$$

c. **Determine the level of significance, α.**
The level of significance is omitted because the question asks for the p-value and not a decision.

Step 3 The Sample Evidence:

a. **Collect the sample information:** $n = 36$, $\bar{x} = 79$.

b. **Calculate the value of the test statistic.**
μ is 82 from H_o; $\sigma = 8$ is a known quantity. We have

$$z\star = \frac{\bar{x} - \mu}{\sigma/\sqrt{n}}: \quad z\star = \frac{79 - 82}{8/\sqrt{36}} = \frac{-3}{1.3333} = -2.25$$

Step 4 The Probability Distribution:

a. **Calculate the p-value for the test statistic.**
Because the alternative hypothesis indicates a two-tailed test, we must find the probability associated with both tails. The p-value is found by doubling the area of one tail (see Table 8.7, p. 377).

$z\star = -2.25$
From Table 3: p-value $= 2 \times P(z < -2.25) = 2(0.0122) = 0.0244$.
or
From Table 5: p-value $= 2 \times P(z > 2.25) = 2(0.0122) = 0.0244$.
or
Use the cumulative probability function on a computer or calculator:
p-value $= 2 \times P(z < -2.25) = 0.0244$.

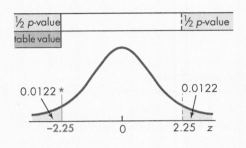

b. **Determine whether or not the *p*-value is smaller than α.** A comparison is not possible; no α value was given in the statement of the question.

Step 5 The Results:

The *p*-value for this hypothesis test is 0.0244. Each individual company now will decide whether to continue to use the Kelley Agency's services or change to the Brown Agency. Each will need to establish the level of significance that best fits its own situation and then make a decision using the decision rule described previously.

FYI See the instructions on pages 375–376.

The *fundamental idea of the p-value* is to express the degree of belief in the null hypothesis:

- When the *p*-value is minuscule (something like 0.0003), the null hypothesis would be rejected by everybody because the sample results are very unlikely for a true H_o.
- When the *p*-value is fairly small (like 0.012), the evidence against H_o is quite strong and H_o will be rejected by many.
- When the *p*-value begins to get larger (say, 0.02 to 0.08), there is too much probability that data like the sample involved could have occurred even if H_o were true, and the rejection of H_o is not an easy decision.
- When the *p*-value gets large (like 0.15 or more), the data are not at all unlikely if the H_o is true, and no one will reject H_o.

The *advantages of the p-value approach* are as follows: (1) The results of the test procedure are expressed in terms of a continuous probability scale from 0.0 to 1.0, rather than simply on a "reject" or "fail to reject" basis. (2) A *p*-value can be reported and the user of the information can decide on the strength of the evidence as it applies to his or her own situation. (3) Computers can do all the calculations and report the *p*-value, thus eliminating the need for tables.

The *disadvantage of the p-value approach* is the tendency for people to put off determining the level of significance. This should not be allowed to happen, because it is then possible for someone to set the level of significance after the fact, leaving open the possibility that the "preferred" decision will result. However, this is probably important only when the reported *p*-value falls in the "hard choice" range (say, 0.02 to 0.08), as described previously.

FYI Do your opponents show you their poker hands before you bet?

E X A M P L E   8 . 1 7

TWO-TAILED HYPOTHESIS TEST WITH SAMPLE DATA

According to the results of Exercise 5.33 (p. 242), the mean of single-digit random numbers is 4.5 and the standard deviation is $\sigma = 2.87$. Draw a random sample of 40 single-digit numbers from Table 1 in Appendix B and test the hypothesis "The mean of the single-digit numbers in Table 1 is 4.5." Use $\alpha = 0.10$.

Solution

Step 1 The Set-Up:

a. **Describe the population parameter of interest.**
The population parameter of interest is the mean μ of the population of single-digit numbers in Table 1 of Appendix B.

b. **State the null hypothesis (H_o) and the alternative hypothesis (H_a).**

$$H_o: \mu = 4.5 \text{ (mean is 4.5)}$$
$$H_a: \mu \neq 4.5 \text{ (mean is not 4.5)}$$

Step 2 The Hypothesis Test Criteria:

a. **Check the assumptions.**
σ is known. Samples of size 40 should be large enough to satisfy the CLT; see the discussion of this issue on page 370.

b. **Identify the probability distribution and the test statistic to be used.**
We use the standard normal probability distribution, and the test statistic is $z\star = \dfrac{\bar{x} - \mu}{\sigma/\sqrt{n}}$; $\sigma = 2.87$.

c. **Determine the level of significance, α.**
$\alpha = 0.10$ (given in the statement of the problem)

Step 3 The Sample Evidence:

a. **Collect the sample information.**
This random sample was drawn from Table 1 in Appendix B [TA08-01]:

2	8	2	1	5	5	4	0	9	1	0	4	6	1
5	1	1	3	8	0	3	6	8	4	8	6	8	
9	5	0	1	4	1	2	1	7	1	7	9	3	

From the sample: $\bar{x} = 3.975$ and $n = 40$.

b. **Calculate the value of the test statistic.**
We use formula (8.4), and μ is 4.5 from H_o, and $\sigma = 2.87$:

$$z\star = \frac{\bar{x} - \mu}{\sigma/\sqrt{n}}: \quad z\star = \frac{3.975 - 4.50}{2.87/\sqrt{40}} = \frac{-0.525}{0.454} = -1.156 = -1.16$$

Step 4 The Probability Distribution:

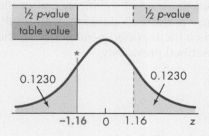

a. **Calculate the p-value for the test statistic.**
Because the alternative hypothesis indicates a two-tailed test, we must find the probability associated with both tails. The p-value is found by doubling the area of one tail. $z\star = -1.16$.
The p-value $= \mathbf{P} = 2 \times P(z < -1.16)$
$= 2(0.1230) = 0.2460$

b. **Determine whether or not the p-value is smaller than α.**
The p-value (0.2460) is greater than α (0.10).

Step 5 The Results:

a. **State the decision about H_o:** Fail to reject H_o.

b. **State the conclusion about H_a.**
The observed sample mean is not significantly different from 4.5 at the 0.10 level of significance.

Suppose we were to take another sample of size 40 from Table 1. Would we obtain the same results? Suppose we took a third sample and a fourth. What results might we expect? What does the p-value in Example 8.17 measure? Table 8.8 lists (1) the means obtained from 50 different random samples of size

---

TABLE 8.8

a. The Means of 50 Random Samples Taken from Table 1 in Appendix B [TA08-08]

3.850	5.075	4.375	4.675	5.200	4.250	3.775	4.075	5.800	4.975
4.225	4.125	4.350	4.925	5.100	4.175	4.300	4.400	4.775	4.525
4.225	5.075	4.325	5.025	4.725	4.600	4.525	4.800	4.550	3.875
4.750	4.675	4.700	4.400	5.150	4.725	4.350	3.950	4.300	4.725
4.975	4.325	4.700	4.325	4.175	3.800	3.775	4.525	5.375	4.225

b. The $z\star$ Values Corresponding to the 50 Means

−1.432	1.267	−0.275	0.386	1.543	−0.551	−1.598	−0.937	2.865	1.047
−0.606	−0.826	−0.331	0.937	1.322	−0.716	−0.441	−0.220	0.606	0.055
−0.606	1.267	−0.386	1.157	0.496	0.220	0.055	0.661	0.110	−1.377
0.551	0.386	0.441	−0.220	1.432	0.496	−0.331	−1.212	−0.441	0.496
1.047	−0.386	0.441	−0.386	−0.716	−1.543	−1.598	0.055	1.928	−0.606

c. The p-Values Corresponding to the 50 Means

0.152	0.205	0.783	0.700	0.123	0.582	0.110	0.349	0.004	0.295
0.545	0.409	0.741	0.349	0.186	0.474	0.659	0.826	0.545	0.956
0.545	0.205	0.700	0.247	0.620	0.826	0.956	0.509	0.912	0.168
0.582	0.700	0.659	0.826	0.152	0.620	0.741	0.226	0.659	0.620
0.295	0.700	0.659	0.700	0.474	0.123	0.110	0.956	0.054	0.545

---

40 that were taken from Table 1 in Appendix B, (2) the 50 values of $z\star$ corresponding to the 50 $\bar{x}$'s, and (3) their 50 corresponding p-values. Figure 8.9 shows a histogram of the 50 $z\star$ values.

FIGURE 8.9

The 50 Values of $z\star$ from Table 8.8

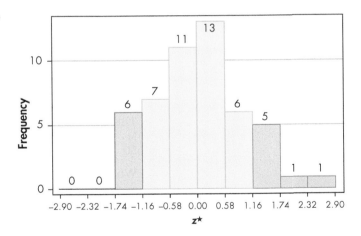

The histogram shows that six values of $z\star$ were less than −1.16 and seven values were greater than 1.16. That means 13 of the 50 samples, or 26%, have mean values more extreme than the mean ($\bar{x} = 3.975$) in Example 8.17. This observed relative frequency of 0.26 represents an empirical look at the p-value. Notice that the empirical value for the p-value (0.26) is very similar to the calculated p-value of 0.2460. Check the list of p-values; do you find that 13 of the 50 p-values are less than 0.2460? Which samples resulted in $|z\star| > 1.16$? Which samples resulted in a p-value greater than 0.2460? How do they compare?

TECHNOLOGY INSTRUCTIONS: HYPOTHESIS TEST FOR MEAN μ WITH A GIVEN σ

MINITAB	Input the data into C1; then continue with:

Choose: **Stat > Basic Statistics > 1-Sample Z**
Enter: Samples in columns: **C1**
            Standard deviation: σ
Select: **Perform hypothesis test**
Enter: Hypothesized mean: μ
Select: **Options**
Select: Alternative: **less than** or **not equal to** or **greater than** > **OK** > **OK**

Excel	Input the data into column A; then continue with:

Choose: **Add-Ins > Data Analysis Plus > Z-Test: Mean > OK**
Enter: Input Range: **(A1:A20 or select cells)**
            Hypothesized Mean: μ
            Standard Deviation (SIGMA): σ > **OK**
Gives p-values for both one-tailed and two-tailed tests.

TI-83/84 Plus	Input the data into L1; then continue with the following, entering the appropriate values and highlighting Calculate:

Choose: **STAT > TESTS > 1:Z-Test**

```
Z-Test
 Inpt:DATA Stats
 μ0:0
 σ:0
 List:L₁
 Freq:1
 μ:≠μ0 <μ0 >μ0
 Calculate Draw
```

FYI The *p*-value approach was "made" for the computer!

The MINITAB solution to the rivet example, used in this section (pp. 371–372, 374–376), is shown here:

One-sample Z: C1
Test of mu = 925.00 vs < 925.00
The assumed standard deviation = 18.0

N	Mean	StDev	SE Mean	Z	P
50	921.18	17.58	2.546	−1.50	0.0668

When the computer is used, all that is left is for you to do is to make the decision and to write the conclusion.

⊡ SECTION 8.4 EXERCISES

8.85 In the example starting on page 371, the aircraft builder who is buying the rivets is concerned that the rivets might not meet the mean-strength spec. State the aircraft manufacturer's null and alternative hypotheses.

8.86 Professor Hart does not believe a statement he heard: "The mean weight of college women is 54.4 kg." State the null and alternative hypotheses he would use to challenge this statement.

8.87 State the null and alternative hypotheses used to test each of the following claims.

a. The mean reaction time is greater than 1.25 seconds.

b. The mean score on that qualifying exam is less than 335.

c. The mean selling price of homes in the area is not $230,000.

d. The mean weight of college football players is no more than 260 lb.

e. The mean hourly wage for a child care giver is at most $15.00.

8.88 State the null hypothesis H_o and the alternative hypothesis H_a that would be used for a hypothesis test related to each of the following statements:

a. The mean age of the students enrolled in evening classes at a certain college is greater than 26 years.

b. The mean weight of packages shipped on Air Express during the past month was less than 36.7 lb.

c. The mean life of fluorescent light bulbs is at least 1600 hours.

d. The mean strength of welds by a new process is different from 570 lb per unit area, the mean strength of welds by the old process.

8.89 Identify the four possible outcomes and describe the situation involved with each with regard to the aircraft manufacturer's testing and buying of rivets. Which is the more serious error: the type I or type II error? Explain.

8.90 A manufacturer wishes to test the hypothesis that "by changing the formula of its toothpaste, it will give its users improved protection." The null hypothesis represents the idea that "the change will not improve the protection," and the alternative hypothesis is "the change will improve the protection." Describe the meaning of the two possible types of errors that can occur in the decision when the test of the hypothesis is conducted.

8.91 Suppose we want to test the hypothesis that the mean hourly charge for automobile repairs is at least $60 per hour at the repair shops in a nearby city. Explain the conditions that would exist if we make an error in decision by committing a type I error. What about a type II error?

8.92 Describe how the null hypothesis, as stated in Example 8.14 (p. 373), is a "starting point" for the decision to be made about the drying time for paint made with the new formula.

8.93 Assume that z is the test statistic and calculate the value of $z\star$ for each of the following:

a. H_o: $\mu = 10$, $\sigma = 3$, $n = 40$, $\bar{x} = 10.6$

b. H_o: $\mu = 120$, $\sigma = 23$, $n = 25$, $\bar{x} = 126.2$

c. H_o: $\mu = 18.2$, $\sigma = 3.7$, $n = 140$, $\bar{x} = 18.93$

d. H_o: $\mu = 81$, $\sigma = 13.3$, $n = 50$, $\bar{x} = 79.6$

8.94 Assume that z is the test statistic and calculate the value of $z\star$ for each of the following:

a. H_o: $\mu = 51$, $\sigma = 4.5$, $n = 40$, $\bar{x} = 49.6$

b. H_o: $\mu = 20$, $\sigma = 4.3$, $n = 75$, $\bar{x} = 21.2$

c. H_o: $\mu = 138.5$, $\sigma = 3.7$, $n = 14$, $\bar{x} = 142.93$

d. H_o: $\mu = 815$, $\sigma = 43.3$, $n = 60$, $\bar{x} = 799.6$

8.95 There are only two possible decisions that can result from a hypothesis test.

a. State the two possible decisions.

b. Describe the conditions that will lead to each of the two decisions identified in part a.

8.96 a. What decision is reached when the p-value is greater than α?

b. What decision is reached when α is greater than the p-value?

8.97 For each of the following pairs of values, state the decision that will occur and why.

a. p-value $= 0.014$, $\alpha = 0.02$

b. p-value $= 0.118$, $\alpha = 0.05$

c. p-value $= 0.048$, $\alpha = 0.05$

d. p-value $= 0.064$, $\alpha = 0.10$

8.98 For each of the following pairs of values, state the decision that will occur and why.

a. p-value $= 0.018$, $\alpha = 0.01$

b. p-value $= 0.033$, $\alpha = 0.05$

c. p-value $= 0.078$, $\alpha = 0.05$

d. p-value $= 0.235$, $\alpha = 0.10$

8.99 The calculated p-value for a hypothesis test is 0.084. What decision about the null hypothesis would occur in the following?

a. The hypothesis test is completed at the 0.05 level of significance.

b. The hypothesis test is completed at the 0.10 level of significance.

8.100 a. A one-tailed hypothesis test is to be completed at the 0.05 level of significance. What calculated values of p will cause a rejection of H_o?

b. A two-tailed hypothesis test is to be completed at the 0.02 level of significance. What calculated values of p will cause a "fail to reject H_o" decision?

8.101 Skillbuilder Applet Exercise estimates the p-value for a one-tailed hypothesis test by simulating

(continue on page 384)

[EX00-000] identifies the filename of an exercise's online dataset—datasets and Skillbuilder Applets available through cengagebrain.com

the taking of many samples. The given hypothesis test is for an $H_o: \mu = 1500$ versus $H_a: \mu < 1500$. A sample of 24 has been taken and the sample mean is 1451.

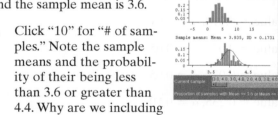

a. Click "10" for "# of samples." Note the sample means and the probability of their being less than 1451 if the true mean is really 1500.

b. Change to "Batch" and simulate 1000 more samples. What is the probability of their being less than 1451? This is your estimated p-value.

c. How does your estimated p-value show on the histogram formed from the taking of many samples? Explain what this p-value means with respect to the test.

d. If the level of significance were 0.01, what would your decision be?

8.102 Skillbuilder Applet Exercise estimates the p-value for a two-tailed hypothesis test by simulating the taking of many samples. The given hypothesis test is for $H_o: \mu = 4$ versus $H_a: \mu \neq 4$. A sample of 100 has been taken and the sample mean is 3.6.

a. Click "10" for "# of samples." Note the sample means and the probability of their being less than 3.6 or greater than 4.4. Why are we including the "greater than 4.4"?

b. Change to "Batch" and simulate 1000 more samples. What is the probability of their being less than 3.6 or greater than 4.4? This is your estimated p-value.

c. How does your estimated p-value show on the histogram formed from the taking of many samples? Explain what this p-value means with respect to the test.

d. If the level of significance were 0.05, what would your decision be?

8.103 Describe in your own words what the p-value measures.

8.104 a. Calculate the p-value, given $H_a: \mu < 45$ and $z\star = -2.3$.

b. Calculate the p-value, given $H_a: \mu > 58$ and $z\star = 1.8$.

8.105 Calculate the p-value, given $H_a: \mu \neq 245$ and $z\star = 1.1$.

8.106 Find the test statistic $z\star$ and the p-value for each of the following situations.

a. $H_o: \mu = 22.5, H_a: \mu > 22.5; \bar{x} = 24.5, \sigma = 6, n = 36$

b. $H_o: \mu = 200, H_a: \mu < 200; \bar{x} = 192.5, \sigma = 40, n = 50$

c. $H_o: \mu = 12.4, H_a: \mu \neq 2.4; \bar{x} = 11.52, \sigma = 2.2, n = 16$

8.107 Calculate the p-value for each of the following:

a. $H_o: \mu = 10, H_a: \mu > 10, z\star = 1.48$

b. $H_o: \mu = 105, H_a: \mu < 105, z\star = -0.85$

c. $H_o: \mu = 13.4, H_a: \mu \neq 13.4, z\star = 1.17$

d. $H_o: \mu = 8.56, H_a: \mu < 8.56, z\star = -2.11$

e. $H_o: \mu = 110, H_a: \mu \neq 110, z\star = -0.93$

8.108 Calculate the p-value for each of the following:

a. $H_o: \mu = 20, H_a: \mu < 20; \bar{x} = 17.8, \sigma = 9, n = 36$

b. $H_o: \mu = 78.5, H_a: \mu > 78.5; \bar{x} = 79.8, \sigma = 15, n = 100$

c. $H_o: \mu = 1.587, H_a: \mu \neq 1.587;$
$\bar{x} = 1.602, \sigma = 0.15, n = 50$

8.109 Find the value of $z\star$ for each of the following:

a. $H_o: \mu = 35$ versus $H_a: \mu > 35$ when p-value $= 0.0582$

b. $H_o: \mu = 35$ versus $H_a: \mu < 35$ when p-value $= 0.0166$

c. $H_o: \mu = 35$ versus $H_a: \mu \neq 35$ when p-value $= 0.0042$

8.110 The null hypothesis, $H_o: \mu = 48$, was tested against the alternative hypothesis, $H_a: \mu > 48$. A sample of 75 resulted in a calculated p-value of 0.102. If $\sigma = 3.5$, find the value of the sample mean, $\bar{x}$.

8.111 The null hypothesis, $H_o: \mu = 16$, was tested against the alternative hypothesis, $H_a: \mu < 16$. A sample of 50 resulted in a calculated p-value of 0.017. If $\bar{x} = 14$, find the value of the population standard deviation.

8.112 Using the MINITAB solution to the rivet example as shown on page 382, describe how MINITAB found each of the six numerical values it reported as results.

8.113 The following computer output was used to complete a hypothesis test.

N	MEAN	STDEV	SE MEAN	Z	P VALUE
TEST OF MU = 525.00 VS MU < 525.00					
THE ASSUMED SIGMA = 60.0					
38	512.14	64.78	9.733	−1.32	0.093

a. State the null and alternative hypotheses.

b. If the test is completed using $\alpha = 0.05$, what decision and conclusion are reached?

c. Verify the value of the standard error of the mean.

8.114 Using the computer output and information in Exercise 8.113, determine the value of the following:

a. Hypothesized value of population mean

b. Sample mean

c. Population standard deviation

d. Test statistic

8.115 The following computer output was used to complete a hypothesis test.

TEST OF MU = 6.250 VS MU not = 6.250
THE ASSUMED SIGMA = 1.40

N	MEAN	STDEV	SE MEAN	Z	P VALUE
78	6.596	1.273	0.1585	2.18	0.029

a. State the null and alternative hypotheses.

b. If the test is completed using $\alpha = 0.05$, what decision and conclusion are reached?

c. Verify the value of the standard error of the mean.

d. Find the values for Σx and Σx^2.

8.116 Using the computer output and information in Exercise 8.115, determine the value of the following:

a. Hypothesized value of population mean

b. Sample mean

c. Population standard deviation

d. Test statistic

8.117 Ponemon Institute, along with Intel, published "The Cost of a Lost Laptop" study in April 2009. With an increasingly mobile workforce carrying around more sensitive data on their laptops, the loss involves much more than the laptop itself. The average cost of a lost laptop based on cases from various industries is $49,246. This figure includes laptop replacement, data breach cost, lost productivity cost, and other legal and forensic costs. A separate study conducted with respect to 30 cases from health care industries produced a mean of $67,873. Assuming that $\sigma = \$25,000$, is there sufficient evidence to support the claim that health care laptop replacement costs are higher in general? Use a 0.001 level of significance.

Source: http://communities.intel.com/

8.118 One of the best indicators of a baby's health is his or her weight at birth. In the United States, mothers who live in poverty generally have babies with lower birth weights than those who do not live in poverty. Although the average birth weight for babies born in the United States is approximately 3300 grams, the birth weight for babies of women living in poverty is 2800 grams with a standard deviation of 500 grams. Recently, a local hospital introduced an innovative new prenatal care program to reduce the number of low-birth-weight babies born in the hospital. At the end of the first year, the birth weights of 25 randomly selected babies were collected; all of the babies were born to women who lived in poverty and participated in the program. Their mean birth weight was 3075 grams. The question posed to you, the researcher, is, "Has there been a significant improvement in the birth weights of babies born to poor women?" Use $\alpha = 0.02$.

Source: http://www.ccnmtl.columbia.edu/

a. Define the parameter.

b. State the null and alternative hypotheses.

c. Specify the hypothesis test criteria.

d. Present the sample evidence.

e. Find the probability distribution information.

f. Determine the results.

8.119 The owner of a local chain of grocery stores is always trying to minimize the time it takes her customers to check out. In the past, she has conducted many studies of the checkout times, and they have displayed a normal distribution with a mean time of 12 minutes and a standard deviation of 2.3 minutes. She has implemented a new schedule for cashiers in hopes of reducing the mean checkout time. A random sample of 28 customers visiting her store this week resulted in a mean of 10.9 minutes. Does she have sufficient evidence to claim the mean checkout time this week was less than 12 minutes? Use $\alpha = 0.02$.

8.120 The average size of a home in 2008 fell to 2343 square feet, according to the National Association of Home Builders and reported in *USA Today* (January 11, 2009). The homebuilders of a northeastern city feel that the average size of homes continues to grow each year. To test their claim, a random sample of 45 new homes was selected and revealed an average of 2490 square feet. Assuming that the population standard deviation is approximately 450 square feet, is there evidence that the average size is larger in the northeast compared to the national 2008 figure? Use a 0.05 level of significance.

8.121 From candy to jewelry to flowers, the average consumer was expected to spend $123.89 for Mother's Day

(continue on page 386)

2009, according to an April 2009 National Retail Federation's survey. Local merchants felt this average was too high for their area and contracted an agency to conduct a study. A random sample of 60 consumers was taken at a local shopping mall the Saturday before Mother's Day and produced a sample mean amount of $106.27. If σ = $39.50, does the sample provide sufficient evidence to support the merchants' claim at the 0.05 level of significance?

Source: http://www.marketingcharts.com

8.122 Imagine that you are a customer living in the shopping area described in Exercise 8.121 and you need to buy a Mother's Day gift. Identify the four possible outcomes and describe the situation involved with each outcome with regard to the average amount spent on a Mother's Day gift. Which is the more serious error: the type I or the type II error? Explain.

8.123 Who says that the more you spend on a wristwatch, the more accurately the watch will keep time? Some say that you can now buy a quartz watch for less than $25 that keeps time just as accurately as watches that cost four times as much. Suppose the average accuracy for all watches being sold today, regardless of price, is within 19.8 seconds per month with a standard deviation of 9.1 seconds. A random sample of 36 quartz watches priced less than $25 is taken, and their accuracy check reveals a sample mean error of 22.7 seconds per month. Based on this evidence, complete the hypothesis test of H_o: μ = 20 vs. H_a: μ > 20 at the 0.05 level of significance using the probability-value approach.

a.   Define the parameter.

b.   State the null and alternative hypotheses.

c.   Specify the hypothesis test criteria.

d.   Present the sample evidence.

e.   Find the probability distribution information.

f.   Determine the results.

8.124 [EX08-124] Major League Baseball games average 2 hours 50.1 minutes with a standard deviation of 21.0 minutes. It has been claimed that St. Louis Cardinals baseball games last, on the average, longer than the games of the other Major League teams. To test the truth of this statement, a sample of 12 Cardinals games was randomly identified and the "time of game" for each obtained.

| 140 | 208 | 187 | 173 | 164 | 195 | 170 | 163 | 187 | 150 | 170 | 208 |

Source: http://mlb.com/

At the 0.05 level of significance, do these data show sufficient evidence to conclude that the mean time of Cardinals baseball games is longer than that of other Major League Baseball teams?

a.   Define the parameter.

b.   State the null and alternative hypotheses.

c.   Specify the hypothesis test criteria.

d.   Present the sample evidence.

e.   Find the probability distribution information.

f.   Determine the results.

8.125 [EX08-125] Nationally, the ratio of nurses to students falls short of the recommended federal standard, according to the *USA Today* article "School nurses in short supply" (August 11, 2009). The recommendation from the Centers for Disease Control and Prevention (CDC) is 1 nurse per 750 students. Use the sample below from 38 randomly selected schools in the state of New York to test the statement "The New York mean number of students per school nurse is significantly higher than the CDC standard of 750." Assume σ = 540.

1062	1070	353	675	1557	1374	459	302	1946	487	295
1047	1751	784	480	377	883	1035	332	330	989	1098
1241	778	1691	963	1645	1594	2125	338	1380	885	707
1267	1412	1037	1603	915						

a.   Describe the parameter of interest.

b.   State the null and alternative hypothesis.

c.   Calculate the value for $z\star$ and find the p-value

d.   State your decision and conclusion using α = 0.01.

8.126 [EX08-126] The National Health and Nutrition Examination Survey (NHANES) indicates that more U.S. adults are becoming either overweight or obese, which is defined as having a body mass index (BMI) of 25 or more. Data from the Centers for Disease Control and Prevention (CDC) indicate that for females aged 35 to 55, the mean BMI is 25.12 with a standard deviation of 5.3. In a similar study that examined female cardiovascular technologists who were registered in the United States and who were within the same age range, the following BMI scores resulted:

| 22 | 28 | 26 | 19 |

*** For remainder of data, logon at cengagebrain.com

Source: "An Assessment of Cardiovascular Risk Behaviors of Registered Cardiovascular Technologists," Dissertation by Dr. Susan Wambold, University of Toledo, 2002. Reprinted with permission.

Test the claim that the cardiovascular technologists have a lower average BMI than the general population. Use α = 0.05.

a.   Describe the parameter of interest.

b.   State the null and alternative hypotheses.

c.   Calculate the value for $z\star$ and find the p-value.

d.   State your decision and conclusion using α = 0.05.

8.127 [EX08-001] In Exercise 8.1, page 344, the heights for a random sample of 50 female American health professionals were given.

a. Determine a 95% confidence interval for the mean height of all female American health professionals. Assume that the standard deviation of female heights is 2.75 inches.

b. The average height of females in the United States is 63.7 inches, according to the National Center for Health Statistics. Does the interval for health professionals contain the mean for all females?

8.128 [EX08-001] According to the National Center for Health Statistics, the average height of females in the United States is 63.7 inches with a standard deviation of 2.75 inches. Using the heights for the random sample of 50 female American health professionals in Exercise 8.1, page 344:

a. Test the claim that the mean height of females in the health profession is different from 63.7 inches, the mean height of all females in the United States. Use a 0.05 level of significance.

b. How is "there is a significant difference" (rejection of the null hypothesis) revealed in the confidence interval formed in Exercise 8.127(a)?

c. Explain how the confidence interval formed in Exercise 8.127(a) could have been used to test the claim (hypothesis test) of part a that the mean height of females in the health profession is different.

d. How would "there is no significant difference" (failure to reject the null hypothesis) be revealed with a confidence interval?

8.129 Use a computer or calculator to select 40 random single-digit numbers. Find the sample mean, $z\star$, and p-value for testing H_o: $\mu = 4.5$ against a two-tailed alternative. Repeat several times as in Table 8.8. Describe your findings.

FYI Use commands for generating integer data on page 91, then continue with hypothesis test commands on page 382.

8.130 Use a computer or calculator to select 36 random numbers from a normal distribution with mean 100 and standard deviation 15. Find the sample mean, $z\star$, and p-value for testing a two-tailed hypothesis test of $\mu = 100$. Repeat several times as in Table 8.8. Describe your findings.

FYI Use commands for generating data on pages 283–284, then continue with hypothesis test commands on page 382.

8.5 Hypothesis Test of Mean μ (σ Known): A Classical Approach (Optional)

In Section 8.3 we surveyed the concepts and much of the reasoning behind a hypothesis test while looking at nonstatistical examples. In this section we are going to formalize the hypothesis test procedure as it applies to statements concerning the mean μ of a population under the restriction that σ, the population standard deviation, is a known value.

The assumption for hypothesis tests about mean μ using a known σ The sampling distribution of $\bar{x}$ has a normal distribution.

The information we need to ensure that this assumption is satisfied is contained in the sampling distribution of sample means and in the central limit theorem.

The sampling distribution of sample means $\bar{x}$ is distributed about a mean equal to μ with a standard error equal to $\sigma/\sqrt{n}$; and (1) if the randomly sampled population is normally distributed, then $\bar{x}$ is normally distributed for all sample sizes, or (2) if the randomly sampled population is not normally distributed, then $\bar{x}$ is approximately normally distributed for sufficiently large sample sizes.

The hypothesis test is a well-organized, step-by-step procedure used to make a decision. Two different formats are commonly used for hypothesis testing. The *classical approach* is the hypothesis test process that has enjoyed popularity for many years. This approach is organized as a five-step procedure.

The Classical Hypothesis Test: A Five-Step Procedure

STEP 1 The Set-Up:

    a. Describe the population parameter of interest.
    b. State the null hypothesis (H_o) and the alternative hypothesis (H_a).

STEP 2 The Hypothesis Test Criteria:

    a. Check the assumptions.
    b. Identify the probability distribution and the test statistic to be used.
    c. Determine the level of significance, α.

STEP 3 The Sample Evidence:

    a. Collect the sample information.
    b. Calculate the value of the test statistic.

STEP 4 The Probability Distribution:

    a. Determine the critical region and critical value(s).
    b. Determine whether or not the calculated test statistic is in the critical region.

STEP 5 The Results:

    a. State the decision about H_o.
    b. State the conclusion about H_a.

A commercial aircraft manufacturer buys rivets to use in assembling airliners. Each rivet supplier that wants to sell rivets to the aircraft manufacturer must demonstrate that its rivets meet the required specifications. One of the specs is "The mean shearing strength of all such rivets, μ, is at least 925 lb." Each time the aircraft manufacturer buys rivets, it is concerned that the mean strength might be less than the 925-lb specification.

Note: Each individual rivet has a shearing strength, which is determined by measuring the force required to shear ("break") the rivet. Clearly, not all the rivets can be tested. Therefore, a sample of rivets will be tested, and a decision about the mean strength of all the untested rivets will be based on the mean of those sampled and tested.

STEP 1 The Set-Up:

a. Describe the population parameter of interest.

    The population parameter of interest is the mean μ, the mean shearing strength of (or mean force required to shear) the rivets being considered for purchase.

b. State the null hypothesis (H_o) and the alternative hypothesis (H_a).

    The null hypothesis and the alternative hypothesis are formulated by inspecting the problem or statement to be investigated and first formulating two opposing statements about the mean μ. For our example, these two opposing statements are: (A) "The mean shearing strength is less than 925" ($\mu < 925$, the aircraft manufacturer's concern), and (B) "The mean shearing strength is at least 925" ($\mu = 925$, the rivet supplier's claim and the aircraft manufacturer's spec).

FYI More specific instructions are given on pages 361–363.

Note: The trichotomy law from algebra states that two numerical values must be related in exactly one of three possible relationships: $<$, $=$, or $>$. All three of these possibilities must be accounted for in the two opposing hypotheses in order for the two hypotheses to be negations of each other. The three possible combinations of signs and hypotheses are shown in Table 8.9. Recall that the null hypothesis assigns a specific value to the parameter in question, and therefore "equals" will always be part of the null hypothesis.

TABLE 8.9 The Three Possible Statements of Null and Alternative Hypotheses

Null Hypothesis	Alternative Hypothesis
1. Greater than or equal to ($\geq$)	Less than ($<$)
2. Less than or equal to ($\leq$)	Greater than ($>$)
3. Equal to ($=$)	Not equal to ($\neq$)

The parameter of interest, the population mean μ, is related to the value 925. Statement (A) becomes the alternative hypothesis:

$$H_a: \mu < 925 \text{ (the mean is less than 925)}$$

This statement represents the aircraft manufacturer's concern and says, "The rivets do not meet the required specs." Statement (B) becomes the null hypothesis:

$$H_o: \mu = 925 \ (\geq) \text{ (the mean is at least 925)}$$

This hypothesis represents the negation of the aircraft manufacturer's concern and says, "The rivets do meet the required specs."

Note: We will write the null hypothesis with just the equal sign, thereby stating the exact value assigned. When "equal" is paired with "less than" or paired with "greater than," the combined symbol is written beside the null hypothesis as a reminder that all three signs have been accounted for in these two opposing statements.

Before continuing with our example, let's look at three examples that demonstrate formulating the statistical null and alternative hypotheses involving population mean μ. Examples 8.18 and 8.19 each demonstrate a "one-tailed" alternative hypothesis.

EXAMPLE 8.18

WRITING NULL AND ALTERNATIVE HYPOTHESES (ONE-TAILED SITUATION)

A consumer advocate group would like to disprove a car manufacturer's claim that a specific model will average 24 miles per gallon of gasoline. Specifically, the group would like to show that the mean miles per gallon is considerably less than 24. State the null and alternative hypotheses.

Solution

To state the two hypotheses, we first need to identify the population parameter in question: the "mean mileage attained by this car model." The parameter μ is being compared with the value 24 miles per gallon, the specific value of interest. The advocates are questioning the value of μ and wish to show it to be less than 24 (i.e., $\mu < 24$). There are three possible relationships: (1) $\mu < 24$, (2) $\mu = 24$, and (3) $\mu > 24$. These three cases must be arranged to form two opposing statements: one states what the advocates are trying to show, "The mean level is less than 24 ($\mu < 24$)," whereas the "negation" is "The mean level is not less than 24 ($\mu \geq 24$)." One of these two statements will become the null hypothesis H_o, and the other will become the alternative hypothesis H_a.

Note: Recall that there are two rules for forming the hypotheses: (1) the null hypothesis states that the parameter in question has a specified value ("H_o must contain the equal sign"), and (2) the consumer advocate group's contention becomes the alternative hypothesis ("less than"). Both rules indicate:

$$H_o: \mu = 24 \ (\geq) \quad \text{and} \quad H_a: \mu < 24$$

EXAMPLE 8.19

WRITING NULL AND ALTERNATIVE HYPOTHESES (ONE-TAILED SITUATION)

Suppose the EPA is suing a large manufacturing company for not meeting federal emissions guidelines. Specifically, the EPA is claiming that the mean amount of sulfur dioxide in the air is dangerously high, higher than 0.09 part per million. State the null and alternative hypotheses for this test situation.

Solution

The parameter of interest is the mean amount of sulfur dioxide in the air, and 0.09 part per million is the specified value. $\mu > 0.09$ corresponds to "The mean amount is greater than 0.09," whereas $\mu \leq 0.09$ corresponds to the negation, "The mean amount is not greater than 0.09." Therefore, the hypotheses are

$$H_o: \mu = 0.09 \ (\leq) \quad \text{and} \quad H_a: \mu > 0.09$$

Example 8.20 demonstrates a "two-tailed" alternative hypothesis.

EXAMPLE 8.20

WRITING NULL AND ALTERNATIVE HYPOTHESES (TWO-TAILED SITUATION)

Job satisfaction is very important to worker productivity. A standard job-satisfaction questionnaire was administered by union officers to a sample of assembly-line workers in a large plant in hopes of showing that the assembly workers' mean score on this questionnaire would be different from the established mean of 68. State the null and alternative hypotheses.

Solution

Either the mean job-satisfaction score is different from 68 ($\mu \neq 68$) or the mean score is equal to 68 ($\mu = 68$). Therefore,

$$H_o: \mu = 68 \quad \text{and} \quad H_a: \mu \neq 68$$

Notes:

1. The alternative hypothesis is referred to as being "two-tailed" when H_a is "not equal."
2. When "less than" is combined with "greater than," they become "not equal to."

The viewpoint of the experimenter greatly affects the way the hypotheses are formed. Generally, the experimenter is trying to show that the parameter value is different from the value specified. Thus, the experimenter is often hoping to be able to reject the null hypothesis so that the experimenter's theory has been substantiated. Examples 8.18, 8.19, and 8.20 also represent the three possible arrangements for the $<$, $=$, and $>$ relationships between the parameter μ and a specified value.

Table 8.10 lists some additional common phrases used in claims and indicates their negations and the hypothesis in which each phrase will be used. Again, notice that "equals" is always in the null hypothesis. Also notice that the negation of "less than" is "not less than," which is equivalent to "greater than or equal to." Think of negation of one sign as the other two signs combined.

TABLE 8.10 Common Phrases and Their Negations

H_o: ($\geq$)	vs.	H_a: ($<$)	H_o: ($\leq$)	vs.	H_a: ($>$)	H_o: ($=$)	vs.	H_a: ($\neq$)
At least		Less than	At most		More than	Is		Is not
No less than		Less than	No more than		More than	Not different from		Different from
Not less than		Less than	Not greater than		Greater than	Same as		Not same as

After the null and alternative hypotheses are established, we will work under the assumption that the null hypothesis is a true statement until there is sufficient evidence to reject it. This situation might be compared with a courtroom trial, where the accused is assumed to be innocent (H_o: Defendant is innocent versus H_a: Defendant is not innocent) until sufficient evidence has been presented to show that innocence is totally unbelievable ("beyond reasonable doubt"). At the conclusion of the hypothesis test, we will make one of two possible decisions. We will decide in opposition to the null hypothesis and say that we "reject H_o" (this corresponds to "conviction" of the accused in a trial), or we will decide in agreement with the null hypothesis and say that we "fail to reject H_o" (this corresponds to "fail to convict" or an "acquittal" of the accused in a trial).

Let's return to the rivet example we interrupted on page 388 and continue with Step 2. Recall that

$$H_o: \mu = 925 \ (\geq) \ (\text{at least } 925) \qquad H_a: \mu < 925 \ (\text{less than } 925)$$

STEP 2 The Hypothesis Test Criteria:

a. Check the assumptions.
Assume that the standard deviation of the shearing strength of rivets is known from past experience to be $\sigma = 18$. Variables like shearing strength typically have a mounded distribution; therefore, a sample of size 50 should be large enough for the CLT to satisfy the assumption; the SDSM is normally distributed.

b. Identify the probability distribution and the test statistic to be used.
The standard normal probability distribution is used because $\bar{x}$ is expected to have a normal or approximately normal distribution.

For a hypothesis test of μ, we want to compare the value of the sample mean with the value of the population mean as stated in the null hypothesis. This comparison is accomplished using the test statistic in formula (8.4):

Test Statistic for Mean

$$z\bigstar = \frac{\bar{x} - \mu}{\sigma/\sqrt{n}} \qquad \text{(8.4)}$$

The resulting calculated value is identified as $z\bigstar$ ("z star") because it is expected to have a standard normal distribution when the null hypothesis is true and the assumptions have been satisfied. The $\bigstar$ ("star") is to remind us that this is the calculated value of the test statistic.

The test statistic to be used is $z\bigstar = \dfrac{\bar{x} - \mu}{\sigma/\sqrt{n}}$.

c. **Determine the level of significance, α.**

Setting α was described as a managerial decision in Section 8.3. To see what is involved in determining α, the probability of the type I error, for our rivet example, we start by identifying the four possible outcomes, their meanings, and the action related to each.

The type I error occurs when a true null hypothesis is rejected. This would occur when the manufacturer tested rivets that did meet the specs and rejected them. Undoubtedly this would lead to the rivets not being purchased even though they do meet the specs. In order for the manager to set a level of significance, related information is needed—namely, how soon is the new supply of rivets needed? If they are needed tomorrow and this is the only vendor with an available supply, waiting a week to find acceptable rivets could be very expensive; therefore, rejecting good rivets could be considered a serious error. On the other hand, if the rivets are not needed until next month, then this error may not be very serious. Only the manager will know all the ramifications, and therefore the manager's input is important here.

After much consideration, the manager assigns the level of significance: $\alpha = 0.05$.

STEP 3 The Sample Evidence:

a. **Collect the sample information.**

We are ready for the data. The sample must be a random sample drawn from the population whose mean μ is being questioned. A random sample of 50 rivets is selected, each rivet is tested, and the sample mean shearing strength is calculated: $\bar{x} = 921.18$ and $n = 50$.

b. **Calculate the value of the test statistic.**

The sample evidence ($\bar{x}$ and n found in Step 3a) is next converted into the calculated value of the test statistic, $z\bigstar$, using formula (8.4). (μ is 925 from H_o, and $\sigma = 18$ is the known quantity.) We have

$$z\bigstar = \frac{\bar{x} - \mu}{\sigma/\sqrt{n}}: \quad z\bigstar = \frac{921.18 - 925.0}{18/\sqrt{50}} = \frac{-3.82}{2.5456} = -1.50$$

STEP 4 The Probability Distribution:

a. **Determine the critical region and critical value(s).**

The standard normal variable z is our test statistic for this hypothesis test.

Critical region The set of values for the test statistic that will cause us to reject the null hypothesis. The set of values that are not in the critical region is called the **noncritical region** (sometimes called the *acceptance region*).

Recall that we are working under the assumption that the null hypothesis is true. Thus, we are assuming that the mean shearing strength of all rivets in the sampled population is 925. If this is the case, then when we select a random sample of 50 rivets, we can expect this sample mean, $\bar{x}$, to be part of a normal distribution that is centered at 925 and to have a standard error of $\sigma/\sqrt{n} = 18/\sqrt{50}$, or approximately 2.55. Approximately 95% of the sample mean values will be greater than 920.8 [a value 1.65 standard errors below the mean: $925 - (1.65)(2.55) = 920.8$]. Thus, if H_o is true and $\mu = 925$, then we expect $\bar{x}$ to be greater than 920.8 approximately 95% of the time and less than 920.8 only 5% of the time.

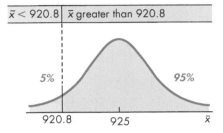

If, however, the value of $\bar{x}$ that we obtain from our sample is less than 920.8 — say, 919.5 — we will have to make a choice. It could be that either: (A) such an $\bar{x}$ value (919.5) is a member of the sampling distribution with mean 925 although it has a very low probability of occurrence (less than 0.05), or (B) $\bar{x} = 919.5$ is a member of a sampling distribution whose mean is less than 925, which would make it a value that is more likely to occur.

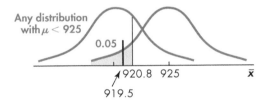

In statistics, we "bet" on the "more probable to occur" and consider the second choice (B) to be the right one. Thus, the left-hand tail of the z distribution becomes the critical region, and the level of significance α becomes the measure of its area.

Critical value(s) The "first" or "boundary" value(s) of the critical region(s).

The critical value for our example is $-z(0.05)$ and has the value of -1.65, as found in Table 4A in Appendix B.

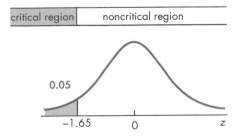

FYI Information about the critical value notation, $z(\alpha)$, is given on pages 292–297.

FYI Shading will be used to identify the critical region.

b. Determine whether or not the calculated test statistic is in the critical region.
Graphically this determination is shown by locating the value for $z\star$ on the sketch in Step 4a.

The calculated value of z, $z\star = -1.50$, is **not in the critical region** (it is in the unshaded portion of the figure).

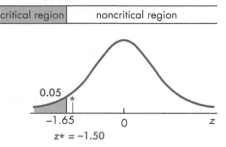

STEP 5 The Result:

a. State the decision about H_o.

In order to make the decision, we need to know the *decision rule*.

> **Decision rule**
>
> a. If the test statistic falls *within the critical region*, then the decision must be **reject H_o**. (The critical value is part of the critical region.)
> b. If the test statistic is *not in the critical region*, then the decision must be **fail to reject H_o**.

The decision is: Fail to reject H_o.

b. State the conclusion about H_a.

FYI Specific information about writing the conclusion is given on page 366.

There is not sufficient evidence at the 0.05 level of significance to show that the rivets have a mean shearing strength less than 925. "We failed to convict" the null hypothesis. In other words, a sample mean as small as 921.18 is not unlikely to occur (as defined by α) when the true population mean value is 925.0. Therefore, the resulting action would be to buy the rivets.

Before we look at another example, let's summarize briefly some of the details we have seen thus far:

1. The null hypothesis specifies a particular value of a population parameter.
2. The alternative hypothesis can take three forms. Each form dictates a specific location of the critical region(s), as shown in the following table.
3. For many hypothesis tests, the sign in the alternative hypothesis "points" in the direction in which the critical region is located. (Think of the not-equal-to sign [$\neq$] as being both less than [$<$] and greater than [$>$], thus pointing in both directions.)

	Sign in the Alternative Hypothesis		
	$<$	$\neq$	$>$
Critical Region	One region Left side One-tailed test	Two regions Half on each side Two-tailed test	One region Right side One-tailed test

The value assigned to α is called the *significance level* of the hypothesis test. Alpha cannot be interpreted to be anything other than the risk (or probability) of rejecting the null hypothesis when it is actually true. We will seldom be able to determine whether the null hypothesis is true or false; we will thus decide only to "reject H_o" or to "fail to reject H_o." The relative frequency with which we reject a true hypothesis is α, but we will never know the relative frequency with which we make an error in decision. The two ideas are quite different; that is, a type I error and an error in decision are two different things altogether. Remember that there are two types of errors: type I and type II.

Let's look at another hypothesis test, one involving the two-tailed procedure.

E X A M P L E   8 . 2 1

TWO-TAILED HYPOTHESIS TEST

It has been claimed that the mean weight of female students at a certain college is 54.4 kg. Professor Hart does not believe the claim and sets out to show that the mean weight is not 54.4 kg. To test the claim, he collects a random sample of 100 weights from among the female students. A sample mean of 53.75 kg results. Is this sufficient evidence for Professor Hart to reject the statement? Use $\alpha = 0.05$ and $\sigma = 5.4$ kg.

Solution

Step 1 The Set-Up:

a. **Describe the population parameter of interest.**
   The population parameter of interest is the mean μ, the mean weight of all female students at the college.

b. **State the null hypothesis (H_o) and the alternative hypothesis (H_a).**
   The mean weight is equal to 54.4 kg, or the mean weight is not equal to 54.4 kg.

$$H_o: \mu = 54.4 \text{ (mean weight is 54.4)}$$
$$H_a: \mu \neq 54.4 \text{ (mean weight is not 54.4)}$$
$$\text{(Remember: } \neq \text{ is } < \text{ and } > \text{ together.)}$$

Step 2 The Hypothesis Test Criteria:

a. **Check the assumptions.**
   σ is known. The weights of an adult group of women are generally approximately normally distributed; therefore, a sample of $n = 100$ is large enough to allow the CLT to apply.

b. **Identify the probability distribution and the test statistic to be used.**
   The standard normal probability distribution and the test statistic
   $$z\star = \frac{\bar{x} - \mu}{\sigma/\sqrt{n}} \text{ will be used; } \sigma = 5.4.$$

c. **Determine the level of significance, α.**
   $\alpha = 0.05$ (given in the statement of problem)

Step 3 The Sample Evidence:

a. **Collect the sample information:** $\bar{x} = 53.75$ and $n = 100$.

b. **Calculate the value of the test statistic.**
   Use formula (8.4), information from H_o: $\mu = 54.4$, and $\sigma = 5.4$ (known):
   $$z\star = \frac{\bar{x} - \mu}{\sigma/\sqrt{n}}: \quad z\star = \frac{53.75 - 54.4}{5.4/\sqrt{100}} = \frac{-0.65}{0.54} = -1.204 = -1.20$$

Step 4 The Probability Distribution:

a. **Determine the critical region and critical value(s).**

   The critical region is both the left tail and the right tail because both smaller and larger values of the sample mean suggest that the null hypothesis is wrong. The level of significance will be split in half, with

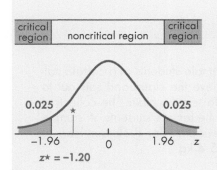

0.025 being the measure in each tail. The critical values are found in Table 4B in Appendix B: $\pm z(0.025) = \pm 1.96$. (Table 4B instructions are on p. 348.)

b. **Determine whether or not the calculated test statistic is in the critical region.**
The calculated value of z, $z\star = -1.20$, is not in the critical region (shown in red on the adjacent figure).

Step 5 The Results:

a. **State the decision about H_o:** Fail to reject H_o.

b. **State the conclusion about H_a.**
There is not sufficient evidence at the 0.05 level of significance to show that the female students have a mean weight different from the 54.4 kg claimed. In other words, there is no statistical evidence to support Professor Hart's contentions.

E X A M P L E   8 . 2 2

TWO-TAILED HYPOTHESIS TEST WITH SAMPLE DATA

According to the results of Exercise 5.33 (p. 242), the mean of single-digit random numbers is 4.5 and the standard deviation is $\sigma = 2.87$. Draw a random sample of 40 single-digit numbers from Table 1 in Appendix B and test the hypothesis "The mean of the single-digit numbers in Table 1 is 4.5." Use $\alpha = 0.10$.

Solution

Step 1 The Set-Up:

a. **Describe the population parameter of interest.**
The parameter of interest is the mean μ of the population of single-digit numbers in Table 1 of Appendix B.

b. **State the null hypothesis (H_o) and the alternative hypothesis (H_a).**

$$H_o: \mu = 4.5 \text{ (mean weight is 4.5)}$$
$$H_a: \mu \neq 4.5 \text{ (mean weight is not 4.5)}$$

Step 2 The Hypothesis Test Criteria:

a. **Check the assumptions.**
σ is known. Samples of size 40 should be large enough to satisfy the CLT; see the discussion of this issue on page 387.

b. **Identify the probability distribution and the test statistic to be used.**

We use the standard normal probability distribution and the test statistic $z\star = \dfrac{\overline{x} - \mu}{\sigma/\sqrt{n}}$; $\sigma = 2.87$.

c. **Determine the level of significance, α.**
$\alpha = 0.10$ (given in the statement of the problem)

Step 3 The Sample Evidence:

a. **Collect the sample information.**
   This random sample was drawn from Table 1 in Appendix B.

TABLE 8.11 Random Sample of Single-Digit Numbers [TA08-01]

2	8	2	1	5	5	4	0	9	1
0	4	6	1	5	1	1	3	8	0
3	6	8	4	8	6	8	9	5	0
1	4	1	2	1	7	1	7	9	3

The sample statistics are $\bar{x} = 3.975$ and $n = 40$.

b. **Calculate the value of the test statistic.**
   Use formula (8.4), information from H_o: $\mu = 4.5$, and $\sigma = 2.87$:

$$z\star = \frac{\bar{x} - \mu}{\sigma/\sqrt{n}}: \quad z\star = \frac{3.975 - 4.50}{2.87/\sqrt{40}} = \frac{-0.525}{0.454} = -1.156 = -1.16$$

Step 4 The Probability Distribution:

a. **Determine the critical region and critical value(s).**
   A two-tailed critical region will be used, and 0.05 will be the area in each tail. The critical values are $\pm z(0.05) = \pm 1.65$.

b. **Determine whether or not the calculated test statistic is in the critical region.**
   The calculated value of z, $z\star = -1.16$, is not in the critical region (shown in red on the figure).

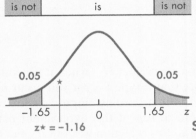

Step 5 The Result:

a. **State the decision about H_o:** Fail to reject H_o.

b. **State the conclusion about H_a.**
   The observed sample mean is not significantly different from 4.5 at the 0.10 significance level.

Suppose we were to take another sample of size 40 from Table 1. Would we obtain the same results? Suppose we took a third sample and a fourth. What results might we expect? What is the level of significance? Yes, its value is 0.10, but what does it measure? Table 8.12 lists the means obtained from 20 different random samples of size 40 that were taken from Table 1 in Appendix B.

TABLE 8.12 Twenty Random Samples of Size 40 Taken from Table 1 in Appendix B [TA08-12]

Sample Number	Sample Mean, $\bar{x}$	Calculated z, $z\star$	Decision Reached	Sample Number	Sample Mean, $\bar{x}$	Calculated z, $z\star$	Decision Reached
1	4.62	+0.26	Fail to reject H_o	11	4.70	+0.44	Fail to reject H_o
2	4.55	+0.11	Fail to reject H_o	12	4.88	+0.83	Fail to reject H_o
3	4.08	−0.93	Fail to reject H_o	13	4.45	−0.11	Fail to reject H_o
4	5.00	+1.10	Fail to reject H_o	14	3.93	−1.27	Fail to reject H_o
5	4.30	−0.44	Fail to reject H_o	15	5.28	+1.71	Reject H_o
6	3.65	−1.87	Reject H_o	16	4.20	−0.66	Fail to reject H_o
7	4.60	+0.22	Fail to reject H_o	17	3.48	−2.26	Reject H_o
8	4.15	−0.77	Fail to reject H_o	18	4.78	+0.61	Fail to reject H_o
9	5.05	+1.21	Fail to reject H_o	19	4.28	−0.50	Fail to reject H_o
10	4.80	+0.66	Fail to reject H_o	20	4.23	−0.61	Fail to reject H_o

The calculated value of $z\star$ that corresponds to each $\bar{x}$ and the decision each would dictate are also listed. The 20 calculated z-scores are shown in Figure 8.10. Note that 3 of the 20 samples (or 15%) caused us to reject the null hypothesis even though we know the null hypothesis is true for this situation. Can you explain this?

FIGURE 8.10
z-Scores from Table 8.12

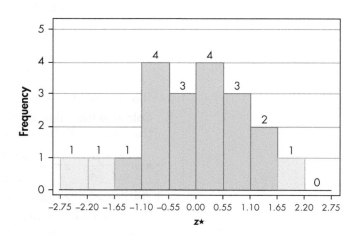

Note: Remember that α is the probability that we "reject H_o" when it is actually a true statement. Therefore, we can anticipate that α type I error will occur α of the time when testing a true null hypothesis. In the preceding empirical situation, we observed a 15% rejection rate. If we were to repeat this experiment many times, the proportion of samples that would lead to a rejection would vary, but the observed relative frequency of rejection should be approximately α or 10%.

SECTION 8.5 EXERCISES

8.131 In the example on page 388, the aircraft builder who is buying the rivets is concerned that the rivets might not meet the mean-strength spec. State the aircraft manufacturer's null and alternative hypotheses.

8.132 Professor Hart does not believe the statement "The mean distance commuted daily by the nonresident students at our college is no more than 9 miles." State the null and alternative hypotheses he would use to challenge this statement.

8.133 State the null and alternative hypotheses used to test each of the following claims:

a. The mean reaction time is less than 1.25 seconds.

b. The mean score on that qualifying exam is different from 335.

c. The mean selling price of homes in the area is no more than $230,000.

8.134 State the null hypothesis, H_o, and the alternative hypothesis, H_a, that would be used for a hypothesis test for each of the following statements:

a. The mean age of the youths who hang out at the mall is less than 16 years.

b. The mean height of professional basketball players is greater than 6 ft 6 in.

c. The mean elevation drop for ski trails at eastern ski centers is at least 285 feet.

d. The mean diameter of the rivets is no more than 0.375 inch.

e. The mean cholesterol level of male college students is different from 200 mg/dL.

8.135 Suppose you want to test the hypothesis that "the mean salt content of frozen 'lite' dinners is more than

350 mg per serving." An average of 350 mg is an acceptable amount of salt per serving; therefore, you use it as the standard. The null hypothesis is "The average content is not more than 350 mg" ($\mu = 350$). The alternative hypothesis is "The average content is more than 350 mg" ($\mu > 350$).

a. Describe the conditions that would exist if your decision results in a type I error.

b. Describe the conditions that would exist if your decision results in a type II error.

8.136 Identify the four possible outcomes and describe the situation involved with each outcome with regard to the aircraft manufacturer's testing and buying of rivets. Which is the more serious error: the type I or type II error? Explain.

8.137 Suppose you wanted to test the hypothesis that the mean minimum home service call charge for plumbers is at most $95 in your area. Explain the conditions that would exist if you made an error in decision by committing a

a. type I error. b. type II error.

8.138 Describe how the null hypothesis in Example 8.21 is a "starting point" for the decision to be made about the mean weight of all female students at the college.

8.139 a. What is the critical region?

b. What is the critical value?

8.140 a. What decision is reached when the test statistic falls in the critical region?

b. What decision is reached when the test statistic falls in the noncritical region?

8.141 Because the size of the type I error can always be made smaller by reducing the size of the critical region, why don't we always choose critical regions that make α extremely small?

8.142 Calculate the test statistic $z\star$, given H_o: $\mu = 356$, $\sigma = 17$, $\bar{x} = 354.3$, and $n = 120$.

8.143 Find the critical region and value(s) for H_a: $\mu < 19$ and $\alpha = 0.01$.

8.144 Find the critical region and value(s) for H_a: $\mu > 34$ and $\alpha = 0.02$.

8.145 Determine the critical region and critical values for z that would be used to test the null hypothesis at the given level of significance, as described in each of the following:

a. H_o: $\mu = 20$, H_a: $\mu \neq 20$, $\alpha = 0.10$

b. H_o: $\mu = 24$ ($\leq$), H_a: $\mu > 24$, $\alpha = 0.01$

c. H_o: $\mu = 10.5$ ($\geq$), H_a: $\mu < 10.5$, $\alpha = 0.05$

d. H_o: $\mu = 35$, H_a: $\mu \neq 35$, $\alpha = 0.01$

8.146 Determine the critical region and the critical values used to test the following null hypotheses:

a. H_o: $\mu = 55$ ($\geq$), H_a: $\mu < 55$, $\alpha = 0.02$

b. H_o: $\mu = -86$ ($\geq$), H_a: $\mu < -86$, $\alpha = 0.01$

c. H_o: $\mu = 107$, H_a: $\mu \neq 107$, $\alpha = 0.05$

d. H_o: $\mu = 17.4$ ($\leq$), H_a: $\mu > 17.4$, $\alpha = 0.10$

8.147 The null hypothesis, H_o: $\mu = 250$, was tested against the alternative hypothesis, H_a: $\mu < 250$. A sample of $n = 85$ resulted in a calculated test statistic of $z\star = -1.18$. If $\sigma = 22.6$, find the value of the sample mean, $\bar{x}$. Find the sum of the sample data, Σx.

8.148 Find the value of $\bar{x}$ for each of the following:

a. H_o: $\mu = 580$, $z\star = 2.10$, $\sigma = 26$, $n = 55$

b. H_o: $\mu = 75$, $z\star = -0.87$, $\sigma = 9.2$, $n = 35$

8.149 The calculated value of the test statistic is actually the number of standard errors that the sample mean differs from the hypothesized value of μ in the null hypothesis. Suppose that the null hypothesis is H_o: $\mu = 4.5$, σ is known to be 1.0, and a sample of size 100 results in $\bar{x} = 4.8$.

a. How many standard errors is $\bar{x}$ above 4.5?

b. If the alternative hypothesis is H_a: $\mu > 4.5$ and $\alpha = 0.01$, would you reject H_o?

8.150 Consider the hypothesis test where the hypotheses are H_o: $\mu = 26.4$ and H_a: $\mu < 26.4$. A sample of size 64 is randomly selected and yields a sample mean of 23.6.

a. If it is known that $\sigma = 12$, how many standard errors below $\mu = 26.4$ is the sample mean, $\bar{x} = 23.6$?

b. If $\alpha = 0.05$, would you reject H_o? Explain.

8.151 There are only two possible decisions as a result of a hypothesis test.

a. State the two possible decisions.

b. Describe the conditions that will lead to each of the two decisions identified in part a.

8.152 a. What proportion of the probability distribution is in the critical region, provided the null hypothesis is correct?

b. What error could be made if the test statistic falls in the critical region?

c. What proportion of the probability distribution is in the noncritical region, provided the null hypothesis is not correct?

d. What error could be made if the test statistic falls in the noncritical region?

8.153 The following computer output was used to complete a hypothesis test.

TEST OF MU = 15.0000 VS MU not = 15.0000				
THE ASSUMED SIGMA = 0.50				
N	MEAN	STDEV	SE MEAN	Z
30	15.6333	0.4270	0.0913	6.94

a. State the null and alternative hypotheses.

b. If the test is completed using $\alpha = 0.01$, what decision and conclusion are reached?

c. Verify the value of the standard error of the mean.

8.154 Using the computer output and information in Exercise 8.153, determine the value of the following:

a. Hypothesized value of population mean

b. Sample mean

c. Population standard deviation

d. Test statistic

8.155 The following computer output was used to complete a hypothesis test.

TEST OF MU = 72.00 VS MU > 72.00				
THE ASSUMED SIGMA = 12.0				
N	MEAN	STDEV	SE MEAN	Z
36	75.2	11.87	2.00	1.60

a. State the null and alternative hypotheses.

b. If the test is completed using $\alpha = 0.05$, what decision and conclusion are reached?

c. Verify the value of the standard error of the mean.

8.156 Using the computer output and information in Exercise 8.155, determine the value of the following:

a. Hypothesized value of population mean

b. Sample mean

c. Population standard deviation

d. Test statistic

8.157 The Texas Department of Health published the statewide results for the Emergency Medical Services Certification Examination. Data for those taking the paramedic exam for the first time gave an average score of 79.68 (out of a possible 100) with a standard deviation of 9.06. Suppose a random sample of 50 individuals taking the exam yielded a mean score of 81.05. Is there sufficient evidence to conclude that "the population from which this random sample was taken, on the average, scored higher than the state average"? Use $\alpha = 0.05$.

8.158 According to the Center on Budget and Policy Priorities' article "Curbing Flexible Spending Accounts Could Help Pay for Health Care Reform" (revised June 10, 2009), flexible-spending accounts encourage the overconsumption of health care. People buy things they do not need; otherwise they lose the money. In 2007, for those who did not use all of their account (about one out of every seven), the average amount lost was $723.

Source: http://www.cbpp.org/

Suppose a random sample of 150 employees who did not use all of their funds in 2009 is taken and an average amount of $683 was lost. Test the hypothesis that there is no significant difference in the average amount forfeited. Assume that $\sigma = \$307$ per year. Use $\alpha = 0.05$.

a. Define the parameter.

b. State the null and alternative hypotheses.

c. Specify the hypothesis test criteria.

d. Present the sample evidence.

e. Find the probability distribution information.

f. Determine the results.

8.159 Women own an average of 15 pairs of shoes. This is based on a survey of female adults by Kelton Research for Eneslow, the New York City–based Foot Comfort Center. Suppose a random sample of 35 newly hired female college graduates was taken and the sample mean was 18.37 pairs of shoes. If $\sigma = 6.12$, does this sample provide sufficient evidence that young female college graduates' mean number of shoes is greater than the overall mean number for all female adults? Use a 0.10 level of significance.

8.160 A fire insurance company thought that the mean distance from a home to the nearest fire department in a suburb of Chicago was at least 4.7 miles. It set its fire insurance rates accordingly. Members of the community set out to show that the mean distance was less than 4.7 miles. This, they thought, would convince the insurance company to lower its rates. They randomly identified 64 homes and measured the distance to the nearest fire department from each. The resulting sample mean was 4.4. If $\sigma = 2.4$ miles, does the sample show sufficient evidence to support the community's claim at the $\alpha = 0.05$ level of significance?

8.161 [EX08-161] The length of Major League Baseball games is approximately normally distributed and averages 2 hours and 50.1 minutes, with a standard deviation of 21.0 minutes. It has been claimed that New York Yankees baseball games last, on the average, longer than the games of the other Major League teams. To test the truth of this statement, a sample of eight Yankees games was randomly identified and the "time of game" (in minutes) for each obtained:

| 199 | 196 | 202 | 213 | 187 | 169 | 169 | 188 |

Source: http://mlb.com/

At the 0.05 level of significance, do these data show sufficient evidence to conclude that the mean time of Yankees baseball games is longer than that of other Major League baseball teams?

8.162 [EX08-162] The manager at Air Express believes that the weights of packages shipped recently are less than those in the past. Records show that in the past, packages have had a mean weight of 36.5 lb and a standard deviation of 14.2 lb. A random sample of last month's shipping records yielded the following 64 data values:

32.1	41.5	16.1	8.9	36.2	12.3	28.4	40.4
45.5	15.2	26.5	13.3	23.5	33.7	18.3	16.3
15.4	39.7	50.3	14.8	44.4	47.7	45.8	52.3
48.4	10.4	59.9	5.5	6.7	17.1	20.0	28.1
48.1	29.5	22.9	47.8	24.8	20.1	40.1	12.6
24.3	43.3	32.4	57.7	42.9	36.7	15.5	46.4
51.3	38.6	39.4	27.1	55.7	37.7	39.4	55.5
26.9	15.7	32.3	47.8	33.2	29.1	31.1	34.5

Is this sufficient evidence to reject the null hypothesis in favor of the manager's claim? Use $\alpha = 0.01$.

8.163 Do you drink the recommended amount of water each day? Most Americans don't! On average, Americans drink 4.6 eight-oz servings of water a day.

Source: http://www.bottledwater.org

A sample of 42 education professionals was randomly selected and their water consumption for a 24-hour period was monitored; the mean amount consumed was 39.3 oz. Assuming the amount of water consumed daily by adults is normally distributed and the standard deviation is 11.2 oz, is there sufficient evidence to show that education professionals consume, on average, more water daily than the national average? Use $\alpha = 0.05$.

8.164 The recommended amount of water a person should drink is eight 8-oz servings per day.

a. Does the sample of educational professionals in Exercise 8.163 show sufficient evidence that the education professionals consume, on average, significantly less water daily than the recommended amount? Use $\alpha = 0.05$.

b. The value of the calculated z-score in part a is unusual. In what way is it unusual, and what does that mean?

8.165 Use a computer or calculator to select 40 random single-digit numbers. Find the sample mean and $z\star$. Using $\alpha = 0.05$, state the decision for testing H_o: $\mu = 4.5$ against a two-tailed alternative. Repeat it several times as in Table 8.12. Describe your findings after several tries.

FYI Use commands for generating integer data on page 91; then continue with the hypothesis test commands on page 382

8.166 Use a computer or calculator to select 36 random numbers from a normal distribution with mean 100 and standard deviation 15. Find the sample mean and $z\star$ for testing a two-tailed hypothesis test of $\mu = 100$. Using $\alpha = 0.05$, state the decision. Repeat several times as in Table 8.12. Describe your findings.

FYI Use commands for generating data on pages 283–284; then continue with the hypothesis test commands on page 382.

© 2010 Image Source/Jupiterimages Corporation

Chapter Review

In Retrospect

Two forms of inference were presented in this chapter: estimation and hypothesis testing. They may be, and often are, used separately. It seems natural, however, for the rejection of a null hypothesis to be followed by a confidence interval. (If the value claimed is wrong, we often want an estimate for the true value.)

These two forms of inference are quite different, but they are related. There is a certain amount of crossover between the use of the two inferences. For example, suppose that you had sampled and calculated a 90% confidence interval for the mean of a population. The interval was 10.5 to 15.6. Then someone claims that the true mean is 15.2. Your confidence interval can be compared with this claim. If the claimed value falls within your interval estimate, you would fail to reject the null hypothesis that $\mu = 15.2$ at a 10% level of significance in a two-tailed test. If the claimed value (say, 16.0) falls outside the interval, you would reject the null hypothesis that $\mu = 16.0$ at $\alpha = 0.10$ in a two-tailed test. If a

one-tailed test is required, or if you prefer a different value of α, a separate hypothesis test must be used.

Many users of statistics (especially those marketing a product) will claim that their statistical results prove that their product is superior. But remember, the hypothesis test does not *prove* or *disprove* anything. The decision reached in a hypothesis test has probabilities associated with the four various situations. If "fail to reject H_o" is the decision, it is possible that an error has occurred. Furthermore, if "reject H_o" is the decision reached, it is possible for this to be an error. Both errors have probabilities greater than zero.

In this chapter we have restricted our discussion of inferences to the mean of a population for which the standard deviation is known. In Chapters 9 and 10 we will discuss inferences about the population mean and remove the restriction about the known value for standard deviation. We will also look at inferences about the parameters proportion, variance, and standard deviation.

CourseMate The **Statistics CourseMate** site for this text brings chapter topics to life with interactive learning, study, and exam preparation tools, including quizzes and flashcards for the Vocabulary and Key Concepts that follow. The site also provides an **eBook** version of the text with highlighting and note taking capabilities. Throughout chapters, the CourseMate icon [icon] flags concepts and examples that have

corresponding interactive resources such as **video** and **animated tutorials** that demonstrate, step by step, how to solve problems; **datasets** for exercises and examples; **Skillbuilder Applets** to help you better understand concepts; **technology manuals**; and software to download including **Data Analysis Plus** (a suite of statistical macros for Excel) and **TI-83/84 Plus** programs—logon at **www.cengagebrain.com**.

Vocabulary and Key Concepts

alpha (α) (p. 365)
alternative hypothesis (pp. 361, 371, 388)
assumptions (pp. 347, 370, 387)
beta (β) (p. 365)
biased statistics (p. 342)
calculated value ($z\star$) (pp. 375, 392)
conclusion (p. 366)
confidence coefficient (p. 348)
confidence interval (pp. 343, 348)

confidence interval procedure (pp. 348, 349)
critical region (p. 392)
critical value (p. 393)
decision rule (pp. 366, 376, 394)
estimation (p. 341)
estimation question (p. 340)
hypothesis (p. 361)
hypothesis test, classical procedure (p. 388)

hypothesis test, p-value procedure (p. 371)
hypothesis-testing question (p. 340)
interval estimate (p. 343)
level of confidence (p. 343)
level of significance (p. 366)
lower confidence limit (p. 348)
maximum error of estimate (pp. 348, 355)
noncritical region (p. 392)

null hypothesis (pp. 361, 371 388)	sample statistic (p. 342)	type B correct decision (pp. 363, 364)
parameter (p. 342)	standard error of mean (p. 348)	type I error (pp. 363, 364)
point estimate for a parameter (p. 342)	statistical hypothesis test (p. 361)	type II error (p. 364)
	test criteria (p. 371)	unbiased statistic (p. 342)
p-value (p. 375)	test statistic (pp. 366, 374, 392)	upper confidence limit (p. 348)
sample size (p. 356)	type A correct decision (pp. 363, 364)	$z(\alpha)$ (pp. 348, 393)

Learning Outcomes

- Understand the difference between descriptive statistics and inferential statistics. p. 1, Ex. 1.8, p. 341
- Understand that an unbiased statistic has a sampling distribution with a mean that is equal to the population parameter being estimated. pp. 342–343

With respect to confidence intervals:
- Understand that a confidence interval is an interval estimate of a population parameter, with a degree of certainty, used when the population parameter is unknown. p. 343
- Understand that a point estimate for a population parameter is the value of the corresponding sample statistic. p. 342, Ex. 8.5
- Understand that the level of confidence is the long-run proportion of the intervals, which will contain the true population parameters, based on repeated sampling. EXP 8.4
- Understand and be able to describe the key components for a confidence interval: point estimate, level of confidence, confidence coefficient, maximum error of estimate, lower confidence limit, and upper confidence limit. p. 348, Ex. 8.27, 8.29, 8.171
- Understand that the assumption for a confidence interval for μ using a known σ is that the sampling distribution of $\bar{x}$ has a normal distribution. Based on this assumption, the standard normal z distribution will be utilized. pp. 347–348
- Compute, describe, and interpret a confidence interval for the population mean, μ. EXP 8.2, Ex. 8.35
- Compute sample sizes required for constructing confidence intervals with varying levels of confidence and acceptable errors. pp. 355–357, Ex. 8.47, 8.51

With respect to hypothesis tests:
- Understand that a hypothesis test is used to make a decision about the value of a population parameter. p. 361
- Understand and be able to define null and alternative hypotheses. p. 361
- Understand and be able to describe the two types of errors in a hypothesis test, type I and type II. Understand that the probability of these errors are α and β, respectively. pp. 363–365, Ex. 8.61
- Understand and be able to describe the two types of correct decisions in a hypothesis test, type A and type B. pp. 363–365, Ex. 8.61
- Understand and be able to describe the relationship between the four possible outcomes of a hypothesis test—the two types of errors and the two types of correct decisions. pp. 363–365, Ex. 8.61
- Demonstrate and understand the three possible combinations for the null and alternative hypotheses. pp. 371–374, Ex. 8.87, pp. 388–391, Ex. 8.133
- Understand that the assumption for a hypothesis test for μ using a known σ is that the sampling distribution of $\bar{x}$ has a normal distribution. Based on this assumption, the standard normal z distribution will be utilized. pp. 370, 387
- Compute and understand the value of the test statistic. Compute the p-value for the test statistic and/or determine the critical region and critical value(s). pp. 375–377, Ex. 8.106, pp. 392–394, Ex. 8.142, 8.145
- Understand and be able to describe what a p-value and/or a critical region is with respect to a hypothesis test. pp. 375–377, 379, 392–394
- Determine and know the proper format for stating a decision in a hypothesis test. pp. 366, 376, 394
- Understand and be able to state the conclusion for a hypothesis test. pp. 366, 376, 394, Ex. 8.185, 8.190, 8.191

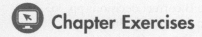

Chapter Exercises

8.167 A sample of 64 measurements is taken from a continuous population, and the sample mean is found to be 32.0. The standard deviation of the population is known to be 2.4. An interval estimation is to be made of the mean with a level of confidence of 90%. State or calculate the following items.

a. $\bar{x}$ b. σ

c. n d. $1 - \alpha$

e. $z(\alpha/2)$ f. $\sigma_{\bar{x}}$

g. E (maximum error of estimate)

h. Upper confidence limit

i. Lower confidence limit

8.168 Suppose that a confidence interval is assigned a level of confidence of $1 - \alpha = 95\%$. How is the 95% used in constructing the confidence interval? If $1 - \alpha$ was changed to 90%, what effect would this have on the confidence interval?

8.169 The average volunteer ambulance member is 45 years old and has 8 years of service, according to the *Democrat & Chronicle* article "Unpaid ambulance workers could get 'pension'" (January 23, 2005). The quoted statistics were based on the Penfield Volunteer Ambulance Squad of 80 members. If the Penfield Volunteer Ambulance Squad is considered representative of all upstate New York volunteer ambulance squads, determine a 95% confidence interval for the mean age of all volunteer ambulance members in upstate New York. Assume the population standard deviation is 7.8 years.

8.170 The standard deviation of a normally distributed population is equal to 10. A sample size of 25 is selected, and its mean is found to be 95.

a. Find an 80% confidence interval for μ.

b. What would the 80% confidence interval be for a sample of size 100?

c. What would be the 80% confidence interval for a sample of size 25 with a standard deviation of 5 (instead of 10)?

8.171 The weights of full boxes of a certain kind of cereal are normally distributed with a standard deviation of 0.27 oz. A sample of 18 randomly selected boxes produced a mean weight of 9.87 oz.

a. Find the 95% confidence interval for the true mean weight of a box of this cereal.

b. Find the 99% confidence interval for the true mean weight of a box of this cereal.

c. What effect did the increase in the level of confidence have on the width of the confidence interval?

8.172 Waiting times (in hours) at a popular restaurant are believed to be approximately normally distributed with a variance of 2.25 during busy periods.

a. A sample of 20 customers revealed a mean waiting time of 1.52 hours. Construct the 95% confidence interval for the population mean.

b. Suppose that the mean of 1.52 hours had resulted from a sample of 32 customers. Find the 95% confidence interval.

c. What effect does a larger sample size have on the confidence interval?

8.173 A random sample of the scores of 100 applicants for clerk-typist positions at a large insurance company showed a mean score of 72.6. The preparer of the test maintained that qualified applicants should average 75.0.

a. Determine the 99% confidence interval for the mean score of all applicants at the insurance company. Assume that the standard deviation of test scores is 10.5.

b. Can the insurance company conclude that it is getting qualified applicants (as measured by this test)?

8.174 The length of time it takes to play a Major League Baseball game is of interest to many fans. To estimate the mean "time of game," a random sample of 48 National League games was identified and the "time of game" (in minutes) obtained for each. The resulting sample mean was 2 hours and 49.1 minutes, and the history of baseball indicates that the time of game variable has a standard deviation of 21 minutes. Construct the 98% confidence interval for the mean time for all National League games.

8.175 [EX08-175] A large order of the no. 9 corks described in Applied Example 6.13 (p. 285) is about to be shipped. The final quality-control inspection includes an estimation of the mean ovality (ovalization; out-of-roundness) of the corks. The diameter of each cork is measured in several places, and the difference between the maximum and minimum diameters is the measure of ovality for each cork. After years of measuring corks, the manufacturer is sure that ovality has a mounded distribution with a standard deviation of 0.10 mm. A random sample of 36 corks is taken from the batch and the ovality is determined for each.

0.32	0.27	0.24	0.31	0.20	0.38	0.32	0.11	0.25
0.22	0.35	0.20	0.28	0.17	0.36	0.28	0.38	0.17
0.34	0.06	0.43	0.13	0.39	0.15	0.18	0.13	0.25
0.20	0.16	0.26	0.47	0.21	0.19	0.34	0.24	0.20

a. The out-of-round spec is "less than 1.0 mm." Does it appear this order meets the spec on an individual cork basis? Explain.

b. The certification sheet that accompanies the shipment includes a 95% confidence interval for the mean ovality. Construct the confidence interval.

c. Explain what the confidence interval found in part b tells about this shipment of corks.

8.176 Use a computer and generate 50 random samples, each of size $n = 25$, from a normal probability distribution with $\mu = 130$ and $\sigma = 10$.

a. Calculate the 95% confidence interval based on each sample mean.

b. What proportion of these confidence intervals contains $\mu = 130$?

c. Explain what the proportion found in part b represents.

8.177 A pharmaceutical company wants to estimate the mean response time for a supplement to reduce blood pressure. How large of a sample should be taken to estimate the mean response time to within 1 week at 99% confidence. Assume $\sigma = 3.7$ weeks.

8.178 An automobile manufacturer wants to estimate the mean gasoline mileage of its new compact model. How many sample runs must be performed to ensure that the estimate is accurate to within 0.3 mpg at 95% confidence? (Assume $\sigma = 1.5$.)

8.179 A fish hatchery manager wants to estimate the mean length of her 3-year-old hatchery-raised trout. She wants to make a 99% confidence interval accurate to within $\frac{1}{3}$ of a standard deviation. How large a sample does she need to take?

8.180 We are interested in estimating the mean life of a new product. How large a sample do we need to take to estimate the mean to within $\frac{1}{10}$ of a standard deviation with 90% confidence?

8.181 Suppose a hypothesis test is conducted using the p-value approach and assigned a level of significance of $\alpha = 0.01$.

a. How is the 0.01 used in completing the hypothesis test?

b. If α is changed to 0.05, what effect would this have on the test procedure?

8.182 Suppose a hypothesis test is conducted using the classical approach and assigned a level of significance of $\alpha = 0.01$.

a. How is the 0.01 used in completing the hypothesis test?

b. If α is changed to 0.05, what effect would this have on the test procedure?

8.183 The expected mean of a continuous population is 100, and its standard deviation is 12. A sample of 50 measurements gives a sample mean of 96. Using a 0.01 level of significance, a test is to be made to decide between "the population mean is 100" and "the population mean is different from 100." State or find each of the following:

a. H_o

b. H_a

c. α

d. μ (based on H_o)

e. $\bar{x}$

f. σ

g. $\sigma_{\bar{x}}$

h. $z\star$, z-score for $\bar{x}$

i. p-value

j. Decision

k. Sketch the standard normal curve and locate $z\star$ and p-value.

8.184 The expected mean of a continuous population is 200, and its standard deviation is 15. A sample of 80 measurements gives a sample mean of 205. Using a 0.01 level of significance, a test is to be made to decide between "the population mean is 200" and "the population mean is different from 200." State or find each of the following:

a. H_o

b. H_a

c. α

d. $z(\alpha/2)$

e. μ (based on H_o)

f. $\bar{x}$

g. σ

h. $\sigma_{\bar{x}}$

i. $z\star$, z-score for $\bar{x}$

j. decision

k. Sketch the standard normal curve and locate $\alpha/2$, $z(\alpha/2)$, the critical region, and $z\star$.

8.185 A lawn and garden sprinkler system is designed to have a delayed start; that is, there is a delay from the moment it is turned on until the water starts. The delay times form a normal distribution with mean 45 seconds and standard deviation 8 seconds. Several customers have complained that the delay time is considerably longer than claimed. The system engineer has selected a random sample of 15 installed systems and has obtained one delay time from each system. The sample mean is 50.1 seconds. Using $\alpha = 0.02$, is there significant evidence to show that the customers might be correct that the mean delay time is more than 45 seconds?

a. Solve using the p-value approach.

b. Solve using the classical approach.

8.186 The college bookstore tells prospective students that the average cost of its textbooks is $90 per book with a standard deviation of $15. The engineering science students think that the average cost of their books is higher than the average for all students. To test the bookstore's claim against their alternative, the engineering students collect a random sample of size 45.

a. If they use $\alpha = 0.05$, what is the critical value of the test statistic?

b. The engineering students' sample data are summarized by $n = 45$ and $\Sigma x = 4380.30$. Is this sufficient evidence to support their contention?

8.187 A manufacturing process produces ball bearings with diameters having a normal distribution and a standard deviation of $\sigma = 0.04$ cm. Ball bearings that have diameters that are too small or too large are undesirable. To test the null hypothesis that $\mu = 0.50$ cm, a sample of 25 is randomly selected and the sample mean is found to be 0.51.

a. Design null and alternative hypotheses such that rejection of the null hypothesis will imply that the ball bearings are undesirable.

b. Using the decision rule established in part a, what is the p-value for the sample results?

c. If the decision rule in part a is used with $\alpha = 0.02$, what is the critical value for the test statistic?

8.188 After conducting a large number of tests over a long period, a rope manufacturer has found that its rope has a mean breaking strength of 300 lb and a standard deviation of 24 lb. Assume that these values are μ and σ. It is believed that by using a recently developed high-speed process, the mean breaking strength has been decreased.

a. Design null and alternative hypotheses such that rejection of the null hypothesis will imply that the mean breaking strength has decreased.

b. Using the decision rule established in part a, what is the p-value associated with rejecting the null hypothesis when 45 tests result in a sample mean of 295?

c. If the decision rule in part a is used with $\alpha = 0.01$, what is the critical value for the test statistic and what value of $\bar{x}$ corresponds to it if a sample of size 45 is used?

8.189 A worker honeybee leaves the hive on a regular basis and travels to flowers and other sources of pollen and nectar before returning to the hive to deliver its cargo. The process is repeated several times each day in order to feed younger bees and support the hive's production of honey and wax. The worker bee can carry an average of 0.0113 gram of pollen and nectar per trip, with a standard deviation of 0.0063 gram. Fuzzy Drone is entering the honey and beeswax business with a new strain of Italian bees that are reportedly capable of carrying larger loads of pollen and nectar than the typical honeybee. After installing three hives, Fuzzy isolated 200 bees before and after their return trip and carefully weighed their cargoes. The sample mean weight of the pollen and nectar was 0.0124 gram. Can Fuzzy's bees carry a greater load of pollen and nectar than the rest of the honeybee population? Complete the appropriate hypothesis test at the 0.01 level of significance.

a. Solve using the p-value approach.

b. Solve using the classical approach.

8.190 In a large supermarket the customer's waiting time to check out is approximately normally distributed with a standard deviation of 2.5 minutes. A sample of 24 customer waiting times produced a mean of 10.6 minutes. Is this evidence sufficient to reject the supermarket's claim that its customer checkout time averages no more than 9 minutes? Complete this hypothesis test using the 0.02 level of significance.

a. Solve using the p-value approach.

b. Solve using the classical approach.

8.191 At a very large firm, the clerk-typists were sampled to see whether salaries differed among departments for workers in similar categories. In a sample of 50 of the firm's accounting clerks, the average annual salary was $16,010. The firm's personnel office insists that the average salary paid to all clerk-typists in the firm is $15,650 and that the standard deviation is $1800. At the 0.05 level of significance, can we conclude that the accounting clerks receive, on average, a different salary from that of the clerk-typists?

a. Solve using the p-value approach.

b. Solve using the classical approach.

8.192 Jack Williams is vice president of marketing for one of the largest natural gas companies in the nation. During the past 4 years, he has watched two major factors erode the profits and sales of the company. First, the average price of crude oil has been virtually flat, and many of his industrial customers are burning heavy oil rather than natural gas to fire their furnaces, regardless of added smokestack emissions. Second, both residential and commercial customers are still pursuing energy-conservation techniques (e.g., adding extra insulation, installing clock-drive thermostats, and sealing cracks around doors and windows to eliminate cold air infiltration). In previous years, residential customers bought an average of 129.2

mcf of natural gas from Jack's company ($\sigma = 18$ mcf), based on internal company billing records, but environmentalists have claimed that conservation is cutting fuel consumption up to 3% per year. Jack has commissioned you to conduct a spot check to see if any change in annual usage has transpired before his next meeting with the officers of the corporation. A sample of 300 customers selected randomly from the billing records reveals an average of 127.1 mcf during the past 12 months. Is there a significant decline in consumption?

a. Complete the appropriate hypothesis test at the 0.01 level of significance using the p-value approach so that you can properly advise Jack before his meeting.

b. Because you are Jack's assistant, why is it best for you to use the p-value approach?

8.193 With a nationwide average drive time of about 24.3 minutes, Americans now spend more than 100 hours a year commuting to work, according to the U.S. Census Bureau's American Community Survey. Yes, that's more than the average 2 weeks of vacation time (80 hours) taken by many workers during a year.

Source: http://usgovinfo.about.com/

A random sample of 150 workers at a large nearby industry was polled about their commute time. If the standard deviation is known to be 10.7 minutes, is the resulting sample mean of 21.7 minutes significantly lower than the nationwide average? Use $\alpha = 0.01$.

a. Solve using the p-value approach.

b. Solve using the classical approach.

8.194 A manufacturer of automobile tires believes it has developed a new rubber compound that has superior antiwearing qualities. It produced a test run of tires made with this new compound and had them road-tested. The data values recorded were the amount of tread wear per 10,000 miles. In the past, the mean amount of tread wear per 10,000 miles, for tires of this quality, has been 0.0625 inch.

The null hypothesis to be tested here is "The mean amount of wear on the tires made with the new compound is the same mean amount of wear with the old compound, 0.0625 inch per 10,000 miles," $H_o: \mu = 0.0625$. Three possible alternative hypotheses could be used: (1) $H_a: \mu < 0.0625$, (2) $H_a: \mu \neq 0.0625$, (3) $H_a: \mu > 0.0625$.

(continue on page 408)

a. Explain the meaning of each of these three alternatives.

b. Which one of the possible alternative hypotheses should the manufacturer use if it hopes to conclude that "use of the new compound does yield superior wear"?

8.195 From a population of unknown mean μ and a standard deviation $\sigma = 5.0$, a sample of $n = 100$ is selected and the sample mean 40.6 is found. Compare the concepts of estimation and hypothesis testing by completing the following:

a. Determine the 95% confidence interval for μ.

b. Complete the hypothesis test involving $H_a: \mu \neq 40$ using the p-value approach and $\alpha = 0.05$.

c. Complete the hypothesis test involving $H_a: \mu \neq 40$ using the classical approach and $\alpha = 0.05$.

d. On one sketch of the standard normal curve, locate the interval representing the confidence interval from part a; the $z\star$, p-value, and α from part b; and the $z\star$ and critical regions from part c. Describe the relationship between these three separate procedures.

8.196 From a population of unknown mean μ and a standard deviation $\sigma = 5.0$, a sample of $n = 100$ is selected and the sample mean 41.5 is found. Compare the concepts of estimation and hypothesis testing by completing the following:

a. Determine the 95% confidence interval for μ.

b. Complete the hypothesis test involving $H_a: \mu \neq 40$ using the p-value approach and $\alpha = 0.05$.

c. Complete the hypothesis test involving $H_a: \mu \neq 40$ using the classical approach and $\alpha = 0.05$.

d. On one sketch of the standard normal curve, locate the interval representing the confidence interval from part a; the $z\star$, p-value, and α from part b; and the $z\star$ and critical regions from part c. Describe the relationship between these three separate procedures.

8.197 From a population of unknown mean μ and a standard deviation $\sigma = 5.0$, a sample of $n = 100$ is selected and the sample mean 40.9 is found. Compare the concepts of estimation and hypothesis testing by completing the following:

a. Determine the 95% confidence interval for μ.

b. Complete the hypothesis test involving $H_a: \mu > 40$ using the p-value approach and $\alpha = 0.05$.

c. Complete the hypothesis test involving $H_a: \mu > 40$ using the classical approach and $\alpha = 0.05$.

d. On one sketch of the standard normal curve, locate the interval representing the confidence interval from part a; the $z\star$, p-value, and α from part b; and the $z\star$ and critical regions from part c. Describe the relationship between these three separate procedures.

8.198 A manufacturer of stone-ground, deli-style mustard uses a high-speed machine to fill jars. The amount of mustard dispensed into the jars forms a normal distribution with a mean 290 grams and a standard deviation 4 grams. Each hour a random sample of 12 jars is taken from that hour's production. If the sample mean is between 287.74 and 292.26, that hour's production is accepted; otherwise, it is rejected and the machine is recalibrated before continuing.

a. What is the probability of the type I error by rejecting the previous hour's production when the mean jar weight is 290 grams?

b. What is the probability of the type II error by accepting the previous hour's production when the mean jar weight is actually 288 grams?

8.199 All drugs must be approved by the U.S. Food and Drug Administration (FDA) before they can be marketed by a drug company. The FDA must weigh the error of marketing an ineffective drug, with the usual risks of side effects, against the consequences of not allowing an effective drug to be sold. Suppose, using standard medical treatment, that the mortality rate (r) of a certain disease is known to be A. A manufacturer submits for approval a drug that is supposed to treat this disease. The FDA sets up the hypothesis to test the mortality rate for the drug as (1) $H_o: r = A$, $H_a: r < A$, $\alpha = 0.005$ or (2) $H_o: r = A$, $H_a: r > A$, $\alpha = 0.005$.

a. If $A = 0.95$, which test do you think the FDA should use? Explain.

b. If $A = 0.05$, which test do you think the FDA should use? Explain.

8.200 The drug manufacturer in Exercise 8.199 has a different viewpoint on the matter. It wants to market the new drug starting as soon as possible so that it can beat its competitors to the marketplace and make lots of money. Its position is, "Market the drug unless the drug is totally ineffective."

a. How would the drug company set up the alternative hypothesis if it were doing the testing: $H_o: r < A$, $H_a: r \neq A$, or $H_a: r > A$? Explain.

b. Does the mortality rate ($A = 0.95$ or $A = 0.05$) of the existing treatment affect the alternative? Explain.

8.201 [EX08-201] This computer output shows a simulated sample of size 28 randomly generated from a normal population with $\mu = 18$ and $\sigma = 4$. Computer commands were then used to complete a hypothesis test for $\mu = 18$ against a two-tailed alternative.

a. State the alternative hypothesis, the decision, and the conclusion that resulted.

b. Verify the values reported for the standard error of mean, $z\star$, and the p-value.

18.7734	21.4352	15.5438	20.2764	23.2434	15.7222	13.9368
14.4112	15.7403	19.0970	19.0032	20.0688	12.2466	10.4158
8.9755	18.0094	20.0112	23.2721	16.6458	24.6146	17.8078
16.5922	16.1385	12.3115	12.5674	18.9141	22.9315	13.3658

TEST OF MU = 18.000 VS MU not = 18.000					
THE ASSUMED STANDARD DEVIATION = 4.00					
N	MEAN	STDEV	SE MEAN	Z	P VALUE
28	17.217	4.053	0.756	−1.04	0.30

8.202. Use a computer and generate 50 random samples, each of size $n = 28$, from a normal probability distribution with $\mu = 18$ and $\sigma = 4$.

a. Calculate the $z\star$ corresponding to each sample mean.

b. In regard to the p-value approach, find the proportion of 50 $z\star$ values that are "more extreme" than the $z = -1.04$ that occurred in Exercise 8.201 ($H_a: \mu \neq 18$). Explain what this proportion represents.

c. In regard to the classical approach, find the critical values for a two-tailed test using $\alpha = 0.01$; find the proportion of 50 $z\star$ values that fall in the critical region. Explain what this proportion represents.

8.203 Use a computer and generate 50 random samples, each of size $n = 28$, from a normal probability distribution with $\mu = 19$ and $\sigma = 4$.

a. Calculate the $z\star$ corresponding to each sample mean that would result when testing the null hypothesis $\mu = 18$.

b. In regard to the p-value approach, find the proportion of 50 $z\star$ values that are "more extreme" than the $z = -1.04$ that occurred in Exercise 8.201 ($H_a: \mu \neq 18$). Explain what this proportion represents.

c. In regard to the classical approach, find the critical values for a two-tailed test using $\alpha = 0.01$; find the proportion of 50 $z\star$ values that fall in the noncritical region. Explain what this proportion represents.

Chapter Practice Test

PART I: Knowing the Definitions

Answer "True" if the statement is always true. If the statement is not always true, replace the words shown in bold with words that make the statement always true.

8.1 **Beta** is the probability of a type I error.

8.2 $1 - \alpha$ is known as the level of significance of a hypothesis test.

8.3 The standard error of the mean is the standard deviation of the **sample selected**.

8.4 The maximum error of estimate is controlled by three factors: **level of confidence**, **sample size**, and **standard deviation**.

8.5 Alpha is the measure of the area under the curve of the standard score that lies in the **rejection region** for H_o.

8.6 The risk of making a **type I error** is directly controlled in a hypothesis test by establishing a level for α.

8.7 Failing to reject the null hypothesis when it is false is a **correct decision**.

8.8 If the noncritical region in a hypothesis test is made wider (assuming σ and n remain fixed), α becomes larger.

8.9 Rejection of a null hypothesis that is false is a **type II error**.

8.10 To conclude that the mean is greater (or less) than a claimed value, the value of the test statistic must fall in the **acceptance region**.

PART II: Applying the Concepts

Answer all questions, showing all formulas, substitutions, and work.

8.11 An unhappy post office customer is frustrated with the waiting time to buy stamps. Upon registering his complaint, he was told, "The average waiting time in the past has been about 4 minutes with a standard deviation of 2 minutes." The customer collected a sample of $n = 45$ customers and found the mean wait was 5.3 minutes. Find the 95% confidence interval for the mean waiting time.

8.12 State the null (H_o) and the alternative (H_a) hypotheses that would be used to test each of these claims:
a. The mean weight of professional football players is more than 245 lb.
b. The mean monthly amount of rainfall in Monroe County is less than 4.5 inches.
c. The mean weight of the baseball bats used by Major League players is not equal to 35 oz.

8.13 Determine the level of significance, test statistic, critical region, and critical value(s) that would be used in completing each hypothesis test using $\alpha = 0.05$:
a. $H_o: \mu = 43$ b. $H_o: \mu = 0.80$ c. $H_o: \mu = 95$
   $H_a: \mu < 43$ $H_a: \mu > 0.80$ $H_a: \mu \neq 95$
   (given $\sigma = 6$) (given $\sigma = 0.13$) (given $\sigma = 12$)

8.14 Find each value:
   a. $z(0.05)$ b. $z(0.01)$ c. $z(0.12)$

8.15 In the past, the grapefruits grown in a particular orchard have had a mean diameter of 5.50 inches and a standard deviation of 0.6 inch. The owner believes this year's crop is larger than those in the past. He collected a random sample of 100 grapefruits and found a sample mean diameter of 5.65 inches.
a. Find the value of the test statistic, $z\star$, that corresponds to $\bar{x} = 5.65$.
b. Calculate the p-value for the owner's hypothesis.

8.16 A manufacturer claims that its light bulbs have a mean lifetime of 1520 hours with a standard deviation of 85 hours. A random sample of 40 such bulbs is selected for testing. If the sample produces a mean value of 1498.3 hours, is there sufficient evidence to claim that the mean lifetime is less than the manufacturer claims? Use $\alpha = 0.01$.

PART III: Understanding the Concepts

8.17 Sugar Creek Convenience Stores has commissioned a statistics firm to survey its customers in order to estimate the mean amount spent per customer. From previous records the standard deviation is believed to be $\sigma = \$5$. In its proposal to Sugar Creek, the statistics firm states that it plans to base the estimate for the mean amount spent on a sample of size 100 and use the 95% confidence level. Sugar Creek's president has suggested that the sample size be increased to 400. If nothing else changes, what effect will this increase in the sample size have on the following?
a. The point estimate for the mean
b. The maximum error of estimation
c. The confidence interval
The CEO wants the level of confidence increased to 99%. If nothing else changes, what effect will this increase in level of confidence have on the following?
d. The point estimate for the mean
e. The maximum error of estimation
f. The confidence interval

8.18 The noise level in a hospital may be a critical factor influencing a patient's speed of recovery. Suppose for the sake of discussion that a research commission has recommended a maximum mean noise level of 30 decibels (db) with a standard deviation of 10 db. The staff of a hospital intend to sample one of its wards to determine whether the noise level is significantly higher than the recommended level. The following hypothesis test will be completed:

$$H_o: \mu = 30 \ (\leq) \text{ versus } H_a: \mu > 30, \alpha = 0.05$$

a. Identify the correct interpretation for each hypothesis with regard to the recommendation and justify your choice.
   H_o: (1) Noise level is not significantly higher than the recommended level, or (2) Noise level is significantly higher than the recommended level.

H_a: (1) Noise level is not significantly higher than the recommended level, or (2) Noise level is significantly higher than the recommended level.

b. Which statement best describes the type I error?

(i) Decision reached was that noise level is within the recommended level when, in fact, it actually is within.

(ii) Decision reached was that noise level is within the recommended level when, in fact, it actually exceeds it.

(iii) Decision reached was that noise level exceeds the recommended level when, in fact, it actually is within.

(iv) Decision reached was that noise level is within the recommended level when, in fact, it actually exceeds it.

c. Which statement in part b best describes the type II error?

d. If α were changed from 0.05 to 0.01, identify and justify the effect (increases, decreases, or remains the same) on P(type I error) and on P(type II error).

8.19 The alternative hypothesis is sometimes called the *research hypothesis*. The conclusion is a statement written about the alternative hypothesis. Explain why these two statements are compatible.

9 Inferences Involving One Population

Courtesy of the author

Courtesy of the author

9.1 Inferences about the Mean μ (σ Unknown)

***Student's t-distribution** is used in inferences about the mean when the population is unknown*

9.2 Inferences about the Binomial Probability of Success

*Sample proportion p' is **approximately normally distributed** under certain conditions*

9.3 Inferences about the Variance and Standard Deviation

*The **chi-square distribution** is employed to test the variance or standard deviation*

9.1 Inferences about the Mean μ (σ Unknown)

Floor to Door

Think about how long it takes you to get ready in the morning—that is, from the time your feet hit the floor until you are going out the door, having showered, groomed, eaten breakfast, and fully dressed.

Some will say they get ready in as little as 5 minutes, but when timed, it is hard to do everything in less than 15 minutes, and even in that case, only if your routine is very well orchestrated. Here is one morning routine: up at 6:55, in shower by 7:05, out of shower by 7:15, makeup on and dressed by 7:30, bookbag packed and breakfast grabbed by 7:45, out the door by 7:46, and at class by 8:00. That is a total of 51 minutes "floor to door."

If you were given the job of estimating the "floor-to-door" time for the typical college woman, what information would you need and how would you use it to determine the estimate?

Inferences about the population mean μ are based on the sample mean $\bar{x}$ and information obtained from the sampling distribution of sample means. Recall that the sampling distribution of sample means has a mean μ and a **standard error** of $\sigma/\sqrt{n}$ for all samples of size n, and it is normally distributed when the sampled population has a normal distribution or approximately normally distributed when the **sample size** is sufficiently large. This means the **test statistic** $z\star = \dfrac{\bar{x} - \mu}{\sigma\sqrt{n}}$ has a standard normal distribution. However, when σ **is unknown**, the standard error $\sigma/\sqrt{n}$ is also unknown. Therefore, the sample standard deviation s will be used as the point estimate for σ. As a result, an estimated standard error of the mean, $s/\sqrt{n}$, will be used and our test statistic will become $\dfrac{\bar{x} - \mu}{\sigma\sqrt{n}}$.

When a **known** σ is being used to make an inference about the mean μ, a sample provides one value for use in the formulas; that one value is $\bar{x}$. When the sample standard deviation s is also used, the sample provides two values: the sample mean $\bar{x}$ and the estimated standard error $s/\sqrt{n}$. As a result, the z-statistic will be replaced with a statistic that accounts for the use of an estimated standard error. This new statistic is known as **Student's t-statistic**.

In 1908, W. S. Gosset, an Irish brewery employee, published a paper about this t-distribution under the pseudonym "Student." In deriving the t-distribution, Gosset assumed that the samples were taken from normal populations. Although this might seem to be restrictive, satisfactory results are obtained when large samples are selected from many nonnormal populations.

Figure 9.1 presents a diagrammatic organization for the inferences about the population mean as discussed in Chapter 8 and in this section of Chapter 9. Two situations exist: σ is known, or σ is unknown. As stated before, σ is almost never a known quantity in real-world problems; therefore, the standard error will almost always be estimated by $s/\sqrt{n}$. The use of an estimated standard error of the mean requires the use of the t-distribution. Almost all real-world inferences about the population mean will be made with Student's t-statistic.

FIGURE 9.1

Do I Use the z-Statistic or the t-Statistic?

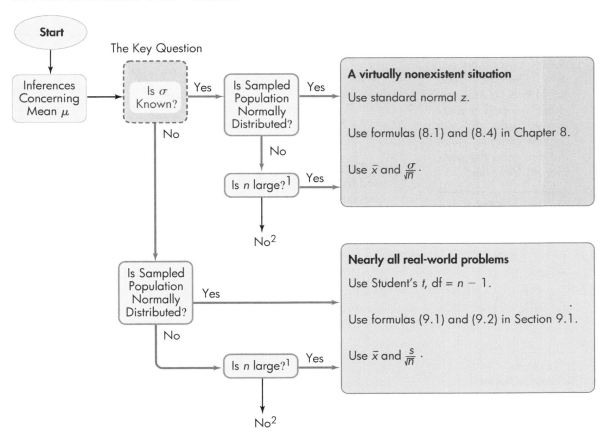

1. Is n large? Samples as small as $n = 15$ or 20 may be considered large enough for the central limit theorem to hold if the sample data are unimodal, nearly symmetrical, short-tailed, and without outliers. Samples that are not symmetrical require larger sample sizes, with 50 sufficing except for extremely skewed samples. See the discussion on page 347.
2. Requires the use of a nonparametric technique; see Chapter 14.

The *t*-distribution has the following properties (see also Figure 9.2):

Properties of the *t*-distribution (df > 2)*

1. *t* is distributed with a mean of zero.
2. *t* is distributed symmetrically about its mean.
3. *t* is distributed so as to form a family of distributions, a separate distribution for each different number of degrees of freedom (df $\geq$ 1).
4. The *t*-distribution approaches the **standard normal distribution** as the number of degrees of freedom increases.
5. *t* is distributed with a variance greater than 1, but as the degrees of freedom increase, the variance approaches 1.
6. *t* is distributed so as to be less peaked at the mean and thicker at the tails than is the normal distribution.

Degrees of freedom, df

A **parameter** that identifies each different distribution of Student's *t*-distribution. For the methods presented in this chapter, the value of df will be the sample size minus 1: df = $n - 1$.

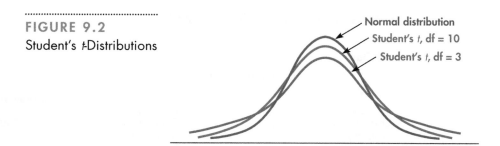

FIGURE 9.2
Student's *t*-Distributions

The number of degrees of freedom associated with s^2 is the divisor $(n - 1)$ used to calculate the sample variance s^2 [formula 2.5, p. 75]; that is, df = $n - 1$. The sample variance is the mean of the squared deviations. The number of degrees of freedom is the "number of unrelated deviations" available for use in estimating σ^2. Recall that the sum of the deviations, $\Sigma(x - \bar{x})$, must be zero. From a sample of size n, only the first $n - 1$ of these deviations has freedom of value. That is, the last, or nth, value of $(x - \bar{x})$ must make the sum of the n deviations total exactly zero. As a result, variance is said to average $n - 1$ unrelated squared deviation values, and this number, $n - 1$, was named "degrees of freedom."

Although there is a separate *t*-distribution for each degree of freedom, df = 1, df = 2, . . . , df = 20, . . . , df = 40, and so on, only certain key **critical values of *t*** will be necessary for our work. Consequently, the table for Student's *t*-distribution (Table 6 in Appendix B) is a table of critical values rather than a complete table, such as Table 3 is for the standard normal distribution for *z*. As you look at Table 6, you will note that the left side of the table is identified by "df," degrees of freedom. This left-hand column starts at 3 at the top and lists consecutive df values to 30, then jumps to 35, . . . , to "df > 100" at the bottom. As we stated, as the degrees of freedom increase, the *t*-distribution

---

*Not all of the properties hold for df = 1 and df = 2. Since we will not encounter situations where df = 1 or 2, these special cases are not discussed further.

approaches the characteristics of the standard normal z-distribution. Once df is "greater than 100," the critical values of the t-distribution are the same as the corresponding critical values of the standard normal distribution as given in Table 4A in Appendix B.

Using the *t*-Distribution Table (Table 6, Appendix B)

The critical values of Student's t-distribution that are to be used both for constructing a confidence interval and for hypothesis testing will be obtained from Table 6 in Appendix B. To find the value of t, you will need to know two identifying values: (1) df, the number of degrees of freedom (identifying the distribution of interest), and (2) α, the area under the curve to the right of the right-hand critical value. A notation much like that used with z will be used to identify a critical value. $t(\text{df}, \alpha)$, read as "t of df, α," is the symbol for the value of t with df degrees of freedom and an area of α in the right-hand tail, as shown in Figure 9.3.

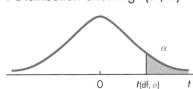

FIGURE 9.3

t-Distribution Showing $t(\text{df}, \alpha)$

E X A M P L E   9 . 1

t ON THE RIGHT SIDE OF THE MEAN

Find the value of $t(10, 0.05)$ (see the diagram).

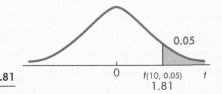

Solution

There are 10 degrees of freedom, and 0.05 is to be the area to the right of the critical value. In Table 6 of Appendix B, we look for the row df = 10 and the column marked "Area in One Tail," $\alpha = 0.05$. At their intersection, we see that $t(10, 0.05) = $ **1.81**.

Portion of Table 6

	Area in One Tail	
df	...	**0.05**	...
10		**1.81**	

$\longrightarrow$ $t(10, 0.05) = $ **1.81**

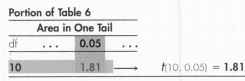

For the values of t on the left side of the mean, we can use one of two notations. The t-value shown in Figure 9.4 could be named $t(\text{df}, 0.95)$, because the area to the right of it is 0.95, or it could be identified by $-t(\text{df}, 0.05)$, because the t-distribution is symmetrical about its mean, zero.

FIGURE 9.4

t-Value on Left Side

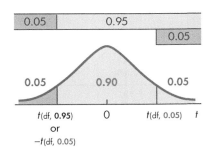

EXAMPLE 9.2

t ON THE LEFT SIDE OF THE MEAN

Find the value of $t(15, 0.95)$.

Solution

There are 15 degrees of freedom. In Table 6 we look for the column marked $\alpha = 0.05$ (one tail) and its intersection with the row df = 15. The table gives us $t(15, 0.95) = 1.75$; therefore, $t(15, 0.95) = -t(15, 0.05) = -1.75$. The value is negative because it is to the left of the mean, zero; see the figure.

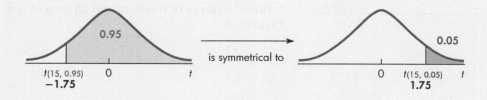

Let's look at another example that connects the t-distribution to percentiles.

EXAMPLE 9.3

t-VALUES THAT BOUND A MIDDLE PERCENTAGE

Find the values of the t-distribution that bound the middle 0.90 of the area under the curve for the distribution with df = 17.

Solution

The middle 0.90 leaves 0.05 for the area of each tail. The value of t that bounds the right-hand tail is $t(17, 0.05) = 1.74$, as found in Table 6. The value that bounds the left-hand tail is -1.74 because the t-distribution is symmetrical about its mean, zero.

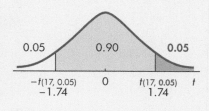

If the df needed is not listed in the left-hand column of Table 6, then use the next smaller value of df that is listed. For example, $t(72, 0.05)$ is estimated using $t(70, 0.05) = 1.67$.

Most computer software packages or statistical calculators will calculate the area related to a specified t-value. The accompanying figure shows the relationship between the cumulative probability and a specific t-value for a t-distribution with df degrees of freedom.

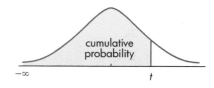

  Video tutorial available—logon and learn more at cengagebrain.com

TECHNOLOGY INSTRUCTIONS: PROBABILITY ASSOCIATED WITH A SPECIFIED VALUE OF t

MINITAB

Cumulative probability for a specified value of t:

Choose: **Calc > Probability Distribution > t**
Select: **Cumulative Probability**
            Noncentrality parameter: 0.0
Enter: **Degrees of freedom: df**
Select: **Input constant***
Enter: **t-value (ex. 1.74) > OK**

*Select Input column if several t-values are stored in C1. Use C2 for optional storage. If the area in the right tail is needed, subtract the calculated probability from 1.

Excel

Probability in one or two tails for a given t-value:
    If several t-values (nonnegative) are to be used, input the values into column A and activate B1; then continue with:

Choose: **Insert function f$_x$ > Statistical > TDIST > OK**
Enter: **X: individual t-value or (A1:A5 or select "t-value" cells)***
            Deg_freedom: df
            Tails: 1 or 2 (one- or two-tailed distributions) > OK
Drag*:   **Bottom right corner of the B1 cell down to give other probabilities**

To find the probability within the two tails or the cumulative probability for one tail, subtract the calculated probability from 1.

TI-83/84 Plus

Cumulative probability for a specified value of t:

Choose: **2nd > DISTR > 5:tcdf(** [†]
Enter: **−1EE99, t-value, df)**

[†] To find the probability between two t-values, enter the two values in place of −1EE99 and t-value.

If the area in the right tail is needed, subtract the calculated probability from 1.

Confidence Interval Procedure

We are now ready to make inferences about the population mean μ using the sample standard deviation. As we mentioned earlier, use of the t-distribution has a condition.

> **The assumption for inferences about the mean μ when σ is unknown** The sampled population is normally distributed.

The procedure to make confidence intervals using the sample standard deviation is very similar to that used when σ is known (see pp. 347–351). The difference is the use of Student's t in place of the standard normal z and the use of s, the sample standard deviation, as an estimate of σ. The central limit theorem (CLT) implies that this technique can also be applied to nonnormal populations when the sample size is sufficiently large.

> **Confidence Interval for Mean**
>
> $$\bar{x} - t(\text{df}, \alpha/2)\left(\frac{s}{\sqrt{n}}\right) \text{ to } \bar{x} + t(\text{df}, \alpha/2)\left(\frac{s}{\sqrt{n}}\right), \text{ with df} = n - 1 \quad (9.1)$$

Example 9.4 will illustrate the formation of a confidence interval utilizing the *t*-distribution.

EXAMPLE 9.4

CONFIDENCE INTERVAL FOR μ WITH σ UNKNOWN

A random sample of 20 weights is taken from babies born at Northside Hospital. A mean of 6.87 lb and a standard deviation of 1.76 lb were found for the sample. Estimate, with 95% confidence, the mean weight of all babies born in this hospital. Based on past information, it is assumed that weights of newborns are normally distributed.

Solution

Step 1 The Set-Up:

Describe the population parameter of interest.

μ, the mean weight of newborns at Northside Hospital

Step 2 The Confidence Interval Criteria:

a. Check the assumptions.
Past information indicates that the sampled population is normal.

b. Identify the probability distribution and the formula to be used.
The value of the population standard deviation, σ, is unknown. Student's *t*-distribution will be used with formula (9.1).

c. State the level of confidence: $1 - \alpha = 0.95$.

Step 3 The Sample Evidence:

Collect the sample information: $n = 20$, $\bar{x} = 6.87$, and $s = 1.76$.

Step 4 The Confidence Interval:

a. Determine the confidence coefficients.
Since $1 - \alpha = 0.95$, $\alpha = 0.05$, and therefore $\alpha/2 = 0.025$. Also, since $n = 20$, df $= 19$. At the intersection of row df $= 19$ and the one-tailed column $\alpha = 0.025$ in Table 6, we find $t(\text{df}, \alpha/2) = t(19, 0.025) = 2.09$. See the figure.
Information about the confidence coefficient and using Table 6 is on pages 415–416.

b. Find the maximum error of estimate.
$$E = t(\text{df}, \alpha/2)\left(\frac{s}{\sqrt{n}}\right): \quad E = t(19, 0.025)\left(\frac{s}{\sqrt{n}}\right)$$
$$= 2.09\left(\frac{1.76}{\sqrt{20}}\right) = (2.09)(0.394) = 0.82$$

c. Find the lower and upper confidence limits.

FYI The five-step confidence interval procedure is given on page 348.

FYI Recall that confidence intervals are two-tailed situations.

FYI df is used to find the confidence coefficient in Table 6; *n* is used in the formula.

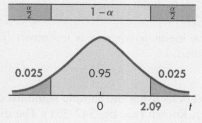

 Video tutorial available—logon and learn more at cengagebrain.com

$$\bar{x} - E \text{ to } \bar{x} + E$$
$$6.87 - 0.82 \text{ to } 6.87 + 0.82$$
$$6.05 \text{ to } 7.69$$

Step 5 The Results:

State the confidence interval.
6.05 to 7.69 is the 95% confidence interval for μ. That is, with 95% confidence we estimate the mean weight of babies born at Northside Hospital to be between 6.05 and 7.69 lb.

TECHNOLOGY INSTRUCTIONS: $1-\alpha$ CONFIDENCE INTERVAL FOR MEAN μ WITH σ UNKNOWN

MINITAB

Input the data into C1; then continue with:

Choose:	**Stat > Basic Statistics > 1-Sample t**
Enter:	Samples in columns: **C1**
Select:	**Options**
Enter:	Confidence level: **$1-\alpha$** (ex. 95.0)
Select:	Alternative: **not equal > OK > OK**

Excel

Input the data into column A; then continue with:

Choose:	**Add-Ins > Data Analysis Plus > t-Estimate: Mean > OK**
Enter:	Input Range: **(A1:A20 or select cells)**
Enter:	Alpha: α (ex. 0.05) **> OK**

TI-83/84 Plus

Input the data into L1; then continue with the following, entering the appropriate values and highlighting Calculate:

Choose: **STAT > TESTS > 8:TInterval**

```
TInterval
 Inpt:DATA Stats
 List:L1
 Freq:1
 C-Level:.95
 Calculate
```

The MINITAB solution to Example 9.4 looks like this:

One-Sample T: C1

Variable	N	Mean	StDev	SE Mean	95% CI
C1	20	6.870	1.760	0.394	(6.047, 7.693)

Hypothesis-Testing Procedure

The *t*-statistic is used to complete a hypothesis test about the population mean μ in much the same manner z was used in Chapter 8. In hypothesis-testing situations, we use formula (9.2) to calculate the value of the **test statistic $t\star$**:

Test Statistic for Mean

$$t\star = \frac{\bar{x} - \mu}{s/\sqrt{n}} \text{ with } df = n - 1 \tag{9.2}$$

The **calculated t** is the number of estimated standard errors that $\bar{x}$ is from the hypothesized mean μ. As with confidence intervals, the CLT indicates that the *t*-distribution can also be applied to nonnormal populations when the **sample size** is sufficiently large.

EXAMPLE 9.5

ONE-TAILED HYPOTHESIS TEST FOR μ WITH σ UNKNOWN

Let's return to the hypothesis of Example 8.13 (p. 372) where the Environmental Protection Agency (EPA) wants to show that the mean carbon monoxide level is higher than 4.9 parts per million. Does a random sample of 22 readings (sample results: $\bar{x} = 5.1$ and $s = 1.17$) present sufficient evidence to support the EPA's claim? Use $\alpha = 0.05$. Previous studies have indicated that such readings have an approximately normal distribution.

Solution

FYI The five-step *p*-value hypothesis test procedure is given on page 371.

FYI Procedures for writing H_o and H_a are discussed on pages 371–373.

Step 1 The Set-Up:
   a. **Describe the population parameter of interest.**
      μ, the mean carbon monoxide level of air in downtown Rochester
   b. **State the null hypothesis (H_o) and the alternative hypothesis (H_a).**

   H_o: $\mu = 4.9(\leq)$ (no higher than)
   H_a: $\mu > 4.9$ (higher than)

Step 2 The Hypothesis Test Criteria:
   a. **Check the assumptions.**
      The assumptions are satisfied because the sampled population is approximately normal and the sample size is large enough for the CLT to apply (see p. 413).
   b. **Identify the probability distribution and the test statistic to be used.**
      σ is unknown; therefore, the *t*-distribution with df $= n - 1 = 21$ will be used, and the test statistic is $t\star$, formula (9.2).
   c. **Determine the level of significance:** $\alpha = 0.05$.

Step 3 The Sample Evidence:
   a. **Collect the sample information:** $n = 22$, $\bar{x} = 5.1$, and $s = 1.17$.

b. Calculate the value of the test statistic.
Use formula (9.2):

$$t\star = \frac{\overline{x} - \mu}{s/\sqrt{n}} : \quad t\star = \frac{5.1 - 4.9}{1.17/\sqrt{22}} = \frac{0.20}{0.2494} = 0.8018 = \mathbf{0.80}$$

Step 4 The Probability Distribution:

| **Using the *p*-value procedure:** | OR | **Using the classical procedure:** |

a. Calculate the *p*-value for the test statistic.
Use the right-hand tail because H_a expresses concern for values related to "higher than."
$\mathbf{P} = P(t\star > 0.80,$ with df = 21) as shown on the figure.

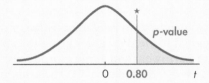

To find the *p*-value, use one of three methods:
1. Use Table 6 in Appendix B to place bounds on the *p*-value: $0.10 < \mathbf{P} < 0.25$.
2. Use Table 7 in Appendix B to read the value directly: $\mathbf{P} = 0.216$.
3. Use a computer or calculator to calculate the *p*-value: $\mathbf{P} = 0.2163$.
Specific details follow this example.

b. Determine whether or not the *p*-value is smaller than α.
The *p*-value is not smaller than α, the level of significance.

a. Determine the critical region and critical value(s).
The critical region is the right-hand tail because H_a expresses concern for values related to "higher than." The critical value is found at the intersection of the df = 21 row and the one-tailed 0.05 column of Table 6: $t(21, 0.05) = 1.72$.

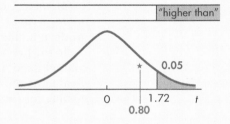

Specific instructions are given on pages 415–417.

b. Determine whether or not the calculated test statistic is in the critical region.
$t\star$ is not in the critical region, as shown in **red** in the figure above.

Step 5 The Results:

a. State the decision about H_o: Fail to reject H_o.

b. State the conclusion about H_a.
At the 0.05 level of significance, the EPA does not have sufficient evidence to show that the mean carbon monoxide level is higher than 4.9.

Calculating the *p*-value when using the *t*-distribution

Method 1: Use Table 6 in Appendix B to place bounds on the p-value. By inspecting the df = 21 row of Table 6, you can determine an interval within which the *p*-value lies. Locate $t\star$ along the row labeled df = 21. If $t\star$ is not listed, locate the two table values it falls between, and read the bounds for the *p*-value from the top of the table. In this case, $t\star = 0.80$ is between 0.686 and 1.32; therefore, $\mathbf{P}$ is between 0.10 and 0.25. Use the one-tailed heading, since H_a is one-tailed in this illustration. (Use the two-tailed heading when H_a is two-tailed.)

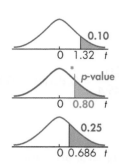

Finding P = $P(t\star > 0.80,$ with df =21)

Portion of Table 6

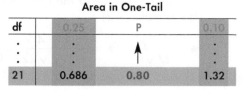

Area in One-Tail

df	0.25	P	0.10
⋮	⋮	⋮	⋮
21	0.686	0.80	1.32

0.10 < P < 0.25

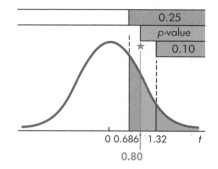

The 0.686 entry in the table tells us that $P(t > 0.686) = 0.25$, as shown on the figure in purple. The 1.32 entry in the table tells us that $P(t > 1.32) = 0.10$, as shown in green. You can see that the *p*-value **P** (shown in blue) is between 0.10 and 0.25. Therefore, $0.10 < \mathbf{P} < 0.25$, and we say that 0.10 and 0.25 are the "bounds" for the *p*-value.

Method 2: Use Table 7 in Appendix B to read the p-value or to "place bounds" on the p-value. Table 7 is designed to yield *p*-values given the $t\star$ and df values or produce bounds on **P** that are narrower than Table 6 produces.

In the preceding example, $t\star = 0.80$ and df = 21. These happen to be row and column headings, so the *p*-value can be read directly from the table. Locate the *p*-value at the intersection of the $t\star = 0.80$ row and the df = 21 column. The *p*-value for $t\star = 0.80$ with df = 21 is **0.216**.

Portion of Table 7

t★	df	. . .	21
⋮			
0.80			0.216

$\longrightarrow$ **P** = $P(t\star < 0.80,$ with df = 21) = **0.216**

To illustrate how to place bounds on the *p*-value when $t\star$ and df are not the heading values, let's consider the situation where $t\star = 2.43$ with df = 16. The $t\star = 2.43$ is between rows $t = 2.4$ and $t = 2.5$, while df = 16 is between columns df = 15 and df = 18. These two rows and two columns intersect a total of four times, namely, at 0.015 and 0.014 in the row $t\star = 2.4$ and at 0.012 and 0.011 in the row $t\star = 2.5$. The *p*-value we are looking for is bounded by the smallest and largest of these four values, namely, 0.011 (lower right) and 0.015 (upper left). Therefore, the bounds for the *p*-value are $0.011 < \mathbf{P} < 0.015$.

Portion of Table 7

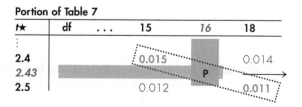

t★	df	. . .	15	16	18
⋮					
2.4			0.015		0.014
2.43				P	
2.5			0.012		0.011

P = $P(t\star > 2.43,$ with df = 16)
0.011 < P < 0.015

Method 3: If you are doing the hypothesis test with the aid of a computer or calculator, most likely it will calculate the *p*-value for you, or you may use the cumulative probability distribution commands described on page 417.

Let's look at a two-tailed hypothesis-testing situation.

EXAMPLE 9.6

TWO-TAILED HYPOTHESIS TEST FOR μ WITH σ UNKNOWN

On a popular self-image test that results in normally distributed scores, the mean score for public-assistance recipients is expected to be 65. A random sample of 28 public-assistance recipients in Emerson County is given the test. They achieve a mean score of 62.1, and their scores have a standard deviation of 5.83. Do the Emerson County public-assistance recipients test differently, on average, than what is expected at the 0.02 level of significance?

Solution

Step 1 The Set-Up:

    a. Describe the population parameter of interest.
      μ, the mean self-image test score for all Emerson County public-assistance recipients

    b. State the null hypothesis (H_o) and the alternative hypothesis (H_a).

$$H_o: \mu = 65 \text{ (mean is 65)}$$
$$H_a: \mu \neq 65 \text{ (mean is different from 65)}$$

Step 2 The Hypothesis Test Criteria:

    a. Check the assumptions.
      The test is expected to produce normally distributed scores; therefore, the assumption has been satisfied; σ is unknown.

    b. Identify the probability distribution and the test statistic to be used.
      The t-distribution with df $= n - 1 = 27$, and the test statistic is $t\star$, formula (9.2).

    c. Determine the level of significance: $\alpha = 0.02$ (given in statement of problem).

Step 3 The Sample Evidence:

    a. Collect the sample information: $n = 28$, $\bar{x} = 62.1$, and $s = 5.83$.

    b. Calculate the value of the test statistic.
      Use formula (9.2):

$$t\star = \frac{\bar{x} - \mu}{s/\sqrt{n}} : \quad t\star = \frac{62.1 - 65.0}{5.83/\sqrt{28}} = \frac{-2.9}{1.1018} = -2.632 = \mathbf{-2.63}$$

Step 4 The Probability Distribution:

Using the *p*-value procedure:	Using the classical procedure:
a. Calculate the *p*-value for the test statistic. Use both tails because H_a expresses concern for values related to "different from." $\mathbf{P} = P(t < -2.63) + P(t > 2.63) = 2 \cdot P(t > 2.63)$, with $df = 27$ as shown in the figure.	**a. Determine the critical region and critical value(s).** The critical region is both tails because H_a expresses concern for values related to "different from." The critical value is found at the

 OR

 Video tutorial available—logon and learn more at cengagebrain.com

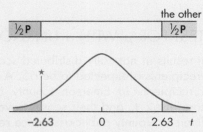

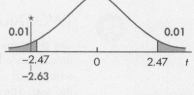

To find the *p*-value, use one of three methods:

1. Use Table 6 in Appendix B to place bounds on the *p*-value: $0.01 < \mathbf{P} < 0.02$.
2. Use Table 7 in Appendix B to place bounds on the *p*-value: $0.012 < \mathbf{P} < 0.016$.
3. Use a computer or calculator to calculate the *p*-value: $\mathbf{P} = 0.0140$.

Specific details follow this example:

b. **Determine whether or not the *p*-value is smaller than α.**

The *p*-value is smaller than the level of significance, α.

intersection of the row and the one-tailed 0.01 column of Table 6: $t(27, 0.01) = 2.47$.

b. **Determine whether or not the calculated test statistic is in the critical region.**

$t\bigstar$ is in the critical region, as shown in **red** in the preceding figure.

Step 5 The Results:

a. **State the decision about H_o:** Reject H_o.

b. **State the conclusion about H_a.**

At the 0.02 level of significance, we do have sufficient evidence to conclude that the Emerson County assistance recipients test significantly different, on average, from the expected 65.

Calculating the *p*-value when using the *t*-distribution

Method 1: Using Table 6, find 2.63 between two entries in the df = 27 row and read the bounds for **P** from the two-tailed heading at the top of the table:

$$0.01 < \mathbf{P} < 0.02$$

Method 2: Generally, bounds found using Table 7 will be narrower than bounds found using Table 6. The following table shows you how to read the bounds from Table 7; find $t\bigstar = 2.63$ between two rows and df = 27 between two columns, and locate the four intersections of these columns and rows. The value of $\frac{1}{2}\mathbf{P}$ is bounded by the upper left and the lower right of these table entries.

Portion of Table 7

$t\bigstar$	Degrees of Freedom		
	25	27	29
⋮	⋮		⋮
2.6	0.008		0.007
2.63		½ P	
2.7	0.006		0.006

$\mathbf{P} = 2P(t\bigstar > 2.63, \text{ with df} = 27)$

$0.006 < \frac{1}{2}\mathbf{P} < 0.008$

$0.012 < \mathbf{P} < 0.016$

Method 3: If you are doing the hypothesis test with the aid of a computer or calculator, most likely it will calculate the *p*-value for you (do not double it). Or you may use the cumulative probability distribution commands described on page 417.

TECHNOLOGY INSTRUCTIONS: HYPOTHESIS TEST FOR MEAN μ WHEN σ UNKNOWN

MINITAB

Input the data into C1; then continue with:

Choose: **Stat > Basic Statistics > 1-Sample t**
Enter: Samples in columns: **C1**
Select: **Perform hypothesis test**
Enter: Hypothesized mean: μ
Select: **Options**
Select: Alternative: **less than** or **not equal** or **greater than > OK > OK**

Excel

Input the data into column A; then continue with:

Choose: **Add-Ins > Data Analysis Plus > t-Test: Mean > OK**
Enter: Input Range: **(A1:A20 or select cells)**
Hypothesized Mean: μ
Alpha: α (ex. 0.05) **> OK**
Gives p-values and critical values for both one-tailed and two-tailed tests.

TI-83/84 Plus

Input the data into L1; then continue with the following, entering the appropriate values and highlighting Calculate:

Choose: **STAT > TESTS > 2:T-Test**

```
T-Test
 Inpt:Data Stats
 μ0:0
 List:L1
 Freq:1
 μ:≠μ0 <μ0 >μ0
 Calculate Draw
```

FYI Compare the MINITAB results to the solution found in Example 9.6.

Here is the MINITAB solution to Example 9.6:

One-Sample T: C1
Test of mu = 65 vs not = 65

Variable	N	Mean	StDev	SE Mean	T	P
C1	28	62.1	5.83	1.102	−2.63	0.0140

APPLIED EXAMPLE 9.7

MOTHERS' USE OF PERSONAL PRONOUNS WHEN TALKING WITH TODDLERS

The calculated t-value and the probability value for five different hypothesis tests are given in the article on the next page. The expression $t(44) = 1.92$ means $t\star = 1.92$ with df = 44 and is significant with p-value < 0.05. Can you verify the p-values? Explain.

ABSTRACT

The verbal interaction of 2-year-old children ($n = 46$; 16 girls, 30 boys) and their mothers was audiotaped, transcribed, and analyzed for the use of personal pronouns, the total number of utterances, the child's mean length of utterance, and the mother's responsiveness to her child's utterances. Mothers' use of the personal pronoun "we" was significantly related to their children's performance on the Stanford-Binet at age 5 and the Wechsler Intelligence Scale for Children at age 8. Mothers' use of "we" in social-vocal interchange, indicating a system for establishing a shared relationship with the child, was closely connected with their verbal responsiveness to their children. The total amount of maternal talking, the number of personal pronouns used by mothers, and their verbal responsiveness to their children were not related to mothers' social class or years of education.

Mothers tended to use more first person singular pronouns (I and me), $t(44) = 1.81$, $p < .10$, and used significantly more first person plural pronouns (we), $t(44) = 1.92$, $p < .05$, with female children than with male children. The mothers also were more verbally responsive to their female children, $t(44) = 2.0$, $p < .06$.

In general, mothers talked more to their first born children, $t(44) = 3.41$, $p < .001$, and were more responsive to their first born children, $t(44) = 3.71$, $p < .001$. Yet, the proportion of personal pronouns used when speaking to first born children was not different from that used when speaking to later born children.

Source: "Mother's Use of Personal Pronouns When Talking to Toddlers," by Dan R. Laks, Leila Beckwith, and Sarale E. Cohen, *The Journal of Genetic Psychology*, 151(1), 25–32, 1990. Reprinted with permission of the publisher (Taylor & Francis Ltd., http://www.informaworld.com).

SECTION 9.1 EXERCISES

9.1 [EX09-001] A random sample of 81 female American college students were each issued a stopwatch and asked to time themselves as they prepared to attend class on the following Thursday morning. The instructions were to start the watch as soon as their feet touched the floor as they got up and to turn it off as they passed through the door of their residence on the way to class.

x = "floor-to-door" time rounded to the nearest minute.

3	4	12	9	12	23	25	25	26	14	17	14	13	17	18
30	28	37	19	18	20	22	38	38	42	38	41	26	23	29
32	23	25	31	29	35	33	37	33	41	42	42	40	46	46
46	46	45	43	44	46	50	48	51	54	55	53	56	53	62
60	59	62	62	60	58	58	16	63	73	71	70	73	78	91
89	98	83	79	75	76									

a. What is the population of interest?

b. Draw a histogram of the "floor-to-door" variable using multiples of 10 for class midpoints. Describe the distribution. Does it appear to be approximately normal? Explain.

c. Redraw the histogram using multiples of 5 for class midpoints. Describe the visible patterns displayed by this second histogram that were not visible in the first one. Explain what causes this strange pattern.

d. Would you say the histogram suggests that the variable, amount of time, is approximately normally distributed? What evidence can you find to support your answer?

9.2 Consider the sample data in exercise 9.1.

a. Find the mean and standard deviation for the "floor-to-door" time.

b. How would you estimate the mean "floor-to-door" time for all female college students?

9.3 Make a list of four numbers that total "zero." How many numbers were you able to pick without restriction? Explain how this demonstrates degrees of freedom.

9.4 Explain the relationship between the critical values found in the bottom row of Table 6 and the critical values of z given in Table 4A.

9.5 Find:

a. $t(12, 0.01)$

b. $t(22, 0.025)$

c. $t(50, 0.10)$

d. $t(8, 0.005)$

9.6 Find these critical values using Table 6 in Appendix B:

a. $t(25, 0.05)$

b. $t(10, 0.10)$

c. $t(15, 0.01)$

d. $t(21, 0.025)$

9.7 Find:

a. $t(18, 0.90)$

b. $t(9, 0.99)$

c. $t(35, 0.975)$

d. $t(14, 0.98)$

9.8 Find these critical values using Table 6 in Appendix B:

a. $t(21, 0.95)$

b. $t(26, 0.975)$

c. $t(27, 0.99)$

d. $t(60, 0.025)$

9.9 Using the notation of Exercise 9.8, name and find the following critical values of t:

a.

b.
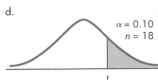

c.

d.
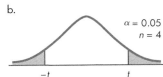

9.10 Using the notation of Exercise 9.8, name and find the following critical values of t:

a.
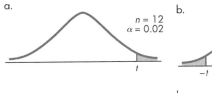

b.

c.

d.

9.11 Find the values of t that bound the middle 0.95 of the distribution for df = 12.

9.12 Find the values of t that bound the middle 0.80 of the distribution for df = 26.

9.13 a. Find the first percentile of Student's t-distribution with 24 degrees of freedom.

b. Find the 95th percentile of Student's t-distribution with 24 degrees of freedom.

c. Find the first quartile of Student's t-distribution with 24 degrees of freedom.

9.14 Find the percent of the Student's t-distribution that lies between the following values:

a. df = 12 and t ranges from -1.36 to 2.68

b. df = 15 and t ranges from -1.75 to 2.95

9.15 Ninety percent of Student's t-distribution lies between $t = -1.89$ and $t = 1.89$ for how many degrees of freedom?

9.16 Ninety percent of Student's t-distribution lies to the right of $t = -1.37$ for how many degrees of freedom?

9.17 Use a computer or calculator to find the area to the left of $t = -2.12$ with df = 18. Draw a sketch showing the question with the answer.

9.18 Use a computer or calculator to find the area to the right of $t = 1.12$ with df = 15. Draw a sketch showing the question with the answer.

9.19 a. State two ways in which the standard normal distribution and Student's t-distribution are alike.

b. State two ways in which they are different.

9.20 The variance for each of Student's t-distributions is equal to df/(df $-$ 2). Find the standard deviation for a Student's t-distribution with each of the following degrees of freedom:

a. 10

b. 20

c. 30

In summary:

d. Explain how this verifies Property 5 of the t-distributions listed on page 414.

9.21 Construct a 95% confidence interval estimate for the mean μ using the sample information $n = 24$, $\bar{x} = 16.7$, and $s = 2.6$.

9.22 Construct a 90% confidence interval estimate for the mean μ using the sample information $n = 53$, $\bar{x} = 87.2$, and $s = 11.9$.

9.23 The National Highway Traffic Safety Administration found the U.S. average EMS response time from EMS notification to arrival at the crash scene in urban areas to be 6.85 minutes. A random sample of 20 reported fatal crashes in South Dakota had a mean notification-to-arrival time of 5.25 minutes with a standard deviation of 2.78 minutes. Find the 98% confidence interval for the true mean notification-to-arrival time in South Dakota if response times are considered nearly symmetrical.

9.24 Based on a survey of 1000 adults by Greenfield Online and reported in a May 2009 *USA Today* Snapshot, adults 24 years of age and under spend a weekly average of $35 on fast food. If 200 of the 1000 adults surveyed who were in the 24 and under age category provided a standard deviation of $14.50, construct a 95% confidence interval for the weekly average expenditure on fast food for adults 24 years of age and under. Assume fast food weekly expenditures are normally distributed.

9.25 The Robertson square drive screw was invented in 1908, but it has gained in popularity with American woodworkers and home craftspeople only within the last 10 years. The advantages of square drives over conventional screws is indeed remarkable—most notably greater strength, increased holding power, and reduced driving resistance and "cam-out." Strength test results published in McFeely's 2005 catalog revealed that the no. 8 Robertson square drive flat head steel screws fail only after an average of 46 inch-pounds of torque is applied, a strength nearly 50% greater than that of the more common slotted- or Phillips-head wood screw.

Source: McFeely's Square Drive Screws, 2005

Suppose an independent testing laboratory randomly selects 22 square drive flat head steel screws from a box of 1000 screws and obtains a mean failure torque of 45.2 inch-pounds and a standard deviation of 5.1 inch-pounds. Estimate with 95% confidence the mean failure torque of the no. 8 wood screws based on the study by the independent laboratory. Specify the population parameter of interest, the criteria, the sample evidence, and the interval limits.

9.26 While writing an article on the high cost of college education, a reporter took a random sample of the cost of new textbooks for a semester. The random variable x is the cost of one book. Her sample data can be summarized by $n = 41$, $\Sigma x = 3582.17$, and $\Sigma(x - \bar{x})^2 = 9960.336$.

a. Find the sample mean, $\bar{x}$.

b. Find the sample standard deviation, s.

c. Find the 90% confidence interval to estimate the true mean textbook cost for the semester based on this sample.

9.27 [EX09-027] The pulse rates for 13 adult women were as follows:

| 83 | 58 | 70 | 56 | 76 | 64 | 80 | 76 | 70 | 97 | 68 | 78 | 108 |

Verify the results shown on the last line of the MINITAB output:

```
MTB > TINTERVAL 90 PERCENT CONFIDENCE
INTERVAL FOR DATA IN C1
      N    MEAN    STDEV   SE MEAN   90% CI
C1   13   75.69   14.54    4.03     (68.50, 82.88)
```

9.28 Using the computer output in Exercise 9.27, determine the value for each of the following:

a. Point estimate

b. Confidence coefficient

c. Standard error of the mean

d. Maximum error of estimate, E

e. Lower confidence limit

f. Upper confidence limit

9.29 [EX02-177] The addition of a new accelerator is claimed to decrease the drying time of latex paint by more than 4%. Several test samples were conducted with the following percentage decrease in drying time.

| 5.2 | 6.4 | 3.8 | 6.3 | 4.1 | 2.8 | 3.2 | 4.7 |

Assume that the percentage decrease in drying time is normally distributed.

a. Find the 95% confidence interval for the true mean decrease in the drying time based on this sample. (The sample mean and standard deviation were found in answering Exercise 2.177, p. 110.)

b. Did the interval estimate reached in part a result in the same conclusion you expressed in answering part c of Exercise 2.177 for these same data?

9.30 Use a computer or calculator to construct a 0.98 confidence interval using the sample data:

| 6 | 7 | 12 | 9 | 10 | 8 | 5 | 9 | 7 | 9 | 6 | 5 |

9.31 [EX09-031] Lunch breaks are often considered too short, and employees frequently develop a habit of "stretching" them. The manager at Giant Mart randomly identified 22 employees and observed the lengths of their lunch breaks (in minutes) for one randomly selected day during the week:

| 30 | 24 | 38 | 35 | 27 | 35 | 23 | 28 | 28 | 22 | 26 |
| 34 | 29 | 25 | 28 | 34 | 24 | 26 | 28 | 32 | 29 | 40 |

a. Show evidence that the normality assumptions are satisfied.

b. Find the 95% confidence interval for "mean length of lunch breaks" at Giant Mart.

9.32 [EX09-032] Many studies have been done telling us that we need to exercise to lower various health risks such as high blood pressure, heart disease, and high cholesterol. But knowing and doing are not the same things. People in the health professions should be even more aware of the need for exercise. The data below are from a study surveying cardiovascular technicians (individuals who perform various cardiovascular diagnostic procedures) as to their own physical exercise per week, measured in minutes.

60	40	50	30	60	50	90	30	60	60
60	80	90	90	60	30	20	120	60	50
20	60	30	120	50	30	90	20	30	40
50	40	30	40	20	30	60	50	60	80

a. Determine whether an assumption of normality is reasonable. Explain.

b. Estimate the mean amount of weekly exercise time for all cardiovascular technicians using a point estimate and a 95% confidence interval.

9.33 [EX09-033] The fuel economy information on a new SUV's window sticker indicates that its new owner can expect 16 mpg (miles per gallon) in city driving and 20 mpg for highway driving and 18 mpg overall. Accurate gasoline records for one such vehicle were kept, and a random sample of mileage per tank of gasoline was collected:

17.6	17.7	18.1	22.0	17.0	19.4	18.9	17.4	21.0	19.2
18.3	19.1	20.7	16.7	19.4	18.2	18.4	17.1	17.4	15.8
17.9	18.0	16.3	17.5	17.3	20.4	19.1	21.0	18.1	19.0
19.6	18.9	16.8	18.2	17.6	19.1	18.0	16.8	20.9	17.9
17.7	20.3	18.6	19.0	16.5	19.4	18.6	18.6	17.3	18.7

a. Determine whether an assumption of normality is reasonable. Explain.

b. Construct a 95% confidence interval for the estimate of the mean mileage per gallon.

c. What does the confidence interval suggest about SUV fuel economy expectations as expressed on the window sticker?

9.34 [EX09-034] James Short (1708–1768), a Scottish optician, constructed the highest-quality reflectors of his time. It was with these reflectors that Short obtained the following measurements of the parallax of the sun (in seconds of a degree), based on the 1761 transit of Venus. The parallax of the sun is the angle α subtended by the Earth, as seen from the surface of the sun. (See accompanying diagram.)

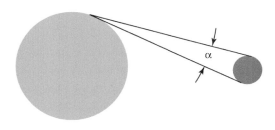

8.50	8.50	7.33	8.64	9.27	9.06	9.25	9.09	8.50
8.06	8.43	8.44	8.14	7.68	10.34	8.07	8.36	9.71
8.65	8.35	8.71	8.31	8.36	8.58	7.80	7.71	8.30
9.71	8.50	8.28	9.87	8.86	5.76	8.44	8.23	8.50
8.80	8.40	8.82	9.02	10.57	9.11	8.66	8.34	8.60
7.99	8.58	8.34	9.64	8.34	8.55	9.54	9.07	

Source: The data and descriptive information are based on material from Stephen M. Stigler. (1977). Do robust estimators work with real data? *Annals of Statistics, 5,* 1055–1098.

a. Determine whether an assumption of normality is reasonable. Explain.

b. Construct a 95% confidence interval for the estimate of the mean parallax of the sun.

c. If the true value is 8.798 seconds of a degree, what does the confidence interval suggest about Short's measurements?

9.35 State the null hypothesis, H_o, and the alternative hypothesis, H_a, that would be used to test each of the following claims:

a. The mean weight of honeybees is at least 11 grams.

b. The mean age of patients at Memorial Hospital is no more than 54 years.

c. The mean amount of salt in granola snack bars is different from 75 mg.

9.36 State the null hypothesis, H_o, and the alternative hypothesis, H_a, that would be used to test each of the following claims:

a. A chicken farmer at Best Broilers claims that his chickens have a mean weight of 56 oz.

b. The mean age of U.S. commercial jets is less than 18 years.

c. The mean monthly unpaid balance on credit card accounts is more than $400.

9.37 Calculate the value of $t\star$ for the hypothesis test: $H_o: \mu = 32, H_a: \mu > 32, n = 16, \bar{x} = 32.93, s = 3.1$.

9.38 Calculate the value of $t\star$ for the following hypothesis test: $H_o: \mu = 73$, $H_a: \mu \neq 73$, $n = 12$, $\bar{x} = 71.46$, $s = 4.1$.

9.39 Determine the p-value for the following hypothesis tests involving Student's t-distribution with 10 degrees of freedom.

a. $H_o: \mu = 15.5, H_a: \mu < 15.5, t\star = -2.01$

b. $H_o: \mu = 15.5, H_a: \mu > 15.5, t\star = 2.01$

c. $H_o: \mu = 15.5, H_a: \mu \neq 15.5, t\star = 2.01$

d. $H_o: \mu = 15.5, H_a: \mu \neq 15.5, t\star = -2.01$

9.40 Determine the critical region and critical value(s) that would be used in the classical approach to test the following null hypotheses:

a. $H_o: \mu = 10, H_a: \mu \neq 10 \ (\alpha = 0.05, n = 15)$

b. $H_o: \mu = 37.2, H_a: \mu > 37.2 \ (\alpha = 0.01, n = 25)$

c. $H_o: \mu = -20.5, H_a: \mu < -20.5 \ (\alpha = 0.05, n = 18)$

d. $H_o: \mu = 32.0, H_a: \mu > 32.0 \ (\alpha = 0.01, n = 42)$

9.41 a.   Find the value of **P** and state the decision for the hypothesis test in Exercise 9.37 using $\alpha = 0.05$.

   b.   Find the critical region and critical value and state the decision for the hypothesis test in Exercise 9.37 using $\alpha = 0.05$.

9.42 a.   Use Table 6 or Table 7 in Appendix B to find the value of **P** for the hypothesis test in Exercise 9.38; state the decision using $\alpha = 0.05$.

   b.   Find the critical region and critical values for the hypothesis test in Exercise 9.38; state the decision using $\alpha = 0.05$.

9.43 Use a computer or calculator to find the p-value for the following hypothesis test: $H_o: \mu = 32, \ H_a: \mu > 32, \ n = 16, \bar{x} = 32.93, s = 3.1$.

9.44 Use a computer or calculator to find the p-value for the following hypothesis test: $H_o: \mu = 73, \ H_a: \mu \neq 73, \ n = 12, \bar{x} = 71.46, s = 4.1$.

9.45 Use both the p-value and the classical approaches to hypothesis testing to reach a decision for each of the following situations. Use $\alpha = 0.05$.

a. $H_o: \mu = 128, H_a: \mu \neq 128, n = 15, t\star = 1.60$

b. $H_o: \mu = 18, H_a: \mu > 18, n = 25, t\star = 2.16$

c. $H_o: \mu = 38, H_a: \mu < 38, n = 45, t\star = -1.73$

d.   Compare the results of the two techniques for each case.

9.46 In reference to Applied Example 9.7 (p. 425):

a.   Verify that $t(44) = 1.92$ is significant at the 0.05 level.

b.   Verify that $t(44) = 3.41$ is significant at the 0.01 level.

c.   Explain why $t(44) = 1.81, p < 0.10$, makes sense only if the hypothesis test is two-tailed.

d.   If the test is one-tailed, what level would be reported?

9.47 A student group maintains that each day, the average student must travel at least 25 minutes one way to reach the college. The college admissions office obtained a random sample of 31 one-way travel times from students. The sample has a mean of 19.4 minutes and a standard deviation of 9.6 minutes. Does the admissions office have sufficient evidence to reject the students' claim? Use $\alpha = 0.01$.

a.   Solve using the p-value approach.

b.   Solve using the classical approach.

9.48 Homes in a nearby college town have a mean value of $88,950. It is assumed that homes in the vicinity of the college have a higher mean value. To test this theory, a random sample of 12 homes is chosen from the college area. Their mean valuation is $92,460, and the standard deviation is $5200. Complete a hypothesis test using $\alpha = 0.05$. Assume prices are normally distributed.

a.   Solve using the p-value approach.

b.   Solve using the classical approach.

9.49 According to the August 2009 *Reader's Digest* article "Where Our Garbage Goes," the average American tosses 4.6 pounds of garbage every day. A small town in Vermont initiated a Going Green campaign and asked residents to work on recycling more and reducing their generation of garbage each day. To estimate the average amount of trash discarded by people in their town, 18 households were randomly selected and all were asked to carefully weigh their trash on the same day. The average amount for the sample was 3.89 pounds, with a standard deviation of 1.322 pounds. Is there sufficient evidence that the Vermont town now has significantly lower average daily garbage amounts than the average American household? Use a 0.05 level of significance and assume weights are normally distributed.

9.50 Up all night? Caffeine cravings may cause long-term health problems. Homework, jobs, and studying all may be causes for teens to consume too much coffee in their everyday lives. Health officials warn that high caffeine intake is not good for anyone, but coffee drinking continues to become more and more popular. However, it is not the coffee that is of concern; it is the amount of caffeine. A moderate amount of caffeine is nothing to worry about, health experts say. There are no health risks drinking three 8-oz cups of regular coffee, which is about 250 mg of caffeine each day, according to the Henry Ford Health System.

Source: New Expressions, http://www.newexpression.org/

A nationwide random sample of college students revealed that 24 students consumed a total of 5428 mg of caffeine each day, with a standard deviation of 48 mg. Assuming that the amount of caffeine consumed per person daily is normally distributed, is there sufficient evidence to conclude that the mean amount of caffeine consumed daily by college students is less than 250 mg, using $\alpha = 0.05$?

a. Complete the test using the *p*-value approach. Include $t\star$, *p*-value, and your conclusion.

b. Complete the test using the classical approach. Include the critical values, $t\star$, and your conclusion.

9.51 [EX09-051] To test the null hypothesis "The mean weight for adult males equals 160 lb" against the alternative, "The mean weight for adult males exceeds 160 lb," the weights of 16 males were obtained:

| 173 | 178 | 145 | 146 | 157 | 175 | 173 | 137 |
| 152 | 171 | 163 | 170 | 135 | 159 | 199 | 131 |

Assume normality and verify the results shown on the following MINITAB analysis by calculating the values yourself.

TEST OF MU = 160.00 VS MU > 160.00

	N	MEAN	STDEV	SE MEAN	T	P
C1	16	160.25	18.49	4.62	0.05	0.48

9.52 Using the computer output in Exercise 9.51, determine the values of the following terms:

a. Hypothesized value of population mean

b. Sample mean

c. Population standard deviation

d. Sample standard deviation

e. Test statistic

9.53 [EX09-053] Use a computer or calculator to complete the hypothesis test H_o: $\mu = 52$, H_a: $\mu < 52$, $\alpha = 0.01$ using the data:

| 45 | 47 | 46 | 58 | 59 | 49 | 46 | 54 | 53 | 52 | 47 | 41 |

9.54 [EX09-054] The recommended number of hours of sleep per night is 8 hours, but everybody "knows" that the average college student sleeps less than 7 hours. The number of hours slept last night by 10 randomly selected college students is listed here:

| 5.2 | 6.8 | 6.2 | 5.5 | 7.8 | 5.8 | 7.1 | 8.1 | 6.9 | 5.6 |

Use a computer or calculator to complete the hypothesis test: H_o: $\mu = 7$, H_a: $\mu < 7$, $\alpha = 0.05$.

9.55 It is claimed that the students at a certain university will score an average of 35 on a given test. Is the claim reasonable if a random sample of test scores from this university yields 33, 42, 38, 37, 30, 42? Complete a hypothesis test using $\alpha = 0.05$. Assume test results are normally distributed.

a. Solve using the *p*-value approach.

b. Solve using the classical approach.

9.56 [EX02-178] Gasoline pumped from a supplier's pipeline is supposed to have an octane rating of 87.5. On 13 consecutive days, a sample was taken and analyzed, with the following results.

| 88.6 | 86.4 | 87.2 | 88.4 | 87.2 | 87.6 | 86.8 | 86.1 | 87.4 | 87.3 | 86.4 | 86.6 | 87.1 |

a. If the octane ratings have a normal distribution, is there sufficient evidence to show that these octane readings were taken from gasoline with a mean octane significantly less than 87.5 at the 0.05 level? (The sample mean and standard deviation were found in answering Exercise 2.178, p. 110.)

b. Did the statistical decision reached in part a result in the same conclusion you expressed in answering part c of Exercise 2.178 for these same data?

9.57 [EX09-032] According to statements from the National Women's Health Information Center and the Centers for Disease Control and Prevention, people should exercise at least 60 minutes per week to lower various health risks.

a. Based on the data from Exercise 9.32, determine if the technicians exercise at least 60 minutes a week. Use a 0.05 level of significance.

b. Did the statistical decision reached in part a result in the same conclusion you expressed in answering part b of Exercise 9.32 for this same data?

9.58 [EX09-001] Consider the "floor-to-door" scenario on page 412, where a random sample of 81 female American college students were each issued a stopwatch and asked to time themselves as they prepared to attend class on the following Thursday morning.

The instructions were to start the watch as soon as their feet touched the floor as they got up and to turn it off as they passed through the door of their residence on the way to class. Use the sample data listed and the results found in Exercises 9.1 and 9.2 (p. 426).

a. What evidence do you have that the assumption of normality is reasonable? Explain.

b. Estimate the mean "floor-to-door" time for all female American college students using a point estimate and a 95% confidence interval.

c. It is assumed that the estimated time of 51 minutes for a typical morning routine, as outlined in the "floor-to-door" scenario on page 412, is a reasonable mean for all American college students. Based on the data from this study, determine if female students are significantly different from the possibly typical student. Use a 0.05 level of significance.

d. Could the statistical decision reached in part c have resulted from your answer in part b? How?

e. Which sample statistic is having an unusually large effect on these results? Explain.

9.59 [EX09-059] The density of the earth relative to the density of water is known to be 5.517 g/cm^3. Henry Cavendish, an English chemist and physicist (1731–1810), was the first scientist to accurately measure the density of the earth. Following are 29 measurements taken by Cavendish in 1798 using a torsion balance.

5.50	5.61	4.88	5.07	5.26	5.55	5.36	5.29	5.58	5.65	5.57
5.53	5.62	5.29	5.44	5.34	5.79	5.10	5.27	5.39	5.42	5.47
5.63	5.34	5.46	5.30	5.75	5.68	5.85				

Source: The data and descriptive information are based on material from "Do robust estimators work with real data?" by Stephen M. Stigler, *Annals of Statistics*, 5 (1977), 1055–1098.

a. What evidence do you have that the assumption of normality is reasonable? Explain.

b. Is the mean of Cavendish's data significantly less than today's recognized standard? Use a 0.05 level of significance.

9.60 [EX09-060] Use a computer or calculator to complete the calculations and the hypothesis test for this exercise. Delco Products, a division of General Motors, produces commutators designed to be 18.810 mm in overall length. (A commutator is a device used in the electrical system of an automobile.) The following data are the lengths of a sample of 35 commutators taken while monitoring the manufacturing process:

18.802	18.810	18.780	18.757	18.824	18.827	18.825
18.809	18.794	18.787	18.844	18.824	18.829	18.817
18.785	18.747	18.802	18.826	18.810	18.802	18.780
18.830	18.874	18.836	18.758	18.813	18.844	18.861
18.824	18.835	18.794	18.853	18.823	18.863	18.808

Source: With permission of Delco Products Division, GMC

Is there sufficient evidence to reject the claim that these parts meet the design requirement "mean length is 18.810" at the $\alpha = 0.01$ level of significance?

9.61 Acetaminophen is an active ingredient found in more than 600 over-the-counter and prescription medicines, such as pain relievers, cough suppressants, and cold medications. It is safe and effective when used correctly, but taking too much can lead to liver damage.

Source: http://www.keepkidshealthy.com/

A researcher believes the mean amount of acetaminophen per tablet in a particular brand of cold tablets is different from the 600 mg claimed by the manufacturer. A random sample of 30 tablets had a mean acetaminophen content of 596.3 mg with a standard deviation of 4.7 mg.

a. Is the assumption of normality reasonable? Explain.

b. Construct a 99% confidence interval for the estimate of the mean acetaminophen content.

c. What does the confidence interval found in part b suggest about the mean acetaminophen content of one pill? Do you believe there is 600 mg per tablet? Explain.

9.62 [EX09-062] A winemaker has placed a large order for the no. 9 corks described in Applied Example 6.13 (p. 285) and is concerned about the number of corks that might have smaller diameters. During the corking process, the corks are squeezed down to 16 to 17 mm in diameter for insertion into bottles with an 18 mm opening. The cork then expands to make the seal. The winemaker wants the corks to be as tight as possible and is therefore concerned about any that might be undersized. The diameter of each cork is measured in several places, and an average diameter is reported for each cork. The cork manufacturer has assured the winemaker that each cork has an average diameter within the specs and that all average diameters have a normal distribution with a mean of 24.0 mm.

a. Why does it make sense for the diameter of the cork to be assigned the average of several different diameter measurements?

A random sample of 18 corks is taken from the batch to be shipped and the diameters (in millimeters) obtained:

23.93	23.91	23.82	24.02	23.93	24.17	23.93	23.84	24.13
24.01	23.83	23.74	23.73	24.10	23.86	23.90	24.32	23.83

b. The average diameter spec is "24 mm + 0.6 mm/ −0.4 mm." Does it appear this order meets the spec on an individual cork basis? Explain.

c. Does the sample in part a show sufficient reason to doubt the truthfulness of the claim, that the mean average diameter is 24.0 mm, at the 0.02 level of significance?

A different sample of 18 corks was randomly selected and the diameters (in millimeters) obtained:

23.90	23.98	24.28	24.22	24.07	23.87	24.05	24.06	23.82
24.03	23.87	24.08	23.98	24.21	24.08	24.06	23.87	23.95

d. Does the preceding sample show sufficient reason to doubt the truthfulness of the claim, that the mean average diameter is 24.0 mm, at the 0.02 level of significance?

e. What effect did the two different sample means have on the calculated test statistic in parts c and d? Explain.

f. What effect did the two different sample standard deviations have on the calculated test statistic in parts c and d? Explain.

9.63 [EX09-063] Length is not very important in evaluating the quality of corks because it has little to do with the effectiveness of a cork in preserving wine. Winemakers have several lengths to choose from and order the length of cork they prefer (long corks tend to make a louder pop when the bottle is uncorked). Length is monitored very closely, though, because it is a specified quality of the cork. The lengths of no. 9 natural corks (24 mm diameter by 45 mm length) have a normal distribution. Twelve randomly selected corks were measured to the nearest hundredth of a millimeter.

44.95	44.95	44.80	44.93	45.22	44.82
45.12	44.62	45.17	44.60	44.60	44.75

a. Does the preceding sample give sufficient reason to show that the mean length is different from 45.0 mm, at the 0.02 level of significance?

A different random sample of 18 corks is taken from the same batch.

45.17	45.02	45.30	45.14	45.35	45.50	45.26	44.88	44.71
44.07	45.10	45.01	44.83	45.13	44.69	44.89	45.15	45.13

b. Does the preceding sample give sufficient reason to show that the mean length is different from 45.0 mm, at the 0.02 level of significance?

c. What effect did the two different sample means have on the calculated test statistic in parts a and b? Explain.

d. What effect did the two different sample sizes have on the calculated test statistic in parts a and b? Explain.

e. What effect did the two different sample standard deviations have on the calculated test statistic in parts a and b? Explain.

9.64 How important is the assumption "The sampled population is normally distributed" to the use of Student's t-distribution? Using a computer, simulate drawing 100 samples of size 10 from each of three different types of population distributions, namely, a normal, a uniform, and an exponential. First generate 1000 data values from the population and construct a histogram to see what the population looks like. Then generate 100 samples of size 10 from the same population; each row represents a sample. Calculate the mean and standard deviation for each of the 100 samples. Calculate $t\star$ for each of the 100 samples. Construct histograms of the 100 sample means and the 100 $t\star$ values. (Additional details can be found in the *Student Solutions Manual*.)

For the samples from the normal population:

a. Does the $\bar{x}$ distribution appear to be normal? Find percentages for intervals and compare them with the normal distribution.

b. Does the distribution of $t\star$ appear to have a t-distribution with df = 9? Find percentages for intervals and compare them with the t-distribution.

For the samples from the rectangular or uniform population:

c. Does the $\bar{x}$ distribution appear to be normal? Find percentages for intervals and compare them with the normal distribution.

d. Does the distribution of $t\star$ appear to have a t-distribution with df = 9? Find percentages for intervals and compare them with the t-distribution.

For the samples from the skewed (exponential) population:

e. Does the $\bar{x}$ distribution appear to be normal? Find percentages for intervals and compare them with the normal distribution.

f. Does the distribution of $t\star$ appear to have a t-distribution with df = 9? Find percentages for intervals and compare them with the t-distribution.

In summary:

g. In each of the preceding three situations, the sampling distribution for $\bar{x}$ appears to be slightly different from the distribution of $t\star$. Explain why.

h. Does the normality condition appear to be necessary in order for the calculated test statistic $t\star$ to have a Student's t-distribution? Explain.

9.2 Inferences about the Binomial Probability of Success

Perhaps the most common inference involves the **binomial parameter** *p*, the "probability of success." Yes, every one of us uses this inference, even if only casually. In thousands of situations we are concerned about something either "happening" or "not happening." There are only two possible outcomes of concern, and that is the fundamental property of a **binomial experiment**. The other necessary ingredient is multiple independent trials. Asking five people whether they are "for" or "against" some issue can create five independent trials; if 200 people are asked the same question, 200 independent trials may be involved; if 30 items are inspected to see if each "exhibits a particular property" or "not," there will be 30 repeated trials; these are the makings of a binomial inference.

The binomial parameter *p* is defined to be the probability of success on a single trial in a binomial experiment.

> ### Sample Binomial Probability
>
> $$p' = \frac{x}{n} \qquad (9.3)$$
>
> where the **random variable** *x* represents the number of successes that occur in a sample consisting of *n* trials

FYI Complete details about binomial experimentation can be found on pages 246–249.

Recall that the mean and standard deviation of the binomial random variable *x* are found by using formula (5.7), $\mu = np$, and formula (5.8), $\sigma = \sqrt{npq}$, where $q = 1 - p$. The distribution of *x* is considered to be approximately normal if *n* is greater than 20 and if *np* and *nq* are both greater than 5. This commonly accepted **rule of thumb** allows us to use the **standard normal distribution** to estimate probabilities for the binomial random variable *x*, the number of successes in *n* trials, and to make inferences concerning the binomial parameter *p*, the probability of success on an individual trial.

Generally, it is easier and more meaningful to work with the distribution of p' (the observed probability of occurrence) than with *x* (the number of occurrences). Consequently, we will convert formulas (5.7) and (5.8) from units of *x* (integers) to units of **proportions** (percentages expressed as decimals) by dividing each formula by *n*, as shown in Table 9.1.

TABLE 9.1 Formulas (9.4) and (9.5)

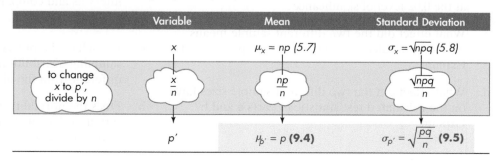

	Variable	Mean	Standard Deviation
	x	$\mu_x = np$ (5.7)	$\sigma_x = \sqrt{npq}$ (5.8)
to change *x* to *p'*, divide by *n*	$\dfrac{x}{n}$	$\dfrac{np}{n}$	$\dfrac{\sqrt{npq}}{n}$
	p'	$\mu_{p'} = p$ **(9.4)**	$\sigma_{p'} = \sqrt{\dfrac{pq}{n}}$ **(9.5)**

Recall that $\mu_{p'} = p$ and that the *sample statistic p'* is an **unbiased estimator for p.** Therefore, the information about the sampling distribution of p' is summarized as follows:

If a random sample of size n is selected from a large population with $p = P(\text{success})$, then the sampling distribution of p' has:

1. A mean $\mu_{p'}$ equal to p
2. A standard error $\sigma_{p'}$ equal to $\sqrt{\dfrac{pq}{n}}$
3. An approximately normal distribution if n is sufficiently large

In practice, using these guidelines will ensure normality:

1.   The sample size is greater than 20.
2.   The products np and nq are both greater than 5.
3.   The sample consists of less than 10% of the population.

We are now ready to make inferences about the population parameter p. Use of the z distribution involves an assumption.

The assumptions for inferences about the binomial parameter p The n random observations that form the sample are selected independently from a population that does not change during the sampling.

Confidence Interval Procedure

Inferences concerning the population binomial parameter p, $P(\text{success})$, are made using procedures that closely parallel the inference procedures used for the population mean μ. When we estimate the **population proportion p,** we will base our estimations on the **unbiased estimator p'.** The point estimate, the sample statistic p', becomes the center of the confidence interval, and the maximum error of estimate is a multiple of the **standard error.** The **level of confidence** determines the confidence coefficient, the number of multiples of the standard error.

Confidence Interval for a Proportion

$$p' - z(\alpha/2)\left(\sqrt{\frac{p'q'}{n}}\right) \quad \text{to} \quad p' + z(\alpha/2)\left(\sqrt{\frac{p'q'}{n}}\right) \qquad (9.6)$$

where $p' = \dfrac{x}{n}$ and $q' = 1 - p'$

Notice that the standard error, $\sqrt{\dfrac{pq}{n}}$, has been replaced by $\sqrt{\dfrac{p'q'}{n}}$. Since we are estimating p, we do not know its value and therefore we must use the best replacement available. That replacement is p', the observed value or the point estimate for p. This replacement will cause little change in the standard error or the width of our confidence interval provided n is sufficiently large.

Example 9.8 will illustrate the formation of a confidence interval for the binomial parameter, p.

EXAMPLE 9.8

CONFIDENCE INTERVAL FOR p

In a discussion about the cars that fellow students drive, several statements were made about types, ages, makes, colors, and so on. Dana decided he wanted to estimate the proportion of cars students drive that are convertibles, so he randomly identified 200 cars in the student parking lot and found 17 to be convertibles. Find the 90% confidence interval for the proportion of cars driven by students that are convertibles.

Solution

FYI The five-step confidence interval procedure is given on page 348.

Step 1 The Set-Up:
Describe the population parameter of interest.
p, the proportion (percentage) of students' cars that are convertibles

Step 2 The Confidence Interval Criteria:
   a. **Check the assumptions.**
      The sample was randomly selected, and each student's response is independent of those of the others surveyed.
   b. **Identify the probability distribution and the formula to be used.**
      The standard normal distribution will be used with formula (9.6) as the test statistic. p' is expected to be approximately normal because:
      (1) $n = 200$ is greater than 20, and
      (2) both np [approximated by $np' = 200(17/200) = 17$] and nq [approximated by $nq' = 200(183/200) = 183$] are greater than 5.
   c. **State the level of confidence:** $1 - \alpha = 0.90$.

Step 3 The Sample Evidence:
Collect the sample information.
$n = 200$ cars were identified, and $x = 17$ were convertibles:

$$p' = \frac{x}{n} = \frac{17}{200} = 0.085$$

Step 4 The Confidence Interval:

FYI Specific instructions are given on pages 348–350.

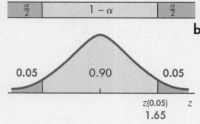

   a. **Determine the confidence coefficient.**
      This is the z-score [$z(\alpha/2)$, "z of one-half of alpha"] identifying the number of standard errors needed to attain the level of confidence and is found using Table 4 in Appendix B; $z(\alpha/2) = z(0.05) = 1.65$ (see the diagram).
   b. **Find the maximum error of estimate.**
      Use the maximum error part of formula (9.6):

$$E = z(\alpha/2)\left(\sqrt{\frac{p'q'}{n}}\right) = 1.65\left(\sqrt{\frac{(0.085)(0.915)}{200}}\right)$$
$$= (1.65)\sqrt{0.000389} = (1.65)(0.020) = \mathbf{0.033}$$

   c. **Find the lower and upper confidence limits.**

$$p' - E \text{ to } p' + E$$
$$0.085 - 0.033 \text{ to } 0.085 + 0.033$$
$$0.052 \text{ to } 0.118$$

Video tutorial available—logon and learn more at cengagebrain.com

Step 5 The Results:

State the confidence interval.

0.052 to 0.118 is the 90% confidence interval for $p = P$(drives convertible).

That is, the true proportion of students who drive convertibles is between 0.052 and 0.118, with 90% confidence.

TECHNOLOGY INSTRUCTIONS: $1 - \alpha$ CONFIDENCE INTERVAL FOR A PROPORTION p

MINITAB	
Choose:	**Stat > Basic Statistics > 1 Proportion**
Select:	**Summarized Data**
Enter:	**Number of events: x**
	Number of trials: n
Select:	**Options**
Enter:	**Confidence level: $1 - \alpha$** (ex. 95.0)
Select:	**Alternative: not equal**
	Use test and interval based on normal distribution. > OK > OK

Excel

Input the data into column A using 0s for failures (or no's) and 1s for successes (or yes's); then continue with:

Choose:	**Add-Ins > Data Analysis Plus > Z-Estimate: Proportion > OK**
Enter:	**Input Range: (A2:A20 or select cells)**
	Code for success: 1
	Alpha: α (ex. 0.05) **> OK**

TI-83/84 Plus

Choose:	**STAT > TESTS > A:1-PropZint**
	Enter the appropriate values and highlight Calculate.

```
1-PropZInt
x:0
n:0
C-Level:.95
Calculate
```

APPLIED EXAMPLE 9.9

MYTH AND REALITY IN REPORTING SAMPLING ERROR

On almost every occasion when we release a new survey, someone in the media will ask, "What is the margin of error for this survey?" When the media print sentences such as "the margin of error is plus or minus three percentage points," they strongly suggest that the results are accurate to within the percentage stated. They want to warn people about sampling error. But they might be better off assuming that all surveys, all opinion polls are estimates, which may be wrong.

In the real world, "random sampling error"—or the likelihood that a pure probability sample would produce replies within a certain band of percentages only because of the sample size—is one of the least of our measurement problems.

For this reason, we (Harris) include a strong warning in all of the surveys that we publish. Typically, it goes as follows: In theory, with a sample of this size, one can say with 95 percent certainty that the results have a statistical precision of plus or minus _ percentage points of what they would be if the entire adult population had been polled with complete accuracy. Unfortunately, there are several other possible sources of error in all polls or surveys that are probably more serious than theoretical calculations of sampling error. They include refusals to be interviewed (non-response), question wording and question order, interviewer bias, weighting by demographic control data, and screening. It is difficult or impossible to quantify the errors that may result from these factors.

If journalists are the least bit interested in all of this they may well ask, "If there are so many sources of error in surveys, why should we bother to read or report any poll results?" To which I normally give two replies:

1. Well-designed, well-conducted surveys work. Their record overall is pretty good. Most social, and marketing, researchers would be very happy with the average forecasting errors of the polls. However, there are enough disasters in the history of election predictions for readers to be cautious about interpreting the results.

2. (And this is more effective.) I re-word Winston Churchill's famous remarks about democracy and say, "Polls are the worst way of measuring public opinion and public behavior, or of predicting elections—except for all of the others."

Source: The Polling Report, May 4, 1998, by Humphrey Taylor, Chairman, Louis Harris & Assoc., Inc. http://www.pollingreport.com/sampling.htm.

Determining the Sample Size

By using the maximum error part of the confidence interval formula, it is possible to determine the **size of the sample** that must be taken in order to estimate p with a desired accuracy. Here is the formula for the **maximum error of estimate for a proportion:**

$$E = z_{(\alpha/2)}\left(\sqrt{\frac{pq}{n}}\right) \tag{9.7}$$

To determine the sample size from this formula, we must decide on the quality we want for our final confidence interval. This quality is measured in two ways: the level of confidence and the preciseness (narrowness) of the interval. The level of confidence we establish will in turn determine the confidence coefficient, $z_{(\alpha/2)}$. The desired preciseness will determine the maximum error of estimate, E. (Remember that we are estimating p, the binomial probability; therefore, E will typically be expressed in hundredths.)

For ease of use, we can solve formula (9.7) for n as follows:

Sample Size for $1 - \alpha$ Confidence Interval of p

$$n = \frac{[z(\alpha/2)]^2 \cdot p^* \cdot q^*}{E^2} \tag{9.8}$$

where p^* and q^* are provisional values of p and q used for planning

By inspecting formula (9.8), we can observe that three components determine the sample size:

1. The level of confidence [$1 - \alpha$, which determines the confidence coefficient, $z(\alpha/2)$]
2. The provisional value of p (p^* determines the value of q^*)
3. The maximum error, E

An increase or decrease in one of these three components affects the sample size. If the level of confidence is increased or decreased (while the other components are held constant), then the sample size will increase or decrease, respectively. If the product of p^* and q^* is increased or decreased (with other components held constant), then the sample size will increase or decrease, respectively. (The product $p^* \cdot q^*$ is largest when $p^* = 0.5$ and decreases as the value of p^* becomes further from 0.5.) An increase or decrease in the desired maximum error will have the opposite effect on the sample size, since E appears in the denominator of the formula. If no provisional values for p and q are available, then use $p^* = 0.5$ and $q^* = 0.5$. Using $p^* = 0.5$ is safe because it gives the largest sample size of any possible value of p. Using $p^* = 0.5$ works reasonably well when the true value is "near 0.5" (say, between 0.3 and 0.7); however, as p gets nearer to either 0 or 1, a sizable overestimate in sample size will occur.

E X A M P L E   9 . 1 0

SAMPLE SIZE FOR ESTIMATING p (NO PRIOR INFORMATION)

Determine the sample size that is required to estimate the true proportion of community college students who are blue-eyed if you want your estimate to be within 0.02 with 90% confidence.

Solution

Step 1 The level of confidence is $1 - \alpha = 0.90$; therefore, the confidence coefficient is $z(\alpha/2) = z(0.05) = 1.65$ from Table 4 in Appendix B; see the diagram.

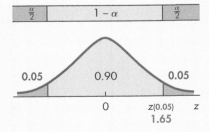

Step 2 The desired maximum error is $E = 0.02$.

Step 3 No estimate was given for p, so use $p^* = 0.5$ and $q^* = 1 - p^* = 0.5$.

Video tutorial available—logon and learn more at cengagebrain.com

FYI When finding the sample size n, always round up to the next larger integer, no matter how small the decimal.

Step 4   Use formula (9.8) to find n:

$$n = \frac{[z(\alpha/2)]^2 \cdot p^* \cdot q^*}{E^2} : n = \frac{(1.65)^2 \cdot 0.5 \cdot 0.5}{(0.02)^2} = \frac{0.680625}{0.0004} = 1701.56 = \mathbf{1702}$$

E X A M P L E   9 . 1 1

SAMPLE SIZE FOR ESTIMATING p (PRIOR INFORMATION)

An automobile manufacturer purchases bolts from a supplier who claims the bolts are approximately 5% defective. Determine the sample size that will be required to estimate the true proportion of defective bolts if you want your estimate to be within ±0.02 with 90% confidence.

Solution

Step 1   The level of confidence is $1 - \alpha = 0.90$; the confidence coefficient is $z(\alpha/2) = z(0.05) = 1.65$.

Step 2   The desired maximum error is $E = 0.02$.

Step 3   There is an estimate for p (supplier's claim is "5% defective"), so use $p^* = 0.05$ and $q^* = 1 - p^* = 0.95$.

FYI Yes, sample size calculations are always rounded up to the next larger integer!

Step 4   Use formula (9.8) to find n:

$$n = \frac{[z(\alpha/2)]^2 \cdot p^* \cdot q^*}{E^2} : n = \frac{(1.65)^2 \cdot 0.05 \cdot 0.95}{(0.02)^2} = \frac{0.12931875}{0.0004} = 323.3 = \mathbf{324}$$

Notice the difference in the sample sizes required in Examples 9.10 and 9.11. The only mathematical difference between the problems is the value used for p^*. In Example 9.10 you used $p^* = 0.5$, and in Example 9.11 you used $p^* = 0.05$. Recall that the use of the provisional value $p^* = 0.5$ gives the maximum sample size. As you can see, it will be an advantage to have some indication of the value expected for p, especially as p becomes increasingly further from 0.5.

Hypothesis-Testing Procedure

When the binomial parameter p is to be tested using a hypothesis-testing procedure, we will use a test statistic that represents the difference between the observed proportion and the hypothesized proportion, divided by the standard error. This test statistic is assumed to be normally distributed when the null hypothesis is true, when the assumptions for the test have been satisfied, and when n is sufficiently large ($n > 20$, $np > 5$, and $nq > 5$).

FYI p' is from the sample, p is from H_o, and $q = 1 - p$.

Test Statistic for a Proportion

$$z\bigstar = \frac{p' - p}{\sqrt{\dfrac{pq}{n}}} \text{ with } p' = \frac{x}{n} \qquad (9.9)$$

E X A M P L E   9 . 1 2

ONE-TAILED HYPOTHESIS TEST FOR PROPORTION p

Many people sleep late on the weekends to make up for "short nights" during the workweek. The Better Sleep Council reports that 61% of us get more than 7 hours of sleep per night on the weekend. A random sample of 350 adults found that 235 had more than 7 hours of sleep each night last weekend. At the 0.05 level of significance, does this evidence show that more than 61% sleep 7 hours or more per night on the weekend?

Solution

Step 1 The Set-Up:

    a. Describe the population parameter of interest.

        p, the proportion of adults who get more than 7 hours of sleep per night on weekends

    b. State the null hypothesis (H_o) and the alternative hypothesis (H_a).

        H_o: $p = P(7 + \text{hours of sleep}) = 0.61 (\leq)$ (no more than 61%)

        H_a: $p > 0.61$ (more than 61%)

Step 2 The Hypothesis Test Criteria:

    a. Check the assumptions.

        The random sample of 350 adults was independently surveyed.

    b. Identify the probability distribution and the test statistic to be used.

        The standard normal z will be used with formula (9.9). Since $n = 350$ is greater than 20 and both $np = (350)(0.61) = 213.5$ and $nq = (350)(0.39) = 136.5$ are greater than 5, p' is expected to be approximately normally distributed.

    c. Determine the level of significance: $\alpha = 0.05$.

Step 3 The Sample Evidence:

    a. Collect the sample information: $n = 350$ and $x = 235$:

$$p' = \frac{x}{n} = \frac{235}{350} = 0.671$$

    b. Calculate the value of the test statistic.

        Use formula (9.9):

$$z\bigstar = \frac{p' - p}{\sqrt{\dfrac{pq}{n}}} : z\bigstar = \frac{0.671 - 0.61}{\sqrt{\dfrac{(0.61)(0.39)}{350}}} = \frac{0.061}{\sqrt{0.0006797}} = \frac{0.061}{0.0261} = \mathbf{2.34}$$

Video tutorial available—logon and learn more at cengagebrain.com

Step 4 The Probability Distribution:

Using the *p*-value procedure:	OR	Using the classical procedure:

a.   Calculate the *p*-value for the test statistic.
Use the right-hand tail because H_a expresses concern for values related to "more than." $\mathbf{P} = p\text{-value} = P(z > 2.34)$ as shown in the figure.

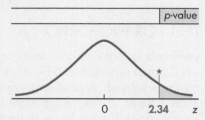

To find the *p*-value, use one of three methods:

1.   Use Table 3 in Appendix B to calculate the *p*-value: $\mathbf{P} = 1.0000 - 0.9904 = \mathbf{0.0096}$.
2.   Use Table 5 in Appendix B to place bounds on the *p*-value: $0.0094 < \mathbf{P} < 0.0107$.
3.   Use a computer or calculator to calculate the *p*-value: $\mathbf{P} = 0.0096$.

For specific instructions, see Method 3 below.

b.   Determine whether or not the *p*-value is smaller than α.

The *p*-value is smaller than α.

a.   Determine the critical region and critical value(s).
The critical region is the right-hand tail because H_a expresses concern for values related to "more than." The critical value is obtained from Table 4A: $z(0.05) = \mathbf{1.65}$.

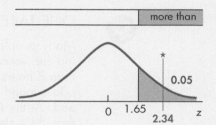

Specific instructions for finding critical values are given on pages 392–394.

b.   Determine whether or not the calculated test statistic is in the critical region.

$z\star$ is in the critical region, as shown in **red** in the accompanying figure.

Step 5 The Results:

a.   State the decision about H_o: Reject H_o.

b.   State the conclusion about H_a.
There is sufficient reason to conclude that the proportion of adults in the sampled population who are getting more than 7 hours of sleep nightly on weekends is significantly higher than 61% at the 0.05 level of significance.

Method 3: If you are doing the hypothesis test with the aid of a computer or calculator, most likely it will calculate the *p*-value for you, or you may use the cumulative probability distribution commands described on page 285.

E X A M P L E   9 . 1 3

TWO-TAILED HYPOTHESIS TEST FOR PROPORTION *p*

While talking about the cars that fellow students drive (see Example 9.8, p. 436), Tom claimed that 15% of the students drive convertibles. Jody finds this hard to believe, and she wants to check the validity of Tom's claim using Dana's random sample. At a level of significance of 0.10, is there sufficient evidence to reject Tom's claim if there are 17 convertibles in his sample of 200 cars?

Solution

Step 1 The Set-Up:

a. **Describe the population parameter of interest.**

$p = P$(student drives convertible)

b. **State the null hypothesis (H_o) and the alternative hypothesis (H_a).**

H_o: $p = 0.15$ (15% do drive convertibles)
H_a: $p \neq 0.15$ (the percentage is different from 15%)

Step 2 The Hypothesis Test Criteria:

a. **Check the assumptions.**
The sample was randomly selected, and each subject's response was independent of other responses.

b. **Identify the probability distribution and the test statistic to be used.**
The standard normal z and formula (9.9) will be used. Since $n = 200$ is greater than 20 and both np and nq are greater than 5, p' is expected to be approximately normally distributed.

c. **Determine the level of significance:** $\alpha = 0.10$.

Step 3 The Sample Evidence:

a. **Collect the sample information:** $n = 200$ and $x = 17$:

$$p' = \frac{x}{n} = \frac{17}{200} = 0.085$$

b. **Calculate the value of the test statistic.**
Use formula (9.9):

$$z\star = \frac{p' - p}{\sqrt{\dfrac{pq}{n}}} : z\star = \frac{0.085 - 0.150}{\sqrt{\dfrac{(0.15)(0.85)}{200}}}$$

$$= \frac{-0.065}{\sqrt{0.00064}} = \frac{-0.065}{0.02525} = -2.57$$

Step 4 The Probability Distribution:

| **Using the *p*-value procedure:** | **OR** | **Using the classical procedure:** |

Using the *p*-value procedure:

a. **Calculate the *p*-value for the test statistic.**
Use both tails because H_a expresses concern for values related to "different from."
$$\mathbf{P} = p\text{-value} = P(z < -2.57) + P(z > 2.57)$$
$$= 2 \times P(z < -2.57) \text{ as shown in the figure}$$

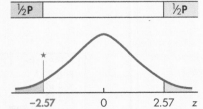

To find the *p*-value, use one of three methods:
1. Use Table 3 in Appendix B to calculate the *p*-value: $\mathbf{P} = 2 \times 0.0051 = 0.0102$.

Using the classical procedure:

a. **Determine the critical region and critical value(s).**
The critical region is two-tailed because H_a expresses concern for values related to "different from." The critical value is obtained from Table 4B: $z(0.05) = 1.65$.

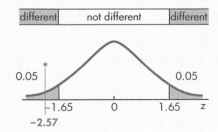

For specific instructions, see pages 395–396.

2. Use Table 5 in Appendix B to place bounds on the p-value: $0.0094 < \mathbf{P} < 0.0108$.

3. Use a computer or calculator to calculate the p-value: $\mathbf{P} = 0.0102$.

For specific instructions, see pages 376–377.

b. Determine whether or not the p-value is smaller than α.

The p-value is smaller than α.

b. Determine whether or not the calculated test statistic is in the critical region.

$z\star$ is in the critical region, as shown in **red** in the accompanying figure.

Step 5 The Results:

a. State the decision about H_o: Reject H_o.

b. State the conclusion about H_a.

There is sufficient evidence to reject Tom's claim and conclude that the percentage of students who drive convertibles is different from 15% at the 0.10 level of significance.

TECHNOLOGY INSTRUCTIONS: HYPOTHESIS TEST FOR A PROPORTION p

MINITAB		
Choose:	Stat > Basic Statistics > 1 Proportion	
Select:	Summarized Data	
Enter:	Number of events: **x**	
	Number of trials: **n**	
Select:	Perform hypothesis test	
Enter:	Hypothesized proportion: **p**	
Select:	Options	
Select:	Alternative: **less than** or **not equal** or **greater than**	
	Use test and interval based on normal distribution > OK > OK	

Excel

Input the data into column A using 0s for failures (or no's) and 1s for successes (or yes's); then continue with:

Choose:	Add-Ins > Data Analysis Plus > Z-Test: Proportion > OK	
Enter:	Input Range: (A2:A20 or select cells)	
	Code for success: 1	
	Hypothesized Proportion: p	
	Alpha: α (ex. 0.05) > OK	
	Gives p-values and critical values for both one-tailed and two-tailed tests.	

TI-83/84 Plus

Choose:	STAT > TESTS > 5:1-PropZTest	
	Enter the appropriate values and highlight Calculate.	

```
1-PropZTest
p₀:0
x:0
n:0
Prop≠p₀ <p₀ >p₀
Calculate Draw
```

Relationship between Confidence Intervals and Hypothesis Tests

There is a relationship between confidence intervals and two-tailed hypothesis tests when the level of confidence and the level of significance add up to 1. The confidence coefficients and the critical values are the same, which means the width of the confidence interval and the width of the noncritical region are the same. The point estimate is the center of the confidence interval, and the hypothesized mean is the center of the noncritical region. Therefore, if the hypothesized value of p is contained in the confidence interval, then the test statistic will be in the noncritical region (see Figure 9.5).

FIGURE 9.5
Confidence Interval Contains p

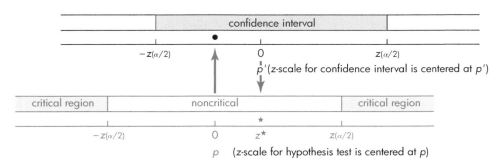

Furthermore, if the hypothesized probability p does not fall within the confidence interval, then the test statistic will be in the critical region (see Figure 9.6).

FIGURE 9.6
Confidence Interval Does Not Contain p

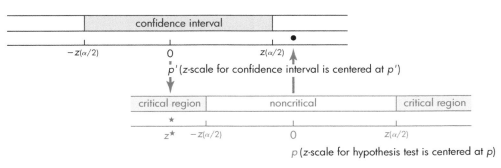

FYI Explore Skillbuilder Applet "$z\star$ & Confidence Level"—available through cengage-brain.com

This comparison should be used only when the hypothesis test is two-tailed and when the same value of α is used in both procedures.

APPLIED EXAMPLE 9.14

Courtesy of Chard the Jeweller

HEADS OR TAILS?

Euro coin accused of unfair flipping

The introduction of the Euro, the largest currency switch in history, has proceeded with few problems - until now. Polish statisticians say the one Euro coin, at least in Belgium, does not have an equal chance of landing "heads" or "tails". . . . The observation is not to be taken lightly on a sports-mad continent where important decisions can turn on the flip of a coin. . . .

Tomasz Gliszczynski and Waclaw Zawadowski, statistics teachers at the Akademia Podlaska in Siedlce, received Belgian Euro coins from Poles returning from jobs in Belgium and immediately set their students spinning them. . . . they have so far managed only 250 spins. Of these, 140, or 56.0 percent, came up heads. Glyszczynski attributes such asymmetry to a heavier embossed image on one side of the coin. All Euros have a national image on the "heads" side and a common design on the "tails". Belgium portrays its portly King Albert, on the heads side.

Not significant But Howard Grubb, an applied statistician at the University of Reading, notes that, "with a sample of only 250, anything between 43.8 percent and 56.2 percent on one side or the other cannot be said to be biased". This is because random variation can produce such scatter even if the coin is truly unbiased. . . . The range of 6.2 percent on either side of 50 percent is expected to cover the results, even with a fair coin, in 95 of every 100 experiments. Nonetheless, Grubb cautions, the Polish result is at the outside of this range, and would be expected in only about 7 of every 100 experiments with a fair coin, leaving a glimmer of hope for their hypothesis. Clearly, more research is needed.

New Scientist carried out its own experiments with the Belgian Euro in its Brussels office. Heads came up five percent less often than tails. This looks like the opposite of the Polish result but in fact - in terms of statistical significance - it is the same one.

Source: Charlotte Denny and Sarah Dennis, *The Guardian*, Friday, January 4, 2002, http://www.guardian.co.uk/euro/story/0,11306,627496,00.html. Copyright © Guardian News & Media Ltd, 2002. Reprinted with permission.

S E C T I O N   9 . 2   E X E R C I S E S

9.65 Of the 150 elements in a random sample, 45 are classified as "success."

a. Explain why x and n are assigned the values 45 and 150, respectively.

b. Determine the value of p'. Explain how p' is found and the meaning of p'.

For each of the following situations, find p'.

c. $x = 24$ and $n = 250$

d. $x = 640$ and $n = 2050$

e. 892 of 1280 responded "Yes"

9.66 a. What is the relationship between $p = P(\text{success})$ and $q = P(\text{failure})$? Explain.

b. Explain why the relationship between p and q can be expressed by the formula $q = 1 - p$.

c. If $p = 0.6$, what is the value of q?

d. If the value of $q' = 0.273$, what is the value of p'?

9.67 a. Does it seem reasonable that the mean of the sampling distribution of observed values of p' should be p, the true proportion? Explain.

b. Explain why p' is an unbiased estimator for the population p.

9.68 Show that $\frac{\sqrt{npq}}{n}$ simplifies to $\sqrt{\frac{pq}{n}}$.

9.69 Find α, the area of one tail, and the confidence coefficients of z that are used with each of the following levels of confidence.

a. $1 - \alpha = 0.90$ b. $1 - \alpha = 0.95$ c. $1 - \alpha = 0.99$

9.70 Find α, the area of one tail, and the confidence coefficients of z that are used with each of the following levels of confidence.

a. $1 - \alpha = 0.80$ b. $1 - \alpha = 0.98$ c. $1 - \alpha = 0.75$

9.71 Refer back to Example 9.8, page 436. Another sample is taken to estimate the proportion of convertibles. Results are $n = 400$ and $x = 92$. Find:

a. the estimate for the standard error.

b. the 95% confidence interval.

9.72 "You say tomato, burger lovers say ketchup!" According to a recent T.G.I. Friday's restaurants' random survey of 1027 Americans, approximately half (47%) said that ketchup is their preferred burger condiment. The survey quoted a margin of error of plus or minus 3.1%.

Source: Harris Interactive/Yankelovich Partners for T.G.I. Friday's restaurants, http://www.knoxville3.com/

a. Describe how this survey of 1027 Americans fits the properties of a binomial experiment. Specifically identify n, a trial, success, p, and x.

b. What is the point estimate for the proportion of all Americans who prefer ketchup on their burger? Is it a parameter or a statistic?

c. Calculate the 95% confidence maximum error of estimate for a binomial experiment of 1027 trials that results in an observed proportion of 0.47.

d. How is the maximum error, found in part c, related to the 3.1% margin of error quoted in the survey report?

e. Find the 95% confidence interval for the true proportion p based on a binomial experiment of 1027 trials that results in an observed proportion of 0.47.

9.73 Although most people are aware of minor dehydration symptoms such as dry skin and headaches, many are less knowledgeable about the causes of dehydration. According to a poll done for the Nutrition Information Center, the results of a random sample of 3003 American adults showed that 20% did not know that caffeine dehydrates. The survey listed a margin of error of plus or minus 1.8%.

Source: Yankelovich Partners for the Nutrition Information Center of the New York Hospital–Cornell Medical Center and the International Bottled Water Association

a. Describe how this survey of 3003 American adults fits the properties of a binomial experiment. Specifically identify n, a trial, success, p, and x.

b. What is the point estimate for the proportion of all Americans who did not know that caffeine dehydrates? Is it a parameter or a statistic?

c. Calculate the 95% confidence maximum error of estimate for a binomial experiment of 3003 trials that result in an observed proportion of 0.20.

d. How is the maximum error, found in part c, related to the 1.8% margin of error quoted in the survey report?

e. Find the 95% confidence interval for the true proportion p based on a binomial experiment of 3003 trials that results in an observed proportion of 0.20.

9.74 A bank randomly selected 250 checking account customers and found that 110 of them also had savings accounts at the same bank. Construct a 95% confidence interval for the true proportion of checking account customers who also have savings accounts.

9.75 In a sample of 60 randomly selected students, only 22 favored the amount budgeted for next year's intramural

and interscholastic sports. Construct the 99% confidence interval for the proportion of all students who support the proposed budget amount.

9.76 The National Highway Traffic Safety Administration found that, among the crashes with recorded times, EMS notification times exceeded 10 minutes in 19.4% of rural fatal crashes. A random sample of 500 reported fatal crashes in Kentucky showed 21.8% of the notification times exceeded 10 minutes. Construct the 95% confidence interval for the true proportion of fatal crashes in Kentucky whose elapsed notification time exceeded 10 minutes.

9.77 In a poll conducted by Harris Interactive of 1179 video-gaming U.S. youngsters, 8.5% displayed behavioral signs that may indicate addiction. Using a 99% confidence interval for the true binomial proportion based on this random sample of 1179 binomial trials and an observed proportion of 0.085, estimate the proportion of video-gaming youngsters that may go on to have an addiction.

Source: USA Today, April 21, 2009, "Kids show addiction symptoms"

9.78 Just one serving a month of kale or collard greens or more than two servings of carrots a week can reduce the risk of glaucoma by more than 60%, according to a UCLA study of 1000 women. Using a 90% confidence interval for the true binomial proportion based on this random sample of 1000 binomial trials and an observed proportion of 0.60, estimate the proportion of glaucoma risk reduction in women who eat the recommended servings of kale, collard greens, or carrots.

Source: Reader's Digest, February 2009, "Tasty Sight Savers"

9.79 Adverse drug reactions to legally prescribed medicines are among the leading causes of drug-related death in the United States. Suppose you investigate drug-related deaths in your city and find that 223 out of 250 incidences were caused by legally prescribed drugs and the rest were the result of illicit drug use. MINITAB was then used to form the 98% confidence interval for the proportion of drug-related deaths that were caused by legally prescribed drugs. Verify the MINITAB results that follow.

CI for One Proportion				
Sample	X	N	Sample p	98% CI
1	223	250	0.892000	(0.846333, 0.937667)

9.80 Using the MINITAB output and information in Exercise 9.79, determine the values of the following terms:

a. Point estimate

b. Confidence coefficient

(continue on page 448)

c. Standard error of the mean

d. Maximum error of estimate, E

e. Lower confidence limit

f. Upper confidence limit

9.81 A Cambridge Consumer Credit Index nationwide telephone survey of 1000 people found that most Americans are not easily swayed by the lure of reward points or rebates when deciding to use a credit card or pay by cash or check. The survey found that 2 out of 3 consumers do not even have credit cards offering reward points or rebates. Explain why you would be reluctant to use this information to construct a confidence interval estimating the true proportion of consumers who do not have credit cards offering reward points or rebates.

9.82 Construct 90% confidence intervals for the binomial parameter p for each of the following pairs of values. Write your answers on the chart.

Observed Proportion $p' = x/n$	Sample Size	Lower Limit	Upper Limit
a. $p' = 0.3$	$n = 30$		
b. $p' = 0.7$	$n = 30$		
c. $p' = 0.5$	$n = 10$		
d. $p' = 0.5$	$n = 100$		
e. $p' = 0.5$	$n = 1000$		

f. Explain the relationship between the answers to parts a and b.

g. Explain the relationship among the answers to parts c–e.

9.83 Three nationwide poll results are described below.

> *USA Today* Snapshot/Rent.com, August 18, 2009; $N = 1000$ adults 18 and over; MoE ± 3. (MoE is margin of error.)
> "What renters look for the most when seeking an apartment:" Washer/dryer—39%, Air Conditioning—30%, Fitness Center—10%, Pool—10%
>
> *USA Today*/Harris Interactive Poll, February 10–15, 2009; $N = 1010$ adults; MoE ± 3.
> "Americans who say people on Wall Street are 'as honest and moral as other people.'" Disagree—70%, Agree—26%, Not sure/refuse to answer—4%
>
> American Association of Retired Persons Bulletin/AARP survey, July 22–August 2, 2009; $N = 1006$ adults age 50 and older; MoE ± 3.
> The American Association of Retired Persons Bulletin Survey reported that 16% of adults, 50 and older, said they are likely to return to school.

Each of the polls is based on approximately 1005 randomly selected adults.

a. Calculate the 95% confidence maximum error of estimate for the true binomial proportion based on binomial experiments with the same sample size and observed proportion as listed first in each article.

b. Explain what caused the values of the maximum errors to vary.

c. The margin of error being reported is typically the value of the maximum error rounded to the next larger whole percentage. Do your results in part a verify this?

d. Explain why the round-up practice is considered "conservative."

e. What value of p should be used to calculate the standard error if the most conservative margin of error is desired?

9.84 a. If x successes result from a binomial experiment with $n = 1000$ and $p = P(\text{success})$, and the 95% confidence interval for the true probability of success is determined, what is the maximum value possible for the "maximum error of estimate"?

b. Compare the numerical value of the "maximum error of estimate" found in part a with "the margin of error" discussed in Applied Example 9.9.

c. Under what conditions are they the same? Not the same?

d. Explain how the results of national polls, like those of Harris and Gallup, are related (similarities and differences) to the confidence interval technique studied in this section.

e. The theoretical sampling error with a level of confidence can be calculated, but the polls typically report only a "margin of error" with no probability (level of confidence). Why is that?

9.85 Karl Pearson once tossed a coin 24,000 times and recorded 12,012 heads.

a. Calculate the point estimate for $p = P(\text{head})$ based on Pearson's results.

b. Determine the standard error of proportion.

c. Determine the 95% confidence interval estimate for $p = P(\text{head})$.

d. It must have taken Mr. Pearson many hours to toss a coin 24,000 times. You can simulate 24,000 coin tosses using the computer and calculator commands that follow. (*Note:* A Bernoulli experiment is like a "single" trial binomial experiment. That is, one toss of a coin is one Bernoulli experiment with $p = 0.5$; and 24,000 tosses of a coin either is a binomial experiment with $n = 24,000$ or is 24,000 Bernoulli experiments. Code: $0 = $ tail, $1 = $ head. The sum of the 1s will be the number of heads in the 24,000 tosses.)

MINITAB
Choose **Calc > Random Data > Bernoulli**, entering 24000 for generate, C1 for Store in column(s) and 0.5 for Probability of success. Sum the data and divide by 24,000.

Excel
Choose **Data > Data Analysis > Random Number Generation > Bernoulli**, entering 1 for Number of Variables, 24000 for Number of Random Numbers and 0.5 for p Value. Sum the data and divide by 24,000.

TI-83/84 PLUS
Choose **MATH > PRB > 5:randInt**, then enter 0, 1, number of trials. The maximum number of elements (trials) in a list is 999 (slow process for large n's). Sum the data and divide by n.

e. How do your simulated results compare with Pearson's?

f. Use the commands (part d) and generate another set of 24,000 coin tosses. Compare these results to those obtained by Pearson. Also, compare the two simulated samples to each other. Explain what you can conclude from these results.

9.86 When a single die is rolled, the probability of a one is 1/6, or 0.167. Let's simulate 3000 rolls of a die. (*Note:* A Bernoulli experiment is like a "single" trial binomial experiment. That is, one roll of a die is one Bernoulli experiment with $p = 1/6$; and 3000 rolls of a die either is a binomial experiment with $n = 3000$ or is 3000 Bernoulli experiments. Code: $0 = 2, 3, 4, 5,$ or 6, and $1 = 1$. The sum of the 1s will be the number of ones in the 3000 tosses.)

a. Use the commands given in Exercise 9.85 and a calculator or computer to simulate the rolling of a single die 3000 times.

Using the results from the simulation:

b. Sum the data and divide by 3000. Explain what this value represents.

c. Determine the standard error of proportion.

d. Determine the 95% confidence interval for $p = P(\text{one})$.

e. How do the results from the simulation compare with your expectations? Explain.

9.87 The "rule of thumb" stated on page 434 indicated that we would expect the sampling distribution of p' to be approximately normal when "$n > 20$ and both np and nq are greater than 5." What happens when these guidelines are not followed?

a. Use the following set of computer or calculator commands to see what happens. Try $n = 15$ and $p = 0.1 (K1 = n$ and $K2 = p)$. Do the distributions look normal? Explain what causes the "gaps." Why do the histograms look alike? Try some different combinations of n (K1) and p (K2):

MINITAB
Choose **Calc > Random Data > Binomial** to simulate 1000 trials for an n of 15 and a p of 0.5. Divide each generated value by n, forming a column of sample p's. Calculate a z value for each sample p by using $z = (p' - p)/\sqrt{p(1 - p)/n}$. Construct a histogram for the sample p's and another histogram for the z's.

Excel
Choose **Data > Data Analysis > Random Number Generation > Binomial** to simulate 1000 trials for an n of 15 and a p of 0.5. Divide each generated value by n, forming a column of sample p's. Calculate a z value for each sample p by using $z = (p' - p)/\sqrt{p(1 - p)/n}$. Construct a histogram for the sample p's and another histogram for the z's.

TI-83/84 Plus
Choose **MATH > PRB > 7:randBin**, then enter n, p, number of trials. The maximum number of elements (trials) in a list is 999 (slow process for large n's). Divide each generated value by n, forming a list of sample p's. Calculate a z value for each sample p by using $z = (p' - p)/\sqrt{p(1 - p)/n}$. Construct a histogram for the sample p's and another histogram for the z's.

b. Try $n = 15$ and $p = 0.01$.

c. Try $n = 50$ and $p = 0.03$.

d. Try $n = 20$ and $p = 0.2$.

e. Try $n = 20$ and $p = 0.8$.

f. What happens when the rule of thumb is not followed?

9.88 Has the law requiring bike helmet use failed? Yankelovich Partners conducted a survey of bicycle riders in the United States. Only 60% of the nationally representative sample of 1020 bike riders reported owning a bike helmet.

Source: http://www.cpsc.gov/

a.   Find the 95% confidence interval for the true proportion p for a binomial experiment of 1020 trials that resulted in an observed proportion of 0.60. Use this to estimate the percentage of bike riders who reported owning a helmet.

b.   Based on the survey results, would you say there is compliance with the law requiring bike helmet use? Explain.

Suppose you wish to conduct a survey in your city to determine what percent of bicyclists own helmets. Use the national figure of 60% for your initial estimate of p.

c.   Find the sample size if you want your estimate to be within 0.02 with 95% confidence.

d.   Find the sample size if you want your estimate to be within 0.04 with 95% confidence.

e.   Find the sample size if you want your estimate to be within 0.02 with 90% confidence.

f.   What effect does changing the maximum error have on the sample size? Explain.

g.   What effect does changing the level of confidence have on the sample size? Explain.

9.89 Find the sample size n needed for a 95% interval estimate in Example 9.10.

9.90 Find n for a 90% confidence interval for p with $E = 0.02$ using an estimate of $p = 0.25$.

9.91 According to a May 2009 Harris Poll, 72% of those who drive and own cell phones say they use them to talk while they are driving. You wish to conduct a survey in your city to determine what percent of the drivers with cell phones use them to talk while driving. Use the national figure of 72% for your initial estimate of p.

a.   Find the sample size if you want your estimate to be within 0.02 with 90% confidence.

b.   Find the sample size if you want your estimate to be within 0.04 with 90% confidence.

c.   Find the sample size if you want your estimate to be within 0.02 with 98% confidence.

d.   What effect does changing the maximum error have on the sample size? Explain.

e.   What effect does changing the level of confidence have on the sample size? Explain.

9.92 Lung cancer is the leading cause of cancer deaths in both women and men in the United States. According to Centers for Disease Control and Prevention 2005 statistics, lung cancer accounts for more deaths than breast cancer, prostate cancer, and colon cancer combined. Overall, only about 16% of all people who develop lung cancer survive 5 years.

Source: http://www.cdc.gov/

Suppose you want to see if this survival rate is still true. How large a sample would you need to take to estimate the true proportion surviving 5 years after diagnosis to within 1% with 95% confidence? (Use the 16% as the initial value of p.)

9.93 State the null hypothesis, H_o, and the alternative hypothesis, H_a, that would be used to test these claims:

a.   More than 60% of all students at our college work part-time jobs during the academic year.

b.   No more than one-third of cigarette smokers are interested in quitting.

c.   A majority of the voters will vote for the school budget this year.

d   At least three-fourths of the trees in our county were seriously damaged by the storm.

e.   The results show the coin was not tossed fairly.

9.94 State the null hypothesis, H_o, and the alternative hypothesis, H_a, that would be used to test these claims:

a.   The probability of our team winning tonight is less than 0.50.

b.   At least 50% of all parents believe in spanking their children when appropriate.

c.   At most, 80% of the invited guests will attend the wedding.

d.   The single-digit numbers generated by the computer do not seem to be equally likely with regard to being odd or even.

e.   Less than half of the customers like the new pizza.

9.95 Calculate the test statistic $z\star$ used in testing the following:

a. $H_o: p = 0.70$ vs. $H_a: p > 0.70$, with the sample $n = 300$ and $x = 224$

b. $H_o: p = 0.50$ vs. $H_a: p < 0.50$, with the sample $n = 450$ and $x = 207$

c. H_o: $p = 0.35$ vs. H_a: $p \neq 0.35$, with the sample $n = 280$ and $x = 94$

d. H_o: $p = 0.90$ vs. H_a: $p > 0.90$, with the sample $n = 550$ and $x = 508$

9.96 Find the value of **P** for each of the hypothesis tests in Exercise 9.95; state the decision using $\alpha = 0.05$.

9.97 Determine the p-value for each of the following hypothesis-testing situations.

a. H_o: $p = 0.5$, H_a: $p \neq 0.5$, $z\star = 1.48$

b. H_o: $p = 0.7$, H_a: $p \neq 0.7$, $z\star = -2.26$

c. H_o: $p = 0.4$, H_a: $p > 0.4$, $z\star = 0.98$

d. H_o: $p = 0.2$, H_a: $p < 0.2$, $z\star = -1.59$

9.98 Find the critical region and critical values for each of the hypothesis tests in Exercise 9.95; state the decision using $\alpha = 0.05$.

9.99 Determine the test criteria that would be used to test the following hypotheses when z is used as the test statistic and the classical approach is used.

a. H_o: $p = 0.5$ and H_a: $p > 0.5$, with $\alpha = 0.05$

b. H_o: $p = 0.5$ and H_a: $p \neq 0.5$, with $\alpha = 0.05$

c. H_o: $p = 0.4$ and H_a: $p < 0.4$, with $\alpha = 0.10$

d. H_o: $p = 0.7$ and H_a: $p > 0.7$, with $\alpha = 0.01$

9.100 The binomial random variable, x, may be used as the test statistic when testing hypotheses about the binomial parameter, p, when n is small (say, 15 or less). Use Table 2 in Appendix B and determine the p-value for each of the following situations.

a. H_o: $p = 0.5$, H_a: $p \neq 0.5$, where $n = 15$ and $x = 12$

b. H_o: $p = 0.8$, H_a: $p \neq 0.8$, where $n = 12$ and $x = 4$

c. H_o: $p = 0.3$, H_a: $p > 0.3$, where $n = 14$ and $x = 7$

d. H_o: $p = 0.9$, H_a: $p < 0.9$, where $n = 13$ and $x = 9$

9.101 The binomial random variable, x, may be used as the test statistic when testing hypotheses about the binomial parameter, p. When n is small (say, 15 or less), Table 2 in Appendix B provides the probabilities for each value of x separately, thereby making it unnecessary to estimate probabilities of the discrete binomial random variable with the continuous standard normal variable z. Use Table 2 to determine the value of α for each of the following:

a. H_o: $p = 0.5$ and H_a: $p > 0.5$, where $n = 15$ and the critical region is $x = 12, 13, 14, 15$

b. H_o: $p = 0.3$ and H_a: $p < 0.3$, where $n = 12$ and the critical region is $x = 0, 1$

c. H_o: $p = 0.6$ and H_a: $p \neq 0.6$, where $n = 10$ and the critical region is $x = 0, 1, 2, 3, 9, 10$

d. H_o: $p = 0.05$ and H_a: $p > 0.05$, where $n = 14$ and the critical region is $x = 4, 5, 6, 7, \ldots, 14$

9.102 Use Table 2 in Appendix B to determine the critical region used in testing each of the following hypotheses. (*Note:* Since x is discrete, choose critical regions that do not exceed the value of α given.)

a. H_o: $p = 0.5$ and H_a: $p > 0.5$, where $n = 15$ and $\alpha = 0.05$

b. H_o: $p = 0.5$ and H_a: $p \neq 0.5$, where $n = 14$ and $\alpha = 0.05$

c. H_o: $p = 0.4$ and H_a: $p < 0.4$, where $n = 10$ and $\alpha = 0.10$

d. H_o: $p = 0.7$ and H_a: $p > 0.7$, where $n = 13$ and $\alpha = 0.01$

9.103 You are testing the hypothesis $p = 0.7$ and have decided to reject this hypothesis if after 15 trials you observe 14 or more successes.

a.   If the null hypothesis is true and you observe 13 successes, which of the following will you do? (1) Correctly fail to reject H_o. (2) Correctly reject H_o. (3) Commit a type I error. (4) Commit a type II error.

b.   Find the significance level of your test.

c.   If the true probability of success is 1/2 and you observe 13 successes, which of the following will you do? (1) Correctly fail to reject H_o. (2) Correctly reject H_o. (3) Commit a type I error. (4) Commit a type II error.

d.   Calculate the p-value for your hypothesis test after 13 successes are observed.

9.104 You are testing the null hypothesis $p = 0.4$ and will reject this hypothesis if $z\star$ is less than -2.05.

a.   If the null hypothesis is true and you observe $z\star$ equal to -2.12, which of the following will you do? (1) Correctly fail to reject H_o. (2) Correctly reject H_o. (3) Commit a type I error. (4) Commit a type II error.

b.   What is the significance level for this test?

c.   What is the p-value for $z\star = -2.12$?

9.105 An insurance company states that 90% of its claims are settled within 30 days. A consumer group selected a random sample of 75 of the company's claims to test this statement. If the consumer group found that 55 of the claims were settled within 30 days, does it have sufficient reason to support the contention that less than 90% of the claims are settled within 30 days? Use $\alpha = 0.05$.

a. Solve using the p-value approach.

b. Solve using the classical approach.

9.106 The full-time student body of a college is composed of 50% males and 50% females. Does a random sample of students (30 male, 20 female) from an introductory chemistry course show sufficient evidence to reject the hypothesis that the proportion of male and of female students who take this course is the same as that of the whole student body? Use $\alpha = 0.05$.

a. Solve using the p-value approach.

b. Solve using the classical approach.

9.107 A politician claims that she will receive 60% of the vote in an upcoming election. The results of a properly designed random sample of 100 voters showed that 50 of those sampled will vote for her. Is it likely that her assertion is correct at the 0.05 level of significance?

a. Solve using the p-value approach.

b. Solve using the classical approach.

9.108 The popularity of personal watercraft (PWCs, also known as jet skis) continues to increase, despite the apparent danger associated with their use. In fact, a sample of 54 watercraft accidents reported to the Game and Parks Commission in the state of Nebraska revealed that 85% of them involved PWCs even though only 8% of the motorized boats registered in the state are PWCs.

Source: *Nebraskaland*, "Officer's Notebook: The Personal Problem"

Suppose the national average proportion of watercraft accidents involving PWCs was 78%. Does the watercraft accident rate for PWCs in the state of Nebraska exceed that of the nation as a whole? Use a 0.01 level of significance.

a. Solve using the p-value approach.

b. Solve using the classical approach.

9.109 An April 21, 2009, *USA Today* article titled "On road, it's do as I say, not as I do" reported that 58% of U.S. adults speed up to beat a yellow light. Suppose you conduct a survey in your hometown of 150 randomly selected adults and find that 71 out of the 150 admit to speeding up to beat a yellow light. Does your hometown have a lower rate for speeding up to beat a yellow light than the nation as a whole? Use a 0.05 level of significance.

9.110 A recent survey conducted by Lieberman Research Worldwide and Charles Schwab reported that the "High Cost of Living" was the top concern that most surprised young adults as they began life on their own. Twenty-six percent reported "High Cost of Living" as their top concern. A disbeliever of this information took his own random sample of 500 young adults starting out on their own in an attempt to show that the true percentage for this top concern is actually higher.

a. Find the p-value if 148 of the young adults surveyed put down "High Cost of Living" as their top concern.

b. Explain why it is important for the level of significance to be established before the sample results are known.

9.111 September is Library Card Sign-up Month. According to a nationwide Harris Poll during August 2008, 68% of American adults own a library card. Suppose you conduct a survey of 1000 randomly chosen adults in order to test $H_o: p = 0.68$ versus $H_a: p < 0.68$, where p represents the proportion of adults who currently have a library card; 651 of the 1000 sampled had a library card. Use $\alpha = 0.01$.

a. Calculate the value of the test statistic.

b. Solve using the p-value approach.

c. Solve using the classical approach.

9.112 Show that the hypothesis test completed as Example 9.13 was unnecessary because the confidence interval had already been completed in Example 9.8.

9.113 The following computer output was used to complete a hypothesis test.

Test for One Proportion
Test of p = 0.225 vs p > 0.225

Sample	X	N	Sample p	95% Lower Bound	Z-Value	P-Value
1	61	200	0.305000	0.251451	2.71	0.003

a. State the null and alternative hypotheses.

b. If the test is completed using $\alpha = 0.05$, what decision and conclusion are reached?

c. Verify the "Sample p."

9.114 Using the computer output and information in Exercise 9.113, determine the value of the following:

a. Hypothesized value of population proportion

b. Sample proportion

c. Test statistic

9.115 Reliable Equipment has developed a machine, *The Flipper,* that will flip a coin with predictable results. They claim that a coin flipped by The Flipper will land heads up at least 88% of the time. What conclusion would result in a hypothesis test, using $\alpha = 0.05$, when 200 coins are flipped and the following results are achieved?

a. 181 heads   b. 172 heads

c. 168 heads   d. 153 heads

9.116 Refer to Applied Example 9.14.

a. State the Polish statisticians' hypothesis.

b. Is their hypothesis the null or alternative hypothesis? Explain.

c. State the null hypothesis. State the alternative hypothesis.

d. Explain the meaning of, "6.2 percent on either side of 50 percent."

e. What statistical term does the phrase, "would be expected in only about 7 of every 100 experiments with a fair coin" represent? Express it using symbols.

f. If the null hypothesis is tested using $\alpha = 0.05$, what decision and conclusion is reached using the results obtained by the students?

g. When New Scientist carried out its own experiment, what value did it obtain for the observed probability of heads?

h. Does New Scientist have evidence that the coin is biased? Explain.

i. Should we be upset by the fact that the results obtained by Gliszczynski's students and the New Scientist results are quite different but lead to the same conclusion? Explain.

9.3 Inferences about the Variance and Standard Deviation

Problems often arise that require us to make inferences about variability. For example, a soft drink bottling company has a machine that fills 16-oz bottles. The company needs to control the standard deviation σ (or variance σ^2) in the amount of soft drink, x, put into each bottle. The mean amount placed into each bottle is important, but a correct mean amount does not ensure that the filling machine is working correctly. If the variance is too large, many bottles will be overfilled and many underfilled. Thus, the bottling company wants to maintain as small a standard deviation (or variance) as possible.

When discussing inferences about the spread of data, we usually talk about variance instead of standard deviation because the techniques (the formulas used) employ the sample variance rather than the standard deviation. However, remember that the standard deviation is the positive square root of the variance; thus, talking about the variance of a population is comparable to talking about the standard deviation.

Inferences about the variance of a normally distributed population use the **chi-square**, χ^2, distributions ("*ki-square*": that's "ki" as in "kite," and χ is the

Greek lowercase letter chi). The chi-square distributions, like Student's t-distributions, are a family of probability distributions, each of which is identified by the **parameter** number of **degrees of freedom**. To use the chi-square distribution, we must be aware of its properties (also see Figure 9.7).

Properties of the chi-square distribution

1. χ^2 is nonnegative in value; it is zero or positively valued.
2. χ^2 is not symmetrical; it is skewed to the right.
3. χ^2 is distributed so as to form a family of distributions, a separate distribution for each different number of degrees of freedom.

FIGURE 9.7

Various Chi-Square Distributions

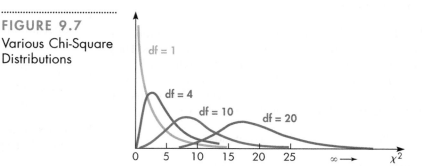

FIGURE 9.8

Location of Mean, Median, and Mode for χ^2 Distribution

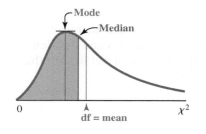

Note: When df = 2, the mean value of the chi-square distribution is df. The mean is located to the right of the mode (the value where the curve reaches its high point) and just to the right of the median (the value that splits the distribution, 50% on each side). By locating zero at the left extreme and the value of df on your sketch of the χ^2 distribution, you will establish an approximate scale so that other values can be located in their respective positions. See Figure 9.8.

For values of χ^2 on the left side of the median, the area to the right will be greater than 0.50.

FIGURE 9.9

Chi-Square Distribution Showing χ^2(df, α)

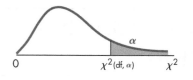

The **critical values for chi-square** are obtained from Table 8 in Appendix B. Each critical value is identified by two pieces of information: df and area under the curve to the right of the critical value being sought. Thus, χ^2(df, α) (read "chi-square of df, alpha") is the symbol used to identify the critical value of chi-square with df degrees of freedom and with α area to the right, as shown in Figure 9.9. Since the chi-square distribution is not symmetrical, the critical values associated with the right and left tails are given separately in Table 8.

EXAMPLE 9.15

χ^2 ASSOCIATED WITH THE RIGHT TAIL

Find χ^2(20, 0.05).

Solution

See the figure. Use Table 8 in Appendix B to find the value of χ^2(20, 0.05) at the intersection of row df = 20 and the column for an area of 0.05 to the right, as shown in the portion of the table that follows:

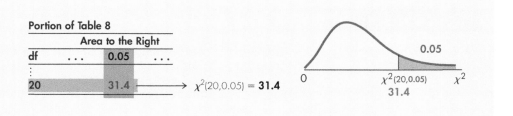

EXAMPLE 9.16

χ^2 ASSOCIATED WITH THE LEFT TAIL

Find $\chi^2(14, 0.90)$.

Solution

See the figure that follows. Use Table 8 in Appendix B to find the value of $\chi^2(14, 0.90)$ at the intersection of row df $= 14$ and the column for an area of 0.90 to the right, as shown in the portion of the table that follows:

Portion of Table 8		
Area to the Right		
df	...	**0.90**	...
⋮			
14		7.79	→ $\chi^2(14,0.90) = $ **7.79**

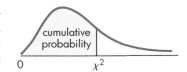

FYI Explore Skillbuilder Applet "Chi-Square Probabilities"— available through cengagebrain.com

Most computer software packages or statistical calculators will calculate the area related to a specified χ^2-value. The accompanying figure shows the relationship between the cumulative probability and a specific χ^2-value for a χ^2-distribution with df degrees of freedom.

TECHNOLOGY INSTRUCTIONS: CUMULATIVE PROBABILITIES FOR χ^2

MINITAB

Input the data into C1; then continue with:

Choose:	**Calc > Probability Distributions > Chi-Square**
Select:	**Cumulative Probability**
	Noncentrality Parameter: 0.0
Enter:	**Degrees of freedom: df**
Select:	**Input constant***
Enter:	χ^2**-value (ex. 47.25) > OK**

*Select Input column if several χ^2-values are stored in C1. Use C2 for optional storage. If the area in the right tail is needed, subtract the calculated probability from 1.

Excel	If several χ^2-values are to be used, input the values into column A and activate B1; then continue with:
	Choose: Insert function f_x > Statistical > CHIDIST > OK
	Enter: X: individual χ^2-value or (A1:A5 or select "χ^2-value" cells)* Deg_freedom: df > OK
	Drag*: Bottom right corner of the B1 cell down to give other probabilities

TI-83/84 Plus	**Choose:** 2nd > DISTR > 7: χ^2cdf(
	Enter: 0, χ^2-value, df)
	If the area in the right tail is needed, subtract the calculated probability from 1.

We are now ready to use chi-square to make inferences about the population variance or standard deviation.

> **The assumptions for inferences about the variance σ^2 or standard deviation σ**
> The sampled population is normally distributed.

The t procedures for inferences about the mean (see Section 9.1) were based on the assumption of normality, but the t procedures are generally useful even when the sampled population is nonnormal, especially for larger samples. However, the same is not true about the inference procedures for the standard deviation. The statistical procedures for the standard deviation are very sensitive to nonnormal distributions (skewness, in particular), and this makes it difficult to determine whether an apparent significant result is the result of the sample evidence or a violation of the assumptions. Therefore, the only inference procedure to be presented here is the hypothesis test for the standard deviation of a normal population.

The **test statistic** that will be used in testing hypotheses about the population variance or standard deviation is obtained by using the following formula:

> **Test Statistic for Variance and Standard Deviation**
>
> $$\chi^2\bigstar = \frac{(n-1)s^2}{\sigma^2} \text{, with df} = n - 1 \tag{9.10}$$

When random samples are drawn from a normal population with a known variance σ^2, the quantity $\frac{(n-1)s^2}{\sigma^2}$ possesses a probability distribution that is known as the chi-square distribution with $n - 1$ degrees of freedom.

Hypothesis-Testing Procedure

Let's return to the example about the bottling company that wishes to detect when the variability in the amount of soft drink placed into each bottle gets out of control. A variance of 0.0004 is considered acceptable, and the company wants to adjust the bottle-filling machine when the variance, σ^2, becomes larger than this value. The decision will be made using the hypothesis-testing procedure.

EXAMPLE 9.17

ONE-TAILED HYPOTHESIS TEST FOR VARIANCE, σ^2

The soft drink bottling company wants to control the variability in the amount of fill by not allowing the variance to exceed 0.0004. Does a sample of size 28 with a variance of 0.0007 indicate that the bottling process is out of control (with regard to variance) at the 0.05 level of significance?

Solution

Step 1 The Set-Up:

    a. Describe the population parameter of interest.
    σ^2, the variance in the amount of fill of a soft drink during a bottling process

    b. State the null hypothesis (H_o) and the alternative hypothesis (H_a).

        H_o: $\sigma^2 = 0.0004$ ($\leq$) (variance is not larger than 0.0004)

        H_a: $\sigma^2 > 0.0004$ (variance is larger than 0.0004)

Step 2 The Hypothesis Test Criteria:

    a. Check the assumptions.
    The amount of fill put into a bottle is generally normally distributed. By checking the distribution of the sample, we could verify this.

    b. Identify the probability distribution and the test statistic to be used.
    The chi-square distribution and formula (9.10), with $df = n - 1 = 28 - 1 = 27$, will be used.

    c. Determine the level of significance: $\alpha = 0.05$.

Step 3 The Sample Evidence:

    a. Collect the sample information: $n = 28$ and $s^2 = 0.0007$.

    b. Calculate the value of the test statistic.
    Use formula (9.10):

$$\chi^2\bigstar = \frac{(n-1)s^2}{\sigma^2} : \chi^2\bigstar = \frac{(28-1)(0.0007)}{0.0004} = \frac{(27)(0.0007)}{0.0004} = \mathbf{47.25}$$

Step 4 The Probability Distribution:

Using the *p*-value procedure:		**Using the classical procedure:**
a. Calculate the *p*-value for the test statistic. Use the right-hand tail because H_a expresses concern for values related to "larger than." $\mathbf{P} = P(\chi^2\bigstar > 47.25$, with df = 27) as shown in the figure.	**OR**	**a. Determine the critical region and critical value(s).** The critical region is the right-hand tail because H_a expresses concern for values related to "larger than." The critical value is obtained from

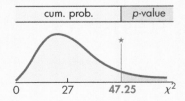

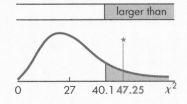

To find the *p*-value, use one of two methods:

1. Use Table 8 in Appendix B to place bounds on the *p*-value: $0.005 < P < 0.01$.
2. Use a computer or calculator to calculate the *p*-value: $P = 0.0093$.

Specific instructions follow this example.

b. Determine whether or not the *p*-value is smaller than α.

The *p*-value is smaller than the level of significance, α (0.05).

Table 8, at the intersection of row df = 27 and column $\alpha = 0.05$: $\chi^2(27, 0.05) = 40.1$. For specific instructions, see page 455.

b. Determine whether or not the calculated test statistic is in the critical region.

$\chi^2\star$ is in the critical region, as shown in **red** in the accompanying figure.

Step 5 The Results:

    a. State the decision about H_o: Reject H_o.

    b. State the conclusion about H_a.

    At the 0.05 level of significance, we conclude that the bottling process is out of control with regard to the variance.

Calculating the *p*-value when using the χ^2-distribution

Method 1: Use Table 8 in Appendix B to place bounds on the p-value. By inspecting the df = 27 row of Table 8, you can determine an interval within which the *p*-value lies. Locate $\chi^2\star$ along the row labeled df = 27. If $\chi^2\star$ is not listed, locate the two values that $\chi^2\star$ falls between, and then read the bounds for the *p*-value from the top of the table. In this case, $\chi^2\star = 47.25$ is between 47.0 and 49.6; therefore, **P** is between 0.005 and 0.01.

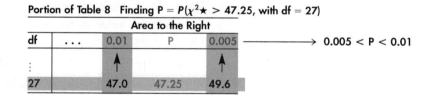

Portion of Table 8   Finding P = P($\chi^2\star$ > 47.25, with df = 27)

| df | ... | 0.01 | P | 0.005 | → | 0.005 < P < 0.01 |

Method 2: Use a computer or calculator. Use the χ^2 probability distribution commands on pages 455–456 to find the *p*-value associated with $\chi^2\star = 47.25$.

EXAMPLE 9.18

ONE-TAILED *p*-VALUE HYPOTHESIS TEST FOR VARIANCE, σ^2

Find the *p*-value for this hypothesis test:

$$H_o: \sigma^2 = 12$$
$$H_a: \sigma^2 < 12 \text{ with df} = 15 \text{ and } \chi^2\star = 7.88$$

Solution

Since the concern is for "smaller" values (the alternative hypothesis is "less than"), the p-value is the area to the left of $\chi^2 \star = 7.88$, as shown in the figure:

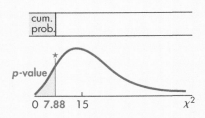

$$\mathbf{P} = P(\chi^2 \star < 7.88 \text{ with df} = 15)$$

To find the p-value, use one of two methods:

Method 1: Use Table 8 in Appendix B to place bounds on the p-value. Inspect the df = 15 row to find $\chi^2 \star = 7.88$. The $\chi^2 \star$ value is between entries, so the interval that bounds **P** is read from the Area to the Left heading at the top of the table.

Portion of Table 8   Finding P = P($\chi^2 \star$ < 7.88, with df = 15)

df	...	0.05	P	0.10	
⋮		↑		↑	→ 0.05 < P < 0.10
15		7.26	7.88	8.55	

Method 2: Use a computer or calculator. Use the χ^2 probability distribution commands on pages 455–456 to find the p-value associated with $\chi^2 \star = 7.88$.

E X A M P L E   9 . 1 9

TWO-TAILED HYPOTHESIS TEST FOR STANDARD DEVIATION, σ

A manufacturer claims that a photographic chemical has a shelf life that is normally distributed about a mean of 180 days with a standard deviation of no more than 10 days. As a user of this chemical, Fast Photo is concerned that the standard deviation might be different from 10 days; otherwise, it will buy a larger quantity while the chemical is part of a special promotion. Twelve random samples were selected and tested, with a standard deviation of 14 days resulting. At the 0.05 level of significance, does this sample present sufficient evidence to show that the standard deviation is different from 10 days?

Solution

Step 1 The Set-Up:

    a. Describe the population parameter of interest.

        σ, the standard deviation for the shelf life of the chemical

    b. State the null hypothesis (H_o) and the alternative hypothesis (H_a).

        H_o: $\sigma = 10$ (standard deviation is 10 days)

        H_a: $\sigma \neq 10$ (standard deviation is different from 10 days)

Step 2 The Hypothesis Test Criteria:

a. **Check the assumptions.**
   The manufacturer claims shelf life is normally distributed; this could be verified by checking the distribution of the sample.

b. **Identify the probability distribution and the test statistic to be used.**
   The chi-square distribution and formula (9.10), with $df = n - 1 = 12 - 1 = 11$, will be used.

c. **Determine the level of significance:** $\alpha = 0.05$.

Step 3 The Sample Evidence:

a. **Collect the sample information:** $n = 12$ and $s = 14$.

b. **Calculate the value of the test statistic.**
   Use formula (9.10):

$$\chi^2\star = \frac{(n-1)s^2}{\sigma^2} : \chi^2\star = \frac{(12-1)(14)^2}{(10)^2} = \frac{2156}{100} = \mathbf{21.56}$$

Step 4 The Probability Distribution:

<table>
<tr><td>

Using the *p*-value procedure:

a. **Calculate the *p*-value for the test statistic.**
Since the concern is for values "different from" 10, the *p*-value is the area of both tails. The area of each tail will represent $\frac{1}{2}\mathbf{P}$. Since $\chi^2\star = 21.56$ is in the right tail, the area of the right tail is $\frac{1}{2}\mathbf{P}$:
$\frac{1}{2}\mathbf{P} = P(\chi^2 > 21.56$, with df = 11), as shown in the figure.

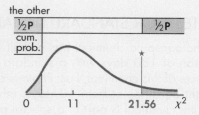

To find $\frac{1}{2}\mathbf{P}$, use one of two methods:
1. Use Table 8 in Appendix B to place bounds on $\frac{1}{2}\mathbf{P}$: $0.025 < \frac{1}{2}\mathbf{P} < 0.05$. Double both bounds to find the bounds for **P**: $2 \times (0.025 < \frac{1}{2}\mathbf{P} < 0.05)$ becomes $0.05 < \mathbf{P} < 0.10$.
2. Use a computer or calculator to find $\frac{1}{2}\mathbf{P}$: $\frac{1}{2}\mathbf{P} = 0.0280$; therefore, $\mathbf{P} = 0.0560$.
Specific instructions follow this example.

b. **Determine whether or not the *p*-value is smaller than α.**
The *p*-value is not smaller than the level of significance, α (0.05).

</td><td>

Using the classical procedure:

a. **Determine the critical region and critical value(s).**
The critical region is split into two equal parts because H_a expresses concern for values related to "different from." The critical values are obtained from Table 8 at the intersections of row df = 11 with columns 0.975 and 0.025 for the area to right: $\chi^2(11, 0.975) = 3.82$ and $\chi^2(11, 0.025) = 21.9$.

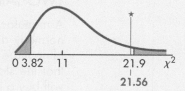

For specific instructions, see page 455.

b. **Determine whether or not the calculated test statistic is in the critical region.**
$\chi^2\star$ is not in the critical region; see the accompanying figure.

</td></tr>
</table>

Step 5 The Results:

    a. State the decision about H_o: Fail to reject H_o.

    b. State the conclusion about H_a.

       There is not sufficient evidence at the 0.05 significance level to conclude that the shelf life of this chemical has a standard deviation different from 10 days. Therefore, Fast Photo should purchase the chemical accordingly.

Calculating the *p*-value when using the χ^2-distribution

Method 1: Use Table 8 in Appendix B to place bounds on the p-value. Inspect the df = 11 row to locate $\chi^2\star = 21.56$. Notice that 21.56 is between two table entries. The bounds for $\frac{1}{2}P$ are read from the Area to the Right heading at the top of the table.

Portion of Table 8		Finding $P = 2 \cdot P(\chi^2\star > 21.56,$ with df $= 11)$		
		Area to the Right		
df	...	0.05	$\frac{1}{2}$P	0.025
11		19.7	21.56	21.9

$\longrightarrow$ $0.025 < \frac{1}{2}P < 0.05$
$0.05 < P < 0.10$

Double both bounds to find the bounds for **P:** $2 \times (0.025 < \frac{1}{2}P < 0.05)$ becomes **$0.05 < P < 0.10$**.

Method 2: Use a computer or calculator. Use the χ^2 probability distribution commands on pages 455–456 to find the *p*-value associated with $\chi^2\star = 21.56$. Remember to double the probability.

Note: When sample data are skewed, just one outlier can greatly affect the standard deviation. It is very important, especially when using small samples, that the sampled population be normal; otherwise, these procedures are not reliable.

APPLIED EXAMPLE 9.20

A textured ceramic floor tile and

its uneven top surface as shown by the varying amount of light showing under the ruler

Courtesy of the author

CERAMIC FLOOR TILE

Ceramic floor tiles come in all sorts of colors, finishes, and textures. One reason for making the surface textured is to create a natural stone look. In nature, the layers within stone vary greatly. For ceramic tiles there must be enough variation that the tiles resemble real stone, yet not so much as to create a safety problem.

    This variation can be measured as surface height, *x*, the distance between the surface and the plane of the "highest" points of the surface. See the figure below.

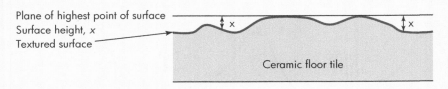

Plane of highest point of surface
Surface height, *x*
Textured surface

Ceramic floor tile

The manufacturing specification calls for the mean surface height to be no greater than 0.025 inch. The manufacturing process is under control when the standard deviation is no greater than 0.01 inch.

Twenty-six randomly located points were measured and the following data resulted.

Surface height, x [EX09-145]

0.000	0.017	0.007	0.011	0.027	0.041	0.010	0.033	0.023
0.008	0.004	0.026	0.025	0.025	0.028	0.017	0.025	0.042
0.015	0.020	0.012	0.024	0.019	0.028	0.022	0.006	

SECTION 9.3 EXERCISES

9.117 a.   Calculate the standard deviation for each set.

A: 5, 6, 7, 7, 8, 10   **B:** 5, 6, 7, 7, 8, 15

   b.   What effect did the largest value changing from 10 to 15 have on the standard deviation?

   c.   Why do you think 15 might be called an outlier?

9.118 The variance of shoe sizes for all manufacturers is 0.1024. What is the standard deviation?

9.119 Find:

a. $\chi^2(10, 0.01)$ b. $\chi^2(12, 0.025)$

c. $\chi^2(10, 0.95)$ d. $\chi^2(22, 0.995)$

9.120 Find these critical values by using Table 8 of Appendix B.

a. $\chi^2(18, 0.01)$ b. $\chi^2(16, 0.025)$

c. $\chi^2(8, 0.10)$ d. $\chi^2(28, 0.01)$

e. $\chi^2(22, 0.95)$ f. $\chi^2(10, 0.975)$

g. $\chi^2(50, 0.90)$ h. $\chi^2(24, 0.99)$

9.121 Using the notation of Exercise 9.120, name and find the critical values of χ^2.

a.
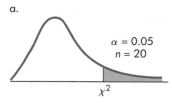
$\alpha = 0.05$
$n = 20$

b.
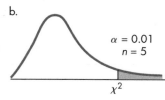
$\alpha = 0.01$
$n = 5$

c.
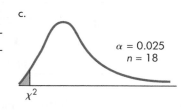
$\alpha = 0.025$
$n = 18$

d.
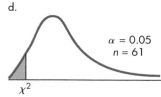
$\alpha = 0.05$
$n = 61$

e.
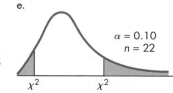
$\alpha = 0.10$
$n = 22$

f.
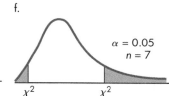
$\alpha = 0.05$
$n = 7$

9.122 Using the notation of Exercise 9.120, name and find the critical values of χ^2.

a.
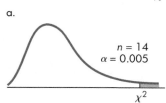
$n = 14$
$\alpha = 0.005$

b.
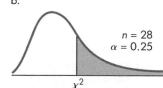
$n = 28$
$\alpha = 0.25$

c.
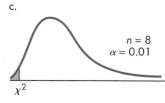
$n = 8$
$\alpha = 0.01$

d.
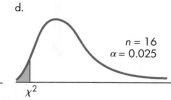
$n = 16$
$\alpha = 0.025$

e.
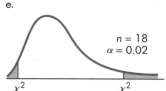
$n = 18$
$\alpha = 0.02$

f.
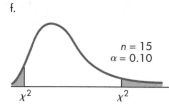
$n = 15$
$\alpha = 0.10$

9.123 a. What value of chi-square for 5 degrees of freedom subdivides the area under the distribution curve such that 5% is to the right and 95% is to the left?

b. What is the value of the 95th percentile for the chi-square distribution with 5 degrees of freedom?

c. What is the value of the 90th percentile for the chi-square distribution with 5 degrees of freedom?

9.124 a. The central 90% of the chi-square distribution with 11 degrees of freedom lies between what values?

b. The central 95% of the chi-square distribution with 11 degrees of freedom lies between what values?

c. The central 99% of the chi-square distribution with 11 degrees of freedom lies between what values?

9.125 For a chi-square distribution having 12 degrees of freedom, find the area under the curve for chi-square values ranging from 3.57 to 21.0.

9.126 For a chi-square distribution having 35 degrees of freedom, find the area under the curve between $\chi^2(35, 0.96)$ and $\chi^2(35, 0.15)$.

9.127 Use a computer or calculator to find the area (a) to the left, and (b) to the right of $\chi^2\star = 20.2$ with df = 15.

9.128 Use a computer or calculator to find the area (a) to the left, and (b) to the right of $\chi^2\star = 14.7$ with df = 24.

9.129 State the null hypothesis, H_o, and the alternative hypothesis, H_a, that would be used to test these claims:

a. The standard deviation has increased from its previous value of 24.

b. The standard deviation is no larger than 0.5 oz.

c. The standard deviation is not equal to 10.

d. The variance is no less than 18.

e. The variance is different from the value of 0.025, the value called for in the specs.

9.130 State the null hypothesis, H_o, and the alternative hypothesis, H_a, that would be used to test these claims:

a. The variance has decreased from 34.5.

b. The standard deviation of shoe size is more than 0.32.

c. The standard deviation is at least 5.5.

d. The variance is at most 35.

e. The variance has shrunk from the value of 0.34 since the assembly lines were retooled.

9.131 Find the test statistic for the hypothesis test:

a. H_o: $\sigma^2 = 532$ versus H_a: $\sigma^2 > 532$ using sample information $n = 18$ and $s^2 = 785$

b. H_o: $\sigma^2 = 52$ versus H_a: $\sigma^2 \neq 52$ using sample information $n = 41$ and $s^2 = 78.2$

9.132 Calculate the value for the test statistic, $\chi^2\star$, for each of these situations:

a. H_o: $\sigma^2 = 20, n = 15, s^2 = 17.8$

b. H_o: $\sigma^2 = 30, n = 18, s = 5.7$

c. H_o: $\sigma = 42, n = 25, s = 37.8$

d. H_o: $\sigma = 12, n = 37, s^2 = 163$

9.133 Calculate the p-value for each of the following hypothesis tests.

a. H_a: $\sigma^2 \neq 20, n = 15, \chi^2\star = 27.8$

b. H_a: $\sigma^2 > 30, n = 18, \chi^2\star = 33.4$

c. H_a: $\sigma^2 \neq 42, \text{df} = 25, \chi^2\star = 37.9$

d. H_a: $\sigma^2 < 12, \text{df} = 40, \chi^2\star = 26.3$

9.134 Determine the critical region and critical value(s) that would be used to test the following using the classical approach:

a. H_o: $\sigma = 0.5$ and H_a: $\sigma > 0.5$, with $n = 18$ and $\alpha = 0.05$

b. H_o: $\sigma^2 = 8.5$ and H_a: $\sigma^2 < 8.5$, with $n = 15$ and $\alpha = 0.01$

c. H_o: $\sigma = 20.3$ and H_a: $\sigma \neq 20.3$, with $n = 10$ and $\alpha = 0.10$

d. H_o: $\sigma^2 = 0.05$ and H_a: $\sigma^2 \neq 0.05$, with $n = 8$ and $\alpha = 0.02$

e. H_o: $\sigma = 0.5$ and H_a: $\sigma < 0.5$, with $n = 12$ and $\alpha = 0.10$

9.135 Complete the hypothesis test in Exercise 9.131a using the following:

a.   The *p*-value method and $\alpha = 0.01$

b.   The classical method and $\alpha = 0.01$

9.136 Complete the hypothesis test in Exercise 9.131b using the following:

a.   The *p*-value method and $\alpha = 0.05$

b.   The classical method and $\alpha = 0.05$

9.137 In the past the standard deviation of weights of certain 32.0-oz packages filled by a machine was 0.25 oz. A random sample of 20 packages showed a standard deviation of 0.35 oz. Is the apparent increase in variability significant at the 0.10 level of significance? Assume package weight is normally distributed.

a.   Solve using the *p*-value approach.

b.   Solve using the classical approach.

9.138 Variation in the life of a battery is expected, but too much variation would be of concern to the consumer, who would never know if the purchased battery might have a very short life. A random sample of 30 AA batteries of a particular brand produced a standard deviation of 350 hours. If a standard deviation of 288 hours (12 days) is considered acceptable, does this sample provide sufficient evidence that this brand of battery has greater variation than what is acceptable at the 0.05 level of significance? Assume battery life is normally distributed.

9.139 A random sample of 51 observations was selected from a normally distributed population. The sample mean was $\bar{x} = 98.2$, and the sample variance was $s^2 = 37.5$. Does this sample show sufficient reason to conclude that the population standard deviation is not equal to 8 at the 0.05 level of significance?

a.   Solve using the *p*-value approach.

b.   Solve using the classical approach.

9.140 A commercial farmer harvests his entire field of a vegetable crop at one time. Therefore, he would like to plant a variety of green beans that mature all at one time (small standard deviation between maturity times of individual plants). A seed company has developed a new hybrid strain of green beans that it believes to be better for the commercial farmer. The maturity time of the standard variety has an average of 50 days and a standard deviation of 2.1 days. A random sample of 30 plants of the new hybrid showed a standard deviation of 1.65 days. Does this sample show a significant lowering of the standard deviation at the 0.05 level of significance? Assume that maturity time is normally distributed.

a.   Solve using the *p*-value approach.

b.   Solve using the classical approach.

9.141 Farm real estate values in rural America fluctuate substantially from state to state and county to county, thus making it difficult for buyers purchasing land or landowners to know precisely what the property is actually worth. For example, the average value of ranchland in Missouri was $548 per acre, whereas the same average in three nearby states (Kansas, Nebraska, and Oklahoma) was more than $200 less.

Source: *Regional Economic Digest,* "Survey of Agricultural Credit Conditions"

This discrepancy could be caused by an exaggerated variability in the value of ranchland acreage in the state of Missouri. Assume that the combined four-state region yields a standard deviation of $85 per acre. Suppose a sample was taken of 31 landowners in Missouri who recently sold their property and a sample standard deviation of $125 per acre resulted. Is the variability in ranchland value in Missouri, at the 0.05 level of significance, greater than the variability for the region as a whole? Using the MINITAB output below, complete the hypothesis test.

Null hypothesis			Sigma = 85		
Alternative hypothesis			Sigma > 85		
N	StDev	Variance	Chi-Square	DF	P-Value
31	125	15625	64.88	30	0.000

9.142 Using the computer output in Exercise 9.141, determine the values of the following terms:

a.   Hypothesized value of population standard deviation

b.   Sample standard deviation

c.   Degrees of freedom—how were they calculated?

d.   Relationship between sample variance and standard deviation

e.   Test statistic.

9.143 [EX09-143] Maybe even more important than how much they weigh is that the plates used in weightlifting be the same weight. When one of each weight is hanging on opposite ends of a bar, they need to balance. A random sample of twenty-four 25-lb weights used for weightlifting were randomly selected and their weights (in pounds) determined:

25.3	22.1	25.7	24.2	25.7	23.9	23.1	21.9
24.7	26.3	26.5	22.2	25.9	23.5	25.8	27.1
25.4	22.0	25.2	21.1	27.9	22.9	27.3	25.7

There have been complaints about the excessive variability in the weights of these 25-lb plates. Does the sample show sufficient evidence to conclude that the variability in the weights is greater than the acceptable 1-lb standard deviation? Use $\alpha = 0.01$.

a. What role does the assumption of normality play in this solution? Explain.

b. What evidence do you have that the assumption of normality is reasonable? Explain.

c. Solve using the *p*-value approach.

d. Solve using the classical approach.

9.144 [EX09-144] A car manufacturer claims that the miles per gallon for a certain model has a mean equal to 40.5 miles with a standard deviation equal to 3.5 miles. Use the following data, obtained from a random sample of 15 such cars, to test the hypothesis that the standard deviation differs from 3.5. Use $\alpha = 0.05$. Assume normality.

37.0	38.0	42.5	45.0	34.0	32.0	36.0	35.5
38.0	42.5	40.0	42.5	35.0	30.0	37.5	

a. Solve using the *p*-value approach.

b. Solve using the classical approach.

9.145 [EX09-145] Refer to Applied Example 9.20, "Ceramic Floor Tile," on page 461. First you need to complete a preliminary investigation of the 26 randomly selected surface heights:

a. Present and describe the sample of surface heights using a histogram, the mean, and the standard deviation.

b. Check the surface heights for a normal distribution. State what you believe to be the case based on the results found in part a. Further, find additional statistical evidence. Very precisely state your conclusion regarding normality for the distribution of this variable.

Statistical testing of the manufacturing process:

c. Is there statistical evidence that the process used to manufacture these floor tiles has produced a textured surface that has a mean surface height no greater than 0.025 inch? State the *p*-value.

d. Complete the hypothesis test at the 0.01 level of significance; be sure to state your decision and conclusion.

9.146 [EX09-145] Refer to Applied Example 9.20, "Ceramic Floor Tile," and Exercise 9.145 to continue the investigation of the floor tile manufacturing process.

a. What are the assumptions for a chi-square test of standard deviation? Do any of the answers in Exercise 9.145 help resolve the assumption requirements?

b. Is there statistical evidence that the process used to manufacture these floor tiles has produced a textured surface that has a standard deviation of surface height no greater than 0.01 inch? State the *p*-value.

c. Complete the hypothesis test at the 0.01 level of significance; be sure to state your decision and conclusion.

d. What conclusions can be reached about the manufacturing process?

9.147 [EX09-147] The dry weight of a cork is another quality that does not affect the ability of the cork to seal a bottle, but it is a variable that is monitored regularly. The weights of the no. 9 natural corks (24 mm in diameter by 45 mm in length) have a normal distribution. Ten randomly selected corks were weighed to the nearest hundredth of a gram.

Dry Weight (in grams)

3.26	3.58	3.07	3.09	3.16	3.02	3.64	3.61	3.02	2.79

a. Does the preceding sample present sufficient reason to show that the standard deviation of the dry weights is different from 0.3275 gram at the 0.02 level of significance?

A different random sample of 20 is taken from the same batch.

Dry Weight (in grams)

3.53	3.77	3.49	3.24	3.00	3.41	3.33	3.51	3.02	3.46
2.80	3.58	3.05	3.51	3.61	2.90	3.69	3.62	3.26	3.58

b. Does the preceding sample present sufficient reason to show that the standard deviation of the dry weights is different from 0.3275 gram at the 0.02 level of significance?

c. What effect did the two different sample standard deviations have on the calculated test statistic in parts a and b? What effect did they have on the *p*-value or critical value? Explain.

d. What effect did the two different sample sizes have on the calculated test statistic in parts a and b? What effect did they have on the *p*-value or critical value? Explain.

9.148 Use a computer or calculator to find the *p*-value for the following hypothesis test: $H_o: \sigma^2 = 7$ versus $H_a: \sigma^2 \neq 7$, if $\chi^2\star = 6.87$ for a sample of $n = 15$.

9.149 Use a computer or calculator to find the *p*-value for the following hypothesis test: $H_o: \sigma = 12.4$ versus $H_a: \sigma > 12.4$, if $\chi^2\star = 36.59$ for a sample of $n = 24$.

[EX00-000] identifies the filename of an exercise's online dataset—available through cengagebrain.com

9.150 The chi-square distribution was described on page 454 as a family of distributions. Let's investigate these distributions and observe some of their properties.

a. Use the MINITAB commands that follow and generate several large random samples of data from various chi-square distributions. Use df values of 1, 2, 3, 5, 10, 20, and 80 (and others if you wish).

Choose: **Calc > Random Data > ChiSquare**
Enter: Number of rows of data to generate: **1000**
           Store in column(s): **C1**
           Degrees of freedom: **df**
Use **Stat > Basic Statistics > Display Descriptive Statistics** to calculate the mean and median of the data in C1. Use **Graph > Histogram** to construct a histogram of the data in C1.

b. What appears to be the relationship between the mean of the sample and the number of degrees of freedom?

c. How do the values of the mean, median, and mode appear to be related? Do your results agree with the information on page 454?

d. Have the computer generate samples for two additional degrees of freedom df = 120 and 150. Describe how these distributions seem to be changing as df increases.

9.151 How important is the assumption "the sampled population is normally distributed" for the use of the chi-square distributions? Use a computer and the two sets of MINITAB commands that can be found in the *Student Solutions Manual* to simulate drawing 200 samples of size 10 from each of two different types of population distributions. The first commands will generate 2000 data values

and construct a histogram so that you can see what the population looks like. The next commands will generate 200 samples of size 10 from the same population; each row represents a sample. The following commands will calculate the standard deviation and $\chi^2\star$ for each of the 200 samples. The last commands will construct histograms of the 200 sample standard deviations and the 200 $\chi^2\star$-values. (Additional details can be found in the *Student Solutions Manual*.) For the samples from the normal population:

a. Does the sampling distribution of sample standard deviations appear to be normal? Describe the distribution.

b. Does the χ^2-distribution appear to have a chi-square distribution with df = 9? Find percentages for intervals (less than 2, less than 4, ..., more than 15, more than 20, etc.), and compare them with the percentages expected as estimated using Table 8 in Appendix B.

For the samples from the skewed population:

c. Does the sampling distribution of sample standard deviations appear to be normal? Describe the distribution.

d. Does the χ^2-distribution appear to have a chi-square distribution with df = 9? Find percentages for intervals (less than 2, less than 4, ..., more than 15, more than 20, etc.), and compare them with the percentages expected as estimated using Table 8.

In summary:

e. Does the normality condition appear to be necessary in order for the calculated test statistic $\chi^2\star$ to have a χ^2-distribution? Explain.

Courtesy of the author

Chapter Review

In Retrospect

We have been studying inferences, both confidence intervals and hypothesis tests, for the three basic population parameters (mean μ, proportion p, and standard deviation σ) of a single population. Most inferences about a single population are concerned with one of these three parameters. Figure 9.10 (p. 467) presents a

visual organization of the techniques presented in Chapters 8 and 9 along with the key questions that you must ask as you are deciding which test statistic and formula to use.

In this chapter we also used the maximum error of estimate, formula (9.7), to determine the size of the sample required to make estimations about the population proportion with the desired accuracy. In Applied Example 9.9 the margin of error reported by the media is described, and its relationship to the maximum error of estimate, as presented in this chapter, is discussed. By combining the reported point estimate and the sample size, we can determine the corresponding binomial proportion maximum error of estimate. Most polls and surveys use the 95% confidence level and then use the maximum error as an estimate for margin of error and do not report a level of confidence, as Humphrey Taylor explained.

In the next chapter we will discuss inferences about two populations whose respective means, proportions, and standard deviations are to be compared.

FIGURE 9.10
Choosing the Right Inference Technique

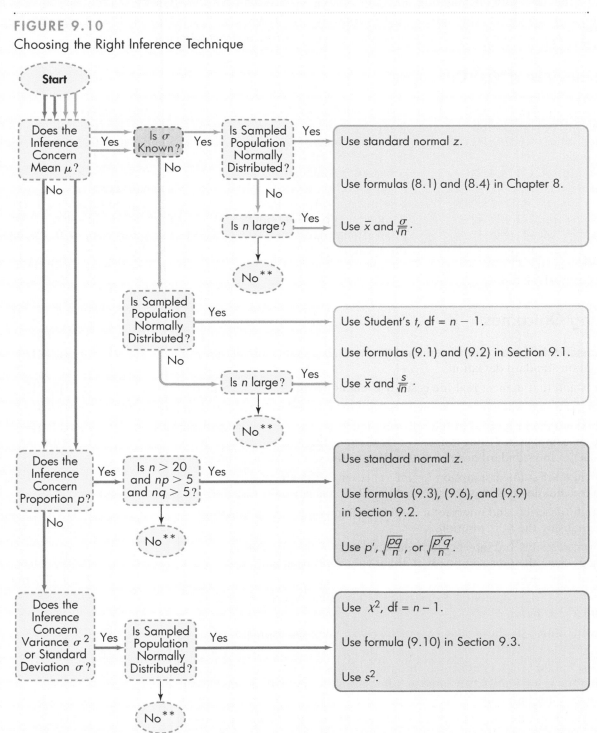

No** means that a nonparametric technique (normal distribution not required) is used; see Chapter 14.

CourseMate   The **Statistics CourseMate** site for this text brings chapter topics to life with interactive learning, study, and exam preparation tools, including quizzes and flashcards for the Vocabulary and Key Concepts that follow. The site also provides an **eBook** version of the text with highlighting and note taking capabilities. Throughout chapters, the CourseMate icon 🖥 flags concepts and examples that have corresponding interactive resources such as **video** and **animated tutorials** that demonstrate, step by step, how to solve problems; **datasets** for exercises and examples; **Skillbuilder Applets** to help you better understand concepts; **technology manuals**; and software to download including **Data Analysis Plus** (a suite of statistical macros for Excel) and **TI-83/84 Plus** programs—logon at **www.cengagebrain.com**.

Vocabulary and Key Concepts

assumptions (pp. 417, 435, 456)
binomial experiment (p. 434)
calculated value (pp. 420, 441)
chi-square (p. 453)
conclusion (pp. 421, 424, 442)
confidence interval (pp. 417, 435, 436)
critical region (pp. 421, 423, 442)
critical value (pp. 414, 421, 423)
decision (pp. 424, 442, 444)
degrees of freedom (pp. 414, 454)
hypothesis test (pp. 420, 423, 441)
inference (pp. 417, 435, 456)

level of confidence (p. 435)
level of significance (pp. 420, 423)
maximum error of estimate (pp. 418, 436, 438)
observed binomial probability, p' (p. 434)
p-value (pp. 424, 457, 458)
parameter (pp. 414, 418, 420, 423)
population proportion (p. 435)
proportions (p. 434)
random variable (p. 434)

rule of thumb (p. 434)
sample size (pp. 438, 439)
sample statistic (p. 435)
σ known (p. 413)
σ unknown (p. 412)
standard error (p. 412)
standard normal, z (p. 414)
Student's t-statistic (p. 413)
test statistic (pp. 420, 421, 423, 441, 456)
unbiased estimator (p. 435)

Learning Outcomes

• Understand that s, the sample standard deviation, is a point estimate for σ, the population standard deviation.	pp. 412–413
• Understand that in most real-life cases, σ is unknown and s is used as its best estimate.	pp. 412–413
• Understand that when σ is unknown, the z-statistic is replaced with Student's t-statistic.	pp. 412–413
• Understand the properties of the t-distribution, how it is a series of distributions based on sample size (using degrees of freedom as the index), and how it approaches the standard normal distribution as the sample size increases.	pp. 413–414, EXP 9.1, 9.2, 9.3, Ex. 9.4, 9.19
• Understand that the assumption for inferences about the mean μ when σ is unknown is that the sampled population is normally distributed.	p. 417, Ex. 9.33a
• Compute, describe, and interpret a confidence interval for the population mean, μ, using the t-distribution.	p. 418, EXP 9.4, Ex. 9.24, 9.31, 9.153
• Perform, describe, and interpret a hypothesis test for the population mean, μ, using the t-distribution with the p-value approach and/or classical approach.	pp. 420–421, EXP 9.5, 9.6, Ex. 9.47, 9.160
• Understand the fundamental properties of a binomial experiment and the binomial parameter, p.	pp. 434–435, Ex. 9.65, 9.66
• Understand that p', the sample proportion, is an unbiased estimator of the population proportion, p.	pp. 434–435, Ex. 9.67
• Understand that the sampling distribution of p' has an approximately normal distribution if n is sufficiently large and therefore the standard normal distribution can be used for inferences.	pp. 434–435, Ex. 9.87
• Understand that the assumption for inferences about the binomial parameter, p, is that the n random observations forming the sample are selected independently from a population that is not changing during the sampling.	p. 435

- Compute, describe, and interpret a confidence interval for the population proportion, p, using the z-distribution. EXP 9.8, Ex. 9.71, 9.75, 9.167
- Compute and describe the required sample size for a confidence interval of p, the population proportion. pp. 438–440, EXP 9.10, 9.11, Ex. 9.88
- Perform, describe, and interpret a hypothesis test for the population proportion, p, using the z-distribution with the p-value approach and/or classical approach. EXP 9.12, 9.13, Ex. 9.105, 9.108
- Understand the properties of the chi-square distribution and how it is a series of distributions based on sample size (using degrees of freedom as the index). pp. 453–454
- Understand that the assumption for inferences about the variance, σ^2, or standard deviation, σ, is that the sampled population is normally distributed. p. 456, Ex. 9.151
- Perform, describe, and interpret a hypothesis test for the population variance, σ^2, or standard deviation, σ, using the χ^2-distribution with the p-value approach and classical approach. EXP 9.17, 9.19, Ex. 9.137, 9.183

Chapter Exercises

9.152 You hurry to the local emergency department in hopes of immediate, urgent care, only to find yourself waiting for what seems like hours. The manager of the large emergency department believes that his new procedures have substantially reduced the wait time for the average urgent care patient. He initiates a study to evaluate the wait time. The records of 18 randomly selected patients seen since the new procedures were put in place are checked, and the time between entering the emergency department and being seen by urgent care personnel was observed. The mean wait time was 17.82 minutes with a standard deviation of 5.68 minutes. Estimate the mean wait time using a 99% confidence interval. Assume that wait times are normally distributed.

9.153 A natural gas utility is considering a contract for purchasing tires for its fleet of service trucks. The decision will be based on expected mileage. For a sample of 100 tires tested, the mean mileage was 36,000 and the standard deviation was 2000 miles. Estimate the mean mileage that the utility should expect from these tires using a 98% confidence interval.

9.154 One of the objectives of a large medical study was to estimate the mean physician fee for cataract removal. For 25 randomly selected cases, the mean fee was found to be $3550 with a standard deviation of $275. Set a 99% confidence interval on μ, the mean fee for all physicians. Assume fees are normally distributed.

9.155 Oranges are selected at random from a large shipment that just arrived. The sample is taken to estimate the size (circumference, in inches) of the oranges. The sample data are summarized as follows: $n = 100$, $\Sigma x = 878.2$, and $\Sigma (x - \bar{x})^2 = 49.91$.

a. Determine the sample mean and standard deviation.

b. What is the point estimate for μ, the mean circumference of all oranges in the shipment?

c. Find the 95% confidence interval for μ.

9.156 [EX09-156] Molds are used in the manufacture of contact lenses so that the lens material for proper preparation and curing will be consistent and meet designated dimensional criteria. Molds were fabricated and a critical dimension measured for 15 randomly selected molds. (Data have been doubly coded to ensure propriety.)

140	130	15	180	95	135	220	105
195	110	150	150	130	120	120	

Source: Courtesy of Bausch & Lomb.

a. Construct a histogram and find the mean and standard deviation.

b. Demonstrate how this set of data satisfies the assumptions for inference.

c. Find the 95% confidence interval for μ.

d. Interpret the meaning of the confidence interval.

9.157 A company claims that its battery lasts no less than 42.5 hours in continuous use in a specified toy. A simple random sample of batteries yields a sample mean life of 41.89 hours with a standard deviation of 4.75 hours. A computer calculates a test statistic of $t = -1.09$ and a p-value of 0.139. If the test uses df = 71, what is the best estimate of the sample size?

9.158 [EX09-158] Getting a college education today is almost as important as breathing and it's expensive! It is

(continue on page 470)

[EX00-000] identifies the filename of an exercise's online dataset — available through cengagebrain.com

not just the tuition, room, and board; textbooks are expensive too. It is very important for students, and their parents, to have an accurate estimate of total textbook costs. The total cost of required textbooks for nine freshman- or sophomore-level classes at 10 randomly selected New York public colleges was collected:

582.19	806.40	913.44	915.75	932.35
957.45	960.92	996.24	1070.44	1223.44

a. Construct a histogram and find the mean and standard deviation.

b. Demonstrate how this set of data satisfies the assumptions for inference.

c. Find the 95% confidence interval for μ, the mean total cost of required textbooks.

d. Interpret the meaning of the confidence interval.

9.159 [EX09-159] The total cost of required textbooks for nine freshman- or sophomore-level classes at 10 randomly selected New York private colleges was collected:

639.00	865.75	868.20	874.25	887.06
890.50	970.13	1013.22	1026.00	1048.96

a. Construct a histogram and find the mean and standard deviation.

b. Demonstrate how this set of data satisfies the assumptions for inference.

c. Find the 95% confidence interval for μ, the mean total cost of required textbooks.

d. Interpret the meaning of the confidence interval.

e. Is there a difference in the mean total cost of the nine required textbooks between the public colleges in Exercise 9.158 and in the private colleges in this exercise? Explain.

f. Explain why the confidence interval for the public colleges is so much wider than the corresponding interval for the private colleges. Be exact and detailed.

9.160 A manufacturer of television sets claims that the maintenance expenditures for its product will average no more than $110 during the first year following the expiration of the warranty. A consumer group has asked you to substantiate or discredit the claim. The results of a random sample of 50 owners of

such television sets showed that the mean expenditure was $131.60 and the standard deviation was $42.46. At the 0.01 level of significance, should you conclude that the manufacturer's claim is true or not likely to be true?

9.161 College students throw out an average of 640 pounds of trash each year, 30% of that in the month before graduation, according to the *Reader's Digest* article "Campus Castoffs." To estimate the amount of trash discarded by the students at State University, 18 students were randomly selected and carefully monitored for one year. The amounts of trash discarded had a mean of 559.9 lbs and a standard deviation of 158.6 lbs. Does State University have sufficient evidence that their students' mean amount of trash is significantly lower than the mean amount from all colleges? Assume normality and use a 0.05 level of significance.

9.162 [EX09-162] The water pollution readings at State Park Beach seem to be lower than those of the prior year. A sample of 12 readings (measured in coliform/100 mL) was randomly selected from the records of this year's daily readings:

3.5	3.9	2.8	3.1	3.1	3.4	4.8	3.2	2.5	3.5	4.4	3.1

Does this sample provide sufficient evidence to conclude that the mean of this year's pollution readings is significantly lower than last year's mean of 3.8 at the 0.05 level? Assume that all such readings have a normal distribution.

9.163 [EX09-163] It has been suggested that abnormal male children tend to be born to older-than-average parents. Case histories of 20 abnormal males were obtained, and the ages of the 20 mothers were as follows:

31	21	29	28	34	45	21	41	27	31
43	21	39	38	32	28	37	28	16	39

The mean age at which mothers in the general population give birth is 28.0 years.

a. Calculate the sample mean and standard deviation.

b. Does the sample give sufficient evidence to support the claim that abnormal male children have older-than-average mothers? Use $\alpha = 0.05$. Assume ages have a normal distribution.

9.164 [EX09-164] Twenty-four oat-producing counties were randomly identified from across Minnesota for the purpose of testing the claim "The mean oat crop yield rate is greater than 60 bushels per acre." For each county identified, the yield rate, in bushels of oats per harvested acre, was obtained. The resulting data are listed:

Yield

56	31	80	53	39	59	63	67	56	66	81	61	63	48	53
46	73	85	77	78	72	63	71	77

Source: http://www.nass.usda.gov/

a. Are the test assumptions satisfied? Explain.

b. Complete the test using $\alpha = 0.05$.

9.165 [EX09-165] Presented here are 100 measurements of the velocity of light in air (km/sec) recorded by Albert Michelson, an American physicist, from June 5 to July 2, 1879. The measurements have had 299,000 subtracted from them and then were adjusted for corrections used by Michelson. In this form, the true constant value for the velocity of light in air becomes 734.5 km/sec. Do Michelson's measurements support the true value that he was trying to measure? Use a 0.01 level of significance.

850	740	900	1070	930	850	950	980	980	880	1000
980	930	650	760	810	1000	1000	960	960	960	940
960	940	880	800	850	880	900	840	830	790	810
880	880	830	800	790	760	800	880	880	880	860
720	720	620	860	970	950	880	910	850	870	840
840	850	840	840	840	890	810	810	820	800	770
760	740	750	760	910	920	890	860	880	720	840
850	850	780	890	840	780	810	760	810	790	810
820	850	870	870	810	740	810	940	950	800	810
870

Source: http://lib.stat.cmu.edu/

Note: The currently accepted "true" value is 299,792.5 km/sec (with no adjustments).

9.166 Even with a heightened awareness of beef quality, 82% of Americans indicated their recent burger-eating behavior has remained the same, according to a recent T.G.I. Friday's restaurants random survey of 1027 Americans. In fact, half of Americans eat at least one beef burger each week. That's a minimum of 52 burgers each year.

Source: Harris Interactive/Yankelovich Partners for T.G.I. Friday's restaurants, http://www.knoxville3.com/

a. What is the point estimate for the proportion of all Americans who eat at least one beef burger per week?

b. Find the 98% confidence interval for the true proportion p in the binomial situation where $n = 1027$ and the observed proportion is one-half.

c. Use the results of part b to estimate the percentage of all Americans who eat at least one beef burger per week.

9.167 The marketing research department of an instant-coffee company conducted a survey of married men to determine the proportion of married men who prefer their brand. Of the 100 men in the random sample, 20 prefer the company's brand. Use a 95% confidence interval to estimate the proportion of all married men who prefer this company's brand of instant coffee. Interpret your answer.

9.168 A company is drafting an advertising campaign that will involve endorsements by noted athletes. For the campaign to succeed, the endorser must be both highly respected and easily recognized. A random sample of 100 prospective customers is shown photos of various athletes. If the customer recognizes an athlete, then the customer is asked whether he or she respects the athlete. In the case of a top woman golfer, 16 of the 100 respondents recognized her picture and indicated that they also respected her. At the 95% level of confidence, what is the true proportion with which this woman golfer is both recognized and respected?

9.169 A local auto dealership advertises that 90% of customers whose autos were serviced by the service department are pleased with the results. As a researcher, you take exception to this statement because you are aware that many people are reluctant to express dissatisfaction. A research experiment was set up in which those in the sample had received service by this dealer within the past 2 weeks. During the interview, the individuals were led to believe that the interviewer was new in town and was considering taking his car to this dealer's service department. Of the 60 sampled, 14 said that they were dissatisfied and would not recommend the department.

a. Estimate the proportion of dissatisfied customers using a 95% confidence interval.

b. Given your answer to part a, what can be concluded about the dealer's claim?

9.170 According to a nationwide study by the U.S. Department of Education that was mentioned in "Beating bullies without a fight," an article in the September 22,

(continue on page 472)

2009, *Democrat and Chronicle*, 79% of children between the ages of 12 and 18 were bullied at least once in the past six months. You wish to conduct a study to estimate the percentage of children between the ages of 12 and 18 in your community who were bullied in the last six months. Assume the population proportion is 79%, as reported by the U.S. Department of Education. What sample size must you use if you want your estimate to be within:

a.   0.03 with 90% confidence?

b.   0.06 with 95% confidence?

c.   0.09 with 99% confidence?

9.171 The May 30, 2008, online article "Live with Your Parents After Graduation?" quoted a 2007 survey conducted by Monster-TRAK.com. The survey found that 48% of college students planned to live at home after graduation. How large of a sample size would you need to estimate the true proportion of students that plan to live at home after graduation to within 2% with 98% confidence?

Source: http://www.nomoreramenonline.com/

9.172 The chief executive officer (CEO) of a small business wishes to hire your consulting firm to conduct a simple random sample of its customers. She wants to determine the proportion of her customers who consider her company the primary source of their products. She requests the margin of error in the proportion be no more than 3% with 95% confidence. Earlier studies have indicated that the approximate proportion is 37%.

a.   What is the minimum size of the sample that you would recommend to meet the requirements of your client if you use the earlier results?

b.   What is the minimum size of the sample that you would recommend to meet the requirements of your client if you ignore the earlier results?

c.   Is the approximate proportion of value needed in conducting the survey? Explain.

9.173 In obtaining the sample size to estimate a proportion, the formula $n = [z(\alpha/2)]^2 \, pq/E^2$ is used. If a reasonable estimate of p is not available, it is suggested that $p = 0.5$ be used because this will give the maximum value for n. Calculate the value of $pq = p(1 - p)$ for $p = 0.1, 0.2, 0.3, \ldots, 0.8, 0.9$ in order to obtain some idea about the behavior of the quantity pq.

9.174 A machine is considered to be operating in an acceptable manner if it produces 0.5% or fewer defective parts. It is not performing in an acceptable manner if more than 0.5% of its production is defective. The hypothesis $H_o: p = 0.005$ is tested against the hypothesis $H_a: p > 0.005$ by taking a random sample of 50 parts produced by the machine. The null hypothesis is rejected if two or more defective parts are found in the sample. Find the probability of the type I error.

9.175 You are interested in comparing the null hypothesis $p = 0.8$ against the alternative hypothesis $p < 0.8$. In 100 trials you observe 73 successes. Calculate the p-value associated with this result.

9.176 The Kaiser Family Foundation conducted a national survey in 2003 of 17,685 seniors. The purpose of the survey was to capture detailed information about seniors' prescription drug use, coverage, and experiences.

Source: http://www.kff.org/

a.   If this were a random sample that satisfied all the requirements for an inference about p, what would be the standard error?

b.   What would be the maximum error of estimate for a 95% confidence interval?

c.   Is a sample this size worthwhile? Give reasons to support your answer.

9.177 The Pizza Shack has been experimenting with different recipes for their pizza crust, thinking they might replace their current recipe. They are planning to sample pizza made with the new crust. Before sampling, a strategy is needed so that after the tasting results are in, Pizza Shack will know how to interpret their customers' preferences. The decision is not being taken lightly because there is much to be gained or lost depending on whether or not the decision is a popular one. A one-tailed hypothesis test of $p = P(\text{prefer new crust}) = 0.50$ is being planned.

a.   If $H_a: p > 0.50$ is used, explain the meaning of the four possible outcomes and their resulting actions.

b.   If $H_a: p < 0.50$ is used, explain the meaning of the four possible outcomes and their resulting actions.

c.   Which alternative hypothesis do you recommend be used, $p > 0.5$ or $p < 0.5$? Explain.

9.178 The Pizza Shack in Exercise 9.177 has completed its sampling and the results are in! On Tuesday afternoon, they sampled 15 customers and 9 preferred the new pizza crust. On Friday evening, they sampled 200 customers and 120 preferred the new pizza crust. Help the manager interpret the meaning of these results. Use a one-tailed test with $H_a: p > 0.50$ and $\alpha = 0.02$. Use z as the test statistic.

a. Is there sufficient evidence to conclude a significant preference for the new crust based on Tuesday's customers?

b. Is there sufficient evidence to conclude a significant preference for the new crust based on Friday's customers?

c. Since the percentage of customers preferring the new crust was the same, $p' = 0.60$ in both samplings, explain why the answers in parts a and b are not the same.

9.179 The owner of the Pizza Shack in Exercises 9.177 and 9.178 does not understand the use of the normal distribution and z in Exercise 9.178. Help the manager interpret the meaning of the results by redoing both hypothesis tests using $x =$ number of customers preferring the new crust as the test statistic and its binomial probability distribution. Use a one-tailed test with $H_a: p > 0.50$ and $\alpha = 0.02$.

The results were as follows: on Tuesday afternoon, they sampled 15 customers and 9 preferred the new pizza crust; on Friday evening, they sampled 200 customers and found 120 preferred the new pizza crust.

a. Is there sufficient evidence to conclude a significant preference for the new crust based on Tuesday's customers?

b. Is there sufficient evidence to conclude a significant preference for the new crust based on Friday's customers?

c. Explain the relationship between the solutions obtained in Exercise 9.178 and here.

9.180 An instructor asks each of the 54 members of his class to write down "at random" one of the numbers 1, 2, 3, . . . , 13, 14, 15. Since the instructor believes that students like gambling, he considers 7 and 11 to be lucky numbers. He counts the number of students, x, who selected 7 or 11. How large must x be before the

hypothesis of randomness can be rejected at the 0.05 level?

9.181 Today's newspapers and magazines often report the findings of survey polls about various aspects of life. The Pew Internet & American Life Project (January 13–February 9, 2005) found that "63% of cell phone users ages 18-27 have used text messaging within the past month." Other information obtained from the project included "random telephone survey of 1,460 cell phone users" and "has a margin of sampling error of plus or minus 3 percentage points." Relate this information to the statistical inferences you have been studying in this chapter.

a. Is a percentage of people a population parameter, and if so, how is it related to any of the parameters that we have studied?

b. Based on the information given, find the 95% confidence interval for the true proportion of cell phone users who have used text messaging.

c. Explain how the terms "point estimate," "level of confidence," "maximum error of estimate," and "confidence interval" relate to the values reported in the article and to your answers in part b.

9.182 To test the hypothesis that the standard deviation on a standard test is 12, a sample of 40 randomly selected students' exams was tested. The sample variance was found to be 155. Does this sample provide sufficient evidence to show that the standard deviation differs from 12 at the 0.05 level of significance?

9.183 Bright-Lite claims that its 60-watt light bulb burns with a length of life that is approximately normally distributed with a standard deviation of 81 hours. A sample of 101 bulbs had a variance of 8075. Is this sufficient evidence to reject Bright-Lite's claim in favor of the alternative, "the standard deviation is larger than 81 hours," at the 0.05 level of significance?

9.184 A production process is considered out of control if the produced parts have a mean length different from 27.5 mm or a standard deviation that is greater than 0.5 mm. A sample of 30 parts yields a sample mean of 27.63 mm and a sample standard deviation of 0.87 mm. If we assume part length is a normally distributed variable, does this sample indicate that the process should be

(continue on page 474)

adjusted to correct the standard deviation of the product? Use $\alpha = 0.05$.

9.185 Julia Jackson operates a franchised restaurant that specializes in soft ice cream cones and sundaes. Recently she received a letter from corporate headquarters warning her that her shop is in danger of losing its franchise because the average sales per customer have dropped "substantially below the average for the rest of the corporation." The statement may be true, but Julia is convinced that such a statement is completely invalid to justify threatening a closing. The variation in sales at her restaurant is bound to be larger than most, primarily because she serves more children, elderly, and single adults rather than large families who run up big bills at the other restaurants. Therefore, her average ticket is likely to be smaller and exhibit greater variability. To prove her point, Julia obtained the sales records from the whole company and found that the standard deviation was $2.45 per sales ticket. She then conducted a study of the last 71 sales tickets at her store and found a standard deviation of $2.95 per ticket. Is the variability in sales at Julia's franchise, at the 0.05 level of significance, greater than the variability for the company?

9.186 All tomatoes that a certain supermarket buys from growers must meet the store's specifications of a mean diameter of 6.0 cm and a standard deviation of no more than 0.2 cm. The supermarket's buyer visits a potential new supplier and selects a random sample of 36 tomatoes from the grower's greenhouse. The diameter of each tomato is measured, and the mean is found to be 5.94 and the standard deviation is 0.24. Do the tomatoes meet the supermarket's specs?

a. Determine whether an assumption of normality is reasonable. Explain.

b. Is the sample evidence sufficient to conclude that the tomatoes do not meet the specs with regard to the mean diameter? Use $\alpha = 0.05$.

c. Is the sample evidence sufficient to conclude that the tomatoes do not meet the specs with regard to the standard deviation? Use $\alpha = 0.05$.

d. Write a short report for the buyer outlining the findings and recommendations as to whether or not to use this tomato grower to supply tomatoes for sale in the supermarket.

9.187 The uniform length of nails is very important to a carpenter—the length of the nails being used are matched to the materials being fastened together, thereby making a small standard deviation an important property of the nails. A sample of 35 randomly selected 2-inch nails is taken from a large quantity of Nails, Inc.'s, recent production run. The resulting length measurements have a mean length of 2.025 inches and a standard deviation of 0.048 inch.

a. Determine whether an assumption of normality is reasonable. Explain.

b. Is the sample evidence sufficient to reject the idea that the nails have a mean length of 2 inches? Use $\alpha = 0.05$.

c. Is there sufficient evidence, at the 0.05 level, to show that the length of nails from this production run has a standard deviation greater than the advertised 0.040 inch?

d. Write a short report outlining the findings and recommendations as to whether or not the carpenter should use these nails for an application that requires 2-inch nails.

9.188 [EX09-188] It is important that the force required to extract a cork from a wine bottle not have a large standard deviation. Years of production and testing indicate that the no. 9 corks in Applied Example 6.13 (p. 285) have an extraction force that is normally distributed with a standard deviation of 36 Newtons. Recent changes in the manufacturing process are thought to have reduced the standard deviation.

a. What would be the problem with the standard deviation being relatively large? What would be the advantage of a smaller standard deviation?

A sample of 20 randomly selected bottles is used for testing.

Extraction Force in Newtons

296	338	341	261	250	347	336	297	279	297
259	334	281	284	279	266	300	305	310	253

b. Is the preceding sample sufficient to show that the standard deviation of extraction force is less than 36.0 Newtons, at the 0.02 level of significance?

During a different testing, a sample of eight bottles is randomly selected and tested.

Extraction Force in Newtons

331.9	312.0	289.4	303.6	346.9	308.1	346.9	276.0

c. Is the preceding sample sufficient to show that the standard deviation of extraction force is less than 36.0 Newtons, at the 0.02 level of significance?

d. What effect did the two different sample sizes have on the calculated test statistic in parts b and c? What effect did they have on the *p*-value or critical value? Explain.

e. What effect did the two different sample standard deviations have on the answers in parts b and c? What effect did they have on the *p*-value or critical value? Explain.

9.189 [EX09-189] A box of Corn Flakes that is labeled "NET WT. 14 OZ." should have 14 oz or more of cereal inside. Twenty of these boxes were randomly selected and the weight of the contents (in ounces) determined.

14.52	14.47	14.80	14.60	14.45	14.25	14.15	14.12	14.36	14.39
14.50	14.29	14.28	14.60	13.85	14.18	14.39	14.45	14.69	14.38

a. Draw a histogram of the weight of cereal per box.

b. Find the sample statistics mean and standard deviation.

c. What percent of the sample is below the 14.0 oz weight?

The plant manager is studying the filling process and needs to estimate the mean weight of all boxes being filled.

d. Determine whether an assumption of normality is reasonable. Explain.

e. Find the 95% confidence interval for the mean weight.

f. The filling process is believed to be running with a standard deviation of fill of no more than 0.2 oz. Test this hypothesis at the 0.01 level.

9.190 [EX09-190] The manager in Exercise 9.189 believes that the cereal-filling machine used for Corn Flakes needs to be replaced and that the new one he is considering will pay for the upgrade within a short time,

mainly due to less variability in the fill amount. The new machine is started, and a test run is made. Twenty of these boxes were randomly selected from the run and the contents weighed (in ounces).

14.17	14.25	14.17	14.16	14.18	14.09	14.19	14.17	14.16	14.06
14.11	14.15	14.12	14.19	14.14	14.19	14.13	14.12	14.16	14.15

a. Draw a histogram of the weight of cereal per box.

b. Find the sample statistics mean and standard deviation.

c. What percent of the sample from the new machine is below the 14.0 oz weight?

The manager needs to estimate the mean weight and test the standard deviation of all boxes being filled.

d. Determine whether an assumption of normality is reasonable. Explain.

e. Find the 95% confidence interval for the mean weight.

f. The filling process for the new machine is claimed to be running with a standard deviation of fill of less than 0.1 oz. Test this hypothesis at the 0.01 level.

9.191 The boxes of Corn Flakes in Exercises 9.189 and 9.190 that have more than 14.2 oz of cereal are being considered "too full." Since the weights appear to be normally distributed for both filling machines, use the normal distribution and find the following information for the manager.

a. What proportion of the boxes filled by the current machine fill the boxes with too much cereal?

b. What proportion of the boxes filled by the new machine fill the boxes with too much cereal?

c. For every 1000 boxes of cereal filled by the current machine, how many boxes can be filled by the new machine using the same total amount of cereal?

d. Summarize what you believe should be the manager's pitch to the company for getting the new filling machine.

Chapter Practice Test

PART I: Knowing the Definitions

Answer "True" if the statement is always true. If the statement is not always true, replace the words shown in bold with words that make the statement always true.

9.1 Student's t-distributions have an approximately normal distribution but are **more** dispersed than the standard normal distribution.

9.2 The **chi-square** distribution is used for inferences about the mean when σ is unknown.

9.3 **Student's t-distribution** is used for all inferences about a population's variance.

9.4 If the test statistic falls in the critical region, the null hypothesis has **been proved true.**

9.5 When the test statistic is t and the number of degrees of freedom gets very large, the critical value of t is very close to that of the **standard normal z.**

9.6 When making inferences about one mean when the value of σ is not known, the **z-score** is the test statistic to use.

9.7 The chi-square distribution is a skewed distribution whose mean value is **2** for df > 2.

9.8 Often, the concern with testing the variance (or standard deviation) is keeping its size under control or relatively small. Therefore, many of the hypothesis tests with chi-square are **one-tailed.**

9.9 $\sqrt{npq}$ is the standard error of proportion.

9.10 The sampling distribution of p' is distributed approximately as a **Student's t**-distribution.

PART II: Applying the Concepts

Answer all questions, showing all formulas, substitutions, and work.

9.11 Find each value:
  a. $z(0.02)$
  b. $t(18, 0.95)$
  c. $\chi^2(25, 0.95)$

9.12 A random sample of 25 data values was selected from a normally distributed population for the purpose of estimating the population mean, μ. The sample statistics are $n = 25$, $\bar{x} = 28.6$, and $s = 3.50$.
  a.   Find the point estimate for μ.
  b.   Find the maximum error of estimate for the 0.95 confidence interval estimate.
  c.   Find the lower confidence limit (LCL) and the upper confidence limit (UCL) for the 0.95 confidence interval estimate for μ.

9.13 Thousands of area elementary school students were recently given a nationwide standardized exam to test their composition skills. If 64 of a random sample of 100 students passed this exam, construct the 0.98 confidence interval estimate for the true proportion of all area students who passed the exam.

9.14 State the null (H_o) and the alternative (H_a) hypotheses that would be used to test each of these claims:
  a.   The mean weight of professional basketball players is no more than 225 lb.
  b.   Approximately 40% of daytime students own their own car.
  c.   The standard deviation for the monthly amounts of rainfall in Monroe County is less than 3.7 inches.

9.15 Determine the level of significance, test statistic, critical region, and critical values(s) that would be used in completing each hypothesis test using the classical approach with $\alpha = 0.05$.
  a. H_o: $\mu = 43$ versus H_a: $\mu < 43$, $\sigma = 6$
  b. H_o: $\mu = 95$ versus H_a: $\mu \neq 95$, σ unknown, $n = 22$
  c. H_o: $p = 0.80$ versus H_a: $p > 0.80$
  d. H_o: $\sigma = 12$ versus H_a: $\sigma \neq 12$, $n = 28$

9.16 The automobile manufacturer of the Alero claims that the typical Alero will average 32 mpg of gasoline. An independent consumer group is somewhat skeptical of this claim and thinks the mean gas mileage is less than the 32 claimed. A sample of 24 randomly selected Aleros produced these sample statistics: mean 30.15 and standard deviation 4.87. At the 0.05 level of significance, does the consumer group have sufficient evidence to refute the manufacturer's claim?

9.17 A coffee machine is supposed to dispense 6 fluid ounces of coffee into a paper cup. In reality, the amount dispensed varies from cup to cup. However, if the machine is operating properly, the standard deviation of the amounts dispensed should be 0.1 oz or less. A random sample of 15 cups produced a standard deviation of 0.13 oz. Does this represent sufficient evidence, at the 0.10 level of significance, to conclude that the machine is not operating properly?

9.18 An unhappy customer is frustrated with the waiting time at the post office when buying stamps. Upon registering his complaint, he was told, "You wait more than 1 minute for service no more than half of the time when you buy only stamps." Not believing this to be the case, the customer collected some data from people who had just purchased stamps only. The sample statistics are $n = 60$ and $x = n(\text{wait more than 1 minute}) = 35$. At the 0.02 level of significance, does our unhappy customer have sufficient evidence to refute the post office's claim?

PART III: Understanding the Concepts

9.19 Student B says the range of a set of data may be used to obtain a crude estimate for the standard deviation of a population. Student A is not sure. How will Student B correctly explain how and under what circumstances his statement is true?

9.20 Is it the null hypothesis or the alternative hypothesis that the researcher usually believes to be true? Explain.

9.21 When you reject a null hypothesis, Student A says that you are expressing disbelief in the value of the parameter as claimed in the null hypothesis. Student B says that instead, you are expressing the belief that the sample statistic came from a population other than the one related to the parameter claimed in the null hypothesis. Who is correct? Explain.

9.22 "Student's t-distribution must be used when making inferences about the population mean, μ, when the population standard deviation, σ, is not known" is a true statement. Student A states that the z-score sometimes plays a role when the t-distribution is used. Explain the conditions that exist and the role played by z that make Student A's statement correct.

9.23 Student A says that the percentage of the sample means that fall outside the critical values of the sampling distribution determined by a true null hypothesis is the p-value for the test. Student B says that the percentage Student A is describing is the level of significance. Who is correct? Explain.

9.24 Student A carries out a study in which she is willing to run a 1% risk of making a type I error. She rejects the null hypothesis and claims that her statistic is significant at the 99% level of confidence. Student B argues that Student A's claim is not properly worded. Who is correct? Explain.

9.25 Student A claims that when you employ a 95% confidence interval to determine an estimation, you do not know for sure whether or not your inference is correct (i.e., whether the parameter is contained within the interval). Student B claims that you do know; you have shown that the parameter cannot be less than the lower limit or greater than the upper limit of the interval. Who is right? Explain.

9.26 Student A says that the best way to improve a confidence interval estimate is to increase the level of confidence. Student B argues that using a high confidence level does not really improve the resulting interval estimate. Who is right? Explain.

10 Inferences Involving Two Populations

© 2010/Jupiterimages Corporation

Image copyright Sergey Peterman, 2012. Used under license from Shutterstock.com

10.1 Dependent and Independent Samples

Battle of the Sexes—Commute Time

The battle of the sexes can take on many forms.   When the topic of which gender is the better, faster, or safer driver comes up on campus, the battles could get quite competitive! Once the dust clears one could also ask, "Who drives the longest commute to college?" The length of commute can be measured in distance (miles) or in time (minutes); and there are many factors that play a role for commuting students. Do they live at home? Do they work a part-time or full-time job? Do they have family obligations?

Male students and female students are two populations. In this chapter we are going to study the procedures for making inferences about two populations. When comparing two populations, we need two samples, one from each population. Two basic kinds of samples can be used: independent and dependent. The dependence or independence of two samples is determined by the sources of the data. A **source** can be a person, an object, or anything else that yields a data value. If the same set of sources or related sets are used to obtain the data representing both populations, we have **dependent samples**. If two unrelated sets of sources are used, one set from each population, we have **independent samples**. The following examples should clarify these ideas.

FIGURE 10.1

"Road Map" to Two Population Inferences

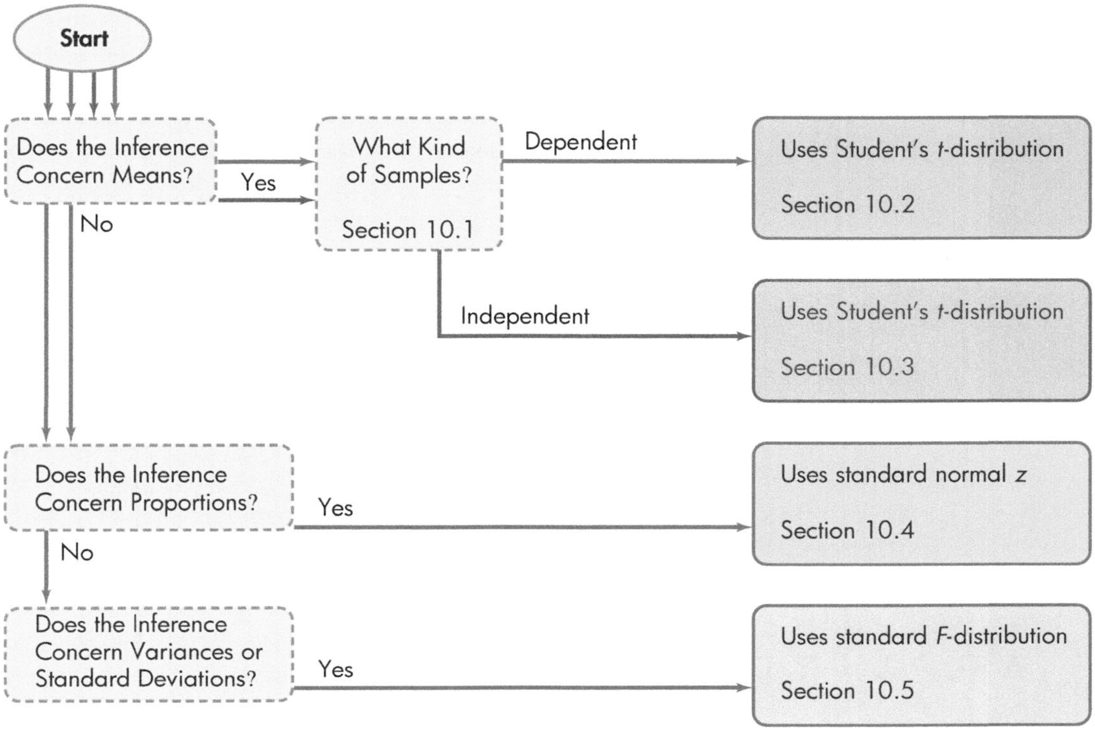

E X A M P L E   1 0 . 1

DEPENDENT VERSUS INDEPENDENT SAMPLES

A test will be conducted to see whether the participants in a physical-fitness class actually improve in their level of fitness. It is anticipated that approximately 500 people will sign up for this course. The instructor decides that she will give 50 of the participants a set of tests before the course begins (a pre-test), and then she will give another set of tests to 50 participants at the end of the course (a post-test). Two sampling procedures are proposed:

Plan A: Randomly select 50 participants from the list of those enrolled and give them the pre-test. At the end of the course, make a second random selection of size 50 and give them the post-test.

Plan B: Randomly select 50 participants and give them the pre-test; give the same set of 50 the post-test when they complete the course.

Plan A illustrates independent sampling; the sources (the class participants) used for each sample (pre-test and post-test) were selected separately. Plan B illustrates dependent sampling; the sources used for both samples (pre-test and post-test) are the same.

Typically, when both a pre-test and a post-test are used, the same subjects participate in the study. Thus, pre-test versus post-test (before versus after) studies usually use dependent samples.

EXAMPLE 10.2

DEPENDENT VERSUS INDEPENDENT SAMPLES

A test is being designed to compare the wearing quality of two brands of automobile tires. The automobiles will be selected and equipped with the new tires and then driven under "normal" conditions for 1 month. Then a measurement will be taken to determine how much wear took place. Two plans are proposed:

Plan C: A sample of cars will be selected randomly, equipped with brand A tires, and driven for 1 month. Another sample of cars will be selected, equipped with brand B tires, and driven for 1 month.

Plan D: A sample of cars will be selected randomly, equipped with one tire of brand A and one tire of brand B (the other two tires are not part of the test), and driven for 1 month.

We suspect that many other factors must be taken into account when testing automobile tires—such as age, weight, and mechanical condition of the car; driving habits of drivers; location of the tire on the car; and where and how much the car is driven. However, at this time we are trying only to illustrate dependent and independent samples. Plan C is independent (unrelated sources), and plan D is dependent (common sources).

APPLIED EXAMPLE 10.3

EXPLORING THE TRAITS OF TWINS

Studies that involve identical twins are a natural for the dependent sampling technique discussed in this section.

A NEW STUDY SHOWS THAT KEY CHARACTERISTICS MAY BE INHERITED

Like many identical twins reared apart, Jim Lewis and Jim Springer found they had been leading eerily similar lives. Separated four weeks after birth in 1940, the Jim twins grew up 45 miles apart in Ohio and were reunited in 1979. Eventually they discovered that both drove the same model blue Chevrolet, chain-smoked Salems, chewed their fingernails, and owned dogs named Toy. Each had spent a good deal of time vacationing at the same three-block strip of beach in Florida. More important, when tested for such personality traits as flexibility, self-control, and sociability, the twins responded almost exactly alike.

The project is considered the most comprehensive of its kind. The Minnesota researchers report the results of six-day tests of their subjects, including 44 pairs of identical twins who were brought up apart. Well-being, alienation, aggression, and the shunning of risk or danger were found to owe as much or more to nature as to nurture. Of eleven key traits or clusters of traits analyzed in the study, researchers estimated that a high of 61 percent of what they call "society potency" (a tendency toward leadership or dominance) is inherited, while "social closeness" (the need for intimacy, comfort, and help) was lowest, at 33 percent.

Source: Behavior: "Exploring the Traits of Twins," By John Leo; Elizabeth Taylor/Chicago, Monday, Jan. 12, 1987. Read more: http://www.time.com/time/magazine/article/0,9171,963211,00.html#ixzz0vw6zjedP. Reprinted by permission of TIME, Inc. All rights reserved.

Independent and dependent samples each have their advantages; these will be emphasized later. Both methods of sampling are often used.

⊡ SECTION 10.1 EXERCISES

10.1 [EX10-001] The data that follow are from two random samples of 37 college males and 42 college females with respect to their commute times to college.

Time—M

15	12	30	15	10	23	20	13	25	20	15	20	23	15	20
15	18	15	20	20	8	10	15	18	20	15	25	20	10	25
18	18	20	27	25	20	7								

Time—F

32	15	20	35	45	20	10	5	35	25	14	25	28	35	30
24	28	15	30	30	30	40	25	20	18	20	15	30	24	30
25	20	10	60	20	25	27	25	40	22	25	25			

a. What are the populations of interest?

b. Describe statistically the distribution of both the male and the female "commute time" data using at least the mean, standard deviation, and a histogram.

c. Do the two sets of data represent dependent or independent samples? Explain why.

d. If the two sample sizes are unequal, does this dictate dependent or independent samples? Explain your answer.

e. If the two sample sizes are equal, does this dictate dependent or independent samples? Explain your answer.

10.2 a. Describe a sampling plan you might use to select two independent samples of male and female commute times to college.

b. Describe a sampling plan you might use to select two dependent samples of male and female commute times to college.

c. Do you foresee any advantages for using one plan over the other?

d. Which of the two plans (from parts a and b) would you prefer to use? Explain your reasons why.

10.3 Explain why studies involving identical twins, as in Applied Example 10.3, result in dependent samples of data.

10.4 a. Describe how you could select two independent samples from among your classmates to compare the heights of female and male students.

b. Describe how you could select two dependent samples from among your classmates to compare their heights when they entered high school with their heights when they entered college.

10.5 The students at a local high school were assigned to do a project for their statistics class. The project involved having sophomores take a timed test on geometric concepts. The statistics students then used these data to determine whether there was a difference between male and female performances. Would the resulting sets of data represent dependent or independent samples? Explain.

10.6 In trying to estimate the amount of growth that took place in the trees recently planted by the County Parks Commission, 36 trees were randomly selected from the 4000 planted. The heights of these trees were measured and recorded. One year later, another set of 42 trees was randomly selected and measured. Do the two sets of data (36 heights, 42 heights) represent dependent or independent samples? Explain.

10.7 Twenty people were selected to participate in a psychology experiment. They answered a short multiple-choice quiz about their attitudes on a particular subject and then viewed a 45-minute film. The following day the same 20 people were asked to answer a follow-up questionnaire about their attitudes. At the completion of the experiment, the experimenter will have had two sets of scores. Do these two samples represent dependent or independent samples? Explain.

10.8 An experiment is designed to study the effect diet has on uric acid level. The study includes 20 white rats. Ten rats are randomly selected and given a junk-food diet; the other 10 rats receive a high-fiber, low-fat diet. Uric acid levels of the two groups are determined. Do the resulting sets of data represent dependent or independent samples? Explain.

10.9 Two different types of disc centrifuges are used to measure the particle size in latex paint. A gallon of paint is randomly selected, and 10 specimens are taken from it for testing on each of the centrifuges. There will be two sets of data, 10 data values each, as a result of the testing. Do the two sets of data represent dependent or independent samples? Explain.

10.10 An insurance company is concerned that garage A charges more for repair work than garage B charges. It plans to send 25 cars to each garage and obtain separate estimates for the repairs needed for each car.

a. How can the company do this and obtain independent samples? Explain in detail.

b. How can the company do this and obtain dependent samples? Explain in detail.

[EX004-000] identifies the filename of an exercise's online dataset—available through cengagebrain.com

10.11 A study is being designed to determine the reasons why adults choose to follow a healthy diet plan. The study will survey 1000 men and 1000 women. Upon completion of the study, the reasons men choose a healthy diet will be compared with the reasons women choose a healthy diet.

a. How can the data be collected if independent samples are to be obtained? Explain in detail.

b. How can the data be collected if dependent samples are to be obtained? Explain in detail.

10.12 With all the media attention from the *Dancing With the Stars* program, local dance studios are seeing an upswing in the numbers of people interested in taking ballroom dancing lessons. Two samples of 15 students are to be judged before taking any lessons and then again after five lessons. The students belonging to the samples are to be randomly selected.

a. How can data be collected if dependent samples are to be obtained? Explain in detail.

b. How can data be collected if independent samples are to be obtained? Explain in detail.

10.2 Inferences Concerning the Mean Difference Using Two Dependent Samples

The procedures for comparing two population means are based on the relationship between two sets of sample data, one sample from each population. When dependent samples are involved, the data are thought of as "paired data." The data may be paired as a result of being obtained from "before" and "after" studies; from pairs of identical twins as in Applied Example 10.3; from a "common" source, as with the amounts of tire wear for each brand in plan D of Example 10.2; or from matching two subjects with similar traits to form "matched pairs." The pairs of data values are compared directly to each other by using the difference in their numerical values. The resulting difference is called a **paired difference**.

> **Paired Difference**
>
> $$d = x_1 - x_2 \qquad (10.1)$$

Using paired data this way has a built-in ability to remove the effect of otherwise uncontrolled factors. The tire-wear problem in Example 10.2 is an excellent example of such additional factors. The wearing ability of a tire is greatly affected by a multitude of factors: the size, weight, age, and condition of the car; the driving habits of the driver; the number of miles driven; the condition and types of roads driven on; the quality of the material used to make the tire; and so on. We create paired data by mounting one tire from each brand on the same car. Since one tire of each brand will be tested under the same conditions, using the same car, same driver, and so on, the extraneous causes of wear are neutralized.

Procedures and Assumptions for Inferences Involving Paired Data

A test was conducted to compare the wearing quality of the tires produced by two tire companies, using plan D as described in Example 10.2. All the aforementioned factors had an equal effect on both brands of tires, car by car. One tire of each brand was placed on each of six test cars. The position (left or right side, front or back) was determined with the aid of a random-number table. Table 10.1 lists the amounts of wear (in thousandths of an inch) that resulted from the test.

TABLE 10.1 Amount of Tire Wear [TA10-01]

Car	1	2	3	4	5	6
Brand A	125	64	94	38	90	106
Brand B	133	65	103	37	102	115

Since the various cars, drivers, and conditions were the same for each tire of a paired set of data, it makes sense to use a third variable, the paired difference *d*. Our two dependent samples of data may be combined into one set of *d* values, where $d = \text{B} - \text{A}$.

Car	1	2	3	4	5	6
$d = \text{B} - \text{A}$	8	1	9	−1	12	9

The difference between the two population means, when dependent samples are used (often called **dependent means**), is equivalent to the **mean of the paired differences**. Therefore, when an inference is to be made about the difference of two means and paired differences are used, the inference will in fact be about the mean of the paired differences. The sample mean of the paired differences will be used as the point estimate for these inferences.

In order to make inferences about the mean of all possible paired differences μ_d, we need to know the *sampling distribution* of $\bar{d}$.

> When paired observations are randomly selected from normal populations, the paired difference, $d = x_1 - x_2$, will be approximately normally distributed about a mean μ_d with a standard deviation of σ_d.

This is another situation in which the *t*-test for one mean is applied; namely, we wish to make inferences about an unknown mean (μ_d) where the random variable (*d*) involved has an approximately normal distribution with an unknown standard deviation (σ_d).

Inferences about the mean of all possible paired differences μ_d are based on samples of *n* dependent pairs of data and the ***t*-distribution** with $n - 1$ degrees of freedom (df), under the following assumption:

> **Assumption for inferences about the mean of paired differences** μ_d The paired data are randomly selected from normally distributed populations.

Confidence Interval Procedure

The $1 - \alpha$ **confidence interval for estimating the mean difference** μ_d is found using this formula:

FYI Formula (10.2) is an adaptation of formula (9.1).

Confidence Interval for Mean Difference (Dependent Samples)

$$\bar{d} - t(\text{df}, \alpha/2) \cdot \frac{s_d}{\sqrt{n}} \quad \text{to} \quad \bar{d} + t(\text{df}, \alpha/2) \cdot \frac{s_d}{\sqrt{n}}, \text{where df} = n - 1 \qquad \textbf{10.2)}$$

where $\overline{d}$ is the mean of the sample differences:

$$\overline{d} = \frac{\Sigma d}{n} \qquad (10.3)$$

FYI Formulas (10.3) and (10.4) are adaptations of formulas (2.1) and (2.9).

and s_d is the standard deviation of the sample differences:

$$s_d = \sqrt{\frac{\Sigma d^2 - \frac{(\Sigma d)^2}{n}}{n-1}} \qquad (10.4)$$

EXAMPLE 10.4

CONSTRUCTING A CONFIDENCE INTERVAL FOR μ_d

Construct the 95% confidence interval for the mean difference in the paired data on tire wear, as reported in Table 10.1. The sample information is $n = 6$ pieces of paired data, $\overline{d} = 6.3$, and $s_d = 5.1$. Assume the amounts of wear are approximately normally distributed for both brands of tires.

Solution

Step 1 **Parameter of interest:** μ_d, the mean difference in the amounts of wear between the two brands of tires

Step 2 **a. Assumptions:** Both sampled populations are approximately normal.
b. Probability distribution: The t-distribution with df $= 6 - 1 = 5$ and formula (10.2) will be used.
c. Level of confidence: $1 - \alpha = 0.95$

Step 3 **Sample information:** $n = 6$, $\overline{d} = 6.3$, and $s_d = 5.1$
The mean:

$$\overline{d} = \frac{\Sigma d}{n}: \qquad \overline{d} = \frac{38}{6} = 6.333 = \mathbf{6.3}$$

The standard deviation:

$$s_d = \sqrt{\frac{\Sigma d^2 - \frac{(\Sigma d)^2}{n}}{n-1}}: \qquad s_d = \sqrt{\frac{372 - \frac{(38)^2}{6}}{6-1}} = \sqrt{26.27} = 5.13 = \mathbf{5.1}$$

Step 4 **a. Confidence coefficient:**
This is a two-tailed situation with $\alpha/2 = 0.025$ in one tail. From Table 6 in Appendix B, $t(df, \alpha/2) = t(5,0.025) = 2.57$.
b. Maximum error of estimate: Using the maximum error part of formula (10.2), we have

$$E = t(df, \alpha/2) \cdot \frac{s_d}{\sqrt{n}}: \qquad E = 2.57 \cdot \left(\frac{5.1}{\sqrt{6}}\right) = (2.57)(2.082) = 5.351 = \mathbf{5.4}$$

For specific instructions about confidence coefficients and Table 6, see pages 415–416.

   Video tutorial available—logon and learn more at cengagebrain.com

       c. **Lower/upper confidence limits:**

$$\bar{d} \pm E$$
$$6.3 \pm 5.4$$
$$6.3 - 5.4 = \textbf{0.9} \quad \text{to} \quad 6.3 + 5.4 = \textbf{11.7}$$

Step 5   a. **Confidence interval:** 0.9 to 11.7 is the 95% confidence interval for μ_d.

        b. That is, with 95% confidence we can say that the mean difference in the amounts of wear is between 0.9 and 11.7 thousandths of an inch. Or, in other words, the population mean tire wear for brand B is between 0.9 and 11.7 thousandths of an inch greater than the population mean tire wear for brand A.

Note: This confidence interval is quite wide, in part because of the small sample size. Recall from the central limit theorem that as the sample size increases, the standard error (estimated by $s_d/\sqrt{n}$) decreases.

TECHNOLOGY INSTRUCTIONS: $1 - \alpha$ CONFIDENCE INTERVAL FOR MEAN μ_d WITH UNKNOWN STANDARD DEVIATION FOR TWO DEPENDENT SETS OF SAMPLE DATA

MINITAB

Input the paired data into C1 and C2; then continue with:

Choose: **Stat > Basic Statistics > Paired t**
Select: **Samples in columns**
Enter: First sample: **C1***
             Second sample: **C2**
Select: **Options**
Enter: Confidence level: **1 − α** (ex. 0.95 or 95.0)
Select: **Alternative: not equal > OK > OK**

*Paired *t* evaluates the first sample minus the second sample.

Excel

Input the paired data into columns A and B; activate C1 or C2 (depending on whether column headings are used or not); then continue with:

Enter: **= A2 − B2*** (if column headings are used)
Drag: **Bottom right corner of C2 down to give other differences**
Choose: **Add-Ins > Data Analysis Plus > t-Estimate: Mean**
Enter: Input range: (**C2:C20** or select cells)
Select: **Labels** (if necessary)
Enter: Alpha: **α** (ex. 0.05) **> OK**

*Enter the expression in the order that is needed: A2 − B2 or B2 − A2.

TI-83/84 Plus	Input the paired data into L1 and L2; then continue with the following, entering the appropriate values and highlighting Calculate:

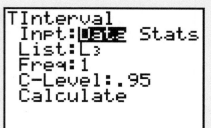

Highlight: **L3**
Enter: **L3 = L1 − L2***
Choose: **STAT > TESTS > 8:TInterval**

*Enter the expression in the order that is needed: L1 − L2 or L2 − L1.

The solution to Example 10.4 looks like this when solved in MINITAB:

Paired T for Brand B − Brand A

	N	Mean	StDev	SE Mean
Brand B	6	92.5	35.2	14.4
Brand A	6	86.2	30.9	12.6
Difference	6	6.33	5.13	2.09

95% CI for mean difference: (0.95, 11.71)

Hypothesis-Testing Procedure

When we test a null **hypothesis about the mean difference**, the test statistic used will be the difference between the sample mean $\overline{d}$ and the hypothesized value of μ_d, divided by the estimated **standard error**. This statistic is assumed to have a t-distribution when the null hypothesis is true and the assumptions for the test are satisfied. The value of the **test statistic $t\star$** is calculated as follows:

FYI Formula (10.5) is an adaptation of formula (9.2).

Test Statistic for Mean Difference (Dependent Samples)

$$t\star = \frac{\overline{d} - \mu_d}{s_d/\sqrt{n}} \text{, where df} = n - 1 \qquad (10.5)$$

Note: A hypothesized mean difference, μ_d, can be any specified value. The most common value specified is zero; however, the difference can be nonzero.

E X A M P L E   1 0 . 5

ONE-TAILED HYPOTHESIS TEST FOR μ_d

In a study on high blood pressure and the drugs used to control it, the effect of calcium channel blockers on pulse rate was one of many specific concerns. Twenty-six patients were randomly selected from a large pool of potential subjects, and their pulse rates were recorded. A calcium channel blocker was administered to each patient for a fixed period of time, and then each patient's pulse rate was again determined. The two resulting sets of data appeared to have approximately normal distributions, and the statistics were $\overline{d} = 1.07$ and $s_d = 1.74$ (d = before − after). Does the sample information provide sufficient evidence to show that the pulse rate is lower after the medication is taken? Use $\alpha = 0.05$.

Video tutorial available—logon and learn more at cengagebrain.com

Solution

FYI "Lower rate" means that "after" is less than "before" and "before − after" is positive.

Step 1   **a. Parameter of interest:** μ_d, the mean difference (reduction) in pulse rate from before to after using the calcium channel blocker for the time period of the test

   b. Statement of hypotheses:

   H_o: $\mu_d = 0$ (≤) (did not lower rate) Remember: d = before − after

   H_a: $\mu_d > 0$ (did lower rate)

Step 2   **a. Assumptions:** Since the data in both sets are approximately normal, it seems reasonable to assume that the two populations are approximately normally distributed.

   b. Test statistic: The t-distribution with df = $n - 1 = 25$, and the test statistic is $t\star$ from formula (10.5).

   c. Level of significance: $\alpha = 0.05$

Step 3   **a. Sample information:** $n = 26$, $\bar{d} = 1.07$, and $s_d = 1.74$

   b. Calculated test statistic:

$$t\star = \frac{\bar{d} - \mu_d}{s_d/\sqrt{n}} : \quad t\star = \frac{1.07 - 0.0}{1.74/\sqrt{26}} = \frac{1.07}{0.34} = \mathbf{3.14}$$

Step 4 The Probability Distribution:

p-**Value:**   **OR**   **Classical:**

a. Use the right-hand tail because H_a expresses concern for values related to "greater than." $\mathbf{P} = P(t\star > 3.14$, with df = 25), as shown in the figure.

To find the *p*-value, you have three options:
1. Use Table 6 (Appendix B): **P < 0.005.**
2. Use Table 7 (Appendix B) to read the value directly: **P = 0.002.**
3. Use a computer or calculator to find the *p*-value: **P = 0.0022.**

Specific instructions are on pages 421–422.

b. The *p*-value is smaller than the level of significance, α.

a. The critical region is the right-hand tail because H_a expresses concern for values related to "greater than." The critical value is obtained from Table 6: $t(25, 0.05) = \mathbf{1.71}$.

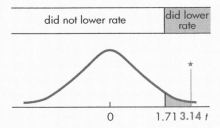

Specific instructions are on pages 415–416.

b. $t\star$ is in the critical region, as shown in **red** in the figure.

Step 5   **a. Decision:** Reject H_o.

   b. Conclusion: At the 0.05 level of significance, we can conclude that the average pulse rate is lower after the administration of the calcium channel blocker.

Statistical significance does not always have the same meaning when the "practical" application of the results is considered. In the preceding detailed hypothesis test, the results showed a statistical significance with a *p*-value of 0.002—that is, 2 chances in 1000. However, a more practical question might be: "Is lowering the pulse rate by this small average amount, estimated to be 1.07 beats per minute, worth the risks of possible side effects of this medication?" Actually, the whole issue is much broader than just this one issue of pulse rate.

TECHNOLOGY INSTRUCTIONS: HYPOTHESIS TEST FOR THE MEAN μ_d WITH UNKNOWN STANDARD DEVIATION FOR TWO DEPENDENT SETS OF SAMPLE DATA

MINITAB

Input the paired data into C1 and C2; then continue with:

Choose: **Stat > Basic Statistics > Paired t**
Select: Samples in columns
Enter: First sample: C1*
Second sample: C2
Select: **Options**
Enter: Test mean: 0.0 or μ_d
Select: Alternative: **less than** or **not equal** or **greater than > OK > OK**

*Paired *t* evaluates the first sample minus the second sample.

Excel

Input the paired data into columns A and B; then continue with:

Choose: **Add-Ins > Data Analysis > t-Test: Paired Two Sample for Means**
Enter: **Variable 1 Range: (A1:A20 or select cells)**
Variable 2 Range: (B1:B20 or select cells)
(subtracts: Var1 – Var2)
Hypothesized Mean Difference: μ_d (usually 0)
Select: **Labels** (if necessary)
Enter: α (ex. 0.05)
Select: **Output Range**
Enter: **(C1 or select cell) > OK**

Use **Home > Cells > Format > Autofit Column Width** to make the output more readable. The output shows *p*-values and critical values for one- and two-tailed tests. The hypothesis test may also be done by first subtracting the two columns and then using the inference about a mean (sigma unknown) commands on page 425 on the differences.

TI-83/84 Plus

Input the paired data into L1 and L2; then continue with the following, entering the appropriate values and highlighting Calculate:

Highlight: L3
Enter: **L3 = L1 − L2***
Choose: **STAT > TESTS > 2:T-Test . . .**

*Enter the expression in the order that is needed: L1 − L2 or L2 − L1.

```
T-Test
 Inpt:Data Stats
 μo:0
 List:L3
 Freq:1
 μ:≠μo <μo >μo
 Calculate Draw
```

The solution to Example 10.5 looks like this when solved in MINITAB:

Paired T for Before – After

	N	Mean	StDev	SE Mean
Difference	26	1.07	1.74	0.34

T-Test of mean difference = 0 (vs > 0): T-Value = 3.14
P-Value = 0.002

E X A M P L E   1 0 . 6

TWO-TAILED HYPOTHESIS TEST FOR μ_d

Suppose the sample data in Table 10.1 (p. 483) were collected with the hope of showing that the two tire brands do not wear equally. Do the data provide sufficient evidence for us to conclude that the two brands show unequal wear, at the 0.05 level of significance? Assume the amounts of wear are approximately normally distributed for both brands of tires.

Solution

Step 1   **a. Parameter of interest:** μ_d, the mean difference in the amounts of wear between the two brands

b. Statement of hypotheses:

H_o: $\mu_d = 0$ (no difference) Remember: $d = B - A$.
H_a: $\mu_d \neq 0$ (difference)

Step 2   **a. Assumptions:** The assumption of normality is included in the statement of this problem.

b. Test statistic: The t-distribution with df = $n - 1 = 6 - 1 = 5$ and $t\star = (\bar{d} - \mu_d)/(s_d/\sqrt{n})$

c. Level of significance: $\alpha = 0.05$

Step 3   **a. Sample information:** $n = 6$, $\bar{d} = 6.3$, and $s_d = 5.1$

b. Calculated test statistic:

$$t\star = \frac{\bar{d} - \mu_d}{s_d/\sqrt{n}}: \quad t\star = \frac{6.3 - 0.0}{5.1/\sqrt{6}} = \frac{6.3}{2.08} = \mathbf{3.03}$$

Step 4 Probability Distribution:

p-Value: (OR)   **Classical:**

a. Use both tails because H_a expresses concern for values related to "different from."

$\mathbf{P} = p\text{-value} = P(t\star < -3.03) + P(t\star > 3.03)$
$= 2 \times P(t\star > 3.03)$, as shown in the figure

a. The critical region is two-tailed because H_a expresses concern for values related to "different from." The critical value is obtained from Table 6: $t(5, 0.025) = \mathbf{2.57}$.

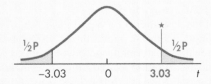

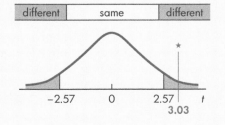

To find the p-value, you have three options:
1.   Use Table 6 (Appendix B): **0.02 < P < 0.05**.

2. Use Table 7 (Appendix B) to place bounds on the p-value: $\mathbf{0.026 < P < 0.030}$.

3. Use a computer or calculator to find the p-value: $\mathbf{P = 2 \times 0.0145 = 0.0290}$.

For specific instructions, see page 424.

b. The p-value is smaller than α.

For specific instructions, see pages 415–416.

b. $t\star$ is in the critical region, as shown in **red** in the figure.

Step 5   **a. Decision:** Reject H_o.

b. Conclusion: There is a significant mean difference in the amounts of wear at the 0.05 level of significance.

APPLIED EXAMPLE 10.7

TESTING ASPHALT SAMPLING PROCEDURES

This Application is an excerpt from a Florida Department of Transportation research report.

COMPARISON OF THE SCOOPING VS. QUARTERING METHODS FOR OBTAINING ASPHALT MIXTURE SAMPLES

Research Report FL/DOT/SMO/00-441
Gregory A. Sholar James A. Musselman Gale C. Page
State Materials Office

ABSTRACT - The standard method of quartering plant produced asphalt mix to obtain samples for maximum specific gravity, gradation and asphalt binder content has been used by the Florida Department of Transportation (FDOT), contractors and independent testing laboratories for many years with great success. This report examines an alternative method for obtaining samples that is somewhat easier and less time consuming than the traditional quartering method. This method, hereafter referred to as the "scooping" method, involves some of the same procedures and techniques that are used with the quartering method. The principle difference is that samples are scooped from the pile of asphalt mix until the desired sample weight is obtained instead of quartering the pile down until the desired sample weight is obtained. Twelve different mixtures were sampled for this study and the following mixture properties were compared for the two different sampling methods: bulk density, maximum specific gravity, % air voids, asphalt binder content and gradation. Analysis of the data indicates that the two sampling methods provide statistically equivalent results for the aforementioned mixture properties. Included in this report is a new version of FM 1-T 168, "Sampling Bituminous Paving Mixtures," which encompasses this new method for sampling asphalt mixtures.

DATA ANALYSIS - Theoretically, if the two sampling methods were identical then the average difference between values obtained for any asphalt property (ex., asphalt binder content) for a particular mix would be zero. A paired difference analysis was performed for each property measured. A paired difference analysis is a t-test performed on the differences between each sampling method.

A 95% confidence interval was used, i.e. $\alpha = 0.05$, to calculate the two-sided t-critical value. The null hypothesis is that the average difference is zero. If t-calculated is less than t-critical, then the null hypothesis cannot be rejected. In the t-test summaries, the important values are the "t-calculated" and the "t-critical" values. For simplicity, all of these "t" values have been summarized

in Table 14. Examination of the statistical results indicate that for all of the properties measured, except for % passing the No. 4 sieve, the null hypothesis cannot be rejected. This indicates that the two methods are statistically equivalent. The one exception is for the % passing the No. 4 sieve. The t-calculated and t-critical values were nearly identical (2.224 vs. 2.228).

TABLE 14
Summary of Paired Difference Analysis

Asphalt Mixture Property	t-Calculated	Absolute Value t-Critical	t-calc. < t-crit.?
Gmb (Nmax)	1.442	2.306	YES
Gmm	0.802	2.201	YES
% Air Voids	1.719	2.306	YES
% AC (ignition)	0.534	2.201	YES
Sieve Size			
1/2"	0.672	2.228	YES
3/8"	0.783	2.228	YES
No. 4	2.224	2.228	Equal
No. 8	1.819	2.228	YES
No. 16	1.047	2.228	YES
No. 30	0.814	2.228	YES
No. 50	0.753	2.228	YES
No. 100	0.387	2.228	YES
No. 200	0.305	2.228	YES

CONCLUSION-Based on the statistical analysis of the data, the two methods of sampling are equivalent with respect to Gmb, Gmm, asphalt binder content and gradation. Since the scooping method is easier and faster it is recommended that the revised Florida method for sampling (FM 1-T 168) be accepted and implemented statewide.

SECTION 10.2 EXERCISES

10.13 Given this set of paired data:

Pairs	1	2	3	4	5
Sample A	3	6	1	4	7
Sample B	2	5	1	2	8

Find:

a. The paired differences, $d = A - B$, for this set of data

b. The mean $\overline{d}$ of the paired differences

c. The standard deviation s_d of the paired differences

10.14 Find $t(15, 0.025)$. Describe the role this number plays when forming a confidence interval for the mean difference.

10.15 a. Find the 95% confidence interval for μ_d given $n = 26$, $\overline{d} = 6.3$, and $s_d = 5.1$. Assume the data are randomly selected from a normal population.

b. Compare your interval to the interval found in Example 10.4 (p. 484).

10.16 [EX10-016] All students who enroll in a certain memory course are given a pre-test before the course

(continue on page 492)

begins. At the completion of the course, 10 students are selected at random and given a post-test; their scores are listed here.

Student	1	2	3	4	5	6	7	8	9	10
Before	93	86	72	54	92	65	80	81	62	73
After	98	92	80	62	91	78	89	78	71	80

MINITAB was used to find the 95% confidence interval for the mean improvement in memory resulting from taking the memory course, as measured by the difference in test scores (d = after − before). Verify the results shown on the output by calculating the values yourself. Assume normality.

Confidence Intervals

Variable	N	Mean	StDev	SE Mean	95% C.I.
C3	10	6.10	4.79	1.52	(2.67, 9.53)

10.17 [EX10-017] Ten subjects with borderline-high cholesterol levels were randomly recruited for a study that involved taking a nutrition education class. Cholesterol readings were taken before the class and 3 months after the class.

Subject	1	2	3	4	5	6	7	8	9	10
Preclass	295	279	250	235	255	290	310	260	275	240
Postclass	265	266	245	240	230	230	235	250	250	215

Let d = preclass cholesterol − postclass cholesterol. Excel was used to find the 95% confidence interval for the mean amount of reduction in cholesterol readings after taking the nutrition education class. Verify the results shown on the output by calculating the values yourself. Assume normality.

t-Estimate: Mean

	$d = pre - post$
Mean	26.3
Standard deviation	24.4997
LCL	8.773924024
UCL	43.82607598

10.18 [EX10-018] Use a computer or calculator to find the 95% confidence interval for estimating μ_d based on these paired data and assuming normality:

Before	75	68	40	30	43	65
After	70	69	32	30	39	63

10.19 [EX10-019] An experiment was designed to estimate the mean difference in weight gain for pigs fed ration A as compared with those fed ration B. Eight pairs of pigs were used. The pigs within each pair were littermates. The rations were assigned at random to the two animals within each pair. The gains (in pounds) after 45 days are shown in the table at top of next column.

Litter	1	2	3	4	5	6	7	8
Ration A	65	37	40	47	49	65	53	59
Ration B	58	39	31	45	47	55	59	51

Assuming weight gain is normal, find the 95% confidence interval estimate for the mean of the differences μ_d, where d = ration A − ration B.

10.20 [EX10-020] Two men, A and B, who usually commute to work together, decide to conduct an experiment to see whether one route is faster than the other. The men believe that their driving habits are approximately the same, and therefore they decide on the following procedure. Each morning for 2 weeks, A will drive to work on one route and B will use the other route. On the first morning, A will toss a coin. If heads appear, he will use route I; if tails appear, he will use route II. On the second morning, B will toss the coin: heads, route I; tails, route II. The times, recorded to the nearest minute, are shown in the following table. Assume commute times are normal and estimate the population mean difference with a 95% confidence interval.

					Day					
Route	M	Tu	W	Th	F	M	Tu	W	Th	F
I	29	26	25	25	25	24	26	26	30	31
II	25	26	25	25	24	23	27	25	29	30

10.21 [EX10-021] An unmarried 19-year-old who has just purchased his/her own 2-year-old Honda Civic might be asking, "Why does auto insurance cost so much?" There are many reasons, according to the insurance agent, one of which is whether the driver is male or female. The insurance rates listed below are for a random sample of 16 ZIP codes within a 50-mile radius of our 19-year-old in question. The data are for a policy whose features are $500 deductible, $25,000/$50,000 bodily injury, $25,000 property, and $25,000/$50,000 uninsured/underinsured motorist.

a. At first glance, does there seem to be a pattern to the relationship between the insurance premiums for males and females? Describe it.

b. Graphically describe each set of data—males, females, and difference—using a histogram and one other graph.

c. Find the mean and the standard deviation for each set of data: males, females, and difference.

d. Are the assumptions for a mean of a paired-difference confidence interval satisfied? Explain.

Table for Exercise 10.21

Male ($)	1215.30	996.30	1179.30	1254.30	1110.30	2086.60	856.30	1298.30
Female ($)	1015.30	812.30	987.30	1045.30	916.30	1804.60	671.30	1132.30
Male ($)	760.30	956.30	1304.30	1548.30	1760.30	1337.30	1037.30	1182.30
Female ($)	606.30	771.30	1095.30	1278.30	1444.30	1095.30	812.30	940.30

e. Using a 95% confidence interval, estimate the mean of the differences. Write a complete confidence interval statement.

f. Do your answers to the above questions suggest any evidence of a difference between the auto insurance rates for male and female 19-year-old drivers? Explain.

10.22 [EX10-022] In evaluating different measuring instruments, one must first determine whether there is a systematic difference between the instruments. Lenses with several different powers were measured once each by two different instruments. The measurement differences (Instrument A – Instrument B) were recorded. The measurement units have been coded for proprietary reasons.

4	5	−2	−3	−7	10	11	−1	3	7	−5	3	−4
−5	−7	4	−1	−18	0	−17	12	9	4	17	−2	

Does there appear to be a systematic difference between the two instruments?

a. Describe the data using a histogram and one other graph.

b. Find the mean and the standard deviation.

c. Are the assumptions required for making inferences satisfied? Explain.

d. Using a 95% confidence interval, estimate the population mean of the differences.

e. Is there any evidence of a difference? Explain.

10.23 State the null hypothesis, H_o, and the alternative hypothesis, H_a, that would be used to test these claims:

a. There is an increase in the mean difference between post-test and pre-test scores.

b. Following a special training session, it is believed that the mean of the difference in performance scores will not be zero.

c. On average, there is no difference between the readings from two inspectors on each of the selected parts.

d. The mean of the differences between pre–self-esteem and post–self-esteem scores showed improvement after involvement in a college learning community.

10.24 State the null hypothesis, H_o, and the alternative hypothesis, H_a, that would be used to test these claims:

a. The mean of the differences between the post-test and the pre-test scores is greater than 15.

b. The mean weight gain, after the change in diet for the laboratory animals, is at least 10 oz.

c. The mean weight loss experienced by people on a new diet plan was no less than 12 lb.

d. The mean difference in the home reassessments from the two town assessors was no more than $200.

10.25 Determine the p-value for each hypothesis test for the mean difference.

a. $H_o: \mu_d = 0$ and $H_a: \mu_d > 0$, with $n = 20$ and $t\bigstar = 1.86$

b. $H_o: \mu_d = 0$ and $H_a: \mu_d \neq 0$, with $n = 20$ and $t\bigstar = -1.86$

c. $H_o: \mu_d = 0$ and $H_a: \mu_d < 0$, with $n = 29$ and $t\bigstar = -2.63$

d. $H_o: \mu_d = 0.75$ and $H_a: \mu_d > 0.75$, with $n = 10$ and $t\bigstar = 3.57$

10.26 Determine the test criteria that would be used with the classical approach to test the following hypotheses when t is used as the test statistic.

a. $H_o: \mu_d = 0$ and $H_a: \mu_d > 0$, with $n = 15$ and $\alpha = 0.05$

b. $H_o: \mu_d = 0$ and $H_a: \mu_d \neq 0$, with $n = 25$ and $\alpha = 0.05$

c. $H_o: \mu_d = 0$ and $H_a: \mu_d < 0$, with $n = 12$ and $\alpha = 0.10$

d. $H_o: \mu_d = 0.75$ and $H_a: \mu_d > 0.75$, with $n = 18$ and $\alpha = 0.01$

10.27 A Bloomberg News poll found that Americans plan to keep spending down over the next six months due to the uncertain economy (*USA Today*, September 17, 2009). Suppose a group of 15 households noted their household spending in March and then noted their household spending six months later in September. The mean monthly difference (former spending – current spending) was calculated to be $75.50 with a standard deviation of $66.20. Does this sample of households show sufficient evidence of increased household savings? Use the 0.05 level of significance and assume normality of spending amounts.

10.28 A random sample of 10 speed skaters, all of the relatively same experience and speed, were selected to try out a new specialty blade. The difference in the short track times were measured as current blade time – specialty blade time, resulting in mean difference of 0.165 second with a standard deviation equal to 0.12 second. Does this sample provide sufficient reason that the specialty blade is beneficial in achieving faster times? Use $\alpha = 0.05$ and assume normality.

10.29 The corrosive effects of various soils on coated and uncoated steel pipe were tested by using a dependent

(continue on page 494)

sampling plan. The data collected are summarized by $n = 40$, $\Sigma d = 220$, and $\Sigma d^2 = 6222$, where d is the amount of corrosion on the coated portion subtracted from the amount of corrosion on the uncoated portion. Does this random sample provide sufficient reason to conclude that the coating is beneficial? Use $\alpha = 0.01$ and assume normality.

a. Solve using the p-value approach.

b. Solve using the classical approach.

10.30 Does a content title help a reader comprehend a piece of writing? Twenty-six participants were given an article to read without a title. They then rated themselves on their comprehension of the information on a scale from 1 to 10, where 10 was complete comprehension. The same 26 participants were then given the article again, this time with an appropriate title, and asked to rate their comprehension. The resulting summarized data were given as $\bar{d} = 4.76$ and $s_d = 2.33$, where $d =$ rating with title – rating without title. Comprehension was generally higher on the second reading than on the first by an average of 3.2 on this scale. Does this sample provide sufficient evidence that a content title does make a difference with respect to comprehension? Use $\alpha = 0.05$.

10.31 Complete the hypothesis test with alternative hypothesis $\mu_d > 0$ based on the paired data that follow and $d = B - A$. Use $\alpha = 0.05$. Assume normality.

A	700	830	860	1080	930
B	720	820	890	1100	960

a. Solve using the p-value approach.

b. Solve using the classical approach.

10.32 Complete the hypothesis test with alternative hypothesis $\mu_d \neq 0$ based on the paired data that follow and $d = O - Y$. Use $\alpha = 0.01$. Assume normality.

Oldest	199	162	174	159	173
Youngest	194	162	167	156	176

a. Solve using the p-value approach.

b. Solve using the classical approach.

10.33 [EX10-033] Ten recently diagnosed diabetics were tested to determine whether an educational program would be effective in increasing their knowledge of diabetes. They were given a test, before and after the educational program, concerning self-care aspects of diabetes. The scores on the test were as follows:

Patient	1	2	3	4	5	6	7	8	9	10
Before	75	62	67	70	55	59	60	64	72	59
After	77	65	68	72	62	61	60	67	75	68

The following MINITAB output may be used to determine whether the scores improved as a result of the program. Verify the values shown on the output [mean difference (MEAN), standard deviation of the difference (STDEV), standard error of the difference (SE MEAN), $t\star$ (T-Value), and p-value] by calculating the values yourself.

Paired T for After – Before

	N	Mean	StDev	SE Mean
After	10	67.50	5.80	1.83
Before	10	64.30	6.50	2.06
Difference	10	3.200	2.741	0.867

T-Test of mean difference = 0 (vs > 0); T-Value = 3.69
P-Value = 0.002

10.34 [EX10-034] Ten subjects with borderline-high cholesterol levels were recruited for a study that involved taking a nutrition education class. Cholesterol readings were taken before the class and 3 months after the class.

Subject	1	2	3	4	5	6	7	8	9	10
Preclass	295	279	250	235	255	290	310	260	275	240
Postclass	265	266	245	240	230	230	235	250	250	215

Let $d =$ preclass cholesterol – postclass cholesterol. Use the following Excel output to test the null hypothesis that the population mean difference equals zero versus the alternative hypothesis that the population mean difference is positive at $\alpha = 0.05$. Rejection of the null hypothesis would indicate that the (population) average cholesterol level after the class is lower than the average level before the class. Assume normality.

t-Test: Paired Two Sample for Means

	Pre-test	Post-test
Mean	268.9	242.6
Variance	618.7666667	256.4888889
Observations	10	10
Hypothesized mean difference	0	
df	9	
t Stat	3.394655392	
P(T ≤ t) one-tail	0.003970146	
t Critical one-tail	1.833113856	

10.35 Use a computer or calculator to complete the hypothesis test with alternative hypothesis $\mu_d < 0$ based on the paired data that follow and $d = M - N$. Use $\alpha = 0.02$. Assume normality.

M	58	78	45	38	49	62
N	62	86	42	39	47	68

10.36 [EX10-036] Ten randomly selected college students who participated in a learning community were given pre–self-esteem and post–self-esteem surveys. A learning community is a group of students who take two or more courses together. Typically, each learning community has a theme, and the faculty involved coordinate assignments linking the courses. Research has shown that higher self-esteem, higher grade point averages (GPAs), and improved satisfaction in courses, as well as better retention rates, result from involvement in a learning community. The scores on the surveys are as follows:

Student	1	2	3	4	5	6	7	8	9	10
Prescore	18	14	11	23	19	21	21	21	11	22
Postscore	17	17	10	25	20	10	24	22	10	24

Does this sample of students show sufficient evidence that self-esteem scores were higher after participation in a learning community? Lower scores indicate higher self-esteem. Use the 0.05 level of significance and assume normality of scores.

10.37 [EX10-037] In reference to the college students who participated in a learning community in Exercise 10.36, a control group of students was also formed for testing and comparison. Ten randomly selected college students who were not involved in the learning community were given pre–self-esteem and post–self-esteem surveys. The scores on the surveys for the control group are as follows:

Student	1	2	3	4	5	6	7	8	9	10
Prescore	19	23	12	20	26	20	15	10	22	12
Postscore	19	21	9	10	23	20	19	10	21	19

Does this sample of students show sufficient evidence that self-esteem scores were higher after participation in a learning community? Lower scores indicate higher self-esteem. Use the 0.05 level of significance and assume normality of scores.

10.38 [EX10-038] To test the effect of a physical-fitness course on one's physical ability, the number of sit-ups that a person could do in 1 minute, both before and after the course, was recorded. Ten randomly selected participants scored as shown in the following table. Can you conclude that a significant amount of improvement took place? Use $\alpha = 0.01$ and assume normality.

Before	29	22	25	29	26	24	31	46	34	28
After	30	26	25	35	33	36	32	54	50	43

a. Solve using the p-value approach.

b. Solve using the classical approach.

10.39 Referring to Applied Example 10.7:

a. What null hypothesis is being tested in each of these 13 tests?

b. Why are the "t-calculated" and the "t-critical" values the important values?

c. Why is it correct to report their absolute values for both t-values in Table 14?

d. What decision is reached for each of these 13 hypotheses tests?

e. What conclusion is reached as a result of these tests?

f. What action is recommended to the State of Florida as a result of the conclusion?

10.40 [EX10-040] A research project was undertaken to evaluate two focimeters. Each of 20 lenses of varying powers was read once on each focimeter. The measurement differences were then calculated, where each difference was Focimeter A – Focimeter B. Assume readings are normally distributed.

-0.016	0.013	0.009	0.000	-0.005	-0.015	-0.006
-0.016	-0.022	-0.006	-0.020	0.015	-0.017	-0.010
-0.003	0.011	-0.012	0.008	-0.005	-0.009	

Source: Courtesy of Bausch & Lomb

a. Using a t-test on these paired differences and an $\alpha = 0.01$, determine whether the corresponding population mean difference is significantly different from zero.

b. Construct a 99% confidence interval for the mean difference in focimeter readings.

c. Explain what both inferential procedures indicate about the differences.

d. If an enterprising experimenter performed this same test using $\alpha = 10\%$, what would the outcome be? Offer comments about proceeding using these ground rules.

10.3 Inferences Concerning the Difference between Means Using Two Independent Samples

When comparing the means of two populations, we typically consider the difference between their means, $\mu_1 - \mu_2$ (often called "**independent means**"). The inferences about $\mu_1 - \mu_2$ will be based on the difference between the observed sample means, $\bar{x}_1 - \bar{x}_2$. This observed difference, $\bar{x}_1 - \bar{x}_2$, belongs to a sampling distribution with the characteristics described in the following statement.

If independent samples of sizes n_1 and n_2 are drawn randomly from large populations with means μ_1 and μ_2 and variances σ_1^2 and σ_2^2, respectively, then the sampling distribution of $\bar{x}_1 - \bar{x}_2$, the difference between the sample means, has

1. mean $\mu_{\bar{x}_1 - \bar{x}_2} = \mu_1 - \mu_2$ and

2. standard error $\sigma_{\bar{x}_1 - \bar{x}_2} = \sqrt{\left(\dfrac{\sigma_1^2}{n_1}\right) + \left(\dfrac{\sigma_2^2}{n_2}\right)}.$ (10.6)

If both populations have normal distributions, then the sampling distribution of $\bar{x}_1 - \bar{x}_2$ will also be normally distributed.

The preceding statement is true for all sample sizes as long as the populations involved are normal and the population variances σ_1^2 and σ_2^2 are known quantities. However, as with inferences about one mean, the variance of a population is generally an unknown quantity. Therefore, it will be necessary to estimate the standard error by replacing the variances, σ_1^2 and σ_2^2, in formula (10.6) with the best estimates available—namely, the sample variances, s_1^2 and s_2^2. The *estimated standard error* will be found using the following formula:

$$\text{estimated standard error} = \sqrt{\left(\frac{s_1^2}{n_1}\right) + \left(\frac{s_2^2}{n_2}\right)}$$ (10.7)

Inferences about the difference between two population means, $\mu_1 - \mu_2$, will be based on the following assumptions.

Assumptions for inferences about the difference between two means, $\mu_1 - \mu_2$ The samples are randomly selected from normally distributed populations, and the samples are selected in an independent manner. NO ASSUMPTIONS ARE MADE ABOUT THE POPULATION VARIANCES.

Since the samples provide the information for determining the standard error, the **t-distribution** will be used as the test statistic. The inferences are divided into two cases.

Case 1: The t-distribution will be used, and the number of degrees of freedom will be calculated.

Case 2: The t-distribution will be used, and the number of degrees of freedom will be approximated.

Case 1 will occur when you are completing the inference *using a computer or statistical calculator and the statistical software or program calculates the number of degrees of freedom* for you. The calculated value for df is a function of both sample sizes and their relative sizes, and both sample variances and their relative sizes. The value of df will be a number between the smaller of $df_1 = n_1 - 1$ or $df_2 = n_2 - 1$, and the sum of the degrees of freedom, $df_1 + df_2 = [(n_1 - 1) + (n_2 - 1)] = n_1 + n_2 - 2$.

Case 2 will occur when you are completing the inference *without the aid of a computer or calculator and its statistical software package.* Use of the t-distribution with the smaller of $df_1 = n_1 - 1$ or $df_2 = n_2 - 1$ will give *conservative* results. Because of this approximation, the true level of confidence for an interval estimate will be slightly higher than the reported level of confidence; or the true p-value and the true level of significance for a hypothesis test will be slightly less than reported. The gap between these reported values and the true values will be quite small,

only man that's ever likely to use them!" Today, Student's *t*-distribution is widely used and respected in statistical research.

unless the sample sizes are quite small and unequal or the sample variances are very different. The gap will decrease as the samples increase in size or as the sample variances become more alike.

Since the only difference between the two cases is the number of degrees of freedom used to identify the *t*-distribution involved, we will study case 2 first.

Note: $A > B$ ("*A* is greater than *B*") is equivalent to $B < A$ ("*B* is less than *A*"). When the difference between *A* and *B* is being discussed, it is customary to express the difference as "larger – smaller" so that the resulting difference is positive: $A - B > 0$. Expressing the difference as "smaller – larger" results in $B - A < 0$ (the difference is negative) and is usually unnecessarily confusing. Therefore, it is recommended that the difference be expressed as "larger – smaller."

Confidence Interval Procedure

FYI Would you say the difference between 5 and 8 is −3? How would you express the difference? Explain.

We will use the following formula to calculate the end points of the **1 − α confidence interval**.

Confidence Interval for the Difference between Two Means (Independent Samples)

$$(\bar{x}_1 - \bar{x}_2) - t(\text{df}, \alpha/2) \cdot \sqrt{\left(\frac{s_1^2}{n_1}\right) + \left(\frac{s_2^2}{n_2}\right)} \quad \text{to} \quad (\bar{x}_1 - \bar{x}_2) + t(\text{df}, \alpha/2) \cdot \sqrt{\left(\frac{s_1^2}{n_1}\right) + \left(\frac{s_2^2}{n_2}\right)} \quad \textbf{(10.8)}$$

where df is either calculated or is the smaller of df_1 or df_2 (see p. 496)

E X A M P L E   1 0 . 8

CONSTRUCTING A CONFIDENCE INTERVAL FOR THE DIFFERENCE BETWEEN TWO MEANS

The heights (in inches) of 20 randomly selected women and 30 randomly selected men were independently obtained from the student body of a certain college in order to estimate the difference in their mean heights. The sample information is given in Table 10.2. Assume that heights are approximately normally distributed for both populations.

TABLE 10.2 Sample Information on Student Heights

Sample	Number	Mean	Standard Deviation
Female (*f*)	20	63.8	2.18
Male (*m*)	30	69.8	1.92

Find the 95% confidence interval for the difference between the mean heights, $\mu_m - \mu_f$.

Solution

Step 1   **Parameter of interest:** $\mu_m - \mu_f$, the difference between the mean height of male students and the mean height of female students

---

 Video tutorial available—logon and learn more at cengagebrain.com

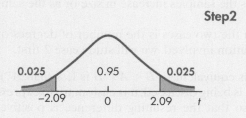

Step 2   **a. Assumptions:** Both populations are approximately normal, and the samples were randomly and independently selected.

   b. Probability distribution: The t-distribution with df = 19, the smaller of $n_m - 1 = 30 - 1 = 29$ or $n_f - 1 = 20 - 1 = 19$, and formula (10.8)

   c. Level of confidence: $1 - \alpha = 0.95$

Step 3   **Sample information:** See Table 10.2.

Step 4   **a. Confidence coefficient:** We have a two-tailed situation with $\alpha/2 = 0.025$ in one tail and df = 19. From Table 6 in Appendix B, $t(df, \alpha/2) = t(19, 0.025) = 2.09$. See the figure. See pages 415–416 for instructions on using Table 6.

   b. Maximum error of estimate: Using the maximum error part of formula (10.8), we have

$$E = t(df, \alpha/2) \cdot \sqrt{\left(\frac{s_1^2}{n_1}\right) + \left(\frac{s_2^2}{n_2}\right)}: \quad E = 2.09 \cdot \sqrt{\left(\frac{1.92^2}{30}\right) + \left(\frac{2.18^2}{20}\right)}$$

$$= (2.09)(0.60) = \mathbf{1.25}$$

   c. Lower and upper confidence limits:

$$(\bar{x}_1 - \bar{x}_2) \pm E$$
$$6.00 \pm 1.25$$
$$6.00 - 1.25 = \mathbf{4.75} \quad \text{to} \quad 6.00 + 1.25 = \mathbf{7.25}$$

Step 5   **a. Confidence interval.**

   4.75 to 7.25 is the 95% confidence interval for $\mu_m - \mu_f$.

   b. That is, with 95% confidence, we can say that the difference between the mean heights of the male and female students is between 4.75 and 7.25 inches; that is, the mean height of male students is between 4.75 and 7.25 inches greater than the mean height of female students.

Hypothesis-Testing Procedure

When we test a null **hypothesis about the difference between two population means,** the test statistic used will be the difference between the observed difference of the sample means and the hypothesized difference of the population means, divided by the estimated standard error. The test statistic is assumed to have approximately a t-distribution when the null hypothesis is true and the normality assumption has been satisfied. The calculated value of the **test statistic** is found using this formula:

Test Statistic for the Difference between Two Means (Independent Samples)

$$t\star = \frac{(\bar{x}_1 - \bar{x}_2) - (\mu_1 - \mu_2)}{\sqrt{\left(\frac{s_1^2}{n_1}\right) + \left(\frac{s_2^2}{n_2}\right)}} \tag{10.9}$$

where df is either calculated or is the smaller of df_1 or df_2 (see p. 496)

Note: A hypothesized difference between the two population means, $\mu_1 - \mu_2$, can be any specified value. The most common value specified is zero; however, the difference can be nonzero.

E X A M P L E   1 0 . 9

ONE-TAILED HYPOTHESIS TEST FOR THE DIFFERENCE BETWEEN TWO MEANS

Suppose that we are interested in comparing the academic success of college students who belong to fraternal organizations with the academic success of those who do not belong to fraternal organizations. The reason for the comparison is the recent concern that fraternity members, on average, are achieving at a lower academic level than nonfraternal students achieve. (Cumulative GPA is used to measure academic success.) Random samples of size 40 are taken from each population. The sample results are listed in Table 10.3.

TABLE 10.3 Sample Information on Academic Success

Sample	Number	Mean	Standard Deviation
Fraternity members (f)	40	2.03	0.68
Nonmembers (n)	40	2.21	0.59

Complete a hypothesis test using $\alpha = 0.05$. Assume that the GPAs for both groups are approximately normally distributed.

Solution

Step 1 **a. Parameter of interest:** $\mu_n - \mu_f$, the difference between the mean GPAs for the nonfraternity members and the fraternity members

FYI Remember: "Larger – smaller" results in a positive difference.

   b. Statement of hypotheses:

   H_o: $\mu_n - \mu_f = 0$ ($\leq$) (fraternity averages are no lower)
   H_a: $\mu_n - \mu_f > 0$ (fraternity averages are lower)

Step 2 **a. Assumptions:** Both populations are approximately normal, and random samples were selected. Since the two populations are separate, the samples are independent.

FYI When df is not in the table, use the next smaller df value.

   b. Test statistic: The t-distribution with df = the smaller of df_n or df_f; since both n's are 40, df = 40 − 1 = **39**; and $t\bigstar$ is calculated using formula (10.9).

   c. Level of significance: $\alpha = 0.05$

Step 3 **a. Sample information:** See Table 10.3.

   b. Calculated test statistic:

$$t\bigstar = \frac{(\bar{x}_1 - \bar{x}_2) - (\mu_1 - \mu_2)}{\sqrt{\left(\frac{s_1^2}{n_1}\right) + \left(\frac{s_2^2}{n_2}\right)}}: \qquad t\bigstar = \frac{(2.21 - 2.03) - (0.00)}{\sqrt{\left(\frac{0.59^2}{40}\right) + \left(\frac{0.68^2}{40}\right)}}$$

$$= \frac{0.18}{\sqrt{0.00870 + 0.01156}} = \frac{0.18}{0.1423} = \mathbf{1.26}$$

Video tutorial available—logon and learn more at cengagebrain.com

Step 4 Probability Distribution:

p-Value: **OR**   Classical:

a. Use the right-hand tail because H_a expresses concern for values related to "greater than." $\mathbf{P} = P(t\star > 1.26$, with df = 39), as shown on the figure.

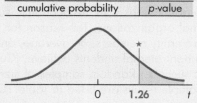

To find the *p*-value, use one of three methods:
1. Use Table 6 (Appendix B) to place bounds on the *p*-value: **0.10 < P < 0.25**.
2. Use Table 7 (Appendix B) to place bounds on the *p*-value: **0.100 < P < 0.119**.
3. Use a computer or calculator to find the *p*-value: **P = 0.1076**.

Specific details follow this example.

b. The *p*-value is not smaller than α.

a. The critical region is the right-hand tail because H_a expresses concern for values related to "greater than." The critical value is obtained from Table 6: $t(39, 0.05) = \mathbf{1.69}$.

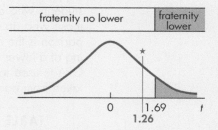

See pages 415–416 for information about critical values.

b. $t\star$ is not in the critical region, as shown in **red** on the figure.

Step 5   **a. Decision:** Fail to reject H_o.

b. Conclusion: At the 0.05 level of significance, the claim that the fraternity members achieve at a lower level than nonmembers is not supported by the sample data.

To find the *p*-value for Example 10.9, use one of three methods:

Method 1: Use Table 6. Find 1.26 between two entries in the df = 39 (use df = 35) row and read the bounds for **P** from the one-tail heading at the top of the table: **0.10 < P < 0.25**.

Method 2: Use Table 7. Find $t\star = 1.26$ between two rows and df = 39 between two columns; read the bounds for $P(t\star > 1.26 \,|\, \mathrm{df} = 39)$; **0.100 < P < 0.119**.

Method 3: If you are doing the hypothesis test with the aid of a computer or calculator, most likely it will calculate the *p*-value for you (see p. 425), or you may use the cumulative probability distribution commands described in Chapter 9 (p. 417).

E X A M P L E   1 0 . 1 0

TWO-TAILED HYPOTHESIS FOR THE DIFFERENCE BETWEEN TWO MEANS

Many students have complained that the soft drink vending machine in the student recreation room (A) dispenses a different amount of soda than the machine in the faculty lounge (B). To test this belief, a student randomly sampled several servings from each machine and carefully measured them, with the results shown in Table 10.4.

TABLE 10.4 Sample Information on Vending Machines

Machine	Number	Mean	Standard Deviation
A	10	5.38	1.59
B	12	5.92	0.83

Does this evidence support the hypothesis that the mean amount dispensed by machine A is different from the mean amount dispensed by machine B? Assume the amounts dispensed by both machines are normally distributed, and complete the test using $\alpha = 0.10$.

Solution

Step 1 **a. Parameter of interest:** $\mu_B - \mu_A$, the difference between the mean amount dispensed by machine B and the mean amount dispensed by machine A

FYI "Larger−smaller" results in a positive difference.

 b. Statement of hypotheses:

H_o: $\mu_B - \mu_A = 0$ (A dispenses a same average amount as B)

H_a: $\mu_B - \mu_A \neq 0$ (A dispenses a different average amount than B)

Step 2 **a. Assumptions:** Both populations are assumed to be approximately normal, and the samples were randomly and independently selected.

 b. Test statistic: The t-distribution with df = the smaller of $n_A - 1 = 10 - 1 = 9$ or $n_B - 1 = 12 - 1 = 11$, df = 9, and $t\bigstar$ calculated using formula (10.9)

 c. Level of significance: $\alpha = 0.10$

Step 3 **a. Sample information:** See Table 10.4.

 b. Calculated test statistic:

$$t\bigstar = \frac{(\bar{x}_B - \bar{x}_A) - (\mu_B - \mu_A)}{\sqrt{\left(\dfrac{s_B^2}{n_B}\right) + \left(\dfrac{s_A^2}{n_A}\right)}}: \quad t\bigstar = \frac{(5.92 - 5.38) - (0.00)}{\sqrt{\left(\dfrac{0.83^2}{12}\right) + \left(\dfrac{1.59^2}{10}\right)}}$$

$$= \frac{0.54}{\sqrt{0.0574 + 0.2528}} = \frac{0.54}{0.557} = \mathbf{0.97}$$

Step 4 Probability Distribution:

p-Value:	OR	Classical:

a. Use both tails because H_a expresses concern for values related to "different than."

$\mathbf{P} = p\text{-value} = P(t\bigstar < -0.97) + P(t\bigstar > 0.97)$
$= 2 \times P(t\bigstar > 0.97 \,|\, \text{df} = 9)$ as in the figure.

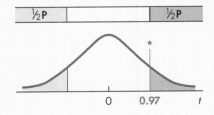

a. The critical region is two-tailed because H_a expresses concern for values related to "different than." The right-hand critical value is obtained from Table 6: $t(9, 0.05) = \mathbf{1.83}$. See the figure.

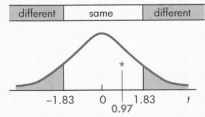

To find the *p*-value, you have three options:
1. Use Table 6 (Appendix B): **0.20 < P < 0.50**.
2. Use Table 7 (Appendix B) to place bounds on the *p*-value: **0.340 < P < 0.394**.
3. Use a computer or calculator to find the *p*-value: **P** = 2 × 0.1787 = **0.3574**.

Specific instructions follow this example.

b. The *p*-value is not smaller than α.

For specific instructions, see pages 415–416.
b. $t\star$ is not in the critical region, as shown in **red** on the figure.

Step 5   **a. Decision:** Fail to reject H_o.

b. Conclusion: The evidence is not sufficient to show that machine A dispenses a different average amount of soft drink than machine B, at the 0.10 level of significance. Thus, for lack of evidence we will proceed as though the two machines dispense, on average, the same amount.

To find the *p*-value for Example 10.10, use one of three methods:

Method 1: Use Table 6. Find 0.97 between two entries in the df = 9 row and read the bounds for **P** from the two-tail heading at the top of the table: **0.20 < P < 0.50**.

Method 2: Use Table 7. Find $t\star$ = 0.97 between two rows and df = 9 between two columns; read the bounds for $P(t\star > 0.97 | df = 9)$: $0.170 < \frac{1}{2} P < 0.197$; therefore, **0.340 < P < 0.394**.

Method 3: If you are doing the hypothesis test with the aid of a computer or calculator, most likely it will calculate the *p*-value (do not double) for you (see p. 425), or you may use the cumulative probability distribution commands described in Chapter 9 (p. 417).

Most computer or calculator statistical packages will complete the inferences for the difference between two means by calculating the number of degrees of freedom.

TECHNOLOGY INSTRUCTIONS: HYPOTHESIS TEST FOR THE DIFFERENCE BETWEEN TWO POPULATION MEANS WITH UNKNOWN STANDARD DEVIATION GIVEN TWO INDEPENDENT SETS OF SAMPLE DATA

MINITAB

MINITAB's 2-Sample *t* (Test and Confidence Interval) command performs both the confidence interval and the hypothesis test at the same time.
Input the two independent sets of data into C1 and C2; then continue with:

Choose: **Stat > Basic Statistics > 2-Sample t**
Select: **Samples in different columns***
Enter: First: **C1**   Second: **C2**
Select: **Assume equal variances** (if known)
Select: **Options**
Enter: Confidence level: **1 − α** (ex. 0.95 or 95.0)
          Test mean: **0.0**
Choose: Alternative: **less than** or **not equal** or **greater than** > **OK** > **OK**

*Note the other possible data formats.

| **Excel** | Input the two independent sets of data into columns A and B; then continue with: |

Choose: **Data > Data Analysis > t-Test: Two-Sample Assuming Unequal Variances**
Enter: Variable 1 Range: **(A1:A20 or select cells)**
       Variable 2 Range: **(B1:B20 or select cells)**
       Hypothesized Mean Difference: $\mu_B - \mu_A$ (usually 0)
Select: **Labels** (if necessary)
Enter: α (ex. 0.05)
Select: **Output Range**
Enter: **(C1 or select cell) > OK**

Use **Home > Cells > Format > AutoFit Column Width** to make the output more readable. The output shows *p*-values and critical values for one- and two-tailed tests.

| **TI-83/84 Plus** | Input the two independent sets of data into L1 and L2.* |

To construct a $1 - \alpha$ confidence interval for the mean difference, continue with the following, entering the appropriate values and highlighting Calculate:

Choose: **STAT > TESTS > 0:2-SampTInt …**

To complete a hypothesis test for the mean difference, continue with the following, entering the appropriate values and highlighting Calculate:

Choose: **STAT > TESTS > 4:2-SampTTest …**

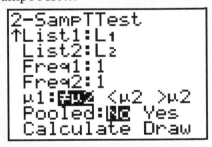

*Enter the data in the order that is needed; the program subtracts as L1 − L2.
Highlight No for Pooled if there are no assumptions about the equality of variances.

Example 10.9 was solved using MINITAB. With 40 cumulative GPAs for non-members in C1 and 40 averages for fraternity members in C2, the preceding commands resulted in the output shown here. Compare these results to the solution of Example 10.9. Notice the difference in **P** and df values. Explain.

Two-Sample T-Test and CI

Sample	N	Mean	StDev	SE Mean
1	40	2.210	0.590	0.093
2	40	2.030	0.680	0.11

Difference = mu(1) − mu(2) Est. diff.: 0.180
95% CI for difference: (−0.10, 0.46)
T-Test diff. = 0 (vs >): T = 1.26 P = 0.105 DF = 76

APPLIED EXAMPLE 10.11

POLISHING A MICROCHIP

Image copyright Joris van den Heuvel, 2012. Used under license from Shutterstock.com

Raul is developing a new technique for polishing the reflective surface of a silicone microchip. This microchip will be used with a laser as part of his research project. The roughness of the surface is measured by the distance, x, between the surface and the plane of the "highest" points on the surface, and is measured in nanometers (nms). See the figure below. (One nanometer is one-billionth of a meter.)

Larger values of this distance, surface height, x, along with a large standard deviation indicate a rougher surface. Typically x ranges in value from 4 to 20 nanometers. To put this into perspective, the human eye cannot see 20 nanometers. Dataset A is a set of measurements taken at random locations on the unpolished surface.

Dataset A (unpolished): Surface Height, x (nms) [EX10-077]

8.651	11.849	7.708	8.184	7.978	4.339	9.194	9.182
5.202	6.309	10.588	8.106	9.877	7.038	9.748	12.049
8.497	7.953	5.641	4.073	7.437	14.824	11.943	8.353
14.730	9.933	7.101	18.570	4.684	8.546	5.216	8.271
10.327	9.748	12.452					

Raul's goal is to make the surface smoother and to statistically show that his new technique does in fact make the surface significantly smoother. This is no easy task, as the microchip is less than 0.25-inch square, and it is thinner than a human hair.

Polished, smoother surface
Plane of highest points
Surface being measured

Dataset B is a set of measurements taken at random locations on the polished surface after the new process had been applied.

Dataset B (polished): Surface Height, x (nms) [EX10-077]

2.077	3.096	2.110	2.264	2.039	2.437	2.181	2.510	2.354
1.732	2.120	2.545	2.054	1.562	2.231	1.480	1.775	2.230
1.465	1.548	1.979	1.993	2.263	1.913	2.177	2.201	2.861
3.241	2.183	1.639	2.342	1.428				

Does it appear that Raul has accomplished his goal? Investigate this question in Exercises 10.77 and 10.78.

DID YOU KNOW

A **nanometer** is a metric unit used to measure things that are very small like atoms and molecules, the smallest pieces of everything around us. It is a unit of measure like inches, feet, and miles—just a whole lot smaller.

1 meter is approximately 39 inches
1 millimeter is 0.001 meter or 10^{-3} m
1 micrometer is 0.000001 meter or 10^{-6} m
1 nanometer is 0.000000001 meter or 10^{-9} m
1 angstrom is 0.0000000001 meter or 10^{-10} m

10.41 Two independent random samples resulted in the following:

Sample 1: $n_1 = 12$, $s_1^2 = 190$
Sample 2: $n_2 = 18$, $s_2^2 = 150$

Find the estimate for the standard error for the difference between two means.

10.42 Two independent random samples resulted in the following:

Sample A: $n_A = 24$, $s_A = 8.5$
Sample B: $n_B = 21$, $s_B = 11.3$

Find the estimate for the standard error for the difference between two means.

10.43 Two independent random samples of sizes 18 and 24 were obtained to make inferences about the difference between two means. What is the number of degrees of freedom? Discuss both cases.

10.44 Find the confidence coefficient, $t(df, \frac{\alpha}{2})$, that would be used to find the maximum error for each of the following situations when estimating the difference between two means, $\mu_1 - \mu_2$.

a. $1 - \alpha = 0.95$, $n_1 = 25$, $n_2 = 15$

b. $1 - \alpha = 0.98$, $n_1 = 43$, $n_2 = 32$

c. $1 - \alpha = 0.99$, $n_1 = 19$, $n_2 = 45$

10.45 Find the 90% confidence interval for the difference between two means based on this information about two samples. Assume independent samples from normal populations.

Sample	Number	Mean	Std. Dev.
1	20	35	22
2	15	30	16

10.46 A study comparing attitudes toward death was conducted in which organ donors (individuals who had signed organ donor cards) were compared with nondonors. The study is reported in the journal *Death Studies*. Templer's Death Anxiety Scale (DAS) was administered to both groups. On this scale, high scores indicate high anxiety concerning death. The results were reported as follows.

	n	Mean	Std. Dev.
Organ Donors	25	5.36	2.91
Nonorgan Donors	69	7.62	3.45

Construct the 95% confidence interval for the difference between the means, $\mu_{non} - \mu_{donor}$.

FYI Results obtained may be noticeably different depending on use of the calculated df or use of the df for a smaller sample.

10.47 The article "Car rental prices can change in a heartbeat" (*USA Today*, March 14, 2007) reported that rates for car rentals can change many times during a day. The national average rate for the January–March quarter of 2007 was $52.71, yet in some cities, car rentals can cost more than $100 a day. A similar study of two major cities found the following results:

City	n	Average Daily Rate	Standard Deviation
Boston	10	95.94	7.50
New York City	16	127.75	15.83

Set a 95% confidence interval on the difference in average daily rates between the two major East Coast cities of Boston and New York City. Assume normality for the sampled populations and that the samples were selected randomly.

10.48 Women on average have 8 more pairs of shoes than men, according to a *USA Today* Snapshot titled "Who owns more shoes?" (July 8, 2009). A recent study at a community college gave the following results:

	n	Mean	Std. Dev.
Males	21	8.48	4.43
Females	30	26.63	21.83

a. Find the 90% confidence interval for the difference between the two mean numbers of pairs of shoes for males and females.

b. Does the confidence interval found in part a agree with the statement made in the *USA Today* Snapshot? Explain.

c. Is the claim "average 8 more" one-sided ("at least 8 more"), or is it two-sided (meaning "not less than 8 or not more than 8")? Explain how you would interpret the claim.

10.49 [EX10-049] A study was designed to estimate the difference in diastolic blood pressure readings between men and women. MINITAB was used to construct a 99% confidence interval for the difference between the means based on the following sample data.

Males	76	76	74	70	80	68	90	70
	90	72	76	80	68	72	96	80

Females	76	70	82	90	68	60	62	68
	80	74	60	62	72			

Two-sample T for Males vs Females

	N	Mean	StDev	SE Mean
Males	16	77.37	8.35	2.1
Females	13	71.08	9.22	2.6

99% C.I. for mu males – mu females: (−2.9, 15.5)

(continue on page 506)

[EX00-000] identifies the filename of an exercise's online dataset—available through cengagebrain.com

Verify the results (the two sample means and standard deviations, and the confidence interval bounds) by calculating the values yourself. Assume normality of blood pressure readings.

10.50 [EX10-050] "Is the length of a steel bar affected by the heat treatment technique used?" This was the question being tested when the following data were collected.

Heat Treatment	Lengths (to the nearest inch)										
1	156	159	151	153	157	159	155	155	151	152	158
	154	156	156	157	155	156	159	153	157	157	159
	158	155	159	152	150	154	156	156	157	160	
2	154	156	150	151	156	155	153	154	149	150	150
	151	154	155	155	154	154	156	150	151	156	154
	153	154	149	150	150	151	154	148	155	158	

a. Find the means and standard deviations for the two sets of data.

b. Find evidence about the sample data (both graphic and numeric) that support the assumption of normality for the two sampled populations.

c. Find the 95% confidence interval for $\mu_1 - \mu_2$.

10.51 [EX10-051] Approximately 95% of the sunflowers raised in the United States are grown in the states of North Dakota, South Dakota, and Minnesota. To compare yield rates between North and South Dakota, 11 sunflower-producing counties were randomly selected from North Dakota and 14 sunflower-producing counties were randomly selected from South Dakota. Their 2008 yields, in pounds per acre, are recorded below.

N. Dakota

1296	1475	1573	1517	1242	1385
1128	1524	1644	1377	1270	

S. Dakota

1551	890	1710	1960	1988	1861	1870
1110	1674	1100	1381	2167	1130	1280

Source: http://www.nass.usda.gov/

Find the 95% confidence interval for the difference between the mean sunflower yield for all North Dakota sunflower-producing counties and the mean sunflower yield for all South Dakota sunflower-producing counties. Assume normality of yield rates.

10.52 [EX10-052] At a large university, a mathematics placement exam is administered to all students. Samples of 36 male and 30 female students are randomly selected from this year's student body and the following scores recorded:

Male	72	68	75	82	81	60	75	85	80	70
	71	84	68	85	82	80	54	81	86	79
	99	90	68	82	60	63	67	72	77	51
	61	71	81	74	79	76				

Female	81	76	94	89	83	78	85	91	83	83
	84	80	84	88	77	74	63	69	80	82
	89	69	74	97	73	79	55	76	78	81

a. Describe each set of data with a histogram (use the same class intervals on both histograms), the mean, and the standard deviation.

b. Construct 95% confidence interval for the mean score for all male students. Do the same for all female students.

c. Do the results found in part b show that the mean scores for males and females could be the same? Justify your answer. Be careful!

d. Construct the 95% confidence interval for the difference between the mean scores for male and female students.

e. Do the results found in part d show that the mean scores for male and female students could be the same? Explain.

f. Explain why the results in part b cannot be used to draw conclusions about the difference between the two means.

10.53 State the null and alternative hypotheses that would be used to test the following claims:

a. There is a difference between the mean age of employees at two different large companies.

b. The mean of population 1 is greater than the mean of population 2.

c. The mean yield of sunflower seeds per county in North Dakota is less than the mean yield per county in South Dakota.

d. There is no difference in the mean number of hours spent studying per week between male and female college students.

10.54 State the null and alternative hypotheses that would be used to test the following claims:

a. The difference between the means of the two populations is more than 20 lb.

b. The mean of population A is less than 50 more than the mean of population B.

c. The difference between the two populations is at least $500.

d. The average size yard for neighborhood A is no more than 30 square yards greater than the average size yard in neighborhood B.

10.55 Calculate the estimate for the standard error of difference between two independent means for each of the following cases:

a. $s_1^2 = 12, s_2^2 = 15, n_1 = 16,$ and $n_2 = 21$

b. $s_1^2 = 0.054, s_2^2 = 0.087, n_1 = 8,$ and $n_2 = 10$

c. $s_1 = 2.8, s_2 = 6.4, n_1 = 16,$ and $n_2 = 21$

10.56 Find the value of $t\star$ for the difference between two means based on an assumption of normality and this information about two samples:

Sample	Number	Mean	Std. Dev.
1	18	38.2	14.2
2	25	43.1	10.6

10.57 Find the value of $t\star$ for the difference between two means based on an assumption of normality and this information about two samples:

Sample	Number	Mean	Std. Dev.
1	21	1.66	0.29
2	9	1.43	0.18

10.58 Determine the p-value for the following hypothesis tests for the difference between two means with population variances unknown.

a. $H_a: \mu_1 - \mu_2 > 0, n_1 = 6, n_2 = 10, t\star = 1.3$

b. $H_a: \mu_1 - \mu_2 < 0, n_1 = 16, n_2 = 9, t\star = -2.8$

c. $H_a: \mu_1 - \mu_2 \neq 0, n_1 = 26, n_2 = 16, t\star = 1.8$

d. $H_a: \mu_1 - \mu_2 \neq 5, n_1 = 26, n_2 = 35, t\star = -1.8$

10.59 Determine the critical values that would be used for the following hypothesis tests (using the classical approach) about the difference between two means with population variances unknown.

a. $H_a: \mu_1 - \mu_2 \neq 0, n_1 = 26, n_2 = 16, \alpha = 0.05$

b. $H_a: \mu_1 - \mu_2 < 0, n_1 = 36, n_2 = 27, \alpha = 0.01$

c. $H_a: \mu_1 - \mu_2 > 0, n_1 = 8, n_2 = 11, \alpha = 0.10$

d. $H_a: \mu_1 - \mu_2 \neq 10, n_1 = 14, n_2 = 15, \alpha = 0.05$

10.60 For the hypothesis test involving $H_a: \mu_B - \mu_A \neq 0$, with df $= 18$ and $t\star = 1.3$:

a.   Find the p-value.

b.   Find the critical values given $\alpha = 0.05$.

10.61 Suppose the calculated $t\star$ had been 1.80 in Example 10.10 (pp. 500–501). Using df $= 9$ or using df $= 20$ results in different answers. Explain how the word *conservative* (p. 496) applies here.

10.62 Is having a long, more complex first name more dignified for a girl? Are girls' names longer than boys' names? With current names like "Alexandra," "Madeleine," and "Savannah," it certainly appears so. To test this theory,

random samples of seventh-grade girls and boys were taken. Let x be the number of letters in each seventh grader's first name.

Boys' Names	$n = 30$	$\bar{x} = 5.767$	$s = 1.870$
Girls' Names	$n = 30$	$\bar{x} = 6.133$	$s = 1.456$

At the 0.05 level of significance, do the data support the contention that the mean length of girls' names is longer than the mean length of boys' names?

10.63 Lauren, a brunette, was tired of hearing, "Blondes have more fun." She set out to "prove" that "brunettes are more intelligent." Lauren randomly (as best she could) selected 40 blondes and 40 brunettes at her high school. The following overall grade statistics were calculated:

Blondes	$n_{Bl} = 40$	$\bar{x}_{Bl} = 88.375$	$s_{Bl} = 6.134$
Brunettes	$n_{Br} = 40$	$\bar{x}_{Br} = 87.600$	$s_{Br} = 6.640$

Upon seeing the sample results, does Lauren have support for her claim that "brunettes are more intelligent than blondes"? Explain. What could Lauren say about blondes' and brunettes' intelligence?

10.64 One could reason that high school seniors would have more money issues than high school juniors. Seniors foresee expenses for college as well as their senior trip and prom. So does this mean that they work more than their junior classmates? Christine, a senior at HFL High School, randomly collected the following data (recorded in hours/week) from students that work:

Seniors	$n_s = 17$	$\bar{x}_s = 16.4$	$s_s = 10.48$
Juniors	$n_j = 20$	$\bar{x}_j = 18.405$	$s_j = 9.69$

Assuming that work hours are normally distributed, do these data suggest that there is a significant difference between the average number of hours that HFL seniors and juniors work per week? Use $\alpha = 0.10$.

10.65 The computer age has allowed teachers to use electronic tutorials to motivate their students to learn. *Issues in Accounting Education* published the results of a study that showed that an electronic tutorial, along with intentionally induced peer pressure, was effective in enhancing preclass preparations and in improving class attendance, test scores, and course evaluations when used by students studying tax accounting.

Suppose a similar study is conducted at your school using an electronic study guide (ESG) as a tutor for students of accounting principles. For one course section, the students were required to use a new ESG computer program that generated and scored chapter review quizzes and practice exams, presented textbook chapter reviews, and tracked progress. Students could use the computer to build, take, and score their own simulated tests and review materials at their own pace before they took their formal in-class quizzes and exams composed of different questions. The same instructor taught the other course section, used the same textbook, and gave the same daily assignments, but he did not require the students to use the ESG. Identical tests were administered to both sections, and the mean scores of all tests and assignments at the end of the year were tabulated:

Section	n	Mean Score	Std. Dev.
ESG (1)	38	79.6	6.9
No ESG (2)	36	72.8	7.6

Do these results show that the mean scores of tests and assignments for students taking accounting principles with an ESG to help them are significantly greater than the mean scores of those not using an ESG? Use a 0.01 level of significance.

a. Solve using the p-value approach.

b. Solve using the classical approach.

10.66 "In a typical month, men spend $178 and women spend $96 on leisure activities," according to the results of an International Communications Research (ICR) for American Express poll, as reported in a *USA Today* Snapshot found on the Internet June 25, 2005.

Suppose random samples were taken from the population of male and female college students. Each student was asked to determine his or her expenditures for leisure activities in the prior month. The sample data results had a standard deviation of $75 for the men and a standard deviation of $50 for the women.

a. If both samples were of size 20, what is the standard error for the difference of two means?

b. Assuming normality in leisure activity expenditures, is the difference found in the ICR poll significant at $\alpha = 0.05$ if the samples in part a are used? Explain.

10.67 Many cheeses are produced in the shape of a wheel, and due to manufacturing inconsistencies, the amount of cheese, measured by weight, varies from wheel to wheel. Heidi Cembert wishes to determine if there is a significant difference, at the 10% level, between the weight per wheel of Gouda and Brie cheese. She randomly samples 16 wheels of Gouda and finds the mean to be 1.2 pounds with a standard deviation of 0.32 pound and then samples 14 wheels of Brie and finds the mean to be 1.05 pounds with a standard deviation of 0.25 pound. At the 0.05 level of significance, is there sufficient evidence to support Heidi's contention that there is a difference in the mean weights of the two types of cheese?

10.68 If a random sample of 18 homes south of Center Street in Provo has a mean selling price of $145,200 and a standard deviation of $4700, and a random sample of 18 homes north of Center Street has a mean selling price of $148,600 and a standard deviation of $5800, can you

conclude that there is a significant difference between the selling prices of homes in these two areas of Provo at the 0.05 level? Assume normality.

a. Solve using the p-value approach.

b. Solve using the classical approach.

10.69 [EX10-069] MINITAB was used to complete a t-test of the difference between two means using the following two independent samples.

Sample 1	33.7	21.6	32.1	38.2	33.2	35.9	34.1	39.8
	23.5	21.2	23.3	18.9	30.3			

Sample 2	28.0	59.9	22.3	43.3	43.6	24.1	6.9	14.1
	30.2	3.1	13.9	19.7	16.6	13.8	62.1	28.1

Two-sample T for sample 1 vs sample 2

	N	Mean	StDev	SE Mean
sample1	13	29.68	7.07	2.0
sample2	16	26.9	17.4	4.4

T-Test mu sample1 = mu sample2 (vs not =):
T = 0.59 P = 0.56 DF = 20

a. Assuming normality, verify the results (two sample means and standard deviations, and the calculated $t\star$) by calculating the values yourself.

b. Use Table 7 in Appendix B to verify the p-value based on the calculated df.

c. Find the p-value using the smaller number of degrees of freedom. Compare the two p-values.

10.70 [EX10-070] It is a known fact that private colleges cost more than public colleges. In fact, according to the College Board [http://www.collegeboard.com/], the average 2008–2009 cost (tuition, fees, room and board) for a public college was $14,333 versus $34,132 for a private college. Does this difference hold when it comes to the average cost of required textbooks per class? The following samples of size 10 were taken.

Public	Private
64.69	71.00
89.60	96.19
101.49	96.47
101.75	97.14
103.59	98.56
106.38	98.94
106.77	107.79
110.69	112.58
118.94	114.00
135.94	116.55

Using the Excel output on the next page and $\alpha = 0.05$, determine if the average cost of required textbooks per class is different between public and private colleges.

a. Solve using the p-value approach.

b. Solve using the classical approach.

t-Test: Two-Sample Assuming Unequal Variances

	Public	Private
Mean	103.984	100.922
Variance	340.6249822	173.2995511
Observations	10	10
Hypothesized Mean Difference	0	
df	16	
t Stat	0.427125511	
$P(T \le t)$ two-tail	0.674980208	
t Critical two-tail	2.119904821	

10.71 [EX10-071] Are females as serious about golf as men are? If so, would the price of a driver for a man be the same as the price of a driver for a female? It was contended that women's drivers would be cheaper. Random samples of drivers were taken from the Golflink.com website. The prices were:

Male

149.99	299.99	49.99	499.99	167.97	299.99
399.99	199.99	99.99	149.99		

Female

199.99	79.99	499.99	199.97	299.99	99.99

At the 0.05 level of significance, is there sufficient evidence to support the contention that men's drivers are more expensive than women's drivers? Assume normality of golf driver prices.

10.72 [EX10-072] Twenty laboratory mice were randomly divided into two groups of 10. Each group was fed according to a prescribed diet. At the end of 3 weeks, the weight gained by each animal was recorded. Do the data in the following table justify the conclusion that the mean weight gained on diet B was greater than the mean weight gained on diet A, at the $\alpha = 0.05$ level of significance? Assume normality.

Diet A	5	14	7	9	11	7	13	14	12	8
Diet B	5	21	16	23	4	16	13	19	9	21

a. Solve using the *p*-value approach.

b. Solve using the classical approach.

10.73 [EX10-073] Many people who are involved with Major League Baseball believe that Yankees' baseball games tend to last longer than games played by other teams. In order to test this theory, one other MLB team, the St. Louis Cardinals, was picked at random. The time of the game (in minutes) for 12 randomly selected Cardinals' games and 14 randomly selected Yankees' games was obtained.

Yankees	Cardinals
155	208
205	135
190	161
193	170
232	150
208	187
174	200
188	143
229	154
158	193
202	128
189	212
232	
211	

Source: MLB.com

Do these samples provide significant evidence to conclude that the mean time of Yankees' baseball games is significantly greater than the mean time of Cardinals' games? Use $\alpha = 0.05$.

10.74 [EX10-074] Penfield and Perinton are two adjacent eastside suburbs of Rochester, New York. They have both always been considered on equal ground with respect to quality of life, housing, and education. Although many new housing developments are going up in Penfield, Perinton offers the added bonus of cheaper energy rates. It is thought that Perinton is able to have higher asking prices because of this bonus. To test this theory, random samples of real estate transactions were taken in each suburb during the week of October 17, 2009. Do the data support the theory for this time frame? Use $\alpha = 0.10$.

Penfield	Perinton
195,700	154,900
137,500	429,000
117,000	272,000
115,000	160,609
176,000	265,000
149,013	144,000
130,000	152,000
266,490	390,000
152,000	130,300
262,765	149,900

10.75 Consider the "commute time" data for college males and females given in Exercise 10.1 on page 481.

a. Are the assumptions of normality satisfied for each sample? Explain.

b. Test the hypothesis that there is no difference between the mean commute times for male and female college students. Use a 0.05 level of significance.

c. If the difference is significant in part b, what factors could be contributing to the difference?

10.76 [EX10-076] When evaluating different measuring instruments, one must first determine whether there is a systematic difference between the instruments. Lenses from two different groups (1 and 2) were measured once each by two different instruments. The measurement differences (Instrument A – Instrument B) were recorded. Measurement units have been coded for proprietary reasons.

Group 1	4	5	−2	−3	−7	10	11	−1	3	7	−5	3	−4
	−5	−7	4	−1	−18	0	−17	12	9	4	17	−2	

Group 2	−13	−12	−5	11	15	7	−33	−10	−6	−2	−16	2
	0	−19	6	−17	−4	−19	−22	−4	8	10	−6	

Does there appear to be a systematic difference between the two instruments?

a. Describe each set of data separately using a histogram and comparatively using one side-by-side graph.

(continue on page 510)

b. Find the mean and the standard deviation for each set of data.

c. Are the assumptions satisfied? Explain.

d. Test the hypothesis that there is no difference between the means of the two differences. Use $\alpha = 0.05$.

e. Is there any evidence of a difference between the two instruments? Explain.

10.77 [EX10-077] Consider the "surface height" data for Raul's unpolished and polished reflective surfaces of a silicone microchip given in Applied Example 10.11 on page 504.

a. Present and describe each set of data (unpolished and polished) using a histogram, the mean, and the standard deviation.

b. Check each set of data for a normal distribution: (1) State what you believe to be the case based on the results found in part a. (2) Further, find additional statistical evidence. (3) Very precisely state your conclusions regarding normality.

10.78 [EX10-077] Consider the "surface height" data for Raul's unpolished and polished reflective surfaces of a silicone microchip given in Applied Example 10.11 on page 504 and initially investigated in Exercise 10.77.

a. Do the two sets of data, unpolished and polished, represent independent or dependent samples? Explain your answer.

b. Produce at least three graphical statistics demonstrating that the new polishing process does in fact produce a smoother reflecting surface. Explain how each graph demonstrates that the goal has been attained.

c. Is there statistical evidence that the process has produced a surface that is significantly smoother? State the p-value.

d. Complete the hypothesis test at the 0.01 level of significance; be sure to state your decision and conclusion.

10.79 Use a computer to demonstrate the truth of the statement describing the sampling distribution of $\bar{x}_1 - \bar{x}_2$. Use two theoretical normal populations: $N_1(100, 20)$ and $N_2(120, 20)$.

a. To get acquainted with the two theoretical populations, randomly select a very large sample from each. Generate 2000 data values, calculate mean and standard deviation, and construct a histogram

using class boundaries that are multiples of one-half of a standard deviation (10) starting at the mean for each population.

b. If samples of size 8 are randomly selected from each population, what do you expect the distribution of $\bar{x}_1 - \bar{x}_2$ to be like (shape of distribution, mean, standard error)?

c. Randomly draw a sample of size 8 from each population, and find the mean of each sample. Find the difference between the sample means. Repeat 99 more times.

d. The set of 100 $(\bar{x}_1 - \bar{x}_2)$ values forms an empirical sampling distribution of $\bar{x}_1 - \bar{x}_2$. Describe the empirical distribution: shape (histogram), mean, and standard error. (Use class boundaries that are multiples of standard error from mean for easy comparison to the expected.)

e. Using the information found in parts a–d, verify the statement about the $\bar{x}_1 - \bar{x}_2$ sampling distribution made on page 496.

f. Repeat the experiment a few times and compare the results.

FYI See the *Student Solutions Manual* for additional information about commands.

10.80 One reason for being conservative when determining the number of degrees of freedom to use with the t-distribution is the possibility that the population variances might be unequal. Extremely different values cause a lowering in the number of df used. Repeat Exercise 10.79 using theoretical normal distributions of $N(100, 9)$ and $N(120, 27)$ and both sample sizes of 8. Check all three properties of the sampling distribution: normality, its mean value, and its standard error. Describe in detail what you discover. Do you think we should be concerned about the choice of df? Explain.

10.81 Unbalanced sample sizes are a factor in determining the number of degrees of freedom for inferences about the difference between two means. Repeat Exercise 10.79 using theoretical normal distributions of $N(100, 20)$ and $N(120, 20)$ and sample sizes of 5 and 20. Check all three properties of the sampling distribution: normality, its mean value, and its standard error. Describe in detail what you discover. Do you think we should be concerned when using unbalanced sample sizes? Explain.

10.82 One assumption for the two-sample t-test is that the "sampled populations are to be normally distributed." What happens when they are not normally distributed? Repeat Exercise 10.79 using two theoretical populations that are not normal and using samples of size 10. The exponential distribution uses a continuous random variable, it has a J-shaped distribution, and its mean and standard deviation are the same value. Use two exponential distributions with means of 50 and 80: Exp(50) and Exp(80). Check all three properties of the sampling distribution: normality, its mean value, and its standard error. Describe in detail what you discover. Do you think we should be concerned when sampling nonnormal populations? Explain.

10.4 Inferences Concerning the Difference between Proportions Using Two Independent Samples

FYI The 3 "p" words (*proportion, percentage, probability*) are all the binomial parameter p, P(success).

FYI Binomial experiments are defined in more detail on page 246.

We are often interested in making statistical comparisons between the **proportions, percentages**, or **probabilities** associated with two populations. These questions ask for such comparisons: Is the proportion of homeowners who favor a certain tax proposal different from the proportion of renters who favor it? Did a larger percentage of this semester's class than of last semester's class pass statistics? Is the probability of a Democratic candidate winning in New York greater than the probability of a Republican candidate winning in Texas? Do students' opinions about the new code of conduct differ from those of the faculty? You have probably asked similar questions.

Note: These are the properties of a **binomial experiment:**

1. The observed probability is $p' = x/n$, where x is the number of observed successes in n trials.
2. $q' = 1 - p'$.
3. p is the probability of success on an individual trial in a binomial probability experiment of n repeated independent trials.

In this section, we will compare two population proportions by using the difference between the observed proportions, $p'_1 - p'_2$, of two independent samples. The observed difference, $p'_1 - p'_2$, belongs to a sampling distribution with the characteristics described in the following statement.

If independent samples of sizes n_1 and n_2 are drawn randomly from large populations with $p_1 = P_1$(success) and $p_2 = P_2$(success), respectively, then the sampling distribution of $p'_1 - p'_2$ has these properties:

1. mean $\mu_{p'_1 - p'_2} = p_1 - p_2$

2. standard error $\sigma_{p'_1 - p'_2} = \sqrt{\left(\dfrac{p_1 q_1}{n_1}\right) + \left(\dfrac{p_2 q_2}{n_2}\right)}$ **(10.10)**

3. an approximately normal distribution if n_1 and n_2 are sufficiently large

In practice, we use the following *guidelines to ensure normality:*

1. The sample sizes are both larger than 20.
2. The products $n_1 p_1$, $n_1 q_1$, $n_2 p_2$, and $n_2 q_2$ are all larger than 5.
3. The samples consist of less than 10% of their respective populations.

Note: p_1 and p_2 are unknown; therefore, the products mentioned in guideline 2 will be estimated by $n_1 p'_1$, $n_1 q'_1$, $n_2 p'_2$, and $n_2 q'_2$.

Inferences about the difference between two population proportions, $p_1 - p_2$, will be based on the following assumptions.

Assumptions for inferences about the difference between two proportions $p_1 - p_2$ The n_1 random observations and the n_2 random observations that form the two samples are selected independently from two populations that do not change during the sampling.

Confidence Interval Procedure

When we estimate the **difference between two proportions**, $p_1 - p_2$, we will base our estimates on the **unbiased sample statistic** $p_1' - p_2'$. The point estimate, $p_1' - p_2'$, becomes the center of the confidence interval and the confidence interval limits are found using the following formula:

Confidence Interval for the Difference between Two Proportions

$$(p_1' - p_2') - z(\alpha/2) \cdot \sqrt{\left(\frac{p_1'q_1'}{n_1}\right) + \left(\frac{p_2'q_2'}{n_2}\right)} \quad \text{to}$$

$$(p_1' - p_2') + z(\alpha/2) \cdot \sqrt{\left(\frac{p_1'q_1'}{n_1}\right) + \left(\frac{p_2'q_2'}{n_2}\right)} \quad \textbf{(10.11)}$$

E X A M P L E   1 0 . 1 2

CONSTRUCTING A CONFIDENCE INTERVAL FOR THE DIFFERENCE BETWEEN TWO PROPORTIONS

In studying his campaign plans, Mr. Morris wishes to estimate the difference between men's and women's views regarding his appeal as a candidate. He asks his campaign manager to take two random independent samples and find the 99% confidence interval for the difference between the proportions of women and men voters who plan to vote for him. A sample of 1000 voters was taken from each population, with 388 men and 459 women favoring Mr. Morris.

Solution

Step 1 **Parameter of interest:** $p_w - p_m$, the difference between the proportion of women voters and the proportion of men voters who plan to vote for Mr. Morris

Step 2 **a. Assumptions:** The samples are randomly and independently selected.

b. Probability distribution: The standard normal distribution. The populations are large (all voters); the sample sizes are larger than 20; and the estimated values for $n_m p_m$, $n_m q_m$, $n_w p_w$, and $n_w q_w$ are all larger than 5. Therefore, the sampling distribution of $p_w' - p_m'$ should have an approximately normal distribution. The interval will be calculated using formula (10.11).

c. Level of confidence: $1 - \alpha = 0.99$

FYI It is customary to place the larger value first; that way, the point estimate for the difference is a positive value.

Step 3 Sample information:

We have $n_m = 1000$, $x_m = 388$, $n_w = 1000$, and $x_w = 459$.

$$p'_m = \frac{x_m}{n_m} = \frac{388}{1000} = \mathbf{0.388} \qquad q'_m = 1 - 0.388 = \mathbf{0.612}$$

$$p'_w = \frac{x_w}{n_w} = \frac{459}{1000} = \mathbf{0.459} \qquad q'_w = 1 - 0.459 = \mathbf{0.541}$$

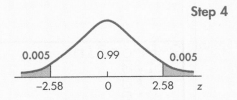

0.005   0.99   0.005

−2.58   0   2.58 z

Step 4   **a. Confidence coefficient:** This is a two-tailed situation, with $\alpha/2$ in each tail. From Table 4B, $z(\alpha/2) = z(0.005) = 2.58$. Instructions for using Table 4B are on page 350.

b. Maximum error of estimate: Using the maximum error part of formula (10.11), we have

$$E = z(\alpha/2) \cdot \sqrt{\left(\frac{p'_w q'_w}{n_w}\right) + \left(\frac{p'_m q'_m}{n_m}\right)}$$

$$E = 2.58 \cdot \sqrt{\left(\frac{(0.459)(0.541)}{1000}\right) + \left(\frac{(0.388)(0.612)}{1000}\right)}$$

$$= 2.58\sqrt{0.000248 + 0.000237} = (2.58)(0.022) = \mathbf{0.057}$$

c. Lower/upper confidence limits:

$$(p'_w - p'_m) \pm E$$
$$0.071 \pm 0.057$$
$$0.071 - 0.057 = \mathbf{0.014} \quad \text{to} \quad 0.071 + 0.057 = \mathbf{0.128}$$

Step 5   **a. Confidence interval:** 0.014 to 0.128 is the 99% confidence interval for $p_w - p_m$. With 99% confidence, we can say that there is a difference of 1.4% to 12.8% in Mr. Morris's voter appeal.

b. That is, a larger proportion of women than men favor Mr. Morris, and the difference in the proportions is between 1.4% and 12.8%.

TECHNOLOGY INSTRUCTIONS: CONFIDENCE INTERVALS FOR THE DIFFERENCE BETWEEN TWO PROPORTIONS GIVEN TWO INDEPENDENT SETS OF SAMPLE DATA

MINITAB		
Choose:	**Stat > Basic Statistics > 2 Proportions**	
Select:	**Summarized data:**	
Enter:	First: **x** (events) **n** (trials)	
	Second: **x** (events) **n** (trials)	
Select:	**Options**	
Enter:	Confidence level: **1 − α** (ex. 0.95 or 95.0)	
Select:	Alternative: **not equal > OK > OK**	

Excel	Input the data for the first sample into column A using 0s for failures (or no's) and 1s for successes (or yes's); then repeat the same procedure for the second sample in column B; then continue with:

Choose: **Add-Ins > Data Analysis Plus > Z-Estimate: Two Proportions**
Enter: Variable 1 Range: (A2:A20 or select cells)
             Variable 2 Range: (B1:B20 or select cells)
             Code for success: 1
Select: **Labels** (if necessary)
Enter: Alpha: α (ex. 0.05) **> OK**

TI-83/84 Plus	Choose: **STAT > TESTS > B:2-PropZInt**

Enter the appropriate values and highlight
Calculate.

```
2-PropZInt
 x1:0
 n1:0
 x2:0
 n2:0
 C-Level:.95
 Calculate
```

Confidence intervals and hypothesis tests can sometimes be interchanged; that is, a confidence interval can be used in place of a hypothesis test. For example, Example 10.12 called for a confidence interval. Now suppose that Mr. Morris asked, "Is there a difference in my voter appeal to men voters as opposed to women voters?" To answer his question, you would not need to complete a hypothesis test if you chose to test at $\alpha = 0.01$ using a two-tailed test. "No difference" would mean a difference of zero, which is not included in the interval from 0.014 to 0.128 (the interval determined in Example 10.12). Therefore, a null hypothesis of "no difference" would be rejected, thereby substantiating the conclusion that a significant difference exists in voter appeal between the two groups.

Hypothesis-Testing Procedure

When the null **hypothesis—there is no difference between two proportions**—is being tested, the **test statistic** will be the difference between the observed proportions divided by the **standard error**; it is found with the following formula:

Test Statistic for the Difference between Two Proportions—Population Proportion Known

$$z\star = \frac{p_1' - p_2'}{\sqrt{pq\left[\left(\frac{1}{n_1}\right) + \left(\frac{1}{n_2}\right)\right]}}$$ (10.12)

Notes:

1. The null hypothesis is $p_1 = p_2$ or $p_1 - p_2 = 0$ (the difference is zero).
2. Nonzero differences between proportions are not discussed in this section.

3. The numerator of formula (10.12) could be written as $(p'_1 - p'_2) - (p_1 - p_2)$, but since the null hypothesis is assumed to be true during the test, $p_1 - p_2 = 0$. By substitution, the numerator becomes simply $p'_1 - p'_2$.

4. Since the null hypothesis is $p_1 = p_2$, the standard error of $p'_1 - p'_2$,
$$\sqrt{\left(\frac{p_1 q_1}{n_1}\right) + \left(\frac{p_2 q_2}{n_2}\right)}, \text{ can be written as } \sqrt{pq\left[\left(\frac{1}{n_1}\right) + \left(\frac{1}{n_2}\right)\right]}, \text{ where } p = p_1 = p_2$$
and $q = 1 - p$.

5. When the null hypothesis states $p_1 = p_2$ and does not specify the value of either p_1 or p_2, the two sets of sample data will be pooled to obtain the estimate for p. This pooled probability (known as p'_p) is the total number of successes divided by the total number of observations with the two samples combined; it is found using the next formula:

$$p'_p = \frac{x_1 + x_2}{n_1 + n_2} \tag{10.13}$$

and q'_p is its complement,

$$q'_p = 1 - p'_p \tag{10.14}$$

When the pooled estimate, p'_p, is being used, formula (10.12) becomes formula (10.15):

Test Statistic for the Difference between Two Proportions—Population Proportion Unknown

$$z\bigstar = \frac{p'_1 - p'_2}{\sqrt{(p'_p)(q'_p)\left[\left(\frac{1}{n_1}\right) + \left(\frac{1}{n_2}\right)\right]}} \tag{10.15}$$

EXAMPLE 10.13

ONE-TAILED HYPOTHESIS TEST FOR THE DIFFERENCE BETWEEN TWO PROPORTIONS

A salesperson for a new manufacturer of cellular phones claims not only that they cost the retailer less but also that the percentage of defective cellular phones found among her products will be no higher than the percentage of defectives found in a competitor's line. To test this statement, a retailer took random samples of each manufacturer's product. The sample summaries are given in Table 10.5. Can we reject the salesperson's claim at the 0.05 level of significance?

TABLE 10.5 Cellular Phone Sample Information

Product	Number Defective	Number Checked
Salesperson's	15	150
Competitor's	6	150

Video tutorial available—logon and learn more at cengagebrain.com

Solution

Step 1 **a. Parameter of interest:** $p_s - p_c$, the difference between the proportion of defectives in the salesperson's product and the proportion of defectives in the competitor's product

b. Statement of hypotheses: The concern of the retailer is that the salesperson's less expensive product may be of a poorer quality, meaning a greater proportion of defectives. If we use the difference "suspected larger proportion – smaller proportion," then the alternative hypothesis is "The difference is positive (greater than zero)."

$H_o: p_s - p_c = 0 \ (\leq)$ (salesperson's defective rate is no higher
than competitor's)

$H_a: p_s - p_c > 0$ (salesperson's defective rate is higher than competitor's)

Step 2 **a. Assumptions:** Random samples were selected from the products of two different manufacturers.

b. The test statistic to be used: The standard normal distribution. Populations are very large (all cellular phones produced); the samples are larger than 20; and the estimated products $n_s p'_s$, $n_s q'_s$, $n_c p'_c$, and $n_c q'_c$ are all larger than 5. Therefore, the sampling distribution should have an approximately normal distribution. $z\star$ will be calculated using formula (10.15).

c. Level of significance: $\alpha = 0.05$

Step 3 **a. Sample information:**

$p'_s = \dfrac{x_s}{n_s} = \dfrac{15}{150} = \mathbf{0.10}$ $\qquad\qquad$ $p'_c = \dfrac{x_c}{n_c} = \dfrac{6}{150} = \mathbf{0.04}$

$p'_p = \dfrac{x_1 + x_2}{n_1 + n_2} = \dfrac{15 + 6}{150 + 150} = \dfrac{21}{300} = \mathbf{0.07}$ $\qquad$ $q'_p = 1 - p'_p = 1 - 0.07 = \mathbf{0.93}$

b. Calculated test statistic:

$$z\star = \frac{p'_s - p'_c}{\sqrt{(p'_p)(q'_p)\left[\left(\dfrac{1}{n_s}\right) + \left(\dfrac{1}{n_c}\right)\right]}} : \quad z\star = \frac{0.10 - 0.04}{\sqrt{(0.07)(0.93)\left[\left(\dfrac{1}{150}\right) + \left(\dfrac{1}{150}\right)\right]}}$$

$$= \frac{0.06}{\sqrt{0.000868}} = \frac{0.06}{0.02946} = \mathbf{2.04}$$

Step 4 Probability Distribution:

p-Value:	**OR**	Classical:

a. Use the right-hand tail because H_a expresses concern for values related to "higher than." **P** = *p*-value = $P(z\star > 2.04)$, as shown in the figure.

To find the *p*-value, you have three options:

1. Use Table 3 (Appendix B) to calculate the *p*-value: **P** = $1.0000 - 0.9793 = \mathbf{0.0207}$.

a. The critical region is the right-hand tail because H_a expresses concern for values related to "higher than." The critical value is obtained from Table 4A: $z(0.05) = \mathbf{1.65}$.

For specific instructions, see pages 393–394.

b. $z\star$ is in the critical region, as shown in **red** in the figure.

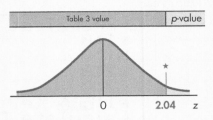

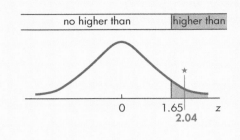

2. Use Table 5 (Appendix B) to place bounds on the *p*-value: **0.0202 < P < 0.0228**.
3. Use a computer or calculator: **P = 0.0207**.

For specific instructions, see page 376.

b. The *p*-value is smaller than α.

Step 5   **a.** **Decision:** Reject H_o.

   b. **Conclusion:** At the 0.05 level of significance, there is sufficient evidence to reject the salesperson's claim; the proportion of her company's cellular phones that are defective is higher than the proportion of her competitor's cellular phones that are defective.

TECHNOLOGY INSTRUCTIONS: HYPOTHESIS TEST FOR THE DIFFERENCE BETWEEN TWO PROPORTIONS, $p_1 - p_2$, FOR TWO INDEPENDENT SETS OF SAMPLE DATA

MINITAB

Choose: **Stat > Basic Statistics > 2 Proportions**
Select: **Summarized data:**
Enter: First: **x (events) n (trials)**
         Second: **x (events) n (trials)**
Select: **Options**
Enter: **Test difference: 0.0**
Select: **Alternative: less than** or **not equal** or **greater than**
Select: **Use pooled estimate of p for test > OK > OK**

Excel

Input the data for the first sample into column A using 0s for failures (or no's) and 1s for successes (or yes's); then repeat the same procedure for the second sample in column B; then continue with:

Choose: **Add-Ins > Data Analysis Plus > Z-Test: Two Proportions**
Enter: **Variable 1 Range: (A1:A20 or select cells)**
         Variable 2 Range: (B1:B20 or select cells)
         Code for success: 1
         Hypothesized difference: 0
Select: **Labels (if necessary)**
Enter: **Alpha: α (ex. 0.05) > OK**

TI-83/84 Plus

Choose: STAT > TESTS > 6:2-PropZTest . . .

Enter the appropriate values and highlight Calculate.

```
2-PropZTest
 x1:0
 n1:0
 x2:0
 n2:0
 p1:≠p2 <p2 >p2
Calculate Draw
```

APPLIED EXAMPLE 10.14

CADAVER KIDNEYS ARE GOOD FOR TRANSPLANTS

In a discovery that could ease the severe shortage of donor organs, Swiss researchers found that kidneys transplanted from cadavers keep working just as long as those from a patient whose heart is still beating. Most transplant organs are taken from brain-dead patients whose hearts have not stopped because doctors have long believed that if they wait until the heart stops, the organs will become damaged from lack of oxygen.

But in the first long-term study comparing the two approaches, doctors at University Hospital Zurich followed nearly 250 transplant patients for up to 15 years and found nearly identical survival rates. At 10 years, 79 percent of patients whose kidney came from a donor with no heartbeat were alive, as were 77 percent of patients whose organ came from a brain-dead donor whose heart was beating. The study, published in Thursday's *New England Journal of Medicine*, could prove especially influential because it was a head-to-head comparison of the two approaches and was the first to follow patients for many years.

Doctors believe similar results may be found for transplants of the liver, pancreas, and lungs. By using organs from "cardiac death" donors, the number of kidneys available could increase up to 30 percent, meaning some 1000 or more extra U.S. donors a year, experts estimate.

Source: Reprinted with permission of The Associated Press.

SECTION 10.4 EXERCISES

10.83 Only 75 of the 250 people interviewed were able to name the vice president of the United States. Find the values for x, n, p', and q'.

10.84 If $n_1 = 40, p'_1 = 0.9, n_2 = 50$, and $p'_2 = 0.9$:

a. Find the estimated values for both np's and both nq's.

b. Would this situation satisfy the guidelines for approximately normal? Explain.

10.85 Calculate the estimate for the standard error of the difference between two proportions for each of the following cases:

a. $n_1 = 40, p'_1 = 0.8, n_2 = 50$, and $p'_2 = 0.8$

b. $n_1 = 33, p'_1 = 0.6, n_2 = 38$, and $p'_2 = 0.65$

10.86 Calculate the maximum error of estimate for a 90% confidence interval for the difference between two proportions for the following cases:

a. $n_1 = 40, p_1' = 0.7, n_2 = 44$, and $p_2' = 0.75$

b. $n_1 = 36, p_1' = 0.33, n_2 = 38$, and $p_2' = 0.42$

10.87 A *Nursing Economics* article titled "Nurse Executive Turnover" compared two groups of nurse executives. One group had participated in a unique program for nurse executives called the Wharton Fellows Program, and the other group had not participated in the program. Of 341 Wharton Fellows, 87 had experienced one change in position; of 40 non-Wharton Fellows, 9 had experienced one change in position. MINITAB was used to construct a 99% confidence interval for the difference in population proportions. Verify the results that follow by calculating them yourself.

Test and CI for Two Proportions

Sample	X	N	Sample p
1	87	341	0.255132
2	9	40	0.225000

Difference = p (1) − p (2)
Estimate for difference: 0.0301320
99% CI for difference: (−0.150483, 0.210747)

10.88 Find the 95% confidence interval for $p_A - p_B$.

Sample	n	x
A	125	45
B	150	48

10.89 The proportions of defective parts produced by two machines were compared, and the following data were collected:

Machine 1: $n = 150$; number of defective parts = 12
Machine 2: $n = 150$: number of defective parts = 6

Determine a 90% confidence interval for $p_1 - p_2$.

10.90 In a random sample of 40 brown-haired individuals, 22 indicated that they use hair coloring. In another random sample of 40 blond individuals, 26 indicated that they use hair coloring. Use a 92% confidence interval to estimate the difference in the population proportions of brunettes and blondes that use hair coloring.

10.91 The Soap and Detergent Association issued its fifth annual Clean Hands Report Card survey for 2009. From the answers to a series of hygiene-related questions posed to American adults, it was found that 62% of 442 women washed their hands more than 10 times per day, while 37% of 446 men did the same. Find the 95% confidence interval for the difference in proportions of women and men that washed their hands more than 10 times a day.

10.92 In a survey of 300 people from city A, 128 prefer New Spring soap to all other brands of deodorant soap. In city B, 149 of 400 people prefer New Spring soap. Find the 98% confidence interval for the difference in the proportions of people from the two cities who prefer New Spring soap.

10.93 State the null hypothesis, H_o, and the alternative hypothesis, H_a, that would be used to test these claims:

a. There is no difference between the proportions of men and women who will vote for the incumbent in next month's election.

b. The percentage of boys who cut classes is greater than the percentage of girls who cut classes.

c. The percentage of college students who drive old cars is higher than the percentage of noncollege people of the same age who drive old cars.

10.94 Show that the standard error of $p_1' - p_2'$, which is

$$\sqrt{\left(\frac{p_1 q_1}{n_1}\right) + \left(\frac{p_2 q_2}{n_2}\right)}, \text{ reduces to } \sqrt{pq\left[\left(\frac{1}{n_1}\right) + \left(\frac{1}{n_2}\right)\right]} \text{ when}$$

$p_1 = p_2 = p$.

10.95 Find the values of p_p' and q_p' for these samples:

Sample	x	n
E	15	250
R	25	275

10.96 Find the value of $z\star$ that would be used to test the difference between the proportions, given the following:

Sample	n	x
G	380	323
H	420	332

10.97 Find the *p*-value for the test with alternative hypothesis $p_E < p_R$ using the data in Exercise 10.95.

10.98 Determine the *p*-value that would be used to test the following hypotheses when z is used as the test statistic.

a. $H_o: p_1 = p_2$ versus $H_a: p_1 > p_2$, with $z\star = 2.47$

b. $H_o: p_A = p_B$ versus, $H_a: p_A \neq p_B$, with $z\star = -1.33$

c. $H_o: p_1 - p_2 = 0$ versus $H_a: p_1 - p_2 < 0$, with $z\star = -0.85$

d. $H_o: p_m - p_f = 0$ versus $H_a: p_m - p_f > 0$, with $z\star = 3.04$

10.99 Determine the critical region and critical value(s) that would be used to test (classical procedure) the following hypotheses when z is used as the test statistic.

a. $H_o: p_1 = p_2$ versus $H_a: p_1 > p_2$, with $\alpha = 0.05$

b. $H_o: p_A = p_B$ versus $H_a: p_A \neq p_B$, with $\alpha = 0.05$

c. $H_o: p_1 - p_2 = 0$ versus $H_a: p_1 - p_2 < 0$, with $\alpha = 0.04$

d. $H_o: p_m - p_f = 0$ versus $H_a: p_m - p_f > 0$, with $\alpha = 0.01$

10.100 PC users are often victimized by hardware problems. A study revealed that hardware problems reported to manufacturers could not be fixed by one in three owners of personal computers. Home PC owners fared even worse than those with work PCs, facing longer waits for service and getting even fewer problems resolved. Relatively few owners gave service technicians high marks for having adequate knowledge or for exerting sincere efforts to help solve the problems with the hardware.

Source: *PC World, "Which PC Makers Can You Trust?"*

Suppose a study is conducted to compare the service provided by manufacturers to both home PC owners and work PC owners. Of 220 home PC owners who had trouble, 98 reported that their problem was not resolved satisfactorily. When the same question was asked of 180 work PC owners who experienced difficulty, 52 reported that the problem was not resolved. Did the home PC owners experience a greater proportion of problems that could not be solved with help from the manufacturer? Use the 0.05 level of significance and the MINITAB output that follows to answer the question.

a.   Solve using the *p*-value approach.

b.   Solve using the classical approach.

Test and CI for Two Proportions			
Sample	X	N	Sample p
1	98	220	0.445455
2	52	180	0.288889
Difference = p (1) − p (2)			
Estimate for difference: 0.156566			
Test for difference = 0 (vs > 0): z = 3.22			
P-Value = 0.001			

10.101 Two randomly selected groups of citizens were exposed to different media campaigns that dealt with the image of a political candidate. One week later, the citizen groups were surveyed to see whether they would vote for the candidate. The results were as follows:

	Exposed to Conservative Image	Exposed to Moderate Image
Number in Sample	100	100
Proportion for the Candidate	0.40	0.50

Is there sufficient evidence to show a difference in the effectiveness of the two image campaigns at the 0.05 level of significance?

a.   Solve using the *p*-value approach.

b.   Solve using the classical approach.

10.102 In a survey of families in which both parents work, one of the questions asked was, "Have you refused a job, promotion, or transfer because it would mean less time with your family?" A total of 200 men and 200 women were asked this question. "Yes" was the response given by 29% of the men and 24% of the women. Based on this survey, can we conclude that there is a difference in the proportion of men and women responding "Yes" at the 0.05 level of significance?

10.103 *Consumer Reports* conducted a survey of 1000 adults concerning wearing bicycle helmets. One of the questions presented to 25- to 54-year olds was whether they wear a helmet most of the time while biking or cycling. This was further broken down to whether or not they had a child at home. Eighty-seven percent of the age group that had a child at home reported that they wear a helmet most of the time, while 74% of those without a child at home reported they wear a helmet most of the time. If the sample size is 340 for both age groups, is the proportion of helmet use significantly greater when there is a child in the home, at the 0.01 level of significance?

10.104 A Harris Interactive poll found that 50% of Democrats follow professional football while 59% of Republicans follow the sport. If the poll results were based on samples of 875 Democrats and 749 Republicans, determine, at the 0.05 level of significance, if the viewpoint of more Republicans following professional football is substantiated.

10.105 The Committee of 200, a professional organization of preeminent women entrepreneurs and corporate leaders, reported the following: 60% of women MBA students say, "Businesses pay their executives too much money" and 50% of the men MBA students agreed.

a.   Does there appear to be a difference in the proportion of women and men who say, "Executives are paid too much"? Explain the meaning of your answer.

b.   If the preceding percentages resulted from two samples of size 20 each, is the difference statistically significant at a 0.05 level of significance? Justify your answer.

c.   If the preceding percentages resulted from two samples of size 500 each, is the difference statistically significant at a 0.05 level of significance? Justify your answer.

d.   Explain how your answers to parts b and c affect your thoughts about your answer to part a.

10.106 Both parents and students have many concerns when considering colleges. One of the top three concerns, based on a College Partnership study, is "Choosing best major/career." Nineteen percent of the parents reported

"Choosing best major/career" as a major concern whereas 15% of students reported it as a major concern.
Source: http://www.collegepartnership.com/

If the study was conducted with a sample of 1750 students and their parents, test the hypothesis that "Choosing best major/career" was a bigger concern for the parents, at the 0.05 level of significance.

10.107 A 2005 Harris Interactive for Korbel poll found that 63% of men and 55% of women think that it is okay for women to make marriage proposals to men. An 8% difference may or may not be statistically significant. What size sample is required to make this difference significant?

Source: *USA Today* Snapshot found on Internet, June 25, 2005

a. If the preceding sample statistics had resulted from a sample of 250 men and a sample of 250 women, would the difference be significant, using $\alpha = 0.05$? Explain.

b. If the samples had each been of size 500, would the difference be significant, using $\alpha = 0.05$? Explain.

c. Determine the size sample that would have the difference of 0.08 corresponding to $p = 0.05$.

10.108 Adverse side effects are always a concern when testing and trying new medicines. Placebo-controlled clinical studies were conducted in patients 12 years of age and older who were receiving "once-a-day" doses of Allegra, a seasonal allergy drug. The following results were published in the April 2005 edition of *Reader's Digest*.

	Allegra (dose once a day)	Placebo (dose once a day)
Side effects	$n = 283$	$n = 293$
Number reporting headaches	30	22

Determine, at the 0.05 level of significance, whether there is a difference in the proportion of patients reporting headaches between the two groups.

10.109 Forty-one small lots of experimental product were manufactured and tested for the occurrence of a particular indication that is an attribute in nature yet causes rejection of the part. Thirty-one lots were made using one particular processing method, and ten lots were made using a second processing method. Each lot was equally sampled ($n = 32$) for the presence of this indication. In practice, optimal processing conditions show little or no occurrence of the indication. Method 1, involving the ten lots, was run before Method 2.

Methods	n	Number of Rejects
Method 1	320	4
Method 2	992	26

Source: Courtesy of Bausch & Lomb

Determine, at the 0.05 level of significance, whether there is a difference in the proportion of reject product between the two methods. (Save your answer for comparison with Exercise 11.72 on p. 575.)

10.110 The guidelines to ensure the sampling distribution of $p_1' - p_2'$ is normal include several conditions about the size of several values. The two binomial distributions $B(100, 0.3)$ and $B(100, 0.4)$ satisfy all of those guidelines.

a. Verify that $B(100, 0.3)$ and $B(100, 0.4)$ satisfy all guidelines.

b. Use a computer to randomly generate 200 samples from each of the binomial populations. Find the observed proportion for each sample and the value of the 200 differences between two proportions.

c. Describe the observed sampling distribution using both graphic and numerical statistics.

d. Does the empirical sampling distribution appear to have an approximately normal distribution? Explain.

FYI See the *Student Solutions Manual* for additional information about commands.

10.5 Inferences Concerning the Ratio of Variances Using Two Independent Samples

When comparing two populations, we naturally compare their two most fundamental distribution characteristics, their "center" and their "spread," by comparing their means and standard deviations. We have learned, in two of the previous sections, how to use the *t*-distribution to make inferences comparing two population means with either dependent or independent samples. These procedures were intended to be used with normal populations, but they work quite well even when the populations are not exactly normally distributed.

The next logical step in comparing two populations is to compare their standard deviations, the most often used measure of spread. However, sampling distributions that deal with sample standard deviations (or variances) are very sensitive to slight departures from the assumptions. Therefore, the only inference procedure to be presented here will be the **hypothesis test for the equality of standard deviations (or variances)** for two normal populations.

The soft drink bottling company discussed in Section 9.3 (pp. 453, 457) is trying to decide whether to install a modern, high-speed bottling machine. There are, of course, many concerns in making this decision, and one of them is that the increased speed may result in increased variability in the amount of fill placed into each bottle; such an increase would not be acceptable. To this concern, the manufacturer of the new system responded that the variance in fills will be no greater with the new machine than with the old. (The new system will fill several bottles in the same amount of time that the old system fills one bottle; this is the reason the change is being considered.) A test is set up to statistically test the bottling company's concern, "Standard deviation of new machine is greater than standard deviation of old," against the manufacturer's claim, "Standard deviation of new is no greater than standard deviation of old."

EXAMPLE 10.15

WRITING HYPOTHESES FOR THE EQUALITY OF VARIANCES

State the null and alternative hypotheses to be used for comparing the variances of the two soft drink bottling machines.

Solution

There are several equivalent ways to express the null and alternative hypotheses, but because the test procedure uses the ratio of variances, the recommended convention is to express the null and alternative hypotheses as ratios of the population variances. Furthermore, it is recommended that the "larger" or "expected to be larger" variance be the numerator. The concern of the soft drink company is that the new modern machine (m) will result in a larger standard deviation in the amounts of fill than its present machine (p); $\sigma_m > \sigma_p$ or equivalently $\sigma_m^2 > \sigma_p^2$, which becomes $\frac{\sigma_m^2}{\sigma_p^2} > 1$. We want to test the manufacturer's claim (the null hypothesis) against the company's concern (the alternative hypothesis):

$$H_o: \frac{\sigma_m^2}{\sigma_p^2} = 1 \quad (m \text{ is no more variable})$$

$$H_a: \frac{\sigma_m^2}{\sigma_p^2} > 1 \quad (m \text{ is more variable})$$

Inferences about the ratio of variances for two normally distributed populations use the **F-distribution**. The F-distribution, similar to Student's *t*- distribution and the χ^2-distribution, is a family of probability distributions. Each F-distribution is identified by two numbers of degrees of freedom, one for each of the two samples involved.

Before continuing with the details of the hypothesis-testing procedure, let's learn about the F-distribution.

Properties of the F-distribution
1. F is nonnegative; it is zero or positive.
2. F is nonsymmetrical; it is skewed to the right.
3. F is distributed so as to form a family of distributions; there is a separate distribution for each pair of numbers of degrees of freedom.

FIGURE 10.2

F-Distribution

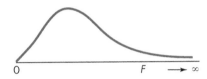

For inferences discussed in this section, the number of degrees of freedom for each sample is $df_1 = n_1 - 1$ and $df_2 = n_2 - 1$. Each different combination of degrees of freedom results in a different F-distribution, and each F-distribution looks approximately like the distribution shown in Figure 10.2.

The critical values for the F-distribution are identified using three values:

df_n, the degrees of freedom associated with the sample whose variance is in the numerator of the calculated F

df_d, the degrees of freedom associated with the sample whose variance is in the denominator

α, the area under the distribution curve to the right of the critical value being sought

FIGURE 10.3

A Critical Value of F

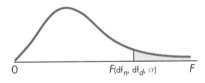

Therefore, the symbolic name for a critical value of F will be $F_{(df_n, df_d, \alpha)}$, as shown in Figure 10.3.

Because it takes three values to identify a single critical value of F, making tables for F is not as simple as it was with previously studied distributions. The tables presented in this textbook are organized so as to have a different table for each different value of α, the "area to the right." Table 9A in Appendix B shows the critical values for $F_{(df_n, df_d, \alpha)}$, when $\alpha = 0.05$; Table 9B gives the critical values when $\alpha = 0.025$; Table 9C gives the values when $\alpha = 0.01$.

EXAMPLE 10.16

FINDING CRITICAL F-VALUES

Find $F_{(5, 8, 0.05)}$, the critical F-value for samples of size 6 and size 9 with 5% of the area in the right-hand tail.

Solution

Using Table 9A ($\alpha = 0.05$), find the intersection of column df = 5 (for the numerator) and row df = 8 (for the denominator) and read the value: $F_{(5, 8, 0.05)} = $ **3.69**. See the accompanying partial table.

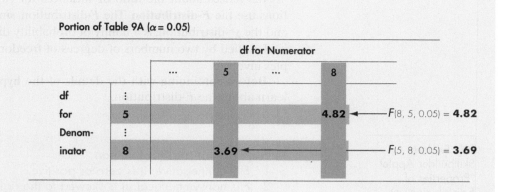

Portion of Table 9A ($\alpha = 0.05$)

Notice that $F(8, 5, 0.05)$ is **4.82**. The degrees of freedom associated with the numerator and with the denominator must be kept in the correct order; 3.69 is different from 4.82. Check some other pairs to verify that interchanging the degrees of freedom numbers will result in different F-values.

TECHNOLOGY INSTRUCTIONS: CUMULATIVE PROBABILITY ASSOCIATED WITH A SPECIFIED VALUE OF F

MINITAB

Choose: **Calc > Probability Distributions > F**
Select: Cumulative Probability Noncentrality parameter: **0.0**
Enter: Numerator degrees of freedom: df_n
            Denominator degrees of freedom: df_d
Select: **Input constant***
Enter: **F-value (ex.1.74) > OK**

*Select the Input column if several F-values are stored in C1. Use C2 for optional storage. If the area in the right tail is needed, subtract the calculated probability from 1.

Excel

If several F-values are to be used, input the values into column A and activate B1; then continue with:

Choose: **Insert function f_x > Statistical > FDIST > OK**
Enter: X: **individual F-value** or **(A1:A5** or select "F-value" cells)*
            Deg_freedom 1: df_n
            Deg_freedom 2: df_d **> OK**

*Drag: Bottom right corner of the B1 cell down to give other probabilities

To find the probability for the left tail (the cumulative probability up to the F-value), subtract the calculated probability from 1.

TI-83/84 Plus

Choose: **2nd > DISTR > 9:Fcdf(**
Enter: 0, F-value, df_n, df_d)

Note: To find the probability between two F-values, enter the two values in place of 0 and the F-value.

If the area in the right tail is needed, subtract the calculated probability from 1.

Use of the *F*-distribution has a condition.

> **Assumptions for inferences about the ratio of two variance:** The samples are randomly selected from normally distributed populations, and the two samples are selected in an independent manner.
>
> **Test Statistic for Equality of Variances**
>
> $$F\star = \frac{s_n^2}{s_d^2}, \text{ with } \text{df}_n = n_n - 1 \text{ and } \text{df}_d = n_d - 1 \qquad (10.16)$$

The sample variances are assigned to the numerator and denominator in the order established by the null and alternative hypotheses for one-tailed tests. The calculated ratio, $F\star$, will have an F-distribution with $\text{df}_n = n_n - 1$ (numerator) and $\text{df}_d = n_d - 1$ (denominator) when the assumptions are met and the null hypothesis is true.

We are ready to use F to complete a hypothesis test about the ratio of two population variances.

E X A M P L E 1 0 . 1 7

ONE-TAILED HYPOTHESIS TEST FOR THE EQUALITY OF VARIANCES

Recall that our soft drink bottling company was to make a decision about the equality of the variances of amounts of fill between its present machine and a modern, high-speed machine. Does the sample information in Table 10.6 present sufficient evidence to reject the null hypothesis (the manufacturer's claim) that the modern, high-speed bottle-filling machine fills bottles with no greater variance than the company's present machine? Assume the amounts of fill are normally distributed for both machines, and complete the test using $\alpha = 0.01$.

TABLE 10.6 Sample Information on Variances of Fills

Sample	n	s^2
Present machine (p)	22	0.0008
Modern, high-speed machine (m)	25	0.0018

Solution

Step 1   **a. Parameter of interest:** $\frac{\sigma_m^2}{\sigma_p^2}$, the ratio of the variances in the amounts of fill placed in bottles for the modern machine versus the company's present machine

b. Statement of hypotheses: The hypotheses were established in Example 10.15 (p. 522):

$$H_o: \frac{\sigma_m^2}{\sigma_p^2} = 1 \ (\leq) \quad (m \text{ is no more variable})$$

$$H_a: \frac{\sigma_m^2}{\sigma_p^2} > 1 \quad (m \text{ is more variable})$$

Note: When the "expected to be larger" variance is in the numerator for a one-tailed test, the alternative hypothesis states, "The ratio of the variances is greater than 1."

Step 2   **a. Assumptions:** The sampled populations are normally distributed (given in the statement of the problem), and the samples are independently selected (drawn from two separate populations).

   b. Test statistic: The F-distribution with the ratio of the sample variances and formula (10.16)

   c. Level of significance: $\alpha = 0.01$

Step 3   **a. Sample information:** See Table 10.6.

   b. Calculated test statistic:
   Using formula (10.16), we have

$$F\star = \frac{s_m^2}{s_p^2}: \qquad F\star = \frac{0.0018}{0.0008} = 2.25$$

   The number of degrees of freedom for the numerator is $df_n = 24$ (or $25 - 1$) because the sample from the modern, high-speed machine is associated with the numerator, as specified by the null hypothesis. Also, $df_d = 21$ because the sample associated with the denominator has size 22.

Step 4 Probability Distribution:

p-Value:	**OR**	**Classical:**

p-Value:

a. Use the right-hand tail because H_a expresses concern for values related to "more than." $\mathbf{P} = P(F\star > 2.25$, with $df_n = 24$ and $df_d = 21)$, as shown in the figure.

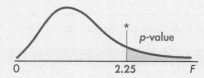

To find the *p*-value, you have two options:
1. Use Tables 9A and 9B (Appendix B) to place bounds on the *p*-value: **0.025 < P < 0.05**.
2. Use a computer or calculator to find the *p*-value: **P = 0.0323**.

Specific instructions follow this example.

b. The *p*-value is not smaller than the level of significance, α (0.01).

Classical:

a. The critical region is the right-hand tail because H_a expresses concern for values related to "more than." $df_n = 24$ and $df_d = 21$. The critical value is obtained from Table 9C: $F(24, 21, 0.01) = \mathbf{2.80}$.

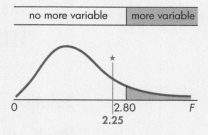

For additional instructions, see page 524.

b. $F\star$ is not in the critical region, as shown in **red** in the figure.

Step 5.   **a. Decision:** Fail to reject H_o.

   b. Conclusion: At the 0.01 level of significance, the samples do not present sufficient evidence to indicate an increase in variance with the new machine.

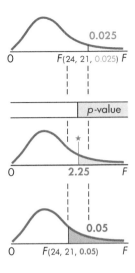

Calculating the *p*-value when using the *F*-distribution

Method 1: Use Table 9 in Appendix B to place bounds on the p-value. Using Tables 9A, 9B, and 9C in Appendix B to estimate the *p*-value is very limited. However, for Example 10.17, the *p*-value can be estimated. By inspecting Tables 9A and 9B, you will find that $F_{(24, 21, 0.025)} = 2.37$ and $F_{(24, 21, 0.005)} = 2.05$. $F\star = 2.25$ is between the values 2.37 and 2.05; therefore, the *p*-value is between 0.025 and 0.05: **0.025 < P < 0.05**. (See figure in margin.)

Method 2: If you are doing the hypothesis test with the aid of a computer or calculator, most likely it will calculate the *p*-value for you, or you may use the cumulative probability distribution commands described on page 524.

Critical *F*-Values for One- and Two-Tailed Tests

The tables of critical values for the *F*-distribution give only the right-hand critical values. This will not be a problem because the right-hand critical value is the only critical value that will be needed. You can adjust the numerator–denominator order so that all the "activity" is in the right-hand tail. There are two cases: one-tailed tests and two-tailed tests.

> *One-tailed tests:* Arrange the null and alternative hypotheses so that the alternative is always "greater than." The $F\star$-value is calculated using the same order as specified in the null hypothesis (as in Example 10.17; also see Example 10.18).

> *Two-tailed tests:* When the value of $F\star$ is calculated, always use the sample with the larger variance for the numerator; this will make $F\star$ greater than 1 and place it in the right-hand tail of the distribution. Thus, you will need only the critical value for the right-hand tail (see Example 10.19).

All hypothesis tests about two variances can be formulated and completed in a way that both the critical value of F and the calculated value of $F\star$ will be in the right-hand tail of the distribution. Since Tables 9A, 9B, and 9C contain only critical values for the right-hand tail, this will be convenient and you will never need critical values for the left-hand tail. The following two examples will demonstrate how this is accomplished.

E X A M P L E   1 0 . 1 8

FORMAT FOR WRITING HYPOTHESES FOR THE EQUALITY OF VARIANCES

Reorganize the alternative hypothesis so that the critical region will be the right-hand tail:

$$H_a: \sigma_1^2 < \sigma_2^2 \quad \text{or} \quad \frac{\sigma_1^2}{\sigma_2^2} < 1 \quad \text{(population 1 is less variable)}$$

Solution

Reverse the direction of the inequality, and reverse the roles of the numerator and denominator.

$$H_a: \sigma_2^2 > \sigma_1^2 \quad \text{or} \quad \frac{\sigma_2^2}{\sigma_1^2} > 1 \quad \text{(population 2 is more variable)}$$

The calculated test statistic $F\star$ will be $\dfrac{s_2^2}{s_1^2}$.

EXAMPLE 10.19

TWO-TAILED HYPOTHESIS TEST FOR THE EQUALITY OF VARIANCES

Find $F\star$ and the critical values for the following hypothesis test so that only the right-hand critical value is needed. Use $\alpha = 0.05$ and the sample information $n_1 = 10$, $n_2 = 8$, $s_1 = 5.4$, and $s_2 = 3.8$.

$$H_o: \sigma_2^2 = \sigma_1^2 \quad \text{or} \quad \frac{\sigma_2^2}{\sigma_1^2} = 1$$

$$H_a: \sigma_2^2 \neq \sigma_1^2 \quad \text{or} \quad \frac{\sigma_2^2}{\sigma_1^2} \neq 1$$

Solution

When the alternative hypothesis is two-tailed ($\neq$), the calculated $F\star$ can be either $F\star = \dfrac{s_1^2}{s_2^2}$ or $F\star = \dfrac{s_2^2}{s_1^2}$. The choice is ours; we need only make sure that we keep df_n and df_d in the correct order. We make the choice by looking at the sample information and using the sample with the larger standard deviation or variance as the numerator. Therefore, in this illustration,

$$F\star = \frac{s_1^2}{s_2^2} = \frac{5.4^2}{3.8^2} = \frac{29.16}{14.44} = \mathbf{2.02}$$

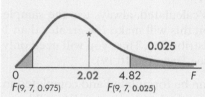

The critical values for this test are left tail, $F(9, 7, 0.975)$, and right tail, $F(9, 7, 0.025)$, as shown in the figure.

Since we chose the sample with the larger standard deviation (or variance) for the numerator, the value of $F\star$ will be greater than 1 and will be in the right-hand tail; therefore, only the right-hand critical value is needed. (All critical values for left-hand tails will be values between 0 and 1.)

TECHNOLOGY INSTRUCTIONS: HYPOTHESIS TEST FOR THE RATIO BETWEEN TWO POPULATION VARIANCES, σ_1^2/σ_2^2, FOR TWO INDEPENDENT SETS OF SAMPLE DATA

MINITAB		
Choose:	**Stat > Basic Statistics > 2 Variances**[*]	
Select:	Data: **Samples in one column:**	
Enter:	Samples: **C1** Subscripts: **C2**	
Or	Select: Data: **Samples in different columns:**	
	Enter: First: **C1** Second: **C2**	
Or	Select: Data: **Sample standard deviations** or **Sample variances**	
	Enter: **Sample size and std dev** or **variance for each sample**	
Select:	**Options**	
Select:	Hypothesized ratio: **StDev 1/StDev 2** or **Variance 1/Variance 2**	
Select:	Alternative: **less than** or **not equal to** or **greater than > OK > OK**	

[*]The 2 Variances procedure evaluates the first sample divided by the second sample.

Excel	Input the data for the numerator (larger spread) into column A and the data for the denominator (smaller spread) into column B; then continue with:

Choose: **Data > Data Analysis > F-Test**: Two-Sample for Variances
Enter: **Variable 1 Range: (A1:A20 or select cells)**
              Variable 2 Range: (B1:B20 or select cells)
Select: **Labels** (if necessary)
Enter: α (ex. 0.05)
Select: **Output Range**
Enter: **(C1 or select cell) > OK**

Use **Home > Cells > Format > AutoFit Column Width** to make the output more readable. The output shows the *p*-value and the critical value for a one-tailed test.

TI-83/84 Plus	Input the data for the numerator (larger spread) into L1 and the data for the denominator (smaller spread) into L2; then continue with the following, entering the appropriate values and highlighting Calculate:

Choose: **STAT > TESTS > D:2-SampFTest…**

```
2-SampFTest
 Inpt:DATA Stats
 List1:L1
 List2:L2
 Freq1:1
 Freq2:1
 σ1:≠σ2 <σ2 >σ2
 Calculate Draw
```

APPLIED EXAMPLE 10.20

PERSONALITY CHARACTERISTICS OF POLICE ACADEMY APPLICANTS

Bruce N. Carpenter and Susan M. Raza concluded that "police applicants are somewhat more like each other than are those in the normative population" when the *F*-test of homogeneity of variance resulted in a *p*-value of less than 0.005. *Homogeneity* means that the group's scores are less variable than the scores for the normative population.

COMPARISONS ACROSS SUBGROUPS AND WITH OTHER POPULATIONS

To determine whether police applicants are a more homogeneous group than the normative population, the *F*-test of homogeneity of variance was used. With the exception of scales F, K, and 6, where the differences are nonsignificant, the results indicate that the police applicants form a somewhat more homogeneous group than the normative population $[F(237, 305) = 1.36, p < 0.005]$. Thus, police applicants are somewhat more like each other than are individuals in the normative population.

Source: Reprinted from the *Journal of Police Science and Administration*, Vol. 15, no. 1, pp. 10–17, copyright held by the International Association of Chiefs of Police, Inc. 515 N. Washington Street, Alexandria, VA 22314. Further reproduction without the express written permission from IACP is strictly prohibited.

⬛ S E C T I O N   1 0 . 5   E X E R C I S E S

10.111 State the null hypothesis, H_o, and the alternative hypothesis, H_a, that would be used to test the following claims:

a. The variances of populations A and B are not equal.

b. The standard deviation of population I is larger than the standard deviation of population II.

c. The ratio of the variances for populations A and B is different from 1.

d. The variability within population C is less than the variability within population D.

10.112 State the null hypothesis, H_o, and the alternative hypothesis, H_a, that would be used to test the following claims:

a. The standard deviation of population X is smaller than the standard deviation of population Y.

b. The ratio of the variances of population A over population B is greater than 1.

c. The standard deviation of population Q_1 is at most that of population Q_2.

d. The variability within population I is more than the variability within population II.

10.113 Explain why the inequality $\sigma_m^2 > \sigma_p^2$ is equivalent to $\dfrac{\sigma_m^2}{\sigma_p^2} > 1$.

10.114 Express the H_o and H_a of Example 10.17 (p. 525) equivalently in terms of standard deviations.

10.115 Using the $F_{(df_1, df_2, \alpha)}$ notation, name each of the critical values shown on the following figures.

a.

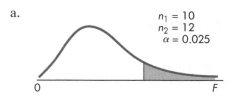
$n_1 = 10$
$n_2 = 12$
$\alpha = 0.025$

b.

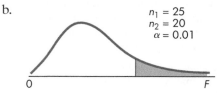
$n_1 = 25$
$n_2 = 20$
$\alpha = 0.01$

c.

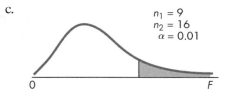
$n_1 = 9$
$n_2 = 16$
$\alpha = 0.01$

d.

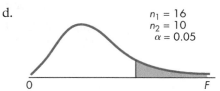
$n_1 = 16$
$n_2 = 10$
$\alpha = 0.05$

10.116 Find the values of $F_{(12, 24, 0.01)}$ and $F_{(24, 12, 0.01)}$.

10.117 Find the following critical values for F from Tables 9A, 9B, and 9C in Appendix B.

a. $F_{(24, 12, 0.05)}$ e. $F_{(15, 18, 0.025)}$

b. $F_{(30, 40, 0.01)}$ f. $F_{(15, 9, 0.025)}$

c. $F_{(12, 10, 0.05)}$ g. $F_{(40, 30, 0.05)}$

d. $F_{(5, 20, 0.01)}$ h. $F_{(8, 40, 0.01)}$

10.118 Determine the p-value that would be used to test the following hypotheses when F is used as the test statistic:

a. $H_o: \sigma_1 = \sigma_2$ versus $H_a: \sigma_1 > \sigma_2$, with $n_1 = 10, n_2 = 16$, and $F\star = 2.47$

b. $H_o: \sigma_1^2 = \sigma_2^2$ versus $H_a: \sigma_1^2 > \sigma_2^2$, with $n_1 = 25, n_2 = 21$, and $F\star = 2.31$

c. $H_o: \dfrac{\sigma_1^2}{\sigma_2^2} = 1$ versus $H_a: \dfrac{\sigma_1^2}{\sigma_2^2} \neq 1$, with $n_1 = 41, n_2 = 61$, and $F\star = 4.78$

d. $H_o: \sigma_1 = \sigma_2$ versus $H_a: \sigma_1 < \sigma_2$, with $n_1 = 10, n_2 = 16$, and $F\star = 2.47$

10.119 Find the critical value for the hypothesis test with $H_a: \sigma_1 > \sigma_2$, with $n_1 = 7, n_2 = 10$, and $\alpha = 0.05$.

10.120 Determine the critical region and critical value(s) that would be used to test the following hypotheses using the classical approach when $F\star$ is used as the test statistic.

a. $H_o: \sigma_1^2 = \sigma_2^2$ versus $H_a: \sigma_1^2 > \sigma_2^2$, with $n_1 = 10, n_2 = 16$, and $\alpha = 0.05$

b. $H_o: \dfrac{\sigma_1^2}{\sigma_2^2} = 1$ versus $H_a: \dfrac{\sigma_1^2}{\sigma_2^2} \neq 1$, with $n_1 = 25$, $n_2 = 31$, and $\alpha = 0.05$

c. $H_o: \dfrac{\sigma_1^2}{\sigma_2^2} = 1$ versus $H_a: \dfrac{\sigma_1^2}{\sigma_2^2} > 1$, with $n_1 = 10$, $n_2 = 10$, and $\alpha = 0.01$

d. $H_o: \sigma_1 = \sigma_2$ versus $H_a: \sigma_1 < \sigma_2$, with $n_1 = 25$, $n_2 = 16$, and $\alpha = 0.01$

10.121 Calculate $F\star$ given $s_1 = 3.2$ and $s_2 = 2.6$.

10.122 Calculate $F\star$ given $s_1^2 = 3.2$ and $s_2^2 = 2.6$.

10.123 What would be the value of $F\star$ in Example 10.19 if $F\star = \dfrac{s_2^2}{s_1^2}$ were used? Why is it less than 1?

10.124 Do the lengths of boys' names have more variation than the lengths of girls' names? With current names like "Nathaniel" and "Christopher" versus "Ian" and "Jack," it certainly appears that boys' names cover a wide range with respect to their length. To test this theory, random samples of seventh-grade girls and boys were taken.

Boys' Names	$n = 30$	$s = 1.870$
Girls' Names	$n = 30$	$s = 1.456$

At the 0.05 level of significance, do the data support the contention that the lengths of boys' names have more variation than the lengths of girls' names?

10.125 Two independent random samples of size 25 were taken from an English class and a chemistry class at a local community college. Students in both classes were asked to draw a 3-inch line to the best of their ability without any measuring device (ruler, etc.). The following data resulted:

English	$n = 25$	$\bar{x} = 2.660$	$s = 0.617$
Chemistry	$n = 25$	$\bar{x} = 2.750$	$s = 0.522$

At the 0.05 level of significance, is there a difference between the standard deviations for 3-inch-line measurements from the English and chemistry classes?

10.126 A study in *Pediatric Emergency Care* compared the injury severity between younger and older children. One measure reported was the Injury Severity Score (ISS). The standard deviation of ISS's for 37 children 8 years or younger was 23.9, and the standard deviation for 36 children older than 8 years was 6.8. Assume that ISS's are normally distributed for both age groups. At the 0.01 level of significance, is there sufficient reason to conclude that the standard deviation of ISS's for younger children is larger than the standard deviation of ISS's for older children?

10.127 A bakery is considering buying one of two gas ovens. The bakery requires that the temperature remain constant during a baking operation. A study was conducted to measure the variance in temperature of the ovens during the baking process. The variance in temperature before the thermostat restarted the flame for the Monarch oven was 2.4 for 16 measurements. The variance for the Kraft oven was 3.2 for 12 measurements. Does this information provide sufficient reason to conclude that there is a difference in the variances for the two ovens? Assume measurements are normally distributed and use a 0.02 level of significance.

10.128 [EX10-128] A study was conducted to determine whether or not there was equal variability in male and female systolic blood pressure readings. Random samples of 16 men and 13 women were used to test the experimenter's claim that the variances were unequal. MINITAB was used to calculate the standard deviations, $F\star$, and the p-value. Assume normality.

Men	120	120	118	112	120	114	130	114	124	125
	130	100	120	108	112	122				

Women	122	102	118	126	108	130	104	116	102	122
	120	118	130							

Standard deviation of Men = 7.8864
Standard deviation of Women = 9.9176
F-Test (normal distribution)
Test Statistic: 1.581
P-Value: 0.398

Verify these results by calculating the values yourself.

10.129 When a hypothesis test is two-tailed and Excel is used to calculate the p-value, what additional step must be taken?

10.130 Referring to Applied Example 10.20 (p. 529):

a.   What null and alternative hypotheses did Carpenter and Raza test?

b.   What does "$p < 0.005$" mean?

c.   Use a computer or calculator to calculate the p-value for $F(237, 305) = 1.36$.

10.131 [EX10-131] The quality of an end product is somewhat determined by the quality of the materials used. Textile mills monitor the tensile strength of the fibers used in weaving their yard goods. The following independent random samples are tensile strengths of cotton fibers from two suppliers.

Supplier A	78	82	85	83	77	84	90	82	93	82
	80	82	77	80	80					

Supplier B	76	79	83	78	72	73	69	80	74	77
	78	78	73	76	78	79				

Calculate the observed value of F, $F\star$, for comparing the variances of these two sets of data.

10.132 [EX10-132] Several counties in Minnesota and Wisconsin were randomly selected and information about the 2007 sweet corn crop was collected. The following yield rates in tons of sweet corn per acre harvested resulted.

Yield, MN	7.38	6.27	6.26	6.02	7.38	6.32	6.74	5.13	6.54	5.71
Yield, WI	7.0	6.9	7.7	8.0	6.5	5.8	6.8	7.5		

Source: http://www.usda.gov/

a.   Is there a difference in the variability of county yields as measured by the standard deviation of the yield rate? Use $\alpha = 0.05$.

b.   Is the mean yield rate in Wisconsin significantly higher than the mean yield rate in Minnesota? Use $\alpha = 0.05$.

10.133 [EX10-133] Professional athletes' salaries are often criticized for being "over the top." Many also earn even more money through endorsements. Various NBA (National Basketball Association) and MLB (Major League Baseball) players can be seen in many high-profile endorsements. Two random samples from each sport's player endorsements were selected and yielded the following amounts in millions of dollars.

NBA	16.0	15.0	21.7	12.0	9.5	0.5	15.5	0.8	2.5	5.0	16.0	15.5
MLB	6.0	8.0	2.5	0.3	3.5	0.5	0.5	0.3				

a.   At the 0.05 level of significance, is there a difference in the variability of endorsement amounts between NBA and MLB players? Assume normality of endorsement amounts.

b.   At the 0.05 level of significance, is the mean endorsement amount for NBA players significantly higher than the mean endorsement amount for MLB players?

10.134 [EX10-134] The data that follow are from two random samples of 37 college males and 42 college females with respect to their commute times to college.

Time—M

15	12	30	15	10	23	20	13	25	20	15	20	23	15
20	15	18	15	20	20	8	10	15	18	20	15	25	20
10	25	18	18	20	27	25	20	7					

Time—F

32	15	20	35	45	20	10	5	35	25	14	25	28	35
30	24	28	15	30	30	30	40	25	20	18	20	15	30
24	30	25	20	10	60	20	25	27	25	40	22	25	25

a.   Do the distributions of "commute time" for males and females appear to be similar in distribution? Center? Spread? Discuss your responses.

b.   Construct a histogram and find the mean and standard deviation for each set of data.

c.   Is it possible that both of the samples were drawn from normal populations? Justify your answer.

d.   Is the mean commute time for females statistically greater than the mean commute time for males? Use $\alpha = 0.05$.

e.   Is there sufficient evidence to show that the standard deviations of these two samples are statistically different? Use $\alpha = 0.05$.

10.135 [EX10-135] A constant objective in the manufacture of contact lenses is improving the level and variation for those features that affect lens power and visual acuity. One such feature involves the tooling from which lenses are ultimately manufactured.

The results of two initial process development runs were examined for Critical Feature A. Two distinct product lots were manufactured with slight differences designed to affect the feature in question. Each lot was then sampled. Lot 1 had a smaller run time than Lot 2, and therefore more samples were taken from Lot 2.

a.   Calculate the mean and standard deviation of Critical Feature A for Lots 1 and 2.

b.   Is there evidence of a difference in the variability of Critical Feature A between Lots 1 and 2? Use an alpha value of 5% to make a determination.

c.   Is there evidence of a difference in the mean levels of Critical Feature A between Lots 1 and 2? Use an alpha value of 5% to make a determination.

Lot 1 Sample	Critical Feature A	Lot 2 Sample	Critical Feature A	Lot 2 Sample	Critical Feature A
1	0.017	1	0.026	14	0.041
2	0.021	2	0.027	15	0.021
3	0.006	3	0.024	16	0.022
4	0.009	4	0.023	17	0.027
5	0.018	5	0.034	18	0.032
6	0.021	6	0.035	19	0.023
7	0.013	7	0.035	20	0.023
8	0.017	8	0.033	21	0.024
		9	0.034	22	0.017
		10	0.033	23	0.023
		11	0.032	24	0.019
		12	0.038	25	0.027
		13	0.041		

10.136 [EX10-136] Americans snooze more on the weekends, according to a poll of 1506 adults for the National Sleep Foundation and reported in a *USA Today* Snapshot during April 2005.

Hours of Sleep	Weekdays	Weekends
Less than 6	0.16	0.10
6–6.9	0.24	0.15
7–7.9	0.31	0.24
8 or more	0.26	0.49

Two independent random samples were taken at a large industrial complex. The workers selected in one sample were asked, "How many hours, to the nearest 15 minutes, did you sleep on Tuesday night this week?" The workers selected for the second sample were asked, "How many hours, to nearest 15 minutes, did you sleep on Saturday night last weekend?"

	Weekday			Weekend	
5.00	7.75	7.25	9.00	7.25	8.75	7.50
9.25	7.25	8.75	6.25	5.25	9.25	9.25
7.00	7.75	6.75	7.50	8.50	8.75	6.50
9.25	7.00	7.75	8.00	8.75	9.50	8.00
9.25	9.25	6.00	8.75	7.75	8.75	7.50

a. Construct a histogram and find the mean and standard deviation for each set of data.

b. Do the distributions of "hours of sleep on weekday" and "hours of sleep on weekend" resulting from the poll appear to be similar in shape? Center? Spread? Discuss your responses.

c. Is it possible that both of the samples were drawn from normal populations? Justify your answer.

d. Is the mean number of hours slept on the weekend statistically greater than the mean number of hours slept on the weekdays? Use $\alpha = 0.05$.

e. Is there sufficient evidence to show that the standard deviations of these two samples are statistically different? Use $\alpha = 0.05$.

f. Explain how the answers to parts b–e now affect your thoughts about your answer to part a.

10.137 [EX10-137] What dollar amount should someone spend on a Valentine's Day gift for you? The results found by a Greenfield Online survey of 653 respondents are pictured in the accompanying graphic.

Random samples were selected in central New York State with the following results.

Men	103	100	100	67	77	63	55	43	139	2	51	5	100	52	139
	86	23	56	40	84	15	32	157	35	4	24	102	52	43	75
	128	206	16	13	98										

Women	36	5	77	97	25	62	91	170	108	112	198	161	54	40	111
	107	241	89	37	10	175	10	84	102	17	32	25	1	38	126
	121	30	147	135	45	230	29	88							

a. Construct a histogram and find the mean and standard deviation for each set of central New York State data.

What Dollar Amount Should Someone Spend on a Valentine's Day Gift for You?

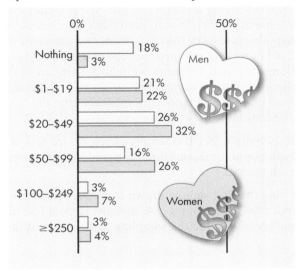

Source: Data from Darryl Haralson and Karl Gelles, USA TODAY; Greenfield Online survey of 653 respondents. Margin of error ±3 percentage points.

b. Do the shapes of "amounts spent" suggested by males and females in central New York State appear to be similar in shape? Center? Spread? Discuss your responses.

c. Is it possible that both of the samples were drawn from normal populations? Justify your answer.

d. Is the mean amount stated by women statistically greater than the mean amount stated by men? Use $\alpha = 0.05$.

e. Is there sufficient evidence to show that the standard deviations of these two samples are statistically different? Use $\alpha = 0.05$.

f. Explain how your answers to parts b–e now affect your thoughts about your answer to part a.

10.138 a. Two independent samples, each of size 3, are drawn from a normally distributed population. Find the probability that one of the sample variances is at least 19 times larger than the other one.

b. Two independent samples, each of size 6, are drawn from a normally distributed population. Find the probability that one of the sample variances is no more than 11 times larger than the other one.

10.139 Use a computer to demonstrate the truth of the theory presented in this section.

a.   The underlying assumptions are "the populations are normally distributed," and while conducting a hypothesis test for the equality of two standard deviations, it is assumed that the standard deviations are equal. Generate very large samples of two theoretical populations: $N(100, 20)$ and $N(120, 20)$. Find graphic and numerical evidence that the populations satisfy the assumptions.

b.   Randomly select 100 samples, each of size 8, from both populations and find the standard deviation of each sample.

c.   Using the first sample drawn from each population as a pair, calculate the $F\star$-statistic. Repeat for all samples. Describe the sampling distribution of the

100 $F\star$-values using both graphic and numerical statistics.

d.   Generate the probability distribution for $F_{(7, 7)}$, and compare it with the observed distribution of $F\star$. Do the two graphs agree? Explain.

FYI See the *Student Solutions Manual* for additional information about commands.

10.140 It was stated in this section that the F-test is very sensitive to minor departures from the assumptions. Repeat Exercise 10.139 using $N(100, 20)$ and $N(120, 30)$. Notice that the only change from Exercise 10.139 is the seemingly slight increase in the standard deviation of the second population. Answer the same questions using the same kind of information and you will see very different results.

Chapter Review

In Retrospect

In this chapter we began the comparisons of two populations by distinguishing between independent and dependent samples, which are statistically important and useful sampling procedures. We then proceeded to examine the inferences concerning the comparison of means, proportions, and variances for two populations.

We are always making comparisons between two groups. We compare means and we compare proportions. In this chapter we have learned how to statistically compare two populations by making inferences about

their means, proportions, or variances. For convenience, Table 10.7 identifies the formulas to use when making inferences about comparisons between two populations.

In Chapters 8, 9, and 10 we have learned how to use confidence intervals and hypothesis tests to answer questions about means, proportions, and standard deviations for one or two populations. From here we can expand our techniques to include inferences about more than two populations as well as inferences of different types.

TABLE 10.7 Formulas to Use for Inferences Involving Two Populations

Situations	Test Statistic	Formula to Be Used — Confidence Interval	Formula to Be Used — Hypothesis Test
Difference between two means			
Dependent samples	t	Formula (10.2) (p. 483)	Formula (10.5) (p. 486)
Independent samples	t	Formula (10.8) (p. 497)	Formula (10.9) (p. 498)
Difference between two proportions	z	Formula (10.11) (p. 512)	Formula (10.15) (p. 515)
Difference between two variances	F		Formula (10.16) (p. 525)

CourseMate The **Statistics CourseMate** site for this text brings chapter topics to life with interactive learning, study, and exam preparation tools, including quizzes and flashcards for the Vocabulary and Key Concepts that follow. The site also provides an **eBook** version of the text with highlighting and note taking capabilities. Throughout chapters, the CourseMate icon 🖥 flags concepts and examples that have corre- sponding interactive resources such as **video** and **animated tutorials** that demonstrate, step by step, how to solve problems; **datasets** for exercises and examples; **Skillbuilder Applets** to help you better understand concepts; **technology manuals**; and software to download including **Data Analysis Plus** (a suite of statistical macros for Excel) and **TI-83/84 Plus** programs—logon at **www.cengagebrain.com**.

Vocabulary and Key Concepts

assumptions (pp. 482, 483, 496, 512, 525)
binomial experiment (p. 511)
binomial p (p. 511)
confidence interval (pp. 483, 497, 512)
dependent means (p. 483)
dependent samples (pp. 478, 479, 480)
F-distribution (p. 523)
F-statistic (p. 525)
hypothesis test (pp. 486, 498, 515, 525)

independent means (p. 495)
independent samples (pp. 478, 497, 498, 511)
mean difference (pp. 483, 486)
mean of the paired differences (p. 483)
paired difference (p. 482)
percentage (p. 511)
pooled observed probability (p. 515)
probability (p. 511)

proportion (p. 511)
p-value (pp. 487, 489, 500, 501)
source (of data) (p. 478)
standard error (pp. 486, 496, 511, 514)
t-distribution (pp. 483, 496)
test statistic (pp. 486, 498, 514, 515, 525)
t-statistic (pp. 486, 498)
unbiased sample statistic (p. 512)
z-statistic (p. 514)

Learning Outcomes

• Understand the difference between dependent and independent samples.	EXP 10.1, 10.2, Ex. 10.5, 10.7
• Understand that the mean difference (mean of the paired differences) should be used to analyze dependent samples.	pp. 482–483
• Compute and/or understand how to compute the mean difference and standard deviation for paired data.	p. 484, Ex. 10.13
• Compute, describe, and interpret a confidence interval for the population mean difference.	EXP 10.4, Ex. 10.19
• Perform, describe, and interpret a hypothesis test for the population mean difference, μ_d, using the p-value approach and/or classical approach.	EXP 10.5, 10.6, Ex. 10.37, 10.148
• Understand that the difference between two means should be used to analyze independent samples.	pp. 495–497
• Understand how to determine the t-distribution's degrees of freedom for the difference between means using two independent samples.	p. 496, Ex. 10.43
• Compute, describe, and interpret a confidence interval for the difference between two means using independent samples.	EXP 10.8, Ex. 10.45, 10.47
• Perform, describe, and interpret a hypothesis test for the difference between two population means, $\mu_1 - \mu_2$, using the p-value approach and/or classical approach.	EXP 10.9, 10.10, Ex. 10.68, 10.72
• Understand that the z-distribution will be used to analyze the difference between two proportions using independent samples, provided the guidelines to ensure normality are met.	pp. 511–512
• Compute sample proportions based on sample size and number of successes.	Ex. 10.83
• Compute, describe, and interpret a confidence interval for the difference between two proportions using independent samples.	EXP 10.12, Ex. 10.89, 10.163

- Perform, describe, and interpret a hypothesis test for the difference between two population proportions, $p_1 - p_2$, using the p-value approach and/or classical approach.   EXP 10.13, Ex. 10.109, 10.168

- Understand the properties of the F-distribution and how it is a series of distributions based on sample sizes (using pairs of numbers for degrees of freedom as the index).   pp. 523–524, EXP 10.15, 10.16, 10.18, Ex. 10.116, 10.117

- Understand that the assumption for inferences about the ratio of two variances is that the sampled populations are normally distributed and that the two samples are selected independently.   p. 525

- Perform, describe, and interpret a hypothesis test for the ratio of two population variances, $\dfrac{\sigma_1^2}{\sigma_2^2}$, using the F-distribution with the p-value approach and/or classical approach.   EXP 10.17, Ex. 10.127, 10.169

 Chapter Exercises

10.141 Eighteen diamonds are identified and each is appraised for retail sale value by two qualified, licensed appraisers. Do the resulting two sets of data represent dependent or independent samples? Explain.

10.142 A chemist is testing a proposed analytical method and has no established standards to compare it with, so she decides to use the currently accepted method for comparison. She takes a specimen of unknown concentrate and determines its concentration 12 times using the proposed method. She then takes another specimen of same unknown concentrate and determines its concentration 12 times using the current method. Do these two samples represent dependent or independent samples? Explain.

10.143 [EX10-143] Using a 95% confidence interval, estimate the mean difference in IQ between the oldest and the youngest members (brothers and sisters) of a family based on the following random sample of IQs. Assume normality.

Oldest	145	133	116	128	85	100	105	150	97	110	120	130
Youngest	131	119	103	93	108	100	111	130	135	113	108	125

10.144 [EX10-144] The diastolic blood pressure readings for 15 patients were determined using two techniques: the standard method used by medical personnel and a method using an electronic device with a digital readout. The results were as follows:

Patient	1	2	3	4	5	6	7	8	9	10	11	12	13	14	15
Standard method	72	80	88	80	80	75	92	77	80	65	69	96	77	75	60
Digital method	70	76	87	77	81	75	90	75	82	64	72	95	80	70	61

Assuming blood pressure is normally distributed, determine the 90% confidence interval for the mean difference in the two readings, where d = standard method − digital readout.

10.145 [EX10-145] We want to know which of two types of filters should be used. A test was designed in which the strength of a signal could be varied from zero to the point where the operator first detects the image. At that point, the intensity setting is recorded. Lower settings are better. Twenty operators were asked to make one reading for each filter.

Operator	1	2	3	4	5	6	7	8	9	10
Filter 1	96	83	97	93	99	95	97	91	100	92
Filter 2	92	84	92	90	93	91	92	90	93	90

Operator	11	12	13	14	15	16	17	18	19	20
Filter 1	88	89	85	94	90	92	91	78	77	93
Filter 2	88	89	86	91	89	90	90	80	80	90

Assuming the intensity readings are normally distributed, estimate the mean difference between the two readings using a 90% confidence interval.

10.146 [EX10-146] At the end of their first day at training camp, 10 new recruits participated in a rifle-shooting competition. The same 10 competed again at the end of a full week of training and practice. Their resulting scores are shown in the following table.

	Recruit									
Time of Competition	1	2	3	4	5	6	7	8	9	10
First day	72	29	62	60	68	59	61	73	38	48
One week later	75	43	63	63	61	72	73	82	47	43

Does this set of 10 pairs of data show that there was a significant amount of improvement in the recruits' shooting abilities during the week? Use $\alpha = 0.05$ and assume normality.

10.147 [EX10-147] Salt-free diets are often prescribed to people with high blood pressure. The following data were obtained from an experiment designed to estimate the reduction in diastolic blood pressure as a result of following a salt-free diet for two weeks. Assume diastolic readings are normally distributed.

Before	93	106	87	92	102	95	88	110
After	92	102	89	92	101	96	88	105

a. What is the point estimate for the mean reduction in the diastolic reading after two weeks on this diet?

b. Can you conclude that the salt-free diet produced a significant mean reduction in diastolic blood pressure? Use the 0.01 level of significance.

10.148 [EX10-148] Immediate-release medications quickly deliver their drug content, with the maximum concentration reached in a short time; sustained-release medications, on the other hand, take longer to reach maximum concentration. As part of a study, immediate-release codeine (irc) was compared with sustained-release codeine (src) using 13 healthy patients. The patients were randomly assigned to one of the two types of codeine and treated for 2.5 days; after a 7-day washout period, each patient was given the other type of codeine. Thus, each patient received both types. The total amount (A) of drug available over the life of the treatment in (ng·mL)/hr follows:

Patient	1	2	3	4	5	6	7
Airc	1091.3	1064.5	1281.1	1921.4	1649.9	1423.6	1308.4
Asrc	1308.5	1494.2	1382.2	1978.3	2004.6	*	1211.1

Patient	8	9	10	11	12	13
Airc	1192.1	766.2	978.6	1618.9	582.9	972.1
Asrc	1002.4	866.6	1345.8	979.2	576.3	999.1

Source: http://exploringdata.cqu.edu.au/

a. Explain why this is a paired-difference design.

b. What adjustment is needed since there is no Asrc for patient 6?

Is there a significant difference in the total amount of drug available over the life of the treatment?

c. Check the test assumptions and describe your findings.

d. Test the claim using $\alpha = 0.05$.

10.149 A test that measures math anxiety was given to 50 male and 50 female students. The results were as follows:

Males: $\bar{x} = 70.5, s = 13.2$
Females: $\bar{x} = 75.7, s = 13.6$

Construct a 95% confidence interval for the difference between the mean anxiety scores.

10.150 The same achievement test is given to soldiers selected at random from two units. The scores they attained are summarized as follows:

Unit 1: $n_1 = 70, \bar{x}_1 = 73.2, s_1 = 6.1$
Unit 2: $n_2 = 60, \bar{x}_2 = 70.5, s_2 = 5.5$

Construct a 90% confidence interval for the difference in the mean level of the two units.

10.151 Experimentation with a new rocket nozzle has led to two slightly different designs. The following data summaries resulted from testing these two designs.

	n	Σx	Σx^2
Design 1	36	278.4	2163.76
Design 2	42	310.8	2332.26

Determine the 99% confidence interval for the difference in the means for these two rocket nozzles.

10.152 [EX10-152] An achievement test in a beginning computer science course was administered to two groups. One group had had a previous computer science course in high school; the other group had not. The test results follow. Assuming test scores are normal, construct a 98% confidence interval for the difference between the two population means.

Group 1 (had high school course)	17	18	27	19	24	36	27	26	35	22	18
	29	29	26	33							
Group 2 (no high school course)	19	25	28	27	21	24	18	14	28	21	22
	20	21	14	29	28	25	17	20	28	31	27

10.153 [EX10-153] Two methods were used to study the latent heat of ice fusion. Both method A (an electrical method) and method B (a method of mixtures) were conducted with the specimens cooled to $-0.72°C$. The data in the following table represent the change in total heat from $0.72°C$ to water at $0°C$ in calories per gram of mass.

Method A	79.98	80.04	80.02	80.04	80.03	80.03	80.04
	79.97	80.05	80.03	80.02	80.00	80.02	
Method B	80.02	79.94	79.98	79.97	79.97	80.03	79.95
	79.97						

Assuming normality, construct a 95% confidence interval for the difference between the means.

10.154 [EX10-154] Sucrose, ordinary table sugar, is probably the single most abundant pure organic chemical in the world and the one most widely known to nonchemists. Whether from sugar cane (20% by weight) or sugar beets (15% by weight), and whether raw or refined, common sugar is still sucrose. Fifteen U.S. sugar

(continue on page 538)

beet–producing counties were randomly selected and their sucrose percentages recorded. Similarly, 12 U.S. sugar cane–producing counties were randomly selected and their sucrose percentages recorded.

SBsucrose	17.30	16.46	16.20	17.53	17.00	18.53	16.77	16.11
	15.30	17.90	15.98	17.30	17.94	17.30	16.60	
SCsucrose	14.1	13.5	15.2	15.0	13.6	13.6	11.7	14.3	13.8	13.8
	14.8	13.7								

Source: http://www.usda.gov/

Find the 95% confidence interval for the difference between the mean sucrose percentage for all U.S. sugar beet–producing counties and all U.S. sugar cane–producing counties.

10.155 George Johnson is the head coach of a college football team that trains and competes at home on artificial turf. George is concerned that the 40-yard sprint time recorded by his players and others increases substantially when running on natural turf as opposed to artificial turf. If so, there is little comparison between his players' speed and that of their opponents whenever his team plays on grass. The next opponent of George's team plays on grass, so he surveyed all the starters in the next game and obtained their best 40-yard sprint times. He then compared them with the best times turned in by his own players. The results are shown in the table that follows:

Player Group	n	Mean (sec)	Std. Dev.
Artificial turf	22	4.85	0.31
Grass	22	4.96	0.42

Do Coach Johnson's players have a lower mean sprint time? Assuming normality, test at the 0.05 level of significance to advise Coach Johnson.

a. Solve using the *p*-value approach.

b. Solve using the classical approach.

10.156 A test concerning some of the fundamental facts about acquired immunodeficiency syndrome (AIDS) was administered to two groups, one consisting of college graduates and the other consisting of high school graduates. A summary of the test results follows:

College graduates: $n = 75, \bar{x} = 77.5, s = 6.2$
High school graduates: $n = 75, \bar{x} = 50.4, s = 9.4$

Do these data show that the college graduates, on average, scored significantly higher on the test? Use $\alpha = 0.05$.

10.157 Some 20 million Americans visit chiropractors annually, and the number of practitioners in the United States is 55,000, nearly double the number two decades ago, according to the American Chiropractic Association. The *New England Journal of Medicine* released a report showing the results of a study that compared chiropractic spinal manipulation (CSM) with physical therapy for treatment of acute lower back pain. After 2 years of treatment, CSM was found to be no more effective at either reducing missed work or preventing a relapse.

Suppose a similar study of 60 patients is made by dividing the sample into two groups. For 1 year, one group is given CSM and the other, physical therapy. During the 1-year period, the number of missed days at work as a result of lower back pain is measured:

Group	n	Mean	Std. Dev.
CSM (1)	32	10.6	4.8
Therapy (2)	28	12.5	6.3

Do these results show that the mean number of missed days of work for people suffering from acute back pain is significantly less for those receiving CSM than for those undergoing physical therapy? Assume normality and use a 0.01 level of significance.

a. Solve using the *p*-value approach.

b. Solve using the classical approach.

10.158 To compare the merits of two short-range rockets, 8 of the first kind and 10 of the second kind are fired at a target. If the first kind has a mean target error of 36 feet and a standard deviation of 15 feet and the second kind has a mean target error of 52 feet and a standard deviation of 18 feet, does this indicate that the second kind of rocket is less accurate than the first? Use $\alpha = 0.01$ and assume normal distribution for target error.

10.159 [EX10-159] The material used in making parts affects not only how long the part lasts but also how difficult it is to repair. The following measurements are for screw torque removal for a specific screw after several operations of use. The first row lists the part number, the second row lists the screw torque removal measurements for assemblies made with material A, and the third row lists the screw torque removal measurements for assemblies made with material B. Assume torque measurements are normally distributed.

Removal Torque (NM, Newton-meters)

Part Number	1	2	3	4	5	6	7	8	9	10	11	12	13	14	15
Material A	16	14	13	17	18	15	17	16	14	16	15	17	14	16	15
Material B	11	14	13	13	10	15	14	12	11	14	13	12	11	13	12

Source: Problem data provided by AC Rochester Division, General Motors, Rochester, NY

a. Find the sample mean, variance, and standard deviation for the material A data.

b. Find the sample mean, variance, and standard deviation for the material B data.

c. At the 0.01 level, do these data show a significant difference in the mean torque required to remove the screws from the two different materials?

10.160 [EX10-160] A group of 17 students participated in an evaluation of a special training session that claimed to improve memory. The students were randomly assigned to two groups: group A, the test group, and group B, the control group. All 17 students were tested for their ability to remember certain material. Group A was given the special training; group B was not. After 1 month, both groups were tested again, with the results as shown in the following table. Do these data support the alternative hypothesis that the special training is effective at the $\alpha = 0.01$ level of significance? Assume normality.

	Group A Students									Group B Students							
Time of Test	1	2	3	4	5	6	7	8	9	10	11	12	13	14	15	16	17
Before	23	22	20	21	23	18	17	20	23	22	20	23	17	21	19	20	20
After	28	29	26	23	31	25	22	26	26	23	25	26	18	21	17	18	20

10.161 [EX10-161] At a large university, a mathematics placement exam is administered to all students. This exam has a history of producing scores with a mean of 77. Samples of 36 male students and 30 female students are randomly selected from this year's student body and the following scores recorded.

Male	72	68	75	82	81	60	75	85	80	70
	71	84	68	85	82	80	54	81	86	79
	99	90	68	82	60	63	67	72	77	51
	61	71	81	74	79	76				

Female	81	76	94	89	83	78	85	91	83	83
	84	80	84	88	77	74	63	69	80	82
	89	69	74	97	73	79	55	76	78	81

a. Describe each set of data with a histogram (use the same class intervals on both histograms), mean, and standard deviation.

b. Test the hypotheses, "Mean score for all males is 77" and "Mean score for all females is 77," using $\alpha = 0.05$.

c. Do the preceding results show that the mean scores for males and females are the same? Justify your answer. Be careful!

d. Test the hypothesis, "There is no difference between the mean scores for male and female students" using $\alpha = 0.05$.

e. Do the results found in part d show that the mean scores for males and females are the same? Explain.

f. Explain why the results found in part b cannot be used to conclude, "The two means are the same."

10.162 A survey was conducted to determine the proportion of Democrats as well as Republicans who support a "get tough" policy in South America. The results of the survey were as follows:

Democrats: $n = 250$, number in support = 120
Republicans: $n = 200$, number in support = 105

Construct the 98% confidence interval for the difference between the proportions of support.

10.163 A consumer group compared the reliability of two comparable microcomputers from two different manufacturers. The proportion requiring service within the first year after purchase was determined for samples from each of two manufacturers.

Manufacturer	Sample Size	Proportion Needing Service
1	75	0.15
2	75	0.09

Find a 0.95 confidence interval for $p_1 - p_2$.

10.164 "It's a draw," according to two Australian researchers. "By age 25, up to 29% of all men and up to 34% of all women have some gray hair, but this difference is so small that it's considered insignificant" ("Silver Threads Among the Gold: Who'll Find Them First, a Man or a Woman?" *Family Circle*).

a. If 1000 men and 1000 women were involved in this research, construct a 95% confidence interval to estimate the true difference.

b. Does the confidence interval found in part a indicate that the difference is significant? Explain.

10.165 According to "Venus vs. Mars" in the May/June 2005 issue of *Arthritis Today,* men and women may be more alike than we think. The Boomers Wellness Lifestyle Survey of men and women aged 35 to 65 found that 88% of

(continue on page 540)

women considered "managing stress" important for maintaining overall well-being. The same survey found that 75% of men consider "managing stress" important. Answer the following and give details to support each of your answers.

a. If these statistics came from samples of 100 men and 100 women, is the difference significant?

b. If these statistics came from samples of 150 men and 150 women, is the difference significant?

c. If these statistics came from samples of 200 men and 200 women, is the difference significant?

d. What effects did the increase in sample size have on the solutions in parts a–c?

10.166 A study in the *New England Journal of Medicine* reported that based on 987 deaths in southern California, right-handers died at an average age of 75 and left-handers died at an average age of 66. In addition, it was found that 7.9% of the lefties died from accident-related injuries, excluding vehicles, versus 1.5% for the right-handers; and 5.3% of the left-handers died while driving vehicles versus 1.4% of the right-handers.

Suppose you examine 1000 randomly selected death certificates, of which 100 were left-handers and 900 were right-handers. If you found that 5 of the left-handers and 18 of the right-handers died while driving a vehicle, would you have evidence to show that the proportion of left-handers who die at the wheel is significantly higher than the proportion of right-handers who die while driving? Calculate the *p*-value and interpret its meaning.

10.167 Who wins disputed cases whenever there is a change made to the tax laws, the taxpayer or the Internal Revenue Service (IRS)? The latest trend indicates that the burden of proof in all court cases has shifted from taxpayers to the IRS, which tax experts predict could set off more intrusive questioning. Of the accountants, lawyers, and other tax professionals surveyed by RIA Group, a tax-information publisher, 55% expect at least a slight increase in taxpayer wins.

Source: *Fortune*, "Tax Reform?"

Suppose samples of 175 accountants and 165 lawyers were asked, "Do you expect taxpayers to win more court cases because of the new burden of proof rules?" Of those surveyed, 101 accountants replied "yes" and 84 lawyers said

"yes." Do the two expert groups differ in their opinions? Use a 0.01 level of significance to answer the question.

a. Solve using the *p*-value approach.

b. Solve using the classical approach.

10.168 In determining the "goodness" of a test question, a teacher will often compare the percentage of better students who answer it correctly with the percentage of poorer students who answer it correctly. One expects that the proportion of better students who will answer the question correctly is greater than the proportion of poorer students who will answer it correctly. On the last test, 35 of the students with the top 60 grades and 27 of the students with the bottom 60 grades answered a certain question correctly. Did the students with the top grades do significantly better on this question? Use $\alpha = 0.05$.

a. Solve using the *p*-value approach.

b. Solve using the classical approach.

10.169 A manufacturer designed an experiment to compare the differences between men and women with respect to the times they require to assemble a product. A total of 15 men and 15 women were tested to determine the time they required, on average, to assemble the product. The time required by the men had a standard deviation of 4.5 minutes, and the time required by the women had a standard deviation of 2.8 minutes. Do these data show that the amount of time needed by men is more variable than the amount of time needed by women? Use $\alpha = 0.05$ and assume the times are approximately normally distributed.

a. Solve using the *p*-value approach.

b. Solve using the classical approach.

10.170 A soft drink distributor is considering two new models of dispensing machines. Both the Harvard Company machine and the Fizzit machine can be adjusted to fill the cups to a certain mean amount. However, the variation in the amount dispensed from cup to cup is a primary concern. Ten cups dispensed from the Harvard machine showed a variance of 0.065, whereas 15 cups dispensed from the Fizzit machine showed a variance of 0.033. The factory representative from the Harvard Company maintains that his machine had no more variability than the Fizzit machine. Assume the amount dispensed is normally distributed.

At the 0.05 level of significance, does the sample refute the representative's assertion?

a. Solve using the *p*-value approach.

b. Solve using the classical approach.

10.171 Mindy Fernandez is in charge of production at the new sport-utility vehicle (SUV) assembly plant that just opened in her town. Lately she has been concerned that the wheel lug studs do not match the chrome lug nuts close enough to keep the assembly of the wheels operating smoothly. Workers are complaining that cross-threading is happening so often that threads are being stripped by the air wrenches and that torque settings also have to be adjusted downward to prevent stripped threads even if the parts match up. In an effort to determine whether the fault lies with the lug nuts or the studs, Mindy decided to ask the quality-control department to test a random sample of 60 lug nuts and 40 studs to see if the variances in threads are the same for both parts. The report from the technician indicated that the thread variance of the sampled lug nuts was 0.00213 and that the thread variance for the sampled studs was 0.00166. What can Mindy conclude about the equality of the variances at the 0.05 level of significance?

a. Solve using the *p*-value approach.

b. Solve using the classical approach.

10.172 [EX10-172] Do students who register for early morning classes typically do better than those who sign up for later day classes? Random samples from two elementary statistics classes were taken to determine if a significant difference exists in the variability and means of grades for a probability test given to both classes.

Early morning grades

78	92	86	78	89	100	97	53	86	58	78	100
92	83										

Later day grades

89	100	92	78	100	67	58	100	78	100	89	83

a. Are the assumptions of normality satisfied? Explain.

b. Is there sufficient evidence to reject the hypothesis of equal variances of grades for these two classes? Use $\alpha = 0.05$.

c. Is there sufficient evidence to reject the hypothesis that there is no difference between the mean probability test grades for these two classes? Use $\alpha = 0.05$.

FYI When using MINITAB's or TI's 2-sample *t*-test, you have the option of selecting "assume equal variances" according to the result in part b.

10.173 [EX10-173] A research project was undertaken to evaluate the amount of force needed to elicit a designated response on equipment made from two distinct designs: the existing design and an improved design. The expectation was that new design equipment would require less force than the current equipment. Fifty units of each design were tested and the required force recorded. A lower force level and reduced variability are both considered desirable.

Control or Existing	Test or New Design
0.003562	0.002477
0.005216	0.002725

*** For remainder of data, logon at cengagebrain.com

a. Describe both sets of data using means, standard deviations, and histograms.

b. Check the assumptions for comparing the variances and means of two independent samples. Describe your findings.

c. Will one- or two-tailed tests be appropriate for testing the expectations for the new design? Why?

d. Is there significant evidence to show that the new design has reduced the variability in the required force? Use $\alpha = 0.05$.

e. Is there significant evidence to show that the new design has reduced the mean amount of force? Use $\alpha = 0.05$.

f. Did the new design live up to expectations? Explain.

Chapter Practice Test

PART I: Knowing the Definitions

Answer "True" if the statement is always true. If the statement is not always true, replace the words shown in bold with words that make the statement always true.

10.1 When the means of two unrelated samples are used to compare two populations, we are dealing with **two dependent means**.

10.2 The use of **paired data (dependent means)** often allows for the control of unmeasurable or confounding variables because each pair is subjected to these confounding effects equally.

10.3 The **chi-square distribution** is used for making inferences about the ratio of the variances of two populations.

10.4 The **z distribution** is used when two dependent means are to be compared.

10.5 In comparing two independent means when the σ's are unknown, we need to use the **standard normal** distribution.

10.6 The **standard normal score** is used for all inferences concerning population proportions.

10.7 The F-distribution is a **symmetrical** distribution.

10.8 The number of degrees of freedom for the critical value of t is equal to **the smaller of $n_1 - 1$ or $n_2 - 1$** when inferences are made about the difference between two independent means in the case when the degrees of freedom are estimated.

10.9 In a confidence interval for the mean difference in paired data, the interval **increases** in width when the sample size is increased.

10.10 A **pooled estimate** for any statistic in a problem dealing with two populations is a value arrived at by combining the two separate sample statistics so as to achieve the best possible point estimate.

PART II: Applying the Concepts

Answer all questions, showing all formulas, substitutions, and work.

10.11 State the null (H_o) and the alternative (H_a) hypotheses that would be used to test each of these claims:

a. There is no significant difference in the mean batting averages for the baseball players of the two major leagues.

b. The standard deviation for the monthly amounts of rainfall in Monroe County is less than the standard deviation for the monthly amounts of rainfall in Orange County.

c. There is a significant difference between the percentages of male and female college students who own their own car.

10.12 Determine the test statistic, critical region, and critical value(s) that would be used in completing each hypothesis test using the classical procedure with $\alpha = 0.05$.

a. $H_o: p_1 - p_2 = 0$
   $H_a: p_1 - p_2 \neq 0$

b. $H_o: \mu_d = 12$
   $H_a: \mu_d \neq 12$
   $(n = 28)$

c. $H_o: \mu_1 - \mu_2 = 17$
   $H_a: \mu_1 - \mu_2 > 17$
   $(n_1 = 8, n_2 = 10)$

d. $H_o: \mu_1 - \mu_2 = 37$
   $H_a: \mu_1 - \mu_2 < 37$
   $(n_1 = 38, n_2 = 50)$

e. $H_o: \sigma_m^2 = \sigma_p^2$
   $H_a: \sigma_m^2 > \sigma_p^2$
   $(n_m = 16, n_p = 25)$

10.13 Find each of the following:

a. $z(0.02)$ e. $z(0.04)$

b. $t(15, 0.025)$ f. $t(38, 0.05)$

c. $F(24, 12, 0.05)$ g. $t(23, 0.99)$

d. $F(12, 24, 0.05)$ h. $z(0.90)$

10.14 [PT10-14] Twenty college freshmen were randomly divided into two groups. Members of one group were assigned to a statistics section that used programmed materials only. Members of the other group were assigned to a section in which the professor lectured. At the end of the semester, all were given the same final exam. Here are the results:

Programmed	76	60	85	58	91
	44	82	64	79	88
Lecture	81	62	87	70	86
	77	90	63	85	83

At the 5% level of significance, do these data provide sufficient evidence to conclude that on the average, the students in the lecture sections performed significantly better on the final exam? Assume normality.

10.15 [PT10-15] The weights of eight people before they stopped smoking and 5 weeks after they stopped smoking are as follows:

	1	2	3	4	5	6	7	8
Before	148	176	153	116	129	128	120	132
After	154	179	151	121	130	136	125	128

At the 0.05 level of significance, does this sample present enough evidence to justify the conclusion that weight increases if one quits smoking? Assume normality.

10.16 In a nationwide sample of 600 school-age boys and 500 school-age girls, 288 boys and 175 girls admitted to having committed a destruction-of-property offense. Use these sample data to construct a 95% confidence interval for the difference between the proportions of boys and girls who have committed this offense.

PART III: Understanding The Concepts

10.17 To compare the accuracy of two short-range missiles, 8 of the first kind and 10 of the second kind were fired at a target. Let x be the distance by which the missile missed the target. Do these two sets of data (8 distances and 10 distances) represent dependent or independent samples? Explain.

10.18 Let's assume that 400 students in our college are taking elementary statistics this semester. Describe how you could obtain two dependent samples of size 20 from these students to test some precourse skill against the same skill after students complete the course. Be very specific.

10.19 Student A says, "I don't see what all the fuss is about the difference between independent and dependent means; the results are almost the same regardless of the method used." Professor C suggests Student A compare the procedures a bit more carefully. Help Student A discover that there is a substantial difference between the procedures.

10.20 Suppose you are testing H_o: $\mu_d = 0$ versus H_a: $\mu_d < 0$ and the sample paired differences are all negative. Does this mean there is sufficient evidence to reject the null hypothesis? How can it not be significant? Explain.

10.21 Truancy is very disruptive to the educational system. A group of high school teachers and counselors have developed a group counseling program that they hope will help improve the truancy situation in their school. They have selected the 80 students in their school with the worst truancy records and have randomly assigned half of them to the group counseling program. At the end of the school year, the 80 students will be rated with regard to their truancy. When the scores have been collected, they will be turned over to you for evaluation. Explain what you will do to complete the study.

10.22 You wish to estimate and compare the proportion of Catholic families whose children attend a private school to the proportion of non-Catholic families whose children attend private schools. How would you go about estimating the two proportions and the difference between them?

11 Applications of Chi-Square

© James Schwabel / Alamy

11.1 Chi-Square Statistic
*Used to test hypotheses concerning **enumerated data***

11.2 Inferences Concerning Multinomial Experiments
*Differs from a binomial experiment in that **each trial has many outcomes***

11.3 Inferences Concerning Contingency Tables
*Tabular representations of frequency counts for data in a **two-way classification***

© 2010 Jupiterimages Corporation/Getty Images

© 2010 Masterfile/Radius Images/Jupiterimages Corporation

© iStockphoto.com/Irina Tischenko

11.1 Chi-Square Statistic

Cooling a Great Hot Taste

If you like hot foods, you probably have a favorite hot sauce and preferred way to "cool" your mouth after eating a mind-blowing spicy morsel. Some of the more common methods used by people are drinking water, milk, soda, or beer or eating bread or other food. There are even a few who prefer not to cool their mouth on such occasions and therefore do nothing.

Putting Out The Fire

Top six ways American adults say they cool their mouths after eating hot sauce:

Water	Bread	Milk	Beer	Soda	Don't
43%	19%	15%	7%	7%	6%

Source: Data from Anne R. Carey and Suzy Parker, © 1995 *USA Today*.

Recently a sample of two hundred adults professing to love hot spicy food were asked to name their favorite way to cool their mouth after eating food with hot sauce. The table summarizes the responses. [EX11-01]

Method	Water	Bread	Milk	Beer	Soda	Nothing	Other
Number	73	29	35	19	20	13	11

Count data like these are often referred to as enumerative data.

There are many problems for which **enumerative data** are categorized and the results shown by way of counts. For example, a set of final exam scores can be displayed as a frequency distribution. These frequency numbers are counts, the number of data that fall in each cell. A survey asks voters whether they are registered as Republican, Democrat, or Other, and whether or not they support a particular candidate. The results are usually displayed on a chart that shows the number of voters in each possible category. Numerous illustrations of this way of presenting data have been given throughout the previous 10 chapters.

Data Set-Up

Suppose that we have a number of **cells** into which n observations have been sorted. (The term *cell* is synonymous with the term *class*; the terms *class* and *frequency* were defined and first used in earlier chapters. Before you continue, a brief review of Sections 2.1, 2.2, and 3.1 might be beneficial.) The **observed frequencies** in each cell are denoted by $O_1, O_2, O_3, \ldots, O_k$ (see Table 11.1). Note that the sum of all the observed frequencies is

$$O_1 + O_2 + \cdots + O_k = n$$

where n is the sample size. What we would like to do is compare the observed frequencies with some **expected**, or theoretical, **frequencies**, denoted by $E_1, E_2, E_3, \ldots, E_k$ (see Table 11.1), for each of these cells. Again, the sum of these expected frequencies must be exactly n:

$$E_1 + E_2 + \cdots + E_k = n$$

TABLE 11.1 Observed Frequencies

	k Categories					Total
	1st	2nd	3rd	. . .	*k*th	
Observed frequencies	O_1	O_2	O_3	. . .	O_k	n
Expected frequencies	E_1	E_2	E_3	. . .	E_k	n

We will then decide whether the observed frequencies seem to agree or disagree with the expected frequencies. We will do this by using a **hypothesis test** with **chi-square**, χ^2 ("ki-square"; that's "ki" as in *kite*; χ is the Greek lowercase letter chi).

Outline of Test Procedure

Test Statistic for Chi-Square

$$\chi^2\star = \sum_{\text{all cells}} \frac{(O - E)^2}{E} \qquad\qquad (11.1)$$

This calculated value for chi-square is the sum of several nonnegative numbers, one from each cell (or category). The numerator of each term in the formula for $\chi^2\bigstar$ is the square of the difference between the values of the observed and the expected frequencies. The closer together these values are, the smaller the value of $(O - E)^2$; the farther apart, the larger the value of $(O - E)^2$. The denominator for each cell puts the size of the numerator into perspective; that is, a difference $(O - E)$ of 10 resulting from frequencies of 110 (O) and 100 (E) is quite different from a difference of 10 resulting from 15 (O) and 5 (E).

These ideas suggest that small values of chi-square indicate agreement between the two sets of frequencies, whereas larger values indicate disagreement. Therefore, it is customary for these tests to be one-tailed, with the critical region on the right.

In repeated sampling, the calculated value of $\chi^2\bigstar$ in formula (11.1) will have a sampling distribution that can be approximated by the chi-square probability distribution when n is large. This approximation is generally considered adequate when all the expected frequencies are equal to or greater than 5. Recall that the chi-square distributions, like Student's t-distributions, are a family of probability distributions, each one being identified by the parameter number of **degrees of freedom**, df. The appropriate value of df will be described with each specific test. In order to use the chi-square distribution, we must be aware of its properties, which were listed in Section 9.3 on page 454. (Also see Figure 9.7.) The critical values for chi-square are obtained from Table 8 in Appendix B. (Specific instructions were given in Section 9.3; see pp. 454–455.)

> **Assumption for using chi-square to make inferences based on enumerative data** The sample information is obtained using a random sample drawn from a population in which each individual is classified according to the categorical variable(s) involved in the test.

A *categorical variable* is a variable that classifies or categorizes each individual into exactly one of several cells or classes; these cells or classes are all-inclusive and mutually exclusive. The side facing up on a rolled die is a categorical variable: the list of outcomes $\{1, 2, 3, 4, 5, 6\}$ is a set of all-inclusive and mutually exclusive categories.

In this chapter we permit a certain amount of "liberalization" with respect to the null hypothesis and its testing. In previous chapters the null hypothesis was always a statement about a population parameter ($\mu, \sigma,$ or p). However, there are other types of hypotheses that can be tested, such as "This die is fair" or "The height and weight of individuals are independent." Notice that these hypotheses are not claims about a parameter, although sometimes they could be stated with parameter values specified.

Suppose that we claim, "This die is fair," $p = P(\text{any one number}) = \frac{1}{6}$, and you want to test the claim. What would you do? Was your answer something like: Roll this die many times and record the results? Suppose that you decide to roll the die 60 times. If the die is fair, what do you expect will happen? Each number $(1, 2, \ldots, 6)$ should appear approximately $\frac{1}{6}$ of the time (that is, 10 times). If it happens that approximately 10 of each number appear, you will certainly accept the claim of fairness ($p = \frac{1}{6}$ for each value). If it happens that the die seems to favor some particular numbers, you will reject the claim. (The calculated test statistic $\chi^2\bigstar$ will have a large value in this case, as we will soon see.)

⊡ S E C T I O N   1 1 . 1   E X E R C I S E S

11.1 [EX11-001] Referring to the sample of 200 adults surveyed in Section 11.1's Cooling a Great Hot Taste (p. 544):

a. What information was collected from each adult in the sample?

b. Define the population and the variable involved in the sample.

c. Using the sample data, calculate percentages for the various methods of cooling one's mouth.

11.2 Referring to the sample of 200 adults surveyed in Section 11.1's Cooling a Great Hot Taste and the accompanying "Putting Out the Fire" graphic:

a. How do the sample percentages calculated in part c of Exercise 11.1 compare to the percentages on the "Putting Out the Fire" graphic?

b. Construct a vertical bar graph of the 200 adults using relative frequency for the vertical scale. (Treat the missing 3% in "Putting Out the Fire" as category "Other.")

c. Superimpose the bar graph from "Putting Out the Fire" on the bar graph in part b.

d. Would you say the sample's distribution looks "similar to" or "quite different from" the distribution shown in the "Putting Out the Fire" graphic? Explain your answer.

11.3 Using Table 8 of Appendix B, find the following:

a. $\chi^2(10, 0.01)$

b. $\chi^2(12, 0.025)$

c. $\chi^2(10, 0.95)$

d. $\chi^2(22, 0.995)$

11.4 Find these critical values by using Table 8 of Appendix B.

a. $\chi^2(18, 0.01)$

b. $\chi^2(16, 0.025)$

c. $\chi^2(40, 0.10)$

d. $\chi^2(45, 0.01)$

11.5 Using the notation seen in Exercise 11.4, name and find the critical values of χ^2.

a.
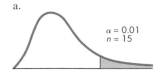

b.
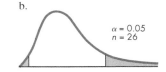

11.6 Using the notation seen in Exercise 11.4, name and find the critical values of χ^2.

a.
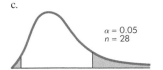

b.
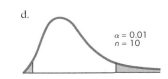

c.
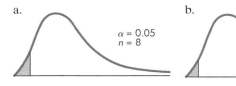

d.

[EX00-000] identifies the filename of an exercise's online dataset—available through cengagebrain.com

11.2 Inferences Concerning Multinomial Experiments

The preceding die problem is a good illustration of a **multinomial experiment**. Let's consider this problem again. Suppose that we want to test this die (at $\alpha = 0.05$) and decide whether to fail to reject or reject the claim "This die is fair." (The probability of each number is $\frac{1}{6}$.) The die is rolled from a cup onto a smooth, flat surface 60 times, with the following observed frequencies:

Number	1	2	3	4	5	6
Observed frequency	7	12	10	12	8	11

The null hypothesis that the die is fair is assumed to be true. This allows us to calculate the expected frequencies. If the die is fair, we certainly expect 10 occurrences of each number.

Now let's calculate an observed value of χ^2. These calculations are shown in Table 11.2. The calculated value is $\chi^2\bigstar = \mathbf{2.2}$.

TABLE 11.2 Computations for Calculating χ^2

Number	Observed (O)	Expected (E)	O − E	(O − E)²	$\dfrac{(O - E)^2}{E}$
1	7	10	−3	9	0.9
2	12	10	2	4	0.4
3	10	10	0	0	0.0
4	12	10	2	4	0.4
5	8	10	−2	4	0.4
6	11	10	1	1	0.1
Total	60	60	0 ⓒⓚ		2.2

Note: $\Sigma(O - E)$ must equal zero because $\Sigma O = \Sigma E = n$. You can use this fact as a check, as shown in Table 11.2.

Now let's use our familiar hypothesis-testing format.

STEP 1   **a. Parameter of interest:** The probability with which each side faces up: $P(1), P(2), P(3), P(4), P(5), P(6)$

b. Statement of hypotheses:

H_o: The die is fair $\left(\text{each } p = \frac{1}{6}\right)$.
H_a: The die is not fair (at least one p is different from the others).

STEP 2   **a. Assumptions:** The data were collected in a random manner, and each outcome is one of the six numbers.

b. Test statistic: The chi-square distribution and formula (11.1), with df $= k - 1 = 6 - 1 = 5$

In a multinomial experiment, df $= k - 1$, where k is the number of cells.

c. Level of significance: $\alpha = 0.05$

STEP 3   **a. Sample information:** See Table 11.2.

b. Calculated test statistic: Using formula (11.1), we have

$$\chi^2\star = \sum_{\text{all cells}} \frac{(O - E)^2}{E}: \quad \chi^2\star = 2.2 \quad \text{(calculations are shown in Table 11.2)}$$

STEP 4 Probability Distribution:

p-Value:

a. Use the right-hand tail because "larger" values of chi-square disagree with the null hypothesis: $\mathbf{P} = P(\chi^2\star > 2.2 \,|\, \text{df} = 5)$, as shown in the figure.

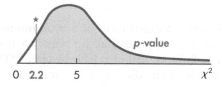

To find the p-value, you have two options:

1. Use Table 8 (Appendix B) to place bounds on the p-value: **0.75 < P < 0.90.**
2. Use a computer or calculator to find the p-value: **P = 0.821.**

For specific instructions, see page 458.

b. The p-value is not smaller than the level of significance, α.

OR

Classical:

a. The critical region is the right-hand tail because "larger" values of chi-square disagree with the null hypothesis. The critical value is obtained from Table 8, at the intersection of row df $= 5$ and column $\alpha = 0.05$:

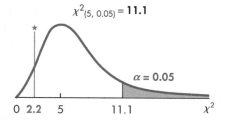

For specific instructions, see pages 454–455.

b. $\chi^2\star$ is not in the critical region, as shown in **red** in the figure.

FYI Computer and calculator commands to find the probability associated with a specified chi-square value can be found in Chapter 9 (pp. 455–456).

STEP 5   **a. Decision:** Fail to reject H_o.
   b. Conclusion: At the 0.05 level of significance, the observed frequencies are not significantly different from those expected of a fair die.

Before we look at other examples, we must define the term *multinomial experiment* and state the guidelines for completing the chi-square test for it.

Multinomial experiment A multinomial experiment has the following characteristics:

1. It consists of n identical independent trials.
2. The outcome of each trial fits into exactly one of k possible cells.
3. There is a probability associated with each particular cell, and these individual probabilities remain constant during the experiment. (It must be the case that $p_1 + p_2 + \cdots + p_k = 1$.)
4. The experiment will result in a set of k observed frequencies, $O_1, O_2, \ldots, O_k$, where each O_i is the number of times a trial outcome falls into that particular cell. (It must be the case that $O_1 + O_2 + \cdots + O_k = n$.)

The die example meets the definition of a multinomial experiment because it has all four of the characteristics described in the definition.

1. The die was rolled n (60) times in an identical fashion, and these trials were independent of each other. (The result of each trial was unaffected by the results of other trials.)
2. Each time the die was rolled, one of six numbers resulted, and each number was associated with a cell.
3. The probability associated with each cell was $\frac{1}{6}$, and this was constant from trial to trial. (Six values of $\frac{1}{6}$ sum to 1.0.)
4. When the experiment was complete, we had a list of six frequencies (7, 12, 10, 12, 8, and 11) that summed to 60, indicating that each of the outcomes was taken into account.

The testing procedure for multinomial experiments is very similar to the testing procedure described in previous chapters. The biggest change comes with the statement of the null hypothesis. It may be a verbal statement, such as in the die example: "This die is fair." Often the alternative to the null hypothesis is not stated. However, in this book the alternative hypothesis will be shown, because it aids in organizing and understanding the problem. It will not be used to determine the location of the critical region, though, as was the case in previous chapters. For multinomial experiments we will always use a one-tailed critical region, and it will be the right-hand tail of the χ^2-distribution because larger deviations (positive or negative) from the expected values lead to an increase in the calculated $\chi^2\bigstar$-value.

The critical value will be determined by the level of significance assigned (α) and the number of degrees of freedom. The number of degrees of freedom (df) will be 1 less than the number of cells (k) into which the data are divided:

Degrees of Freedom for Multinomial Experiments

$$\text{df} = k - 1 \qquad (11.2)$$

Each expected frequency, E_i, will be determined by multiplying the total number of trials n by the corresponding probability (p_i) for that cell; that is,

Expected Value for Multinomial Experiments

$$E_i = n \cdot p_i \tag{11.3}$$

One guideline should be met to ensure a good approximation to the chi-square distribution: each expected frequency should be at least 5 (i.e., each $E_i \geq 5$). Sometimes it is possible to combine "smaller" cells to meet this guideline. If this guideline cannot be met, then corrective measures to ensure a good approximation should be used. These corrective measures are not covered in this book but are discussed in many other sources.

EXAMPLE 11.1

A MULTINOMIAL HYPOTHESIS TEST WITH EQUAL EXPECTED FREQUENCIES

College students have regularly insisted on freedom of choice when they register for courses. This semester there were seven sections of a particular mathematics course. The sections were scheduled to meet at various times with a variety of instructors. Table 11.3 shows the number of students who selected each of the seven sections. Do the data indicate that the students had a preference for certain sections, or do they indicate that each section was equally likely to be chosen?

TABLE 11.3 Data on Section Enrollments

| | Section | | | | | | | |
	1	2	3	4	5	6	7	Total
Number of students	18	12	25	23	8	19	14	119

Solution

If no preference were shown in the selection of sections, then we would expect the 119 students to be equally distributed among the seven classes: We would expect 17 students to register for each section. The hypothesis test is completed at the 5% level of significance.

Step 1   a. **Parameter of interest:** Preference for each section, the probability that a particular section is selected at registration

   b. **Statement of hypotheses:**

   H_o: There was no preference shown (equally distributed).

   H_a: There was a preference shown (not equally distributed).

Step 2   a. **Assumptions:** The 119 students represent a random sample of the population of all students who register for this particular course. Since no new regulations were introduced in the selection of courses and registration seemed to proceed in its usual pattern, there is no reason to believe this is other than a random sample.

   b. Test statistic: The chi-square distribution and formula (11.1), with df = 6

   c. Level of significance: $\alpha = 0.05$

Step 3   **a. Sample information:** See Table 11.3 (p. 550).

   b. Calculated test statistic: Using formula (11.1), we have

$$x^2\star = \sum_{\text{all cells}} \frac{(O - E)^2}{E} : x^2\star = \frac{(18 - 17)^2}{17} + \frac{(12 - 17)^2}{17} + \frac{(25 - 17)^2}{17}$$
$$+ \frac{(23 - 17)^2}{17} + \frac{(8 - 17)^2}{17} + \frac{(19 - 17)^2}{17} + \frac{(14 - 17)^2}{17}$$
$$= \frac{(1)^2 + (-5)^2 + (8)^2 + (6)^2 + (-9)^2 + (2)^2 + (-3)^2}{17}$$
$$= \frac{1 + 25 + 64 + 36 + 81 + 4 + 9}{17} = \frac{220}{17} = 12.9411$$
$$= \mathbf{12.94}$$

Step 4 Probability Distribution:

| *p*-Value: | **OR** | Classical: |

p-Value:

a. Use the right-hand tail because "larger" values of chi-square disagree with the null hypothesis: $\mathbf{P} = P(x^2\star > 12.94 \mid df = 6)$, as shown in the figure:

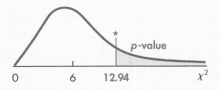

To find the *p*-value, you have two options:
1. Use Table 8 (Appendix B) to place bounds on the *p*-value: **0.025 < P < 0.05.**
2. Use a computer or calculator to find the *p*-value: **P = 0.044.**

For specific instructions, see page 458.

b. The *p*-value is smaller than the level of significance, α.

Classical:

a. The critical region is the right-hand tail because "larger" values of chi-square disagree with the null hypothesis. The critical value is obtained from Table 8, at the intersection of row df = 6 and column $\alpha = 0.05$:

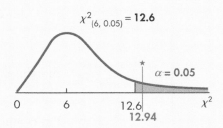

For specific instructions, see pages 454–455.

b. $x^2\star$ is in the critical region, as shown in **red** in the figure.

Step 5   **a. Decision:** Reject H_o.

   b. Conclusion: At the 0.05 level of significance, there does seem to be a preference shown. We cannot determine from the given information what the preference is. It could be teacher preference, time preference, or a schedule conflict.

   Conclusions must be worded carefully to avoid suggesting conclusions that the data cannot support.

   Not all multinomial experiments result in equal expected frequencies, as we will see in Example 11.2.

EXAMPLE 11.2

A MULTINOMIAL HYPOTHESIS TEST WITH UNEQUAL EXPECTED FREQUENCIES

The Mendelian theory of inheritance claims that the frequencies of round and yellow, wrinkled and yellow, round and green, and wrinkled and green peas will occur in the ratio $9:3:3:1$ when two specific varieties of peas are crossed. In testing this theory, Mendel obtained frequencies of 315, 101, 108, and 32, respectively. Do these sample data provide sufficient evidence to reject the theory at the 0.05 level of significance?

Solution

Step 1   **a. Parameter of interest:** The proportions: P(round and yellow), P(wrinkled and yellow), P(round and green), P(wrinkled and green)

        b. Statement of hypotheses:

H_o: $9:3:3:1$ is the ratio of inheritance.
H_a: $9:3:3:1$ is not the ratio of inheritance.

Step 2   **a. Assumptions:** We will assume that Mendel's results form a random sample.

        b. Test statistic: The chi-square distribution and formula (11.1), with df = 3

        c. Level of significance: $\alpha = 0.05$

Step 3   **a. Sample information:** The observed frequencies were: 315, 101, 108, and 32.

        b. Calculated test statistic: The ratio $9:3:3:1$ indicates probabilities of $\frac{9}{16}, \frac{3}{16}, \frac{3}{16},$ and $\frac{1}{16}$.

Therefore, the expected frequencies are $\frac{9n}{16}, \frac{3n}{16}, \frac{3n}{16},$ and $\frac{1n}{16}$. We have

$$n = \Sigma O_i = 315 + 101 + 108 + 32 = 556$$

The computations for calculating $\chi^2\star$ are shown in Table 11.4.

TABLE 11.4 Computations Needed to Calculate $\chi^2\star$

O	E	$O - E$	$\dfrac{(O - E)^2}{E}$
315	312.75	2.25	0.0162
101	104.25	−3.25	0.1013
108	104.25	3.75	0.1349
32	34.75	−2.75	0.2176
556	556.00	0 ✓	0.4700 → $\chi^2\star = \sum\limits_{\text{all cells}} \dfrac{(O - E)^2}{E} = \mathbf{0.47}$

Step 4 Probability Distribution:

p-Value:		Classical:
a. Use the right-hand tail because "larger" values of chi-square disagree with the null hypothesis:		**a.** The critical region is the right-hand tail because "larger" values of chi-square disagree with the

$\mathbf{P} = \mathbf{P(\chi^2\star > 0.47 \mid df = 3)}$, as shown in the figure. To find the p-value, you have two options:

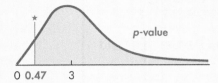

1. Use Table 8 (Appendix B) to place bounds on the p-value: $\mathbf{0.90 < P < 0.95}$.
2. Use a computer or calculator to find the p-value: $\mathbf{P = 0.925}$.

For specific instructions, see page 458.

b. The p-value is not smaller than the level of significance, α.

null hypothesis. The critical value is obtained from Table 8, at the intersection of row df $= 3$ and column $\alpha = 0.05$:

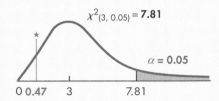

For specific instructions, see pages 454–455.

b. $\chi^2\star$ is not in the critical region, as shown in **red** in the figure.

Step 5 **a. Decision:** Fail to reject H_o.

b. Conclusion: At the 0.05 level of significance, there is not sufficient evidence to reject Mendel's theory.

TECHNOLOGY INSTRUCTIONS: GOODNESS OF FIT TEST

Minitab

Input the observed frequencies into C1. If performing a test with unequal expected frequencies, input the specific proportions into C2. Then continue with:

Choose: Stat > Tables > Chi-Square Goodness-of-Fit Test (One Variable) . . .
Enter: Observed counts: C1
Select: Equal Proportions > OK
         or
         Specific Proportions
         Enter: C2 > OK

Excel

Input the observed frequencies into column A and the corresponding expected frequencies into column B. (Can use Excel to convert probabilities into expected frequencies.) Then continue with:

Choose: Insert function, f_x > Statistical > CHITEST > OK
Enter: Actual range: (A1:A6 or select cells)
         Expected range: (B1:B6 or select cells) > OK

Excel output only provides the p-value for the test.

TI-84 Plus*

Input the observed frequencies into L1 and the expected frequencies into L2; then continue with:

Choose: STAT > TESTS > D:χ^2 GOF-Test . . .
Enter: Observed: L1
         Expected: L2
         df: k − 1

Highlight: Calculate > ENTER

```
X²GOF-Test
 Observed:L₁
 Expected:L₂
 df:5
 Calculate Draw
```

*Goodness of Fit Test is only available on the TI-84 Plus.

APPLIED EXAMPLE 11.3

Most Popular Days for Babies

The most popular day of the week for U.S. babies to enter the world is Tuesday, with almost 13,000 births on average.
Slowest day: Sunday.

Source: Census Bureau by Anne R. Carey and Ron Coddington, *USA TODAY*

BIRTH DAYS

The Census Bureau collects data for many variables. The information provided with the accompanying graphic is based on the U.S. census and fits the format of a multinomial experiment. Verify that these data qualify as a multinomial experiment (see Exercise 11.7).

APPLIED EXAMPLE 11.4

Teens and Downloading

For the 33% of Americans ages 8 to 18 who own cell phones, special features are a plus. When downloading extras, they choose:

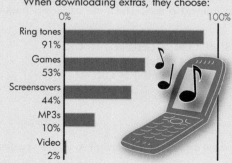

	0%	100%
Ring tones	91%	
Games	53%	
Screensavers	44%	
MP3s	10%	
Video	2%	

Source: Data from Justin Dickerson and Adrienne Lewis, © 2005 *USA Today*.

DOWNLOADING WHAT?

The graphic "Teens and Downloading" displays the results of surveying 8- to 18-year-olds about what they download using their cell phones. This information does not qualify as a multinomial experiment. What property is violated? (See Exercise 11.8.)

SECTION 11.2 EXERCISES

11.7 Verify that Applied Example 11.3, "Birth Days" (p. 554), is a multinomial experiment. Be specific.

a. What is one trial?

b. What is the variable?

c. What are the possible levels of results from each trial?

11.8 Why is the information shown in Applied Example 11.4, "Downloading What?", on page 554 not that of a multinomial experiment? Be specific.

11.9 State the null hypothesis, H_o, and the alternative hypothesis, H_a, that would be used to test the following statements:

a. The five numbers 1, 2, 3, 4, and 5 are equally likely to be drawn.

b. The multiple-choice question has a history of students selecting answers in the ratio of 2:3:2:1.

c. The poll will show a distribution of 16%, 38%, 41%, and 5% for the possible ratings of excellent, good, fair, and poor on that issue.

11.10 State the null hypothesis H_o and the alternative hypothesis H_a that would be used to test the following statements.

a. The four choices are all equally likely.

b. The poll showed the political party distributions of 23%, 36%, and 41% for Republicans, Democrats, and Independents, respectively.

c. Favoring responses with respect to sustainability and the four designated generation intervals were in the ratio of 11:15:8:6.

11.11 Determine the p-value for the following hypotheses tests involving the χ^2-distribution.

a. H_o: $P(1) = P(2) = P(3) = P(4) = 0.25$, with $\chi^2\star = 12.25$

b. H_o: $P(I) = 0.25$, $P(II) = 0.40$, $P(III) = 0.35$, with $\chi^2\star = 5.98$

11.12 Determine the critical value and critical region that would be used in the classical approach to test the null hypothesis for each of the following multinomial experiments.

a. H_o: $P(1) = P(2) = P(3) = P(4) = 0.25$, with $\alpha = 0.05$

b. H_o: $P(I) = 0.25$, $P(II) = 0.40$, $P(III) = 0.35$, with $\alpha = 0.01$

11.13 Explain how 9:3:3:1 becomes $\frac{9}{16}, \frac{3}{16}, \frac{3}{16}$, and $\frac{1}{16}$ in Example 11.2 on pages 552–553.

11.14 Explain how 312.75, 2.25, and 0.0162 were obtained in the first row of Table 11.4 on page 552.

11.15 A manufacturer of floor polish conducted a consumer-preference experiment to determine which of five different floor polishes was the most appealing in appearance. A sample of 100 consumers viewed five patches of flooring that had each received one of the five polishes. Each consumer indicated the patch he or she preferred. The lighting and background were approximately the same for all patches. The results were as follows:

Polish	A	B	C	D	E	Total
Frequency	27	17	15	22	19	100

Solve the following using the p-value approach and the classical approach:

a. State the hypothesis for "no preference" in statistical terminology.

b. What test statistic will be used in testing this null hypothesis?

c. Complete the hypothesis test using $\alpha = 0.10$.

11.16 Skittles Original Fruit bite-size candies are multiple-colored candies in a bag, and you can "Taste the Rainbow" with their five colors and flavors: green—lime, purple—grape, yellow—lemon, orange—orange, red—strawberry. Unlike some of the other multicolored candies available, Skittles claims that their five colors are equally likely. In an attempt to reject this claim, a 4-oz bag of Skittles was purchased and the colors counted:

Red	Orange	Yellow	Green	Purple
18	21	23	17	27

Does this sample contradict Skittles' claim at the 0.05 level?

a. Solve using the p-value approach.

b. Solve using the classical approach.

11.17 An October 16, 2009, *USA Today* Snapshot titled "Are public cell phone conversations rude?" reported the following results from a Fox TV/Rasmussen Reports poll:

Poll Response	Percent
Yes	51
No	37
Not sure	12

(continue on page 556)

As a member of the Civility Committee at your college, you decide to conduct a survey of students with respect to this issue. The following table shows the 300 student responses:

Poll Response	Number
Yes	126
No	118
Not sure	56

Does the distribution of responses from the college students differ significantly from the published survey results? Use a 0.01 level of significance.

11.18 National health care is currently a big issue for Americans. The October 21, 2009, *USA Today* article "Poll: Americans skittish over health care changes" reported the following percentages with respect to "Insurance company requirements you have to meet to get certain treatments covered" if a health care bill passes:

Viewpoint Sept. 11–13	Percentage
Get better	22%
Not change	35%
Get worse	38%
Unknown	5%

One month later, during October 16–19, another poll was taken of 1521 adults. Those viewpoints are categorized in the table below.

Viewpoint Oct. 16–19	Number
Get better	380
Not change	380
Get worse	700
Unknown	61

At the 0.05 level of significance, did the distributions of viewpoints change significantly from September 2009 to October 2009?

11.19 A certain type of flower seed will produce magenta, chartreuse, and ochre flowers in the ratio 6:3:1 (one flower per seed). A total of 100 seeds are planted and all germinate, yielding the following results.

Magenta	Chartreuse	Ochre
52	36	12

Solve the following using the *p*-value approach and the classical approach:

a. If the null hypothesis (6:3:1) is true, what is the expected number of magenta flowers?

b. How many degrees of freedom are associated with chi-square?

c. Complete the hypothesis test using $\alpha = 0.10$.

11.20 Bird foraging behavior is being studied in a managed forest that is made up of Douglas fir (52% of canopy volume), ponderosa pine (36%), and grand fir (12%). Two hundred thirty-eight red-breasted nuthatches were observed, with 105 in Douglas fir, 92 in ponderosa pine, and 41 in grand fir. The null hypothesis being tested is: the birds forage randomly without regard to the species of tree.

a. State the alternative hypothesis.

b. Determine the expected values for the number of birds foraging each species of tree.

c. Complete the hypothesis test using $\alpha = 0.05$ and carefully state the conclusion.

11.21 A large supermarket carries four qualities of ground beef. Customers are believed to purchase these four varieties with probabilities of 0.10, 0.30, 0.35, and 0.25, respectively, from the least to most expensive variety. A sample of 500 purchases resulted in sales of 46, 162, 191, and 101 of the respective qualities. Does this sample contradict the expected proportions? Use $\alpha = 0.05$.

a. Solve using the *p*-value approach.

b. Solve using the classical approach.

11.22 [EX11-22] One of the major benefits of e-mail is that it makes it possible to communicate rapidly without getting a busy signal or no answer, two major criticisms of telephone calls. But does e-mail succeed in helping solve the problems people have trying to run computer software? A study polled the opinions of consumers who had tried to use e-mail to obtain help by posting a message online to their PC manufacturer or authorized representative. Results are shown in the following table.

Result of Online Query	Percent
Never got a response	14
Got a response, but it didn't help	30
Response helped, but didn't solve problem	34
Response solved problem	22

Source: *PC World*, "PC World's Reliability and Service Survey"

As marketing manager for a large PC manufacturer, you decide to conduct a survey of your customers comparing your e-mail records against the published results. To ensure a fair comparison, you elect to use the same questionnaire and examine returns from 500 customers who attempted to use e-mail to get help from your technical support staff. The results follow:

Result of Online Query	Number Responding
Never got a response	35
Got a response, but it didn't help	102
Response helped, but didn't solve problem	125
Response solved problem	238
Total	500

Does the distribution of responses differ from the distribution obtained from the published survey? Test at the 0.01 level of significance.

a. Solve using the *p*-value approach.

b. Solve using the classical approach.

11.23 [EX11-23] *Nursing Magazine* reported results of a survey of more than 1800 nurses across the country concerning job satisfaction and retention. Nurses from magnet hospitals (hospitals that successfully attract and retain nurses) describe the staffing situation in their units as follows:

Staffing Situation	Percent
1. Desperately short of help—patient care has suffered	12
2. Short, but patient care hasn't suffered	32
3. Adequate	38
4. More than adequate	12
5. Excellent	6

A survey of 500 nurses from nonmagnet hospitals gave the following responses to the staffing situation.

Staffing Situation	1	2	3	4	5
Number	165	140	125	50	20

Do the data indicate that the nurses from the nonmagnet hospitals have a different distribution of opinions? Use $\alpha = 0.05$.

a. Solve using the *p*-value approach.

b. Solve using the classical approach.

11.24 [EX11-24] A program for generating random numbers on a computer is to be tested. The program is instructed to generate 100 single-digit integers between 0 and 9. The frequencies of the observed integers are as follows:

Integer	0	1	2	3	4	5	6	7	8	9
Frequency	11	8	7	7	10	10	8	11	14	14

At the 0.05 level of significance, is there sufficient reason to believe that the integers are not being generated uniformly?

a. Solve using the *p*-value approach.

b. Solve using the classical approach.

11.25 [EX11-25] "Climbing out of debt, step by step," an article in the April 29, 2005, *USA Today*, reported results of a survey of 260 members of the Financial Planning Association. Financial planners each reported what he or she considers to be the one most valuable step people can take to improve their financial lives.

Most Valuable Step	Percent
1. Establish goals	30
2. Pay yourself first	21
3. Create and stick to a budget	17

4. Save on a regular basis	12
5. Pay down credit card debt	7
6. Invest the maximum in 401(k)	5
7. Other	8

A survey of 60 financial planners from an upstate metropolitan area gave the following responses to the "one most valuable step" question.

Answer to Question	1	2	3	4	5	6	7
Number	10	13	13	8	9	3	4

Do the data indicate that the financial planners from the upstate metropolitan area have a different distribution of opinions? Use $\alpha = 0.05$.

a. Solve using the *p*-value approach.

b. Solve using the classical approach.

11.26 [EX11-26] The U.S. census found that babies enter the world on the days of the week in the proportions that follow.

Weekday	P(Day)	Weekday	P(Day)
Sunday	0.098	Thursday	0.160
Monday	0.149	Friday	0.159
Tuesday	0.166	Saturday	0.111
Wednesday	0.157		

Source: U.S. Census Bureau

A random sample selected from the birth records for a large metropolitan area resulted in the following data:

Day	Su	M	Tu	W	Th	F	Sa
Observed	10	6	9	13	9	17	11

a. Do these data provide sufficient evidence to reject the claim "Births occur in this metropolitan area in the same daily proportions" as reported by the U.S. Census Bureau? Use $\alpha = 0.05$.

b. Do these data provide sufficient evidence to reject the claim "Births occur in this metropolitan area on all days with the same likeliness"? Use $\alpha = 0.05$.

c. Compare the results obtained in parts a and b. State your conclusions.

11.27 Referring to the sample of 200 adults surveyed in Section 11.1's Cooling a Great Hot Taste and the accompanying "Putting Out the Fire" graphic:

Does the sample of 200 adults show a distribution that is significantly different from the distribution shown in the "Putting Out the Fire" graph (p. 544)? Use $\alpha = 0.05$.

11.28 To demonstrate/explore the effect increased sample size has on the calculated chi-square value, let's consider the Skittles candies in Exercise 11.16 and sample some larger bags of the candy.

(continue on page 558)

a. Suppose we purchase a 16-oz bag of Skittles, count the colors, and observe exactly the same proportion of colors as found in Exercise 11.16:

Red	Orange	Yellow	Green	Purple
72	84	92	68	108

Calculate the value of chi-square for these data. How is the new chi-square value related to the one found in Exercise 11.16? What effect does this new value have on the test results? Explain.

b. To continue this demonstration/exploration, suppose we purchase a 48-oz bag, count the colors, and observe exactly the same proportion of colors as found in Exercise 11.16 and part a of this exercise.

Red	Orange	Yellow	Green	Purple
216	252	276	204	324

Calculate the value of chi-square for these data. How is the new chi-square value related to the one found in Exercise 11.16? Explain.

c. What effect does the size of the sample have on the calculated chi-square value when the proportion of observed frequencies stays the same as the sample size increases?

d. Explain in what way this indicates that if a large enough sample is taken, the hypothesis test will eventually result in a rejection.

11.29 [EX11-29] According to The Harris Poll, the proportion of all adults who live in households with rifles (29%), shotguns (29%), or pistols (23%) has not changed significantly. However, today more people live in households with no guns (61%). The 1014 adults surveyed gave the following results.

	All Adults (%)	All Gun Owners (%)
Have rifle, shotgun, and pistol (3 out of 3)	16	41
Have 2 out of 3 (rifle, shotgun, or pistol)	11	27
Have 1 out of 3 (rifle, shotgun, or pistol)	11	29
Decline to answer/Not sure	1	3
Total	39%	100%

In a survey of 2000 adults in Memphis who said they own guns, 780 said they own all three types, 550 said they own 2 of the 3 types, 560 said they own 1 of the 3 types, and 110 declined to specify what types of guns they own.

a. Test the null hypothesis that the distribution of number of types owned is the same in Memphis as it is nationally as reported by The Harris Poll. Use a level of significance equal to 0.05.

b. What caused the calculated value of $\chi^2\star$ to be so large? Does it seem right that one cell should have this much effect on the results? How could this test be completed differently (hopefully, more meaningfully) so that the results might not be affected as they were in part a? Be specific.

11.30 Why is the chi-square test typically a one-tail test with the critical region in the right tail?

a. What kind of value would result if the observed frequencies and the expected frequencies were very close in value? Explain how you would interpret this situation.

b. Suppose you had to roll a die 60 times as an experiment to test the fairness of the die as discussed in the example on pages 547–549; but instead of rolling the die yourself, you paid your little brother $1 to roll it 60 times and keep a tally of the numbers. He agreed to perform this deed for you and ran off to his room with the die, returning in a few minutes with his resulting frequencies. He demanded his $1. You, of course, paid him before he handed over his results, which were as follows: 10, 10, 10, 10, 10, and 10. The observed results were exactly what you had "expected," right? Explain your reactions. What value of $\chi^2\star$ will result? What do you think happened? What did you demand of your little brother and why? What possible role might the left tail have in the hypothesis test?

c. Why is the left tail not typically of concern?

11.3 Inferences Concerning Contingency Tables

A **contingency table** is an arrangement of data in a two-way classification. The data are sorted into cells, and the count for each cell is reported. The contingency table involves two factors (or variables), and a common question concerning such tables is whether the data indicate that the two variables are independent or dependent (see pp. 121–123, 208–209).

Two different tests use the contingency table format. The first one we will look at is the *test of independence*.

Test of Independence

To illustrate a test of independence, let's consider a random sample that shows the gender of liberal arts college students and their favorite academic area.

E X A M P L E   1 1 . 5

HYPOTHESIS TEST FOR INDEPENDENCE

Each person in a group of 300 students was identified as male or female and then asked whether he or she prefers taking liberal arts courses in the area of math–science, social science, or humanities. Table 11.5 is a contingency table that shows the frequencies found for these categories. Does this sample present sufficient evidence to reject the null hypothesis "Preference for math–science, social science, or humanities is independent of the gender of a college student"? Complete the **hypothesis test** using the 0.05 level of significance.

TABLE 11.5 Sample Results for Gender and Subject Preference

Gender	Favorite Subject Area			Total
	Math–Science (MS)	Social Science (SS)	Humanities (H)	
Male (M)	37	41	44	122
Female (F)	35	72	71	178
Total	72	113	115	300

Solution

Step 1   **a. Parameter of interest:** Determining the independence of the variables "gender" and "favorite subject area" requires us to discuss the probability of the various cases and the effect that answers about one variable have on the probability of answers about the other variable. Independence, as defined in Chapter 4, requires $P(MS|M) = P(MS|F) = P(MS)$; that is, gender has no effect on the probability of a person's choice of subject area.

b. Statement of hypotheses:

H_o: Preference for math–science, social science, or humanities is independent of the gender of a college student.

H_a: Subject area preference is not independent of the gender of the student.

Step 2   **a. Assumptions:** The sample information is obtained using one random sample drawn from one population, with each individual then classified according to gender and favorite subject area.

b. Test statistic.

In the case of contingency tables, the number of degrees of freedom is exactly the same as the number of cells in the table that may be filled in freely when you are given the *marginal totals*. The totals in this example are shown in the following table:

			122
			178
72	113	115	300

Given these totals, you can fill in only two cells before the others are all determined. (The totals must, of course, remain the same.) For example, once we pick two arbitrary values (say, 50 and 60) for the first two cells of the first row, the other four cell values are fixed (see the following table):

50	60	C	122
D	E	F	178
72	113	115	300

The values have to be $C = 12$, $D = 22$, $E = 53$, and $F = 103$. Otherwise, the totals will not be correct. Therefore, for this problem there are two free choices. Each free choice corresponds to 1 degree of freedom. Hence, the number of degrees of freedom for our example is 2 (df = 2).

> The chi-square distribution will be used along with formula (11.1), with df = 2.

> **c. Level of significance:** $\alpha = 0.05$

Step 3 **a. Sample information:** See Table 11.5.

b. Calculated test statistic:

Before we can calculate the value of chi-square, we need to determine the expected values, E, for each cell. To do this we must recall the null hypothesis, which asserts that these factors are independent. Therefore, we would expect the values to be distributed in proportion to the marginal totals. There are 122 males; we would expect them to be distributed among MS, SS, and H proportionally to the 72, 113, and 115 totals. Thus, the expected cell counts for males are

$$\frac{72}{300} \cdot 122 \qquad \frac{113}{300} \cdot 122 \qquad \frac{115}{300} \cdot 122$$

Similarly, we would expect for the females

$$\frac{72}{300} \cdot 178 \qquad \frac{113}{300} \cdot 178 \qquad \frac{115}{300} \cdot 178$$

Thus, the expected values are as shown in Table 11.6. Always compare the marginal totals for the expected values against the marginal totals for the observed values.

TABLE 11.6 Expected Values

	MS	SS	H	Total
Male	29.28	45.95	46.77	122.00
Female	42.72	67.05	68.23	178.00
Total	72.00	113.00	115.00	300.00

Note: We can think of the computation of the expected values in a second way. Recall that we assume the null hypothesis to be true until there is evidence to reject it. Having made this assumption in our example, we are saying in effect that the event that a student picked at random is male and the event that a student picked at random prefers math–science courses are independent. Our point estimate for the probability that a student is male is $\frac{122}{300}$, and the point estimate for the probability that the student prefers math–science courses is $\frac{72}{300}$. Therefore, the probability that both events occur is the product of the probabilities. [Refer to formula (4.7), p. 211.] Thus,

$\left(\frac{122}{300}\right)\left(\frac{72}{300}\right)$ is the probability of a selected student being male and preferring math–science. The number of students out of 300 who are expected to be male and prefer math–science is found by multiplying the probability (or proportion) by the total number of students (300). Thus, the expected number of males who prefer math–science is $\left(\frac{122}{300}\right)\left(\frac{72}{300}\right)300 = \left(\frac{122}{300}\right)(72) = 29.28$. The other expected values can be determined in the same manner.

Typically, the contingency table is written so that it contains all this information (see Table 11.7).

TABLE 11.7 Contingency Table Showing Sample Results and Expected Values

	Favorite Subject Area			
Gender	MS	SS	H	Total
Male	37 (29.28)	41 (45.95)	44 (46.77)	122
Female	35 (42.72)	72 (67.05)	71 (68.23)	178
Total	72	113	115	300

The calculated chi-square is

$$\chi^2\star = \sum_{\text{all cells}} \frac{(O - E)^2}{E} : \chi^2\star = \frac{(37 - 29.28)^2}{29.28} + \frac{(41 - 45.95)^2}{45.95} + \frac{(44 - 46.77)^2}{46.77}$$

$$+ \frac{(35 - 42.72)^2}{42.72} + \frac{(72 - 67.05)^2}{67.05} + \frac{(71 - 68.23)^2}{68.23}$$

$$= 2.035 + 0.533 + 0.164 + 1.395 + 0.365 + 0.112$$

$$= \mathbf{4.604}$$

Step 4 Probability Distribution:

_p_-Value:

OR

Classical:

a. Use the right-hand tail because "larger" values of chi-square disagree with the null hypothesis: **P** = $P(\chi^2\star > 4.604 \mid \mathrm{df} = 2)$, as shown in the figure.

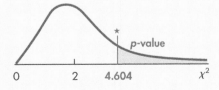

To find the _p_-value, you have two options:
1. Use Table 8 (Appendix B) to place bounds on the _p_-value: **0.10 < P < 0.25.**
2. Use a computer or calculator to find the _p_-value: **P = 0.1001.**

For specific instructions, see page 458.

b. The _p_-value is not smaller than α.

a. The critical region is the right-hand tail because "larger" values of chi-square disagree with the null hypothesis. The critical value is obtained from Table 8, at the intersection of row df = 2 and column $\alpha = 0.05$:

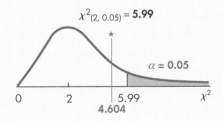

For specific instructions, see pages 454–455.

b. $\chi^2\star$ is not in the critical region, as shown in **red** in the figure.

STEP 5 **a. Decision:** Fail to reject H_o.

  b. Conclusion: At the 0.05 level of significance, the evidence does not allow us to reject independence between the gender of a student and the student's preferred academic subject area.

In general, the **$r \times c$ contingency table** (r is the number of **rows**; c is the number of **columns**) is used to test the independence of the row factor and the column factor. The number of **degrees of freedom** is determined by

Degrees of Freedom for Contingency Tables

$$\text{df} = (r - 1) \cdot (c - 1) \tag{11.4}$$

where r and c are both greater than 1

(This value for df should agree with the number of cells counted, according to the general description on pages 559–560.)

The **expected frequencies** for an $r \times c$ contingency table are found by means of the formulas given in each cell in Table 11.8, where n = grand total. In general, the expected frequency at the intersection of the ith row and the jth column is given by

Expected Frequencies for Contingency Tables

$$E_{ij} = \frac{\text{row total} \times \text{column total}}{\text{grand total}} = \frac{R_i \times C_j}{n} \tag{11.5}$$

TABLE 11.8 Expected Frequencies for an $r \times c$ Contingency Table

Row	Column						Total
	1	2	...	jth column	...	c	
1	$\dfrac{R_1 \times C_1}{n}$	$\dfrac{R_1 \times C_2}{n}$	...	$\dfrac{R_1 \times C_j}{n}$	...	$\dfrac{R_1 \times C_c}{n}$	R_1
2	$\dfrac{R_2 \times C_1}{n}$						R_2
⋮	⋮			⋮			⋮
ith row	$\dfrac{R_i \times C_1}{n}$			$\dfrac{R_i \times C_j}{n}$	...		R_i
⋮	⋮			⋮			⋮
r	$\dfrac{R_r \times C_1}{n}$						
Total	C_1	C_2	...	C_j	...	...	n

We should again observe the previously mentioned guideline: Each $E_{i,j}$ should be at least 5.

Note: The notation used in Table 11.8 and formula (11.5) may be unfamiliar to you. For convenience in referring to cells or entries in a table, we use $E_{i,j}$ to denote the entry in the ith row and the jth column. That is, the first letter in the subscript corresponds to the row number and the second letter corresponds to the column number. Thus, $E_{1,2}$ is the entry in the first row, second column, and $E_{2,1}$ is the entry in the second row, first column. In Table 11.6 (p. 560), $E_{1,2}$ is 45.95 and $E_{2,1}$ is 42.72. The notation used in Table 11.8 is interpreted in a similar manner; that is, R_1 corresponds to the total from row 1, and C_1 corresponds to the total from column 1.

Test of Homogeneity

The second type of contingency table problem is called a *test of homogeneity*. This test is used when one of the two variables is controlled by the experimenter so that the row or column totals are predetermined.

For example, suppose that we want to poll registered voters about a piece of legislation proposed by the governor. In the poll, 200 urban, 200 suburban, and 100 rural residents are randomly selected and asked whether they favor or oppose the governor's proposal. That is, a simple random sample is taken for each of these three groups. A total of 500 voters are polled. But notice that it has been predetermined (before the sample is taken) just how many are to fall within each row category, as shown in Table 11.9, and each category is sampled separately.

TABLE 11.9 Registered Voter Poll with Predetermined Row Totals

Residence	Governor's Proposal		Total
	Favor	Oppose	
Urban			200
Suburban			200
Rural			100
Total			500

In a test of this nature, we are actually testing the hypothesis: The distribution of proportions within the rows is the same for all rows. That is, the distribution of proportions in row 1 is the same as that in row 2, is the same as that in row 3, and so on. The alternative is: The distribution of proportions within the rows is not the same for all rows. This type of example may be thought of as a comparison of several multinomial experiments.

Beyond this conceptual difference, the actual testing for independence and homogeneity with contingency tables is the same. Let's demonstrate this **hypothesis test** by completing the polling illustration.

EXAMPLE 11.6

HYPOTHESIS TEST FOR HOMOGENEITY

Each person in a random sample of 500 registered voters (200 urban, 200 suburban, and 100 rural residents) was asked his or her opinion about the governor's proposed legislation. Does the sample evidence shown in Table 11.10 support the hypothesis "Voters within the different residence groups have different opinions about the governor's proposal"? Use $\alpha = 0.05$.

TABLE 11.10 Sample Results for
Residence and Opinion

Residence	Governor's Proposal		Total
	Favor	Oppose	
Urban	143	57	200
Suburban	98	102	200
Rural	13	87	100
Total	254	246	500

Solution

Step 1   **a. Parameter of interest:** The proportion of voters who favor or oppose (i.e., the proportion of urban voters who favor, the proportion of suburban voters who favor, the proportion of rural voters who favor, and the proportion of all three groups, separately, who oppose)

b. Statement of hypotheses:

H_o: The proportion of voters who favor the proposed legislation is the same in all three residence groups.

H_a: The proportion of voters who favor the proposed legislation is not the same in all three groups. (That is, in at least one group, the proportion is different from the others.)

Step 2   **a. Assumptions:** The sample information is obtained using three random samples drawn from three separate populations in which each individual is classified according to his or her opinion.

b. Test statistic: The chi-square distribution and formula (11.1), with $df = (r - 1)(c - 1) = (3 - 1)(2 - 1) = 2$

c. Level of significance: $\alpha = 0.05$

Step 3   **a. Sample information:** See Table 11.10.

b. Calculated test statistic: The expected values are found by using formula (11.5) (p. 562) and are given in Table 11.11.

TABLE 11.11 Sample Results and Expected Values

Residence	Governor's Proposal		Total
	Favor	Oppose	
Urban	143 (101.6)	57 (98.4)	200
Suburban	98 (101.6)	102 (98.4)	200
Rural	13 (50.8)	87 (49.2)	100
Total	254	246	500

Note: Each expected value is used twice in the calculation of $\chi^2\star$; therefore, it is a good idea to keep extra decimal places while doing the calculations.

The calculated chi-square is

$$\chi^2\star = \sum_{\text{all cells}} \frac{(O - E)^2}{E} : \chi^2\star = \frac{(143 - 101.6)^2}{101.6} + \frac{(57 - 98.4)^2}{98.4} + \frac{(98 - 101.6)^2}{101.6}$$

$$+ \frac{(102 - 98.4)^2}{98.4} + \frac{(13 - 50.8)^2}{50.8} + \frac{(87 - 49.2)^2}{49.2}$$

$$= 16.87 + 17.42 + 0.13 + 0.13 + 28.13 + 29.04$$

$$= \mathbf{91.72}$$

Step 4 Probability Distribution:

***p*-Value:**	**OR**	**Classical:**

p-Value:

a. Use the right-hand tail because "larger" values of chi-square disagree with the null hypothesis: $\mathbf{P} = \mathbf{P(\chi^2\star > 91.72 \mid df = 2)}$, as shown in the figure.

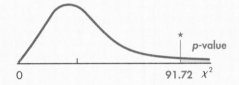

To find the *p*-value, you have two options:
1. Use Table 8 (Appendix B) to place bounds on the *p*-value: **P < 0.005**.
2. Use a computer or calculator to find the *p*-value: **P = 0.000+**.

For specific instructions, see page 458.

b. The *p*-value is smaller than α.

Classical:

a. The critical region is the right-hand tail because "larger" values of chi-square disagree with the null hypothesis. The critical value is obtained from Table 8, at the intersection of row df = 2 and column $\alpha = 0.05$:

$$\chi^2_{(2,\ 0.05)} = \mathbf{5.99}$$

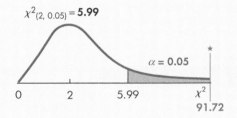

For specific instructions, see pages 454–455.

b. $\chi^2\star$ is in the critical region, as shown in **red** in the figure.

STEP 5   a. **Decision:** Reject H_o.

b. **Conclusion:** The three groups of voters do not all have the same proportions favoring the proposed legislation.

TECHNOLOGY INSTRUCTIONS: HYPOTHESIS TEST OF INDEPENDENCE OR HOMOGENEITY

MINITAB	Input each column of observed frequencies from the contingency table into C1, C2, . . . ; then continue with:

Choose: **Stat > Tables > Chi-Square Test (Two-Way Table in Worksheet)**
Enter: Columns containing the table: **C1 C2 > OK**

COMPUTER SOLUTION MINITAB Printout for Example 11.6:
Chi-square Test: C1, C2
Expected counts are printed below observed counts
Chi-Square contributions are printed below expected counts

	C1	C2	Total
1	143	57	200
	101.60	98.40	
	16.870	17.418	
2	98	102	200
	101.60	98.40	
	0.128	0.132	
3	13	87	100
	50.80	49.20	
	28.127	29.041	
Total	254	246	500

Chi-Sq = 91.715, DF = 2, P-Value = 0.000

Excel

Input each column of observed frequencies from the contingency table into columns A, B, . . . ; then continue with:

Choose: **Add-Ins > Data Analysis Plus > Contingency Table > OK**
Enter: **Input range: (A1:B4 or select cells)**
Select: **Labels** (if necessary)
Enter: **Alpha: α** (ex. 0.05)

TI-83/84 Plus

Input the observed frequencies from the $r \times c$ contingency table into an $r \times c$ matrix A. Set up matrix B as an empty $r \times c$ matrix for the expected frequencies.

Choose: **MATRX > EDIT > 1:[A]**
Enter: **r > ENTER > c > ENTER**
            Each observed frequency with an ENTER afterward
Then continue with:
Choose: **MATRX > EDIT > 2[B]**
Enter: **r > ENTER > c > ENTER**
Choose: **STAT > TESTS > C: χ^2–Test. . .**
Enter: **Observed: [A]** or wherever the contingency table is located
            Expected: [B] place for expected frequencies
Highlight: **Calculate > ENTER**

APPLIED EXAMPLE 11.7

BAKED POTATOES RULE FOR WESTERNERS

Baked Potatoes Rule for Westerners

Americans eat potatoes an average of three times a week and 47% prefer theirs "baked" over mashed (23%) or french fried (16%). Those who preferred baked by region:

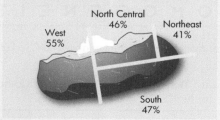

Source: Data from Anne R. Carey and Sam Ward, © 1998 *USA Today.*

The graphic "Baked Potatoes Rule for Westerners" reports the percentage of Americans who prefer to eat baked potatoes by region as well as for the whole country. If the actual number of people in each category were given, we would have a contingency table and we would be able to complete a hypothesis test about the homogeneity of the four regions. (See Exercises 11.46 and 11.55.)

SECTION 11.3 EXERCISES

11.31 State the null hypothesis, H_o, and the alternative hypothesis, H_a, that would be used to test the following statements:

a. The voters expressed preferences that were not independent of their party affiliations.

b. The distribution of opinions is the same for all three communities.

c. The proportion of "yes" responses was the same for all categories surveyed.

11.32 The "test of independence" and the "test of homogeneity" are completed in identical fashion, using the contingency table to display and organize the calculations. Explain how these two hypothesis tests differ.

11.33 Find the expected value for the cell shown.

□	...	50
⋮		
40		200

11.34 Identify these values from Table 11.7:

a. C_2 b. R_1 c. n d. $E_{2,3}$

11.35 MINITAB was used to complete a chi-square test of independence between the number of boat-related manatee deaths and two Florida counties.

County	Boat-Related Deaths	Non–Boat-Related Deaths	Total Deaths
Lee County	23	25	48
Collier County	8	23	31

Chi-Square Test: Boat-Related Deaths, Non-Boat-Related Deaths
Expected counts are printed below observed counts
Chi-Square contributions are printed below expected counts

	Boat-Related Deaths	Non-Boat-Related Deaths	Total
1	23	25	48
	18.84	29.16	
	0.921	0.595	
2	8	23	31
	12.16	18.84	
	1.426	0.921	
Total	31	48	79

Chi-Sq = 3.862, DF = 1, P-Value = 0.049

a. Verify the results (expected values and the calculated $\chi^2\star$) by calculating the values yourself.

b. Use Table 8 to verify the p-value based on the calculated df.

c. Is the proportion of boat-related deaths independent of the county? Use $\alpha = 0.05$.

11.36 Results on seatbelt usage from the 2003 Youth Risk Behavior Survey were published in a *USA Today* Snapshot on January 13, 2005. The following table outlines the results from the high school students who were surveyed in the state of Nebraska. They were asked whether they rarely or never wear seatbelts when riding in someone else's car.

	Female	Male
Rarely or never use seatbelt	208	324
Use seatbelt	1217	1184

Source: http://www.cdc.gov/

Using $\alpha = 0.05$, does this sample present sufficient evidence to reject the hypothesis that gender is independent of seatbelt usage?

a. Solve using the p-value approach.

b. Solve using the classical approach.

11.37 The State Conservation Department used surveillance cameras to study the reaction of white-tailed deer to traffic as they used a wildlife underpass to cross a major highway. When a car or truck passed over while the deer were in the underpass, they recorded "continued" when the deer continued in the original direction or "reversed" when the deer reversed their direction.

	Continued	Reversed
Car	315	73
Truck	84	97

Is the direction the white-tailed deer took independent of the type of vehicle passing over the underpass? Answer using $\alpha = 0.01$.

11.38 A survey of randomly selected travelers who visited the service station restrooms of a large U.S. petroleum distributor showed the following results:

Gender of Respondent	Quality of Restroom Facilities			Totals
	Above Average	Average	Below Average	
Female	7	24	28	59
Male	8	26	7	41
Total	15	50	35	100

Using $\alpha = 0.05$, does the sample present sufficient evidence to reject the hypothesis "Quality of responses is independent of the gender of the respondent"?

a. Solve using the p-value approach.

b. Solve using the classical approach.

11.39 Tourette's syndrome is an inheritable, childhood-onset neurological disorder involving multiple motor tics and at least one vocal tic. A U.S. study that was published

(continue on page 568)

in the June 5, 2009, CDC *Morbidity and Mortality Weekly Report* indicated that the syndrome occurs in 3 out of every 1000 school-age children. Further analysis broke the data into ethnicity/race categories—see the chart below. At the 0.05 level of significance, does this sample indicate that having Tourette's is independent of ethnicity and race?

	Hispanic	Non-Hispanic White	Non-Hispanic Black
Have Tourette's	26	164	18
No Tourette's	7,321	43,602	6,427

11.40 Tourette's syndrome is an inheritable, childhood-onset neurological disorder involving multiple motor tics and at least one vocal tic. A U.S. study that was published in the June 5, 2009, CDC *Morbidity and Mortality Weekly Report* indicated that the syndrome occurs in 3 out of every 1000 school-age children. Further analysis broke the data into household income categories with respect to the federal poverty level—see the chart below. At the 0.05 level of significance, does this sample indicate that having Tourette's is independent of household income?

	Below 200%	200%–400%	Above 400%
Have Tourette's	65	80	80
No Tourette's	17,581	21,795	24,432

11.41 A survey of employees at an insurance firm concerned worker–supervisor relationships. One statement for evaluation was, "I am not sure what my supervisor expects." The results of the survey are presented in the following contingency table.

	I Am Not Sure What My Supervisor Expects		
Years of Employment	True	Not True	Totals
Less than 1 year	18	13	31
1 to 3 years	20	8	28
3 to 10 years	28	9	37
10 years or more	26	8	34
Total	92	38	130

Can we reject the hypothesis that "The responses to the statement and the years of employment are independent" at the 0.10 level of significance?

a. Solve using the *p*-value approach.

b. Solve using the classical approach.

11.42 [EX11-42] The following table is from the publication *Vital and Health Statistics* from the Centers for Disease Control and Prevention/National Center for Health Statistics. The individuals in the following table have an eye irritation, a nose irritation, or a throat irritation. They have only one of the three.

	Age (years)			
Type of Irritation	18–29	30–44	45–64	65 and Older
Eye	440	567	349	59
Nose	924	1311	794	102
Throat	253	311	157	19

Is there sufficient evidence to reject the hypothesis that the type of ear, nose, or throat irritation is independent of the age group at a level of significance equal to 0.05?

a. Solve using the *p*-value approach.

b. Solve using the classical approach.

11.43 [EX11-43] A random sample of 500 married men was taken; each person was cross-classified as to the size community that he was presently residing in and the size community that he was reared in. The results are shown in the following table.

	Size of Community Residing In			
Size of Community Reared In	Less Than 10,000	10,000 to 49,999	50,000 or Over	Total
Less than 10,000	24	45	45	114
10,000 to 49,999	18	64	70	152
50,000 or over	21	54	159	234
Total	63	163	274	500

Does this sample contradict the claim of independence, at the 0.01 level of significance?

a. Solve using the *p*-value approach.

b. Solve using the classical approach.

11.44 It is hypothesized that sick animals receiving a certain drug (the treated group) will survive at a more favorable rate than those that do not receive the drug (the control group). The following results were recorded from the test.

	Survived	Did Not Survive
Treated	46	18
Control	38	35

a. Explain why the hypothesis stated in the exercise cannot be the null hypothesis.

b. Explain why the null hypothesis is correctly stated as "Survival is independent of the drug treatment."

c. Complete the hypothesis test, finding the *p*-value.

d. If the test is completed using $\alpha = 0.02$, state the decision that must be reached.

e. If the test is completed using $\alpha = 0.02$, carefully state the conclusion and its meaning.

11.45 The manager of an assembly process wants to determine whether or not the number of defective articles manufactured depends on the day of the week the articles are produced. She collected the following information.

Day of Week	M	Tu	W	Th	F
Nondefective	85	90	95	95	90
Defective	15	10	5	5	10

Is there sufficient evidence to reject the hypothesis that the number of defective articles is independent of the day of the week on which they are produced? Use $\alpha = 0.05$.

a. Solve using the *p*-value approach.

b. Solve using the classical approach.

11.46 Referring to Applied Example 11.7 (p. 566):

a. Express the percentage of Americans who "prefer baked" to "other" by region as a 2 × 4 contingency table.

b. Explain why the following question could be tested using the chi-square statistic: "Is the preference for baked the same in all four regions of the United States?"

c. Explain why this is a test of homogeneity.

11.47 Blogging is a hot topic nowadays. A "blog" is an Internet log. Blogs are created for personal or professional use. According to the Xtreme Recruiting website (http://www.xtremerecruiting.org/), there is a new blog born every 7 seconds—and quite a few people are reading these blogs. The table that follows shows the number of new blog readers for each of the months listed. Is the distribution of blog creators and readers the same for the months listed? Use $\alpha = 0.05$.

	Blog Creators	Blog Readers
March 2003	74	205
February 2004	93	316
November 2004	130	502

Source: *USA Today*, "Warning: Your clever little blog could get you fired," June 15, 2005

11.48 The November 12, 2009, *USA Today* Snapshot "Rabies in cats on the rise" reported that almost 7000 animals were reported to have rabies in 2008. Utilizing information from the *Journal of the American Veterinary Medical Association*, the following rabies cases were logged for cats and dogs:

	Dogs	Cats
2007	93	274
2008	75	294

At the 0.05 level of significance, is the distribution of rabies cases for dogs and cats the same for the years listed?

11.49 The athletic director at a large high school wants to compare the proportions of different kinds of ankle injuries that occur to his school's basketball players and volleyball players. Inspection of last year's records revealed the following number of ankle injuries for each sport.

	Basketball	Volleyball
Sprains	28	19
Breaks	11	7
Torn ligaments	6	8
Other injuries	10	13

Is there evidence of a significant difference between the two sports? Use $\alpha = 0.05$.

11.50 Students use many kinds of criteria when selecting courses. "Teacher who is a very easy grader" is often one criterion. Three teachers are scheduled to teach statistics next semester. A sample of previous grade distributions for these three teachers is shown here.

	Professor		
Grades	**#1**	**#2**	**#3**
A	12	11	27
B	16	29	25
C	35	30	15
Other	27	40	23

At the 0.01 level of significance, is there sufficient evidence to conclude "The distribution of grades is not the same for all three professors?"

a. Solve using the *p*-value approach.

b. Solve using the classical approach.

c. Which professor is the easiest grader? Explain, citing specific supporting evidence.

11.51 Fear of darkness is a common emotion. The following data were obtained by asking 200 individuals in each age group whether they had serious fears of darkness. At $\alpha = 0.01$, do we have sufficient evidence to reject the hypothesis that "The same proportion of each age group has serious fears of darkness"?

Age Group	Elementary	Jr. High	Sr. High	College	Adult
No. Who Fear Darkness	83	72	49	36	114

a. The above table is an incomplete contingency table even though at first glance it might appear to be multinomial. Explain why. (*Hint:* The contingency table must account for all 1000 people.)

b. Solve using the *p*-value approach.

c. Solve using the classical approach.

Table for Exercise 11.52

Occupation	Construction	Production	Engineering	Politics	Education
# who smoke	43	37	17	17	12

11.52 According to a report from the Substance Abuse and Mental Health Services Administration, food-service workers have the highest rate for smoking cigarettes: 45% of food-service workers reported smoking cigarettes in the past month. Do some careers lend themselves to cigarette smoking more than others? If 100 people in each of the following occupations were asked about smoking in the last month, do the data support that some careers correspond to higher rates of smoking? Use a 0.05 level of significance.

11.53 [EX11-53] All new drugs must go through a drug study before being approved by the U.S. Food and Drug Administration (FDA). A drug study typically includes clinical trials whereby participants are randomized to receive different dosages as well as a placebo but are unaware of which group they are in. To control as many factors as possible, it is best to assign participants randomly yet homogeneously across the treatments. Consider the following arrangement for homogeneity with respect to gender and dosages.

Gender	10-mg Drug	20-mg Drug	Placebo
Female	54	56	60
Male	32	27	26

At the 0.01 level of significance, is the distribution of drug the same for both genders?

Considering the same study, homogeneity of ages would also be an important feature. At the 0.01 level of significance, is the distribution of the drug that follows the same for all age groups?

Age	10-mg Drug	20-mg Drug	Placebo
40–49	18	20	19
50–59	48	41	57
60–69	20	22	10

11.54 [EX11-54] Are younger and younger people able to obtain illegal guns? According to the October 11, 2009, Rochester, NY, *Democrat & Chronicle* article "The Gun Used to Shoot DiPonzio," which cited a 14-year-old shooting a police officer, it appears that the number of people in younger age groups found with illegal guns continues to grow. At the 0.01 level of significance, does it appear that the distribution of ages possessing illegal guns is the same for the years listed?

Year	21 and Less	22–30	31–50	50+
2005	103	93	111	33
2006	119	136	96	31
2007	155	140	130	76
2008	159	160	104	60

11.55 Applied Example 11.7 (p. 566) reports percentages describing people's preferences with regard to how potatoes are prepared. Do you believe there is a significant difference between the four regions of America with regard to the percentage who prefer baked? Notice that the article does not mention the sample size.

a. Assume the percentages reported were based on four samples of size 100 from each region and calculate $\chi^2\bigstar$ and its *p*-value.

b. Repeat part a using sample sizes of 200 and 300.

c. Are the four percentages reported in the graphic of who prefers baked potatoes significantly different? Describe in detail the circumstances for which they are significantly different.

© James Schwabel/Alamy

Chapter Review

In Retrospect

In this chapter we have been concerned with tests of hypotheses using chi-square, with the cell probabilities associated with the multinomial experiment, and with the simple contingency table. In each case the basic assumptions are that a large number of observations have been made and that the resulting test statistic, $\sum \frac{(O - E)^2}{E}$, is approximately distributed as chi-square. In general, if n is large and the minimum allowable expected cell size is 5, then this assumption is satisfied.

The contingency table can be used to test independence and homogeneity. The test for homogeneity and the test for independence look very similar and, in fact, are carried out in exactly the same way. The concepts being tested, however—same distributions and independence, respectively—are quite different. The two tests are easily distinguished because the test of homogeneity has predetermined marginal totals in one direction in the table. That is, before the data are collected, the experimenter determines how many subjects will be observed in each category. The only predetermined number in the test of independence is the grand total.

A few words of caution: The correct number of degrees of freedom is critical if the test results are to be meaningful. The degrees of freedom determine, in part, the critical region, and its size is important. As in other tests of hypothesis, failure to reject H_o does not mean outright acceptance of the null hypothesis.

CourseMate The **Statistics CourseMate** site for this text brings chapter topics to life with interactive learning, study, and exam preparation tools, including quizzes and flashcards for the Vocabulary and Key Concepts that follow. The site also provides an **eBook** version of the text with highlighting and note taking capabilities. Throughout chapters, the CourseMate icon 🖥 flags concepts and examples that have corresponding interactive resources such as **video** and **animated tutorials** that demonstrate, step by step, how to solve problems; **datasets** for exercises and examples; **Skillbuilder Applets** to help you better understand concepts; **technology manuals**; and software to download including **Data Analysis Plus** (a suite of statistical macros for Excel) and **TI-83/84 Plus** programs—logon at **www.cengagebrain.com**.

Vocabulary and Key Concepts

assumptions (p. 546)
cell (p. 545)
chi-square (p. 545)
column (p. 562)
contingency table (pp. 562, 558)
degrees of freedom (pp. 562, 546)

enumerative data (p. 545)
expected frequency (p. 562)
homogeneity (p. 563)
hypothesis test (pp. 545, 559, 563)
independence (p. 559)
marginal totals (p. 559)

multinomial experiment (pp. 547, 549)
observed frequency (p. 545)
$r \times c$ contingency table (pp. 545, 558, 562)
rows (pp. 558, 562)
test statistic (p. 545)

Learning Outcomes

- Understand that enumerative data are data that can be counted and placed into categories. — pp. 545–546
- Understand that the chi-square distribution will be used to test hypotheses involving enumerative data. — pp. 544–546
- Understand the properties of the chi-square distribution and how series of distributions based on sample size (using degrees of freedom as the index). — pp. 544, 545–546, Ex. 11.3, 11.5
- Understand the key elements of a multinomial experiment and be able to define n, k, O_i, and P_i. — pp. 547–549
- Know and be able to calculate expected values using $E = np$. — EXP 11.2, Ex. 11.14
- Know and be able to calculate a chi-square statistic: $\chi^2 = \sum_{\text{all cells}} \frac{(O - E)^2}{E}$. — EXP 11.1
- Know and be able to calculate the degrees of freedom for a multinomial experiment (df $= k - 1$). — EXP 11.1, 11.2
- Perform, describe, and interpret a hypothesis test for a multinomial experiment using the chi-square distribution with the p-value approach and/or the classical approach. — Ex. 11.15, 11.19, 11.21
- Understand and know the definition of independence of two events. — pp. 560–561
- Know and be able to calculate expected values using $E_{ij} = \dfrac{R_i \cdot C_j}{n}$ — pp. 560, 562–563, Ex. 11.33

- Know and be able to calculate the degrees of freedom for a test of independence or homogeneity [df = $(r - 1)(c - 1)$].   p. 562

- Perform, describe, and interpret a hypothesis test for a test of independence or homogeneity using the chi-square distribution with the p-value approach and/or the classical approach.   EXP 11.5, 11.6, Ex. 11.38, 11.51

- Understand the differences and similarities between tests of independence and tests of homogeneity   pp. 570–571

 Chapter Exercises

11.56 The psychology department at a certain college claims that the grades in its introductory course are distributed as follows: 10% A's, 20% B's, 40% C's, 20% D's, and 10% F's. In a poll of 200 randomly selected students who had completed this course, it was found that 16 had received A's; 43, B's; 65, C's; 48, D's; and 28, F's. Does this sample contradict the department's claim at the 0.05 level?

a. Solve using the p-value approach.

b. Solve using the classical approach.

11.57 When interbreeding two strains of roses, we expect the hybrid to appear in three genetic classes in the ratio 1:3:4. If the results of an experiment yield 80 hybrids of the first type, 340 of the second type, and 380 of the third type, do we have sufficient evidence to reject the hypothesized genetic ratio at the 0.05 level of significance?

a. Solve using the p-value approach.

b. Solve using the classical approach.

11.58 A sample of 200 individuals are tested for their blood type, and the results are used to test the hypothesized distribution of blood types:

Blood Type	A	B	O	AB
Percent	0.41	0.09	0.46	0.04

The observed results were as follows:

Blood Type	A	B	O	AB
Number	75	20	95	10

At the 0.05 level of significance, is there sufficient evidence to show that the stated distribution is incorrect?

11.59 [EX11-59] As reported in *USA Today*, about 8.9 million families sent students to college this year and more than half live away from home. Where students live:

Parents' or guardian's home	46%
Campus housing	26%
Off-campus rental	18%
Own off-campus housing	9%
Other arrangements	2%

Note: Exceeds 100% due to rounding error.

A random sample of 1000 college students resulted in the following information:

Parents' or guardian's home	484
Campus housing	230
Off-campus rental	168
Own off-campus housing	96
Other arrangements	22

Is the distribution of this sample significantly different from the distribution reported in the newspaper? Use $\alpha = 0.05$. (To adjust for the rounding error, subtract 2 from each expected frequency.)

a. Solve using the p-value approach.

b. Solve using the classical approach.

11.60 [EX11-60] Over the years, African-American actors in major cinema releases have been more likely than white actors to have major roles in comedies. The table shows the percent of all roles by type of picture.

Type of Picture	Percent of Roles
Action and adventure	13.2
Comedy	31.9
Drama	23.0
Horror and suspense	12.5
Romantic comedy	8.2
Other	11.2

The next table shows a sample of the latest released films with the number of leading roles played by African-Americans for each type of film.

Type of Picture	Number of Roles
Action and adventure	9
Comedy	40
Drama	17
Horror and suspense	11
Romantic comedy	5
Other	7

At the 0.05 level of significance, does the distribution of African-American roles differ from the overall distribution of roles in major cinema releases?

a. Solve using the p-value approach.

b. Solve using the classical approach.

11.61 [EX11-61] Most golfers are probably happy to play 18 holes of golf whenever they get a chance to play. Ben Winter, a club professional, played 306 holes in 1 day at a charity golf marathon in Stevens, Pennsylvania. A nationwide survey conducted by *Golf* magazine over the Internet revealed the following frequency distribution of the most number of holes ever played by the respondents in 1 day:

Most Holes Played in 1 Day	Percent	Most Holes Played in 1 Day	Percent
18	5	37 to 45	20
19 to 27	12	46 to 54	18
28 to 36	28	55 or more	17

Source: *Golf*, "18 Is Not Enough"

Suppose one of your local public golf courses asked 200 golfers who teed off to answer the same question. The following table summarizes their responses:

Most Holes Played in 1 Day	Number	Most Holes Played in 1 Day	Number
18	12	37 to 45	44
19 to 27	35	46 to 54	35
28 to 36	60	55 or more	14

Does the distribution of largest number of holes played by "marathon golfers" at your public course differ from the distribution compiled by *Golf* magazine using responses polled on the Internet? Test at the 0.01 level of significance.

a. Solve using the *p*-value approach.

b. Solve using the classical approach.

11.62 [EX11-62] The U.S. census found that babies enter the world during the various months in the proportions that follow.

Month	P(Month)	Month	P(Month)	Month	P(Month)
January	0.082	May	0.084	September	0.087
February	0.076	June	0.081	October	0.086
March	0.082	July	0.089	November	0.080
April	0.081	August	0.089	December	0.083

Source: U.S. Census Bureau

A random sample selected from the birth records for a large metropolitan area resulted in the following data:

Month	Jan	Feb	Mar	Apr	May	Jun	Jul	Aug	Sep	Oct	Nov	Dec
Observed	14	12	12	10	16	9	16	11	17	7	17	9

a. Do these data provide sufficient evidence to reject the claim "Births occur in this metropolitan area in the same monthly proportions" as those reported by the U.S. Census Bureau? Use $\alpha = 0.05$.

b. Do these data provide sufficient evidence to reject the claim "Births occur in this metropolitan area in all months with the same likeliness"? Use $\alpha = 0.05$.

c. Compare the results obtained in parts a and b. State your conclusions.

11.63 [EX11-63] The weights (x) of 300 adult males were determined and used to test the hypothesis that the weights are normally distributed with a mean of 160 lb and a standard deviation of 15 lb. The data were grouped into the following classes.

Weight (x)	Observed Frequency	Weight (x)	Observed Frequency
$x < 130$	7	$160 \le x < 175$	102
$130 \le x < 145$	38	$175 \le x < 190$	40
$145 \le x < 160$	100	190 and over	13

Using the normal tables, the percentages for the classes are 2.28%, 13.59%, 34.13%, 34.13%, 13.59%, and 2.28%, respectively. Do the observed data show significant reason to discredit the hypothesis that the weights are normally distributed with a mean of 160 lb and a standard deviation of 15 lb? Use $\alpha = 0.05$.

a. Verify the percentages for the classes.

b. Solve using the *p*-value approach.

c. Solve using the classical approach.

11.64 Do you have a favorite "comfort food"? How do you obtain it? The *USA Today* Snapshot lists three methods used by Americans and the percentage who use each method. Which category do you belong to? A random sample of 120 Americans living on the East Coast was asked, "How do you obtain your favorite comfort food?"

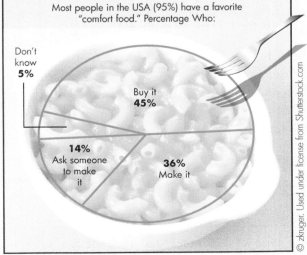

Licking Our Chops

Most people in the USA (95%) have a favorite "comfort food." Percentage Who:

Don't know 5%

Buy it 45%

14% Ask someone to make it

36% Make it

Source: Opinion Research Corp. for Lactaid
By Justin Dickerson and Suzy Parker, USA TODAY

© zkruger. Used under license from Shutterstock.com

(continue on page 574)

Comfort Food	Buy It	Make It	Ask Someone to Make It	Don't Know
East Coast	57	44	12	7

Using the percentages given in the graphic as the national "standard," does the evidence indicate that the East Coast responses are different from those of the nation as a whole? Use $\alpha = 0.05$.

a. Solve using the p-value approach.

b. Solve using the classical approach.

11.65 [EX11-65] The following table gives the color counts for a sample of 30 bags (47.9-gram size) of M&M's.

Case	Red	Gr	Blue	Or	Yel	Br
1	15	9	3	3	9	19
2	9	17	19	3	3	8

*** For remainder of data, logon at cengagebrain.com

Source: http://www.math.uah.edu/, Christine Nickel and Jason York, ST 687 project

Before the Global Color Vote (GCV) of 2002, the target percentage for each color in the six-color mix was as follows: brown, 30%; red and yellow, 20%; blue, green, and orange, 10%.

a. Does case 1 show that bag 1 has a significantly different distribution of colors than the target distribution? Use $\alpha = 0.05$.

b. Combine cases 1 and 2. Does the total of bags 1 and 2 show a significantly different distribution of colors than the target distribution?

c. Combine the results of all 30 bags. Does the total of all 30 bags show a significantly different distribution of colors than the target distribution?

d. Discuss the findings of parts a–c.

11.66 Adults 21 and older volunteer from one to nine hours each week at a center for disabled senior citizens. The program recruits adult community college students, four-year college students, and nonstudents. The table below lists a sample of the volunteers, their number of hours per week, and the volunteer type.

	Comm. College Students	Four-Year College	Nonstudents
1–3 hours	109	115	117
4–6 hours	82	123	138
7–9 hours	34	28	47

Is the type of volunteer and the number of hours volunteered independent at the 0.05 level of significance?

FYI In Exercise 11.67, do not use rounded values.

11.67 [EX11-67] A manufacturer of women's shoes wishes to compare the distribution of defects found in the shoes produced by the three shifts of workers in his plant. A sample of shoes with defects is classified as to type of defect and production shift.

	Type of Defect			
Shift	A	B	C	D
6 a.m.	17	23	43	17
2 p.m.	27	37	33	9
10 p.m.	31	19	53	18

a. Would the proportions of the four defects being the same for all three shifts imply that defects are independent of shift? Explain why or why not.

b. What would it imply if the proportions vary from shift to shift? Explain.

c. Do the data above show that the proportions vary significantly from shift to shift at the $\alpha = 0.01$ level of significance? Explain, stating your decision, conclusion, and evidence very carefully.

11.68 [EX11-68] The following table shows the number of reported crimes committed last year in the inner part of a large city. The crimes were classified according to the type of crime and the district of the inner city where it occurred. Do these data show sufficient evidence to reject the hypothesis that the type of crime and the district in which it occurred are independent? Use $\alpha = 0.01$.

	Crime				
District	Robbery	Assault	Burglary	Larceny	Stolen Vehicle
1	54	331	227	1090	41
2	42	274	220	488	71
3	50	306	206	422	83
4	48	184	148	480	42
5	31	102	94	596	56
6	10	53	92	236	45

a. Solve using the p-value approach.

b. Solve using the classical approach.

11.69 [EX11-69] Based on the results of a survey questionnaire, 400 individuals were classified as politically conservative, moderate, or liberal. In addition, each person was classified by age, as shown in the following table.

	Age Group			
	20–35	36–50	Older Than 50	Totals
Conservative	20	40	20	80
Moderate	80	85	45	210
Liberal	40	25	45	110
Total	140	150	110	400

Is there sufficient evidence to reject the hypothesis that "Political preference is independent of age"? Use $\alpha = 0.01$.

a. Solve using the *p*-value approach.

b. Solve using the classical approach.

11.70 [EX11-70] On May 21, 2004, the National Center for Chronic Disease Prevention and Health Promotion, the Centers for Disease Control and Prevention, reported the results of the Youth Risk Behavior Surveillance—United States, 2003. The report split the sample of 15,184 American teenagers into grade levels as noted in the table that follows. The students admitted to carrying a weapon within the 30 days preceding the survey and to being in a physical fight during the past year. The following table summarizes two portions of the results.

	At Least Once	Never	Total
Carried a weapon			
Grades 9 and 10	1,436	7,008	8,444
Grades 11 and 12	1,140	5,600	6,740
Total	2,576	12,608	15,184
In a physical fight			
Grades 9 and 10	3,057	5,387	8,444
Grades 11 and 12	1,942	4,798	6,740
Total	4,999	10,185	15,184

Source: Data from http://www.cdc.gov/

Does the sample evidence show that students in grades 9 and 10 and grades 11 and 12 have different tendencies to carry weapons to school? Get into a physical fight? Use the 0.01 level of significance in each case.

a. Solve using the *p*-value approach.

b. Solve using the classical approach.

11.71 [EX11-71] Based on data from the U.S. Census Bureau, the National Association of Home Builders forecasted a rise in homeownership rates for this past decade. Part of the forecast predicted new housing starts by region. The following table shows what they forecasted.

Region	Average Housing Starts		
	1996–2000	2001–2005	2006–2010
Northeast	145	161	170
South	710	687	688
Midwest	331	314	313
West	382	385	373

Do the data present sufficient evidence to reject the hypothesis that the distribution of housing starts across the regions was the same for all years? Use $\alpha = 0.05$.

a. Solve using the *p*-value approach.

b. Solve using the classical approach.

11.72 [EX11-72] Forty-one small lots of experimental product were manufactured and tested for the occurrence of a particular indication that is attributed in nature yet causes rejection of the part. Ten lots were made using one particular processing method, and 31 lots were made using a second processing method. Each lot was equally sampled ($n = 32$) for presence of this indication. In practice, optimal processing conditions show little or no occurrence of the indication. Method 1, involving the 10 lots, was run before Method 2.

Methods	n	Number of Rejects
Method 1	320	4
Method 2	992	26

Source: Courtesy of Bausch & Lomb

a. Determine, at the 0.05 level of significance, if there is a difference in the proportion of reject product between the two methods.

b. Compare your finding in part a with your findings in Exercise 10.109 (p. 521). Include all parts of the hypothesis tests in your comparison.

c. Would you say that these two testing procedures are equivalent? Give specific evidence to support your answer.

11.73 Four brands of popcorn were tested for popping. One hundred kernels of each brand were popped, and the number of kernels not popped was recorded in each test (see the following table). Can we reject the null hypothesis that all four brands pop equally? Test at $\alpha = 0.05$.

Brand	A	B	C	D
No. Not Popped	14	8	11	15

a. Solve using the *p*-value approach.

b. Solve using the classical approach.

11.74 [EX11-74] An average of two players per boys' or girls' high school basketball team are injured during a season. The following table shows the distribution of injuries for a random sample of 1000 girls and 1000 boys taken from the season records of all reported injuries.

Injury	Girls	Boys
Ankle/foot	360	383
Hip/thigh/leg	166	147
Knee	130	103
Forearm/wrist/hand	112	115
Face/scalp	88	122
All others	144	130

Does this sample information present sufficient evidence to conclude that the distribution of injuries is different for girls than for boys? Use $\alpha = 0.05$.

a. Solve using the *p*-value approach.

(continue on page 576)

b. Solve using the classical approach.

11.75 [EX11-75] "Have you designated on your driver's license that you are an organ donor?" We have all heard this question, and the word must be getting out, according to the March 30, 2005, article "Organ Transplants Reach New High of Almost 27,000 in 2004." The exact results for types of organ donations are as follows.

	From a Deceased Donor	From a Living Donor
2003	18,650	6,812
2004	20,018	6,966

Source: http://www.seniorjournal.com/

a. What percent of organ donors were deceased for each year? Do you view these percentages as significantly different? Explain.

b. At the 0.05 level of significance, did the rates of deceased donor to living donor change significantly between 2003 and 2004?

c. Compare the decision reached in part b to your answer in part a. Describe any differences and explain what caused them.

11.76 [EX11-76] Last year's work record for absenteeism in each of four categories for 100 randomly selected employees is compiled in the following table. Do these data provide sufficient evidence to reject the hypothesis that the rate of absenteeism is the same for all categories of employees? Use $\alpha = 0.01$ and 240 workdays for the year.

	Married Male	Single Male	Married Female	Single Female
Number of Employees	40	14	16	30
Days Absent	180	110	75	135

a. Solve using the p-value approach.

b. Solve using the classical approach.

11.77 If you were to roll a die 600 times, how different from 100 could the observed frequencies for each face be before the results would become significantly different from equally likely at the 0.05 level?

11.78 Consider the following set of data.

	Response		Total
	Yes	No	
Group 1	75	25	100
Group 2	70	30	100
Total	145	55	200

a. Compute the value of the test statistic $z\star$ that would be used to test the null hypothesis that $p_1 = p_2$, where p_1 and p_2 are the proportions of "yes" responses in the respective groups.

b. Compute the value of the test statistic $\chi^2\star$ that would be used to test the hypothesis that "Response is independent of group."

c. Show that $\chi^2\star = (z\star)^2$.

11.79 Write a paragraph (50+ words) describing the circumstances that would call for the use of the multinomial chi-square method. Include assumptions that would be made when using this method.

11.80 Write a paragraph (50+ words) describing the circumstances that would call for the use of the contingency table/independence chi-square method. Include assumptions that would be made when using this method.

11.81 Write a paragraph (50+ words) describing the circumstances that would call for the use of the contingency table/homogeneity chi-square method. Include assumptions that would be made when using this method.

11.82 Write a paragraph (50+ words) describing the similarities and differences between the multinomial and homogeneity chi-square tests.

11.83 Write a paragraph (50+ words) describing the similarities and differences between the independence and homogeneity chi-square tests.

Chapter Practice Test

Part I: Knowing the Definitions

Answer "True" if the statement is always true. If the statement is not always true, replace the words shown in bold with words that make the statement always true.

11.1 The number of degrees of freedom for a test of a multinomial experiment is **equal to** the number of cells in the experimental data.

11.2 The **expected frequency** in a chi-square test is found by multiplying the hypothesized probability of a cell by the total number of observations in the sample.

11.3 The **observed** frequency of a cell should not be allowed to be smaller than 5 when a chi-square test is being conducted.

11.4 In a **multinomial experiment** we have $(r - 1)(c - 1)$ degrees of freedom (r is the number of rows, and c is the number of columns).

11.5 A multinomial experiment consists of *n* **identical independent trials**.

11.6 A **multinomial experiment** arranges the data in a two-way classification such that the totals in one direction are predetermined.

11.7 The charts for both the multinomial experiment and the contingency table **must** be set in such a way that each piece of data will fall into exactly one of the categories.

11.8 The test statistic $\sum \frac{(O - E)^2}{E}$ has a distribution tha is **approximately normal**.

11.9 The data used in a chi-square multinomial test are always **enumerative**.

11.10 The null hypothesis being tested by a test of **homogeneity** is that the distribution of proportions is the same for each of the subpopulations.

Part II: Applying the Concepts

Answer all questions. Show formulas, substitutions, and work.

11.11 State the null and alternative hypotheses that would be used to test each of these claims:

a. The single-digit numerals generated by a certain random number generator were not equally likely.

b. The results of the last election in our city suggest that the votes cast were not independent of the voter's registered party.

c. The distributions of types of crimes committed against society are the same in the four largest U.S. cities.

11.12 Find each value:

a. $\chi^2(12, 0.975)$ b. $\chi^2(17, 0.005)$

11.13 Three hundred consumers were asked to identify which one of three different items they found to be the most appealing. The table shows the number that preferred each item.

Item	1	2	3
Number	85	103	112

Do these data present sufficient evidence at the 0.05 level of significance to indicate that the three items were not equally preferred?

11.14 To study the effect of the type of soil on the amount of growth attained by a new hybrid plant, saplings were planted in three different types of soil and their subsequent amounts of growth classified into three categories:

Growth	Soil Type		
	Clay	Sand	Loam
Poor	16	8	14
Average	31	16	21
Good	18	36	25
Total	65	60	60

Does the quality of growth appear to be distributed differently for the tested soil types at the 0.05 level?

a. State the null and alternative hypotheses.

b. Find the expected value for the cell containing 36.

c. Calculate the value of chi-square for these data.

d. Find the *p*-value.

e. Find the test criteria [level of significance, test statistic, its distribution, critical region, and critical value(s)].

f. State the decision and the conclusion for this hypothesis test.

Part III: Understanding the Concepts

11.15 Explain how a multinomial experiment and a binomial experiment are similar and also how they are different.

11.16 Explain the distinction between a test for independence and a test for homogeneity.

11.17 Student A says that tests for independence and homogeneity are the same, and Student B says that they are not at all alike because they are tests of different concepts. Both students are partially right and partially wrong. Explain.

11.18 You are interpreting the results of an opinion poll on the role of recycling in your town. A random sample of 400 people was asked to respond strongly in favor, slightly in favor, neutral, slightly against, or strongly against for each of several questions. There are four key questions that concern you, and you plan to analyze their results.

a. How do you calculate the expected probabilities for each answer?

b. How would you decide whether the four questions were answered the same?

Appendix A: Introductory Concepts and Review Lessons

A. Basic Principles of Counting
B. Summation Notation
C. Using the Random Number Table
D. Round-Off Procedure
E. The Coordinate-Axis System and the Equation of a Straight Line
F. Tree Diagrams
G. Venn Diagrams
H. The Use of Factorial Notation
I. Answers to Introductory Concepts and Review Lessons Exercises

Appendix A is available online–logon at cengagebrain.com.

Appendix B: Tables

TABLE 1
Random Numbers

10 09 73 25 33	76 52 01 35 86	34 67 35 48 76	80 95 90 91 17	39 29 27 49 45
37 54 20 48 05	64 89 47 42 96	24 80 52 40 37	20 63 61 04 02	00 82 29 16 65
08 42 26 89 53	19 64 50 93 03	23 20 90 25 60	15 95 33 43 64	35 08 03 36 06
99 01 90 25 29	09 37 67 07 15	38 31 13 11 65	88 67 67 43 97	04 43 62 76 59
12 80 79 99 70	80 15 73 61 47	64 03 23 66 53	98 95 11 68 77	12 17 17 68 33
66 06 57 47 17	34 07 27 68 50	36 69 73 61 70	65 81 33 98 85	11 19 92 91 70
31 06 01 08 05	45 57 18 24 06	35 30 34 26 14	86 79 90 74 39	23 40 30 97 32
85 26 97 76 02	02 05 16 56 92	68 66 57 48 18	73 05 38 52 47	18 62 38 85 79
63 57 33 21 35	05 32 54 70 48	90 55 35 75 48	28 46 82 87 09	83 49 12 56 24
73 79 64 57 53	03 52 96 47 78	35 80 83 42 82	60 93 52 03 44	35 27 38 84 35
98 52 01 77 67	14 90 56 86 07	22 10 94 05 58	60 97 09 34 33	50 50 07 39 98
11 80 50 54 31	39 80 82 77 32	50 72 56 82 48	29 40 52 42 01	52 77 56 78 51
83 45 29 96 34	06 28 89 80 83	13 74 67 00 78	18 47 54 06 10	68 71 17 78 17
88 68 54 02 00	86 50 75 84 01	36 76 66 79 51	90 36 47 64 93	29 60 91 10 62
99 59 46 73 48	87 51 76 49 69	91 82 60 89 28	93 78 56 13 68	23 47 83 41 13
65 48 11 76 74	17 46 85 09 50	58 04 77 69 74	73 03 95 71 86	40 21 81 65 44
80 12 43 56 35	17 72 70 80 15	45 31 82 23 74	21 11 57 82 53	14 38 55 37 63
74 35 09 98 17	77 40 27 72 14	43 23 60 02 10	45 52 16 42 37	96 28 60 26 55
69 91 62 68 03	66 25 22 91 48	36 93 68 72 03	76 62 11 39 90	94 40 05 64 18
09 89 32 05 05	14 22 56 85 14	46 42 75 67 88	96 29 77 88 22	54 38 21 45 98
91 49 91 45 23	68 47 92 76 86	46 16 28 35 54	94 75 08 99 23	37 08 92 00 48
80 33 69 45 98	26 94 03 68 58	70 29 73 41 35	54 14 03 33 40	42 05 08 23 41
44 10 48 19 49	85 15 74 79 54	32 97 92 65 75	57 60 04 08 81	22 22 20 64 13
12 55 07 37 42	11 10 00 20 40	12 86 07 46 97	96 64 48 94 39	28 70 72 58 15
63 60 64 93 29	16 50 53 44 84	40 21 95 25 63	43 65 17 70 82	07 20 73 17 90
61 19 69 04 46	26 45 74 77 74	51 92 43 37 29	65 39 45 95 93	42 58 26 05 27
15 47 44 52 66	95 27 07 99 53	59 36 78 38 48	82 39 61 01 18	33 21 15 94 66
94 55 72 85 73	67 89 75 43 87	54 62 24 44 31	91 19 04 25 92	92 92 74 59 73
42 48 11 62 13	97 34 40 87 21	16 86 84 87 67	03 07 11 20 59	25 70 14 66 70
23 52 37 83 17	73 20 88 98 37	68 93 59 14 16	26 25 22 96 63	05 52 28 25 62
04 49 35 24 94	75 24 63 38 24	45 86 25 10 25	61 96 27 93 35	65 33 71 24 72
00 54 99 76 54	64 05 18 81 59	96 11 96 38 96	54 69 28 23 91	23 28 72 95 29
35 96 31 53 07	26 89 80 93 54	33 35 13 54 62	77 97 45 00 24	90 10 33 93 33
59 80 80 83 91	45 42 72 68 42	83 60 94 97 00	13 02 12 48 92	78 56 52 01 06
46 05 88 52 36	01 39 09 22 86	77 28 14 40 77	93 91 08 36 47	70 61 74 29 41
32 17 90 05 97	87 37 92 52 41	05 56 70 70 07	86 74 31 71 57	85 39 41 18 38
69 23 46 14 06	20 11 74 52 04	15 95 66 00 00	18 74 39 24 23	97 11 89 63 38
19 56 54 14 30	01 75 87 53 79	40 41 92 15 85	66 67 43 68 06	84 96 28 52 07
45 15 51 49 38	19 47 60 72 46	43 66 79 45 43	59 04 79 00 33	20 82 66 95 41
94 86 43 19 94	36 16 81 08 51	34 88 88 15 53	01 54 03 54 56	05 01 45 11 76
98 08 62 48 26	45 24 02 84 04	44 99 90 88 96	39 09 47 34 07	35 44 13 18 80
33 18 51 62 32	41 94 15 09 49	89 43 54 85 81	88 69 54 19 94	37 54 87 30 43
80 95 10 04 06	96 38 27 07 74	20 15 12 33 87	25 01 62 52 98	94 62 46 11 71
79 75 24 91 40	71 96 12 82 96	69 86 10 25 91	74 85 22 05 39	00 38 75 95 79
18 63 33 25 37	98 14 50 65 71	31 01 02 46 74	05 45 56 14 27	77 93 89 19 36

Specific details about using this table can be found on page 20; in Appendix A at cengagebrain.com; or in the Student Solutions Manual.

TABLE 1
Random Numbers (*continued*)

74 02 94 39 02	77 55 73 22 70	97 79 01 71 19	52 52 75 80 21	80 81 45 17 48
54 17 84 56 11	80 99 33 71 43	05 33 51 29 69	56 12 71 92 55	36 04 09 03 24
11 66 44 98 83	52 07 98 48 27	59 38 17 15 39	09 97 33 34 40	88 46 12 33 56
48 32 47 79 28	31 24 96 47 10	02 29 53 68 70	32 30 75 75 46	15 02 00 99 94
69 07 49 41 38	87 63 79 19 76	35 58 40 44 01	10 51 82 16 15	01 84 87 69 38
09 18 82 00 97	32 82 53 95 27	04 22 08 63 04	83 38 98 73 74	64 27 85 80 44
90 04 58 54 97	51 98 15 06 54	94 93 88 19 97	91 87 07 61 50	68 47 66 46 59
73 18 95 02 07	47 67 72 62 69	62 29 06 44 64	27 12 46 70 18	41 36 18 27 60
75 76 87 64 90	20 97 18 17 49	90 42 91 22 72	95 37 50 58 71	93 82 34 31 78
54 01 64 40 56	66 28 13 10 03	00 68 22 73 98	20 71 45 32 95	07 70 61 78 13
08 35 86 99 10	78 54 24 27 85	13 66 15 88 73	04 61 89 75 53	31 22 30 84 20
28 30 60 32 64	81 33 31 05 91	40 51 00 78 93	32 60 46 04 75	94 11 90 18 40
53 84 08 62 33	81 59 41 36 28	51 21 59 02 90	28 46 66 87 95	77 76 22 07 91
91 75 75 37 41	61 61 36 22 69	50 26 39 02 12	55 78 17 65 14	83 48 34 70 55
89 41 59 26 94	00 39 75 83 91	12 60 71 76 46	48 94 97 23 06	94 54 13 74 08
77 51 30 38 20	86 83 42 99 01	68 41 48 27 74	51 90 81 39 80	72 89 35 55 07
19 50 23 71 74	69 97 92 02 88	55 21 02 97 73	74 28 77 52 51	65 34 46 74 15
21 81 85 93 13	93 27 88 17 57	05 68 67 31 56	07 08 28 50 46	31 85 33 84 52
51 47 46 64 99	68 10 72 36 21	94 04 99 13 45	42 83 60 91 91	08 00 74 54 49
99 55 96 83 31	62 53 52 41 70	69 77 71 28 30	74 81 97 81 42	43 86 07 28 34
33 71 34 80 07	93 58 47 28 69	51 92 66 47 21	58 30 32 98 22	93 17 49 39 72
85 27 48 68 93	11 30 32 92 70	28 83 43 41 37	73 51 59 04 00	71 14 84 36 43
84 13 38 96 40	44 03 55 21 66	73 85 27 00 91	61 22 26 05 61	62 32 71 84 23
56 73 21 62 34	17 39 59 61 31	10 12 39 16 22	85 49 65 75 60	81 60 41 88 80
65 13 85 68 06	87 60 88 52 61	34 31 36 58 61	45 87 52 10 69	85 64 44 72 77
38 00 10 21 76	81 71 91 17 11	71 60 29 29 37	74 21 96 40 49	65 58 44 96 98
37 40 29 63 97	01 30 47 75 86	56 27 11 00 86	47 32 46 26 05	40 03 03 74 38
97 12 54 03 48	87 08 33 14 17	21 81 53 92 50	75 23 76 20 47	15 50 12 95 78
21 82 64 11 34	47 14 33 40 72	64 63 88 59 02	49 13 90 64 41	03 85 65 45 52
73 13 54 27 42	95 71 90 90 35	85 79 47 42 96	08 78 98 81 56	64 69 11 92 02
07 63 87 79 29	03 06 11 80 72	96 20 74 41 56	23 82 19 95 38	04 71 36 69 94
60 52 88 34 41	07 95 41 98 14	59 17 52 06 95	05 53 35 21 39	61 21 20 64 55
83 59 63 56 55	06 95 89 29 83	05 12 80 97 19	77 43 35 37 83	92 30 15 04 98
10 85 06 27 46	99 59 91 05 07	13 49 90 63 19	53 07 57 18 39	06 41 01 93 62
39 82 09 89 52	43 62 26 31 47	64 42 18 08 14	43 80 00 93 51	31 02 47 31 67
59 58 00 64 78	75 56 97 88 00	88 83 55 44 86	23 76 80 61 56	04 11 10 84 08
38 50 80 73 41	23 79 34 87 63	90 82 29 70 22	17 71 90 42 07	95 95 44 99 53
30 69 27 06 68	94 68 81 61 27	56 19 68 00 91	82 06 76 34 00	05 46 26 92 00
65 44 39 56 59	18 28 82 74 37	49 63 22 40 41	08 33 76 56 76	96 29 99 08 36
27 26 75 02 64	13 19 27 22 94	07 47 74 46 06	17 98 54 89 11	97 34 13 03 58
91 30 70 69 91	19 07 22 42 10	36 69 95 37 28	28 82 53 57 93	28 97 66 62 52
68 43 49 46 88	84 47 31 36 22	62 12 69 84 08	12 84 38 25 90	09 81 59 31 46
48 90 81 58 77	54 74 52 45 91	35 70 00 47 54	83 82 45 26 92	54 13 05 51 60
06 91 34 51 97	42 67 27 86 01	11 88 30 95 28	63 01 19 89 01	14 97 44 03 44
10 45 51 60 19	14 21 03 37 12	91 34 23 78 21	88 32 58 08 51	43 66 77 08 83
12 88 39 73 43	65 02 76 11 84	04 28 50 13 92	17 97 41 50 77	90 71 22 67 69
21 77 83 09 76	38 80 73 69 61	31 64 94 20 96	63 28 10 20 23	08 81 64 74 49
19 52 35 95 15	65 12 25 96 59	86 28 36 82 58	69 57 21 37 98	16 43 59 15 29
67 24 55 26 70	35 58 31 65 63	79 24 68 66 86	76 46 33 42 22	26 65 59 08 02
60 58 44 73 77	07 50 03 79 92	45 13 42 65 29	26 76 08 36 37	41 32 64 43 44
53 85 34 13 77	36 06 69 48 50	58 83 87 38 59	49 36 47 33 31	96 24 04 36 42
24 63 73 97 36	74 38 48 93 42	52 62 30 79 92	12 36 91 86 01	03 74 28 38 73
83 08 01 24 51	38 99 22 28 15	07 75 95 17 77	97 37 72 75 85	51 97 23 78 67
16 44 42 43 34	36 15 19 90 73	27 49 37 09 39	85 13 03 25 52	54 84 65 47 59
60 79 01 81 57	57 17 86 57 62	11 16 17 85 76	45 81 95 29 79	65 13 00 48 60

From tables of the RAND Corporation. Reprinted from Wilfred J. Dixon and Frank J. Massey, Jr., *Introduction to Statistical Analysis*. 3rd ed. (New York: McGraw-Hill, 1969), pp. 446–447. Reprinted by permission of the RAND Corporation.

TABLE 2

Binomial Probabilities $\left[\binom{n}{x} \cdot p^x \cdot q^{n-x} \right]$

								P							
n	x	0.01	0.05	0.10	0.20	0.30	0.40	0.50	0.60	0.70	0.80	0.90	0.95	0.99	x
2	0	.980	.902	.810	.640	.490	.360	.250	.160	.090	.040	.010	.002	0+	0
	1	.020	.095	.180	.320	.420	.480	.500	.480	.420	.320	.180	.095	.020	1
	2	0+	.002	.010	.040	.090	.160	.250	.360	.490	.640	.810	.902	.980	2
3	0	.970	.857	.729	.512	.343	.216	.125	.064	.027	.008	.001	0+	0+	0
	1	.029	.135	.243	.384	.441	.432	.375	.288	.189	.096	.027	.007	0+	1
	2	0+	.007	.027	.096	.189	.288	.375	.432	.441	.384	.243	.135	.029	2
	3	0+	0+	.001	.008	.027	.064	.125	.216	.343	.512	.729	.857	.970	3
4	0	.961	.815	.656	.410	.240	.130	.062	.026	.008	.002	0+	0+	0+	0
	1	.039	.171	.292	.410	.412	.346	.250	.154	.076	.026	.004	0+	0+	1
	2	.001	.014	.049	.154	.265	.346	.375	.346	.265	.154	.049	.014	.001	2
	3	0+	0+	.004	.026	.076	.154	.250	.346	.412	.410	.292	.171	.039	3
	4	0+	0+	0+	.002	.008	.026	.062	.130	.240	.410	.656	.815	.961	4
5	0	.951	.774	.590	.328	.168	.078	.031	.010	.002	0+	0+	0+	0+	0
	1	.048	.204	.328	.410	.360	.259	.156	.077	.028	.006	0+	0+	0+	1
	2	.001	.021	.073	.205	.309	.346	.312	.230	.132	.051	.008	.001	0+	2
	3	0+	.001	.008	.051	.132	.230	.312	.346	.309	.205	.073	.021	.001	3
	4	0+	0+	0+	.006	.028	.077	.156	.259	.360	.410	.328	.204	.048	4
	5	0+	0+	0+	0+	.002	.010	.031	.078	.168	.328	.590	.774	.951	5
6	0	.941	.735	.531	.262	.118	.047	.016	.004	.001	0+	0+	0+	0+	0
	1	.057	.232	.354	.393	.303	.187	.094	.037	.010	.002	0+	0+	0+	1
	2	.001	.031	.098	.246	.324	.311	.234	.138	.060	.015	.001	0+	0+	2
	3	0+	.002	.015	.082	.185	.276	.312	.276	.185	.082	.015	.002	0+	3
	4	0+	0+	.001	.015	.060	.138	.234	.311	.324	.246	.098	.031	.001	4
	5	0+	0+	0+	.002	.010	.037	.094	.187	.303	.393	.354	.232	.057	5
	6	0+	0+	0+	0+	.001	.004	.016	.047	.118	.262	.531	.735	.941	6
7	0	.932	.698	.478	.210	.082	.028	.008	.002	0+	0+	0+	0+	0+	0
	1	.066	.257	.372	.367	.247	.131	.055	.017	.004	0+	0+	0+	0+	1
	2	.002	.041	.124	.275	.318	.261	.164	.077	.025	.004	0+	0+	0+	2
	3	0+	.004	.023	.115	.227	.290	.273	.194	.097	.029	.003	0+	0+	3
	4	0+	0+	.003	.029	.097	.194	.273	.290	.227	.115	.023	.004	0+	4
	5	0+	0+	0+	.004	.025	.077	.164	.261	.318	.275	.124	.041	.002	5
	6	0+	0+	0+	0+	.004	.017	.055	.131	.247	.367	.372	.257	.066	6
	7	0+	0+	0+	0+	0+	.002	.008	.028	.082	.210	.478	.698	.932	7
8	0	.923	.663	.430	.168	.058	.017	.004	.001	0+	0+	0+	0+	0+	0
	1	.075	.279	.383	.336	.198	.090	.031	.008	.001	0+	0+	0+	0+	1
	2	.003	.051	.149	.294	.296	.209	.109	.041	.010	.001	0+	0+	0+	2
	3	0+	.005	.033	.147	.254	.279	.219	.124	.047	.009	0+	0+	0+	3
	4	0+	0+	.005	.046	.136	.232	.273	.232	.136	.046	.005	0+	0+	4
	5	0+	0+	0+	.009	.047	.124	.219	.279	.254	.147	.033	.005	0+	5
	6	0+	0+	0+	.001	.010	.041	.109	.209	.296	.294	.149	.051	.003	6
	7	0+	0+	0+	0+	.001	.008	.031	.090	.198	.336	.383	.279	.075	7
	8	0+	0+	0+	0+	0+	.001	.004	.017	.058	.168	.430	.663	.923	8

For specific details about using this table, see pages 250–251. Table 2 was generated using Excel.

TABLE 2

Binomial Probabilities $\left[\binom{n}{x} \cdot p^x \cdot q^{n-x}\right]$ (continued)

n	x	0.01	0.05	0.10	0.20	0.30	0.40	0.50	0.60	0.70	0.80	0.90	0.95	0.99	x
9	0	.914	.630	.387	.134	.040	.010	.002	0+	0+	0+	0+	0+	0+	0
	1	.083	.299	.387	.302	.156	.060	.018	.004	0+	0+	0+	0+	0+	1
	2	.003	.063	.172	.302	.267	.161	.070	.021	.004	0+	0+	0+	0+	2
	3	0+	.008	.045	.176	.267	.251	.164	.074	.021	.003	0+	0+	0+	3
	4	0+	.001	.007	.066	.172	.251	.246	.167	.074	.017	.001	0+	0+	4
	5	0+	0+	.001	.017	.074	.167	.246	.251	.172	.066	.007	.001	0+	5
	6	0+	0+	0+	.003	.021	.074	.164	.251	.267	.176	.045	.008	0+	6
	7	0+	0+	0+	0+	.004	.021	.070	.161	.267	.302	.172	.063	.003	7
	8	0+	0+	0+	0+	0+	.004	.018	.060	.156	.302	.387	.299	.083	8
	9	0+	0+	0+	0+	0+	0+	.002	.010	.040	.134	.387	.630	.914	9
10	0	.904	.599	.349	.107	.028	.006	.001	0+	0+	0+	0+	0+	0+	0
	1	.091	.315	.387	.268	.121	.040	.010	.002	0+	0+	0+	0+	0+	1
	2	.004	.075	.194	.302	.233	.121	.044	.011	.001	0+	0+	0+	0+	2
	3	0+	.010	.057	.201	.267	.215	.117	.042	.009	.001	0+	0+	0+	3
	4	0+	.001	.011	.088	.200	.251	.205	.111	.037	.006	0+	0+	0+	4
	5	0+	0+	.001	.026	.103	.201	.246	.201	.103	.026	.001	0+	0+	5
	6	0+	0+	0+	.006	.037	.111	.205	.251	.200	.088	.011	.001	0+	6
	7	0+	0+	0+	.001	.009	.042	.117	.215	.267	.201	.057	.010	0+	7
	8	0+	0+	0+	0+	.001	.011	.044	.121	.233	.302	.194	.075	.004	8
	9	0+	0+	0+	0+	0+	.002	.010	.040	.121	.268	.387	.315	.091	9
	10	0+	0+	0+	0+	0+	0+	.001	.006	.028	.107	.349	.599	.904	10
11	0	.895	.569	.314	.086	.020	.004	0+	0+	0+	0+	0+	0+	0+	0
	1	.099	.329	.384	.236	.093	.027	.005	.001	0+	0+	0+	0+	0+	1
	2	.005	.087	.213	.295	.200	.089	.027	.005	.001	0+	0+	0+	0+	1
	3	0+	.014	.071	.221	.257	.177	.081	.023	.004	0+	0+	0+	0+	3
	4	0+	.001	.016	.111	.220	.236	.161	.070	.017	.002	0+	0+	0+	4
	5	0+	0+	.002	.039	.132	.221	.226	.147	.057	.010	0+	0+	0+	5
	6	0+	0+	0+	.010	.057	.147	.226	.221	.132	.039	.002	0+	0+	6
	7	0+	0+	0+	.002	.017	.070	.161	.236	.220	.111	.016	.001	0+	7
	8	0+	0+	0+	0+	.004	.023	.081	.177	.257	.221	.071	.014	0+	8
	9	0+	0+	0+	0+	.001	.005	.027	.089	.200	.295	.213	.087	.005	9
	10	0+	0+	0+	0+	0+	.001	.005	.027	.093	.236	.384	.329	.099	10
	11	0+	0+	0+	0+	0+	0+	0+	.004	.020	.086	.314	.569	.895	11
12	0	.886	.540	.282	.069	.014	.002	0+	0+	0+	0+	0+	0+	0+	0
	1	.107	.341	.377	.206	.071	.017	.003	0+	0+	0+	0+	0+	0+	1
	2	.006	.099	.230	.283	.168	.064	.016	.002	0+	0+	0+	0+	0+	2
	3	0+	.017	.085	.236	.240	.142	.054	.012	.001	0+	0+	0+	0+	3
	4	0+	.002	.021	.133	.231	.213	.121	.042	.008	.001	0+	0+	0+	4
	5	0+	0+	.004	.053	.158	.227	.193	.101	.029	.003	0+	0+	0+	5
	6	0+	0+	0+	.016	.079	.177	.226	.177	.079	.016	0+	0+	0+	6
	7	0+	0+	0+	.003	.029	.101	.193	.227	.158	.053	.004	0+	0+	7
	8	0+	0+	0+	.001	.008	.042	.121	.213	.231	.133	.021	.002	0+	8
	9	0+	0+	0+	0+	.001	.012	.054	.142	.240	.236	.085	.017	0+	9
	10	0+	0+	0+	0+	0+	.002	.016	.064	.168	.283	.230	.099	.006	10
	11	0+	0+	0+	0+	0+	0+	.003	.017	.071	.206	.377	.341	.107	11
	12	0+	0+	0+	0+	0+	0+	0+	.002	.014	.069	.282	.540	.886	12

Table 2 was generated using Excel.

TABLE 2
Binomial Probabilities $\left[\binom{n}{x} \cdot p^x \cdot q^{n-x}\right]$ (continued)

n	x							P							x
		0.01	0.05	0.10	0.20	0.30	0.40	0.50	0.60	0.70	0.80	0.90	0.95	0.99	
13	0	.878	.513	.254	.055	.010	.001	0+	0+	0+	0+	0+	0+	0+	0
	1	.115	.351	.367	.179	.054	.011	.002	0+	0+	0+	0+	0+	0+	1
	2	.007	.111	.245	.268	.139	.045	.010	.001	0+	0+	0+	0+	0+	2
	3	0+	.021	.100	.246	.218	.111	.035	.006	.001	0+	0+	0+	0+	3
	4	0+	.003	.028	.154	.234	.184	.087	.024	.003	0+	0+	0+	0+	4
	5	0+	0+	.006	.069	.180	.221	.157	.066	.014	.001	0+	0+	0+	5
	6	0+	0+	.001	.023	.103	.197	.209	.131	.044	.006	0+	0+	0+	6
	7	0+	0+	0+	.006	.044	.131	.209	.197	.103	.023	.001	0+	0+	7
	8	0+	0+	0+	.001	.014	.066	.157	.221	.180	.069	.006	0+	0+	8
	9	0+	0+	0+	0+	.003	.024	.087	.184	.234	.154	.028	.003	0+	9
	10	0+	0+	0+	0+	.001	.006	.035	.111	.218	.246	.100	.021	0+	10
	11	0+	0+	0+	0+	0+	.001	.010	.045	.139	.268	.245	.111	.007	11
	12	0+	0+	0+	0+	0+	0+	.002	.011	.054	.179	.367	.351	.115	12
	13	0+	0+	0+	0+	0+	0+	0+	.001	.010	.055	.254	.513	.878	13
14	0	.869	.488	.229	.044	.007	.001	0+	0+	0+	0+	0+	0+	0+	0
	1	.123	.359	.356	.154	.041	.007	.001	0+	0+	0+	0+	0+	0+	1
	2	.008	.123	.257	.250	.113	.032	.006	.001	0+	0+	0+	0+	0+	2
	3	0+	.026	.114	.250	.194	.085	.022	.003	0+	0+	0+	0+	0+	3
	4	0+	.004	.035	.172	.229	.155	.061	.014	.001	0+	0+	0+	0+	4
	5	0+	0+	.008	.086	.196	.207	.122	.041	.007	0+	0+	0+	0+	5
	6	0+	0+	.001	.032	.126	.207	.183	.092	.023	.002	0+	0+	0+	6
	7	0+	0+	0+	.009	.062	.157	.209	.157	.062	.009	0+	0+	0+	7
	8	0+	0+	0+	.002	.023	.092	.183	.207	.126	.032	.001	0+	0+	8
	9	0+	0+	0+	0+	.007	.041	.122	.207	.196	.086	.008	0+	0+	9
	10	0+	0+	0+	0+	.001	.014	.061	.155	.229	.172	.035	.004	0+	10
	11	0+	0+	0+	0+	0+	.003	.022	.085	.194	.250	.114	.026	0+	11
	12	0+	0+	0+	0+	0+	.001	.006	.032	.113	.250	.257	.123	.008	12
	13	0+	0+	0+	0+	0+	0+	.001	.007	.041	.154	.356	.359	.123	13
	14	0+	0+	0+	0+	0+	0+	0+	.001	.007	.044	.229	.488	.869	14
15	0	.860	.463	.206	.035	.005	0+	0+	0+	0+	0+	0+	0+	0+	0
	1	.130	.366	.343	.132	.031	.005	0+	0+	0+	0+	0+	0+	0+	1
	2	.009	.135	.267	.231	.092	.022	.003	0+	0+	0+	0+	0+	0+	2
	3	0+	.031	.129	.250	.170	.063	.014	.002	0+	0+	0+	0+	0+	3
	4	0+	.005	.043	.188	.219	.127	.042	.007	.001	0+	0+	0+	0+	4
	5	0+	.001	.010	.103	.206	.186	.092	.024	.003	0+	0+	0+	0+	5
	6	0+	0+	.002	.043	.147	.207	.153	.061	.012	.001	0+	0+	0+	6
	7	0+	0+	0+	.014	.081	.177	.196	.118	.035	.003	0+	0+	0+	7
	8	0+	0+	0+	.003	.035	.118	.196	.177	.081	.014	0+	0+	0+	8
	9	0+	0+	0+	.001	.012	.061	.153	.207	.147	.043	.002	0+	0+	9
	10	0+	0+	0+	0+	.003	.024	.092	.186	.206	.103	.010	.001	0+	10
	11	0+	0+	0+	0+	.001	.007	.042	.127	.219	.188	.043	.005	0+	11
	12	0+	0+	0+	0+	0+	.002	.014	.063	.170	.250	.129	.031	0+	12
	13	0+	0+	0+	0+	0+	0+	.003	.022	.092	.231	.267	.135	.009	13
	14	0+	0+	0+	0+	0+	0+	0+	.005	.031	.132	.343	.366	.130	14
	15	0+	0+	0+	0+	0+	0+	0+	0+	.005	.035	.206	.463	.860	15

Table 2 was generated using Excel.

TABLE 3

Cumulative Areas of the Standard Normal Distribution

The entries in this table are the cumulative probabilities for the standard normal distribution z (that is, the normal distribution with mean 0 and standard deviation 1). The shaded area under the curve of the standard normal distribution represents the cumulative probability to the left of a z-value in the **left-hand tail**.

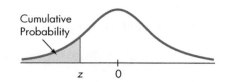

Cumulative Probability

z	0.00	0.01	0.02	0.03	0.04	0.05	0.06	0.07	0.08	0.09
−5.0	0.0000003									
−4.5	0.000003									
−4.0	0.00003	0.00003	0.00003	0.00003	0.00003	0.00003	0.00002	0.00002	0.00002	0.00002
−3.9	0.00005	0.00005	0.00004	0.00004	0.00004	0.00004	0.00004	0.00004	0.00003	0.00003
−3.8	0.00007	0.00007	0.00007	0.00006	0.00006	0.00006	0.00006	0.00005	0.00005	0.00005
−3.7	0.00011	0.00010	0.00010	0.00010	0.00009	0.00009	0.00008	0.00008	0.00008	0.00008
−3.6	0.0002	0.0002	0.0002	0.00014	0.00014	0.00013	0.00013	0.00012	0.00012	0.00011
−3.5	0.0002	0.0002	0.0002	0.0002	0.0002	0.0002	0.0002	0.0002	0.0002	0.0002
−3.4	0.0003	0.0003	0.0003	0.0003	0.0003	0.0003	0.0003	0.0003	0.0003	0.0002
−3.3	0.0005	0.0005	0.0005	0.0004	0.0004	0.0004	0.0004	0.0004	0.0004	0.0004
−3.2	0.0007	0.0007	0.0006	0.0006	0.0006	0.0006	0.0006	0.0005	0.0005	0.0005
−3.1	0.0010	0.0009	0.0009	0.0009	0.0008	0.0008	0.0008	0.0008	0.0007	0.0007
−3.0	0.0014	0.0013	0.0013	0.0012	0.0012	0.0011	0.0011	0.0011	0.0010	0.0010
−2.9	0.0019	0.0018	0.0018	0.0017	0.0016	0.0016	0.0015	0.0015	0.0014	0.0014
−2.8	0.0026	0.0025	0.0024	0.0023	0.0023	0.0022	0.0021	0.0021	0.0020	0.0019
−2.7	0.0035	0.0034	0.0033	0.0032	0.0031	0.0030	0.0029	0.0028	0.0027	0.0026
−2.6	0.0047	0.0045	0.0044	0.0043	0.0042	0.0040	0.0039	0.0038	0.0037	0.0036
−2.5	0.0062	0.0060	0.0059	0.0057	0.0055	0.0054	0.0052	0.0051	0.0049	0.0048
−2.4	0.0082	0.0080	0.0078	0.0076	0.0073	0.0071	0.0070	0.0068	0.0066	0.0064
−2.3	0.0107	0.0104	0.0102	0.0099	0.0096	0.0094	0.0091	0.0089	0.0087	0.0084
−2.2	0.0139	0.0136	0.0132	0.0129	0.0126	0.0122	0.0119	0.0116	0.0113	0.0110
−2.1	0.0179	0.0174	0.0170	0.0166	0.0162	0.0158	0.0154	0.0150	0.0146	0.0143
−2.0	0.0228	0.0222	0.0217	0.0212	0.0207	0.0202	0.0197	0.0192	0.0188	0.0183
−1.9	0.0287	0.0281	0.0274	0.0268	0.0262	0.0256	0.0250	0.0244	0.0239	0.0233
−1.8	0.0359	0.0352	0.0344	0.0336	0.0329	0.0322	0.0314	0.0307	0.0301	0.0294
−1.7	0.0446	0.0436	0.0427	0.0418	0.0409	0.0401	0.0392	0.0384	0.0375	0.0367
−1.6	0.0548	0.0537	0.0526	0.0516	0.0505	0.0495	0.0485	0.0475	0.0465	0.0455
−1.5	0.0668	0.0655	0.0643	0.0630	0.0618	0.0606	0.0594	0.0582	0.0571	0.0559
−1.4	0.0808	0.0793	0.0778	0.0764	0.0749	0.0735	0.0721	0.0708	0.0694	0.0681
−1.3	0.0968	0.0951	0.0934	0.0918	0.0901	0.0885	0.0869	0.0853	0.0838	0.0823
−1.2	0.1151	0.1131	0.1112	0.1094	0.1075	0.1057	0.1038	0.1020	0.1003	0.0985
−1.1	0.1357	0.1335	0.1314	0.1292	0.1271	0.1251	0.1230	0.1210	0.1190	0.1170
−1.0	0.1587	0.1563	0.1539	0.1515	0.1492	0.1469	0.1446	0.1423	0.1401	0.1379
−0.9	0.1841	0.1814	0.1788	0.1762	0.1736	0.1711	0.1685	0.1660	0.1635	0.1611
−0.8	0.2119	0.2090	0.2061	0.2033	0.2005	0.1977	0.1949	0.1922	0.1894	0.1867
−0.7	0.2420	0.2389	0.2358	0.2327	0.2297	0.2266	0.2236	0.2207	0.2177	0.2148
−0.6	0.2743	0.2709	0.2676	0.2643	0.2611	0.2578	0.2546	0.2514	0.2483	0.2451
−0.5	0.3085	0.3050	0.3015	0.2981	0.2946	0.2912	0.2877	0.2843	0.2810	0.2776
−0.4	0.3446	0.3409	0.3372	0.3336	0.3300	0.3264	0.3228	0.3192	0.3156	0.3121
−0.3	0.3821	0.3783	0.3745	0.3707	0.3669	0.3632	0.3594	0.3557	0.3520	0.3483
−0.2	0.4207	0.4168	0.4129	0.4090	0.4052	0.4013	0.3974	0.3936	0.3897	0.3859
−0.1	0.4602	0.4562	0.4522	0.4483	0.4443	0.4404	0.4364	0.4325	0.4286	0.4247
0.0	0.5000	0.4960	0.4920	0.4880	0.4840	0.4801	0.4761	0.4721	0.4681	0.4641

For specific details about using this table to find probabilities, see pages 272–274, 292–294; p-values, pages 375–377. Table 3 was generated using Minitab.

TABLE 3

Cumulative Areas of the Standard Normal Distribution (*continued*)

The entries in this table are the cumulative probabilities for the standard normal distribution z (that is, the normal distribution with mean 0 and standard deviation 1). The shaded area under the curve of the standard normal distribution represents the cumulative probability to the left of a z-value in the **left-hand tail**.

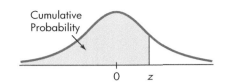

Cumulative Probability

z	0.00	0.01	0.02	0.03	0.04	0.05	0.06	0.07	0.08	0.09
0.0	0.5000	0.5040	0.5080	0.5120	0.5160	0.5199	0.5239	0.5279	0.5319	0.5359
0.1	0.5398	0.5438	0.5478	0.5517	0.5557	0.5596	0.5636	0.5675	0.5714	0.5754
0.2	0.5793	0.5832	0.5871	0.5910	0.5948	0.5987	0.6026	0.6064	0.6103	0.6141
0.3	0.6179	0.6217	0.6255	0.6293	0.6331	0.6368	0.6406	0.6443	0.6480	0.6517
0.4	0.6554	0.6591	0.6628	0.6664	0.6700	0.6736	0.6772	0.6808	0.6844	0.6879
0.5	0.6915	0.6950	0.6985	0.7019	0.7054	0.7088	0.7123	0.7157	0.7190	0.7224
0.6	0.7258	0.7291	0.7324	0.7357	0.7389	0.7422	0.7454	0.7486	0.7518	0.7549
0.7	0.7580	0.7612	0.7642	0.7673	0.7704	0.7734	0.7764	0.7794	0.7823	0.7852
0.8	0.7881	0.7910	0.7939	0.7967	0.7996	0.8023	0.8051	0.8079	0.8106	0.8133
0.9	0.8159	0.8186	0.8212	0.8238	0.8264	0.8289	0.8315	0.8340	0.8365	0.8389
1.0	0.8413	0.8438	0.8461	0.8485	0.8508	0.8531	0.8554	0.8577	0.8599	0.8621
1.1	0.8643	0.8665	0.8686	0.8708	0.8729	0.8749	0.8770	0.8790	0.8810	0.8830
1.2	0.8849	0.8869	0.8888	0.8907	0.8925	0.8944	0.8962	0.8980	0.8997	0.9015
1.3	0.9032	0.9049	0.9066	0.9082	0.9099	0.9115	0.9131	0.9147	0.9162	0.9177
1.4	0.9192	0.9207	0.9222	0.9236	0.9251	0.9265	0.9279	0.9292	0.9306	0.9319
1.5	0.9332	0.9345	0.9357	0.9370	0.9382	0.9394	0.9406	0.9418	0.9430	0.9441
1.6	0.9452	0.9463	0.9474	0.9485	0.9495	0.9505	0.9515	0.9525	0.9535	0.9545
1.7	0.9554	0.9564	0.9573	0.9582	0.9591	0.9599	0.9608	0.9616	0.9625	0.9633
1.8	0.9641	0.9649	0.9656	0.9664	0.9671	0.9678	0.9686	0.9693	0.9700	0.9706
1.9	0.9713	0.9719	0.9726	0.9732	0.9738	0.9744	0.9750	0.9756	0.9762	0.9767
2.0	0.9773	0.9778	0.9783	0.9788	0.9793	0.9798	0.9803	0.9808	0.9812	0.9817
2.1	0.9821	0.9826	0.9830	0.9834	0.9838	0.9842	0.9846	0.9850	0.9854	0.9857
2.2	0.9861	0.9865	0.9868	0.9871	0.9875	0.9878	0.9881	0.9884	0.9887	0.9890
2.3	0.9893	0.9896	0.9898	0.9901	0.9904	0.9906	0.9909	0.9911	0.9913	0.9916
2.4	0.9918	0.9920	0.9922	0.9925	0.9927	0.9929	0.9931	0.9932	0.9934	0.9936
2.5	0.9938	0.9940	0.9941	0.9943	0.9945	0.9946	0.9948	0.9949	0.9951	0.9952
2.6	0.9953	0.9955	0.9956	0.9957	0.9959	0.9960	0.9961	0.9962	0.9963	0.9964
2.7	0.9965	0.9966	0.9967	0.9968	0.9969	0.9970	0.9971	0.9972	0.9973	0.9974
2.8	0.9974	0.9975	0.9976	0.9977	0.9977	0.9978	0.9979	0.9980	0.9980	0.9981
2.9	0.9981	0.9982	0.9983	0.9983	0.9984	0.9984	0.9985	0.9985	0.9986	0.9986
3.0	0.9987	0.9987	0.9987	0.9988	0.9988	0.9989	0.9989	0.9989	0.9990	0.9990
3.1	0.9990	0.9991	0.9991	0.9991	0.9992	0.9992	0.9992	0.9992	0.9993	0.9993
3.2	0.9993	0.9993	0.9994	0.9994	0.9994	0.9994	0.9994	0.9995	0.9995	0.9995
3.3	0.9995	0.9995	0.9996	0.9996	0.9996	0.9996	0.9996	0.9996	0.9996	0.9997
3.4	0.9997	0.9997	0.9997	0.9997	0.9997	0.9997	0.9997	0.9997	0.9998	0.9998
3.5	0.9998	0.9998	0.9998	0.9998	0.9998	0.9998	0.9998	0.9998	0.9998	0.9998
3.6	0.99984	0.99985	0.99985	0.99986	0.99986	0.99987	0.99987	0.99988	0.99988	0.99989
3.7	0.99989	0.99990	0.99990	0.99990	0.99991	0.99991	0.99992	0.99992	0.99992	0.99992
3.8	0.99993	0.99993	0.99993	0.99994	0.99994	0.99994	0.99994	0.99995	0.99995	0.99995
3.9	0.99995	0.99995	0.99996	0.99996	0.99996	0.99996	0.99996	0.99996	0.99997	0.99997
4.0	0.99997	0.99997	0.99997	0.99997	0.99997	0.99997	0.99998	0.99998	0.99998	0.99998
4.5	0.999997									
5.0	0.9999997									

Table 3 was generated using Minitab.

TABLE 4
Critical Values of Standard Normal Distribution

A   ONE-TAILED SITUATIONS

The entries in this table are the critical values for z for which the area under the curve representing α is in the right-hand tail. Critical values for the left-hand tail are found by symmetry.

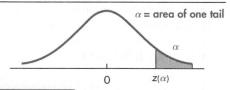

	Amount of α in one tail							
α	0.25	0.10	0.05	0.025	0.02	0.01	0.005
$z(\alpha)$	0.67	1.28	1.65	1.96	2.05	2.33	2.58

One-tailed example:
$\alpha = 0.05$
$z(\alpha) = z(0.05) = 1.65$

B   TWO-TAILED SITUATIONS

The entries in this table are the critical values for z for which the area under the curve representing α is split equally between the two tails.

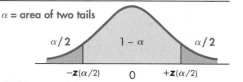

	Amount of α in two tails					
α	0.25	0.20	0.10	0.05	0.02	0.01
$z(\alpha/2)$	1.15	1.28	1.65	1.96	2.33	2.58
$1 - \alpha$	0.75	0.80	0.90	0.95	0.98	0.99
Area in the "center"						

Two-tailed example:
$\alpha = 0.05$ or $1 - \alpha = 0.95$
$\alpha/2 = 0.025$
$z(\alpha/2) = z(0.025) = 1.96$

For specific details about using Table A to find critical values, see page 393.

For specific details about using Table B to find confidence coefficients, see pages 348, 350, 356; for critical values, see pages 393, 395–396.

TABLE 5
p-Values for Standard Normal Distribution

The entries in this table are the p-values related to the right-hand tail for the calculated $z\star$ for the standard normal distribution.

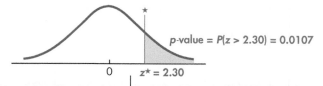

p-value $= P(z > 2.30) = 0.0107$

$z\star$	p-value	$z\star$	p-value	$z\star$	p-value	$z\star$	p-value	$z\star$	p-value
0.00	0.5000	0.80	0.2119	1.60	0.0548	2.40	0.0082	3.20	0.0007
0.05	0.4801	0.85	0.1977	1.65	0.0495	2.45	0.0071	3.25	0.0006
0.10	0.4602	0.90	0.1841	1.70	0.0446	2.50	0.0062	3.30	0.0005
0.15	0.4404	0.95	0.1711	1.75	0.0401	2.55	0.0054	3.35	0.0004
0.20	0.4207	1.00	0.1587	1.80	0.0359	2.60	0.0047	3.40	0.0003
0.25	0.4013	1.05	0.1469	1.85	0.0322	2.65	0.0040	3.45	0.0003
0.30	0.3821	1.10	0.1357	1.90	0.0287	2.70	0.0035	3.50	0.0002
0.35	0.3632	1.15	0.1251	1.95	0.0256	2.75	0.0030	3.55	0.0002
0.40	0.3446	1.20	0.1151	2.00	0.0228	2.80	0.0026	3.60	0.0002
0.45	0.3264	1.25	0.1056	2.05	0.0202	2.85	0.0022	3.65	0.0001
0.50	0.3085	1.30	0.0968	2.10	0.0179	2.90	0.0019	3.70	0.0001
0.55	0.2912	1.35	0.0885	2.15	0.0158	2.95	0.0016	3.75	0.0001
0.60	0.2743	1.40	0.0808	2.20	0.0139	3.00	0.0013	3.80	0.0001
0.65	0.2578	1.45	0.0735	2.25	0.0122	3.05	0.0011	3.85	0.0001
0.70	0.2420	1.50	0.0668	2.30	0.0107	3.10	0.0010	3.90	0+
0.75	0.2266	1.55	0.0606	2.35	0.0094	3.15	0.0008	3.95	0+

For specific details about using this table to find p-values, see pages 376–378.

TABLE 6
Critical Values of Student's *t*-Distribution

The entries in this table are the critical values of the Student's *t*-distribution, for which the area under the curve is: a) in the right-hand tail, or b) in two tails. See the illustrations at the bottom of the page.

Area in One Tail

	0.25	0.10	0.05	0.025	0.01	0.005

Area in Two Tails

df	0.50	0.20	0.10	0.05	0.02	0.01
3	0.765	1.64	2.35	3.18	4.54	5.84
4	0.741	1.53	2.13	2.78	3.75	4.60
5	0.727	1.48	2.02	2.57	3.36	4.03
6	0.718	1.44	1.94	2.45	3.14	3.71
7	0.711	1.41	1.89	2.36	3.00	3.50
8	0.706	1.40	1.86	2.31	2.90	3.36
9	0.703	1.38	1.83	2.26	2.82	3.25
10	0.700	1.37	1.81	2.23	2.76	3.17
11	0.697	1.36	1.80	2.20	2.72	3.11
12	0.695	1.36	1.78	2.18	2.68	3.05
13	0.694	1.35	1.77	2.16	2.65	3.01
14	0.692	1.35	1.76	2.14	2.62	2.98
15	0.691	1.34	1.75	2.13	2.60	2.95
16	0.690	1.34	1.75	2.12	2.58	2.92
17	0.689	1.33	1.74	2.11	2.57	2.90
18	0.688	1.33	1.73	2.10	2.55	2.88
19	0.688	1.33	1.73	2.09	2.54	2.86
20	0.687	1.33	1.72	2.09	2.53	2.85
21	0.686	1.32	1.72	2.08	2.52	2.83
22	0.686	1.32	1.72	2.07	2.51	2.82
23	0.685	1.32	1.71	2.07	2.50	2.81
24	0.685	1.32	1.71	2.06	2.49	2.80
25	0.684	1.32	1.71	2.06	2.49	2.79
26	0.684	1.31	1.70	2.05	2.47	2.77
27	0.684	1.31	1.70	2.05	2.47	2.77
28	0.683	1.31	1.70	2.05	2.47	2.76
29	0.683	1.31	1.70	2.05	2.46	2.76
30	0.683	1.31	1.70	2.04	2.46	2.75
35	0.682	1.31	1.69	2.03	2.44	2.72
40	0.681	1.30	1.68	2.02	2.42	2.70
50	0.679	1.30	1.68	2.01	2.40	2.68
70	0.678	1.29	1.67	1.99	2.38	2.65
100	0.677	1.29	1.66	1.98	2.36	2.63
df > 100	0.675	1.28	1.65	1.96	2.33	2.58

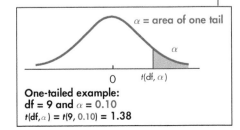

One-tailed example:
df = 9 and $\alpha = 0.10$
$t(df, \alpha) = t(9, 0.10) = 1.38$

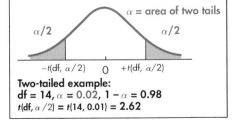

Two-tailed example:
df = 14, $\alpha = 0.02$, $1 - \alpha = 0.98$
$t(df, \alpha/2) = t(14, 0.01) = 2.62$

For specific details about using this table to find confidence coefficients, see pages 415–416, 418; *p*-values, pages 421–422; critical values, pages 415, 421. Table 6 was generated using Minitab.

TABLE 7
Probability-Values for Student's *t*-distribution

The entries in this table are the *p*-values related to the right-hand tail for the calculated $t\star$ value for the *t*-distribution of df degrees of freedom.

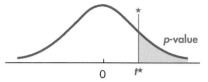

$t\star$						Degrees of Freedom									
	3	4	5	6	7	8	10	12	15	18	21	25	29	35	df $\geq$ 45
0.0	0.500	0.500	0.500	0.500	0.500	0.500	0.500	0.500	0.500	0.500	0.500	0.500	0.500	0.500	0.500
0.1	0.463	0.463	0.462	0.462	0.462	0.461	0.461	0.461	0.461	0.461	0.461	0.461	0.461	0.460	0.460
0.2	0.427	0.426	0.425	0.424	0.424	0.423	0.423	0.422	0.422	0.422	0.422	0.422	0.421	0.421	0.421
0.3	0.392	0.390	0.388	0.387	0.386	0.386	0.385	0.385	0.384	0.384	0.384	0.383	0.383	0.383	0.383
0.4	0.358	0.355	0.353	0.352	0.351	0.350	0.349	0.348	0.347	0.347	0.347	0.346	0.346	0.346	0.346
0.5	0.326	0.322	0.319	0.317	0.316	0.315	0.314	0.313	0.312	0.312	0.311	0.311	0.310	0.310	0.310
0.6	0.295	0.290	0.287	0.285	0.284	0.283	0.281	0.280	0.279	0.278	0.277	0.277	0.277	0.276	0.276
0.7	0.267	0.261	0.258	0.255	0.253	0.252	0.250	0.249	0.247	0.246	0.246	0.245	0.245	0.244	0.244
0.8	0.241	0.234	0.230	0.227	0.225	0.223	0.221	0.220	0.218	0.217	0.216	0.216	0.215	0.215	0.214
0.9	0.217	0.210	0.205	0.201	0.199	0.197	0.195	0.193	0.191	0.190	0.189	0.188	0.188	0.187	0.186
1.0	0.196	0.187	0.182	0.178	0.175	0.173	0.170	0.169	0.167	0.165	0.164	0.163	0.163	0.162	0.161
1.1	0.176	0.167	0.161	0.157	0.154	0.152	0.149	0.146	0.144	0.143	0.142	0.141	0.140	0.139	0.139
1.2	0.158	0.148	0.142	0.138	0.135	0.132	0.129	0.127	0.124	0.123	0.122	0.121	0.120	0.119	0.118
1.3	0.142	0.132	0.125	0.121	0.117	0.115	0.111	0.109	0.107	0.105	0.104	0.103	0.102	0.101	0.100
1.4	0.128	0.117	0.110	0.106	0.102	0.100	0.096	0.093	0.091	0.089	0.088	0.087	0.086	0.085	0.084
1.5	0.115	0.104	0.097	0.092	0.089	0.086	0.082	0.080	0.077	0.075	0.074	0.073	0.072	0.071	0.070
1.6	0.104	0.092	0.085	0.080	0.077	0.074	0.070	0.068	0.065	0.064	0.062	0.061	0.060	0.059	0.058
1.7	0.094	0.082	0.075	0.070	0.066	0.064	0.060	0.057	0.055	0.053	0.052	0.051	0.050	0.049	0.048
1.8	0.085	0.073	0.066	0.061	0.057	0.055	0.051	0.049	0.046	0.044	0.043	0.042	0.041	0.040	0.039
1.9	0.077	0.065	0.058	0.053	0.050	0.047	0.043	0.041	0.038	0.037	0.036	0.035	0.034	0.033	0.032
2.0	0.070	0.058	0.051	0.046	0.043	0.040	0.037	0.034	0.032	0.030	0.029	0.028	0.027	0.027	0.026
2.1	0.063	0.052	0.045	0.040	0.037	0.034	0.031	0.029	0.027	0.025	0.024	0.023	0.022	0.022	0.021
2.2	0.058	0.046	0.040	0.035	0.032	0.029	0.026	0.024	0.022	0.021	0.020	0.019	0.018	0.017	0.016
2.3	0.052	0.041	0.035	0.031	0.027	0.025	0.022	0.020	0.018	0.017	0.016	0.015	0.014	0.014	0.013
2.4	0.048	0.037	0.031	0.027	0.024	0.022	0.019	0.017	0.015	0.014	0.013	0.012	0.012	0.011	0.010
2.5	0.044	0.033	0.027	0.023	0.020	0.018	0.016	0.014	0.012	0.011	0.010	0.010	0.009	0.009	0.008
2.6	0.040	0.030	0.024	0.020	0.018	0.016	0.013	0.012	0.010	0.009	0.008	0.008	0.007	0.007	0.006
2.7	0.037	0.027	0.021	0.018	0.015	0.014	0.011	0.010	0.008	0.007	0.007	0.006	0.006	0.005	0.005
2.8	0.034	0.024	0.019	0.016	0.013	0.012	0.009	0.008	0.007	0.006	0.005	0.005	0.005	0.004	0.004
2.9	0.031	0.022	0.017	0.014	0.011	0.010	0.008	0.007	0.005	0.005	0.004	0.004	0.004	0.003	0.003
3.0	0.029	0.020	0.015	0.012	0.010	0.009	0.007	0.006	0.004	0.004	0.003	0.003	0.003	0.002	0.002
3.1	0.027	0.018	0.013	0.011	0.009	0.007	0.006	0.005	0.004	0.003	0.003	0.002	0.002	0.002	0.002
3.2	0.025	0.016	0.012	0.009	0.008	0.006	0.005	0.004	0.003	0.002	0.002	0.002	0.002	0.001	0.001
3.3	0.023	0.015	0.011	0.008	0.007	0.005	0.004	0.003	0.002	0.002	0.002	0.001	0.001	0.001	0.001
3.4	0.021	0.014	0.010	0.007	0.006	0.005	0.003	0.003	0.002	0.002	0.001	0.001	0.001	0.001	0.001
3.5	0.020	0.012	0.009	0.006	0.005	0.004	0.003	0.002	0.002	0.001	0.001	0.001	0.001	0.001	0.001
3.6	0.018	0.011	0.008	0.006	0.004	0.004	0.002	0.002	0.001	0.001	0.001	0.001	0.001	0+	0+
3.7	0.017	0.010	0.007	0.005	0.004	0.003	0.002	0.002	0.001	0.001	0.001	0.001	0+	0+	0+
3.8	0.016	0.010	0.006	0.004	0.003	0.003	0.002	0.001	0.001	0.001	0.001	0+	0+	0+	0+
3.9	0.015	0.009	0.006	0.004	0.003	0.002	0.001	0.001	0.001	0.001	0+	0+	0+	0+	0+
4.0	0.014	0.008	0.005	0.004	0.003	0.002	0.001	0.001	0.001	0+	0+	0+	0+	0+	0+

For specific details about using this table to find *p*-values, see pages 421–422.

TABLE 8
Critical Values of χ^2 ("Chi-Square") Distribution

The entries in this table are the critical values for the χ^2 distribution for which the area under the curve is: a) to the right-hand tail, or b) to the left-hand tail (the cumulative area). See the illustrations at the bottom of the page.

a) Area to the Right

	0.995	0.99	0.975	0.95	0.90	0.75	0.50	0.25	0.10	0.05	0.025	0.01	0.005

b) Area to the Left (the Cumulative Area)   Median

df	0.005	0.01	0.025	0.05	0.10	0.25	0.50	0.75	0.90	0.95	0.975	0.99	0.995
1	0.0000393	0.000157	0.000982	0.00393	0.0158	0.102	0.455	1.32	2.71	3.84	5.02	6.63	7.88
2	0.0100	0.0201	0.0506	0.103	0.211	0.575	1.39	2.77	4.61	5.99	7.38	9.21	10.6
3	0.0717	0.115	0.216	0.352	0.584	1.21	2.37	4.11	6.25	7.81	9.35	11.3	12.8
4	0.207	0.297	0.484	0.711	1.06	1.92	3.36	5.39	7.78	9.49	11.1	13.3	14.9
5	0.412	0.554	0.831	1.15	1.61	2.67	4.35	6.63	9.24	11.1	12.8	15.1	16.7
6	0.676	0.872	1.24	1.64	2.20	3.45	5.35	7.84	10.6	12.6	14.4	16.8	18.5
7	0.989	1.24	1.69	2.17	2.83	4.25	6.35	9.04	12.0	14.1	16.0	18.5	20.3
8	1.34	1.65	2.18	2.73	3.49	5.07	7.34	10.2	13.4	15.5	17.5	20.1	22.0
9	1.73	2.09	2.70	3.33	4.17	5.90	8.34	11.4	14.7	16.9	19.0	21.7	23.6
10	2.16	2.56	3.25	3.94	4.87	6.74	9.34	12.5	16.0	18.3	20.5	23.2	25.2
11	2.60	3.05	3.82	4.57	5.58	7.58	10.34	13.7	17.3	19.7	21.9	24.7	26.8
12	3.07	3.57	4.40	5.23	6.30	8.44	11.34	14.8	18.5	21.0	23.3	26.2	28.3
13	3.57	4.11	5.01	5.89	7.04	9.30	12.34	16.0	19.8	22.4	24.7	27.7	29.8
14	4.07	4.66	5.63	6.57	7.79	10.2	13.34	17.1	21.1	23.7	26.1	29.1	31.3
15	4.60	5.23	6.26	7.26	8.55	11.0	14.34	18.2	22.3	25.0	27.5	30.6	32.8
16	5.14	5.81	6.91	7.96	9.31	11.9	15.34	19.4	23.5	26.3	28.8	32.0	34.3
17	5.70	6.41	7.56	8.67	10.1	12.8	16.34	20.5	24.8	27.6	30.2	33.4	35.7
18	6.26	7.01	8.23	9.39	10.9	13.7	17.34	21.6	26.0	28.9	31.5	34.8	37.2
19	6.84	7.63	8.91	10.1	11.7	14.6	18.34	22.7	27.2	30.1	32.9	36.2	38.6
20	7.43	8.26	9.59	10.9	12.4	15.5	19.34	23.8	28.4	31.4	34.2	37.6	40.0
21	8.03	8.90	10.3	11.6	13.2	16.3	20.34	24.9	29.6	32.7	35.5	38.9	41.4
22	8.64	9.54	11.0	12.3	14.0	17.2	21.34	26.0	30.8	33.9	36.8	40.3	42.8
23	9.26	10.2	11.7	13.1	14.8	18.1	22.34	27.1	32.0	35.2	38.1	41.6	44.2
24	9.89	10.9	12.4	13.8	15.7	19.0	23.34	28.2	33.2	36.4	39.4	43.0	45.6
25	10.5	11.5	13.1	14.6	16.5	19.9	24.34	29.3	34.4	37.7	40.6	44.3	46.9
26	11.2	12.2	13.8	15.4	17.3	20.8	25.34	30.4	35.6	38.9	41.9	45.6	48.3
27	11.8	12.9	14.6	16.2	18.1	21.7	26.34	31.5	36.7	40.1	43.2	47.0	49.6
28	12.5	13.6	15.3	16.9	18.9	22.7	27.34	32.6	37.9	41.3	44.5	48.3	51.0
29	13.1	14.3	16.0	17.7	19.8	23.6	28.34	33.7	39.1	42.6	45.7	49.6	52.3
30	13.8	15.0	16.8	18.5	20.6	24.5	29.34	34.8	40.3	43.8	47.0	50.9	53.7
40	20.7	22.2	24.4	26.5	29.1	33.7	39.34	45.6	51.8	55.8	59.3	63.7	66.8
50	28.0	29.7	32.4	34.8	37.7	42.9	49.33	56.3	63.2	67.5	71.4	76.2	79.5
60	35.5	37.5	40.5	43.2	46.5	52.3	59.33	67.0	74.4	79.1	83.3	88.4	92.0
70	43.3	45.4	48.8	51.7	55.3	61.7	69.33	77.6	85.5	90.5	95.0	100.4	104.2
80	51.2	53.5	57.2	60.4	64.3	71.1	79.33	88.1	96.6	101.9	106.6	112.3	116.3
90	59.2	61.8	65.6	69.1	73.3	80.6	89.33	98.6	107.6	113.1	118.1	124.1	128.3
100	67.3	70.1	74.2	77.9	82.4	90.1	99.33	109.1	118.5	124.3	129.6	135.8	140.2

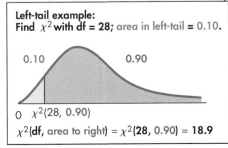

Left-tail example:
Find χ^2 with df = 28; area in left-tail = 0.10.

0.10   0.90

0   $\chi^2(28, 0.90)$

χ^2(df, area to right) = χ^2(28, 0.90) = 18.9

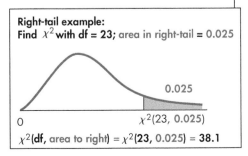

Right-tail example:
Find χ^2 with df = 23; area in right-tail = 0.025

0.025

0                  $\chi^2(23, 0.025)$

χ^2(df, area to right) = χ^2(23, 0.025) = 38.1

For specific details about using this table to find p-values, see pages 458–461; critical values, pages 454–455.
Table 8 was generated using Minitab.

TABLE 9A
Critical Values of the F Distribution
($\alpha = 0.05$)

The entries in this table are critical values of F for which the area under the curve to the right is equal to 0.05.

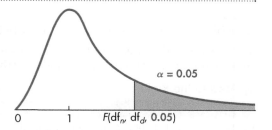

$\alpha = 0.05$

$0 \qquad 1 \qquad F(df_n, df_d, 0.05)$

Degrees of Freedom for Numerator

		1	2	3	4	5	6	7	8	9	10
	1	161.	200.	216.	225.	230.	234.	237.	239.	241.	242.
	2	18.5	19.0	19.2	19.2	19.3	19.3	19.4	19.4	19.4	19.4
	3	10.1	9.55	9.28	9.12	9.01	8.94	8.89	8.85	8.81	8.79
	4	7.71	6.94	6.59	6.39	6.26	6.16	6.09	6.04	6.00	5.96
	5	6.61	5.79	5.41	5.19	5.05	4.95	4.88	4.82	4.77	4.74
	6	5.99	5.14	4.76	4.53	4.39	4.28	4.21	4.15	4.10	4.06
	7	5.59	4.74	4.35	4.12	3.97	3.87	3.79	3.73	3.68	3.64
	8	5.32	4.46	4.07	3.84	3.69	3.58	3.50	3.44	3.39	3.35
	9	5.12	4.26	3.86	3.63	3.48	3.37	3.29	3.23	3.18	3.14
	10	4.96	4.10	3.71	3.48	3.33	3.22	3.14	3.07	3.02	2.98
	11	4.84	3.98	3.59	3.36	3.20	3.09	3.01	2.95	2.90	2.85
	12	4.75	3.89	3.49	3.26	3.11	3.00	2.91	2.85	2.80	2.75
	13	4.67	3.81	3.41	3.18	3.03	2.92	2.83	2.77	2.71	2.67
	14	4.60	3.74	3.34	3.11	2.96	2.85	2.76	2.70	2.65	2.60
	15	4.54	3.68	3.29	3.06	2.90	2.79	2.71	2.64	2.59	2.54
	16	4.49	3.63	3.24	3.01	2.85	2.74	2.66	2.59	2.54	2.49
	17	4.45	3.59	3.20	2.96	2.81	2.70	2.61	2.55	2.49	2.45
	18	4.41	3.55	3.16	2.93	2.77	2.66	2.58	2.51	2.46	2.41
	19	4.38	3.52	3.13	2.90	2.74	2.63	2.54	2.48	2.42	2.38
	20	4.35	3.49	3.10	2.87	2.71	2.60	2.51	2.45	2.39	2.35
	21	4.32	3.47	3.07	2.84	2.68	2.57	2.49	2.42	2.37	2.32
	22	4.30	3.44	3.05	2.82	2.66	2.55	2.46	2.40	2.34	2.30
	23	4.28	3.42	3.03	2.80	2.64	2.53	2.44	2.37	2.32	2.27
	24	4.26	3.40	3.01	2.78	2.62	2.51	2.42	2.36	2.30	2.25
	25	4.24	3.39	2.99	2.76	2.60	2.49	2.40	2.34	2.28	2.24
	30	4.17	3.32	2.92	2.69	2.53	2.42	2.33	2.27	2.21	2.16
	40	4.08	3.23	2.84	2.61	2.45	2.34	2.25	2.18	2.12	2.08
	60	4.00	3.15	2.76	2.53	2.37	2.25	2.17	2.10	2.04	1.99
	120	3.92	3.07	2.68	2.45	2.29	2.18	2.09	2.02	1.96	1.91
	10,000	3.84	3.00	2.61	2.37	2.21	2.10	2.01	1.94	1.88	1.83

Degrees of Freedom for Denominator (left vertical label)

For specific details about using this table to find *p*-values, see page 527; critical values, pages 523–524. Table 9A was generated using Minitab.

TABLE 9A

Critical Values of the F Distribution ($\alpha = 0.05$) (*continued*)

				Degrees of Freedom for Numerator					
	12	15	20	24	30	40	60	120	10,000
1	244.	246.	248.	249.	250.	251.	252.	253.	254.
2	19.4	19.4	19.4	19.5	19.5	19.5	19.5	19.5	19.5
3	8.74	8.70	8.66	8.64	8.62	8.59	8.57	8.55	8.53
4	5.91	5.86	5.80	5.77	5.75	5.72	5.69	5.66	5.63
5	4.68	4.62	4.56	4.53	4.50	4.46	4.43	4.40	4.37
6	4.00	3.94	3.87	3.84	3.81	3.77	3.74	3.70	3.67
7	3.57	3.51	3.44	3.41	3.38	3.34	3.30	3.27	3.23
8	3.28	3.22	3.15	3.12	3.08	3.04	3.01	2.97	2.93
9	3.07	3.01	2.94	2.90	2.86	2.83	2.79	2.75	2.71
10	2.91	2.85	2.77	2.74	2.70	2.66	2.62	2.58	2.54
11	2.79	2.72	2.65	2.61	2.57	2.53	2.49	2.45	2.41
12	2.69	2.62	2.54	2.51	2.47	2.43	2.38	2.34	2.30
13	2.60	2.53	2.46	2.42	2.38	2.34	2.30	2.25	2.21
14	2.53	2.46	2.39	2.35	2.31	2.27	2.22	2.18	2.13
15	2.48	2.40	2.33	2.29	2.25	2.20	2.16	2.11	2.07
16	2.42	2.35	2.28	2.24	2.19	2.15	2.11	2.06	2.01
17	2.38	2.31	2.23	2.19	2.15	2.10	2.06	2.01	1.96
18	2.34	2.27	2.19	2.15	2.11	2.06	2.02	1.97	1.92
19	2.31	2.23	2.16	2.11	2.07	2.03	1.98	1.93	1.88
20	2.28	2.20	2.12	2.08	2.04	1.99	1.95	1.90	1.84
21	2.25	2.18	2.10	2.05	2.01	1.96	1.92	1.87	1.81
22	2.23	2.15	2.07	2.03	1.98	1.94	1.89	1.84	1.78
23	2.20	2.13	2.05	2.01	1.96	1.91	1.86	1.81	1.76
24	2.18	2.11	2.03	1.98	1.94	1.89	1.84	1.79	1.73
25	2.16	2.09	2.01	1.96	1.92	1.87	1.82	1.77	1.71
30	2.09	2.01	1.93	1.89	1.84	1.79	1.74	1.68	1.62
40	2.00	1.92	1.84	1.79	1.74	1.69	1.64	1.58	1.51
60	1.92	1.84	1.75	1.70	1.65	1.59	1.53	1.47	1.39
120	1.83	1.75	1.66	1.61	1.55	1.50	1.43	1.35	1.26
10,000	1.75	1.67	1.57	1.52	1.46	1.40	1.32	1.22	1.03

Degrees of Freedom for Denominator (row labels)

Table 9A was generated using Minitab.

TABLE 9B
Critical Values of the F Distribution
$(\alpha = 0.025)$
The entries in this table are critical values of F for which the area under the curve to the right is equal to 0.025.

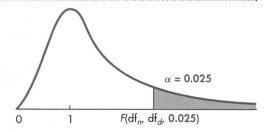

$\alpha = 0.025$

$0 \qquad 1 \qquad F(df_n, df_d, 0.025)$

Degrees of Freedom for Numerator

		1	2	3	4	5	6	7	8	9	10
Degrees of Freedom for Denominator	1	648.	800.	864.	900.	922.	937.	948.	957.	963.	969.
	2	38.5	39.0	39.2	39.2	39.3	39.3	39.4	39.4	39.4	39.4
	3	17.4	16.0	15.4	15.1	14.9	14.7	14.6	14.5	14.5	14.4
	4	12.2	10.6	9.98	9.60	9.36	9.20	9.07	8.98	8.90	8.84
	5	10.0	8.43	7.76	7.39	7.15	6.98	6.85	6.76	6.68	6.62
	6	8.81	7.26	6.60	6.23	5.99	5.82	5.70	5.60	5.52	5.46
	7	8.07	6.54	5.89	5.52	5.29	5.12	4.99	4.90	4.82	4.76
	8	7.57	6.06	5.42	5.05	4.82	4.65	4.53	4.43	4.36	4.30
	9	7.21	5.71	5.08	4.72	4.48	4.32	4.20	4.10	4.03	3.96
	10	6.94	5.46	4.83	4.47	4.24	4.07	3.95	3.85	3.78	3.72
	11	6.72	5.26	4.63	4.28	4.04	3.88	3.76	3.66	3.59	3.53
	12	6.55	5.10	4.47	4.12	3.89	3.73	3.61	3.51	3.44	3.37
	13	6.41	4.97	4.35	4.00	3.77	3.60	3.48	3.39	3.31	3.25
	14	6.30	4.86	4.24	3.89	3.66	3.50	3.38	3.28	3.21	3.15
	15	6.20	4.77	4.15	3.80	3.58	3.41	3.29	3.20	3.12	3.06
	16	6.12	4.69	4.08	3.73	3.50	3.34	3.22	3.12	3.05	2.99
	17	6.04	4.62	4.01	3.66	3.44	3.28	3.16	3.06	2.98	2.92
	18	5.98	4.56	3.95	3.61	3.38	3.22	3.10	3.01	2.93	2.87
	19	5.92	4.51	3.90	3.56	3.33	3.17	3.05	2.96	2.88	2.82
	20	5.87	4.46	3.86	3.51	3.29	3.13	3.01	2.91	2.84	2.77
	21	5.83	4.42	3.82	3.48	3.25	3.09	2.97	2.87	2.80	2.73
	22	5.79	4.38	3.78	3.44	3.22	3.05	2.93	2.84	2.76	2.70
	23	5.75	4.35	3.75	3.41	3.18	3.02	2.90	2.81	2.73	2.67
	24	5.72	4.32	3.72	3.38	3.15	2.99	2.87	2.78	2.70	2.64
	25	5.69	4.29	3.69	3.35	3.13	2.97	2.85	2.75	2.68	2.61
	30	5.57	4.18	3.59	3.25	3.03	2.87	2.75	2.65	2.57	2.51
	40	5.42	4.05	3.46	3.13	2.90	2.74	2.62	2.53	2.45	2.39
	60	5.29	3.93	3.34	3.01	2.79	2.63	2.51	2.41	2.33	2.27
	120	5.15	3.80	3.23	2.89	2.67	2.52	2.39	2.30	2.22	2.16
	10,000	5.03	3.69	3.12	2.79	2.57	2.41	2.29	2.19	2.11	2.05

For specific details about using this table to find p-values, see page 527; critical values, pages 523–524. Table 9B was generated using Minitab.

TABLE 9B

Critical Values of the F Distribution ($\alpha = 0.025$) (*continued*)

	12	15	20	24	30	40	60	120	10,000
1	977.	985.	993.	997.	1001.	1006.	1010.	1014.	1018.
2	39.4	39.4	39.4	39.5	39.5	39.5	39.5	39.5	39.5
3	14.3	14.3	14.2	14.1	14.1	14.0	14.0	13.9	13.9
4	8.75	8.66	8.56	8.51	8.46	8.41	8.36	8.31	8.26
5	6.52	6.43	6.33	6.28	6.23	6.18	6.12	6.07	6.02
6	5.37	5.27	5.17	5.12	5.07	5.01	4.96	4.90	4.85
7	4.67	4.57	4.47	4.42	4.36	4.31	4.25	4.20	4.14
8	4.20	4.10	4.00	3.95	3.89	3.84	3.78	3.73	3.67
9	3.87	3.77	3.67	3.61	3.56	3.51	3.45	3.39	3.33
10	3.62	3.52	3.42	3.37	3.31	3.26	3.20	3.14	3.08
11	3.43	3.33	3.23	3.17	3.12	3.06	3.00	2.94	2.88
12	3.28	3.18	3.07	3.02	2.96	2.91	2.85	2.79	2.73
13	3.15	3.05	2.95	2.89	2.84	2.78	2.72	2.66	2.60
14	3.05	2.95	2.84	2.79	2.73	2.67	2.61	2.55	2.49
15	2.96	2.86	2.76	2.70	2.64	2.59	2.52	2.46	2.40
16	2.89	2.79	2.68	2.63	2.57	2.51	2.45	2.38	2.32
17	2.82	2.72	2.62	2.56	2.50	2.44	2.38	2.32	2.25
18	2.77	2.67	2.56	2.50	2.44	2.38	2.32	2.26	2.19
19	2.72	2.62	2.51	2.45	2.39	2.33	2.27	2.20	2.13
20	2.68	2.57	2.46	2.41	2.35	2.29	2.22	2.16	2.09
21	2.64	2.53	2.42	2.37	2.31	2.25	2.18	2.11	2.04
22	2.60	2.50	2.39	2.33	2.27	2.21	2.14	2.08	2.00
23	2.57	2.47	2.36	2.30	2.24	2.18	2.11	2.04	1.97
24	2.54	2.44	2.33	2.27	2.21	2.15	2.08	2.01	1.94
25	2.51	2.41	2.30	2.24	2.18	2.12	2.05	1.98	1.91
30	2.41	2.31	2.20	2.14	2.07	2.01	1.94	1.87	1.79
40	2.29	2.18	2.07	2.01	1.94	1.88	1.80	1.72	1.64
60	2.17	2.06	1.94	1.88	1.82	1.74	1.67	1.58	1.48
120	2.05	1.95	1.82	1.76	1.69	1.61	1.53	1.43	1.31
10,000	1.95	1.83	1.71	1.64	1.57	1.49	1.39	1.27	1.04

Degrees of Freedom for Numerator (column header)
Degrees of Freedom for Denominator (row header)

Table 9B was generated using Minitab.

TABLE 9C
Critical Values of the F Distribution
($\alpha = 0.01$)
The entries in the table are critical values of F for which the area under the curve to the right is equal to 0.01

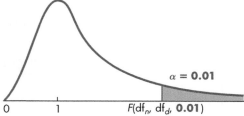

$\alpha = 0.01$

$0 \qquad 1 \qquad F(df_n, df_d, 0.01)$

	Degrees of Freedom for Numerator									
	1	2	3	4	5	6	7	8	9	10
1	4052.	5000.	5403.	5625.	5764.	5859.	5928.	5981.	6022.	6056.
2	98.5	99.0	99.2	99.2	99.3	99.3	99.4	99.4	99.4	99.4
3	34.1	30.8	29.5	28.7	28.2	27.9	27.7	27.5	27.3	27.2
4	21.2	18.0	16.7	16.0	15.5	15.2	15.0	14.8	14.7	14.5
5	16.3	13.3	12.1	11.4	11.0	10.7	10.5	10.3	10.2	10.1
6	13.7	10.9	9.78	9.15	8.75	8.47	8.26	8.10	7.98	7.87
7	12.2	9.55	8.45	7.85	7.46	7.19	6.99	6.84	6.72	6.62
8	11.3	8.65	7.59	7.01	6.63	6.37	6.18	6.03	5.91	5.81
9	10.6	8.02	6.99	6.42	6.06	5.80	5.61	5.47	5.35	5.26
10	10.0	7.56	6.55	5.99	5.64	5.39	5.20	5.06	4.94	4.85
11	9.65	7.21	6.22	5.67	5.32	5.07	4.89	4.74	4.63	4.54
12	9.33	6.93	5.95	5.41	5.06	4.82	4.64	4.50	4.39	4.30
13	9.07	6.70	5.74	5.21	4.86	4.62	4.44	4.30	4.19	4.10
14	8.86	6.51	5.56	5.04	4.70	4.46	4.28	4.14	4.03	3.94
15	8.68	6.36	5.42	4.89	4.56	4.32	4.14	4.00	3.89	3.80
16	8.53	6.23	5.29	4.77	4.44	4.20	4.03	3.89	3.78	3.69
17	8.40	6.11	5.19	4.67	4.34	4.10	3.93	3.79	3.68	3.59
18	8.29	6.01	5.09	4.58	4.25	4.01	3.84	3.71	3.60	3.51
19	8.18	5.93	5.01	4.50	4.17	3.94	3.77	3.63	3.52	3.43
20	8.10	5.85	4.94	4.43	4.10	3.87	3.70	3.56	3.46	3.37
21	8.02	5.78	4.87	4.37	4.04	3.81	3.64	3.51	3.40	3.31
22	7.95	5.72	4.82	4.31	3.99	3.76	3.59	3.45	3.35	3.26
23	7.88	5.66	4.76	4.26	3.94	3.71	3.54	3.41	3.30	3.21
24	7.82	5.61	4.72	4.22	3.90	3.67	3.50	3.36	3.26	3.17
25	7.77	5.57	4.68	4.18	3.86	3.63	3.46	3.32	3.22	3.13
30	7.56	5.39	4.51	4.02	3.70	3.47	3.30	3.17	3.07	2.98
40	7.31	5.18	4.31	3.83	3.51	3.29	3.12	2.99	2.89	2.80
60	7.08	4.98	4.13	3.65	3.34	3.12	2.95	2.82	2.72	2.63
120	6.85	4.79	3.95	3.48	3.17	2.96	2.79	2.66	2.56	2.47
10,000	6.64	4.61	3.78	3.32	3.02	2.80	2.64	2.51	2.41	2.32

Degrees of Freedom for Denominator

For specific details about using this table to find *p*-values, see page 527; critical values, pages 523–524. Table 9C was generated using Minitab.

TABLE 9C

Critical Values of the F Distribution ($\alpha = 0.01$) (*continued*)

Degrees of Freedom for Numerator

	12	15	20	24	30	40	60	120	10,000
1	6106.	6157.	6209.	6235.	6261.	6287.	6313.	6339.	6366.
2	99.4	99.4	99.4	99.5	99.5	99.5	99.5	99.5	99.5
3	27.1	26.9	26.7	26.6	26.5	26.4	26.3	26.2	26.1
4	14.4	14.2	14.0	13.9	13.8	13.7	13.7	13.6	13.5
5	9.89	9.72	9.55	9.47	9.38	9.29	9.20	9.11	9.02
6	7.72	7.56	7.40	7.31	7.23	7.14	7.06	6.97	6.88
7	6.47	6.31	6.16	6.07	5.99	5.91	5.82	5.74	5.65
8	5.67	5.52	5.36	5.28	5.20	5.12	5.03	4.95	4.86
9	5.11	4.96	4.81	4.73	4.65	4.57	4.48	4.40	4.31
10	4.71	4.56	4.41	4.33	4.25	4.17	4.08	4.00	3.91
11	4.40	4.25	4.10	4.02	3.94	3.86	3.78	3.69	3.60
12	4.16	4.01	3.86	3.78	3.70	3.62	3.54	3.45	3.36
13	3.96	3.82	3.66	3.59	3.51	3.43	3.34	3.25	3.17
14	3.80	3.66	3.51	3.43	3.35	3.27	3.18	3.09	3.01
15	3.67	3.52	3.37	3.29	3.21	3.13	3.05	2.96	2.87
16	3.55	3.41	3.26	3.18	3.10	3.02	2.93	2.84	2.75
17	3.46	3.31	3.16	3.08	3.00	2.92	2.83	2.75	2.65
18	3.37	3.23	3.08	3.00	2.92	2.84	2.75	2.66	2.57
19	3.30	3.15	3.00	2.92	2.84	2.76	2.67	2.58	2.49
20	3.23	3.09	2.94	2.86	2.78	2.69	2.61	2.52	2.42
21	3.17	3.03	2.88	2.80	2.72	2.64	2.55	2.46	2.36
22	3.12	2.98	2.83	2.75	2.67	2.58	2.50	2.40	2.31
23	3.07	2.93	2.78	2.70	2.62	2.54	2.45	2.35	2.26
24	3.03	2.89	2.74	2.66	2.58	2.49	2.40	2.31	2.21
25	2.99	2.85	2.70	2.62	2.54	2.45	2.36	2.27	2.17
30	2.84	2.70	2.55	2.47	2.39	2.30	2.21	2.11	2.01
40	2.66	2.52	2.37	2.29	2.20	2.11	2.02	1.92	1.80
60	2.50	2.35	2.20	2.12	2.03	1.94	1.84	1.73	1.60
120	2.34	2.19	2.03	1.95	1.86	1.76	1.66	1.53	1.38
10,000	2.19	2.04	1.88	1.79	1.70	1.59	1.48	1.33	1.05

Degrees of Freedom for Denominator (row labels)

Table 9C was generated using Minitab.

TABLE 10
Confidence Belts for the Correlation Coefficient $(1 - \alpha) = 0.95$
The numbers on the curves are sample sizes.

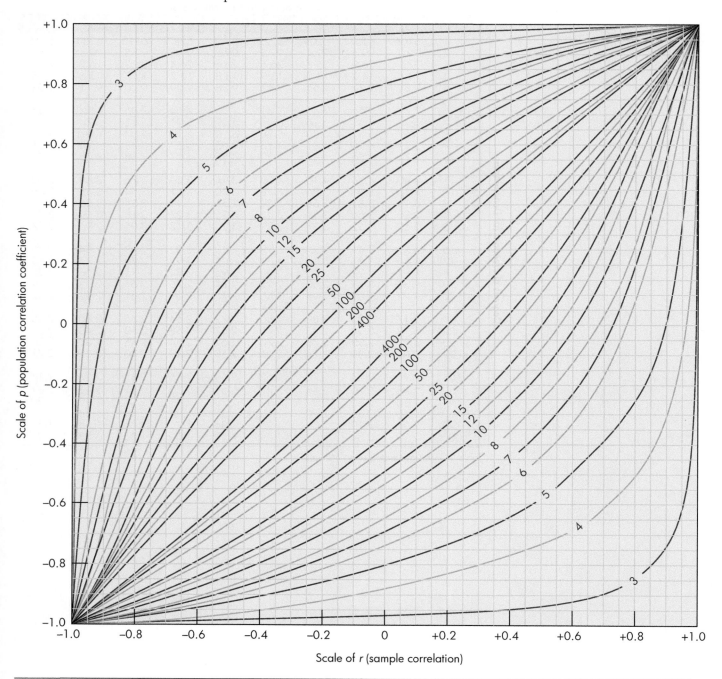

For specific details about using this table to find confidence intervals, see page 620.

TABLE 11
Critical Values of r When $\rho = 0$

The entries in this table are the critical values of r for a two-tailed test at α. For simple correlation, df $= n - 2$, where n is the number of pairs of data in the sample. For a one-tailed test, the value of α shown at the top of the table is double the value of α being used in the hypothesis test.

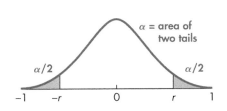

df	0.10	0.05	0.02	0.01
1	0.988	0.997	1.000	1.000
2	0.900	0.950	0.980	0.990
3	0.805	0.878	0.934	0.959
4	0.729	0.811	0.882	0.917
5	0.669	0.754	0.833	0.875
6	0.621	0.707	0.789	0.834
7	0.582	0.666	0.750	0.798
8	0.549	0.632	0.715	0.765
9	0.521	0.602	0.685	0.735
10	0.497	0.576	0.658	0.708
11	0.476	0.553	0.634	0.684
12	0.458	0.532	0.612	0.661
13	0.441	0.514	0.592	0.641
14	0.426	0.497	0.574	0.623
15	0.412	0.482	0.558	0.606
16	0.400	0.468	0.543	0.590
17	0.389	0.456	0.529	0.575
18	0.378	0.444	0.516	0.561
19	0.369	0.433	0.503	0.549
20	0.360	0.423	0.492	0.537
25	0.323	0.381	0.445	0.487
30	0.296	0.349	0.409	0.449
35	0.275	0.325	0.381	0.418
40	0.257	0.304	0.358	0.393
45	0.243	0.288	0.338	0.372
50	0.231	0.273	0.322	0.354
60	0.211	0.250	0.295	0.325
70	0.195	0.232	0.274	0.302
80	0.183	0.217	0.256	0.283
90	0.173	0.205	0.242	0.267
100	0.164	0.195	0.230	0.254

For specific details about using this table to find *p*-values and critical values, see pages 621–623.

TABLE 12
Critical Values of the Sign Test

The entries in this table are the critical values for the number of the least frequent sign for a two-tailed test at α for the binomial $p = 0.5$. For a one-tailed test, the value of α shown at the top of the table is double the value of α being used in the hypothesis test.

n	α 0.01	0.05	0.10	0.25	n	α 0.01	0.05	0.10	0.25
1					51	15	18	19	20
2					52	16	18	19	21
3				0	53	16	18	20	21
4				0	54	17	19	20	22
5			0	0	55	17	19	20	22
6		0	0	1	56	17	20	21	23
7		0	0	1	57	18	20	21	23
8	0	0	1	1	58	18	21	22	24
9	0	1	1	2	59	19	21	22	24
10	0	1	1	2	60	19	21	23	25
11	0	1	2	3	61	20	22	23	25
12	1	2	2	3	62	20	22	24	25
13	1	2	3	3	63	20	23	24	26
14	1	2	3	4	64	21	23	24	26
15	2	3	3	4	65	21	24	25	27
16	2	3	4	5	66	22	24	25	27
17	2	4	4	5	67	22	25	26	28
18	3	4	5	6	68	22	25	26	28
19	3	4	5	6	69	23	25	27	29
20	3	5	5	6	70	23	26	27	29
21	4	5	6	7	71	24	26	28	30
22	4	5	6	7	72	24	27	28	30
23	4	6	7	8	73	25	27	28	31
24	5	6	7	8	74	25	28	29	31
25	5	7	7	9	75	25	28	29	32
26	6	7	8	9	76	26	28	30	32
27	6	7	8	10	77	26	29	30	32
28	6	8	9	10	78	27	29	31	33
29	7	8	9	10	79	27	30	31	33
30	7	9	10	11	80	28	30	32	34
31	7	9	10	11	81	28	31	32	34
32	8	9	10	12	82	28	31	33	35
33	8	10	11	12	83	29	32	33	35
34	9	10	11	13	84	29	32	33	36
35	9	11	12	13	85	30	32	34	36
36	9	11	12	14	86	30	33	34	37
37	10	12	13	14	87	31	33	35	37
38	10	12	13	14	88	31	34	35	38
39	11	12	13	15	89	31	34	36	38
40	11	13	14	15	90	32	35	36	39
41	11	13	14	16	91	32	35	37	39
42	12	14	15	16	92	33	36	37	39
43	12	14	15	17	93	33	36	38	40
44	13	15	16	17	94	34	37	38	40
45	13	15	16	18	95	34	37	38	41
46	13	15	16	18	96	34	37	39	41
47	14	16	17	19	97	35	38	39	42
48	14	16	17	19	98	35	38	40	42
49	15	17	18	19	99	36	39	40	43
50	15	17	18	20	100	36	39	41	44

From Wilfred J. Dixon and Frank J. Massey, Jr., *Introduction to Statistical Analysis*, 3d ed. (New York: McGraw-Hill, 1969), p. 509. Reprinted by permission.

For specific details about using this table: confidence intervals, see pages 664–665; p-values, pages 666–667; critical values, page 666.

TABLE 13
Critical Values of *U* in the Mann-Whitney Test

A. The entries are the critical values of U for a one-tailed test at 0.025 or for a two-tailed test at 0.05.

n2 \ n1	1	2	3	4	5	6	7	8	9	10	11	12	13	14	15	16	17	18	19	20
1																				
2								0	0	0	0	1	1	1	1	1	2	2	2	2
3					0	1	1	2	2	3	3	4	4	5	5	6	6	7	7	8
4				0	1	2	3	4	4	5	6	7	8	9	10	11	11	12	13	13
5			0	1	2	3	5	6	7	8	9	11	12	13	14	15	17	18	19	20
6			1	2	3	5	6	8	10	11	13	14	16	17	19	21	22	24	25	27
7			1	3	5	6	8	10	12	14	16	18	20	22	24	26	28	30	32	34
8		0	2	4	6	8	10	13	15	17	19	22	24	26	29	31	34	36	38	41
9		0	2	4	7	10	12	15	17	20	23	26	28	31	34	37	39	42	45	48
10		0	3	5	8	11	14	17	20	23	26	29	33	36	39	42	45	48	52	55
11		0	3	6	9	13	16	19	23	26	30	33	37	40	44	47	51	55	58	62
12		1	4	7	11	14	18	22	26	29	33	37	41	45	49	53	57	61	65	69
13		1	4	8	12	16	20	24	28	33	37	41	45	50	54	59	63	67	72	76
14		1	5	9	13	17	22	26	31	36	40	45	50	55	59	64	67	74	78	83
15		1	5	10	14	19	24	29	34	39	44	49	54	59	64	70	75	80	85	90
16		1	6	11	15	21	26	31	37	42	47	53	59	64	70	75	81	86	92	98
17		2	6	11	17	22	28	34	39	45	51	57	63	67	75	81	87	93	99	105
18		2	7	12	18	24	30	36	42	48	55	61	67	74	80	86	93	99	106	112
19		2	7	13	19	25	32	38	45	52	58	65	72	78	85	92	99	106	113	119
20		2	8	13	20	27	34	41	48	55	62	69	76	83	90	98	105	112	119	127

B. The entries are the critical values of U for a one-tailed test at 0.05 or for a two-tailed test at 0.10.

n2 \ n1	1	2	3	4	5	6	7	8	9	10	11	12	13	14	15	16	17	18	19	20
1																			0	0
2					0	0	0	1	1	1	1	2	2	2	3	3	3	4	4	4
3			0	0	1	2	2	3	3	4	5	5	6	7	7	8	9	9	10	11
4			0	1	2	3	4	5	6	7	8	9	10	11	12	14	15	16	17	18
5		0	1	2	4	5	6	8	9	11	12	13	15	16	18	19	20	22	23	25
6		0	2	3	5	7	8	10	12	14	16	17	19	21	23	25	26	28	30	32
7		0	2	4	6	8	11	13	15	17	19	21	24	26	28	30	33	35	37	39
8		1	3	5	8	10	13	15	18	20	23	26	28	31	33	36	39	41	44	47
9		1	3	6	9	12	15	18	21	24	27	30	33	36	39	42	45	48	51	54
10		1	4	7	11	14	17	20	24	27	31	34	37	41	44	48	51	55	58	62
11		1	5	8	12	16	19	23	27	31	34	38	42	46	50	54	57	61	65	69
12		2	5	9	13	17	21	26	30	34	38	42	47	51	55	60	64	68	72	77
13		2	6	10	15	19	24	28	33	37	42	47	51	56	61	65	70	75	80	84
14		2	7	11	16	21	26	31	36	41	46	51	56	61	66	71	77	82	87	92
15		3	7	12	18	23	28	33	39	44	50	55	61	66	72	77	83	88	94	100
16		3	8	14	19	25	30	36	42	48	54	60	65	71	77	83	89	95	101	107
17		3	9	15	20	26	33	39	45	51	57	64	70	77	83	89	96	102	109	115
18		4	9	16	22	28	35	41	48	55	61	68	75	82	88	95	102	109	116	123
19	0	4	10	17	23	30	37	44	51	58	65	72	80	87	94	101	109	116	123	130
20	0	4	11	18	25	32	39	47	54	62	69	77	84	92	100	107	115	123	130	138

Reproduced from the *Bulletin of the Institute of Educational Research at Indiana University,* vol. 1, no. 2; with the permission of the author and the publisher. For specific details about using this table to find *p*-values, see pages 679–680; critical values, page 679.

TABLE 14

Critical Values for Total Number of Runs (V)

The entries in this table are the critical values for a two-tailed test using $\alpha = 0.05$. For a one-tailed test at $\alpha = 0.025$, use only one of the critical values: the smaller critical value for a left-hand critical region, the larger for a right-hand critical region.

The larger of n_1 and n_2

Each cell lists two critical values (lower critical value / upper critical value).

The smaller of n_1 and n_2	5	6	7	8	9	10	11	12	13	14	15	16	17	18	19	20
2								2/6	2/6	2/6	2/6	2/6	2/6	2/6	2/6	2/6
3		2/8	2/8	2/8	2/8	2/8	2/8	2/8	2/8	2/8	3/8	3/8	3/8	3/8	3/8	3/8
4	2/9	2/9	2/10	3/10	3/10	3/10	3/10	3/10	3/10	3/10	3/10	4/10	4/10	4/10	4/10	4/10
5	2/10	3/10	3/11	3/11	3/12	3/12	4/12	4/12	4/12	4/12	4/12	4/12	4/12	5/12	5/12	5/12
6		3/11	3/12	3/12	4/13	4/13	4/13	4/13	5/14	5/14	5/14	5/14	5/14	5/14	6/14	6/14
7			3/13	4/13	4/14	5/14	5/14	5/14	5/15	5/15	6/15	6/16	6/16	6/16	6/16	6/16
8				4/14	5/14	5/15	5/15	6/16	6/16	6/16	6/16	6/17	7/17	7/17	7/17	7/17
9					5/15	5/16	6/16	6/16	6/17	7/17	7/18	7/18	7/18	8/18	8/18	8/18
10						6/16	6/17	7/17	7/18	7/18	7/18	8/19	8/19	8/19	8/20	9/20
11							7/17	7/18	7/19	8/19	8/19	8/20	9/20	9/20	9/21	9/21
12								7/19	8/19	8/20	8/20	9/21	9/21	9/21	10/22	10/22
13									8/20	9/20	9/21	9/21	10/22	10/22	10/23	10/23
14										9/21	9/22	10/22	10/23	10/23	11/23	11/24
15											10/22	10/23	11/23	11/24	11/24	12/25
16												11/23	11/24	11/25	12/25	12/25
17													11/25	12/25	12/26	13/26
18														12/26	13/26	13/27
19															13/27	13/27
20																14/28

From C. Eisenhart and F. Swed, "Tables for testing randomness of grouping in a sequence of alternatives," *Annals of Statistics*, vol. 14 (1943): 66–87. Reprinted by permission.

For specific details about using this table to find *p*-values, see pages 688–689; critical values, page 688.

TABLE 15
Critical Values of Spearman's Rank Correlation Coefficient

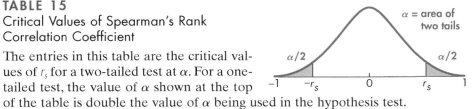

α = area of two tails

The entries in this table are the critical values of r_s for a two-tailed test at α. For a one-tailed test, the value of α shown at the top of the table is double the value of α being used in the hypothesis test.

n	$\alpha = 0.10$	$\alpha = 0.05$	$\alpha = 0.02$	$\alpha = 0.01$
5	0.900	—	—	—
6	0.829	0.886	0.943	—
7	0.714	0.786	0.893	0.929
8	0.643	0.738	0.833	0.881
9	0.600	0.700	0.783	0.833
10	0.564	0.648	0.745	0.794
11	0.536	0.618	0.709	0.755
12	0.503	0.587	0.678	0.727
13	0.484	0.560	0.648	0.703
14	0.464	0.538	0.626	0.679
15	0.446	0.521	0.604	0.654
16	0.429	0.503	0.582	0.635
17	0.414	0.485	0.566	0.615
18	0.401	0.472	0.550	0.600
19	0.391	0.460	0.535	0.584
20	0.380	0.447	0.520	0.570
21	0.370	0.435	0.508	0.556
22	0.361	0.425	0.496	0.544
23	0.353	0.415	0.486	0.532
24	0.344	0.406	0.476	0.521
25	0.337	0.398	0.466	0.511
26	0.331	0.390	0.457	0.501
27	0.324	0.382	0.448	0.491
28	0.317	0.375	0.440	0.483
29	0.312	0.368	0.433	0.475
30	0.306	0.362	0.425	0.467

From *Non Parametrics Statistical Methods*, Hollander & Wolfe, 2d ed. Adapted, in part, from J. H. Zar, Significance testing of the Spearman rank correlation coefficient, *Journal of the American Statistical Association* 67 (1972): 578–580. Reprinted with permission from the *Journal of the American Statistical Association*. Copyright © 1972 by the American Statistical Association. All rights reserved, and, in part, from A. Otten. Note on the Spearman rank correlation coefficient, *Journal of the American Statistical Association* 68 (1973): 585. Reprinted with permission from the *Journal of the American Statistical Association*. Copyright © 1973 by the American Statistical Association. All rights reserved.

For specific details about using this table to find *p*-values, see page 698; critical values, page 698.

Answers to Selected Exercises

Chapter 1

1.1 a. How often do you eat fruit?
b. Internet visitors at the Postyour.info website.
c. 63
d. $1/63 = 0.01587, 11/63 = 0.1746, 16/63 = 0.25397$
e. No. Only people who visited the site and wanted to answer the question did so.

1.3 a. Americans
b. length of time before a WiFi user gets antsy and needs to check his or her messages
c. 47% of those people surveyed say they get antsy within 1 hour about checking email, etc,

1.7 a. inferential   b. descriptive

1.9 a. married women, ages 25–50, who have 2 or more children
b. 1170
c. how often moms sad they have a date night with their spouse
d. 18% of those surveyed say they have a date night every 4–6 months
e. $(0.18)(1170) = 211$

1.11 a. U.S. teens
b. 501
c. what everyday invention they thought would be obsolete in 5 years
d. $501(0.21) = 105$
e. actual percentage could be 4.3% lower or 4.3% higher than quoted
f. between 16.7% and 25.3%

1.13 a. 45% (100%–55%)
b. Percentages are from different groups

1.15 a. all U.S. adults
b. 1200 randomly selected adults
c. "allergy status" for each adult
d. 33.2% based on the sampled adults
e. percent of all U.S. adults with an allergy, 36%

1.19 a. ZIP code, gender, highest level of education
b. annual income, age, distance to store

1.21 a. gender, nominal
b. height, continuous

1.23 a. severity of side effects
b. attribute (ordinal)

1.25 a. weight of books and supplies
b. numerical (continuous)
c. 5.67 lbs, 15.2 lbs

1.27 a. average cost of textbooks for the semester
b. all students enrolled for the semester
c. cost of textbooks for this semester
d. the 100 students
e. average cost of textbooks; add 100 values, divide by 100

1.29 a. all students currently enrolled at the college
b. finite
c. the 10 students selected
d. discrete, continuous (cost rounded to the nearest cent), nominal

1.31 a. all 2009 pickup trucks listed on MPGoMatic.com
b. 165, sample = 6 trucks
c. 8 variables
d. Manufacturer, Model, Drive, Transmission
e. all nominal
f. Engine size, Engine Size Displacement, City MPG, Hwy MPG
g. discrete: Engine Size; continuous: Engine Size Displacement, City MPG, Hwy MPG

1.33 a. numerical   b. attribute   c. numerical
d. attribute   e. numerical   f. numerical

1.35 a. Population contains all objects of interest, sample contains only those actually studied.
b. convenience, availability, practicality

1.37 football players, wider range

1.39 Price/standard unit makes price the only variable.

1.41 too easy, too hard, can't distinquish among the students' knowledge

1.43 a. volunteer   b. yes

1.45 volunteer; bias

1.47 convenience sampling

1.51 probability samples

1.53 Statistical methods assume the use of random samples.

1.55 randomly select first item between 1 and 25; select every 25th thereafter

1.57 A proportional sample would work best.

1.59 Only people with telephones and listed phone numbers will be considered, possibly eliminating those with only cell phones.

1.61 a. Fluorescent bulbs use up to 75% less energy than incandescent light bulbs; the average life of compact fluorescent bulbs is up to 10 times as long as that of incandescent light bulbs.
  b. yes
  c. no
  d. yes, so one would know which bulb is best
  e. part d
  f. collect data on the amount of energy used in each type of bulb for a certain amount of time
  g. collect data on the lifetimes of a sample of each bulb

1.63 draw graphs, print charts, calculate statistics

1.65 a. color of hair, major, gender, marital status
  b. number of courses taken, height, distance from hometown to college

1.67 a. data value
  b. What is the average of the sample?
  c. What is the average for all people?

1.69 a. credit card holders
  b. type/name of credit card, number of months past due on payment, debt amount
  c. name – attribute; number of months, debt amount – numerical

1.71 qualitative, ordinal; responses were descriptive and could be ranked

1.73 a. observational study
  b. percent or proportion of sunglass use
  c. proportion of sample that wore sunglasses, 4 out of 10 adults

Chapter 2

2.3 b. relative proportions as a whole
  c. relative proportions between the individual answers

2.5 a. **The current dress code at my company is . . .**

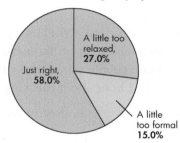

  b.
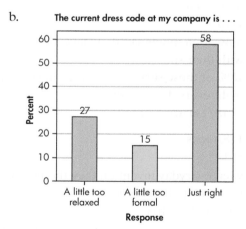

2.7 a.
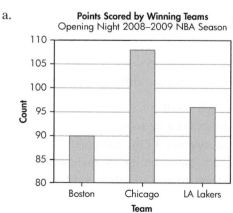

b.

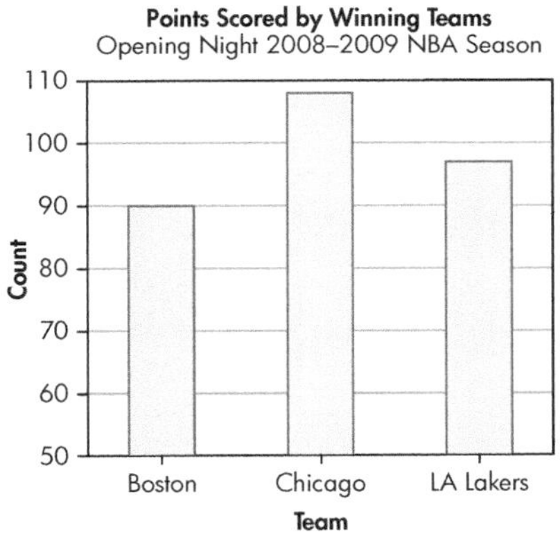
Points Scored by Winning Teams
Opening Night 2008–2009 NBA Season
Count
110
100
90
80
70
60
50
Boston
Chicago
LA Lakers
Team

c. bar graph in 'a'

d. begin the vertical scale at zero

2.9 a.

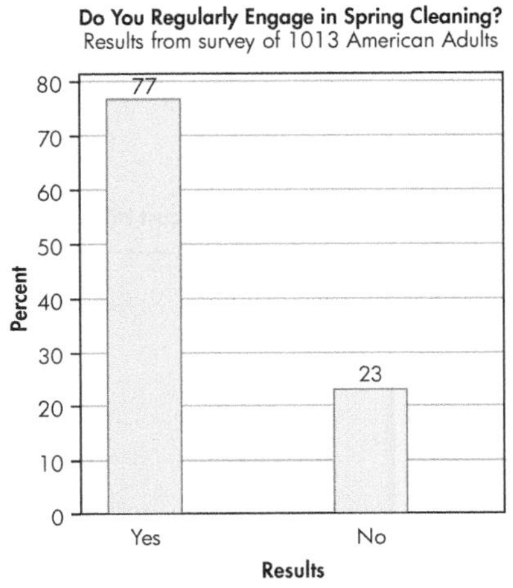
Do You Regularly Engage in Spring Cleaning?
Results from survey of 1013 American Adults
Percent
77
23
Yes
No
Results

b.

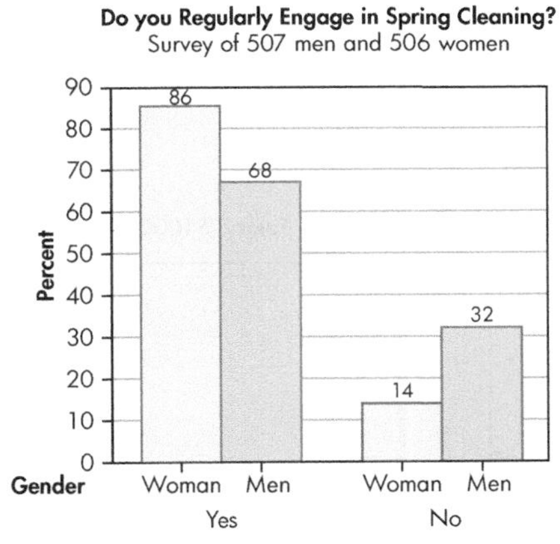
Do you Regularly Engage in Spring Cleaning?
Survey of 507 men and 506 women
Percent
86
68
14
32
Gender
Woman
Men
Woman
Men
Yes
No

2.11

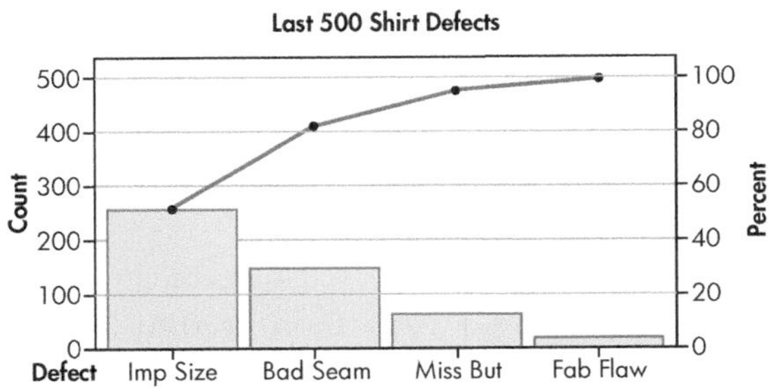
Last 500 Shirt Defects
Count
Percent
Defect
Imp Size
Bad Seam
Miss But
Fab Flaw

Defect	Imp Size	Bad Seam	Miss But	Fab Flaw
Count	258	153	67	22
Percent	51.6	30.6	13.4	4.4
Cum %	51.6	82.2	95.6	100.0

2.13 a.

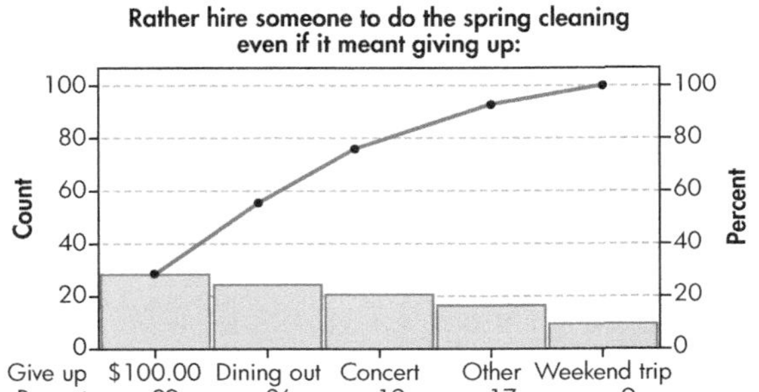
Rather hire someone to do the spring cleaning even if it meant giving up:
Count
Percent

Give up	$100.00	Dining out	Concert	Other	Weekend trip
Percent	29	26	19	17	9
Percent	29.0	26.0	19.0	17.0	9.0
Cum %	29.0	55.0	74.0	91.0	100.0

b. it is a collection of several answers; needs to be broken down

2.15 a. 150 defects

b. 0.30

c. $(56 + 45 + 23 + 12)/150 = 136/150$

d. Blem and Scratch, total 67.3%

2.17 a.

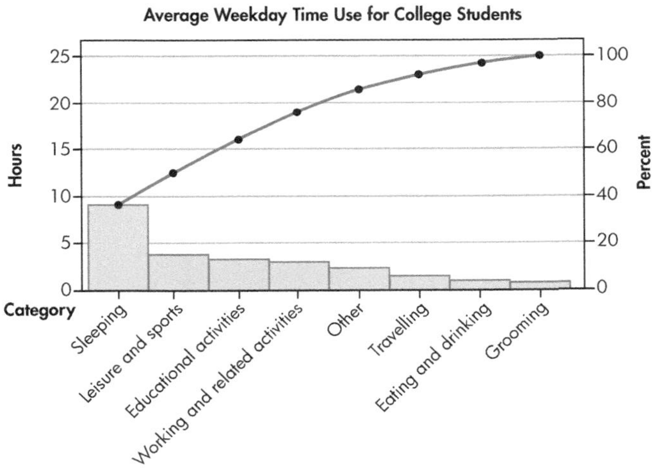
Average Weekday Time Use for College Students
Hours
Percent
Category
Sleeping
Leisure and sports
Educational activities
Working and related activities
Other
Travelling
Eating and drinking
Grooming

Hours	8.3	3.9	3.2	3.0	2.3	1.5	1.0	0.8
Percent	34.6	16.2	13.3	12.5	9.6	6.2	4.2	3.3
Cum%	34.6	50.8	64.2	76.7	86.3	92.5	96.7	100.0

b. sleeping, leisure and sports, educational activities, working and related activities

2.19

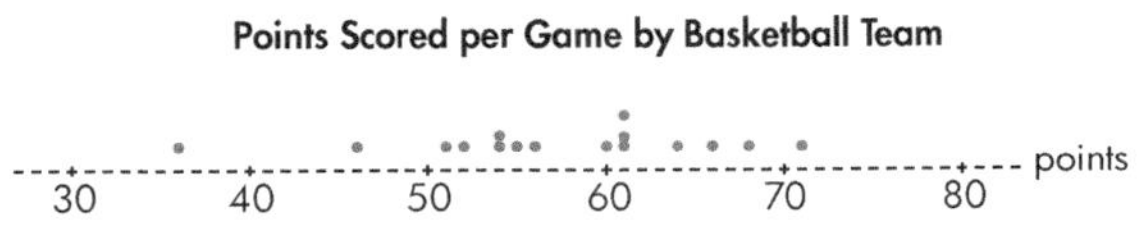
Points Scored per Game by Basketball Team
points
30
40
50
60
70
80

2.21 a.

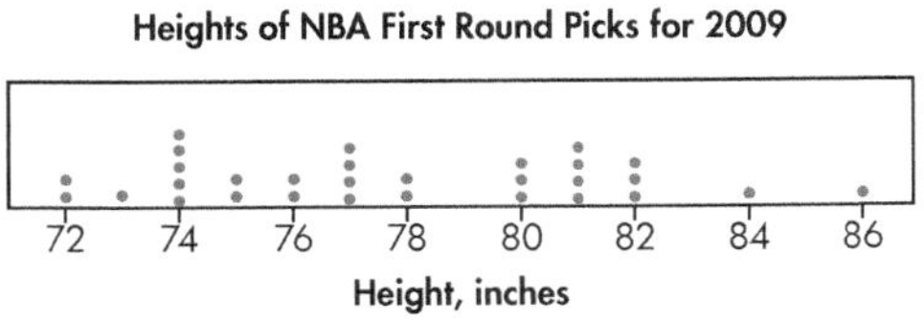
Heights of NBA First Round Picks for 2009
72
74
76
78
80
82
84
86
Height, inches

b. 72 inches, 86 inches

c. 74 inches, 5 players

d. tallest column

2.23

Overall Length of Commutators

```
    .  ..     :.. :  ::!:.  !!.: :. .. .
--+----+----+----+----+----+----+----+--
 18.740   18.780   18.820   18.860   length
```

2.25 Points scored per game

```
3 | 6
4 | 6
5 | 6 4 5 4 2 1
6 | 1 1 8 0 6 1 4
7 | 1
```

2.27 a. **Quik Delivery's delivery charges**

```
2. | 0
2. | 9 8 8 9
3. | 1 1
3. | 5 8 8 5 8 6 6 8 7 7 8
4. | 0 3 1 0 0
4. | 5 5 9 6 8 6
5. | 0 4 0 2 4
5. | 6 7
6. | 0 1
6. | 8
7. |
7. | 8
```

b. skewed right

2.29 a. place value of the leaves is hundredths

b. 16

c. 5.97, 6.01, 6.04, 6.08

d. Cumulative frequencies starting at the top and bottom

2.31 a.

x	f
0	2
1	5
2	3
3	0
4	2

b. f is frequency, value of 1 occurred 5 times

c. 12

d. number of data, or sample size

2.33 a. Bargraph   c. Histogram

2.35 b.

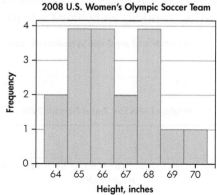

d. 66.7%

2.37 a.

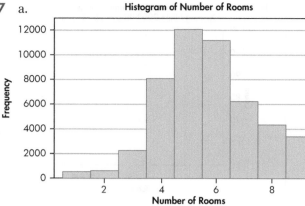

b. Mounded, truncated on the right due to 9+ class

c. Centered on 5 rooms, 4 to 7 account for most

2.39 b.

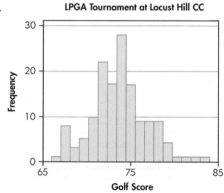

2.41 a. 35–45

d.

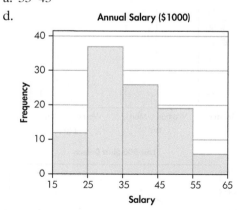

2.43 a. 12 and 16 b. 2, 6, 10, 14, 18, 22, 26

c. 4.0   d. 0.08, 0.16, 0.16, 0.40, 0.12, 0.06, 0.02

e.

KSW Test Scores

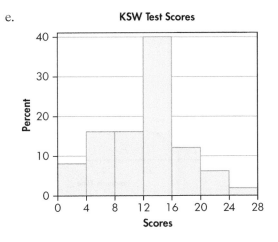

2.45 a. freq: 1, 14, 22, 8, 5, 3, 2
b. 6
c. 27; 24; 30
d.

Speed of 55 Cars on City Street

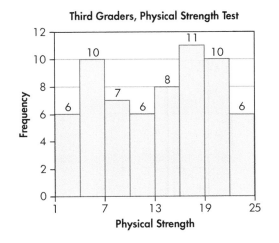

2.47 a.

Third Graders at Roth Elementary School

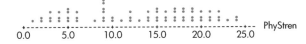

b. freq: 6, 10, 7, 6, 8, 11, 10, 6

Third Graders, Physical Strength Test

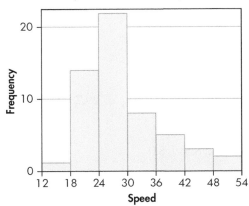

c. freq: 3, 10, 4, 9, 7, 11, 11, 7, 2

Third Graders, Physical Strength Test

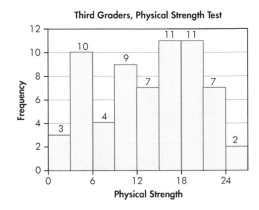

d. freq: 3, 13, 13, 15, 17, 3

Third Graders, Physical Strength Test

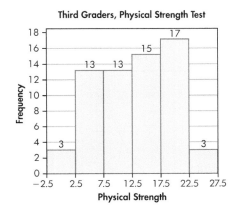

f. b and c, bimodal; d, skewed left;
Dotplot shows mode to be 9; histogram shows
two modal classes to be 4–7 and 16–22; mode
is not in either modal class

2.49 a. 1, 9, 10, 12, 4
b. 1
c. 4.5, 5.5, 6.5, 7.5, 8.5,
d.

Coal, Nuclear, Electric, and Alternate Fuels Report
Average revenue per kilowatt hour

2.53 a. Cum Freq: 12, 49, 75, 94, 100

c.

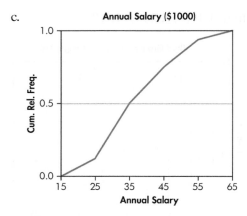

Annual Salary ($1000)

d. $45,000

e. $45,000; they're the same, just asked differently

2.55 a. Cum.Rel.Freq: 0.08, 0.24, 0.40, 0.80, 0.92, 0.98, 1.00

b.

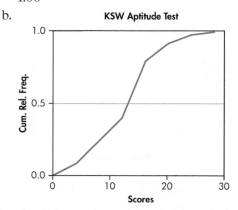

KSW Aptitude Test

c. ≈75–80%

2.57 a. Freq: 2, 6, 2, 11, 13, 9, 4, 2, 2

b. Rel Freq: 0.039, 0.118, 0.039, 0.216, 0.255, 0.176, 0.078, 0.039, 0.039

c.

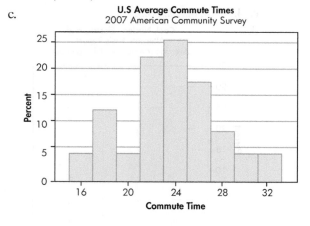

U.S Average Commute Times
2007 American Community Survey

d. Cum Rel Freq: 0.039, 0.157, 0.196, 0.412, 0.667, 0.843, 0.921, 0.960, 0.999

e.

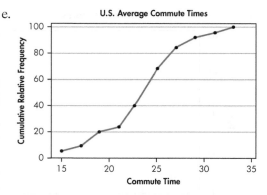

U.S. Average Commute Times

f. ≈25 minutes, approx. 70% of the average commute times are less than 25 minutes

2.59 A quantitative variable results in numbers for which arithmetic is meaningful; a qualitative variable does not

2.61 $102.07

2.63 a. 157.5   b. 94.5

2.65 $635

2.67 3rd; 73

2.69 a. 36.7   b. 32.5   d. 29.7, 30   e. mean

2.71 2

2.73 a. 8.2; 8.5; 9; 8.0

2.75 a. 6.0   b. 3.5th; 6.5   c. 7   d. 5.5

2.77 a. 32.2   b. 5.5th; 30   c. 34.5   d. 21

2.79 a.

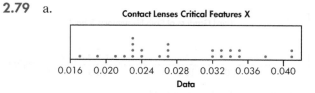

Contact Lenses Critical Features X

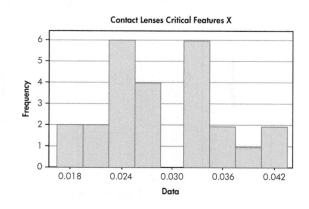

Contact Lenses Critical Features X

b. 0.0286   c. 13th, 0.027   d. 0.029

e. 0.023   f. bimodal

2.81 a.

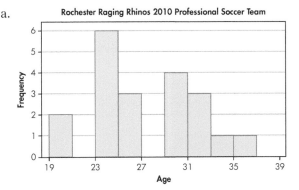

b. skewed right

d. 27.05, 25.5

e. median, mean pulled by few high values

2.83 a.

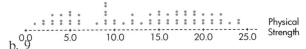

Third Graders at Roth Elementary School

b. 9

c.

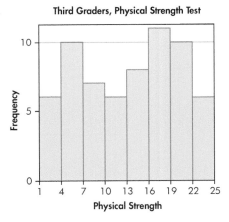

d. appears bimodal; 4–7, 16–19.

f. No

g. mode is single data value; modal class, cluster of data values

2.85 a. & b.

	Runs at Home	Runs Away
Mean	4.775	4.527
Median	4.735	4.625
Maximum	5.99	5.14
Minimum	3.57	4.00
Midrange	4.78	4.57

c. teams score more runs at home

2.87 c. little difference, more male drivers, more female drivers

d.

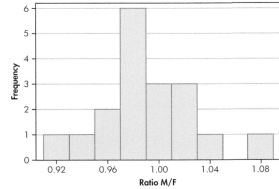

e. mounded

f. 0.990

2.89 a. two different variables

b. $5273.00

c. 12.85%

d. Minimum and maximum values are all that is given, midrange.

2.93 a. $4982   b. 7.5%

2.95 $\Sigma(x - \bar{x}) = \Sigma x - n\bar{x} = \Sigma x - n \cdot (\Sigma x/n) = \Sigma x - \Sigma x = 0$

2.97 a. $\Sigma x/n = 25/5$; $\Sigma(x - \bar{x})^2 = 46$; 11.5

b. $\Sigma x^2 = 171$; 11.5

c. same

2.99 a. 5

b. $n = 6$, $\Sigma x = 36$, $\Sigma(x - \bar{x})^2 = 16$; 3.2

c. 1.8

2.101 a. $\Sigma x/n = 104/15$; $\Sigma(x - \bar{x})^2 = 42.95$; 3.1

b. $\Sigma x^2 = 764$; 3.1

c. 1.8

2.103 a. $n = 6$, $\Sigma x = 37{,}116$, $\Sigma x^2 = 229{,}710{,}344$; 22,153.6

b. $n = 6$, $\Sigma x = 1{,}116$, $\Sigma x^2 = 318{,}344$; 22,153.6

2.105 a.

Police Recruits

b. 30.05   c. 9

d. $\Sigma x^2 = 18{,}209$; 7.8   e. 2.8

g. Except for the value $x = 30$, the distribution looks rectangular.

2.107 a.

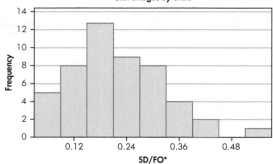

Percent of Structurally Deficient or Functionally Obsolete
U.S. Bridges by State

b. skewed right   c. 0.2176   d. 0.20

e. 0.52   f. 0.1038

2.109 set 1: 0, 54, 9

       set 2: 0, 668, 35

2.111 incorrect; standard deviation is never negative;
error in calculations or typographical error

2.115 a. 44th position from Low value;
7th position from High value

     b. 10.5^{th}; $P_{20} = 64$;
18th; $P_{35} = 70$

     c. 10.5th from H; $P_{80} = 88.5$;
3rd from H; $P_{95} = 95$

2.117 a.

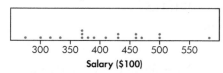

Elementary School Teachers' Salaries

     b. 2nd from the L, 17th from the H

     c. 5th, \$36,700

     d. 14th, \$45,800

2.119 a. 3.8; 5.6

     b. 4.7

     c. 3.5^{th}, 3.5; 7th, 4.0; 18.5^{th}, 6.9

2.121

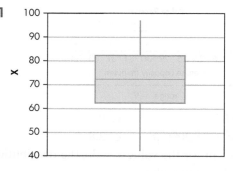

2.123 a.

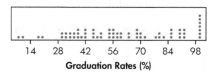

2009 Men's Teams' Graduation Rates
NCAA Division 1 Basketball Tournament

b.

Stem-and-Leaf Display: Graduation Rate, %
Stem-and-leaf of Graduation Rate, %
N = 63
Leaf Unit = 1.0

1	0	8
3	1	07
5	2	09
15	3	0113466788
23	4	01225667
(10)	5	0033355677
30	6	00347779
22	7	017
19	8	002666999
10	9	122
7	10	0000000

c. 5-number summary: 8, 40, 57, 86, 100

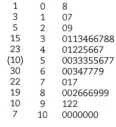

Graduation Rates for 2009 Men's Teams
NCAA Division 1 Basketball Tournament

d. 20, 100

e. Slightly skewed left

2.125 a.

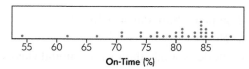

Major Airport On-Time Arrival Performace

b.

Stem-and-leaf of On-Time, % N = 31
Leaf Unit = 1.0

1	5	3
1	5	
2	6	2
3	6	6
7	7	0144
12	7	67779
(16)	8	0011112333333444
3	8	668

c. 53.5, 76.2, 81.3, 83.9, 88.2

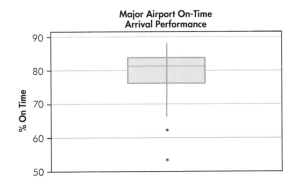

d. 70.7, 74.4

f. only the lowest are poor

g. yes, the ones with the lowest percentages

2.127 symmetric

2.129 1.67, −0.75

2.131 a. −1.76   b. −0.54   c. 0.42   d. 1.63

2.133 a. 120   b. 144.0   c. 92.0   d. 161.0

2.135 b. 0.03, 0.14, 0.20, 0.30, 0.55

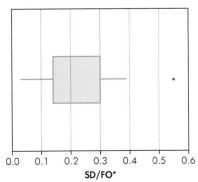

c. 0.22; 0.16

d. −0.75, 1.66, −1.42, 0.22, 3.20

2.137 1.625, 1.2; A

2.139 175 through 225 words, inclusive.

2.141 Nearly all data, 99.7%, lies within 3 standard deviations of the mean

2.143 a. 2.5%   b. 70.4 to 97.6 hours

2.145 a. 50%   b. 0.16   c. 0.84   d. 0.815

2.147 a. at least 75%   b. at least 89%

2.149 a. at most 11%   b. at most 6.25%

2.151 a. at least 75%   b. approximately 95%

2.153 a.

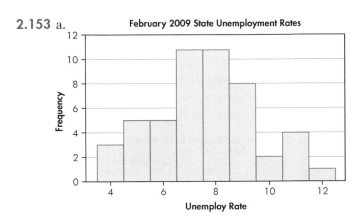

c. 7.649, 1.969

d. 5.68 to 9.618, 3.711 to 11.587, 1.742 to 13.556; 66.7%, 98.0%, 100%

2.155 a.

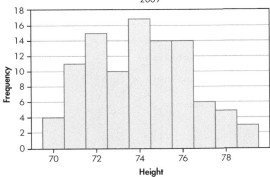

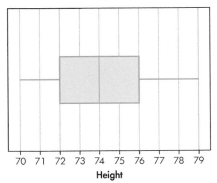

b. 74.09, 2.292

d. 71.798 to 76.382, 70%; 69.506 to 78.674, 97%; 67.214 to 80.966, 100%

e. 70%, 97% and 100% do agree

f. 97%, and 100% satisfy the theorem

2.159 a. bar graph; a person's age is used to identify the age group; age is not used as the variable

b. Smaller groups. 18–19 are different kind of shoppers than 35–39 year olds.

2.167 a.

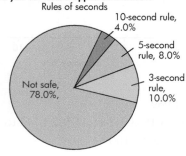

Do you eat food dropped on the floor?
Rules of seconds

10-second rule, 4.0%

5-second rule, 8.0%

3-second rule, 10.0%

Not safe, 78.0%,

b. 12, 24, 30, 234

2.169 a.

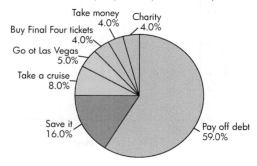

Winning $1 million in a March Madness Basketball Pool
What adults say they would spend the money on first

Take money 4.0% Charity 4.0%

Buy Final Four tickets 4.0%

Go ot Las Vegas 5.0%

Take a cruise 8.0%

Save it 16.0%

Pay off debt 59.0%

2.171 a.

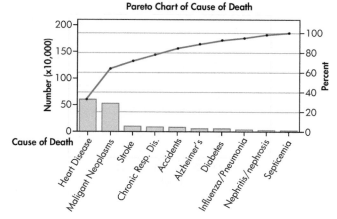

Pareto Chart of Cause of Death

Number (x10,000)	63.2	56.0	13.7	12.5	7.2	7.2	5.6	4.5	4.1	3.4
Percent	34.1	30.2	7.4	6.7	3.9	3.9	3.0	2.4	2.2	1.8
Cum %	34.1	64.3	71.6	78.4	88.8	92.7	95.7	98.2	98.2	100.0

2.173 a. numerical   b. attribute   c. numerical

d. attribute   e. numerical

2.175 a, d, e, f, and g increased; b and c unchanged

2.177 $n = 8$, $\Sigma x = 36.5$, $\Sigma x^2 = 179.11$

a. 4.56   b. 1.34   c. very closely to 4%

2.179 $n = 118$, $\Sigma x = 2364$

a. 20.0   b. 59.5^{th}, 17   c. 16

d. 30th, 15; 89th, 21   e. 12th, 14; 113th, 43

2.181 $n = 25$, $\Sigma x = 1,997$, $\Sigma x^2 = 163,205$; 79.9; 12.4

2.183 a. P: U.S. commercial airline industry; V: 3 are involved - n(reports), n(passengers), n(reports)/1000

b. data, values of variable

c. statistic, average for one month

d. No

2.185 a. 13.15   b. 13.85

c. 15.0   d. 12.95

e. 5.7   f. 25.5^{th}, 10.95; 75.5^{th}, 14.9

g. 12.925   h. 35.5^{th}, 12.05; 64.5^{th}, 14.5

j.

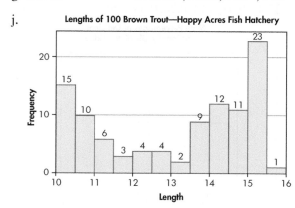

Lengths of 100 Brown Trout—Happy Acres Fish Hatchery

l.

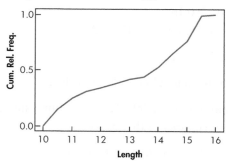

Lengths of 100 Brown Trout — Happy Acres Fish Hatchery

2.187 e. $n = 48$, $\Sigma x = 8503.88$; 177.2; 24.5th, 86.3; no mode; 539.425

f.
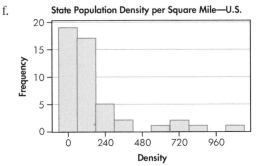
State Population Density per Square Mile—U.S.

g. NJ, RI, MA, CT, MD; WY, MT, ND, SD, NM

2.189 a. Weight

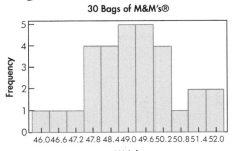

b. $\bar{x} = 49.215$, median = 49.07, s = 1.522, min = 46.22, max = 52.06

c. No

f. $\bar{x} = 57.1$, median = 58, s = 2.383, min = 50, max = 61

g. One bag has "only 50" M&M's in it

2.191 a. ≈ -0.8 or -0.9 b. $\approx +1.6$ or $+1.7$

2.193 z-scores must be changed to percentiles; P_{97}, P_{84}, P_{84}, P_{16}, P_{50}

2.195 $x = 8$, $\Sigma x = 31{,}825$, $\Sigma x^2 = 126{,}894{,}839$
a. 3978.1   b. 203.9   c. 3570.3 to 4385.9

2.197 a.

```
Stem-and-leaf of Time (min) N = 50
Leaf Unit = 1.0

  2    1   89
  5    2   134
 12    2   5777889
 20    3   11122223
(14)   3   55555566888888
 16    4   03333
 11    4   566689
  5    5   01223
```

b. $n = 50$, $\Sigma x = 1810$, $\Sigma x^2 = 69{,}518$; 36.2, 35, 35, 35, 35.5, 81.551, 9.03

c. 18, 31, 35, 43, 53

d. between 18.14 and 54.26 minutes, 98%

e. 40 minutes

Chapter 3

3.1 a. Yes   b. Somewhat

3.3 a.

	On Airplane	Hotel Room	All Other	Marginal total
Business	35.5%	9.5%	5.0%	50%
Leisure	25.0%	16.5%	8.5%	50%

b.

	On Airplane	Hotel Room	All Other	Marginal total
Business	71.0%	19.0%	10.0%	100%
Leisure	50.0%	33.0%	17.0%	100%

Business and leisure are separate distributions.

c.

	On Airplane	Hotel Room	All Other	Marginal total
Business	58.7%	36.5%	37.0%	50%
Leisure	41.3%	63.5%	63.0%	50%

Each category is a separate distribution.

3.5 a. Adults; Gender; Age would like to remain rest of life

b.
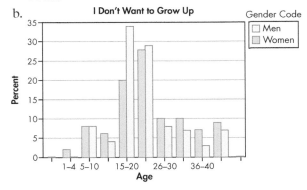

c. No

3.7 a. 3350

b. Two variables, political affiliation and television network; both qualitative

c. 880

d. 46.9%

e. 19.2%

f. 5.9%

3.9   East: $\bar{x} = 9.438$, $\tilde{x} = 9.65$; West: $\bar{x} = 9.287$, $\tilde{x} = 9.00$

3.11   a.

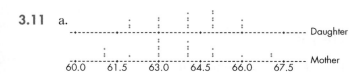

b. mother heights more spread

c.

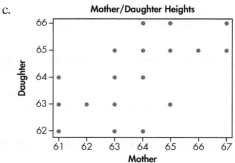

d. As mothers' heights increased, daughters' heights increased.

3.13   height, weight is often predicted

3.15   a.

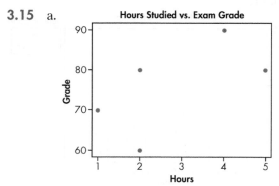

b. As hours studied increased, exam grades increase

3.17   a. age, height
          b. Age = 3 yrs., height = 87 cm.
          c. Growth is above or below normal.

3.19

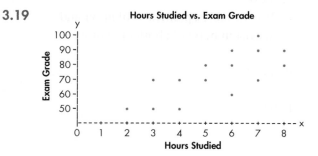

3.21   a. as distance increases, so does commute time

          b.

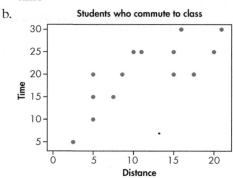

          c. yes

3.23   a. weak relationship

          b.

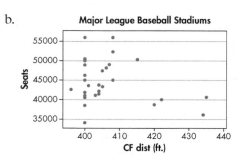

          c. no relationship

3.25   a.

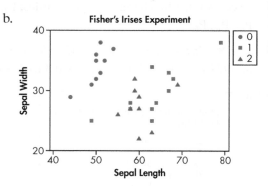

          b.

          c. Type 0 displays a different pattern than types 1 and 2

d.

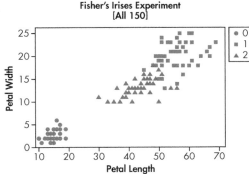

Fisher's Irises Experiment
[All 150]

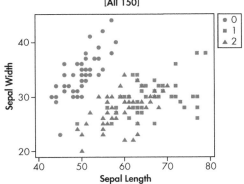

Fisher's Irises Experiment
[All 150]

3.27 a. Become closer to a straight line with a positive slope;
b. Become closer to a straight line with a negative slope

3.29 very little or no linear correlation

3.31 a. $SS(x) = 10.8$; $SS(y) = 520$; $SS(xy) = 46$
b. 0.61

3.33 a.

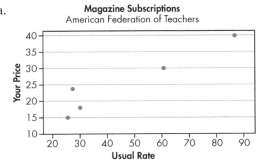

Magazine Subscriptions
American Federation of Teachers

b. $SS(x) = 3028.718$
c. $SS(y) = 398.326$
d. $SS(xy) = 1043.237$
e. 0.95

3.35 a. Manatees, powerboats
b. Number of registrations, manatee deaths
c. As one increases the other does also

3.37 a. near 2/3 of 0.70
b. $SS(x) = 49.6$; $SS(y) = 35.333$; $SS(xy) = 31.0$; 0.74

3.39 a.

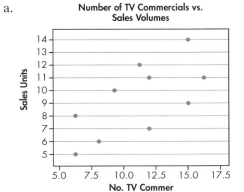

Number of TV Commercials vs.
Sales Volumes

b. from 1/2 to 2/3
c. $SS(x) = 122$; $SS(y) = 72.10$; $SS(xy) = 62.0$; 0.66

3.41 positive vs. negative; nearness to straight line, etc.

3.43 a. 0.95   b. 1.00   d. CO_2 will double

3.45 a.

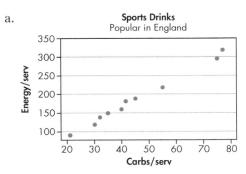

Sports Drinks
Popular in England

b. Yes
c. $SS(x) = 3125.511$; $SS(y) = 48505.6$;
$SS(xy) = 12264.84$; $r = 0.996$
d. strong, positive correlation
e.

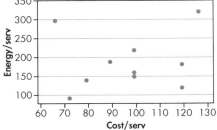

Sports Drinks
Popular in England

no linear relationship
$SS(x) = 3794.1$; $SS(y) = 48505.6$; $SS(xy) = 2044.4$;
$r = 0.15$; little or no correlation

3.47 a.

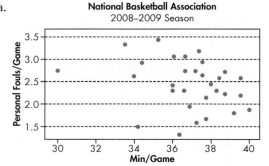

b. no upward or downward; pattern; (29.8, 2.78) from the Knicks is in a low category by itself

c. −0.261

d. Yes

3.49 No, both increase during months of warmer weather

3.51

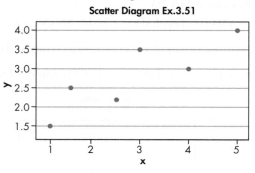

Yes

3.53 a.

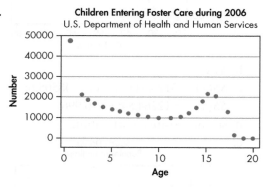

b. The number of children entering foster care increases as they approach the teen year.

c. There appears to be little or no linear correlation.

d. no

e. between the ages of 1 and 10, 11 and 15, 16 and 18

3.55 a. $SS(x) = 10.8$; $SS(xy) = 46$; $\hat{y} = 64.1 + 4.26x$

b.

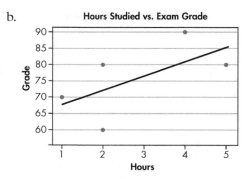

c. Yes, as the hours studied increased, the exam graded increased.

3.57 a. $\hat{y} = 28.1$; $\hat{y} = 47.9$ b. yes

3.59 a.

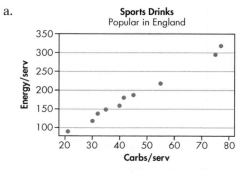

linear

b. $SS(x) = 3125.511$; $SS(xy) = 12264.84$; $\hat{y} = 9.55 + 3.924x$

c. 166.51

d. 264.61

3.61 a. cost when no long-distance calls are made

b. $1.28 is the increase for each additional long-distance call.

3.63 a. For each increase in height of one inch, weight increased by 4.71 pounds

b. The scale for the y-axis starts at $y = 95$ and the scale for the x-axis starts at $x = 60$.

3.65 6.81 or $68,100

3.67 a. $492,411,000

b. $990,241,000

c. $1,488,041,000

3.69 Vertical scale is at $x = 58$ and is not the y-axis.

3.71 a. The data would lie on a straight line with slope 1.618.

b. The data would be scattered about but would generally follow a straight path with slope 1.618.

3.75 a.

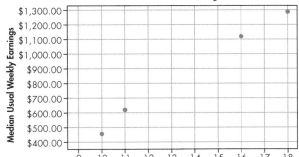

b. Yes, as years of schooling increase, so do median weekly earnings.

c. 0.997

d. yes

e. $\hat{y} = -647.25 + 108.25x$

f. For every additional year of schooling, median weekly earnings increase by $108.25.

h. -647.25, 0 years of schooling is not in the range of data

3.77 a. increasing amounts

b.

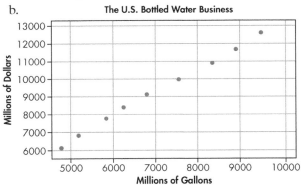

c. yes

d. $\hat{y} = 42.8 + 1.33x$ (Millions of $ = 43 + 1.33$ Millions of Gallons)

e. An increase of $1.33 millions of dollars for each additional million gallons of bottled water sold

3.79 a.

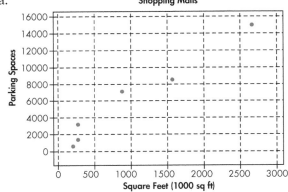

b. Yes, both increase linearly.

c. $\hat{y} = 801 + 5.28x$

d.

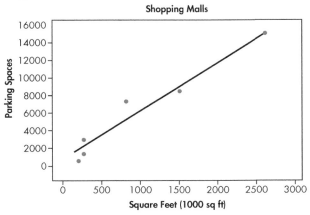

As size of mall increases, so does number of parking spaces.

e. type of store (e.g. Large furniture store – large area but requires less parking)

f.

g. Yes, both increase.

h. $\hat{y} = -23 + 53.13x$

i.

j. type of store

k.

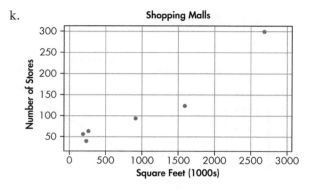

l. Yes, both increase.

m. $\hat{y} = 23.50 + 0.09x$

n.

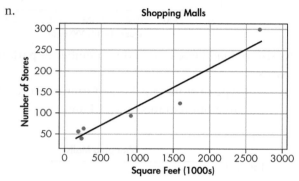

3.83   a. fear: 138; do not: 362

b.

	Elem.	Jr.H.	Sr.H.	Coll.	Adult
Fear	7.4%	5.6%	5.0%	5.4%	4.2%
Do not	12.6%	14.4%	15.0%	14.6%	15.8%

d.

	Elem.	Jr.H.	Sr.H.	Coll.	Adult
Fear	26.8%	20.3%	18.1%	19.6%	15.2%
Do not	17.4%	19.9%	20.7%	20.2%	21.8%

e.

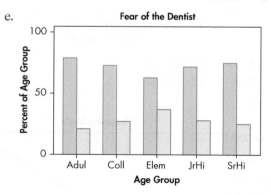

3.89   a. determine if linearly related; result is r

b. determine the equation of the line of best fit; result is the equation

3.91   a.

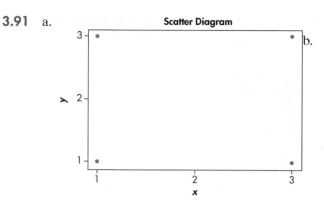

$SS(x) = 4.0$; $SS(y) = 4.0$; $SS(xy) = 0.0$; $r = 0.00$

c. $\hat{y} = 2.0 + 0.0x$

3.95   a.

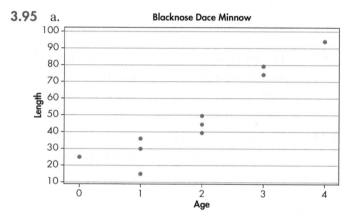

b. $SS(x) = 12.9$; $SS(y) = 6112.9$; $SS(xy) = 263.1$; 0.937

c. $\hat{y} = 10.34 + 20.40x$

3.97   a.

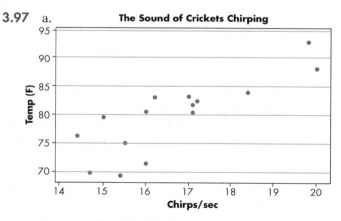

b. strong linearly, increasing

c. $SS(x) = 40.5573$; $SS(xy) = 133.508$;
$\hat{y} = 25.2 + 3.29x$

d. 71°F, 91°F

e. Temperatures range from 70° to 90°F on summer nights.

3.99 a.

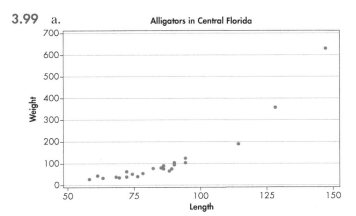

Alligators in Central Florida

b. yes

c. no

d. not a straight line

e. $SS(x) = 9826.96$; $SS(y) = 409,446$;
$SS(xy) = 58,002.2$; $r = 0.914$

f. very elongated

3.101 a. New York City values are about 4 times as large.

c.

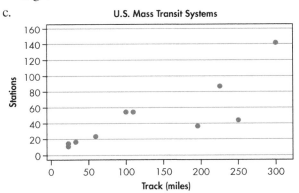

U.S. Mass Transit Systems

d. linear for the most part

e. $\hat{y} = 4.2 + 0.335x$

f. every 10 miles, about 3 stations

g.

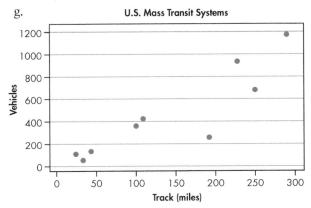

U.S. Mass Transit Systems

h. linear for the most part

i. $\hat{y} = -59 + 3.63x$

j. for every mile of track, about 3 or 4 vehicles

k.

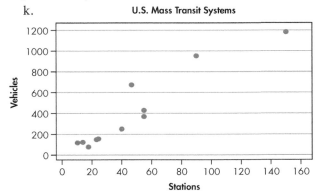

U.S. Mass Transit Systems

l. linear for the most part

m. $\hat{y} = -21.6 + 9.15x$

n. For every additional station, there is an increase of 9 vehicles.

o. 21 stations, 122 to 123 vehicles

p. At varying miles, the y-intercept will have varying degrees of effect on the final answers.

q. 38 stations, 304 vehicles

3.103 a.

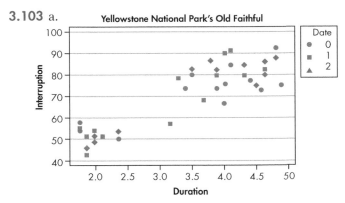

Yellowstone National Park's Old Faithful

b. overall linear pattern, strong positive relationship, with two separate clusters

c. yes

d. between 70 and 90 minutes

e. $SS(x) = 48.3959$; $SS(xy) = 578.274$;
$\hat{y} = 30.0 + 11.9x$

f. approximately 78 minutes

g. very little

3.105 $SS(x) = 9.2$; $SS(y) = 26.8$; $SS(xy) = 14.6$; 0.9298

Chapter 4

4.1 a. Most: yellow, blue and orange Least: brown, red, and green

b. Not exactly, but similar

4.3 5, 6, 6, 9, 8, 6 respectively

4.5 $P'(5) = 0.225$

4.7 a. 15.4%   b. 53.8%   c. 46.2%

4.9 a. 0.150   b. 0.206
         c. 0.326   d. 0.478

4.11 a. 0.10   b. 0.27
          c. 0.49   d. 0.00

4.13 a. {0, 1, 2, 3, 4, 5, 6, 7, 8, 9}   b. 0.1   c. 5/10 = 0.5

4.15 a. 42, 35   b. 77   c.   35/77
          d. 77/77   e. 0/77

4.19 $P(5) = 4/36; P(6) = 5/36; P(7) = 6/36;$
          $P(8) = 5/36; P(9) = 4/36; P(10) = 3/36;$
          $P(11) = 2/36; P(12) = 1/36$

4.23 All are inappropriate

4.25 a.

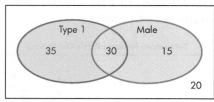

          b. 0.55
          c. 0.35

4.27 0.04; 4%

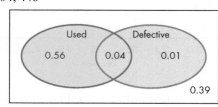

4.29 a. expect a 1 to occur approximately 1/6th of the
             time when you roll a single die
          b. 50% of the tosses are expected to be heads,
             the other 50% tails.

4.31 a.

          b. {(H,1),(H,2),(H,3),(H,4),(H,5),(H,6),
             (T,1),(T,2),(T,3),(T,4),(T,5),(T,6)}

4.35 1/5

4.37 4/5

4.39 a. 1/7   b. 6:1

4.41 a. 39:156   b. 39/195 = 0.20   c. 84:111
          d. 84/195 = 0.43   e. place, twice as likely as 1st place

4.43 a. approx 1:73   b. approx 38:3
          c. 0.0045   d. 0.048

4.45 a. 70   b. M = 2, D = 20, S = 20, W = 10
          c. 0.14   d. 0.43   e. 0.00   f. 0.019
          g. 0.17   h. 0.507

4.47 a. probability   b. statistics

4.49 a. statistics   b. probability
          c. statistics   d. probability

4.51 a. 0.45   b. 0.40   c. 0.55

4.53 a. 0.34   b. 0.38   c. 0.64
          d. 0.03   e. 0.92   f. 0.75
          g. 0.98

4.55 a. 0.59   b. 0.41   c. 0.35
          d. 0.27   e. 0.30   f. 0.60
          g. 0.60   h. different ways of asking
                              same question

4.57 a. Some categories would be counted twice.
b. 0.10   c. 0.77
d. 0.09   e. 0.36

4.59 a. 0.3   b. 0.22

4.61 0.37

4.63 0.8

4.65 0.2

4.67 0.81

4.69 4%

4.71 0.28

4.73 0.5

4.75 0.098

4.77 0.90

4.79 a.

b. 2/5 or 1/5, depending on first pick
c. 0.067
d. 0.067, same probability

4.81 0.62

4.83 0.133

4.85 a. 0.6   b. 0.7   c. 0.5

4.87 a. 0.4   b. 0.4

4.89 a. Not mutually exclusive
b. Not mutually exclusive
c. Not mutually exclusive
d. Mutually exclusive

4.91 there is no intersection

4.93 a. 0.7   b. 0.6   c. 0.7   d. 0.0

4.95 a. Yes   b. No   c. No   d. Yes
e. No   f. Yes   g. No

4.97 a. A & C and A & E are mutually exclusive
b. 12/36, 11/36, 10/36

4.99 a. yes   b. yes
c. no   d. 0.516
e. 0.512   f. 0.558
g. 0.721   h. 0.233
i. 0.552   j. 0.014

4.101 0.54

4.103 a. independent   b. not independent
c. independent   d. independent
e. not independent   f. not independent

4.105 0.28

4.107 0.5

4.109 a. 0.12   b. 0.4   c. 0.3

4.111 a. 0.5   b. 0.667   c. No

4.113 a. independent   b. independent
c. dependent

4.115 a. 0.51   b. 0.15   c. 0.1326

4.117 a. 0.1225   b. 0.4225   c. 0.0150

4.119 0.4565

4.121 a. 0.36   b. 0.16   c. 0.48

4.123 a. 3/5   b. 0.16, 0.48, 0.36

4.127 a. cannot occur at the same time
b. occurrence of one has no effect on the probability of the other
c. Mutually exclusive - whether or not share common elements; independence - effect one event has on the other event's probability

4.129 a. 0.25   b. 0.2   c. 0.6
d. 0.8   e. 0.7   f. No
g. No

4.131 a. 0.0   b. 0.7   c. 0.6
d. 0.0   e. 0.5   f. No

4.133 a. 0.625
 b. 0.25
 c. $P(\text{satisfied}|\text{skilled female}) = 0.25$
  $P(\text{satisfied}|\text{unskilled female}) = 0.667$
  not independent

4.135 a. 0.41   b. 0.007
 c. 0.02   d. Without

4.137 0.300

4.139 a. $S = \{$GGG, GGR, GRG, GRR, RGG, RGR, RRG, RRR$\}$
 b. 3/8
 c. 7/8

4.141 7/8

4.143 a. 0.40   b. 0.49
 c. 0.06   d. 0.82
 e. 0.40   f. 0.45

4.145 a. Brazil, Spain, India, etc.
 b. 'based on countries included'
 c. 44%
 d. 0.44
 e. same question, answer in a different format

4.147 $P(A \text{ or } B) = P(A) + P(B) - P(A \text{ and } B)$
  $= P(A) + P(B) - P(A) = P(B)$

4.149 a. 0.30   b. 0.60
 c. 0.10   d. 0.60
 e. 0.333   f. 0.25

4.151 a. 0.3168   b. 0.4659
 c. No   d. No
 e. "candidate wants job" and "RJB wants candidate" could not both happen.

4.153 a. 0.429   b. 0.476   c. 0.905

4.155 a. 0.531   b. 0.262   c. 0.047

4.157 a. 0.508   b. 0.202   c. 0.334   d. 0.563
 e. 0.194   f. 0.499   g. complements

4.159 a. False   b. True
 c. False   d. False

4.161 8/30

4.163 a. 1/2, 1/4, 1/8   b. 9/16, 9/32, 9/64

4.165 0.592

4.167 a. 26/52   b. 26/52
 c. 32/52   d. 20/52

4.169 a. 0.60   b. 0.648   c. 0.710
 d. (a) 0.70   (b) 0.784   (c) 0.874
 e. (a) 0.90   (b) 0.972   (c) 0.997
 f. "best" team most likely wins-more games, greater difference between teams

4.171 a. 1/7   b. 1/7

Chapter 5

5.1 a. 22%
 b. 1 vehicle
 c. Number of vehicles per household
 d. Yes, events $(1, 2, 3, \ldots 8)$ are non-overlapping

5.3 number of siblings $x = 0, 1, 2, 3, \ldots, n$; length of conversation $x = 0$ to ? minutes

5.5 a. discrete, count; continuous, measurable
 b. count
 c. measurable

5.7 a. number of new jobs
 b. discrete, countable

5.9 distance, $x = 0$ to n, $n =$ radius of the target, continuous

5.11 a. amount of time spent per week on various activities
 b. continuous, measurable

5.13

x	0	1
$P(x)$	1/2	1/2

5.15 a. events never overlap
 b. all outcomes accounted for

5.17 a. $P(x)$ is a probability function

x	$P(x)$
1	0.12
2	0.18
3	0.28
4	0.42

 b.

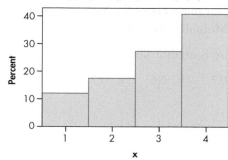

$P(x) = (x^2 + 5)/50$, for $x = 1, 2, 3, 4$

5.19

x	0	1	2	3
P(x)	0.20	0.30	0.40	0.10

5.21 $\sigma^2 = \Sigma[(x - \mu)]^2 \cdot P(x)]$

$= \Sigma[x^2 - 2x\mu + \mu^2) \cdot P(x)]$

$= \Sigma[x^2 \cdot P(x) - 2x\mu \cdot P(x) + \mu^2 \cdot P(x)]$

$= \Sigma[x^2 \cdot P(x)] - 2\mu \cdot \Sigma[x \cdot P(x)] + \mu^2 \cdot [\Sigma P(x)]$

$= \Sigma[x^2 \cdot P(x)] - 2\mu \cdot [\mu] + \mu^2 \cdot [1]$

$= \Sigma[x^2 \cdot P(x)] - 2\mu^2 + \mu^2$

$= \Sigma[x^2 \cdot P(x)] - \mu^2$ or $\Sigma[x^2 \cdot P(x)] - \{\Sigma[x \cdot P(x)]\}^2$

5.23 Nothing of any meaning

5.25 2.0, 1.4

5.27 a. $x = 1, 2, 3, 4, 5; P(x) = 0.209, 0.213, 0.241,$
0.194, 0.143
b. 2.849, 1.34

5.29 a. yes
b.

Number of Pet Dogs
(in United States)

Number of Pet Dogs

c. 1.49, 0.70
d. average number of dogs per household is 1.49
e. Mean is actually higher and standard deviation larger.

5.31 a. 2.0, 1.4
b. −0.8 to 4.8 encompasses the numbers 1, 2, 3 and 4
c. 0.9

5.33 a.

Random Single Digits; P(x) = 0.1, for x = 0, 1,…,9

b. 4.5, 2.87
d. 100%

5.35 No, random variable is an attribute variable, random variables are numerical.

5.39 a. Each question is a separate trial
b. four different ways one correct and three wrong answers can be obtained
c. 1/3 is the probability of success, 4 is the number of independent trials, number of questions

5.41 defective items should be fairly small and easier to count

5.43 a. 24   b. 5,040
c. 1   d. 360
e. 10   f. 15
g. 0.0081   h. 35
i. 10   j. 1
k. 0.4096   l. 0.16807

5.45 $n = 100$ trials (shirts), two outcomes (first quality or irregular), $p = P(\text{irregular})$, $x = n(\text{irregular})$; any integer value from 0 to 100.

5.47 a. trials are not independent
b. $n = 4$; ace, not ace; $p = P(\text{ace}) = 4/52$ and $q = P(\text{not ace}) = 48/52$; $x = n(\text{aces})$, 0, 1, 2, 3 or 4

5.49 a.

e. $P(x) = \binom{3}{x}p^x q^{3-x}$, for $x = 0, 1, 2, 3$

5.51 $P(x) = \binom{3}{x}(0.5)^x(0.5)^{3-x}$, 0.125, 0.375, 0.125

5.53 a. 0.3585   b. 0.0159   c. 0.9245

5.55 a. 0.4116   b. 0.384
c. 0.5625   d. 0.329218
e. 0.375   f. 0.0046296

5.57 $n = 5; p = 1/2, q = 1/2 (p + q = 1)$; exponents add up to 5; x any integer from zero to $n = 5$; binomial

5.59 0.143

5.61 a. 0.088   b. 0.039   c. 0.00000154

5.63 0.0011

5.65 0.410

5.67 a. 0.590   b. 0.918

5.69 a. 0.006   b. 0.215
         c. 0.618   d. 0.167

5.71 0.984

5.73 18, 2.7

5.75 a. $\Sigma[xP(x)] = 0.55, \Sigma[x^2P(x)] = 0.819; 0.55, 0.72$
         b. same

5.77 a. 25.0, 3.5   b. 4.4, 1.98
         c. 24.0, 4.7   d. 44.0, 2.3

5.79 b. 0.4338
         c. 4.92, 1.7
         d.

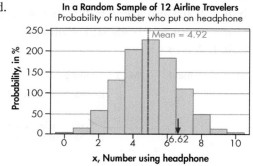

In a Random Sample of 12 Airline Travelers
Probability of number who put on headphone

5.81 a. 0.240   b. 0.240

5.85 0.9666

5.87 a. 0.03132   b. 0.99962

5.89 $p = 0.5, n = 400$

5.91 $\mu = 6, \sigma = 1.9; 0.03383$

5.93 a. Percentage of minorities is "less than would reasonably be expected."
         b. Percentage of minorities is "not less than would be reasonably expected."

5.95 b. 0.886385
         c. 0.99383
         d. 0.12, 0.345

e. $-0.225, 0.465; 0.88638$
f. $-0.57, 0.81; 0.88638$
g. do not agree with the empirical rule; does agree with Chebyshev's

5.97 1. Each $P(x)$, is a value between zero and one inclusive

         2. sum of all the $P(x)$ is exactly one

5.99 a. probability function
         b. probability function
         c. NOT a probability function
         d. NOT a probability function

5.101 a. 0.1   b. 0.4   c. 0.6

5.103 a. 3.3   b. 1.187

5.105 No, variable is attribute

5.107 a. 0.930   b. 0.264

5.109 0.103

5.111 a. 0.999   b. 0.206   c. 0.279

5.113 P(defective) changes, trials are not independent

5.115 a. 0.914
          b. 0.625
          c. Even though the P(defective) changes from trial to trial, if the population is very large, the probabilities will be very similar.

5.117 a. 0.1116   b. 0.645   c. 0.006

5.119 68.8

5.121 a. binomial; n = number of seeds/row;
             $p = P$(germination)
          b. number of seeds planted per row
          c. $B(4, 0.25)$ is a fairly good fit.

5.123 b. 0.001   c. 0.007   d. no

Chapter 6

6.1 a. It's a quotient
        b. I.Q.: 100, 16; SAT: 500, 100; Standard score: 0, 1
        c. $z = (I.Q. - 100)/16; z = (SAT - 500)/100$
        d. 2, 132, 700
        e. same

6.3 a. Proportion   b. Percentage   c. Probability

6.5 a. bell-shaped, mean of 0, standard deviation of 1
b. reference used to determine the probabilities for all other normal distributions

6.7 a. 0.0968   b. 0.0052
c. 0.0007   d. 0.2611

6.9 a. 0.9821   b. 0.8849
c. 0.9994   d. 0.7612

6.11 a. 0.6808   b. 0.8437   c. 0.9996

6.13 a. 0.0007   b. 0.0329   c. 0.2266

6.15 0.4177

6.17 0.8571

6.19 a. 0.4394   b. 0.0606
c. 0.9394   d. 0.8788

6.21 a. 0.5000   b. 0.1469   c. 0.9893
d. 0.9452   e. 0.0548

6.23 a. 0.4906   b. 0.9725
c. 0.4483   d. 0.9306

6.25 a. 0.2704   b. 0.8528
c. 0.1056   d. 0.9599

6.27 0.2144

6.29 a. 0.2978   b. 0.0217
c. 0.0919   d. 0.3630

6.31 a. 1.14   b. 0.47   c. 1.66
d. 0.86   e. 1.74   f. 2.23

6.33 a. 1.65   b. 1.96   c. 2.33

6.35 1.28, 1.65, 2.33

6.37 −0.84

6.39 a. 0.84   b. 1.04   c. −0.67 and +0.67

6.41 a. 0.84   b. −1.15 and +1.15

6.43 a. 0.7620   b. 0.0376   c. 0.2682

6.45 2.88

6.47 a. 0.5000   b. 0.3849   c. 0.6072
d. 0.2946   e. 0.9502   f. 0.0139

6.49 a. 0.3944   b. 0.8943

6.51 a. 0.6826 or 68.26%
b. 0.9545 or 95.45%

c. 0.9973 or 99.73%
d. $0.6828 \approx 68\%$; $0.9545 \approx 95\%$; $0.9973 \approx 99.7\%$

6.53 a. 0.0289 or 2.9%   b. 0.0869 or 8.7%
c. 0.9452 or 94.5%   d. 0.3787 or 37.9%
e. 0.0485 or 4.9%

6.55 a. 0.0869 = 8.7%   b. 0.0668 = 6.7%

6.57 a. 0.4090   b. 0.3821
c. 0.0764   d. 25.42

6.59 a. 0.3557   b. $100(0.3557) \approx 36$ bottles

6.61 a. 89.6   b. 79.2   c. 57.3

6.63 20.26

6.65 7.664

6.67 a. from 150 N to 400 N
b. 0.99377 or 99.4%
c. 0.9525 or 95.3%
d. 0.8354 or 83.5%

6.69 d. 0.2329, 0.2316

Section 6.4 Exercises

6.75 a. $z(0.03)$ b. $z(0.14)$
c. $z(0.75)$ d. $z(0.22)$
e. $z(0.87)$ f. $z(0.98)$

6.77 a. $z(0.01)$ b. $z(0.13)$
c. $z(0.975)$ d. $z(0.90)$

6.81 a. 1.96   b. 1.65   c. 2.33

6.83 a. 1.65   b. 2.33   c. 1.96
d. −1.96   e. −2.05

6.85 ±1.28

6.87 ±1.15, ±1.65, ±1.96, ±2.58,

6.89 a. area, 0.4602.   b. z-score, 1.28.
c. area, 0.5199.   d. z-score, −1.65

6.91 $np = 2$, $nq = 98$; No

6.93 binomial: 0.829; normal approx.: 0.8133

6.95 0.1812; 0.183

6.97 0.6406; 0.655

6.99 $x = n$(survive), $\mu = 245$, $\sigma = 2.21$; 0.999997

6.101 $\mu = 25.1, \sigma = 3.54$
    a. 0.5438   b. 0.0002

6.103 $\mu = 27.63, \sigma = 3.86; 0.6844$

6.105 $\mu = 504, \sigma = 17.10$
    a. 0.4200   b. 0.4182   c. 0.4187

6.107 $\mu = 50, \sigma = 6.12$
    a. 0.0057   b. 0.00006   c. 0.7097

6.109 ±0.84

6.111 a. −0.92   b. −2.03   c. −0.74

6.113 a. 2.07   b. 1.53

6.115 a. 0.9973   b. 0.950   c. 0.09

6.117 a. 0.0091   b. 0.2694   c. 0.2949
    d. 0.8577   e. 0.8888   f. 0.0401

6.119 $\mu = 169; \sigma = 21$
    a. 0.3015   b. 0.1841
    c. 154.9 to 183.1   d. 134.4 to 203.6

6.121 10.033

6.123 a. 0.0143   b. 619.4
    c. 107.2   d. 755.4

6.125 a. $np = 7.5, nq = 17.5$; both greater than 5
    b. 7.5, 2.29

6.127 b. 0.77023   c. 0.751779

6.129 a. $P(0) + P(1) + \cdots + P(75)$
    b. 0.9856
    c. 0.9873

6.131 0.0087

6.133 $\mu = 8, \sigma = 2.6$
    a. 0.0418   b. 0.4247   c. 0.7128

6.135 $\mu = 33.3, \sigma = 4.71$
    a. 0.0307   b. 0.0630

6.137 a. −0.00342, 0.02089
    c. 75.5% vs. 68%, 95.5% vs. 95%, 99.1% vs. 99.7%
    d. 0.8727 = 87.3%

Chapter 7

7.1 a. Histogram

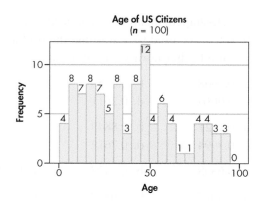

b. Mounded from 0 to 60, skewed right
    c. not exactly, but fairly close

7.3 a. No   b. Variability

7.5 a. distribution formed by means from all
        possible samples of a fixed size taken from
        a population
    b. It is one element in the sampling distribution

7.11 a. not all drawn from the same population, each
        type of vehicle has a different sample size
    b. monitoring transit vehicle populations which
        are continually changing

7.17 b. very close to $\mu = 65.15$
    d. approximately normal
    e. Took many (1001) samples of size 4 from an
        approximately normal population; 1) mean of
        the $x$bars $\approx \mu$, 2) $s_{\bar{x}} \approx \sigma/\sqrt{n}$, 3) approximately
        normal distribution

7.19 a. 1.0
    b. $\sigma_{\bar{x}} = \sigma/\sqrt{n}$; as n increases the value of this
        fraction gets smaller

7.21 a. 500
    b. 5
    c. approximately normal

7.23 a. Approximately normal
    b. 4.58 hours
    c. 0.133

7.25 a. 86.5 pounds/person
    b. 2.392
    c. approximately normal

7.29 2.69

7.31 $z = +2.00$ corresponds to 0.9773 cum prob., $z = -2.00$ corresponds to 0.0228 cum prob.; subtract to find area in between

7.33 a. approximately normal   b. 50   c. 1.667
d. 0.9973   e. 0.8849   f. 0.9282

7.35 a. approximately normal, $\mu = 69$, $\sigma = 4$.
b. 0.4013
c. approximately normal
d. $\mu_{\bar{x}} = 69$; $\sigma_{\bar{x}} = 1.0$
e. 0.1587
f. 0.0228

7.37 a. 0.3830   b. 0.9938
c. 0.3085   d. 0.0031

7.39 a. 0.2033
b. 0.0064
c. No, especially for (a); wind speed will be skewed to the right, not normal
d. Actual probabilities are most likely not as high.

7.41 a. 0.9821
b. 0.00006
c. Normal distribution should allow for reasonable estimates since $n > 30$.

7.43 38.73 inches

7.45 a. computer: 0.68269
Table 3: 0.6826

7.47 6.067, 3.64, 2.6, 1.82

7.49 a. Normally distributed with a *mean* = $775 and a standard *deviation* = $115.
b. 0.5696
c. Approximately normally distributed with a *mean* = $775 and a standard *error* = $23
d. 0.6641

7.51 a. $e = 0.49$ b. $E = 0.049$

7.53 a. 0.0060   b. 0.1635
c. skewed distribution

7.55 a. 0.1498   b. 0.0089

7.57 0.0228

7.59 0.0023

7.61 a. Σx total weight; approximately 1.000
b. 0.9773

7.65 a. $\mu = 60$; $\sigma = 6.48$

Chapter 8

8.1 a. female health professionals
b. $\bar{x} = 64.8$, $s = 3.5$
c. mounded about center, approximately symmetrical
d. approx. 65 inches; 62 to 69, the heights most frequent in this data.
e. A narrower interval would be very desirable and/or larger sample size.

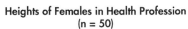

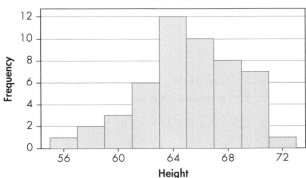

Heights of Females in Health Profession
(n = 50)

8.3 Point estimate is a single number; interval estimate is an interval of some width.

8.5 $n = 15$, $\Sigma x = 271$, $\Sigma x^2 = 5015$
a. 18.1 dollars
b. 8.5
c. 2.9 dollars

8.7 a. II has lower variability
b. II has a mean value equal to the parameter
c. Neither is a good choice, II is better

8.9 difficulty, collector fatigue; cost of sampling; destruction of product

8.11 a. U.S. married-couple families; family income
b. mean; $90,835
c. same
d. $101
e. 0.90
f. $90,734 to $90,936; an interval of values that is 0.90 likely to include the population true mean value

8.13   3

8.15   a. $1 - 2(0.1003) = 0.7994$
           b. $1 - 2(0.0749) = 0.8502$
           c. $1 - 2(0.0250) = 0.9500$
           d. $1 - 2(0.0099) = 0.9802$

8.17   a. an interval of values, 101 to 113, that is 0.95
              likely to include the population true number
              per 1000 showing a prevalence of self-reported
              hip pain among the men
           b. 3.06
           c. 3.57

8.19   a. Between 3:09 p.m. and 3:29 p.m.
           b. yes
           c. 90% occur within predicted interval

8.21   Sampling distribution of sample means must be
           normal.

8.23   a. $z(0.01) = 2.33$ b. $z(0.005) = 2.58$

8.25   a. 25.76 to 31.64
           b. Yes, population is normal

8.27   a. 125.58 to 131.42   b. Yes, CLT

8.29   a. 128.5   b. $z(0.05) = 1.65$ c. 1.76845
           d. 2.92   e. 125.58   f. 131.42

8.33   a. 15.9; $\approx 68\%$
           b. 31.4; $\approx 95\%$
           c. 41.2; $\approx 99\%$
           d. Higher level makes for a wider width.

8.35   a. 75.92   b. 0.368   c. 75.552 to 76.288

8.37   a. 14.01 to 14.59   b. 13.89 to 14.71

8.39   a. speed reading   b. 176.99 to 191.01
           c. 175.67 to 192.33

8.41   a. mean length   b. 75.92
           c. 75.512 to 76.328

8.43   b. 450.6, 173.4
           c. 107.86814506, 0.00001734
           d. approximately normal
           e. both apply
           f. no
           g. Sample standard deviation will be used.
           h. 401.5 to 499.7 or 107.86814015 to
              107.86814997

8.45   a. $33,257.74 to $34,782.26
           b. $13,525.28 to $14,564.72
           c. basically the same

8.47   49

8.49   27

8.51   25

8.55 H_o: system is reliable
           H_a: system is not reliable

8.57   a. H_a: Special delivery mail takes too much time
           b. H_a: New design is more comfortable
           c. H_a: Cigarette smoke has an effect
           d. H_a: Hair conditioner is effective on "split
              ends"

8.59   A: party will be a dud; did not go
           B: party will be a great time, did go
           I: party will be a dud, did go
           II: party will be a great time, did not go

8.61   a. H_a: The victim is not alive
           b. A: alive, treated as though alive
              I: alive, treated as though dead
              II: dead, treated as if alive
              B: dead, treated as dead
           c. I very serious, victim may very well be dead
              shortly without attention
              II not serious, victim is receiving attention that
              is of no value

8.63   You missed a great time.

8.69   a. Type I   b. Type II
           c. Type I   d. Type II

8.71   a. Commercial is not effective.
           b. Commercial is effective.

8.73   a. very serious   b. somewhat serious
           c. not at all serious

8.75   a. α b. β

8.77 α is probability rejecting a *true* null hypothesis; $1-\beta$
           is probability of rejecting a *false* null hypothesis

8.79   a. "See I told you so"
           b. "Okay not significant, I'll try again tomorrow"

8.81   a. 0.1151   b. 0.2119

8.83   a. 29 corks pass Part 1 inspection
           b. Batch refused, 3 corks that do not meet the
              specification.

8.85 H_o: The mean shearing strength is at least 925 lbs.
           H_a: The mean shearing strength is less than
           925 lbs.

8.87 a. $H_a: \mu > 1.25$ b. $H_a: \mu < 335$
c. $H_a: \mu \neq 230{,}000$ d. $H_a: \mu > 260$
e. $H_a: \mu > 15.00$

8.89 II; buy and use weak rivets

8.91 I: reject H_o interpreted as 'mean hourly charge is less than \$60, when in fact is at least \$60
II: fail to reject H_o interpreted as 'mean hourly charge is at least \$60, when in fact is less than \$60

8.93 a. 1.26   b. 1.35
c. 2.33   d. -0.74

8.95 a. Reject H_o, Fail to reject H_o
b. p-value is smaller than or equal to α, reject H_o. p-value is larger than α, fail to reject H_o.

8.97 a. Reject H_o, $P < \alpha$
b. Fail to reject H_o, $P > \alpha$
c. Reject H_o, $P < \alpha$
d. Reject H_o, $P < \alpha$

8.99 a. Fail to reject H_o
b. Reject H_o

8.101 b. ≈ 0.0000 d. Reject H_o.

8.105 0.2714

8.107 a. 0.0694   b. 0.1977   c. 0.2420
d. 0.0174   e. 0.3524

8.109 a. 1.57   b. -2.13 c. $-2.87, +2.87$

8.111 6.67

8.113 a. $H_a: \mu < 525$ b. Fail to reject H_o
c. $60.0\sqrt{38}$

8.115 a. $H_a: \mu \neq 6.25$ b. Reject H_o
c. $1.4\sqrt{78}$ d. 514.488, 3518.3437

8.117 $H_a: \mu > \$49{,}246$; $z^* = 4.08$; $P = 0.00002$; Reject H_o

8.119 $H_a: \mu < 12$; $z^* = -2.53$; $P = 0.0057$; Reject H_o

8.121 $H_a: \mu < \$123.89$; $z^* = -3.46$; $P = 0.0003$; Reject H_o

8.123 a. mean accuracy of quartz watches
b. $H_a: \mu > 20$
c. normality assumed, $n = 36$; $\sigma = 9.1$
d. $n = 36, \bar{x} = 22.7$
e. $z^* = 1.78$; $P = 0.0375$
f. $P < \alpha$; Reject H_o

8.125 a. Mean r
New Y
b. $H_o: \mu$
c. $z^* = \,'$
f. $P <$

8.127 a. 64.(

8.131 $H_o: r$
$H_a:$ ...

8.133 a. $H_a: \mu < 1.25$ b. ...
c. $H_a: \mu > 230{,}000$

8.135 a. decided average salt content is more than 350 mg when in fact, it is not
b. decided average salt content is less than or equal to 350 mg when in fact it is greater

8.137 I: decided mean minimum is greater than \$95, when in fact it is not
II: decided mean minimum is at most \$95, when in fact it is greater

8.139 a. set of all values of test statistic that will cause us to reject H_o.
b. critical value(s) is value(s) of the test statistic that form boundary between critical region non-critical region, the critical value is in the critical region.

8.141 If one is reduced, the other one becomes larger.

8.143 $z \leq -2.33$

8.145 a. $z \leq -1.65, z \geq 1.65$
b. $z \geq 2.33$
c. $z \leq -1.65$
d. $z \leq -2.58, z \geq 2.58$

8.147 $\bar{x} = 247.1$; $\Sigma x = 21{,}004.133$

8.149 a. 3.0 standard errors
b. critical region is $z \geq 2.33$, reject H_o

8.151 a. Reject H_o or Fail to reject H_o
b. calculated test statistic falls in the critical region, reject H_o
calculated test statistic falls in the non-critical region, fail to reject H_o

8.153 a. $H_o: \mu = 15.0$ vs. $H_a: \mu \neq 15.0$
b. ± 2.58; reject H_o
c. $0.5\sqrt{30}$

762

8.155 a. $H_o: \mu = 72$(
b. Fail to reje
c. $12.0\sqrt{3}$

8.157 $H_a: \mu$
$z^* =$

8.159

) vs. $H_a: \mu > 72$

ct H_o

79.68; normality is assumed, $n = 50$;
1.07; $z(0.05) = 1.65$; Fail to reject H_o

$H_a: \mu > 15$; normality assumed, $n = 35$; $z^* = 3.26$;
$z \geq 1.28$; Reject H_o

.161 $H_a: \mu > 170.1$; normality indicated; $z^* = 2.73$;
$z \geq 1.65$; Reject H_o

8.163 $H_a: \mu > 36.8$; normality assumed, $n = 42$;
$z^* = 1.45$; $z \geq 1.65$; Fail to reject H_o

8.167 a. 32.0   b. 2.4   c. 64
d. 0.90   e. 1.65   f. 0.3
g. 0.495   h. 32.495   i. 31.505

8.169 43.3 to 46.7

8.171 a. 9.75 to 9.99
b. 9.71 to 10.03
c. widened the interval

8.173 a. 69.89 to 75.31   b. Yes

8.175 a. Yes, all measurements are < 1.0 mm.
b. 0.221 to 0.287

8.177 92

8.179 60

8.181 a. "boundary" for decision
b. none

8.183 a. $H_o: \mu = 100$ b. $H_a: \mu \neq 100$
c. 0.01   d. 100
e. 96   f. 12
g. 1.70   h. −2.35
i. 0.0188   j. Fail to reject H_o

8.185 $H_a: \mu > 45$; $z^* = 2.47$; $P = 0.0068$; $z \geq 2.05$;
Reject H_o

8.187 a. $H_a: \mu \neq 0.50$
b. 0.2112
c. $z \leq -2.33$, $z \geq 2.33$

8.189 $H_a: \mu > 0.0113$; $z^* = 2.47$; $P = 0.0068$; $z \geq 2.33$;
Reject H_o

8.191 $H_a: \mu \neq 15{,}650$; $z^* = 1.41$; $P = 0.1586$; $z \leq -1.96$,
$z \geq 1.96$; Fail to reject H_o

8.193 $H_a: \mu < 24.3$; $z^* = -2.98$; $P = 0.0014$;
$z \leq -2.33$; Reject H_o

8.195 a. 39.6 to 41.6
b. $H_a: \mu \neq 40$; $z^* = 1.20$; $P = 0.2302$; Fail to
reject H_o
c. $H_a: \mu \neq 40$; $z^* = 1.20$; $z \leq -1.96$, $z \geq 1.96$; Fail
to reject H_o

8.197 a. 39.9 to 41.9
b. $H_a: \mu > 40$; $z^* = 1.80$; $P = 0.0359$; Reject H_o
c. $H_a: \mu > 40$; $z^* = 1.80$; $z \geq 1.65$; Reject H_o

8.199 a. $H_a: r > A$, burden on old drug
b. $H_a: r < A$, burden on new drug

8.201 a. $H_a: \mu \neq 18$; Fail to reject H_o; population mean
is not significantly different from 18
b. 0.756; $z^* = -1.04$; p-value = 0.2984

Chapter 9

9.1 a. Female American college students.
b. yes, mounded

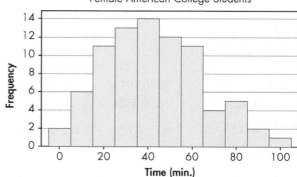

c. mounded but jagged, time of first class

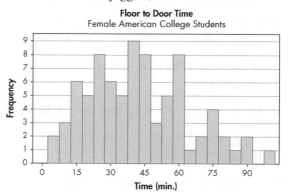

d. yes, approximately; $p = 0.193$ for normality test

9.5 a. 2.68   b. 2.07   c. 1.30   d. 3.36

9.7 a. −1.33   b. −2.82   c. −2.03
d. computer: −2.26; Table 6: $-2.62 < t < -2.14$;
Interpolation: −2.30

9.9 a. 1.73   b. ±3.18   c. −2.55   d. 1.33

9.11 ±2.18

9.13 a. −2.49   b. 1.71   c. −0.685

9.15 df = 7

9.17 0.0241

9.19 a. Symmetric about mean: mean is 0
b. Standard deviation of t is greater than 1; t has df; t-distribution is family of distributions; one z-distribution

9.21 15.60 to 17.8

9.23 3.67 to 6.83 minutes

9.25 42.94 to 47.46

9.29 3.44 to 5.68

9.31 b. 27.138 to 31.502 minutes

9.33 a.

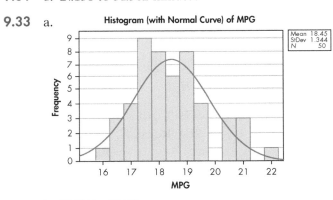

Histogram (with Normal Curve) of MPG

Mean 18.45
StDev 1.344
N 50

b. 18.07 to 18.83

9.35 a. H_o: $\mu = 11 (\geq)$ vs. H_a: $\mu < 11$
b. H_o: $\mu = 54 (\leq)$ vs. H_a: $\mu > 54$
c. H_o: $\mu = 75$ vs. H_a: $\mu \neq 75$

9.37 1.20

9.39 a. $0.025 < P < 0.05$ b. $0.025 < P < 0.05$
c. $0.05 < P < 0.10$ d. $0.05 < P < 0.10$

9.41 a. $0.10 < P < 0.25$; Fail to reject H_o
b. 1.75; Fail to reject H_o

9.43 0.124

9.45 a. $0.10 < P < 0.20$; $t = \pm 2.14$; Fail to reject H_o
b. $0.01 < P < 0.025$; $t \geq 1.71$; Reject H_o
c. $0.025 < P < 0.05$; $t \leq -1.68$; Reject H_o
d. identical

9.47 H_a: $\mu < 25$; $t\bigstar = -3.25$;
$P < 0.005$; $t \leq -2.46$; Reject H_o

9.49 H_a: $\mu < 4.6$; $t\bigstar = -2.28$;
$P = 0.036$; $t \leq -1.74$; Reject H_o

9.53 Fail to reject H_o

9.55 H_a: $\mu \neq 35$; $t\bigstar = 1.02$; $0.20 < P < 0.50$;
$\pm t(5, 0.025) = \pm 2.57$; Fail to reject H_o

9.57 H_a: $\mu < 60$; $t\bigstar = -1.60$;
$0.05 < P < 0.10$; $t \leq -1.69$; Fail to reject H_o

9.59 a.

Histogram of Density, with Normal Curve

b. H_a: $\mu < 5.517$; $t\bigstar = -1.68$;
$0.05 < P < 0.10$; $t \leq -1.70$; Fail to reject H_o

9.61 a. Yes
b. 593.93 to 598.67
c. mean is less than 600 mg

9.63 a. H_a: $\mu \neq 45.0$; $t\bigstar = -1.95$; $0.05 < P < 0.10$;
$\pm t(11, 0.01) = \pm 2.72$; Fail to reject H_o
b. H_a: $\mu \neq 45.0$; $t\bigstar = 0.24$; $P > 0.500$;
$\pm t(17, 0.01) = \pm 2.57$; Fail to reject H_o

9.65 a. number of successes, sample size
b. 0.30   c. 0.096   d. 0.312   e. 0.697

9.67 a. Yes   b. the mean of p' is p

9.69 a. $z(\alpha/2) = 1.65$ b. $z(\alpha/2) = 1.96$
c. $z(\alpha/2) = 2.58$

9.71 a. 0.021   b. 0.189 to 0.271

9.73 a. $p = P$(did not know)   b. 0.20, statistic
c. 0.0143   e. 0.186 to 0.214

9.75 0.206 to 0.528

9.77 0.064 to 0.106

9.83 a. 0.030, 0.028, 0.023
b. differing product of pq
c. Yes
e. 0.5

9.85 a. 0.5005   b. 0.003227
      c. 0.4942 to 0.5068

9.89 2401

9.91 a. 1373   b. 344   c. 2737
      d. Increasing maximum error decreases sample size
      e. Increasing level of confidence increases
         sample size

9.93 a. $H_a: p > 0.60$ b. $H_a: p > 1/3$
      c. $H_a: p > 0.50$ d. $H_a: p < 0.75$
      e. $H_a: p \neq 0.50$

9.95 a. 1.78   b. -1.70 c. -0.49 d. 1.88

9.97 a. 0.1388   b. 0.0238
      c. 0.1635   d. 0.0559

9.99 a. $z \geq 1.65$ b. $z \leq -1.96, z \geq 1.96$
      c. $z \leq -1.28$ d. $z \geq 1.65$

9.101 a. 0.017   b. 0.085
      c. 0.101   d. 0.004

9.103 a. Correctly fail to reject H_o b. 0.036
      c. Commit a type II error   d. 0.128

9.105 $H_a: p < 0.90$; $z\bigstar = -4.82$;
       $\mathbf{P} = 0.000003$; $z \leq -1.65$; Reject H_o

9.107 $H_a: p < 0.60$; $z\bigstar = -2.04$;
       $\mathbf{P} = 0.0207$; $z \leq -1.65$; Reject H_o

9.109 $H_a: p < 0.58$; $z\bigstar = -2.66$; $\mathbf{P} = 0.0039$;
       $z \leq -1.65$; Reject H_o

9.111 a. -1.97
      b & c. $\mathbf{P} = 0.0244$; $z \leq -2.33$; Fail to reject H_o

9.113 b. Reject H_o c. 0.305

9.115 $H_a: p < 0.88$; $z \leq -1.65$
      a. $z\bigstar = 1.09$; $\mathbf{P} = 0.8621$; Fail to Reject H_o
      b. $z\bigstar = -0.87$; $\mathbf{P} = 0.1922$; Fail to Reject H_o
      c. $z\bigstar = -1.74$; $\mathbf{P} = 0.0409$; Reject H_o
      d. $z\bigstar = -5.00$; $\mathbf{P} = 0.0000003$; Reject H_o

9.117 a. A: 1.72; B: 3.58
      b. Increased
      c. quite different from the rest of the data, had a
         big effect on the standard deviation

9.119 a. 23.2   b. 23.3   c. 3.94   d. 8.64

9.121 a. 30.1   b. 13.3
      c. 7.56   d. 43.2
      e. 11.6 and 32.7   f. 1.24 and 14.4

9.123 a. $\chi^2(5, 0.05) = 11.1$
      b. $\chi^2(5, 0.05) = 11.1$
      c. $\chi^2(5, 0.10) = 9.24$

9.125 0.94

9.127 a. 0.8356   b. 0.1644

9.129 a. $H_a: \sigma > 24$ b. $H_a: \sigma > 0.5$
      c. $H_a: \sigma \neq 10$ d. $H_a: \sigma^2 < 18$
      e. $H_a: \sigma^2 \neq 0.025$

9.131 a. 25.08   b. 60.15

9.133 a. $0.02 < \mathbf{P} < 0.05$ b. 0.01
      c. $0.05 < \mathbf{P} < 0.10$ d. $0.025 < \mathbf{P} < 0.05$

9.135 a. $0.05 < \mathbf{P} < 0.10$; Fail to reject H_o
      b. $\chi^2 \geq 33.4$; Fail to reject H_o

9.137 $H_a: \sigma > 0.25$; $\chi^2\bigstar = 37.24$; $0.005 < \mathbf{P} < 0.01$;
       $\chi^2 \geq 27.2$; Reject H_o

9.139 $H_a: \sigma \neq 8$; $\chi^2\bigstar = 29.3$; $0.01 < \mathbf{P} < 0.02$;
       $\chi^2 \leq 32.4$, $\chi^2 \geq 71.4$; Reject H_o

9.141 $H_a: \sigma > 85$; $\chi^2\bigstar = 64.88$; $\mathbf{P} < 0.005$;
       $\chi^2 \geq 43.8$; Reject H_o

9.143 b.

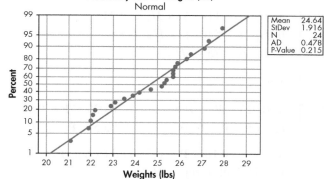

      c. & d. $H_a: \sigma > 1.0$; $\chi^2\bigstar = 84.43$
              $\mathbf{P} < 0.005$; $\chi^2 \geq 41.6$; Reject H_o

9.145 a. mounded, $\bar{x} = 0.01981$, $s = 0.01070$
      b. assume normal based on normality test
      c. $H_a: \mu > 0.025$; $t\bigstar = -2.47$; $\mathbf{P} > 0.50$ or
         $\mathbf{P} = 0.990$
      d. $t \geq 2.49$; Fail to Reject H_o

9.147 a. H_a: $\sigma \neq 0.3275$; $\chi^2 \star = 7.15$; $0.50 < \mathbf{P} < 1.00$;
$\chi^2 \leq 2.09$, $\chi^2 \geq 21.7$; Fail to reject H_o

b. H_a: $\sigma \neq 0.3275$; $\chi^2 \star = 13.97$; $0.20 < \mathbf{P} < 0.50$;
$\chi^2 \leq 2.09$, $\chi^2 \geq 21.7$; Fail to reject H_o

c. Larger sample standard deviations increase chi-square value

9.149 0.0359

9.153 35,524 to 36,476 miles

9.155 a. 8.782, 0.710

b. 8.78

c. 8.64 to 8.92 inches

9.157 72

9.159 a. $\overline{x} = \$908.30$, $s = \$118.50$

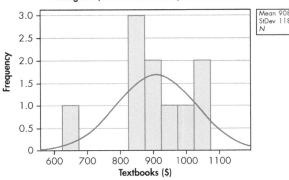

c. $823.61 to $992.99

9.161 H_a: $\mu < 640$; $t\star = -2.14$;
$0.01 < \mathbf{P} < 0.025$; $t \leq -1.74$; Reject H_o

9.163 a. 31.45, 8.049

b. H_a: $\mu > 28.0$; $t\star = 1.92$;
$0.025 < \mathbf{P} < 0.05$; $t \geq 1.73$; Reject H_o

9.165

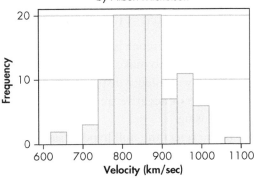

H_a: $\mu \neq 734.5$; $t\star = 14.92$;
$\mathbf{P} = 0.00+$; $t \leq -2.65$, $t \geq 2.65$; Reject H_o

9.167 0.122 to 0.278

9.169 a. 0.126 to 0.340

b. overestimate percent of satisfied customers

9.171 3388

9.173

$p =$	0.1	0.2	0.3	0.4	0.5	0.6	0.7	0.8	0.9
$pq =$	0.09	0.16	0.21	0.24	0.25	0.24	0.21	0.16	0.09

9.175 0.0401

9.181 a. parameter; binomial p, P(success)

b. 0.60 to 0.66

9.183 H_a: $\sigma > 81$; $\chi^2 \star = 123.1$;
$0.05 < \mathbf{P} < 0.10$; $\chi^2 \geq 124.0$; Fail to reject H_o

9.185 H_a: $\sigma > \$2.45$; $\chi^2 \star = 101.5$;
$0.005 < \mathbf{P} < 0.01$; $\chi^2 \geq 90.5$; Reject H_o

9.187 b. H_a: $\mu \neq 2.0$; $t\star = 3.08$;
$\mathbf{P} < 0.010$; $t \leq -2.04$, $t \geq 2.04$; Reject H_o

c. H_a: $\sigma > 0.040$; $\chi^2 \star = 48.96$;
$0.01 < \mathbf{P} < 0.25$; $\chi^2 \geq 43.8$; Reject H_o

9.189 a.

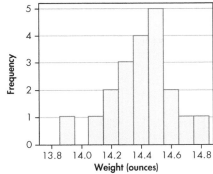

b. $\overline{x} = 14.386$, $s = 0.217$ c. 5%

d.

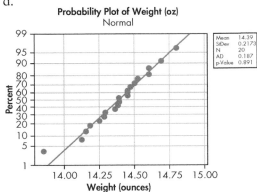

e. $\bar{x} = 14.386$, $s = 0.217$; 14.285 to 14.487

f. $H_a: \sigma > 0.2$; $\chi^2\star = 22.37$;
$\mathbf{P} = 0.2662$; $\chi^2 \geq 36.2$; Fail to reject H_o

9.191 a. 0.8051   b. 0.1271

c. 1016.46 boxes

Chapter 10

10.1 a. male commute time to college, female commute time to college

b. male: $\bar{x} = 17.97$, $s = 5.42$; female: $\bar{x} = 25.64$, $s = 9.95$

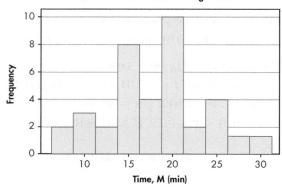

c. independent, separate, unrelated samples, no pairings

d. yes, independent; cannot be paired

e. no: if paired, then dependent, if not paired then independent

10.5 Independent, samples are separate sets of students

10.7 Dependent, each person provides one data for each sample

10.9 Independent, separate samples

10.13 a. $d = A - B$: 1 1 0 2 −1

b. 0.6   c. 1.14

10.15 a. 4.24 to 8.36;

b. narrower confidence interval

10.19 $d = A - B$, $n = 8$, $\bar{d} = 3.75$, $s_d = 5.726$
−1.03 to 8.53

10.21 a. Male insurance costs appear higher than female insurance costs.

c. males: $\bar{x} = \$1242.70$, $s = \$334.90$; females: $\bar{x} = \$1026.80$, $s = \$299.50$; difference: $\bar{x} = \$215.90$, $s = \$44.30$

d. male and female cost distributions, approx normal

e. $192.31 to $239.49

f. yes, entire confidence interval above zero

10.23 a. $H_a: \mu_d > 0$; $d =$ posttest − pretest

b. $H_a: \mu_d \neq 0$; $d =$ after − before

c. $H_a: \mu_d \neq 0$; $d =$ reading1 − reading2

d. $H_a: \mu_d > 0$; $d =$ postscore − prescore

10.25 a. $P(t > 1.86 | \mathrm{df} = 19)$; $0.025 < \mathbf{P} < 0.05$

b. $2P(t < -1.86 | \mathrm{df} = 19)$; $0.05 < \mathbf{P} < 0.10$

c. $P(t < -2.63 | \mathrm{df} = 28)$; $0.005 < \mathbf{P} < 0.01$

d. $P(t > 3.57 | \mathrm{df} = 9)$; $< \mathbf{P} < 0.005$

10.27 $H_a: \mu_d > 0$ (increased); $t\star = 4.42$;
$\mathbf{P} < 0.005$; $t(14, 0.05) = 1.76$; Reject H_o

10.29 $H_a: \mu_d > 0$ (beneficial); $t\star = 3.067$;
$\mathbf{P} < 0.005$; $t(39, 0.01) = 2.44$; Reject H_o

10.31 $t\star = 2.45$; $0.025 < \mathbf{P} < 0.05$;
$t(4, 0.05) = 2.13$; Reject H_o

10.35 $t\star = -1.30$; $\mathbf{P} = 0.13$; Fail to reject H_o

10.37 $H_a: \mu_d > 0$ (improvement); $t\star = 0.56$;
$\mathbf{P} > 0.25$; $t(9, 0.05) = 1.83$; Fail to reject H_o

10.39 a. average difference is zero

b. values used to make the decision

c. test is two-tailed; t-distribution is symmetric; absence of negative numbers makes less confusing

d. fail to reject the null hypothesis in 12 of them

e. The two methods are equivalent.

f. revised Florida method for sampling be accepted and implemented

10.41 4.92

10.43 Case I: between 17 and 40; Case II: 17

10.45 −6.3 to 16.3

10.47 $21.40 to $42.22

10.51 S. Dakota: $n = 14, \bar{x} = 1548, s = 401$;
N. Dakota: $n = 11, \bar{x} = 1403, s = 159$;
−116.8 to 406.8

10.53 a. $H_a: \mu_1 - \mu_2 \neq 0$ b. $H_a: \mu_1 - \mu_2 > 0$
c. $H_a: \mu_S - \mu_N > 0$ d. $H_a: \mu_M - \mu_F \neq 0$

10.55 a. 1.21   b. 0.1243   c. 1.56

10.57 2.64

10.59 a. ±2.13   b. −2.47   c. 1.41   d. −2.16

10.63 No

10.65 $H_a: \mu_1 - \mu_2 > 0; t\star = 4.02$;
$P < 0.005; t \geq 2.44$; Reject H_o

10.67 $H_a: \mu_G - \mu_B \neq 0; t\star = 1.44$;
$0.10 < P < 0.20; t \leq -2.16, t \geq 2.16$; Fail to
reject H_o

10.69 b. $0.554 < P < 0.624$
c. $0.560 < P < 0.626$

10.71 $H_a: \mu_M - \mu_F > 0; t\star = 0.02$;
$P > 0.25; t \geq 2.02$; Fail to reject H_o

10.73 $H_a: \mu_Y - \mu_C > 0; t\star = 2.57$;
$0.01 < P < 0.025; t \geq 1.80$; reject H_o

10.75 a. yes, p-values from normality tests > 0.05
b. $H_a: \mu_F - \mu_M \neq 0; t\star = 4.32$;
$P < 0.01; t \leq -2.03, t \geq 2.03$; Reject H_o
c. living at home or not, having children, having a
job, method of commuting

10.77 a. unpolished: $\bar{x} = 8.98, s = 3.12$, polished:
$\bar{x} = 2.126, s = 0.437$
b. Histograms show mounded, slightly skewed
distributions; normality tests demonstrate
normality in both distributions.

10.83 75, 250, 0.30, 0.70

10.85 a. 0.085   b. 0.115

10.89 0.000 to 0.080

10.91 0.186 to 0.314

10.93 a. $H_a: p_m - p_w \neq 0$ b. $H_a: p_b - p_a > 0$
c. $H_a: p_c - p_{nc} > 0$

10.95 0.076; 0.924

10.97 1.34; 0.0901

10.99 a. $z \geq 1.65$ b. $z \leq -196, z \geq 1.96$
c. $z \leq -1.75$ d. $z \geq 2.33$

10.101 $H_a: p_m - p_c \neq 0$; $z\star = 1.42$; $P = 0.1556$;
$z \leq -1.96$ and $z \geq 1.96$; Fail to reject H_o

10.103 $H_a: p_c - p_{nc} > 0$; $z\star = 4.32$;
$P < 0.00002$; $z \geq 2.33$; Reject H_o

10.105 b. $H_a: p_w - p_m \neq 0$; $z\star = 0.64$; $P = 0.5222$;
$z \leq -1.96$ and $z \geq 1.96$; Fail to reject H_o
c. $z\star = 3.18$; $P = 0.0014$; $z \leq -1.96$ and
$z \geq 1.96$; Reject H_o
d. takes a reasonably large sample size to
show significance

10.107 a. $H_a: p_M - p_W \neq 0$; $z\star = 1.82$; $P = 0.0688$;
$z \leq -1.96$ and $z \geq 1.96$; Fail to reject H_o
b. $z\star = 2.57$; $P = 0.0102$;
$z \leq -1.96$ and $z \geq 1.96$; Reject H_o
c. 291

10.109 $H_a: p_2 - p_1 \neq 0$; $z\star = 1.43$; $P = 0.1528$;
$z \leq -1.96$ and $z \geq 1.96$; Fail to reject H_o

10.111 a. $H_a: \sigma_A^2 \neq \sigma_B^2$ b. $H_a: \sigma_I/\sigma_{II} > 1$
c. $H_a: \sigma_A^2/\sigma_B^2 \neq 1$ d. $H_a: \sigma_D^2/\sigma_C^2 > 1$

10.113 Divide inequality by σ_p^2

10.115 a. $F_{(9, 11, 0.025)}$ b. $F_{(24, 19, 0.01)}$
c. $F_{(8, 15, 0.01)}$ d. $F_{(15, 9, 0.05)}$

10.117 a. 2.51   b. 2.20   c. 2.91   d. 4.10
e. 2.67   f. 3.77   g. 1.79   h. 2.99

10.119 3.37

10.121 1.51

10.123 0.495; smaller variance in numerator

10.125 $H_a: \sigma_e \neq \sigma_c; F\star = 1.40; P > 0.05$;
$F \geq 2.27$; Fail to reject H_o

10.127 $H_a: \sigma_k^2 \neq \sigma_m^2; F\star = 1.33; P > 0.10$;
$F \geq 3.73$; Fail to reject H_o

10.129 Multiply by 2

10.131 $(4.43)^2/(3.50)^2 = 1.60$

10.133 a. $H_a: \sigma_{NBA}^2 \neq \sigma_{MLB}^2$;
$F\star = 5.72; 0.025 < P$
$F \geq 4.76$; Reject H_o

b. H_a: $\mu_{NBA} - \mu_{MLB} > 0$;
   $t\star = 3.56$; $\mathbf{P} < 0.005$;
   $t \geq 1.89$; Reject H_o

10.135 a. $\bar{x}_1 = 0.01525$, $s_1 = 0.00547$;
        $\bar{x}_2 = 0.02856$, $s_2 = 0.00680$
     b. H_a: $\sigma_1^2 \neq \sigma_2^2$; $F\star = 1.55$; $P > 0.10$; $F \geq 4.42$;
        Fail to reject H_o
     c. H_a: $\mu_2 - \mu_1 \neq 0$; $t\star = 5.64$; $\mathbf{P} < 0.01$;
        $t \leq -2.36$, $t \geq 2.36$; Reject H_o

10.137 a. Men: $\bar{x}_m = \$68.14$, $s_m = \$47.95$
        Women: $\bar{x}_W = \$85.90$, $s_w = \$63.50$

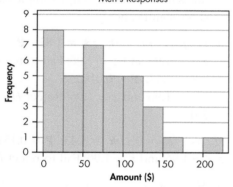

How much should someone spend on you for Valentine's Day?
Men's Responses

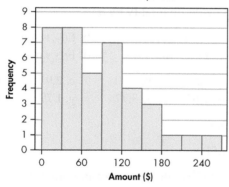

How much should someone spend on you for Valentine's Day?
Women's Responses

     d. H_a: $\mu_w - \mu_m > 0$;
        $t\star = 1.36$; $0.05 < \mathbf{P} < 0.10$
        $t \geq 1.70$; Fail to reject H_o
     e. H_a: $\sigma_w \neq \sigma_m$; $F\star = 1.75$; $\mathbf{P} \approx 0.10$;
        $F \geq 2.07$; Fail to reject H_o
     f. difference was not significant

'0.141 dependent; same set of 18 diamonds is
        appraised by two appraisers

'3 -8.85 to 16.02

      0.95 to 3.05

10.147 a. 1.0
     b. H_a: $\mu_d > 0$ (before $-$ after); $t\star = 1.18$;
        $0.10 < \mathbf{P} < 0.25$; $t \geq 3.00$; Fail to reject H_o

10.149 -0.21 to 10.61

10.151 -0.12 to 0.78

10.153 0.012 to 0.072

10.155 H_a: $\mu_2 - \mu_1 > 0$; $t\star = 0.988$;
        $0.10 < \mathbf{P} < 0.25$; $t \geq 1.72$; Fail to reject H_o

10.157 H_a: $\mu_2 - \mu_1 > 0$; $t\star = 1.30$; $0.10 < \mathbf{P} < 0.25$;
        $t \geq 2.47$; Fail to reject H_o

10.159 H_a: $\mu_A - \mu_B \neq 0$; $t\star = 5.84$; $\mathbf{P} < 0.01$;
        $t \leq -2.98$, $t \geq 2.98$; Reject H_o

10.161 a. M: $\bar{x} = 74.69$, $s = 10.19$,
        F: $\bar{x} = 79.83$, $s = 8.80$,

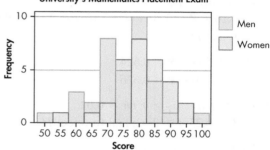

University's Mathematics Placement Exam

     b. M: H_a: $\mu \neq 77$;
        $t\star = -1.36$; $0.10 < \mathbf{P} < 0.20$;
        $t \leq -2.03$, $t \geq 2.03$; Fail to reject H_o
        F: H_a: $\mu \neq 77$; $t\star = 1.76$; $0.05 < \mathbf{P} < 0.10$;
        $t \leq -2.05$, $t \geq 2.05$; Fail to reject H_o
     c. both not significantly different than 77
     d. H_a: $\mu_F - \mu_M \neq 0$;
        $t\star = 2.19$; $0.02 < \mathbf{P} < 0.05$;
        $t \leq -2.05$, $t \geq 2.05$; Reject H_o
     e. & f.   asking different questions

10.163 -0.044 to 0.164

10.165 a. $z\star = 2.37$; $\mathbf{P} = 0.0178$;
        significant difference for $\alpha \geq 0.02$
     b. $z\star = 2.90$; $\mathbf{P} = 0.0038$;
        significant difference for $\alpha \geq 0.01$
     c. $z\star = 3.35$; $\mathbf{P} = 0.0008$;
        significant difference for $\alpha \geq 0.001$
     d. standard error became smaller

10.167 $H_a: p_a - p_1 \neq 0$; $z\star = 1.26$; $\mathbf{P} = 0.2076$; $z \leq -2.58$ and $z \geq 2.58$; Fail to reject H_o

10.169 $H_a: \sigma_m^2 > \sigma_f^2$; $F\star = 2.58$; $0.025 < \mathbf{P} < 0.05$; $F \geq 2.53$; Reject H_o

10.171 $H_a: \sigma_n^2 \neq \sigma_s^2$; $F\star = 1.28$; $\mathbf{P} > 0.10$; $F \geq 2.01$; Fail to reject H_o

10.173 a.

	N	Mean	StDev
Cont	50	0.005459	0.000763
Test	50	0.003507	0.000683

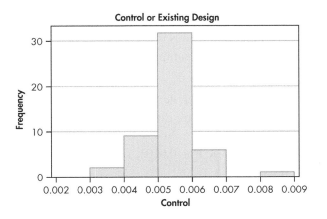

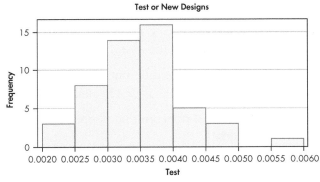

b. Both approximately normal

c. One-tailed – looking for a reduction

d. $H_a: \sigma_c^2 > \sigma_t^2$; $F\star = 1.248$; $\mathbf{P} > 0.05$; $F \geq 1.69$; Fail to reject H_o

e. $H_a: \mu_c - \mu_t > 0$; $t\star = 13.48$; $\mathbf{P} = +0.000$; $t \leq -1.68, t \geq 1.68$; Reject H_o

f. mean force has been reduced, but not the variability

Chapter 11

11.1 a. preferred way to cool mouth after hot sauce

b. U.S. adults professing to love eating hot, spicy food; method of cooling

c. 36.5%, 14.5%, 17.5%, 9.5%, 10%, 6.5%, 5.5%

11.3 a. 23.2   b. 23.3   c. 3.94   d. 8.64

11.5 a. $\chi^2(14, 0.01) = 29.1$ b. $\chi^2(25, 0.025) = 40.6$, $\chi^2(25, 0.025) = 13.1$

11.7 a. Asking one person

b. Birth day of the week

c. the 7 days of the week

11.9 a. $H_o: P(1) = P(2) = P(3) = P(4) = P(5) = 0.2$
H_a: not equally likely

b. $H_o: P(1) = 2/8, P(2) = 3/8$, $P(3) = 2/8, P(4) = 1/8$
H_a: at least one is different

c. $H_o: P(E) = 0.16, P(G) = 0.38$, $P(F) = 0.41, P(P) = 0.05$
H_a: percentages are different from specified

11.11 a. $P(\chi^2 > 12.25|\mathrm{df} = 3)$; $0.005 < P < 0.01$

b. $P(\chi^2 > 5.98|\mathrm{df} = 2)$; $0.05 < P < 0.10$

11.15 a. $H_o: P(A) = P(B) = P(C) = P(D) = P(E) = 0.2$

b. χ^2

c. H_a: preferences not all equal; $\chi^2\star = 4.40$; $\mathbf{P} = 0.355$; $\chi^2 \geq 7.78$; fail to reject H_o

11.17 H_a: proportions different; $\chi^2\star = 16.317$; $\mathbf{P} < 0.005$; $\chi^2 \geq 9.21$; reject H_o

11.19 a. 60

b. 2

c. H_a: ratio other than 6:3:1
$\chi^2\star = 2.67$; $\mathbf{P} = 0.263$; $\chi^2 \geq 4.61$; fail to reject H_o

11.21 H_a: proportions are different
$\chi^2\star = 7.35$; $\mathbf{P} = 0.062$; $\chi^2 \geq 7.81$; fail to reject H_o

11.23 H_a: Opinions distributed differently
$\chi^2\star = 213.49$; $\mathbf{P} < 0.005$; $\chi^2 \geq 9.49$; reject H_o

11.25 H_a: Opinions distributed differently
$\chi^2\star = 10.05$; $\mathbf{P} = 0.123$; $\chi^2 \geq 12.6$; fail to reject H_o

11.27 H_a: proportions different
$\chi^2\star = 13.537$; $\mathbf{P} = 0.0352$; $\chi^2 \geq 12.6$; reject H_o

11.29 a. H_a: proportions different from listed
$\chi^2\star = 44.4928$; $\mathbf{P} < 0.005$; $\chi^2 \geq 7.81$; reject H_o

b. 4th cell

11.31 a. H_a: Voters preference and party affiliation are not independent.

b. H_a: The distribution is not the same for all three

c. H_a: The proportion of yes's is not the same in all categories

11.33 10

11.37 H_a: Direction was not independent of vehicle type, $\chi^2\star = 71.249$; $\mathbf{P} = 0.000$; $\chi^2 \geq 6.63$; reject H_o

11.39 H_a: Having Tourette's is not independent of ethnicity and race; $\chi^2\star = 1.434$; $\mathbf{P} = 0.488$; $\chi^2 \geq 5.99$; fail to reject H_o

11.41 H_a: Response is not independent of years $\chi^2\star = 3.390$; $\mathbf{P} = 0.335$; $\chi^2 \geq 6.25$; fail to reject H_o

11.43 H_a: Size of community of residence is not independent of size of community reared in $\chi^2\star = 35.741$; $\mathbf{P} < 0.005$; $\chi^2 \geq 13.3$; reject H_o

11.45 H_a: number of defectives is not independent of day $\chi^2\star = 8.548$; $\mathbf{P} = 0.074$; $\chi^2 \geq 9.49$; fail to reject H_o

11.47 H_a: Blog creators and readers are not proportioned the same $\chi^2\star = 3.954$; $\mathbf{P} = 0.138$; $\chi^2 \geq 5.99$; fail to reject H_o

11.49 H_a: Distributions different $\chi^2\star = 2.678$; $\mathbf{P} = 0.444$; $\chi^2 \geq 7.81$; fail to reject H_o

11.51 H_a: Fear and Do Not Fear are not proportioned the same $\chi^2\star = 80.959$; $\mathbf{P} < 0.005$; $\chi^2 \geq 13.3$; reject H_o

11.53 Gender:
H_a: Females and males are not proportioned the same for each dosage $\chi^2\star = 0.978$; $\mathbf{P} = 0.613$; $\chi^2 \geq 9.21$; fail to reject H_o

Dosage:
H_a: Age groups are not proportioned the same for each dosage $\chi^2\star = 7.449$; $\mathbf{P} = 0.114$; $\chi^2 \geq 13.3$; fail to reject H_o

11.55 a. $\chi^2\star = 4.043$; $\mathbf{P} = 0.257$
b. $\chi^2\star = 8.083$, $\mathbf{P} = 0.044$; $\chi^2\star = 12.127$, $\mathbf{P} = 0.007$
c. Yes

11.57 H_a: proportions are other than 1:3:4 $\chi^2\star = 10.33$; $\mathbf{P} = 0.006$; $\chi^2 \geq 5.99$; reject H_o

11.59 H_a: percentages different from listed $\chi^2\star = 6.693$; $\mathbf{P} = 0.153$; $\chi^2 \geq 9.49$; fail to reject H_o

11.61 H_a: percentages different from listed $\chi^2\star = 17.92$; $\mathbf{P} = 0.003$; $\chi^2 \geq 15.1$; reject H_o

11.63 $P(x < 130) = 0.0228$,
$P(130 < x < 145) = 0.1359$,
$P(145 < x < 160) = 0.3413$,
$P(160 < x < 175) = 0.3413$,
$P(175 < x < 190) = 0.1359$,
$P(x > 190) = 0.0228$
H_a: The weights are not N(160, 15); $\chi^2\star = 5.812$; $\mathbf{P} = 0.325$; $\chi^2 \geq 11.1$; fail to reject H_o

11.65 a. H_a: Distributions are different $\chi^2\star = 6.1954$; $\mathbf{P} = 0.2877$; $\chi^2 \geq 11.1$; fail to reject H_o
b. $\chi^2\star = 36.761$; $\mathbf{P} < 0.005$; $\chi^2 \geq 11.1$; reject H_o
c. $\chi^2\star = 92.93$; $\mathbf{P} < 0.005$; $\chi^2 \geq 11.1$; reject H_o
d. Chi-square becomes more sensitive to variations as the sample size gets larger

1.67 a. yes
b. shift had an effect, defects depend on shifts
c. H_a: proportions are different from shift to shift; $\chi^2\star = 16.734$; $\mathbf{P} = 0.0103$, $\chi^2 \geq 16.8$, fail to reject H_o
Note – use of a rounded p-value will lead to the opposite decision of what the classical method produces

11.69 H_a: Political preference is not independent of age; $\chi^2\star = 23.339$; $\mathbf{P} < 0.005$; $\chi^2 \geq 13.3$; reject H_o

11.71 H_a: Distributions were different; $\chi^2\star = 3.123$; $\mathbf{P} = 0.793$; $\chi^2 \geq 12.6$; fail to reject H_o

11.73 H_o: Proportion of popcorn that popped is same for all brands; $\chi^2\star = 2.839$; $\mathbf{P} = 0.417$; $\chi^2 \geq 7.81$; fail to reject H_o

11.75 a. 2003: 73.2%; 2004: 74.2%
b. H_a: Ratio of organ donors is not the same; $\chi^2 = 5.955$; $\mathbf{P} = 0.015$; $\chi^2 \geq 3.84$; reject H_o
c. With very large sample sizes, differences must be very small to be considered non-existent.

Answers to Chapter Practice Tests

Part 1: Only the replacement for the word(s) in boldface type is given. (If the statement is true, no answer is shown. If the statement is false, a replacement is given.)

Chapter 1, Page 30

Part I

1.1 descriptive
1.2 inferential
1.4 sample
1.5 population
1.6 attribute or qualitative
1.7 quantitative
1.9 random

Part II

1.11 **a.** Nominal   **b.** Ordinal   **c.** Continuous
     d. Discrete   **e.** Nominal

1.12 c, g, h, b, e, a, d, f

Part III

1.13 See definitions; examples will vary. Note: *population* is set of ALL possible, whereas *sample* is the actual set of subjects studied.

1.14 See definitions; examples will vary. Note: *variable* is the idea of interest, whereas *data* are the actual values obtained.

1.15 See definitions; examples will vary. Note: *data value* is the value describing one source, the *statistic* is a value (usually calculated) describing all the data in the sample, the *parameter* is a value describing the entire population (usually unknown).

1.16 Every element of the population has an equal chance of being selected.

Chapter 2, Page 116

Part I

2.1 median
2.2 dispersion
2.3 never
2.5 zero
2.6 higher than

Part II

2.11 **a.** 30   **b.** 46   **c.** 91   **d.** 15   **e.** 1   **f.** 61
     g. 75   **h.** 76   **i.** 91   **j.** 106 or 114

2.12 **a.** two items purchased
     b. Nine people purchased 3 items each.
     c. 40   **d.** 120   **e.** 5   **f.** 2   **g.** 3   **h.** 3

2.13 **a.** 6.7   **b.** 7   **c.** 8   **d.** 6.5   **e.** 5   **f.** 6
     g. 3.0   **h.** 1.7   **i.** 5

2.14 **a.** -1.5 **b.** 153

Part III

2.15 **a.** 98   **b.** 50
     c. 121   **d.** 100

2.16 **a.** $32,000, $26,500, $20,000, $50,000
     b.

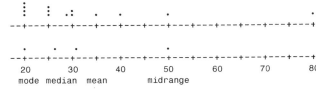

     c. Mr. VanCott—midrange; business manager—mean; foreman—median; new worker—mode
     d. The distribution is J-shaped.

2.17 There is more than one possible answer for these.

    a. 12, 12, 12   **b.** 15, 20, 25

    c. 12, 15, 15, 18   **d.** 12, 15, 16, 25 25

    e. 12, 12, 15, 16, 17   **f.** 20, 25, 30, 32, 32, 80

2.18 A is right; B is wrong; standard deviation will not change.

2.19 B is correct. For example, if standard deviation is $5, then the variance, (standard deviation)^2, is "25 dollars squared." Who knows what "dollars squared" are?

Chapter 3, Page 169

Part I

3.1 regression

3.2 strength of the

3.3 +1 or −1

3.5 positive

3.7 positive

3.8 −1 and +1

3.9 output or predicted value

Part II

3.11 **a.** b, d, a, c   **b.** 12   **c.** 10   **d.** 175 **e.** N

      f. (125, 13)   **g.** N   **h.** P

3.12 Someone made a mistake in arithmetic; r must be between −1 and +1.

3.13 **a.** 12   **b.** 10   **c.** 8   **d.** 0.73   **e.** 0.67

      f. 4.33   **g.** $\hat{y} = 4.33 + 0.67x$

Part III

3.14 Young children have small feet and probably tend to have less mathematics ability, whereas adults have larger feet and would tend to have more ability.

3.15 Student B is correct. −1.78 can occur only as a result of faulty arithmetic.

3.16 These answers will vary but should somehow include the basic thought:

    a. strong negative   **b.** strong positive

    c. no correlation   **d.** no correlation

    e. impossible value, bad arithmetic

3.17 There is more than one possible answer for these.

    a. (1, 1), (2, 1), (3, 1)   **b.** (1, 1), (3, 3), (5, 5)

    c. (1, 5), (3, 3), (5, 1)   **d.** (1, 1), (5, 1), (1, 5), (5, 5)

Chapter 4, Page 229

Part I

4.1 any number value between 0 and 1, inclusive

4.4 simple

4.5 seldom

4.6 sum to 1.0

4.7 dependent

4.8 complementary

4.9 mutually exclusive or dependent

4.10 multiplication rule

Part II

4.11 **a.** $\dfrac{4}{8}$ **b.** $\dfrac{4}{8}$ **c.** $\dfrac{2}{8}$ **d.** $\dfrac{6}{8}$ **e.** $\dfrac{2}{8}$

      f. $\dfrac{6}{8}$ **g.** 0   **h.** $\dfrac{6}{8}$ **i.** $\dfrac{1}{8}$ **j.** $\dfrac{5}{8}$

      k. $\dfrac{2}{4}$ **l.** 0   **m.** $\dfrac{1}{2}$ **n.** no (e)

      o. yes (g)   **p.** no (i)   **q.** yes (a, k)

      r. no (b, 1)   **s.** yes (a, m)

4.12 **a.** 0   **b.** 0.7   **c.** 0   **d.** no (c)

4.13 **a.** 0.7   **b.** 0.5   **c.** no, $P(E \text{ and } F) = 0.2$

      d. yes, $P(E) = P(E \mid F)$

4.14 0.51

Part III

4.15. Student B is right. *Mutually exclusive* means no intersection, whereas *independence* means one event does not affect the probability of the other.

4.16 These answers will vary but should somehow include the basic thoughts:

    a. no common occurrence

    b. either event has no effect on the probability of the other

    c. the relative frequency with which the event occurs.

    d. probability that an event will occur even though the conditional event has previously occurred

Chapter 5, Page 267

Part I

5.1 continuous

5.3 one

5.5   exactly two
5.6   binomial
5.7   one success occurring on 1 trial
5.8   population
5.9   population parameters

Part II

5.11 **a.** Each $P(x)$ is between 0 and 1, and the sum of all $P(x)$ is exactly 1.
      b. 0.2   **c.** 0   **d.** 0.8   **e.** 3.2   **f.** 1.25

5.12 **a.** 0.230   **b.** 0.085   **c.** 1.2   **d.** 1.04

Part III

5.13 n independent repeated trials of two outcomes; the two outcomes are "success" and "failure"; $p = P(\text{success})$ and $q = P(\text{failure})$ and $p + q = 1$; $x = n(\text{success}) = 0, 1, 2, \ldots, n$.

5.14 Student B is correct. The sample mean and standard deviation are statistics found using formulas studied in Chapter 2. The probability distributions studied in Chapter 5 are theoretical populations and their means and standard deviations are parameters.

5.15 Student B is correct. There are no restrictions on the values of the variable x.

Chapter 6, Page 310

Part I

6.1   its mean
6.4   1 standard deviation
6.6   right
6.7   zero, 1
6.8   some (many)
6.9   mutually exclusive events
6.10 normal

Part II

6.11 **a.** 0.4922   **b.** 0.9162   **c.** 0.1020   **d.** 0.9082

6.12 **a.** 0.63   **b.** −0.95   **c.** 1.75

6.13 **a.** $z(0.8100)$ **b.** $z(0.2830)$

6.14 0.7910
6.15 28.03
6.16 **a.** 0.0569   **b.** 0.9890   **c.** 537
      d. 417   **e.** 605

Part III

6.17 This answer will vary but should somehow include the basic properties: bell-shaped, mean of 0, standard deviation of 1.

6.18 This answer will vary but should somehow include the basic ideas: it is a z-score, a represents the area under the curve and to the right of z.

6.19 All normal distributions have the same shape and probabilities relative to the z-score.

Chapter 7, Page 338

Part I

7.1   is not
7.2   some (many)
7.3   population
7.4   divided by $\sqrt{n}$
7.5   decreases
7.6   approximately normal
7.7   sampling
7.8   means
7.9   random

Part II

7.11 **a.** 0.4364   **b.** 0.2643
7.12 **a.** 0.0918   **b.** 0.9525
7.13 0.6247

Part III

7.14 In this case each head produced one piece of data, the estimated length of the line. The CLT assures us that the mean value of a sample is far less variable than individual values of the variable x.

7.15 All samples must be of one fixed size.

7.16 Student A is correct. A population distribution is a distribution formed by all x values that make up the entire population.

7.17 Student A is correct. The standard error is found by dividing the standard deviation by the square root of the *sample size*.

Chapter 8, Page 409

Part I

8.1   alpha
8.2   alpha

8.3   sample distribution of the mean
8.7   type II error
8.8   beta
8.9   correct decision
8.10 critical (rejection) region

Part II

8.11 4.72 to 5.88

8.12 **a.** $H_o: \mu = 245, H_a: \mu > 245$
　　　　b. $H_o: \mu = 4.5, H_a: \mu < 4.5$
　　　　c. $H_o: \mu = 35, H_a: \mu \neq 35$

8.13 **a.** $0.05, z, z \leq -1.65$
　　　　b. $0.05, z, z \geq +1.65$
　　　　c. $0.05, z, z \leq -1.96$ or $z \geq +1.96$

8.14 **a.** 1.65　　　**b.** 2.33　　　**c.** 1.18

8.15 **a.** $z\star = 2.50$　　**b.** 0.0062

8.16 $H_o: \mu = 1520$ vs. $H_a: \mu < 1520$, crit. reg.
　　　　$z \leq -2.33, z\star = -1.61$, fail to reject H_o

Part III

8.17 **a.** no specific effect
　　　　b. reduces it　　**c.** narrows it　　**d.** no effect
　　　　e. increases it　　**f.** widens it

8.18 **a.** $H_o - $ (a), $H_a - $ (b)　　**b.** 3　　**c.** 2
　　　　d. P(type I errror) is alpha, decreases:
　　　　　　P(type II error) increases

8.19 The alternative hypothesis expresses the concern;
　　　　the conclusion answers the concern.

Chapter 9, Page 476

Part I

9.2   Student's t
9.3   chi-square
9.4   to be rejected
9.6 t score
9.7 $n - 1$
9.9 $\sqrt{pq/n}$
9.10 z(normal)

Part II

9.11 **a.** 2.05　　　**b.** -1.73　　　**c.** 14.6

9.12 **a.** 28.6　　　**b.** 1.44　　　**c.** 27.16 to 30.04

9.13 0.528 to 0.752

9.14 **a.** $H_o: \mu = 255, H_a: \mu > 225$
　　　　b. $H_o: p = 0.40, H_a: p \neq 0.40$
　　　　c. $H_o: \sigma = 3.7, H_a: \sigma < 3.7$

9.15 **a.** $0.05, z, z \leq -1.65$
　　　　b. $0.05, t, t \leq -2.08$ or $t \geq +2.08$
　　　　c. $0.05, z, z \geq +1.65$
　　　　d. $0.05, \chi^2, \chi^2 \leq 14.6$ or $\chi^2 \geq 43.2$

9.16 $H_o: \mu = 26$ vs. $H_a: \mu < 26$, crit. reg, $t \leq -1.71$,
　　　　$t\star = -1.86$, reject H_o

9.17 $H_o: \sigma = 0.1$ vs. $H_a: \sigma > 0.1$, crit. reg. $\chi^2 \geq 21.1$,
　　　　$\chi^2\star = 23.66$, reject H_o

9.18 $H_o: p = 0.50$ vs. $H_a: p > 0.50$, crit. reg. $z \geq 2.05$,
　　　　$z\star = 1.29$, fail to reject H_o

Part III

9.19 If the distribution is normal, 6 standard devia-
　　　　tions is approximately equal to the range.
9.20 Alternative hypothesis
9.21 They are both correct.
9.22 When the sample size, n, is large, the critical value
　　　　of t is estimated by using the critical value from
　　　　the standard normal distribution of z.
9.23 Student A
9.24 Student B is right. It is significant at the 0.01
　　　　level of significance.
9.25 Student A is correct.
9.26 It depends on what it means to improve the
　　　　confidence interval. For most purposes, an
　　　　increased sample size would be the best
　　　　improvement.

Chapter 10, Page 542

Part I

10.1 two independent means
10.3 F-distribution
10.4 Student's t-distribution
10.5 Student's t
10.7 nonsymmetrical (or skewed)
10.9 decreases

Part II

10.11 **a.** $H_o: \mu_N - \mu_A = 0, H_a: \mu_N - \mu_A \neq 0$
　　　　b. $H_o: \sigma_o/\sigma_m = 1.0, H_a: \sigma_o/\sigma_m > 1.0$
　　　　c. $H_o: p_m - p_f = 0, H_a: p_m - p_f \neq 0$

10.12 a. $z, z \leq -1.96$ or $z \geq 1.96$
   b. $t, t \leq -2.05, t \geq 2.05$
   c. t, df = 7, $t \geq 1.89$
   d. t, df = 37, $t \leq -1.69$
   e. $F, F \geq 2.11$

10.13 a. 2.05   **b.** 2.13   **c.** 2.51   **d.** 2.18
   e. 1.75   **f.** 1.69   **g.** -2.50 **h.** -1.28

10.14 $H_o: \mu_L - \mu_P = 0$ vs. $H_a: \mu_L - \mu_P > 0$, crit. reg.
   $t \geq +1.83, t\star = 0.979$, fail to reject H_o

10.15 $H_o: \mu_d = 0$ vs. $H_a: \mu_d > 0$, crit. reg. $t \geq 1.89$,
   $t\star = 1.88$, fail to reject H_o

10.16 0.072 to 0.188

Part III

10.17 independent

10.18 One possibility: Test all students before the course starts, then randomly select 20 of those who finish the course and test them afterward. Use the before scores for these 20 as the before sample.

10.19 For starters, if the two independent samples are of different sizes, the techniques for dependent samples could not be completed. They are testing very different concepts, the "mean of the differences of paired data" and the "difference between two mean values."

10.20 It is only significant if the calculated t-score is in the critical region. The variation among the data and their relative size will play a role.

10.21 The 80 scores actually are two independent samples of size 40. A test to compare the mean scores of the two groups could be completed.

10.22 A fairly large sample of both Catholic and non-Catholic families would need to be taken, and the number of each whose children attended private schools would need to be obtained. The difference between two proportions could then be estimated.

Chapter 11, Page 576

Part I

11.1 one less than
11.3 expected
11.4 contingency table
11.6 test of homogeneity
11.8 approximated by chi-square

Part II

11.11 a. H_o: Digits generated occur with equal probability.
   H_a: Digits do not occur with equal probability.
   b. H_o: Votes were cast independently of party affiliation.
   H_a: Votes were not cast independently of party affiliation.
   c. H_o: The crimes distributions are the same for all four cities.
   H_a: The crimes distributions are not all the same.

11.12 a. 4.40   **b.** 35.7

11.13 $H_o: P(1) = P(2) = P(3) = \dfrac{1}{3}$
   H_a: preferences not all equal, $\chi^2\star = 3.78$;
   $0.10 < \mathbf{P} < 0.25$ or crit. reg. $\chi^2 \geq 5.99$; fail to reject H_o

11.14 a. H_o: The distribution is the same for all types of soil.
   H_a: The distributions are not all the same.
   b. 25.622
   c. 13.746
   d. $0.005 < \mathbf{P} < 0.01$
   e. $\chi^2 \geq 9.49$
   f. Reject H_o: There is sufficient evidence to show that the growth distribution is different for at least one of the three soil types.

Part III

11.15 Similar in that there are n repeated independent trials. Different in that the binomial has two possible outcomes, whereas the multinomial has several. Each possible outcome has a probability and these probabilities sum to 1 for each different experiment, both for binomial and multinomial.

11.16 The test of homogeneity compares several distributions in a side-by-side comparison, whereas the test for independence tests the independence of the two factors that create the rows and columns of the contingency table.

11.17 Student A is right in that the calculations are completed in the same manner. Student B is correct in that the test of independence starts with one large sample and homogeneity has several samples.

11.18 a. If a chi-square test is to be used, the results of the four questions would be pooled to estimate the expected probability.
   b. Use a chi-square test for homogeneity.

Index

Index of Applications

Index for Computer and Calculator Instructions

Cumulative Areas of the Standard Normal Distribution (*continued*)

The entries in this table are the cumulative probabilities for the standard normal distribution z (that is, the normal distribution with mean 0 and standard deviation 1). The shaded area under the curve of the standard normal distribution represents the cumulative probability to the left of a z-value in the **left-hand tail**.

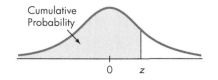

z	0.00	0.01	0.02	0.03	0.04	0.05	0.06	0.07	0.08	0.09
0.0	0.5000	0.5040	0.5080	0.5120	0.5160	0.5199	0.5239	0.5279	0.5319	0.5359
0.1	0.5398	0.5438	0.5478	0.5517	0.5557	0.5596	0.5636	0.5675	0.5714	0.5754
0.2	0.5793	0.5832	0.5871	0.5910	0.5948	0.5987	0.6026	0.6064	0.6103	0.6141
0.3	0.6179	0.6217	0.6255	0.6293	0.6331	0.6368	0.6406	0.6443	0.6480	0.6517
0.4	0.6554	0.6591	0.6628	0.6664	0.6700	0.6736	0.6772	0.6808	0.6844	0.6879
0.5	0.6915	0.6950	0.6985	0.7019	0.7054	0.7088	0.7123	0.7157	0.7190	0.7224
0.6	0.7258	0.7291	0.7324	0.7357	0.7389	0.7422	0.7454	0.7486	0.7518	0.7549
0.7	0.7580	0.7612	0.7642	0.7673	0.7704	0.7734	0.7764	0.7794	0.7823	0.7852
0.8	0.7881	0.7910	0.7939	0.7967	0.7996	0.8023	0.8051	0.8079	0.8106	0.8133
0.9	0.8159	0.8186	0.8212	0.8238	0.8264	0.8289	0.8315	0.8340	0.8365	0.8389
1.0	0.8413	0.8438	0.8461	0.8485	0.8508	0.8531	0.8554	0.8577	0.8599	0.8621
1.1	0.8643	0.8665	0.8686	0.8708	0.8729	0.8749	0.8770	0.8790	0.8810	0.8830
1.2	0.8849	0.8869	0.8888	0.8907	0.8925	0.8944	0.8962	0.8980	0.8997	0.9015
1.3	0.9032	0.9049	0.9066	0.9082	0.9099	0.9115	0.9131	0.9147	0.9162	0.9177
1.4	0.9192	0.9207	0.9222	0.9236	0.9251	0.9265	0.9279	0.9292	0.9306	0.9319
1.5	0.9332	0.9345	0.9357	0.9370	0.9382	0.9394	0.9406	0.9418	0.9430	0.9441
1.6	0.9452	0.9463	0.9474	0.9485	0.9495	0.9505	0.9515	0.9525	0.9535	0.9545
1.7	0.9554	0.9564	0.9573	0.9582	0.9591	0.9599	0.9608	0.9616	0.9625	0.9633
1.8	0.9641	0.9649	0.9656	0.9664	0.9671	0.9678	0.9686	0.9693	0.9700	0.9706
1.9	0.9713	0.9719	0.9726	0.9732	0.9738	0.9744	0.9750	0.9756	0.9762	0.9767
2.0	0.9773	0.9778	0.9783	0.9788	0.9793	0.9798	0.9803	0.9808	0.9812	0.9817
2.1	0.9821	0.9826	0.9830	0.9834	0.9838	0.9842	0.9846	0.9850	0.9854	0.9857
2.2	0.9861	0.9865	0.9868	0.9871	0.9875	0.9878	0.9881	0.9884	0.9887	0.9890
2.3	0.9893	0.9896	0.9898	0.9901	0.9904	0.9906	0.9909	0.9911	0.9913	0.9916
2.4	0.9918	0.9920	0.9922	0.9925	0.9927	0.9929	0.9931	0.9932	0.9934	0.9936
2.5	0.9938	0.9940	0.9941	0.9943	0.9945	0.9946	0.9948	0.9949	0.9951	0.9952
2.6	0.9953	0.9955	0.9956	0.9957	0.9959	0.9960	0.9961	0.9962	0.9963	0.9964
2.7	0.9965	0.9966	0.9967	0.9968	0.9969	0.9970	0.9971	0.9972	0.9973	0.9974
2.8	0.9974	0.9975	0.9976	0.9977	0.9977	0.9978	0.9979	0.9980	0.9980	0.9981
2.9	0.9981	0.9982	0.9983	0.9983	0.9984	0.9984	0.9985	0.9985	0.9986	0.9986
3.0	0.9987	0.9987	0.9987	0.9988	0.9988	0.9989	0.9989	0.9989	0.9990	0.9990
3.1	0.9990	0.9991	0.9991	0.9991	0.9992	0.9992	0.9992	0.9992	0.9993	0.9993
3.2	0.9993	0.9993	0.9994	0.9994	0.9994	0.9994	0.9994	0.9995	0.9995	0.9995
3.3	0.9995	0.9995	0.9996	0.9996	0.9996	0.9996	0.9996	0.9996	0.9996	0.9997
3.4	0.9997	0.9997	0.9997	0.9997	0.9997	0.9997	0.9997	0.9997	0.9998	0.9998
3.5	0.9998	0.9998	0.9998	0.9998	0.9998	0.9998	0.9998	0.9998	0.9998	0.9998
3.6	0.99984	0.99985	0.99985	0.99986	0.99986	0.99987	0.99987	0.99988	0.99988	0.99989
3.7	0.99989	0.99990	0.99990	0.99990	0.99991	0.99991	0.99992	0.99992	0.99992	0.99992
3.8	0.99993	0.99993	0.99993	0.99994	0.99994	0.99994	0.99994	0.99995	0.99995	0.99995
3.9	0.99995	0.99995	0.99996	0.99996	0.99996	0.99996	0.99996	0.99996	0.99997	0.99997
4.0	0.99997	0.99997	0.99997	0.99997	0.99997	0.99997	0.99998	0.99998	0.99998	0.99998
4.5	0.999997									
5.0	0.9999997									

Table 3 was generated using Minitab.